HANDBOOK OF SEMIDEFINITE PROGRAMMING

INTERNATIONAL SERIES IN OPERATIONS RESEARCH & MANAGEMENT SCIENCE

Frederick S. Hillier, Series Editor
Stanford University

Saigal, R. / *LINEAR PROGRAMMING: A Modern Integrated Analysis*

Nagurney, A. & Zhang, D. / *PROJECTED DYNAMICAL SYSTEMS AND VARIATIONAL INEQUALITIES WITH APPLICATIONS*

Padberg, M. & Rijal, M. / *LOCATION, SCHEDULING, DESIGN AND INTEGER PROGRAMMING*

Vanderbei, R. / *LINEAR PROGRAMMING: Foundations and Extensions*

Jaiswal, N.K. / *MILITARY OPERATIONS RESEARCH: Quantitative Decision Making*

Gal, T. & Greenberg, H. / *ADVANCES IN SENSITIVITY ANALYSIS AND PARAMETRIC PROGRAMMING*

Prabhu, N.U. / *FOUNDATIONS OF QUEUEING THEORY*

Fang, S.-C., Rajasekera, J.R. & Tsao, H.-S.J. / *ENTROPY OPTIMIZATION AND MATHEMATICAL PROGRAMMING*

Yu, G. / *OPERATIONS RESEARCH IN THE AIRLINE INDUSTRY*

Ho, T.-H. & Tang, C. S. / *PRODUCT VARIETY MANAGEMENT*

El-Taha, M. & Stidham , S. / *SAMPLE-PATH ANALYSIS OF QUEUEING SYSTEMS*

Miettinen, K. M. / *NONLINEAR MULTIOBJECTIVE OPTIMIZATION*

Chao, H. & Huntington, H. G. / *DESIGNING COMPETITIVE ELECTRICITY MARKETS*

Weglarz, J. / *PROJECT SCHEDULING: Recent Models, Algorithms & Applications*

Sahin, I. & Polatoglu, H. / *QUALITY, WARRANTY AND PREVENTIVE MAINTENANCE*

Tavares, L. V. / *ADVANCED MODELS FOR PROJECT MANAGEMENT*

Tayur, S., Ganeshan, R. & Magazine, M. / *QUANTITATIVE MODELING FOR SUPPLY CHAIN MANAGEMENT*

Weyant, J./ *ENERGY AND ENVIRONMENTAL POLICY MODELING*

Shanthikumar, J.G. & Sumita, U./*APPLIED PROBABILITY AND STOCHASTIC PROCESSES*

Liu, B. & Esogbue, A.O. / *DECISION CRITERIA AND OPTIMAL INVENTORY PROCESSES*

Gal, Stewart & Hanne/ *MULTICRITERIA DECISION MAKING: Advances in MCDM Models, Algorithms, Theory, and Applications*

Fox, B. L./ *STRATEGIES FOR QUASI-MONTE CARLO*

Hall, R.W. / *HANDBOOK OF TRANSPORTATION SCIENCE*

Grassman, W.K./ *COMPUTATIONAL PROBABILITY*

Pomerol & Barba-Romero / *MULTICRITERION DECISION IN MANAGEMENT*

Axsäter / *INVENTORY CONTROL*

HANDBOOK OF SEMIDEFINITE PROGRAMMING

Theory, Algorithms, and Applications

Edited by
Henry Wolkowicz
Department of Combinatorics and Optimization
Faculty of Mathematics
University of Waterloo
Waterloo, Ontario, Canada N2L 3G1
Canada
Romesh Saigal
Department of Industrial and Operations Engineering
University of Michigan
Ann Arbor, Michigan, 48109-2117
USA
Lieven Vandenberghe
Electrical Engineering Department
UCLA
Los Angeles, CA 90095-1594
USA

Kluwer Academic Publishers
Boston/Dordrecht/London

Distributors for North, Central and South America:
Kluwer Academic Publishers
101 Philip Drive
Assinippi Park
Norwell, Massachusetts 02061 USA
Telephone (781) 871-6600
Fax (781) 871-6528
E-Mail <kluwer@wkap.com>

Distributors for all other countries:
Kluwer Academic Publishers Group
Distribution Centre
Post Office Box 322
3300 AH Dordrecht, THE NETHERLANDS
Telephone 31 78 6392 392
Fax 31 78 6546 474
E-Mail <orderdept@wkap.nl>

Electronic Services <http://www.wkap.nl>

Library of Congress Cataloging-in-Publication

Handbook of semidefinite programming : theory, algorithms, and applications / edited by Henry Wolkowicz, Romesh Saigal, Lieven Vandenberghe.
p. cm. -- (International series in operations research & management science ; 27)
Includes bibliographical references and index.
ISBN 0-7923-7771-0
1. Linear programming. 2. Mathematical optimization. I. Wolkowicz, Henry. II. Saigal, Romesh. III. Vandenberghe, Lieven. IV. Series.

T57.74.H355 2000
519.7'2--dc21

99-086382

Printed on acid-free paper.

Printed in the United States of America

This book is dedicated to the memory of Svata Poljak, who made many of the early, major, contributions to this area. He died on April 2, 1995 in a car accident.

Contents

Contributing Authors

Abdo Alfakih is in the Department of Combinatorics and Optimization, University of Waterloo, Waterloo, Ontario, N2L 3G1, Canada.
E-mail: aalfakih@orion.uwaterloo.ca.

Farid Alizadeh is at RUTCOR, Rutgers University, P.O. Box 5062, New Brunswick, NJ 08903-5062.
E-mail alizadeh@rutcor.rutgers.edu.
Home page http://new-rutcor.rutgers.edu/~alizadeh.

Venkataramanan Balakrishnan is Associate Professor of Electrical and Computer Engineering, Purdue University, School of Electrical and Computer Engineering, 1285 Electrical Engineering Building, West Lafayette, Indiana 47907-1285. Research supported in part by the Office of Naval Research under contract no. N00014-97-1-0640, and in part by NEC and General Motors Faculty Fellowships.
E-mail ragu@ecn.purdue.edu.
Home page http://ece.www.ecn.purdue.edu/~ragu.

Dimitris Bertsimas is Boeing Professor of Operations Research, Sloan School of Management, Rm. E53-363, Massachusetts Institute of Technology, Cambridge, Mass. 02142. Research supported by NSF grant DMI-9610486.
E-mail dbertsim@mit.edu
Home page http://web.mit.edu/dbertsim/www.

Aharon Ben-Tal is a professor in the Faculty of Industrial Engineering & Management, Technion Israel Institute of Technology, Haifa 32000, Israel.
E-mail morbt@ie.technion.ac.il.

Etienne de Klerk is an Assistant Professor in the Faculty of Information Technology and Systems, Delft University of Technology, Delft, The Netherlands.
E-mail E.deKlerk@twi.tudelft.nl.

Laurent El Ghaoui is Associate Professor in the Department of Electrical Engineering and Computer Sciences, University of California at Berkeley, Berkeley, CA, 94720.
Email elghaoui@eecs.berkeley.edu
Home page http://www.eecs.berkeley.edu/~elghaoui.

Valerii Fedorov is Director of Research Statistics at Smithkline Beecham Pharmaceuticals, 1250 Collegeville Road, Collegeville, PA 19426 - 0989.
Email Valeri_V_Fedorov@sbphrd.com

Michel Goemans is at MIT, Department of Mathematics, Room 2-351 , 77 Massachusetts Avenue, Cambridge, MA 02139.
Chapter partially written while at C.O.R.E., Louvain-La-Neuve, Belgium, and as part of DONET, a European network supported by the European Community within the frame of the Training and Mobility of Researchers Programme (contract number ERB TMRX-CT98-0202).
E-mail goemans@math.mit.edu.
Home page http://theory.lcs.mit.edu/~goemans/.

Christoph Helmberg is at Konrad-Zuse-Zentrum für Informationstechnik Berlin, Takustrasse 7, D-14195 Berlin, Germany.
E-mail helmberg@zib-berlin.de.
Home page http://www.zib.de/helmberg.

Florian Jarre is at Department of Mathematics, University of Notre Dame, Notre Dame, IN 46556-5683, USA
E-mail jarre.1@nd.edu
Home page http://ifamus.mathematik.uni-wuerzburg.de/~jarre/index.html.

Serge Kruk is in the Department of Combinatorics & Optimization, University of Waterloo, Waterloo, ON Canada N2L 3G1.
E-mail sgkruk@mercator.uwaterloo.ca.

Jon Lee is a Professor in the Department of Mathematics, University of Kentucky, Lexington, Kentucky 40506-0027, U.S.A.
E-mail jlee@ms.uky.edu
Home page http://www.ms.uky.edu/~jlee.

Zhiquan Luo is at McMaster University, Department of Electrical and Computer Engineering, Hamilton, Ontario, Canada L8S 4K1. Supported by the Natural Science and Engineering Research Council of Canada, Grant No. OPG0090391.
Email luozq@mcmail.cis.mcmaster.ca.

Renato Monteiro is an Associate Professor in School of Industrial and Systems Engineering, Georgia Institute of Technology, Atlanta, Georgia 30332, USA. Research supported in part by NSF through grants INT-9600343 and CCR-9700448 and CCR-9902010.
E-mail monteiro@isye.gatech.edu.
Home page http://www.isye.gatech.edu/people/faculty/Renato_Monteiro/.

Yurii Nesterov is a professor at CORE, Catholic University of Louvain, Voie du Roman Pays, 34, B-1348 Louvain-La-Neuve, Belgium.
E-mail nesterov@core.ucl.ac.be.

Arkadi Nemirovski is a professor in the Faculty of Industrial Engineering & Management, Technion Israel Institute of Technology, Haifa 32000, Israel.
E-mail nemirovs@ie.technion.ac.il.

Francois Oustry is at ENSTA, Control and Optimization Group, LMA (Pihce 329), 32, boulevard Victor, 75739 Paris cedex 15, FRANCE
E-mail: Francois.Oustry@inrialpes.fr.

Gábor Pataki is in the Department of IE/OR, Columbia University, New York, NY,
E-mail gabor@ieor.columbia.edu.

Franz Rendl is at Universität Klagenfurt, Institut für Mathematik, A-9020 Klagenfurt, Austria.
E-mail franz.rendl@uni-klu.ac.at.

Kees Roos is a Professor at Delft University of Technology, Faculty of Information Technology and Systems, Delft, and at the University of Leiden, Mathematical Institute, Leiden, both in The Netherlands.
E-mail c.roos@twi.tudelft.nl.

Romesh Saigal is a Professor in Department of Industrial and Operations Engineering, University of Michigan, Ann Arbor, Michigan, 48109-2117, USA.

Katya Scheinberg is in the Department of Mathematical Sciences, IBM T. J. Watson Research Center, Rm. 32-228, Yorktown Heights, NY, 10598.
E-mail katyas@watson.ibm.com.

Stefan Schmieta is in the Department of IE/OR, Columbia University, New York, NY.
E-mail schmieta@rutcor.rutgers.edu

Jay Sethuraman is in the Operations Research Center, MIT, Cambridge, MA 02139.
E-mail jayc@mit.edu.

Alexander Shapiro is a Professor at the School of Industrial and Systems Engineering, Georgia Institute of Technology, Atlanta, Georgia 30332-0205, USA.
E-mail ashapiro@isye.gatech.edu.

Jos Sturm is a lecturer at Maastricht University, Department of Quantitative Economics, P.O.Box 616, NL-6200 MD Maastricht, The Netherlands.
Supported by the Netherlands Organization for Scientific Research (NWO).
Email j.sturm@ke.unimaas.nl.
Home page http://sparky.mcmaster.ca/ASPC/people/sturm/.

Tamás Terlaky is a Professor in the Department of Computing and Software, McMaster University, Hamilton, Ontario, Canada, L8S 4L7.
Email terlaky@cas.mcmaster.ca.
Home page http://www.cas.mcmaster.ca/~terlaky.

Michael J. Todd is a Professor in the School of Operations Research and Industrial Engineering, Cornell University, Ithaca, NY 14853-3801, USA. Research supported in part by NSF through grant DMS-9505155 and ONR through grant N00014-96-1-0050.
E-mail miketodd@cs.cornell.edu.
Home page http://www.orie.cornell.edu/~miketodd/todd.html.

Levent Tuncel is Associate Professor at University of Waterloo, Department of Combinatorics & Optimization, Waterloo, ON Canada N2L 3G1.
E-mail ltuncel@math.uwaterloo.ca.
Home page http://math.uwaterloo.ca/~ltuncel/.

Lieven Vandenberghe is Associate Professor at UCLA, Electrical Engineering Department, Los Angeles, CA 90095-1594, USA.
E-mail vandenbe@ee.ucla.edu.
Home page http://www.ee.ucla.edu/people/vandenbe/.

Fan Wang is in Electrical and Computer Engineering, Purdue University, School of Electrical and Computer Engineering, 1285 Electrical Engineering Building, West Lafayette, Indiana 47907-1285.
E-mail fanw@ecn.purdue.edu.
Home page http://ece.www.ecn.purdue.edu/~fanw.

Henry Wolkowicz is a Professor in the Department of Combinatorics and Optimization, University of Waterloo, Waterloo, Ontario, N2L 3G1, Canada.
Research supported in part by The Natural Sciences Engineering Research Council Canada.
E-mail hwolkowi@orion.uwaterloo.ca
Home page http://orion.math.uwaterloo.ca/~hwolkowi.

Yinyu Ye is a Professor in the Department of Management Sciences, The University of Iowa, Iowa City, Iowa 52242.
E-mail yinyu-ye@uiowa.edu.
Home page http://www.biz.uiowa.edu/faculty/yye/.

List of Figures

List of Tables

Preface

Semidefinite programming (or SDP) has been one of the most exciting and active research areas in optimization during the 1990s. It has attracted researchers with very diverse backgrounds, including experts in convex programming, linear algebra, numerical optimization, combinatorial optimization, control theory, and statistics. This tremendous research activity was spurred by the discovery of important applications in combinatorial optimization and control theory, the development of efficient interior-point algorithms for solving SDP problems, and the depth and elegance of the underlying optimization theory.

This book includes nineteen chapters on the theory, algorithms, and applications of semidefinite programming. Written by the leading experts on the subject, it offers an advanced and broad overview of the current state of the field. The coverage is somewhat less comprehensive, and the overall level more advanced, than we had planned at the start of the project. In order to finish the book in a timely fashion, we have had to abandon hopes for separate chapters on some important topics (such as a discussion of SDP algorithms in the framework of general convex programming using the theory of self-concordant barriers, or an overview of numerical implementation aspects of interior-point methods for SDP). We apologize for any important gaps in the contents, and hope that the historical notes at the end of the book provide a useful guide to the literature on the topics that are not adequately covered in this handbook.

We would like to thank all the authors for their outstanding contributions, their editorial help, and for their patience during the many revisions of the handbook. In addition, we thank Mike Todd for his valuable editorial advice on many occasions. We would also like to thank The Natural Sciences Engineering Research Council Canada and The Fields Institute for their financial support, and Erna Unrau for her help in combining some of the bibliographies into one.

HENRY WOLKOWICZ, ROMESH SAIGAL, LIEVEN VANDENBERGHE

1 INTRODUCTION

Henry Wolkowicz, Romesh Saigal, Lieven Vandenberghe

1.1 SEMIDEFINITE PROGRAMMING

Semidefinite programming refers to optimization problems that can be expressed in the form

$$\begin{array}{lllll} \text{minimize} & C \bullet X & & & \\ \text{subject to} & A_i \bullet X & = & b_i & \text{for all } i = 1, \ldots, m \\ & X & \succeq & 0 & \end{array} \tag{1.1.1}$$

where the variable is $X \in \mathcal{S}^n$, the space of real symmetric $n \times n$ matrices. The vector $b \in I\!R^m$, and the matrices $A_i \in \mathcal{S}^n$ and $C \in \mathcal{S}^n$ are (given) problem parameters. The inequality $X \succeq 0$ means X is positive semidefinite. (More generally, throughout this book, the notation $A \preceq B$ will mean $B - A$ is positive semidefinite.) The notation $C \bullet X$ stands for the inner product of the symmetric matrices C and X:

$$C \bullet X = \sum_{i=1}^{n} \sum_{j=1}^{n} C_{ij} X_{ij}.$$

(Some authors write the inner product as $C \bullet X = \operatorname{Tr} CX$ where $\operatorname{Tr} CX$ is the trace of the matrix CX.) In other words, $C \bullet X$ is a linear function of the elements X_{ij}. In (1.1.1) we minimize a linear function of the matrix variable X, subject to m linear equality constraints $A_i \bullet X = b_i$, and the positive semidefinitess constraint $X \succeq 0$.

We refer to problem (1.1.1) as a semidefinite program or SDP. We can think of an SDP as a generalization of the standard form linear programming problem

$$\begin{array}{lllll} \text{minimize} & c^T x & & & \\ \text{subject to} & a_i^T x & = & b_i & \text{for all } i = 1, \ldots, m \\ & x & \geq & 0 & \end{array} \tag{1.1.2}$$

in which the elementwise nonnegativity constraint $x \geq 0$ is replaced by a generalized inequality with respect to the cone of positive semidefinite matrices. It is therefore not surprising to find many parallels between semidefinite and linear programming. Semidefinite programming, however, is much more general than linear programming, and a wide variety of nonlinear convex optimization problems can be formulated and efficiently solved as SDPs.

Though SDPs have been studied (under various names) as far back as the 1940s, the interest in this problem has grown tremendously in the last decade. This research activity was motivated by the discovery of new applications in several areas, combined with the development of efficient new algorithms. This handbook presents a comprehensive survey of this decade of research.

1.2 OVERVIEW OF THE HANDBOOK

The chapters are organized in three parts: Theory, Algorithms, and Applications and Extensions.

Theory

The theory underlying semidefinite programming can be studied from different viewpoints. First of all, SDPs are convex optimization problems, so they can be studied using the general techniques developed in convex analysis. Semidefinite programming in turn offers a beautiful domain of application of convex analysis, and it is fair to say that the recent popularity of SDP has also brought a resurgence in the interest in convex analysis. Chapter 2 (Florian Jarre) reviews the basic results from convex analysis, applied to convex functions of symmetric matrices and semidefinite programming. The chapter covers the basic definitions, a review of concepts from linear algebra, special results from convex analysis for symmetric matrices, and differential formulae for functions of eigenvalues.

Semidefinite programming can also be studied as an extension of linear programming, i.e., as a linear optimization problem over a nonpolyhedral convex cone (the cone of positive semidefinite matrices). Chapter 3 (Gabor Pataki) studies the geometry of SDP as a 'cone-LP', including a characterization of faces and of the set of optimal solutions, and questions about complementary slackness and strong duality.

The next two chapters cover semidefinite programming duality. Chapter 4 is divided into two parts. The first part 4.1 (Alexander Shapiro) covers the basic duality theory for convex and nonconvex nonlinear cone constrained optimization problems. It also discusses constraint qualifications including Slater type conditions, first order optimality conditions for convex and nonconvex problems, second order optimality conditions with quadratic growth conditions and their equivalence under constraint qualification assumptions. In the second part 4.2 (Katya Scheinberg), the sensitivity analysis of general nonlinear SDPs (presented in the first part) is specialized to the linear SDP case. The quadratic growth condition is shown to be equivalent to nondegeneracy under strict complementarity. A comparison is made with sensitivity analysis in linear programming, and properties of the optimal objective value as a function of a parameter are studied.

In Chapter 5, de Klerk, Terlaky, and Roos discuss self-dual embeddings for SDP. This technique has been highly successful for linear programs and is now extended to SDP.

The last two chapters of part I discuss the related subjects of sensitivity and robustness. Chapter 6 by Ben-Tal, El Ghaoui, and Nemirovski, deals with robustness in seimidefinite programming. In many practical applications, the "data" of the problem are subject to possibly large uncertainties. The authors develop the notion of 'robust counterpart of an SDP' and 'robust solution' to address the problem of uncertainty in the data.

Finally, Chapter 7 by Luo and Sturm considers the solution set of linear conic systems and finds bounds for the distance of a point to the solution set in terms of the constraint violation. This work generalizes many known error bounds.

Algorithms

In the second part we focus on algorithms for semidefinite programming. The first three chapters discuss interior-point methods. As we mentioned, it was the development of efficient polynomial-time algorithms for SDP at the end of the 1980s that spurred the increased interest in the field. In the seminal work by Nesterov-Nemirovski [583], it was shown that we can solve a wide family of convex optimization problems in polynomial time using a *self-concordant barrier function.* They show that $\log\det X^{-1}$ is a self-concordant barrier function for SDP. Thus SDP can be solved in polynomial time using a sequence of barrier subproblems. Independently, Alizadeh [17] presented an elegant approach to extend potential reduction methods from LP to SDP. Simultaneously, strong numerical results for SDP and the special case of the min-max eigenvalue problem appeared in e.g. [339] and [364]. This started a flood of results in SDP from researchers in LP.

The division of this part into three chapters is somewhat arbitrary, since all interior-point methods are closely related, and some overlap between the chapters is inevitable.

The first chapter (chapter 8) by Alizadeh and Schmieta, emphasizes the analogy with linear programming, and discusses algorithms that are 'word by word' translations of interior-point algorithms for linear programming. The treatment in this chapter is more general than SDP, and applies to cone-LP problems over symmetric cones (such as the second-order cone).

Chapter 9 (Levent Tuncel) focuses on the class of symmetric primal-dual potential reduction algorithms for SDPs, and formulates different extensions of the primal-dual search directions used for linear programming.

Chapter 10 (Monteiro and Todd) discusses primal-dual path-following methods. It describes various different families of search directions, and the main convergence results for path-following algorithms based on these families of search directions.

The last chapter of Part II (Chapter 11) presents bundle methods, a class of algorithms from nondifferentiable convex optimization. These methods are quite different from the interior-point methods discussed in the other chapters, and particularly interesting for very large structured SDPs.

Applications and Extensions

Semidefinite programming plays an important role in combinatorial optimization where it is used to solve nonlinear convex relaxations of NP-hard problems. The interest in SDP techniques for combinatorial optimization was motivated by remarkable theoretical results that characterize the quality of the bounds obtained via semidefinite programming. The first two chapters in Part III cover those applications. Chapter 12 surveys applications in combinatorial optimization, while chapter 13 discusses the use of semidefinite programming to derive bounds for nonconvex quadratic optimization problems.

We also find many applications of semidefinite programming in engineering. Chapters 14 and 15 provide introductions to the two most important areas of engineering applications of SDP. In control theory (Chapter 14) SDP has emerged as an important practical tool for the analysis of nonlinear and time-varying systems, and for controller synthesis. Efficient algorithms for semidefinite programming make it possible to numerically solve a wide variety of control problems for which no analytical solution is known. Semidefinite programming is therefore of great interest to computer-aided control system design. Another important area of engineering applications is structural optimization (Chapter 15).

In Chapter 16, Bertsimas and Sethuraman show that semidefinite programming can be used to derive bounds for stochastic optimization problems, and apply the results to queuing networks, finance, and discrete optimization.

Chapter 17 discusses applications of SDP, and more general optimization problems with linear matrix inequality constraints, to the optimal design of experiments. The first part of the chapter states the general problem of experimental design and describes some specialized algorithms. The second part of the chapter (by Jon Lee) describes in more detail the connections between experimental design and semidefinite programming.

The final three chapters discuss application of SDP (and extensions) in linear algebra and nonlinear optimization.

Chapter 18 discusses the two types of matrix completion problems: the *positive semidefinite matrix completion problem* and the *Euclidean distance matrix completion problem.*

Chapter 19 addresses the solution of nonlinear optimization problems with (possibly nonconvex) matrix inequality constraints, and presents a generalization of a primal interior-point method for SDP to this more general problem.

Finally, Chapter 20 presents a study of semidefinite relaxations of quadratic constrained quadratic programming problems applied to sequential quadratic programming algorithms in constrained nonlinear programming.

1.3 NOTATION

Why is "abbreviation" such a long word? *Anonymous*

1.3.1 General comments

As in linear programming, there are various alternative (but equivalent) formulations of the 'standard form' SDP (1.1.1), several of which are used in this handbook. A very common format is the 'inequality form'

$$\begin{array}{ll} \text{minimize} & b^T y \\ \text{subject to} & \sum_{i=1}^m y_i A_i \preceq C, \end{array} \tag{1.3.3}$$

where the variable is $y \in I\!R^n$, and the problem data are the matrices $A_i, C \in \mathcal{S}^n$ and the vector b. The constraint

$$\sum_{i=1}^m y_i A_i \preceq C,$$

is called a *linear matrix inequality.*

Problem (1.3.3) is readily transformed to a problem of the form (1.1.1) and vice-versa. We will give the details for the transformation of (1.3.3) into an SDP of the form (1.1.1).

We can assume without loss of generality that the matrces A_i, $i = 1, \ldots, m$, are linearly independent. Introducing a slack variable S we can express (1.3.3) as

$$\begin{array}{lrcrcl} \text{minimize} & b^T y & & & & \\ \text{subject to} & \sum_{i=1}^m A_i y_i & + & S & = & C \\ & & & S & \succeq & 0, \end{array} \tag{1.3.4}$$

with variables y and S. Since by assumption the matrices A_i are linearly independent, the set

$$\mathcal{V} = \left\{ S = C - \sum_{i=1}^m y_i A_i \;\middle|\; y \in I\!R^m \right\}$$

is an affine subset of $\mathcal{S}^n$ with dimension m. Therefore there exist symmetric matrices G_i, $i = 1, \ldots, n(n+1)/2 - m$, and a vector d such that

$$\mathcal{V} = \{ S \in \mathcal{S}^n \mid G_i \bullet S = d_i,\ i = 1, \ldots, n(n+1)/2 - m \}.$$

We can also compute a matrix G_0 that satisfies

$$G_0 \bullet A_i = -b_i, \quad i = 1, \ldots, m,$$

by solving an underdetermined set of m equations in $n(n+1)/2$ variables, so that for $S = C - \sum_i A_i y_i$ we have

$$b^T y = G_0 \bullet S - G_0 \bullet C.$$

With those definitions we can express (1.3.4) as

$$\begin{array}{ll} \text{minimize} & G_0 \bullet S \\ \text{subject to} & G_i \bullet S = d_i, \quad i = 1, \ldots, n(n+1)/2 - m \\ & S \succeq 0, \end{array}$$

which is an SDP in the form (1.1.1).

As in linear programming, we can also consider SDPs with inequality and equality constraints and mixtures of restricted and free variables.

It is also common to have more than one matrix variable in (1.1.1). More precisely we can consider problems of the form

$$\begin{array}{ll} \text{minimize} & \sum_{j=1}^{L} C_j \bullet X_j \\ \text{subject to} & \sum_{j=1}^{L} A_{i,j} \bullet X_j = b_i, \quad i = 1, \ldots, m \\ & X_j \succeq 0, \quad j = 1, \ldots, L, \end{array} \tag{1.3.5}$$

where the variables are $X_j \in \mathcal{S}^{n_j}$. This problem, however, is not more general than (1.1.1), since it reduces to (1.1.1) if we define block-diagonal matrices

$$X = \text{diag}(X_1, \ldots, X_L), \quad C = \text{diag}(C_1, \ldots, C_L), \quad A_i = \text{diag}(A_{i,1}, \ldots, A_{i,L}),$$

As an extreme case, we can have diagonal blocks of dimension one in (1.3.5), in which case the problem reduces to an LP.

In a similar way, we can consider SDPs in inequality form (1.3.3), with multiple linear matrix inequality constraints

$$\begin{array}{ll} \text{minimize} & b^T y \\ \text{subject to} & \sum_{i=1}^{m} y_i A_{i,j} \preceq C_j, \quad j = 1, \ldots, L. \end{array}$$

Again this is equivalent to (1.3.3) if we define block-diagonal matrices

$$C = \text{diag}(C_1, \ldots, C_L), \quad A_i = \text{diag}(A_{i,1}, \ldots, A_{i,L}).$$

Finally, we mention that some authors prefer the following compact formulation of the SDP (1.1.1):

$$\begin{array}{llll} \text{minimize} & C \bullet X & & \\ \text{subject to} & \mathcal{A}X & = & b \\ & X & \succeq & 0 \end{array}$$

where $\mathcal{A} : \mathcal{S}^n \to I\!R^m$ is a linear operator.

1.3.2 Overview

We conclude this introductory chapter by listing the most important notational conventions used in this book.

1. **Operations**

   - $\succeq$ the Löwner partial order on positive semidefinite matrices, $A \succeq B$ if $B - A$ is positive semidefinite.
   - $\succ$ the order on positive definite matrices, $A \succ B$ if $B - A$ is positive definite.
   - $\otimes$ the Kronecker product.
   - $\bullet$ the trace operator, i.e., $A \bullet B = \text{Trace}\,(A^T B)$. Or

     $$\langle X, Y\rangle = \sum_{i=1}^{m}\sum_{j=1}^{n} X_{ij}Y_{ij}.$$

     the standard inner product on two $(m \times n)$ real matrices.

2. **Sets**

   - $\mathbb{R}^n$ the n dimensional euclidean space.
   - $\mathbb{M}^{m\times n}$ or $\mathbb{R}^{m\times n}$ the space of all $m \times n$ matrices.
   - $\mathcal{S}^m$ the space of all $m \times m$ symmetric matrices.
   - $\mathcal{P} = \{X : X \succeq 0\}$ the convex cone of all symmetric and positive semidefinite matrices.

3. **Vectors**

   - x a vector.
   - x^T the transpose of the vector x.
   - x_i the ith component of the vector x.
   - $x^{(k)}$ the kth vector in a sequence of vectors.
   - $x^T y$ the inner-product of two vectors.
   - e_i the ith coordinate vector of R^n.
   - $1_n = e \in R^n$ the vector of all ones.
   - $I_n = \text{Diag}\,(1_n)$ is a unit matrix.
   - 0_n is the zero vector in R^n.

4. **Matrices**

   - A a matrix.
   - A^T the transpose of the matrix A.

- A^k the kth matrix in a sequence of matrices.
- $(A^k)^T$ the transpose of A^k.
- $((A^k)^T)_{ij}$ the ijth entry of the transpose of A^k.

5. **Operators on matrices**

   (a) $\mathcal{A} : \mathcal{S}^n \longrightarrow I\!R^m$.

   (b) $\mathcal{A}^*$ the adjoint of the linear operator $\mathcal{A}$.

6. **Semidefinite programming problem**

   The matrices C, A_i are in $\mathcal{S}^n$.

$$\begin{array}{lrcll} \text{minimize} & C \bullet X & & & \\ \text{subject to} & A_i \bullet X & = & b_i & \text{for all } i = 1, \cdots, m \\ & X & \succeq & 0 & \end{array}$$

with its dual:

$$\begin{array}{lrcrcl} \text{maximize} & \sum_{i=1}^m b_i y_i & & & & \\ \text{subject to} & \sum_{i=1}^m A_i y_i & + & S & = & C \\ & & & S & \succeq & 0 \end{array}$$

Alternatively, SDP can be formulated as:

$$\begin{array}{lrcl} \text{minimize} & C \bullet X & & \\ \text{subject to} & \mathcal{A}X & = & b \\ & X & \succeq & 0 \end{array}$$

with its dual:

$$\begin{array}{lrcl} \text{maximize} & b^T y & & \\ \text{subject to} & \mathcal{A}^* y & \preceq & C \end{array}$$

I THEORY

The most important fundamental laws and facts of physical science have all been discovered, and these are now so firmly established that the possibility of their ever being supplemented in consequence of new discoveries is exceedingly remote.

Michelson, Albert, Abraham (1852-1931) b. Germany (In 1903)

(With reference to a correspondent)
The young specialist in English Lit, ... lectured me severely on the fact that in every century people have thought they understood the Universe at last, and in every century they were proved to be wrong. It follows that the one thing we can say about our modern "knowledge" is that it is wrong.

... My answer to him was, "... when people thought the Earth was flat, they were wrong. When people thought the Earth was spherical they were wrong. But if you think that thinking the Earth is spherical is just as wrong as thinking the Earth is flat, then your view is wronger than both of them put together."

In: Isaac Asimov,The Relativity of Wrong, Kensington Books, New York, 1996, p 226.

Asimov, Isaac
(1920-1992) b. Petrovichi, Russia.

2 CONVEX ANALYSIS ON SYMMETRIC MATRICES

Florian Jarre

It is more important to have beauty in one's equations than to have them fit experiment... If one is working from the point of view of getting beauty in one's equations, and if one has really a sound insight, one is on a sure line of progress. If there is not complete agreement between the results of one's work and experiment, one should not allow oneself to be too discouraged, because the discrepancy may well be due to minor features that are not properly taken into account and that will get cleared up with further development of the theory.

Paul Adrien Maurice Dirac (1902 - 1984) (taken from Scientific American, May 1963)

2.1 INTRODUCTION

This introductory chapter serves as a reference to keep the handbook self-contained in particular for non-experts. We state basic definitions, review concepts from linear algebra that relate to semi-definite optimization, and summarize special results from convex analysis for symmetric matrices. It is our intention to state each result in a simple language, and therefore we limit our attention to the case of real symmetric matrices. For each result references are given where a discussion can be found, and in many cases also interesting counterparts for complex or nonsymmetric matrices.

Given a basis of $\mathbb{R}^n$, it is well-known that any linear mapping of $\mathbb{R}^n$ to itself can uniquely be represented by a real $n \times n$-matrix. Since the basis will always be clear from the context we do not distinguish between the mapping and the associated matrix in this chapter. The concatenation of two linear mappings is again a linear mapping, and thus, the vector space $\mathbb{R}^{n\times n}$ of real $n \times n$-matrices is naturally equipped with a bilinear operation $(A, B) \mapsto AB$ given by the concatenation of two linear mappings A and B, and resulting in the usual *matrix product* $AB = C$ with $C_{ik} = \sum_{j=1}^{n} A_{ij}B_{jk}$. The space $\mathbb{R}^{n\times n}$ is a Euclidean vector space, equipped with the *inner product* of two matrices $A, B \in \mathbb{R}^{n\times n}$

$$A \bullet B = \text{Trace}\, A^T B = \sum_{i=1}^{n} A_{ij}B_{ij},$$

where A^T is the transpose of a matrix A, and the trace of a matrix is the sum of all its diagonal elements. Because of the Euclidean structure, all results known from convex analysis in some space $\mathbb{R}^p$ also hold for the space of real (or real symmetric) $n \times n$-matrices. For a detailed introduction to convex analysis we refer to Stoer and Witzgall [755], Hiriart-Urruty and Lemarechal [345], or to Rockafellar [679]. The additional structure of $\mathbb{R}^{n\times n}$ induced by the matrix product provides the basis for further results not covered by the general theory of [755, 679], and is thus the "origin" of the present handbook. For special results on convex analysis with symmetric matrices see also the recent papers by Lewis [479] and Lewis and Overton [483].

Well-written standard references to matrix theory include Golub and Van Loan [292] or Horn and Johnson [352]. Important relationships governing the eigenvalues of a matrix are further treated in Stewart and Sun [752], Wilkinson [829], Marshall and Olkin [523], Marcus and Minc [516] or Kato [406], the latter book presents an advanced in depth treatment.

2.2 SYMMETRIC MATRICES

Symmetric matrices are a linear subspace of $\mathbb{R}^{n\times n}$ denoted by $\mathcal{S}^n$. In addition to the characterization $A = A^T$ of symmetric matrices, the spectral theorem [352, Thm 4.1.5] allows a geometrical characterization:

Theorem 2.2.1 (Spectral theorem for symmetric matrices) *The real $n\times n$-matrix A is symmetric if and only if there exists an orthonormal basis with respect to which A is real and diagonal, i.e. if and only if there exists a matrix $U \in \mathbb{R}^{n\times n}$ such that $U^TU = I$ is the identity and $U^TAU = \Lambda$ where Λ is a diagonal matrix.*

The columns u_i of U are the eigenvectors of A, satisfying

$$Au_i = \lambda_i u_i,$$

where λ_i is the i-th diagonal entry of Λ. When λ is a vector we denote the diagonal matrix Λ with entries λ_i by

$$\text{Diag}\,(\lambda),$$

and for a matrix A we write $d = \text{diag}(A)$ for the vector with the components $d_i = A_{ii}$ given by the diagonal elements of A.

2.2.1 *Operations on symmetric matrices*

The following operations on symmetric matrices will be discussed in this and subsequent chapters.

Sum of two symmetric matrices: As a consequence of the spectral theorem, if A and B are linear mappings that are real and diagonal with respect to two possibly different orthogonal bases, i.e. $A = U_1^T \Lambda_1 U_1$ and $B = U_2^T \Lambda_2 U_2$, then also the sum $A+B$ is real and diagonal with respect to a third orthogonal basis, $A + B = U_3^T \Lambda_3 U_3$. The dependence of U_3 and Λ_3 on $U_1, U_2, \Lambda_1, \Lambda_2$ is rather complicated in general, and may lead to surprising results, see e.g. relation (2.3.10) below.

Product of two symmetric matrices: Unlike the sum, the matrix product is not closed on the space of symmetric matrices; if A and B are real symmetric of the same dimension, then AB is symmetric if and only if $AB = BA$ if and only if there is an orthogonal matrix U that simultaneously diagonalizes A and B, i.e. $A = U^T \Lambda_1 U$ and $B = U^T \Lambda_2 U$, see [352, Thm 4.1.6].

Symmetric product of two symmetric matrices: In later chapters of this book, pairs of symmetric matrices X and Z satisfying $XZ \approx \mu I$ will be of special importance. As long as $XZ = \mu I$ is satisfied exactly, the matrices X and Z have the same eigenvectors, and hence $XZ = ZX$ commute. One of the central questions of this book will be how to linearize the equation "$(X + \Delta X)(Z + \Delta Z) = \mu I$" for $X, Z \in \mathcal{S}^n$ when X and Z satisfy $XZ \approx \mu I$ but not $XZ = ZX$. An approach proposed by [23] is to replace the bilinear mapping $(X, Z) \mapsto XZ$ by the symmetric bilinear mapping

$$(X, Z) \mapsto X * Z := \tfrac{1}{2}(XZ + ZX).$$

Hadamard product of two symmetric matrices: Another operation of two $m \times n$-matrices A, B is the "Hadamard product" defined by $A \circ B = C$ with $C_{ij} = A_{ij} B_{ij}$. Clearly, $A \circ B = B \circ A$, and if A and B are square and symmetric, then so is $A \circ B$. The Hadamard product arises naturally with trigonometric moments or integral operators, see, e.g. [352], but also in interior-point methods as componentwise product $x \circ s$ of two vectors $x, s \in \mathbb{R}^n$.

Kronecker product of two matrices: Finally, an operation that will be used in later chapters of this book is the "Kronecker product" or tensor product of an $m \times n$-matrix A and a $p \times q$-matrix B. It is defined as

$$A \otimes B = \begin{pmatrix} A_{11}B & \dots & A_{1n}B \\ \vdots & \ddots & \vdots \\ A_{m1}B & \dots & A_{mn}B \end{pmatrix}.$$

The Kronecker product is used to represent linear operators $\mathbb{R}^{q \times n} \to \mathbb{R}^{p \times m}$ defined by the map:

$$X \mapsto BXA^T.$$

To explain this representation, the mapping $\mathbf{nvec} : \mathbb{R}^{q \times n} \to \mathbb{R}^{qn}$ is used that stacks the columns of a matrix in a vector. It is straightforward to verify that

$$(A \otimes B)\, \mathbf{nvec}(X) = \mathbf{nvec}(BXA^T). \tag{2.2.1}$$

Likewise, one can verify that

$$(A \otimes B)(X \otimes Y) = AX \otimes BY$$

for $n \times n$-matrices A, X, and $p \times p$-matrices B, Y. Hence, for invertible A and B,

$$(A \otimes B)^{-1} = A^{-1} \otimes B^{-1}.$$

Again, if A and B are square and symmetric, then so is $A \otimes B$. A good survey on the Kronecker product can be found in [352, Chap 4].

Symmetric Kronecker product: In analogy to the symmetric product of two matrices, one may also consider a symmetric Kronecker product of two matrices. When $A, B \in \mathbb{R}^{n\times n}$ we use the "symmetrization" of (2.2.1)

$$(A \odot B)\text{svec}\,(X) = \text{svec}\,(\tfrac{1}{2}(BXA^T + AXB^T))$$

to define a symmetric Kronecker product. Here, svec maps $\mathcal{S}^n$ to $\mathbb{R}^{n(n+1)/2}$ by

$$\text{svec}\,(X) = \left[X_{11}, \sqrt{2}X_{12}, \ldots, \sqrt{2}X_{1n}, X_{22}, \ldots, \sqrt{2}X_{2n}, \ \ldots \ , X_{nn}\right]^T.$$

The definition of svec is chosen to guarantee the identity

$$A \bullet B = \text{svec}\,(A)^T \text{svec}\,(B).$$

Further, the identities $A \odot B = B \odot A$ and $(A \odot A)^{-1} = A^{-1} \odot A^{-1}$ hold true, but in general, $(A \odot B)^{-1} \neq A^{-1} \odot B^{-1}$, see e.g. [23].

2.3 ANALYSIS WITH SYMMETRIC MATRICES

2.3.1 *Continuity of eigenvalues*

As just seen, the spectral theorem (Theorem 2.2.1) allows to describe a symmetric matrix by its eigenvalues and an associated orthogonal basis. Understanding the behavior of the eigenvalues of a symmetric matrix with respect to perturbations of the matrix entries is the topic in this subsection.

The norm associated with the scalar product "$\bullet$" is the Frobenius norm $||A||_F = (A \bullet A)^{1/2}$. From the spectral theorem (Theorem 2.2.1) follows that the Frobenius norm of a symmetric matrix A is given by $||A||_F = (\sum_{k=1}^n \lambda_k^2)^{1/2}$, and the 2-norm $||A||_2 = \max_{||x||_2=1} ||Ax||_2$ is given by $||A||_2 = \max_{1\leq k\leq n}\{|\lambda_k|\}$. Both, the Frobenius norm and the 2-norm are *unitarily invariant* in the sense that

$$||A|| = ||UAV|| \qquad (2.3.2)$$

for any unitary matrices U and V. (Since we restrict our attention to real matrices, the name "orthogonally invariant" might be more suitable, the notation "unitarily invariant" is more common, however.) The eigenvalues of a symmetric matrix are a Lipschitz-continuous function of the matrix entries, and the

Lipschitz-constant is one: If $\lambda(A)$ is the vector of all eigenvalues of A arranged in non-increasing order then the eigenvalues of A and $A+B$ satisfy

$$\|\lambda(A+B) - \lambda(A)\|_\infty \le \|\lambda(B)\|_\infty = \|B\|_2 \tag{2.3.3}$$

$$\|\lambda(A+B) - \lambda(A)\|_2 \le \|\lambda(B)\|_2 = \|B\|_F. \tag{2.3.4}$$

The bounds (2.3.3) and (2.3.4) are known as Mirsky's theorem and Hoffman-Wielandt theorem, see e.g. [752, Cor 4.12].

2.3.2 *Smoothness of eigenvalues*

The determinant (see e.g. [292] for a definition) of an $n \times n$-matrix A is zero if and only if A is singular, i.e. if there exists a nonzero vector v such that $Av = 0$. The eigenvalues of the matrix A are thus characterized by the roots of the characteristic polynomial of A, i.e.

$$p(\lambda, A) := \det(A - \lambda I) = 0.$$

Clearly, by the implicit function theorem, as long as λ is a simple solution (of multiplicity one) of $p(\,.\,, A) = 0$, the root λ depends locally analytically on the entries of A. In this case, also the eigenvectors depend smoothly on A. In the case of roots with multiplicity more than one this statement is no longer true. For symmetric matrices the following weaker version is true:

Theorem 2.3.1 (Rellich) *See [406, Chap II, Thm 6.8]. Let two real numbers $a < b$ be given. When $A(\varepsilon)$ is a symmetric matrix and continuously differentiable for all $\varepsilon \in (a, b)$, then there exist n continuously differentiable functions $\lambda_i(\varepsilon)$ $(1 \le i \le n)$ that represent the repeated eigenvalues of $A(\varepsilon)$.*

Hence, if the eigenvalues λ_i are suitably ordered, they always have directional derivatives. Consider for example the matrix

$$A(x, y) := \begin{pmatrix} 1 + x - y & y \\ y & 1 - x - y \end{pmatrix}$$

with eigenvalues $1-y+\sqrt{x^2+y^2}$ and $1-y-\sqrt{x^2+y^2}$. If we take $\varepsilon = x = y$, the eigenvalues are $\lambda_1 = 1-\varepsilon+\sqrt{2}|\varepsilon|$ and $\lambda_2 = 1-\varepsilon-\sqrt{2}|\varepsilon|$. When ordered as above, i.e. $\lambda_1 \ge \lambda_2$, the eigenvalues appear to be nondifferentiable at $\varepsilon = 0$ in contrast to the statement of the theorem. By changing signs at $\varepsilon = 0$ however, these eigenvalues can be written as $\tilde{\lambda}_1(\varepsilon) = 1 + (\sqrt{2} - 1)\varepsilon$ and $\tilde{\lambda}_2(\varepsilon) = 1 - (\sqrt{2} - 1)\varepsilon$ which are both smooth in ε—but no longer satisfy $\tilde{\lambda}_1(\varepsilon) \ge \tilde{\lambda}_2(\varepsilon)$ for all ε.

A slightly more complicated example shows that the eigen*vectors* might not even be continuous functions of the single parameter ε. Moreover, no matter how the eigenvalues $\lambda_i(x, y)$ of $A(x, y)$ are arranged, $\lambda_i(x, y)$ is *not* totally differentiable[1] with respect to the two parameters x and y.

[1] I.e. for $i \in \{1, 2\}$ there does not exist a gradient vector $\nabla\lambda_i$ with

$$\lambda_i(x + \Delta x, y + \Delta y) = \lambda_i(x, y) + \nabla\lambda_i^T \begin{pmatrix} \Delta x \\ \Delta y \end{pmatrix} + o\left(\left\| \begin{pmatrix} \Delta x \\ \Delta y \end{pmatrix} \right\|\right).$$

Even the directional differentiability of the eigenvalues may seem surprising at first sight. The roots (eigenvalues) are *not* differentiable—not even locally Lipschitz continuous—functions of the coefficients α_i of the characteristic polynomial

$$p(\lambda, A) = \alpha_0 + \alpha_1\lambda + \ldots + \alpha_{n-1}\lambda^{n-1} + \lambda^n.$$

However, since symmetric perturbations of the entries of $A = A(0)$ to $A(\varepsilon)$ result in very special perturbations $\Delta\alpha_i(\varepsilon)$ of the coefficients α_i, the roots are nevertheless differentiable functions of ε and do not violate Rellich's theorem.

Rellich's theorem relies on the 'right' order of the eigenvalues. If the order is suppressed, we obtain a special class of functions $F : \mathcal{S}^n \to \mathbb{R}$ that inherits the smoothness of the eigenvalues in a very nice way.

Definition 2.3.1 *A permutation matrix is a matrix* Π *with entries* 0 *and* 1 *and* $\Pi^T\Pi = I$. *We say a set* $\mathcal{C} \subset \mathbb{R}^n$ *is permutation-invariant if* $x \in \mathcal{C}$ *implies* $\Pi x \in \mathcal{C}$ *for all permutation matrices* Π. *In two dimensions, for example, permutation-invariant sets are all sets that are symmetric about the line* $x_1 = x_2$. *For* $n > 2$ *the sets are symmetric about all hyperplanes* $x_i = x_j$ *for* $1 \le i < j \le n$. *A function* $f : \mathcal{C} \to \mathbb{R}$ *is* permutation-invariant *if* $\mathcal{C}$ *is permutation-invariant and* $f(x) = f(\Pi x)$ *for all* x *and all permutation matrices* Π.

We say a set $\mathcal{C} \subset \mathbb{R}^n$ *is convex if for any* $x, y \in \mathcal{C}$ *and any* $\lambda \in [0,1]$ *also* $\lambda x + (1-\lambda)y \in \mathcal{C}$. *The function* f *is* convex *if* $\mathcal{C}$ *is convex, and*

$$f(\lambda x + (1-\lambda)y) \le \lambda f(x) + (1-\lambda)f(y) \qquad (2.3.5)$$

for all $x, y \in \mathcal{C}$ *and any* $\lambda \in [0,1]$. *The function* f *is* absolute *if* $f(x)$ *is invariant under sign changes of the components* x_i. *The function* f *is a* symmetric gauge function *if* f *is a permutation-invariant absolute norm on* $\mathbb{R}^n$.

For a symmetric $n \times n$-matrix X we denote by $\lambda(X)$ the vector of eigenvalues of X arranged in non-increasing order. This order is chosen to allow for a unique definition of λ. As just seen, it does not preserve smoothness of λ.

Theorem 2.3.2 (Lewis) *Let* f *be a permutation-invariant function. Then* $F : \mathcal{S}^n \to \mathbb{R}$ *defined by*

$$F(X) = f(\lambda(X)) \qquad (2.3.6)$$

is totally (Fréchet-) differentiable at X *if and only if* f *is totally differentiable at* $\lambda(X)$. *The derivative of* F *with respect to the scalar product "* $\bullet$ *" is then given by*

$$DF(X) = U^T \mathrm{Diag}\,(f'(\lambda(X)))U,$$

where $f'(y) = Df(y)$ *is the derivative (row vector) of* f *at point* y, *and* U *is any orthogonal matrix that diagonalizes* X, *i.e.* $X = U^T \mathrm{Diag}\,(\lambda(X))U$.

For a proof and a discussion of related results see for example [479]. In addition to differentiability, also convexity is inherited by F, see Section 2.3.8 below. First, we review some standard results from linear algebra that are needed throughout this book.

2.3.3 *The Courant-Fischer-Theorem and its consequences*

Due to the orthogonality of the eigenvectors, symmetric matrices allow special min-max characterizations of the eigenvalues that are important also in the context of semidefinite programming. We begin with a simple and well-known characterization of the extremal eigenvalues of a symmetric $n \times n$-matrix A. As before we denote the eigenvalues of A by $\lambda_1(A) \geq \ldots \geq \lambda_n(A)$.

Theorem 2.3.3 (Raleigh-Ritz) *Let A be a symmetric $n \times n$-matrix. Then*

$$\lambda_1(A) = \max_{x \neq 0} \frac{x^T Ax}{x^T x} \qquad \text{and} \qquad \lambda_n(A) = \min_{x \neq 0} \frac{x^T Ax}{x^T x}.$$

For a proof and modifications of this and further results in this subsection, see e.g. [352, Chap 4.2] and [292, Chap 8.1]. The Theorem by Raleigh-Ritz is a special case of the Courant-Fischer Theorem.

Theorem 2.3.4 (Courant-Fischer) *Let A be a symmetric $n \times n$-matrix and k with $1 \leq k \leq n$ be fixed. Then*

$$\lambda_k(A) = \min_{u_1, u_2, \ldots, u_{k-1} \in \mathbb{R}^n} \max_{\substack{x \in \mathbb{R}^n,\ x \neq 0 \\ x \perp u_1, u_2, \ldots, u_{k-1}}} \frac{x^T Ax}{x^T x}$$

and

$$\lambda_k(A) = \max_{u_1, u_2, \ldots, u_{n-k} \in \mathbb{R}^n} \min_{\substack{x \in \mathbb{R}^n,\ x \neq 0 \\ x \perp u_1, u_2, \ldots, u_{n-k}}} \frac{x^T Ax}{x^T x}$$

This theorem has many consequences some of which are listed below.

Theorem 2.3.5 (Weyl) *Let A, and B be symmetric $n \times n$-matrices and k with $1 \leq k \leq n$ be fixed. Then*

$$\lambda_k(A) + \lambda_1(B) \leq \lambda_k(A + B) \leq \lambda_k(A) + \lambda_n(B).$$

The Weyl theorem sharpens the continuity bound in (2.3.3) above.

Theorem 2.3.6 (Interlace property) *Let A be a symmetric $n \times n$-matrix and $a \in \mathbb{R}$, $b \in \mathbb{R}^n$ be fixed. Let*

$$\hat{A} = \begin{bmatrix} A & b \\ b^T & a \end{bmatrix}. \tag{2.3.7}$$

Denote the eigenvalues of A by λ_i and those of $\hat{A}$ by $\hat{\lambda}_i$, then

$$\hat{\lambda}_1 \geq \lambda_1 \geq \hat{\lambda}_2 \geq \lambda_2 \geq \ldots \geq \lambda_n \geq \hat{\lambda}_{n+1}. \tag{2.3.8}$$

Conversely, if some numbers λ_i *and* $\hat{\lambda}_i$ *are given that satisfy the interlace property (2.3.8), then there exist a vector* $b \in \mathbb{R}^n$ *and a number* $a \in \mathbb{R}$ *such that* $\hat{A}$ *in (2.3.7) with* $A = \text{Diag}(\lambda)$ *has exactly the eigenvalues* $\lambda_i(\hat{A}) = \hat{\lambda}_i$.

The *inertia* of a symmetric matrix A is the triplet (m, z, p) where m, z, p are the number of negative, zero, and positive eigenvalues of A.

Theorem 2.3.7 (Sylvester law of inertia) *If* $X \in \mathbb{R}^{n\times n}$ *is nonsingular and* $A \in \mathcal{S}^n$ *is symmetric, then the inertia of* A *and of* X^TAX *coincide.*

Theorem 2.3.8 (Sum of k **largest eigenvalues)** *Let* A *be a symmetric* $n\times n$*-matrix and* k *with* $1 \le k \le n$ *be fixed. Then*

$$\lambda_1 + \ldots + \lambda_k(A) = \max_{U\in\mathbb{R}^{n\times k}:\ U^TU=I} \text{Trace}\,(U^TAU)$$

and

$$\lambda_{n-k+1} + \ldots + \lambda_n(A) = \min_{U\in\mathbb{R}^{n\times k}:\ U^TU=I} \text{Trace}\,(U^TAU)$$

Here, I *is the* $k \times k$ *identity matrix.*

This relation is important in establishing a duality theory for certain eigenvalue optimization problems.

More generally, the sum of certain k eigenvalues in any order (for example the second plus the fifth plus the tenth eigenvalue) can be stated by a min-max characterization somewhat more complicated than the above one. The resulting theorem is called Wielandt's theorem, and in fact, the Courant-Fischer theorem is a just special case of Wielandt's theorem, see e.g. [752, Thm 4.5]. The eigenvalue bounds (2.3.3) and (2.3.4) follow from this theorem as well.

2.3.4 Positive definite matrices

A real symmetric matrix A is positive definite (positive semidefinite), if all its eigenvalues are positive (nonnegative), or, equivalently, if $x^TAx > 0$ ($x^TAx \ge 0$) for all $x \in \mathbb{R}^n$ with $x \ne 0$. We then write $A \succ 0$ or $A \succeq 0$. Since $x^TAx > 0$ and $x^TBx > 0$ imply

$$x^T(\lambda A + (1-\lambda)B)x = \lambda x^TAx + (1-\lambda)x^TBx > 0$$

for $\lambda \in [0, 1]$ it follows that the set of positive definite symmetric $n\times n$-matrices is a convex cone denoted by $\mathcal{S}^n_{++}$ in the sequel. Likewise, the set of positive semi-definite symmetric $n \times n$-matrices is a convex cone denoted by $\mathcal{S}^n_+$. (Note that the sum of two nonsymmetric diagonalizable matrices A and B with positive eigenvalues may have negative eigenvalues. By Weyl's theorem (Theorem 2.3.5), this cannot happen when A and B are both symmetric.)

Since $\mathcal{S}^n_+$ is a convex cone it defines a partial order for $A, B \in \mathcal{S}^n$ by $A \succeq B$ whenever $A - B$ is positive semi-definite, $A - B \succeq 0$. The order is called the *Löwner partial order.* (It is transitive; and $A \succeq B$ and $B \succeq A$ imply $A = B$.)

We write $A \succ B$ when $A - B$ is positive definite. A few simple rules for the Löwner partial order are given next.

When $A \succeq B$ then $\lambda_i(A) \geq \lambda_i(B)$ for $1 \leq i \leq n$ where λ_i denotes the i-th largest eigenvalue. The converse is not true, i.e. when $\lambda_i(A) \geq \lambda_i(B)$ for all i it may be that $A \not\succeq B$.

Let A, B be symmetric $n \times n$-matrices. When

$$A \succeq B \quad \text{then} \quad S^T AS \succeq S^T BS$$

for all $n \times m$-Matrices S, and when

$$A \succ B \quad \text{then} \quad S^T AS \succ S^T BS$$

for all $n \times m$-Matrices S with rank m. The converse of both implications is true when S is an invertible $n \times n$-matrix, see also Theorem 2.3.7.

Let $A \succ 0$ and $B \succ 0$. Then

$$A \succeq B \quad \text{if and only if} \quad B^{-1} \succeq A^{-1}.$$

Theorem 2.3.9 (Schur complement) *Let*

$$M = \begin{pmatrix} A & B \\ B^T & C \end{pmatrix},$$

where A is a symmetric, positive definite $n \times n$-matrix, and C is a symmetric, positive definite $m \times m$-matrix. Then the following are equivalent:

a) M is positive semi-definite,

b) $C \succeq B^T A^{-1} B$,

c) $\rho(B^T A^{-1} B C^{-1}) \leq 1$. Here, $\rho(S)$ denotes the spectral radius *of a $p \times p$-matrix S, and is $\rho(S) = \max_{1 \leq i \leq p} |\lambda_i(S)|$, see e.g. [292, Chap 10.1.2].*

d) $(x^T Ax)(y^T Cy) \geq (x^T By)^2$ for all $x \in \mathbb{R}^n$ and $y \in \mathbb{R}^m$.

e) $x^T Ax + y^T Cy \geq 2(x^T By)$ for all $x \in \mathbb{R}^n$ and $y \in \mathbb{R}^m$.

The theorem is also true when "positive semi-definite" is replaced by "positive definite" and all inequalities are replaced by strict inequalities, (and $x \neq 0$, $y \neq 0$ for d) and e)). For a proof see e.g. [352, Thm 7.7.7].

The Schur complement is a very useful tool for writing certain constraints that arise for example in control theory as semi-definiteness constraints.

Theorem 2.3.10 (Hadamard products) *If A and B are symmetric positive (semi-) definite $n \times n$-matrices, then so is $A \circ B$, and if A, B are positive definite then*

$$\begin{array}{lcccc} & A^{-1} \circ B^{-1} & \succeq & (A \circ B)^{-1}, & \\ \text{and} & A^{-1} \circ A & \succeq & I & \succeq (A^{-1} \circ A)^{-1}. \end{array}$$

The first part of the theorem is obvious; if $A = \sum_i a_i a_i^T$, and $B = \sum_j b_j b_j^T$, then $A \circ B = \sum_{i,j}(a_i \circ b_j)(a_i \circ b_j)^T \succeq 0$. For the second part and further inequalities see e.g. [352, Chap 7.7]. Finally, as a corollary of Theorem 2.3.10 we obtain Fejer's theorem which provides an important characterization of positive definite matrices.

Theorem 2.3.11 (Fejer) *A symmetric matrix A is positive semi-definite if and only if $A \bullet B \geq 0$ for all positive semi-definite matrices B.*

This theorem relates to an important concept used in convex programming, namely the polar cone of a set S. Let E be a finite dimensional Euclidean space (typically, $E = \mathbb{R}^n$ or $E = \mathcal{S}^n$), and $S \subset E$. We then define the polar cone S^{P} to S by

$$S^{\mathrm{P}} := \{y \in E \mid \langle x, y\rangle \leq 0 \quad \text{for all} \quad x \in S\}.$$

Obviously, for any set S, the polar cone S^{P} is a closed convex cone. It is well-known, see e.g. [755, Thm 2.7.7], that $(S^{\mathrm{P}})^{\mathrm{P}} = S$ if and only if S is a closed convex cone. When $E = \mathcal{S}^n$ with the scalar product $\langle A, B\rangle = A \bullet B$, and S is the cone of positive semidefinite matrices $S = \mathcal{S}^n_+$, Theorem 2.3.11 implies that $(\mathcal{S}^n_+)^{\mathrm{P}} = -\mathcal{S}^n_+$. We say the cone $\mathcal{S}^n_+$ is *self-polar.*

Let us briefly consider the facial structure of $\mathcal{S}^n_+$. A convex subset S of a convex set C is called a *face* or *extreme subset* of C if $x \in S$ and $x = \lambda a + (1-\lambda) b$ with $\lambda \in (0, 1)$ and $a, b \in C$ implies $a, b \in S$. A subset S of a convex set C is called an *exposed subset* if there exists a linear form f that assumes precisely on S its minimum over C,

$$S = \{y \in C \mid f(y) \leq f(x) \;\; \forall\, x \in C\}.$$

Every exposed subset is an extreme subset, see e.g. [755, Thm 2.4.12]. For $C = \mathcal{S}^n$ the nonempty extreme subsets and the exposed subsets coincide, and are given by

$$\mathcal{F} = \left\{ \begin{pmatrix} W & 0 \\ 0 & 0 \end{pmatrix} \mid W \in \mathcal{S}^m_+ \right\} \qquad \text{(with } m \leq n\text{)}$$

and their rotations $U^T \mathcal{F} U$ for orthogonal U, see e.g. [482].

2.3.5 Monotonicity of the Löwner partial order

For 1×1-matrices $A \succ B$, the Löwner partial order reduces to the order of the real numbers $A > B$. As explained above, when A, B, C are symmetric $n \times n$-matrices with $n > 1$ transitivity of "$\succ$" is "inherited" from "$>$" in the sense that $A \succ B$ and $B \succ C$ imply $A \succ C$. Likewise $A \succ B$ is equivalent to $A + C \succ B + C$. Nevertheless we need to be careful when generalizing other rules for the real order "$>$" to the Löwner partial order. We start this subsection with some examples that show how several anticipated monotonicity properties of the Löwner partial order fail.

- Recall the operation "$*$" of Section 2.2.1, $A * B := \frac{1}{2}(AB + BA)$. Assume that A and B are positive definite $n \times n$-matrices. (This is the situation considered in [23].) Then, by [352, Thm 7.6.3] the product AB has n real positive eigenvalues. The same is true, of course, for BA which is in fact similar to AB. Nevertheless, as the example

$$A = \begin{pmatrix} 20 & 2 \\ 2 & 1 \end{pmatrix} \succ 0 \quad \text{and} \quad B = \begin{pmatrix} 2 & -1 \\ -1 & 1 \end{pmatrix} \succ 0 \tag{2.3.9}$$

shows, the (2,2)-entry of $A * B$ is -1. In particular, $A * B$ has at least one negative eigenvalue implying $A * B \not\succeq 0$. We note that the converse implication is true, namely if $A \succ 0$ and $A * B \succ 0$ then $B \succ 0$, see e.g. [523, Lemma E.1, Chap 16].

- Note that $A \pm B \succ 0$ in the previous example. Therefore, when $C := A + B$ and $D := A - B$ we have $C \succ D \succ 0$. Nevertheless $C^2 \not\succeq D^2$ because $C^2 - D^2 = 2(AB + BA) = 4A * B \not\succeq 0$. On the other hand, it is easy to verify that whenever $C \succeq D \succeq 0$ then also $C^{1/2} \succeq D^{1/2} \succeq 0$. In other words, the square root is a *monotone matrix function*, but the square is not.

- For a third example we generalize the absolute value function and apply it to a symmetric matrix A. We define $|A| = \sqrt{A^2}$ as the matrix with the same eigenvectors as A and the eigenvalues given by the absolute value of the eigenvalues of A. It is straightforward to show that

$$|A| = \operatorname{argmin}\{\operatorname{Trace} B \mid B \succeq \pm A\} = U|\Lambda|U^T$$

if $A = U\Lambda U^T$ with an orthogonal matrix U and a diagonal matrix Λ. One might conjecture that $|\ .\ |$ satisfies the triangle inequality. It does not! As pointed out by [567], even for any fixed $\epsilon > 0$ there are examples with

$$|A| + |B| \not\succeq \epsilon |A + B|. \tag{2.3.10}$$

To understand why (2.3.10) is true, we consider the following example: Let $u = (1,0)^T$ and $v = (1,\epsilon)^T$ for some small positive ϵ. Define $A = uu^T$ and $B = -vv^T$. Then

$$|A| + |B| = \begin{pmatrix} 2 & \epsilon \\ \epsilon & \epsilon^2 \end{pmatrix} \quad \text{and} \quad A + B = -\begin{pmatrix} 0 & \epsilon \\ \epsilon & \epsilon^2 \end{pmatrix}.$$

Simple calculations show that the eigenvalues of $|A| + |B|$ are approximately 2 and $\epsilon^2/2$, while the eigenvalues of $A + B$ are approximately $\pm\epsilon$. Hence, $|A + B| \approx \epsilon I$, the eigenvalues of which are about $2/\epsilon$-times larger than the smallest eigenvalue of $|A| + |B|$. (!)

For further properties of the absolute value function we refer to [352, Chap 3.5, Problem 6-10].

The second example above states that the square root

$$\sqrt{\ }: \quad \mathcal{S}_+^n \to \mathcal{S}_+^n, \quad X \mapsto X^{1/2}$$

is a monotone matrix function [352] (with respect to the Löwner partial order). In [523], such functions are also called matrix-increasing. Further monotone matrix functions have been characterized in [498]: All real functions that have an analytic continuation on the upper complex half plane and have function values with nonnegative imaginary part there. Hence, the square root function or, more generally, the functions

$$X \mapsto X^\alpha \qquad \text{and} \qquad X \mapsto -X^{-\alpha} \tag{2.3.11}$$

with $0 \leq \alpha \leq 1$ on the domain of positive (semi-) definite matrices X are monotone. Also the function $X \mapsto \ln X$ is matrix monotone, see [352, Chap 6.6, Prob 20].

In (2.3.11), we have used the natural extension of a real-valued function to real symmetric matrices. If $f : \mathbb{R} \to \mathbb{R}$ is a function, then f defines a function $F : \mathcal{S}^n \to \mathcal{S}^n$ by

$$F(X) = U^T \text{Diag}\,(f(\lambda(X)))U \tag{2.3.12}$$

where $X = U^T \text{Diag}\,(\lambda(X))U$ is the spectral decomposition of X (Theorem 2.2.1), and the function f is applied componentwise to the vector $\lambda(X)$. The above extension F of a real function f to a symmetric matrix is also called a "primary matrix function". When f is given by a power series $f(x) = \sum a_k(x - \bar{x})^k$ with radius of convergence θ, and X is a symmetric matrix such that the spectral radius of $X - \bar{x}I$ satisfies $\rho(X - \bar{x}I) < \theta$, then the primary matrix function F coincides with the power series

$$F(X) = \sum a_k(X - \bar{x}I)^k.$$

The primary matrix function induced by $x \mapsto \max\{0, x\}$, or by $x \mapsto x^\alpha$ for $\alpha > 1$ or by $x \mapsto e^x$ are not monotone, see [523, Chap 16, E4].

The Hadamard Product with a positive semi-definite matrix A: $X \mapsto A \circ X$ (see [352, Thm 7.5.3]), and the Kronecker products with a positive definite matrix A: $X \mapsto A \otimes X$ and $X \mapsto X \otimes A$ are also monotone matrix functions, see e.g. [523, Chap 16 E] and [352, Chap 6] for this and further examples.

2.3.6 Majorization

The results in this subsection are based on the observation that the eigenvalues of a symmetric matrix are more "spread out" than the diagonal elements of the matrix. More precisely, let

$$\mu = \frac{1}{n} \sum_{i=1}^n A_{ii}$$

be the mean value of the diagonal entries of a symmetric matrix A (and hence also the mean value of the eigenvalues). Further, let $A = U^T \Lambda U$ be the spectral decomposition of A, (Theorem 2.2.1). Then, from unitary invariance of the Frobenius norm we obtain

$$\sum_{i=1}^{n}(A_{ii}-\mu)^2 \le \|A-\mu I\|_F^2 = \|U^T(\Lambda-\mu I)U\|_F^2 = \|\Lambda-\mu I\|_F^2 = \sum_{i=1}^{n}(\lambda_i-\mu)^2,$$

i.e. the eigenvalues vary stronger about their mean value than the diagonal elements of A. A mathematically precise statement how to formulate "vary more" is given next.

For a vector $x \in \mathbb{R}^n$ we denote by $x_{[1]} \ge x_{[2]} \ge \ldots \ge x_{[n]}$ the components of x arranged in non-increasing order. We say the vector $x \in \mathbb{R}^n$ majorizes the vector $y \in \mathbb{R}^n$ if

$$\sum_{i=1}^{n} x_i = \sum_{i=1}^{n} y_i, \qquad \text{and} \qquad \sum_{i=1}^{k} x_{[i]} \ge \sum_{i=1}^{k} y_{[i]} \quad \text{for} \quad 1 \le k \le n. \tag{2.3.13}$$

We then write $x \succ y$. When the symbol "$\succ$" is used between vectors it means majorization in this chapter; for symmetric matrices it still denotes the Löwner partial order. Intuitively, when $x \succ y$ then x and y have the same mean value, but since the large components of x are larger than the large components of y, the components of x "vary more". Note that this interpretation is also true if some or all the x_i or y_i are negative. In fact, by adding a large constant to all x_i and all y_i, the relation (2.3.13) remains unchanged, and thus, we can "shift" x and y to the positive orthant without changing (2.3.13).

The above notation was introduced by Hardy, Littlewood and Polya, and is also used in [523]. (Note, however, that the definition of majorization is exactly the opposite in [352].)

As the next two theorems show, the concept of majorization is intimately related to both, convex analysis and symmetric matrices.

Theorem 2.3.12 (Hardy, Littlewood, Polya) *The condition $x \succ y$ is necessary and sufficient in order that the inequality*

$$\sum_{i=1}^{n} f(x_i) \ge \sum_{i=1}^{n} f(y_i) \tag{2.3.14}$$

holds for all convex functions $f : \mathbb{R} \to \mathbb{R}$.

When we apply the definition (2.3.5) of convexity to $f : \mathbb{R} \to \mathbb{R}$ and restrict ourselves to $\lambda = \frac{1}{2}$, we obtain the inequality

$$f(a) + f(b) \ge 2f(\tfrac{1}{2}(a+b)) \qquad \text{for all} \quad a, b \in \mathbb{R}. \tag{2.3.15}$$

For continuous functions f, (2.3.15) is in fact equivalent to (2.3.5). Note that

$$\begin{pmatrix} x_1 \\ x_2 \end{pmatrix} := \begin{pmatrix} a \\ b \end{pmatrix} \succ \begin{pmatrix} (a+b)/2 \\ (a+b)/2 \end{pmatrix} =: \begin{pmatrix} y_1 \\ y_2 \end{pmatrix}.$$

Hence, inequality (2.3.14) can be viewed as a generalization of inequality (2.3.15) from $n = 2$ to values $n \geq 2$. For a further discussion see [523, Chap 1].

Theorem 2.3.13 (Horn) *Let $d = \text{diag}(A)$ denote the vector of diagonal elements of a symmetric matrix A, and $\lambda = \lambda(A)$ the vector of eigenvalues. Then $\lambda \succ d$.*

Conversely, the condition $x \succ y$ is also sufficient in order that there exists a matrix A with $x = \lambda(A)$ and $y = \text{diag}(A)$.

The first part of Horn's theorem complements the observation of the beginning of this subsection. For a proof and further references, see [352, Thm 4.3.26-4.3.32].

2.3.7 Convex matrix functions

Definition 2.3.2 *A mapping $\Phi : \mathcal{C} \to \mathcal{S}^n$ is convex if $\mathcal{C}$ is convex, and*

$$(1-\alpha)\Phi(A) + \alpha\Phi(B) \succeq \Phi((1-\alpha)A + \alpha B), \qquad (2.3.16)$$

holds for all $\alpha \in (0,1)$. Φ is strictly convex if "$\succeq$" in (2.3.16) can be replaced by "$\succ$" for $A \neq B$.

A function $f : \mathbb{R} \to \mathbb{R}$ is called a (strictly) convex matrix function if the primary matrix function (2.3.12) is (strictly) convex.

The functions $x \mapsto x^2$, $x \mapsto x^{-1}$ for $x > 0$, $x \mapsto -x^\alpha$ for $\alpha \in (0,1)$ and $x > 0$, are all convex matrix functions, see [352, Chap 6.6].

2.3.8 Convex real-valued functions of matrices

We now return to the definition of a function $F : \mathcal{S}^n \to \mathbb{R}$ via a function $f : \mathbb{R}^n \to \mathbb{R}$. Our first result is due to Davis [180].

Theorem 2.3.14 (Davis) *Let $F : \mathcal{S}^n \to \mathbb{R}$ be weakly orthogonally invariant in the sense that $F(U^TAU) = F(A)$ for all orthogonal matrices U. Then F is convex if and only if the restriction of F to diagonal matrices is convex.*

Note that permutation matrices are orthogonal, and hence the function F restricted to diagonal matrices is permutation-invariant (Definition 2.3.1). Thus F is precisely of the form (2.3.6).

As a corollary we can categorize convex, unitarily invariant subsets $\mathcal{C}$ of $\mathcal{S}^n$ as sets for which there exists a permutation invariant subset $C \subset \mathbb{R}^n$ such that

$$\mathcal{C} = \{X \in \mathcal{S}^n \mid \lambda(X) \in C\}.$$

Moreover, the exposed faces of $\mathcal{C}$ are naturally associated with the exposed faces of C, see [479].

Closely related to Theorem 2.3.14 is the following result by Von Neumann [590] which relates matrix norms and symmetric gauge functions (named "Minkowsky gauge functions" in [590]), see Definition 2.3.1:

Theorem 2.3.15 (Von Neumann) *The unitarily invariant norms* $\| \,.\, \|$ *on* $\mathbb{R}^{n\times n}$ *(see (2.3.2)) are precisely the norms induced by symmetric gauge functions* f *via* $\|X\| = f(\sigma(X))$. *Here,* $\sigma(X) = \lambda(X^T X)^{1/2}$.

One of the key elements of convex analysis is the construction of the Fenchel conjugate of a function. Let E be a (real and finite-dimensional) Euclidean vector space (typically $\mathbb{R}^n$ or $\mathcal{S}^n$), and for $x, y \in E$ let $\langle x, y\rangle$ denote the scalar product of x and y — i.e. $\langle x, y\rangle = x^T y$ for $x, y \in \mathbb{R}^n$ and $\langle x, y\rangle = x \bullet y$ if $x, y \in \mathbb{R}^{n\times n}$. Let $f : \mathbb{R}^n \to [-\infty, \infty]$ be a function that takes values in $\mathbb{R} \cup \{\pm\infty\}$. We say f is closed if $\{(x, r) \subset E \times \mathbb{R} \mid f(x) \le r\}$ is closed [483]. Continuous functions are obviously closed. The function f^* defined by

$$f^*(y) = \sup\{\langle x, y\rangle - f(x) \mid x \in E\}$$

is closed and convex, and is called the *Fenchel conjugate* of f. Roughly speaking, for convex differentiable functions f, the gradient of f^* is just the inverse function of the gradient of f. More precisely, we denote by

$$\partial f(x) = \{y \in E \mid \langle y, z - x\rangle \le f(z) - f(x) \ \text{ for all } \ z \in E\}$$

the *subdifferential* of a convex function f. When f is convex and differentiable, the subdifferential coincides with the derivative, $\partial f(x) = \{\nabla f(x)\}$; for further properties of the subdifferential see e.g. [679]. (Above, we write ∇f rather than Df to indicate that ∇f is an element of E; when $E = \mathbb{R}^n$ is the space of real column vectors of dimension n, then Df denotes a row vector, while ∇f denotes a column vector.) The next theorem relates the subdifferential of f and f^*.

Theorem 2.3.16 *If* $f : E \to (-\infty, \infty]$ *is closed and convex then*

$$y \in \partial f(x) \iff x \in \partial f^*(y).$$

We close this chapter with a useful result for finding the Fenchel conjugate of functions of the form (2.3.6).

Theorem 2.3.17 (Lewis) *If* $f : \mathbb{R}^n \to (-\infty, \infty]$ *is permutation-invariant the conjugate function of* F *defined by* $F(X) = f(\lambda(X))$ *is* F^* *defined by*

$$F^*(Y) = \sup\{X \bullet Y - F(X) \mid X \in \mathcal{S}^n\} = f^*(\lambda(Y)).$$

For a proof and discussion see [479].

Acknowledgements

The author would like to thank T. Huckle and H. Wolkowicz for helpful discussions and references and M. Wechs and R. Lepenis for improving the presentation of the paper.

3 THE GEOMETRY OF SEMIDEFINITE PROGRAMMING

Gábor Pataki

My work has always tried to unite the true with the beautiful and when I had to choose one or the other, I usually chose the beautiful.

Hermann Weyl (1885 - 1955) quoted in an obituary by Freeman J. Dyson in Nature, March 10, 1956

[Inscription above Plato's Academy:]
Let no one ignorant of geometry enter here.

Plato

3.1 INTRODUCTION

Consider the primal-dual pair of optimization problems

$$(P) \quad \begin{array}{ll} Min & \langle c, x \rangle \\ s.t. & x \in K \\ & Ax = b \end{array} \qquad \begin{array}{ll} Max & \langle b, y \rangle \\ s.t. & z \in K^* \\ & A^* y + z = c \end{array} \quad (D)$$

where

- X and Y are Euclidean spaces with $\dim X \geq \dim Y$.
- $A : X \to Y$ is a linear operator, assumed to be onto.

- $A^* : Y \to X$ is its adjoint.
- K is a closed, convex, facially exposed cone in X.
- $K^* := \{ z \mid \langle z, x \rangle \geq 0 \ \forall x \in K \}$ is the dual of K, also a closed, convex, facially exposed cone.

The problems (P) and (D) are called a primal-dual pair of conic linear programs, cone programs, or cone-LP's. With the appropriate choice of X, Y and K, they include: ordinary LP's; semidefinite programs (SDP's); programs over p-cones, in particular over second order cones.

Besides the wide applicability of cone-LP's, their main attraction is their elegance: both their duality theory, and the algorithmic approaches to solve them are natural extensions of their counterparts in linear programming. As we shall see in this chapter, the situation is similar regarding their geometry. By the "geometry" of a cone-LP we mean the characterization of the

(1) Set of optimal solutions, in particular, of whether this set is a singleton (the question of uniqueness).

(2) Tangent cone and tangent space of the feasible set at an optimal solution (thus through polarity, also of the normal cone).

(3) Analogous sets in the dual problem.

In this chapter we give the overview of a theory that describes the geometry of cone-LP's. It is reminiscent of how the geometry of LP is usually described: through the *facial structure* of the feasible set. Since the solution set of a cone-LP is always a face of the feasible set, regardless of what the underlying cone is, this approach is quite natural. It generalizes the notions, and corresponding theorems known in LP about the facial structure of the feasible set, on nondegeneracy, and strict complementarity.

Its essence: given a feasible solution to a cone program, the minimal face of the *feasible set* containing it is the intersection of the minimal face of the *cone* containing it with the affine constraints. Whether this solution is an extreme point of the feasible set can be characterized using these two latter sets. (E.g. when the cone program is an LP, the extremity of a feasible solution depends only on the *position* of the nonzeros in it; in other words on the minimal face of the nonnegative orthant that contains it). Moreover, its nondegeneracy is defined by imposing the extremity condition on the *dual* with the *complementary* face of the cone; and strict complementarity of a solution-pair is imposed by requiring them to be in the relative interior of complementary faces of K and K^*. The other objects we want to study (the tangent cone and tangent space of the feasible set at a solution) will have similar descriptions.

This theory can be specialized to various classes of cone-LP's by using the description of the faces of the underlying cones. In all the interesting cases this description is quite handy; for the nonnegative orthant and p-cones, it is trivial; for the semidefinite cone it is given by a classical result (see e.g. [73]).

The chapter is structured as follows: section 3.2 collects the notation, and necessary basic results that will be used later on. Section 3.3 presents the theory on the geometry of cone programs. In subsection 3.3.1 we describe their facial structure, the notions of nondegeneracy and strict complementarity, and prove the generalizations of the results connecting them in LP. We also show how several previous results on the geometry of SDP are subsumed by this framework. Subsection 3.3.2 describes the tangent spaces of the feasible sets. In subsection 3.3.3 we derive the family of *boundary structure inequalities* that relate the dimensions of

- Minimal faces in the primal and dual *cones* that contain a given optimal solution.
- Minimal faces in the primal and dual *feasible sets* that contain a given optimal solution.
- Tangent spaces at the primal and dual feasible sets at a given optimal solution.

These inequalities provide a surprising amount of information about the boundary structure of (P) and (D). For example: at a strictly complementary optimal solution pair in an SDP, it is impossible to have full-dimensional normal cones at both the primal and dual optima; i.e. both the primal and dual optimal solutions cannot be "kinky".

Subsection 3.3.4 translates the previous results for equivalent cone programs formulated with different variables (e.g. when the dual slack is eliminated), and subsection 3.3.5 gives a detailed example. In section 3.4 we present several examples of "semidefinite combinatorics", ie. apply the results on the geometry of cone programs to deduce some instructive structural results about problems that can be formulated as an SDP.

Finally, in section 3.5 we study two algorithmic aspects; converting a feasible solution of a cone-lp into one, which is also an extreme point of the feasible set, and performing sensitivity analysis.

3.2 PRELIMINARIES

Spaces and cones of interest. The space of n by n symmetric, and the cone of n by n symmetric, positive semidefinite matrices are denoted by $\mathcal{S}^n$, and $\mathcal{S}^n_+$, respectively. The space $\mathcal{S}^n$ is equipped with the inner product

$$\langle x, z \rangle \quad := \quad \sum_{i,j=1}^{n} x_{ij} z_{ij},$$

and it is a well-known fact, that $\mathcal{S}^n_+$ is self-dual with respect to it.

If $1 < p < +\infty$, then the p-cone is defined as

$$K_p \quad = \quad \{ (x_0, x) \mid x_0 \geq ||x||_p \}.$$

If $K = K_p$, then $K^* = K_q$, where $\frac{1}{p} + \frac{1}{q} = 1$.

Operators and matrices. Linear operators are denoted by capital letters; when a matrix is considered to be an element of a euclidean space, and not a linear operator, it is usually denoted by a small letter. The ith row of matrix a is denoted by $a_{i\,.}$ and the jth column by $a_{.\,j}$.
The range space of an operator A [of a matrix x] is denoted by $\mathcal{R}(A)$ $[\mathcal{R}(x)]$. If $x \in \mathcal{S}^n$, then $\lambda_i(x)$ denotes its ith largest eigenvalue, and $\lambda(x)$ the vector $(\lambda_1(x), \ldots, \lambda_n(x))^T$.
The identity linear operator, and the identity matrix are denoted by I, and the vector of all ones by e.
The positive part of the vector $x \in \mathcal{R}^n$ is denoted by x_+, i.e. $(x_+)_i = x_i$, if $x_i > 0$, and $(x_+)_i = 0$, otherwise.
The inner product of $x^1, x^2 \in X$ is denoted by $\langle x^1, x^2 \rangle$. Even if the inner products in X and Y are different (say if $X = \mathcal{S}^n$ and $Y = \mathcal{R}^m$), we still use the notation $\langle , \rangle$ for both; the context should make it clear, which one is meant. The matrix product of matrices x^1 and x^2 is denoted by $x^1 x^2$. The block diagonal matrix obtained by placing x^1 and x^2 on the main diagonal is denoted by $x^1 \oplus x^2$.

The dimension of the sum of subspaces. The following simple proposition will be used many times, hence we state, and prove it for convenience.

Proposition 3.2.1 *Suppose that L_1 and L_2 are subspaces. Then*

$$\dim\,[\,L_1 + L_2\,] \quad = \quad \dim\,L_1 + \dim\,L_2 - \dim\,[\,L_1 \cap L_2\,].$$

Proof **Case 1** If $L_1 \cap L_2 = \{0\}$, then the claim is obvious.
Case 2 $L_1 \cap L_2 \neq \{0\}$. Let L_3 be a subspace that satisfies

$$L_3 \subseteq L_1, \quad L_3 \cap L_2 = \{0\}, \quad L_1 = L_3 + [\,L_1 \cap L_2\,].$$

Then using Case 1 with L_3 in place of L_1, and $L_1 \cap L_2$ in place of L_2, we obtain

$$\dim\,L_1 \quad = \quad \dim\,L_3 + \dim\,[\,L_1 \cap L_2\,]. \tag{3.2.1}$$

Also,

$$\begin{aligned} L_1 + L_2 &= [\,L_3 + [L_1 \cap L_2]\,] + L_2 \\ &= L_3 + L_2. \end{aligned}$$

Again using Case 1, we get

$$\dim\,[\,L_1 + L_2\,] \quad = \quad \dim\,L_3 + \dim\,L_2. \tag{3.2.2}$$

Subtracting (3.2.2) from (3.2.1) and rearranging yields the required formula.

■

Faces, feasible directions, tangent cones and tangent spaces in convex sets. For vectors y and z, we denote the open line-segment between y and z by

$$(y,z) \quad = \quad \{\, \mu y + (1-\mu)z \,|\, 0 < \mu < 1 \,\}.$$

Let C be a closed convex set. A convex subset F of C is called a *face* of C, and this fact is denoted by $F \lhd C$, if

$$x \in F,\; y,z \in C,\; x \in (y,z) \text{ implies } y,z \in F. \tag{3.2.3}$$

An *extreme point* of C is a face consisting of a single element. If S is a subset of C, then we denote by face (S,C) the *minimal face of* C containing S, and if $x \in C$, then we write face (x,C) for face $(\{x\},C)$.
A *supporting hyperplane* of C is a set of the form

$$H \quad = \quad \{\, x \,|\, \langle a, x\rangle = \alpha \,\},$$

where

$$\alpha \quad = \quad \max\{\, \langle a,x\rangle \,|\, x \in C \,\}.$$

The set $C \cap H$ is called an *exposed face* of C. An exposed face is always a face, whereas the opposite may not be true. For a discussion of the difference between faces and exposed faces, see e.g. Section 5 in [142].

Proposition 3.2.2 *Let C be a convex set, $C' \lhd C$, and D a convex subset of C.*

(i) If $\mathrm{ri}\, D \cap C' \neq \emptyset$, *then* $D \subseteq C'$.

(ii) If $D \subseteq C'$, *and* $D \cap \mathrm{ri}\, C' \neq \emptyset$, *then* $C' = \mathrm{face}\,(D,C)$.

(iii) $C' = \mathrm{face}\,(D,C)$, *iff* $\mathrm{ri}\, D \cap \mathrm{ri}\, C' \neq \emptyset$.

(iv) $C' = C \cap \mathrm{aff}\, C'$.

Proof The statement (i) is [679, Theorem 18.1]; (ii) follows from [679, Theorem 18.2], and (iii) by putting (i) and (ii) together. Statement (iv) is Exercise 5.4 in [142]. ■

For $x \in C$, the cone of feasible directions, the tangent cone and the tangent space at x in C are defined as

$$\begin{aligned}
\mathrm{dir}\,(x,C) \quad &= \quad \{\, y \,|\, x + ty \in C \text{ for some } t > 0 \,\}, \\
\mathrm{tcone}\,(x,C) \quad &= \quad \mathrm{Cl}\ \mathrm{dir}\,(x,C) \;=\; \{\, y \,|\, \mathrm{dist}(x+ty, C) = o(t) \,\}, \\
\tan(x,C) \quad &= \quad \mathrm{tcone}\,(x,C) \cap -\mathrm{tcone}\,(x,C) \\
&= \quad \{\, y \,|\, \mathrm{dist}(x \pm ty, C) = o(t) \,\}.
\end{aligned}$$

The equivalence of the alternative expressions for $\operatorname{tcone}(x, C)$ follows e.g. from [345], page 135.
An important fact ([345, Proposition 5.3.1]), that we state for the ease of reference is

Proposition 3.2.3 *If C_1 and C_2 are nonempty closed convex sets, $x \in C_1 \cap C_2$, then*

$$\operatorname{tcone}(x, C_1 \cap C_2) \quad \subseteq \quad \operatorname{tcone}(x, C_1) \cap \operatorname{tcone}(x, C_2),$$

with equality holding, if $\operatorname{ri} C_1 \cap \operatorname{ri} C_2 \neq \emptyset$.

■

Faces, complementary faces, feasible directions and tangent spaces in cones. A convex set K is a *cone*, if $\mu K \subseteq K$ holds for all $\mu \geq 0$. If K is a cone, then a simple argument shows that (3.2.3) is equivalent to

$$x \in F, \; y, z \in K, \; x = y + z \; \text{ implies } \; y, z \in F. \tag{3.2.4}$$

The *dual* of the cone K is

$$K^* \quad = \quad \{ z \,|\, \langle z, x \rangle \geq 0 \text{ for all } x \in K \}.$$

If K, K_1 and K_2 are convex cones, then

$$\begin{aligned} K^{**} &= \operatorname{Cl} K, \\ (K_1 + K_2)^* &= K_1^* \cap K_2^*, \\ (K_1 \cap K_2)^* &= \operatorname{Cl}(K_1^* + K_2^*). \end{aligned}$$

If $F \triangleleft K$, and $\bar{x} \in \operatorname{ri} F$ is fixed, then the *complementary* (or *conjugate*) face of F is defined alternatively as

$$\begin{aligned} F^{\triangle} &= \{ z \in K^* \,|\, \langle z, x \rangle = 0 \;\text{ for all } x \in F \} \\ &= \{ z \in K^* \,|\, \langle z, \bar{x} \rangle = 0 \}. \end{aligned}$$

The equivalence of the two definitions is straightforward. The complementary face of $G \triangleleft K^*$ is defined analogously, and is denoted by $G^{\triangle}$. K is facially exposed, i.e. all faces of K arise as the intersection of K with a supporting hyperplane, iff for all $F \triangleleft K$, $F^{\triangle\triangle} = F$, see [142, Theorem 6.7]. For brevity, if $F \triangleleft K$, then we write $F^{\triangle *}$ for $(F^{\triangle})^*$, and $F^{\triangle\perp}$ for $(F^{\triangle})^{\perp}$.
It is well known that if K is a polyhedral cone, then for all $F \triangleleft K$, then $\operatorname{lin} F + \operatorname{lin} F^{\triangle}$ is the whole space. The *residual subspace* of $F \triangleleft K$ is meant to measure, "to what extent F is nonpolyhedral". It is defined as

$$\operatorname{res} F \quad = \quad (\operatorname{lin} F + \operatorname{lin} F^{\triangle})^{\perp}.$$

We say, that K is *nice*, if one of the following equivalent statements holds

$$\begin{array}{rll} K^* + F^\perp & \text{is closed} & \forall \; F \triangleleft K, \\ \operatorname{Proj}_{\,\operatorname{lin} F}(K^*) & \text{is closed} & \forall \; F \triangleleft K. \end{array} \tag{3.2.5}$$

Next, we list several examples of cones, along with the description of their faces. The corresponding complementary faces, and residual subspaces can be found in Table 3.2.

Example 3.2.1 (The nonnegative orthant) If $\bar{x} \in K = \mathcal{R}^n_+$, then

$$\operatorname{face}(\bar{x}, \mathcal{R}^n_+) \;=\; \{ x \in \mathcal{R}^n_+ \mid x_i = 0 \;\; \forall i \; s.t. \; \bar{x}_i = 0 \}.$$

This face, (after permuting components) can be brought to the form

$$\operatorname{face}((e, 0)^T, \mathcal{R}^n_+),$$

for an e of appropriate size. ■

Example 3.2.2 (The semidefinite cone) If $\bar{x} \in K = \mathcal{S}^n_+$, then

$$\operatorname{face}(\bar{x}, \mathcal{S}^n_+) \;=\; \{ x \in \mathcal{S}^n_+ \mid \mathcal{R}(x) \subseteq \mathcal{R}(\bar{x}) \}, \tag{3.2.6}$$

$$\operatorname{face}(\bar{x}, \mathcal{S}^n_+)^\triangle \;=\; \{ x \in \mathcal{S}^n_+ \mid \mathcal{R}(x) \subseteq \mathcal{R}(\bar{x})^\perp \}, \tag{3.2.7}$$

([73], for a simple proof, see Appendix 3.7.1.) Let q be an orthonormal matrix such that

$$\bar{x} \;=\; q \left(\begin{smallmatrix} \Lambda & 0 \\ 0 & 0 \end{smallmatrix} \right) q^T,$$

where Λ is a diagonal matrix with positive diagonal. All transformations $v^T(.)v$, where v is an invertible matrix, are one-to-one mappings of $\mathcal{S}^n_+$ to itself. Therefore $\operatorname{face}(\bar{x}, \mathcal{S}^n_+)$ can be brought to the form

$$q^T(\operatorname{face}(\bar{x}, \mathcal{S}^n_+))q \;=\; \operatorname{face}(q^T \bar{x} q, \mathcal{S}^n_+) \;=\; \operatorname{face}(\left(\begin{smallmatrix} \Lambda & 0 \\ 0 & 0 \end{smallmatrix} \right), \mathcal{S}^n_+) \;=\; \operatorname{face}(\left(\begin{smallmatrix} I & 0 \\ 0 & 0 \end{smallmatrix} \right), \mathcal{S}^n_+).$$

If the rank of x is r, then

$$\dim \operatorname{face}(\bar{x}, \mathcal{S}^n_+) \;=\; t(r) \;:=\; r(r+1)/2,$$

where $t(r)$ denotes the rth triangular number. ■

Example 3.2.3 (The p-cones) Let $1 < p < +\infty$. Since K_p is obtained by "lifting" the unit ball of the norm $||\,.\,||_p$, all of its nontrivial faces (i.e. apart from the origin and itself) are of the form

$$\operatorname{cone}\{ (||x||_p, x)^T \}$$

K	A typical F	$F^{\triangle}$	res F
$\mathcal{R}^n_+$	face $((e,0)^T, \mathcal{R}^n_+)$	face $((0,e)^T, \mathcal{R}^n_+)$	$\{0\}$
$\mathcal{S}^n_+$	face $(\begin{pmatrix} I & 0 \\ 0 & 0 \end{pmatrix}, \mathcal{S}^n_+)$	face $(\begin{pmatrix} 0 & 0 \\ 0 & I \end{pmatrix}, \mathcal{S}^n_+)$	$\{ y \in \mathcal{S}^n \mid y = \begin{pmatrix} 0 & v \\ v^T & 0 \end{pmatrix} \}$
K_p	cone $\{ (\|\|x\|\|_p, x)^T \}$	cone $\{ (\|\|x\|\|_q, -x)^T \}$	$\{ (0,y)^T \mid \langle y, x \rangle = 0 \}$

Table 3.1 The faces, complementary faces, and residual subspaces in $\mathcal{R}^n_+$, $\mathcal{S}^n_+$ and K_p

for some x. ∎

It is not hard to see, that all these cones are facially exposed. They are also nice, by using the second criterion in (3.2.5). In the case of $\mathcal{R}^n_+$ and $\mathcal{S}^n_+$, the projection in question is just a smaller copy of the original cone. In the case of K_p the linear span of any nontrivial face is a line, and all cones contained in a line are closed.

Next, we show that the set of feasible directions, and several related sets for an $x \in K$ can be conveniently described in terms of face (x, K).

Lemma 3.2.1 *Let $x \in K$, and write $F =$ face (x,K). Then the following relations hold.*

$$\text{dir } (x,K) = K + \operatorname{lin} F, \tag{3.2.8}$$

$$\text{dir } (x,K)^* = K^* \cap F^{\perp} = F^{\triangle} = K^* \cap \operatorname{lin} F^{\triangle}, \tag{3.2.9}$$

$$\text{Cl dir } (x,K) = \text{Cl}\,(K + \operatorname{lin} F) = F^{\triangle *} = \text{Cl}\,(K + \operatorname{lin} F + \text{res } F) \tag{3.2.10}$$

Furthermore, if K is nice, then

$$\tan(x,K) = \operatorname{lin} F + \text{res } F. \tag{3.2.11}$$

Proof of (3.2.8) "$\supseteq$:" Let $v \in \operatorname{lin} F$, $z \in K$. Then $x + \alpha v \in K$ for some $\alpha > 0$ $x + \alpha v \in K$, hence $x + \alpha(v + z) \in K$. "$\subseteq$:" Let $y \in$ dir (x,K), $\alpha > 0$, $x' := x + \alpha y \in K$. Then $y = \frac{1}{\alpha}(x' - x) \in K + \operatorname{lin} F$.

Proof of (3.2.9) The first equality follows from (3.2.8) by taking the dual, the second by the definition of $F^{\triangle}$, and the third by Proposition 3.2.2 (iii), since $F^{\triangle}$ is a face.

Proof of (3.2.10) The first and third equalities follow from (3.2.9).

Proof of (3.2.11) From the definition of $\tan(x,K)$ and since K is nice,

$$\tan(x,K) = (K + \operatorname{lin} F + \text{res } F) \cap -(K + \operatorname{lin} F + \text{res } F).$$

Therefore "$\supseteq$" in (3.2.11) is obvious. For "$\subseteq$", let $x^1 \in K$, $y^1 \in \operatorname{lin} F + \operatorname{res}\ F$ such that

$$x^1 + y^1 \in (K + \operatorname{lin} F + \operatorname{res}\ F) \cap -(K + \operatorname{lin} F + \operatorname{res}\ F).$$

That is, for some $x^2 \in K, y^2 \in \operatorname{lin} F + \operatorname{res}\ F$,

$$\begin{array}{lcllccl} x^1 + y^1 & = & -(x^2 + y^2) & \Rightarrow & x^1 + x^2 & = & y^1 + y^2 \quad \Rightarrow \\ x^1 + x^2 & \in & K \cap (\operatorname{lin} F + \operatorname{res}\ F) = F & \Rightarrow & x^1, x^2 & \in & F \quad \Rightarrow \\ x^1 + y^1 & \in & \operatorname{lin} F + \operatorname{res}\ F. & & & & \end{array}$$

■

3.3 THE GEOMETRY OF CONE-LP'S: MAIN RESULTS

3.3.1 Facial structure, nondegeneracy and strict complementarity

We say that (P) satisfies the Slater condition, if there is an $\bar{x} \in \operatorname{ri} \mathrm{K}$ feasible for (P), and that (D) satisfies the Slater condition, if there is a $(\bar{y}, \bar{z}) \in Y \times \operatorname{ri} \mathrm{K}^*$ feasible for (D). We assume that the optimal values of (P) and (D) are equal, both attained, (this is ensured if both satisfy the Slater condition) and denote:

- their feasible sets by *Feas(P)* and *Feas(D)*, resp.
- the set of their optimal solutions by *Opt(P)* and *Opt(D)*, resp.

Theorem 3.3.1 (Primal Faces) *Let $\bar{x} \in Feas(P)$, and let*

$$\begin{array}{rcl} \mathcal{F} & = & \operatorname{face}(\bar{x}, Feas(P)), \\ F & = & \operatorname{face}(\bar{x}, K). \end{array}$$

Then

(1) $F = \operatorname{face}(\mathcal{F}, K)$,

(2) *(a)* $\begin{array}{rcl} \operatorname{aff} \mathcal{F} & = & \operatorname{lin} F \cap \{ x \mid Ax = b \} \\ & = & \bar{x} + [\operatorname{lin} F \cap \mathcal{N}(A)], \end{array}$

(b) $\mathcal{F} = F \cap \{ x \mid Ax = b \}$,

(3) $\dim \mathcal{F} = \dim [\operatorname{lin} F \cap \mathcal{N}(A)] = \dim F - \dim Y + \dim [F^\perp \cap \mathcal{R}(A^*)]$.

(4) *$\mathcal{F}$ is a singleton set, i.e. $\mathcal{F}$ (or equivalently $\bar{x}$) is an extreme point of Feas(P), if and only if*

$$\operatorname{lin} F \cap \mathcal{N}(A) \quad = \quad \{0\}.$$

Proof
(1): Follows from (iii) in Proposition 3.2.2.

(2)(a) "$\subseteq$" : Follows from (1).
(2)(a) "$\supseteq$" : Let $v \in \operatorname{lin} F \cap \{x \mid Ax = b\}$. As $\bar{x} \in \mathcal{F} \cap \operatorname{ri} \mathrm{F}$, there exists $\epsilon > 0$ such that

$$\begin{aligned} x^1 &= \bar{x} + \epsilon(\bar{x} - v), \\ x^2 &= \bar{x} - \epsilon(\bar{x} - v), \\ x^1, x^2 &\in F. \end{aligned}$$

Clearly, x^1 and x^2 also satisfy the affine constraint, so they are in *Feas(P)*. As $\bar{x} \in \mathcal{F} \lhd Feas(P)$, and $\bar{x} \in (x^1, x^2)$, we get

$$\begin{aligned} x^1, x^2 &\in \mathcal{F} \Longrightarrow \\ v &\in \operatorname{aff}\{x^1, x^2\} \subseteq \operatorname{aff} \mathcal{F}, \end{aligned}$$

as required.
(2)(b): We have

$$\begin{aligned} \mathcal{F} &= Feas(P) \cap \operatorname{aff} \mathcal{F} \\ &= (K \cap \{x \mid Ax = b\}) \cap (\operatorname{lin} F \cap \{x \mid Ax = b\}) \\ &= (K \cap \operatorname{lin} F) \cap \{x \mid Ax = b\} \\ &= F \cap \{x \mid Ax = b\}, \end{aligned}$$

where the first equality follows by $\mathcal{F} \lhd$ *Feas(P)*, and the last by $F \lhd K$.
(3): By (2)(a),

$$\begin{aligned} \operatorname{aff} \mathcal{F} - \bar{x} &= \mathcal{N}(A) \cap \operatorname{lin} F & \Rightarrow \\ \dim \mathcal{F} &= \dim\,[\operatorname{lin} F \cap \mathcal{N}(A)] \\ &= \dim F + \dim \mathcal{N}(A) - \dim\,[\mathcal{N}(A) + \operatorname{lin} F] \\ &= \dim F + \dim X - \dim Y - \dim\,[\mathcal{N}(A) + \operatorname{lin} F] \\ &= \dim F - \dim Y + \dim\,[\mathcal{R}(A^*) \cap F^\perp], \end{aligned}$$

with the third equality following from Proposition 3.2.1.
(4): $\mathcal{F}$ is a singleton set, iff $\operatorname{aff} \mathcal{F}$ is, so the equivalence follows from (2)(a).
■

Theorem 3.3.2 (Dual Faces) *Let* $(\bar{y}, \bar{z}) \in Feas(D)$, *and let*

$$\begin{aligned} \mathcal{G} &= \operatorname{face}((\bar{y}, \bar{z}), Feas(D)), \\ Y \times G &= \operatorname{face}((\bar{y}, \bar{z}), (Y \times K^*)) \quad (\Leftrightarrow\ G = \operatorname{face}(\bar{z}, K^*)). \end{aligned}$$

Then

(1) $Y \times G = \operatorname{face}(\mathcal{G}, (Y \times K^*))$,

(2) *(a)* $$\begin{aligned} \operatorname{aff} \mathcal{G} &= (Y \times \operatorname{lin} G) \cap \{(y, z) \mid A^* y + z = c\} \\ &= (\bar{y}, \bar{z}) + [(Y \times \operatorname{lin} G) \cap \mathcal{N}(A^*, I)], \end{aligned}$$

(b) $\mathcal{G} = (Y \times G) \cap \{(y, z) \mid A^* y + z = c\}$,

(3) $$\begin{aligned} \dim \mathcal{G} &= \dim\,[\operatorname{lin} G \cap \mathcal{R}(A^*)] \\ &= \dim G - (\dim X - \dim Y) + \dim\,[G^\perp \cap \mathcal{N}(A)]. \end{aligned}$$

(4) $\mathcal{G}$ is a singleton set, i.e. $\mathcal{G}$ (or equivalently $(\bar{y}, \bar{z})$) is an extreme point of Feas(D), if and only if

$$\operatorname{lin} G \cap \mathcal{R}(A^*) = \{0\}.$$

Proof (1) and (2) follow along the same lines as their counterparts in the Primal Faces Theorem, by noting that the relative interior [affine hull], of $Y \times G$ is the the direct product of Y with the relative interior [affine hull] of G.
(3): From (2)(a)

$$\begin{aligned} \operatorname{aff} \mathcal{G} - (\bar{y}, \bar{z}) &= (Y \times \operatorname{lin} G) \cap \mathcal{N}(A^*, I), \\ \dim \mathcal{G} &= \dim [\operatorname{lin} G \cap \mathcal{R}(A^*)] \\ &= \dim \mathcal{R}(A^*) + \dim G - \dim [\operatorname{lin} G + \mathcal{R}(A^*)] \\ &= \dim G - (\dim X - \dim Y) + \dim [G^{\perp} \cap \mathcal{N}(A)]. \end{aligned}$$

(4): $\mathcal{G}$ is a singleton set, iff $\operatorname{aff} \mathcal{G}$ is, so the equivalence follows from (2)(a). ■

Remark 3.3.1 If A is not onto, then $\dim Y$ in all the previous results should be replaced by $\operatorname{rank} A$.

Given a feasible solution of a cone-LP, and the minimal face of the *cone* containing it, its nondegeneracy is defined by imposing the extremity condition in the dual problem, with the complementary face in the dual cone; and strict complementarity of a solution pair by requiring them to be in the relative interior of complementary faces.

Definition 3.3.1 *Let $\bar{x}$ be feasible for (P), $(\bar{y}, \bar{z})$ feasible for (D),* $F = \operatorname{face}(\bar{x}, K)$*, and* $G = \operatorname{face}(\bar{z}, K^*)$*. We say that*

- $\bar{x}$ *is* (primal) nondegenerate *if*

$$\operatorname{lin} F^{\triangle} \cap \mathcal{R}(A^*) = \{0\},$$

- $(\bar{y}, \bar{z})$ *is* (dual) nondegenerate *if*

$$\operatorname{lin} G^{\triangle} \cap \mathcal{N}(A) = \{0\}.$$

Furthermore, if $\bar{x}$ and $(\bar{y}, \bar{z})$ are optimal solutions, then we say that

- *They are* strictly complementary *if* $F^{\triangle} = G$.

It turns out, that all results well known from LP that connect nondegeneracy in either (P) or (D), strict complementarity, and uniqueness of the optimal solution in the "opposite" problem carry over word by word to our more general framework.

Theorem 3.3.3 *Let $\bar{x}$, $(\bar{y}, \bar{z})$, F and G be as in Definition 3.3.1. Then the following hold.*

(1) (a) If $(\bar{y}, \bar{z})$ is nondegenerate, then $\bar{x}$ is a unique primal optimal solution.

(b) The converse of (1)(a) holds, assuming that they are strictly complementary.

(c) If $(\bar{y}, \bar{z})$ is nondegenerate, then

$$\dim G^{\triangle\perp} \;\geq\; \dim X - \dim Y.$$

(2) (a) If $\bar{x}$ is nondegenerate, then $(\bar{y}, \bar{z})$ is a unique dual optimal solution.

(b) The converse holds, assuming that they are strictly complementary.

(c) If $\bar{x}$ is nondegenerate, then

$$\dim F^{\triangle\perp} \;\geq\; \dim Y.$$

Proof of (1) The duality gap between arbitrary primal and dual feasible solutions x and (y, z) is $\langle x, z\rangle$. Hence

$$\begin{aligned} Opt(P) &= Feas(P) \cap \{\, x \in K^* \mid \langle x, \bar{z}\rangle = 0 \,\} \\ &= Feas(P) \cap G^{\triangle} \implies \\ E := \text{face}\,(Opt(P), K) &\subseteq G^{\triangle}. \end{aligned}$$

By the Primal Faces Theorem (3), *Opt(P)* is a singleton, iff

$$\text{lin}\, E \cap \mathcal{N}(A) \;=\; \{\, 0 \,\}.$$

From this (1)(a) is immediate. If strict complementarity holds, i.e. $\bar{x} \in Opt(P) \cap \text{ri}\, \text{G}^{\triangle}$, then by Proposition 3.2.2, (ii)

$$E \;=\; G^{\triangle}.$$

So, (1)(b) also follows. The proof of (1)(c) is immediate from the definition.
Proof of (2) Analogous. ■

Remark 3.3.2 With the sole exception of the (b) parts of Theorem 3.3.3, all results derived so far hold true, even in the case when K is not facially exposed. The only difficulty arising in this case is that (using the notation there), strict complementarity could be stated in 2 different ways:

$$\text{(i) } F^{\triangle} = G, \quad \text{or} \quad \text{(ii) } F = G^{\triangle}.$$

Neither of (i) or (ii) implies the other, unless K is facially exposed, in which case $F = F^{\triangle\triangle}$, and $G = G^{\triangle\triangle}$, hence (1) and (2) are equivalent.
However, the proof of the (b) parts in Theorem 3.3.3 implies, that if (1) (which may be called "primal strict complementarity") holds, and $(\bar{y}, \bar{z})$ is unique, then $\bar{x}$ is (primal) nondegenerate. Similarly, if (2) (which may be called "dual strict complementarity") holds, and $\bar{x}$ is unique, then $(\bar{y}, \bar{z})$ is (dual) nondegenerate.

Since all cones known so far that occur in practice are facially exposed, and the results for the non-exposed case are simple generalizations of the exposed case, we restrict ourselves to the latter. ■

Remark 3.3.3 Note that nondegeneracy simply requires the *extremity condition* given in parts (4) of the Primal and Dual Faces Theorems to hold for a *strictly complementary* solution in the "opposite" program (even though a strictly complementary solution pair may not always exist; for SDP, see the discussion at the end of this subsection). ■

Remark 3.3.4 Our definition of nondegeneracy is a generalization of the one used by Alizadeh, Haeberly and Overton for SDP in [24], and of the one by Alizadeh and Schmieta in [26]. In Definition 3.3.1 primal nondegeneracy is defined by

$$\begin{aligned} \mathcal{R}(A^*) \cap \operatorname{lin} F^{\triangle} &= \{0\} \Leftrightarrow \\ \mathcal{N}(A) + F^{\triangle\perp} &= X, \end{aligned} \tag{3.3.12}$$

and in [26] by

$$\mathcal{N}(A) + \tan(\bar{x}, K) \;=\; X, \tag{3.3.13}$$

"assuming that $\tan(\bar{x}, K)$ exists". If we define $\tan(\bar{x}, K)$ as in Section 3.2, and assume that K (such as $\mathcal{R}^n_+$, $\mathcal{S}^n_+$ and the p-cone) is nice, then by Lemma 3.2.1 $F^{\triangle\perp} = \tan(\bar{x}, K)$. Therefore (3.3.13) and (3.3.12) are equivalent. The case of dual nondegeneracy is similar, assuming that K^* is nice.
On the other hand, our theory covers not only nondegeneracy and strict complementarity, but also the characterization of the faces of the feasible sets (and tangent spaces and other sets, see the discussions below), which is of independent interest. ■

In the remainder of Subsection 3.3.1 we specialize these results for linear and semidefinite programs.

Example 3.3.1 (Linear programming) If $X = \mathcal{R}^n, Y = \mathcal{R}^m, K = K^* = \mathcal{R}^n_+$, then (P) and (D) are a standard pair of primal and dual linear programs. For $\bar{x} \in K$,

$$\text{lin face}\,(\bar{x}, K) \;=\; \{\, y \,|\, y_i = 0 \;\; \forall i \; s.t. \; \bar{x}_i = 0 \,\}.$$

In particular, (3) in Theorem 3.3.1 specializes to: $\bar{x}$ is an extreme point iff the columns of A corresponding to its nonzero components are linearly independent. Similarly, $\bar{x}$ is nondegenerate, iff the *rows* of this submatrix are linearly independent.

If $\bar{x}$ and $(\bar{y}, \bar{z})$ are a strictly complementary pair of solutions, then

$$\begin{aligned} \dim \mathcal{F} &= \dim F - \dim Y + \dim [\, F^{\perp} \cap \mathcal{R}(A^*)\,] \\ &= \dim F - m + \dim [\, \mathrm{lin}\, G \cap \mathcal{R}(A^*)\,] \\ &= \dim F - m + \dim \mathcal{G}, \end{aligned} \tag{3.3.14}$$

where the first equality follows from (3) in the Primal Faces Theorem, the second by $F^{\perp} = \mathrm{lin}\, G$, and the third by (3) in the Dual Faces Theorem. The formula (3.3.14) reduces to a result of Tijssen and Sierksma [777], as follows. They define

$$\sigma(\mathcal{F}) = \dim \mathcal{F} + \mathrm{bnd}\, \mathcal{F} - n,$$

as the *degree of degeneracy* of face $\mathcal{F}$, with $\mathrm{bnd}\, \mathcal{F}$ being the number of hyperplanes in the description of the polyhedron *Feas(P)*, which are tight at $\bar{x}$ (equivalently at every point in $\mathcal{F}$). (The number $\sigma(\mathcal{F})$ can depend on the representation of *Feas(P)*.) In words, the degeneracy degree is the number of "superfluous" hyperplanes at $\mathcal{F}$, i.e. of hyperplanes which are tight at $\mathcal{F}$, but not necessary to define its affine hull. They prove

$$\dim \mathcal{G} = \sigma(\mathcal{F}). \tag{3.3.15}$$

Rewriting (3.3.14) yields

$$\dim \mathcal{G} = \dim \mathcal{F} + m - \dim F,$$

and it is not hard to see, that

$$m - \dim F = \mathrm{bnd}\, \mathcal{F} - n.$$

Therefore (3.3.14) and (3.3.15) are indeed equivalent. ■

Example 3.3.2 (Semidefinite programming) If $X = \mathcal{S}^n, Y = \mathcal{R}^m, K = K^* = \mathcal{S}^n_+$, A and A^* are defined via $a^1, \ldots, a^m \in \mathcal{S}^n$ as

$$Ax = \begin{pmatrix} \langle a^1, x\rangle \\ \vdots \\ \langle a^m, x\rangle \end{pmatrix}, \quad A^* y = \sum_{i=1}^m y_i a^i,$$

then (P) and (D) are a pair of semidefinite programs. By the characterization of the faces of $\mathcal{S}^n_+$ given in Example 3.2.2,

$$\mathrm{lin}\,(\mathrm{face}\,(\bar{x}, \mathcal{S}^n_+)) = \{\, x \in \mathcal{S}^n \,|\, \mathcal{R}(x) \subseteq \mathcal{R}(\bar{x})\,\}.$$

Therefore specializing the results in the Primal and Dual Faces Theorems is straightforward. We obtain

Corollary 3.3.1 *Suppose that $\bar{x} \in Feas(P), (\bar{y}, \bar{z}) \in Feas(D)$, where (P) and (D) are a pair of SDP's defined by the operator A above, and $b \in \mathcal{R}^m$, $c \in \mathcal{S}^n$. Let*

$$\begin{array}{lllllll} \mathcal{F} & = & \text{face}\,(\bar{x}, Feas(P)), & & r & = & \text{rank}\,\bar{x}, \\ \mathcal{G} & = & \text{face}\,((\bar{y}, \bar{z}), Feas(D)), & & s & = & \text{rank}\,\bar{z}. \end{array}$$

Then the following hold.

(1) $t(r) \leq m + \dim \mathcal{F}$.

(2) If $\bar{x}$ is nondegenerate, then $t(n-r) \leq t(n) - m$.

(3) $t(s) \leq (t(n) - m) + \dim \mathcal{G}$.

(4) If $(\bar{y}, \bar{z})$ is nondegenerate, then $t(n-s) \leq m$.

(5) $\bar{x}$ and $(\bar{y}, \bar{z})$ are strictly complementary, iff $r + s = n$.

Proof Note that $\dim X = t(n)$, $\dim Y = m$, and by Example 3.2.2 if $x \in \mathcal{S}^n_+$ then

$$\text{rank}\, x \leq r \quad \Leftrightarrow \quad \dim\ \text{face}\,(x, \mathcal{S}^n_+) \leq t(r).$$

Then, (1) follows by (3) in the Primal Faces Theorem, (2) by (1)(c) in Theorem 3.3.3, (3) by (3) in the Dual Faces Theorem, and (4) by (2)(c) in Theorem 3.3.3. ■

Moreover, just as in the LP case, one can obtain and elegant characterization of extreme and nondegenerate solutions. Let $\bar{x}$ be a feasible solution of (P), and q an orthonormal matrix such that

$$\bar{x} \quad = \quad q\left(\begin{smallmatrix} \Lambda & 0 \\ 0 & 0 \end{smallmatrix}\right)q^{\mathrm{T}},$$

where Λ is a diagonal matrix with positive diagonal. Partition q as $q = [q^1, q^2]$, where $\mathcal{R}(q^1) = \mathcal{R}(\bar{x})$ and $\mathcal{R}(q^2)$ is its orthogonal complement. Since

$$A\bar{x} \quad = \quad \begin{pmatrix} \langle a^1, \bar{x} \rangle \\ \vdots \\ \langle a^m, \bar{x} \rangle \end{pmatrix} \quad = \quad \begin{pmatrix} \langle q^T a^1 q, \left(\begin{smallmatrix} \Lambda & 0 \\ 0 & 0 \end{smallmatrix}\right) \rangle \\ \vdots \\ \langle q^T a^m q, \left(\begin{smallmatrix} \Lambda & 0 \\ 0 & 0 \end{smallmatrix}\right) \rangle \end{pmatrix},$$

we obtain that $\bar{x}$ is an extreme [nondegenerate] point in *Feas(P)* iff $\left(\begin{smallmatrix} \Lambda & 0 \\ 0 & 0 \end{smallmatrix}\right)$ is such a point in the primal SDP defined by rhs b, and linear operator A_q, where

$$A_q x \quad = \quad \begin{pmatrix} \langle q^T a^1 q, x \rangle \\ \vdots \\ \langle q^T a^m q, x \rangle \end{pmatrix}.$$

Therefore,

Corollary 3.3.2 *Using the notation of Corollary 3.3.1, the following hold.*

(1) $\bar{x}$ is an extreme point of Feas(P) $\Leftrightarrow$ the matrices

$$(q^1)^T a^1 q^1, \ldots, (q^1)^T a^m q^1$$

span $\mathcal{S}^r$.

(2) $\bar{x}$ is a nondegenerate point of Feas(P) $\Leftrightarrow$ the matrices

$$\begin{pmatrix} (q^1)^T a^1 q^1 & (q^1)^T a^1 q^2 \\ (q^2)^T a^1 q^1 & 0 \end{pmatrix}, \ldots, \begin{pmatrix} (q^1)^T a^m q^1 & (q^1)^T a^m q^2 \\ (q^2)^T a^m q^1 & 0 \end{pmatrix}$$

are linearly independent.

■

To characterize the extremity, and nondegeneracy of a *dual* feasible solution $(\bar{y}, \bar{z})$, one does not need to repeat the calculations. Recall from Remark 3.3.3 that nondegeneracy requires the extremity condition to hold for a strictly complementary solution in the "opposite" problem. Suppose that $(\bar{y}, \bar{z})$ is a feasible solution of (D), and write

$$\bar{x} = \widetilde{q} \left(\begin{smallmatrix} 0 & 0 \\ 0 & \Omega \end{smallmatrix} \right) \widetilde{q}^T,$$

where Ω is a diagonal matrix with positive diagonal, $s = \operatorname{rank} \bar{z}$, $\widetilde{q} = [\widetilde{q}^1, \widetilde{q}^2]$, where $\mathcal{R}(\widetilde{q}^2) = \mathcal{R}(\bar{z})$ and $\mathcal{R}(\widetilde{q}^1)$ is its orthogonal complement. Then we obtain

Corollary 3.3.3 *Using the notation of Corollary 3.3.1, the following hold.*

(1') *$(\bar{y}, \bar{z})$ is an extreme point of Feas(D) iff the matrices in (2) of Corollary 3.3.2 with $\widetilde{q}$ in place of q are linearly independent.*

(2') *$(\bar{y}, \bar{z})$ is a nondegenerate point of Feas(D) iff the matrices in (1) of Corollary 3.3.2 with $\widetilde{q}$ in place of q span $\mathcal{S}^{n-s}$.*

■

In ordinary linear programs, a strictly complementary solution-pair always exists, as was shown by Goldman and Tucker [290]. This is not the case for SDP. An example was given in [24], which we reproduce here. Let $n = m = 3$,

$$b = (1 \quad 0 \quad 0)^T, \quad c = \begin{pmatrix} 0 & 0 & 0 \\ 0 & 0 & 0 \\ 0 & 0 & 1 \end{pmatrix},$$

$$a^1 = \begin{pmatrix} 1 & 0 & 0 \\ 0 & 0 & 0 \\ 0 & 0 & 0 \end{pmatrix}, \quad a^2 = \begin{pmatrix} 0 & 0 & 1 \\ 0 & 1 & 0 \\ 1 & 0 & 0 \end{pmatrix}, \quad a^3 = \begin{pmatrix} 0 & 1 & 0 \\ 1 & 0 & 0 \\ 0 & 0 & 1 \end{pmatrix}.$$

Then it is straightforward to see that $\bar{x}$ and $(\bar{y}, \bar{z})$ given by

$$\bar{x} = \begin{pmatrix} 1 & 0 & 0 \\ 0 & 0 & 0 \\ 0 & 0 & 0 \end{pmatrix}, \quad \bar{y} = \begin{pmatrix} 0 & 0 & 0 \end{pmatrix}^{\mathrm{T}}, \quad \bar{z} = \begin{pmatrix} 0 & 0 & 0 \\ 0 & 0 & 0 \\ 0 & 0 & 1 \end{pmatrix}$$

are both nondegenerate, therefore both unique optimal solutions, which do not satisfy strict complementarity.
An instructive family of SDP examples where strict complementarity fails can be created from (convex) quadratically constrained quadratic programs (QCQP's). All such problems can be written in the form

$$\begin{array}{ll} \max & \langle b, y \rangle \\ s.t. & \|d^i y - f^i\|^2 \;\leq\; \gamma_i \;\; (i = 1, \ldots, m) \end{array} \qquad \text{(QCQP)}$$

with the d^i's being appropriate symmetric matrices and the f^i's vectors. The problem $(QCQP)$ then has an SDP representation

$$\begin{array}{ll} \max & \langle b, y \rangle \\ s.t. & \begin{pmatrix} I & d^i y - f^i \\ (d^i y - f^i)^T & \gamma_i \end{pmatrix} \succeq 0 \;\; (i = 1, \ldots, m) \end{array}$$

which is equivalent to

$$\begin{array}{ll} \max & \langle b, y \rangle \\ s.t. & z^i \succeq 0 \\ & \sum_{j=1}^{m} \begin{pmatrix} 0 & -d^i_{\cdot j} \\ -d^i_{\cdot j} & 0 \end{pmatrix} y_j + z^i = \begin{pmatrix} I & -f^i \\ -(f^i)^T & \gamma_i \end{pmatrix} \;\; (i = 1, \ldots, m) \end{array}$$

$$\text{(SDP-QCQP)}$$

The proof of the following theorem is straightforward, therefore omitted.

Theorem 3.3.4 *Suppose that*

- $\bar{y}$ *is a unique optimal solution of* $(QCQP)$ *(therefore also of (SDP-*$QCQP$*) with the appropriate* $\bar{z}^1, \ldots, \bar{z}^m$ *slacks), and*
- *in* $(QCQP)$ *constraints* 1 *through* k *are the tight ones.*

Then (1) and (2) below are equivalent.

(1) b *is a strictly positive combination of the gradients of the tight constraints of* $(QCQP)$.

(2) The dual of (SDP-$QCQP$*) has an optimal solution, which is strictly complementary with* $(\bar{y}, \bar{z}^1, \ldots, \bar{z}^m)$.

■

Finally, since the dimension formulas that follow from parts (3) in the Primal and Dual Faces Theorems and Theorem 3.3.3 look quite elegant for *extreme and nondegenerate* solutions, we state them separately in Corollary 3.3.4. More refined and tighter formulas will follow in subsection 3.3.3.

Corollary 3.3.4 *Suppose that $\bar{x}$ is an extreme and nondegenerate point in Feas(P) and $(\bar{y}, \bar{z})$ in Feas(D),*

$$F = \operatorname{face}(\bar{x}, K), \quad G = \operatorname{face}(\bar{z}, K^*).$$

Then

$$\begin{array}{rcccl} \dim F & \leq & \dim Y & \leq & \dim F^{\triangle\perp}, \\ \dim G & \leq & \dim X - \dim Y & \leq & \dim G^{\triangle\perp}. \end{array} \qquad (3.3.18)$$

In particular, if (P) and (D) are ordinary linear programs, with data defined as in Example 3.3.1,

$$r = \#\text{ of nonzeros in } \bar{x}, \quad s = \#\text{ of nonzeros in } \bar{z},$$

then

$$\begin{array}{rcl} r & = & m, \\ s & = & n - m. \end{array} \qquad (3.3.19)$$

If (P) and (D) are semidefinite programs, with data defined as in Example 3.3.2,

$$r = \operatorname{rank} \bar{x}, \quad s = \operatorname{rank} \bar{z},$$

then

$$\begin{array}{rcccl} t(r) & \leq & m & \leq & t(n) - t(n-r), \\ t(s) & \leq & t(n) - m & \leq & t(n) - t(n-s). \end{array} \qquad (3.3.20)$$

■

3.3.2 Tangent spaces

In this subsection we characterize the *tangent spaces* of the feasible sets of (P) and (D) at given feasible solutions. It turns out, that they can described in a manner similar to how the faces were described. The main result is

Theorem 3.3.5 (Tangent Spaces) *Suppose that K and K^* are nice, and both (P) and (D) satisfy the Slater condition. Let $\bar{x}$ and $(\bar{y}, \bar{z})$ be primal and dual feasible solutions, respectively,*

$$\begin{array}{rclcrcl} F & = & \operatorname{face}(\bar{x}, K), & & \mathcal{T} & = & \tan(\bar{x}, Feas(P)), \\ G & = & \operatorname{face}(\bar{z}, K^*), & & \mathcal{U} & = & \tan((\bar{y}, \bar{z}), Feas(D)). \end{array}$$

Then

(1) $\mathcal{T} = F^{\triangle\perp} \cap \mathcal{N}(A)$,

(2) $\dim \mathcal{T} = \dim [F^{\triangle\perp} \cap \mathcal{N}(A)]$
$= \dim F^{\triangle\perp} - \dim Y + \dim [\operatorname{lin} F^{\triangle} \cap \mathcal{R}(A^*)]$,

(3) $\mathcal{U} = (Y \times G^{\triangle\perp}) \cap \mathcal{N}(A^*, I)$,

(4) $\dim \mathcal{U} = \dim [G^{\triangle\perp} \cap \mathcal{R}(A^*)]$
$= \dim G^{\triangle\perp} - (\dim X - \dim Y) + \dim [\operatorname{lin} G^{\triangle} \cap \mathcal{N}(A)]$.

Proof First, from (3.2.11) in Lemma 3.2.1 recall

$$\tan(\bar{x}, K) = F^{\triangle\perp}, \quad \tan(\bar{z}, K^*) = G^{\triangle\perp}.$$

(1): Since (P) satisfies the Slater condition, by Proposition 3.2.3

$$\begin{aligned} \operatorname{tcone}(\bar{x}, Feas(P)) &= \operatorname{tcone}(\bar{x}, K) \cap \mathcal{N}(A) \Rightarrow \\ \mathcal{T} = \tan(\bar{x}, Feas(P)) &= \tan(\bar{x}, K) \cap \mathcal{N}(A) = F^{\triangle\perp} \cap \mathcal{N}(A). \end{aligned}$$

(2): Analogous to the proof of (1).
(3): Since

$$\begin{aligned} \operatorname{aff} \mathcal{F} - \bar{x} &= \operatorname{lin} F \cap \mathcal{N}(A), \\ \mathcal{T} &= F^{\triangle\perp} \cap \mathcal{N}(A), \end{aligned}$$

we can use a similar calculation (with $F^{\triangle\perp}$ in place of $\operatorname{lin} F$) as in computing $\dim \mathcal{F}$ in the proof of the Primal Faces Theorem.
(4): Since

$$\begin{aligned} \operatorname{aff} \mathcal{G} - (\bar{y}, \bar{z}) &= (Y \times \operatorname{lin} G) \cap \mathcal{N}(A^*, I), \\ \mathcal{U} &= (Y \times G^{\triangle\perp}) \cap \mathcal{N}(A^*, I), \end{aligned}$$

we can use a similar calculation (with $G^{\triangle\perp}$ in place of $\operatorname{lin} G$) as in computing $\dim \mathcal{G}$ in the proof of the Primal Faces Theorem. ∎

Remark 3.3.5 If we can compute $\dim \mathcal{T}$ and $\dim F^{\triangle\perp}$, then equation (2) makes it possible to check whether $\bar{x}$ is nondegenerate (with an analogous statement relating dual nondegeneracy to equation (4)).

To keep the presentation relatively short, we do not write out the specializations for LP and SDP; they are quite straightforward.

3.3.3 The boundary structure inequalities

Putting together the results of the previous subsections, we shall now derive several instructive inequalities that relate the dimensions of

- Minimal faces in the primal and dual *cones* that contain a given optimal solution.

- Minimal faces in the primal and dual *feasible sets* that contain a given optimal solution.

- Tangent spaces at the primal and dual feasible sets at a given optimal solution.

We will call these inequalities the *boundary structure inequalities.* Theorem 3.3.6 below contains their statement, and an intuitive explanation follows.

Theorem 3.3.6 (Boundary Structure Inequalities) *Suppose that K and K^* are nice, and both (P) and (D) satisfy the Slater condition. Let $\bar{x}$ and $(\bar{y}, \bar{z})$ be complementary solutions of (P) and (D), respectively,*

$$\begin{array}{lll} F = \text{face}\,(\bar{x}, K), & \mathcal{F} = \text{face}\,(\bar{x}, Feas(P)), & \mathcal{T} = \tan(\bar{x}, Feas(P)), \\ G = \text{face}\,(\bar{z}, K^*), & \mathcal{G} = \text{face}\,((\bar{y}, \bar{z}), Feas(D)), & \mathcal{U} = \tan((\bar{y}, \bar{z}), Feas(D)). \end{array}$$

Then

(1) $\dim F - \dim Y + \dim \mathcal{U} \;\leq\; \dim \mathcal{F} \;\leq\; \dim G^{\triangle} - \dim Y + \dim \mathcal{U}$,

(2) $\dim \mathcal{F} + (\dim X - \dim \mathcal{T}) \;\geq\; \dim F + \dim F^{\triangle} + \dim\,[\,\text{res}\; F \cap \mathcal{R}(A^*)\,]$.

Suppose that strict complementarity holds. Then

(3) $\dim \mathcal{U} + \dim \mathcal{T} \;=\; \dim \mathcal{F} + \dim \mathcal{G} + \dim \text{res}\; F$,

(4) $\dim F^{\triangle\perp} - \dim \mathcal{T} \;\leq\; \dim Y \;\leq\; \dim F + \dim \mathcal{U}$,

with the left inequality at equality iff the primal solution is unique, and the right inequality at equality iff the dual solution is unique.

Proof First, from the Primal and Dual Faces Theorems, and the Tangent Spaces Theorem we recall

$$\begin{aligned} \dim \mathcal{F} &= \dim F - \dim Y + \dim\,[\,F^{\perp} \cap \mathcal{R}(A^*)\,], && (3.3.21) \\ \dim \mathcal{T} &= \dim F^{\triangle\perp} - \dim Y + \dim\,[\,\text{lin}\, F^{\triangle} \cap \mathcal{R}(A^*)\,], && (3.3.22) \\ \dim \mathcal{G} &= \dim\,[\,\text{lin}\, G \cap \mathcal{R}(A^*)\,], && (3.3.23) \\ \dim \mathcal{U} &= \dim\,[\,G^{\triangle\perp} \cap \mathcal{R}(A^*)\,]. && (3.3.24) \end{aligned}$$

Moreover,

$$F \;\subseteq\; G^{\triangle}, \quad F^{\perp} \;\supseteq\; G^{\triangle\perp}.$$

(1): The first inequality:

$$\begin{aligned} \dim \mathcal{F} &= \dim F - \dim Y + \dim\,[\,F^{\perp} \cap \mathcal{R}(A^*)\,] \\ &\geq \dim F - \dim Y + \dim\,[\,G^{\triangle\perp} \cap \mathcal{R}(A^*)\,] \\ &= \dim F - \dim Y + \dim \mathcal{U}. \end{aligned}$$

The second:

$$\begin{aligned} \dim \mathcal{F} &= \dim F - \dim Y + \dim\,[\,F^{\perp} \cap \mathcal{R}(A^*)\,] \\ &\leq \dim G^{\triangle} - \dim Y + \dim\,[\,G^{\triangle\perp} \cap \mathcal{R}(A^*)\,] \\ &= \dim G^{\triangle} - \dim Y + \dim \mathcal{U}. \end{aligned}$$

as the number of linearly independent vectors in $\operatorname{lin} G^{\triangle} \setminus \operatorname{lin} F$ is at least as large as in $[\,F^{\perp} \cap \mathcal{R}(A^*)\,] \setminus [\,G^{\triangle\perp} \cap \mathcal{R}(A^*)\,]$. Also, this inequality is an equality if strict complementarity holds.
(2): Taking (3.3.21) – (3.3.22) yields

$$\begin{array}{rcl} \dim \mathcal{F} - \dim \mathcal{T} & = & \dim F - \dim F^{\triangle\perp} + \dim\,[\,F^{\perp} \cap \mathcal{R}(A^*)\,] \\ & & - \dim\,[\operatorname{lin} F^{\triangle} \cap \mathcal{R}(A^*)\,] \\ & \Leftrightarrow & \\ \dim \mathcal{F} + (\dim X - \dim \mathcal{T}) & = & \dim F + \dim F^{\triangle} + \dim\,[\,F^{\perp} \cap \mathcal{R}(A^*)\,] \\ & & - \dim\,[\operatorname{lin} F^{\triangle} \cap \mathcal{R}(A^*)\,] \\ & \geq & \dim F + \dim F^{\triangle} + \dim\,[\operatorname{res}\, F \cap \mathcal{R}(A^*)\,]. \end{array}$$

(3): By symmetry from (1), or from (3.3.23) we obtain

$$\begin{array}{rcl} \dim G - (\dim X - \dim Y) + \dim \mathcal{T} & \leq & \dim \mathcal{G} \\ & \leq & \dim F^{\triangle} - (\dim X - \dim Y) + \dim \mathcal{T} \quad \Leftrightarrow \\ \dim Y - \dim G^{\perp} + \dim \mathcal{T} & \leq & \dim \mathcal{G} \\ & \leq & \dim Y - \dim F^{\triangle\perp} + \dim \mathcal{T}. \end{array}$$

Adding the second of these inequality chains to (1) yields

$$\begin{array}{rcl} \dim F - \dim G^{\perp} + (\dim \mathcal{U} + \dim \mathcal{T}) & \leq & \dim \mathcal{F} + \dim \mathcal{G} \\ & \leq & \dim G^{\triangle} - \dim F^{\triangle\perp} + (\dim \mathcal{U} + \dim \mathcal{T}), \end{array}$$

therefore, if strict complementarity holds, then

$$\dim \mathcal{U} + \dim \mathcal{T} \quad = \quad \dim \mathcal{F} + \dim \mathcal{G} + \dim \operatorname{res}\, F.$$

(4): From (1) we have

$$\dim \mathcal{F} \quad = \quad \dim F - \dim Y + \dim \mathcal{U} \qquad (3.3.25)$$

By symmetry, and strict complementarity

$$\begin{array}{rcl} \dim \mathcal{G} & = & \dim F^{\triangle} - (\dim X - \dim Y) + \dim \mathcal{T} \\ & = & \dim Y - \dim F^{\triangle\perp} + \dim \mathcal{T}. \end{array} \qquad (3.3.26)$$

Putting (3.3.25) and (3.3.26) together we obtain

$$\begin{array}{rcl} \dim Y & = & \dim F - \dim \mathcal{F} + \dim \mathcal{U} \\ & = & \dim \mathcal{G} + \dim F^{\triangle\perp} - \dim \mathcal{T}, \end{array}$$

from which the two inequalities in (4) readily follow. ■

The geometry behind the inequalities Inequality (1) in the Boundary Structure Inequalities Theorem is a generalization of the Tijssen-Sierksma equality (3.3.15). Inequality (2) proves that "the primal *feasible set* at $\bar{x}$ is at least as nonsmooth as the primal *cone* at $\bar{x}$".

To see what this means, for a convex set $C \subseteq \mathcal{R}^n$, and $x \in C$, the normal cone to C at x is defined as

$$\text{ncone}\,(x, C) \;=\; \{\, v \,|\, \langle v, x\rangle \geq \langle v, y\rangle \text{ for all } y \in C\,\}.$$

It is well known, that

$$\begin{aligned} \text{ncone}\,(x, C)^* &= -\text{tcone}\,(x, C) \;\Rightarrow \\ \text{ncone}\,(x, C)^\perp &= \tan(x, C) \;\Rightarrow \\ \dim\ \text{ncone}\,(x, C) + \dim\ \tan(x, C) &= n. \end{aligned} \tag{3.3.27}$$

Also, if $\bar{x}$ and F are as in the Boundary Structure Inequalities Theorem, then

$$\text{ncone}\,(x, K) \;=\; -F^\triangle.$$

The quantity

$$\dim\ \text{face}\,(x, C) + \dim\ \text{ncone}\,(x, C)$$

is an intuitive measure of the nonsmoothness of C at x. Since face $(x, C) \subseteq \tan(x, C)$, by (3.3.27) this number is always less than or equal than n. If C is a polyhedron, then it is equal to n; and if C is a sphere (that is "as smooth as possible"), and x is on its boundary, then it is 1.
Now, inequality (2) can be rewritten as

$$\begin{aligned} \dim\ \mathcal{F} + \dim\ \text{ncone}\,(\bar{x}, Feas(P)) \;\geq\; & \dim\ F + \dim\ F^\triangle \\ & + \dim\ [\mathcal{R}(A^*) \cap \text{res}\ F].(3.3.28) \end{aligned}$$

Since

- $\dim\ \mathcal{F}$ + $\dim\ \text{ncone}\,(\bar{x}, Feas(P))$ is the measure of nonsmoothness of *Feas(P)* at $\bar{x}$, and
- $\dim\ F + \dim\ F^\triangle$ is the measure of nonsmoothness of K at $\bar{x}$,

inequality (3.3.28) indeed shows that "*Feas(P)* at $\bar{x}$ is at least as nonsmooth as K at $\bar{x}$". Confer this with the case, when $K = \mathcal{R}^n_+$; in this case, res $F = \{0\}$, and inequality (3.3.28) proves that *Feas(P)*, which is a polyhedron, is exactly as nonsmooth as K.
Finally, consider equality (3) in the theorem. This proves, that if strict complementarity holds, then

$$\dim\ \mathcal{T} + \dim\ \mathcal{U} \;\geq\; \dim\ \text{res}\ F. \tag{3.3.29}$$

That is, unless the face F is polyhedral, ie. res $F = \{0\}$, then at least one of the primal and dual feasible sets will have a nontrivial tangent space at the given optimal solutions.

Specializing the Boundary Structure Inequalities Theorem for SDP, and using the substitution for the dimension of the normal cone, as described above, we obtain

Corollary 3.3.5 *If (P) and (D) are semidefinite programs, with data defined as in Example 3.3.2, $\bar{x}$ and $(\bar{y}, \bar{z})$ are complementary solutions of (P) and (D), respectively,*

$$\begin{array}{lll} r = \operatorname{rank} \bar{x}, & \mathcal{F} = \text{face}\,(\bar{x}, Feas(P)), & \mathcal{T} = \tan(\bar{x}, Feas(P)), \\ s = \operatorname{rank} \bar{z}, & \mathcal{G} = \text{face}\,((\bar{y}, \bar{z}), Feas(D)), & \mathcal{U} = \tan((\bar{y}, \bar{z}), Feas(D)). \end{array}$$

Then

(1) $t(r) - m + \dim \mathcal{U} \ \leq\ \dim \mathcal{F} \ \leq\ t(n-s) - m + \dim \mathcal{U}.$

(2) $\dim \mathcal{F} + \dim \text{ncone}\,(\bar{x}, Feas(P)) \ \geq\ t(r) + t(s) + \dim\,[\,\text{res } F \cap \mathcal{R}(A^*)\,].$

Suppose that strict complementarity holds. Then

(3) $\dim \mathcal{U} + \dim \mathcal{T} \ =\ \dim \mathcal{F} + \dim \mathcal{G} + r(n-r),$

(4) $t(n) - t(n-r) - \dim \mathcal{T} \ \leq\ m \ \leq\ t(r) + \dim \mathcal{U},$

with the left inequality at equality iff the primal solution is unique, and the right inequality at equality iff the dual solution is unique.

■

3.3.4 The geometry of the feasible sets expressed with different variables

Frequently, the dual problem is given in a form without the slack variable z, as

$$\begin{array}{lll} Max & \langle b, y \rangle & \\ s.t. & A^* y \leq_{K^*} c & \qquad (D_y) \end{array}$$

The points in *Feas(D)* and *Feas(D)* are obviously in one-to-one correspondence: $y \in Feas(D_y) \Leftrightarrow (y, \phi(y)) \in Feas(D)$, with $\phi(y) = c - A^* y$. It is easy, if somewhat tedious to translate all results proved previously on the geometry of *Feas(D)* to describe the geometry of *Feas(D_y)*. The proof of Lemma 3.3.1 is deferred to Appendix B.

Lemma 3.3.1 *The following hold.*

(1) *(a)* *F is a face of Feas(D), if and only if* $\text{Proj}_{\,y}(F)$ *is a face of Feas(D_y).*

(b) $\dim F = \dim \text{Proj}_{\,y}(F)$.

(2) *If* $(\bar{y}, \bar{z}) \in Feas(D)$*, then*

(a) $\tan(\bar{y}, Feas(D_y)) = \text{Proj}_{\,y}\,[\,\tan((\bar{y}, \bar{z}), Feas(D))\,],$

(b) $\dim\,[\,\tan(\bar{y}, Feas(D_y))\,] = \dim\,[\,\tan((\bar{y}, \bar{z}), Feas(D))\,].$

■

3.3.5 A detailed example

In this subsection we give a detailed example that illustrates the formulas

$$\begin{aligned}
\dim \mathcal{G} &= \dim\,[\,\mathrm{lin}\, G \cap \mathcal{R}(A^*)\,] \\
&= \dim G - (\dim X - \dim Y) + \dim\,[\,G^{\perp} \cap \mathcal{N}(A)\,], \quad (3.3.30)\\
\dim \mathcal{U} &= \dim\,[\,G^{\triangle\perp} \cap \mathcal{R}(A^*)\,] \\
&= \dim G^{\triangle\perp} - (\dim X - \dim Y) + \dim[\,\mathrm{lin}\, G^{\triangle} \cap \mathcal{N}(A)\,] \quad (3.3.31)\\
\dim \mathcal{F} &= \dim\,[\,\mathrm{lin}\, F \cap \mathcal{N}(A)\,], \quad (3.3.32)\\
\dim \mathcal{T} &= \dim\,[\,F^{\triangle\perp} \cap \mathcal{N}(A)\,], \quad (3.3.33)
\end{aligned}$$

from the Dual Faces and Tangent Spaces Theorems, by showing a primal-dual pair of cone program, with 3 different objective functions for the dual. For each objective, we will

- Find a complementary, hence optimal primal-dual solution pair.
- Compute $\dim G$ and $\dim G^{\triangle\perp}$ algebraically.
- Find $\dim \mathcal{U}$ and $\dim \mathcal{G}$ by inspecting the graph of the dual feasible set projected onto the y space.
- Deduce, whether the dual solution is nondegenerate from (3.3.31).
- If strict complementary holds, compute $\dim \mathcal{F}$ from (3.3.32) and $\dim \mathcal{T}$ from (3.3.33).

For each objective, the reader may easily check, that both (P) and (D) satisfy the Slater condition, hence the results of the Tangent Spaces Theorem will hold. The pair of cone programs are defined by $X = S^2 \times S^2$, $Y = \mathcal{R}^2$, $K = K^* = S^2_+ \times S^2_+$ as

$$(P)\quad \begin{array}{lllll}
Min & \langle \begin{pmatrix} 0 & -1 \\ -1 & 1 \end{pmatrix}, x^1 \rangle & + \langle \begin{pmatrix} 0 & 1 \\ 1 & 1 \end{pmatrix}, x^2 \rangle & & \\
st. & x^1 \in S^2_+, & x^2 \in S^2_+ & & \\
& \langle \begin{pmatrix} 0 & -1 \\ -1 & 0 \end{pmatrix}, x^1 \rangle & + \langle \begin{pmatrix} 0 & -1 \\ -1 & 0 \end{pmatrix}, x^2 \rangle & = & b_1 \\
& \langle \begin{pmatrix} -1 & 0 \\ 0 & 0 \end{pmatrix}, x^1 \rangle & + \langle \begin{pmatrix} -1 & 0 \\ 0 & 0 \end{pmatrix}, x^2 \rangle & = & b_2
\end{array}$$

$$(D)\quad \begin{array}{llllll}
Max & b_1 y_1 + b_2 y_2 & & & & \\
st. & z^1 \in S^2_+, & z^2 \in S^2_+ & & & \\
& y_1 \begin{pmatrix} 0 & -1 \\ -1 & 0 \end{pmatrix} & + y_2 \begin{pmatrix} -1 & 0 \\ 0 & 0 \end{pmatrix} & + z^1 & = & \begin{pmatrix} 0 & -1 \\ -1 & 1 \end{pmatrix} \\
& y_1 \begin{pmatrix} 0 & -1 \\ -1 & 0 \end{pmatrix} & + y_2 \begin{pmatrix} -1 & 0 \\ 0 & 0 \end{pmatrix} & + z^2 & = & \begin{pmatrix} 0 & 1 \\ 1 & 1 \end{pmatrix}
\end{array}$$

and we will have 3 different choices for b. A simple calculation shows

$$Feas(D_y) \;=\; \{\,(y_1, y_2) \,|\, y_2 \geq (y_1 - 1)^2,\; y_2 \geq (y_1 + 1)^2\,\}$$

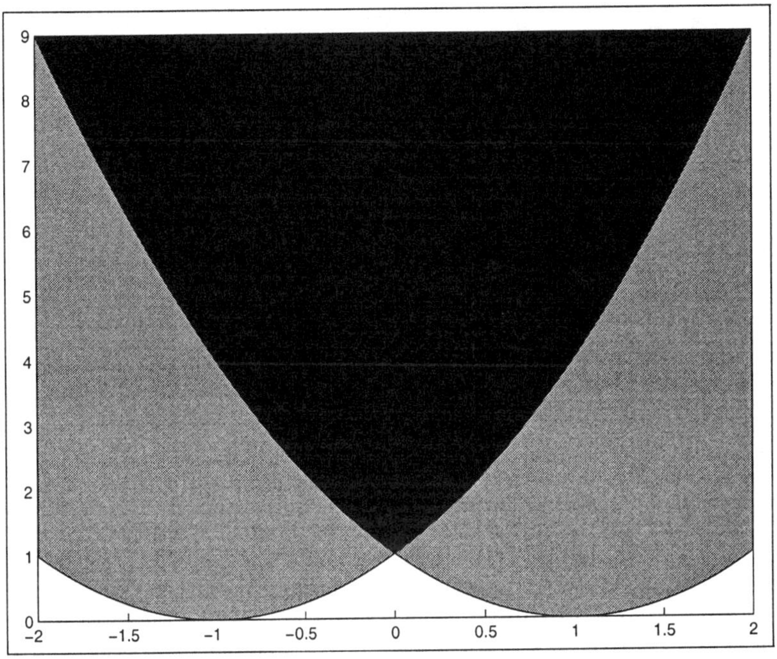

Figure 3.1 $\text{Feas}(D_y)$ for the example of subsection 3.3.5

and $Feas(D_y)$ is shown on Figure 3.1. First, let $(b_1, b_2) = (0, -1)$. An optimal solution pair of (D) and (P) is given by

$$\bar{y} = (0,1)^T, \quad \bar{z}^1 = \begin{pmatrix} 1 & 1 \\ 1 & 1 \end{pmatrix}, \quad \bar{z}^2 = \begin{pmatrix} 1 & -1 \\ -1 & 1 \end{pmatrix},$$
$$\bar{x}^1 = \tfrac{1}{2}\begin{pmatrix} 1 & -1 \\ -1 & 1 \end{pmatrix}, \quad \bar{x}^2 = \tfrac{1}{2}\begin{pmatrix} 1 & 1 \\ 1 & 1 \end{pmatrix}.$$

We find

$$\dim \mathcal{U} = 0, \qquad \dim \mathcal{G} = 0$$

from Figure 3.1, since the dimensions of the optimal face, and tangent space are the same in the (y, z)-space as in the y-space by the results of the previous subsection. Also,

$$\begin{aligned} \dim G &= 2, \qquad \dim G^{\triangle\perp} = 4, \\ \dim X - \dim Y &= 4. \end{aligned}$$

Plugging these numbers into (3.3.31) and (3.3.30) yields

$$\begin{aligned} \dim\,[\operatorname{lin} G^{\triangle} \cap \mathcal{N}(A)\,] &= 0, \\ \dim\,[\,G^{\perp} \cap \mathcal{N}(A)\,] &= 2. \end{aligned}$$

Therefore, $(\bar{y}, \bar{z}^1 \oplus \bar{z}^2)$ is nondegenerate, hence $\bar{x}^1 \oplus \bar{x}^2$ is a unique primal optimal solution. Also, as they are strictly complementary,

$$\dim \mathcal{T} = \dim\,[\,G^{\perp} \cap \mathcal{N}(A)\,] = 2.$$

Next, let $(b_1, b_2) = (-2, -1)$. An optimal solution pair of (D) and (P) is given by the same $(\bar{y}, \bar{z}^1 \oplus \bar{z}^2)$ as before and

$$\bar{x}^1 = 0, \quad \bar{x}^2 = \begin{pmatrix} 1 & 1 \\ 1 & 1 \end{pmatrix}.$$

Note that now strict complementary does not hold; indeed, since $(-2, -1)$ is the gradient of the constraint function $-y_2 + (y_1 - 1)^2$ at $\bar{y}$, so by Theorem 3.3.4 a strictly complementary solution pair cannot exist. Still, by nondegeneracy, the primal optimal solution is unique, and

$$\dim \mathcal{T} \;\leq\; \dim\,[\,G^{\perp} \cap \mathcal{N}(A)\,] \;=\; 2,$$

although the exact dimension of $\mathcal{T}$ now can only be computed algebraically (by finding $\dim\,[\,F^{\triangle\perp} \cap \mathcal{N}(A)\,]$).

Finally, let $(b_1, b_2) = (-4, -1)$. An optimal solution pair of (D) and (P) is now

$$\bar{y} = (-1, 4)^T, \quad \bar{z}^1 = \begin{pmatrix} 4 & -2 \\ -2 & 1 \end{pmatrix}, \quad \bar{z}^2 = \begin{pmatrix} 4 & 0 \\ 0 & 1 \end{pmatrix},$$
$$\bar{x}^1 = \begin{pmatrix} 1 & 2 \\ 2 & 4 \end{pmatrix}, \qquad \bar{x}^2 = 0.$$

We have

$$\begin{aligned} \dim \mathcal{U} &= 1, & \dim \mathcal{G} &= 0, \\ \dim G &= 4, & \dim G^{\triangle\perp} &= 5, \\ \dim X - \dim Y &= 4. \end{aligned}$$

hence

$$\begin{aligned} \dim\,[\,\mathrm{lin}\, G^{\triangle} \cap \mathcal{N}(A)\,] &= 0, \\ \dim\,[\,G^{\perp} \cap \mathcal{N}(A)\,] &= 0. \end{aligned}$$

therefore $(\bar{y}, \bar{z}^1 \oplus \bar{z}^2)$ is nondegenerate, so $\bar{x}$ is a unique primal optimal solution, and by strict complementarity

$$\dim \mathcal{T} \;=\; \dim\,[\,G^{\perp} \cap \mathcal{N}(A)\,] \;=\; 0.$$

3.4 SEMIDEFINITE COMBINATORICS

The subject of polyhedral combinatorics is the study of optimization problems via the polyhedral structure of their linear programming formulations. Combining knowledge on the extremal structure of polyhedra with the specifics of the LP-formulation can give valuable insights about the given optimization problem.

In this section we give three examples to illustrate that similar studies are possible, and worthwhile using convex programming, in particular SDP formulations. In a convex programming formulation one may want to characterize and study

(1) The extreme point optimal solutions in the primal and dual problems.

(2) The nondegenerate optimal solutions in the primal and dual problems.

(3) The solution pairs which lack strict complementarity.

3.4.1 *The Multiplicity of Optimal Eigenvalues*

Let A be an affine function from $\mathcal{R}^m$ to $\mathcal{S}^n$. The *eigenvalue-optimization problem* is

$$Min\ \{ f_k(A(x)) : x \in \mathcal{R}^m \}, \qquad (EV_k)$$

where $f_k(A(x))$ is the sum of the k largest eigenvalues of $A(x)$.
The problem (EV_k) can be formulated as a semidefinite program, as it was shown by Alizadeh [21] and Nesterov and Nemirovskii [583]. In fact, it is the earliest instance of SDP that has been the subject of a computational study; see [178, 196]. In these studies, and in many more recent papers dealing with eigenvalue-optimization, the following phenomenon was observed. At optimal solutions of (EV_k) the eigenvalues of the optimal matrix tend to coalesce; if x^* achieves the minimum, then frequently $\lambda_k(A(x^*)) = \lambda_{k+1}(A(x^*))$, and $\lambda_k(A(x^*))$ often has multiplicity even larger than two.
The clustering phenomenon plays a central role in eigenvalue-optimization. As proven by Cullum, Donath, and Wolfe [178] the function f_k is differentiable at a fixed symmetric matrix a if and only if $\lambda_k(a) > \lambda_{k+1}(a)$. If this condition fails to hold, then the dimension of the subgradient of f_k at a grows quadratically with the multiplicity of $\lambda_k(a)$. Furthermore, if f_k is nonsmooth at $A(x^*)$ then generally the composite function $f_k \circ A$ is also nonsmooth at x^*.
Therefore, clustering tends to cause the nondifferentiability of the objective function $f_k \circ A$ at a solution point, making (EV_k) a "model problem" in nonsmooth optimization. We remark, that $f_k \circ A$ *may* be differentiable at x^* even if f_k is *not* differentiable at $A(x^*)$ (e.g. when $A(x) \equiv I$); this, however is usually not the case.

Example 3.4.1 Consider the problem with $k = 1$, and

$$A(x) = x_1\begin{pmatrix} 1 & 0 \\ 0 & -1 \end{pmatrix} + x_2\begin{pmatrix} 0 & 1 \\ 1 & 0 \end{pmatrix} + \begin{pmatrix} 1 & 0 \\ 0 & 1 \end{pmatrix}.$$

The graph of the function $f_1 \circ A : \mathcal{R}^2 \longrightarrow \mathcal{R}$ is pictured in Figure 2. Clearly, at the unique optimal solution $x^* = (0,0)$ $f_1 \circ A$ is not differentiable.
Somewhat surprisingly, in the case $n = m = 2$ every choice of the coefficient matrices in the definition of $A(x)$ gives rise to at least one point where the objective function is nonsmooth. Another example is given in Figure 3. Here the optimal solution is not unique, still, there is a nonsmooth optimum. ∎

In the rest of the section we outline, how the rankbounds on extreme matrices in SDP's can be used to explain eigenvalue-clustering in extreme point optimal solutions of (EV_k). This material is taken from [623], and proofs are omitted, or only sketched.

Lemma 3.4.1 *Fix $a \in \mathcal{S}^n$, and suppose that*

$$a = q(\mathrm{Diag}\,(\lambda(a)))q^T.$$

Figure 3.2 The graph of $f_1 \circ A$

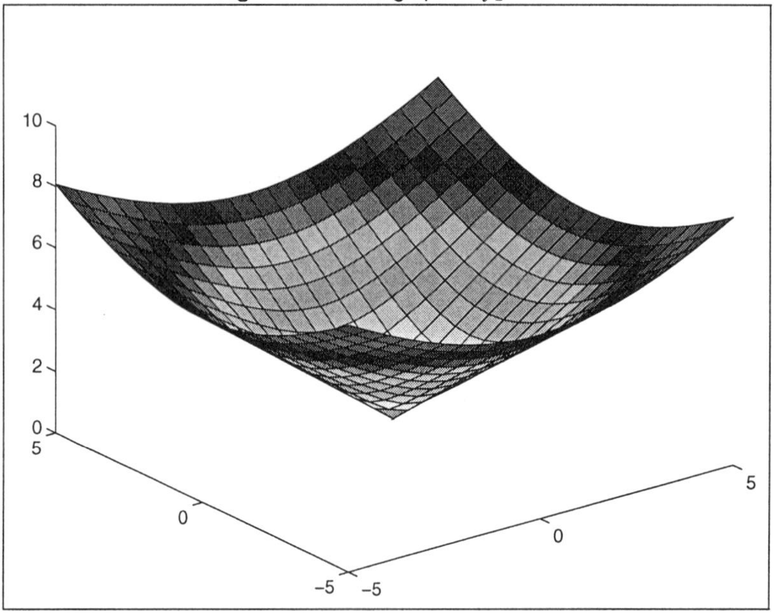

Figure 3.3 The graph of $f_1 \circ A$ for different data

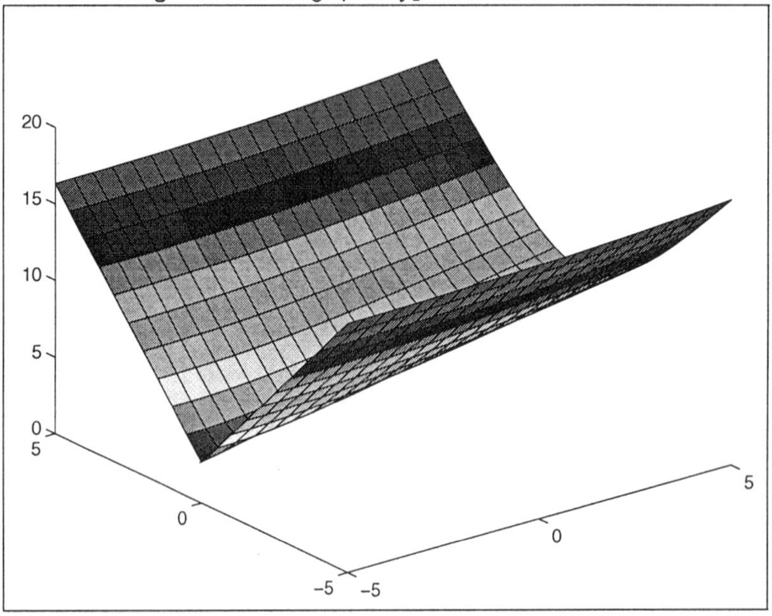

The optimal value of the problem

$$\begin{array}{lll} Min_{z,v,w} & kz + \langle I, v\rangle & \\ s.t. & v \in \mathcal{S}^n_+,\ w \in \mathcal{S}^n_+ & \quad (3.4.35) \\ & zI + v - w = a & \end{array}$$

is $f_k(a)$. The triple (z^*, v^*, w^*) is optimal to (3.4.35) if and only if

$$\begin{array}{rcl} z^* & \in & [\,\lambda_k(a),\ \lambda_{k+1}(a)\,], \\ v^* & = & q\,(\,\mathrm{Diag}\,(\lambda(a) - z^* e\,)_+\,)\,q^T, \\ w^* & = & q\,(\,\mathrm{Diag}\,(\,z^* e - \lambda(a)\,)_+\,)\,q^T. \end{array}$$

■

Remark 3.4.1 The optimal solutions of (3.4.35) do not depend on the choice of a's eigenvectors. Suppose that the distinct eigenvalues of a are

$$\lambda_{i_1}(a), \ldots, \lambda_{i_r}(a)$$

in descending order and $\lambda_k(a) = \lambda_{i_s}(a)$. Then the distinct eigenvalues of v^* and w^* in Lemma 3.4.1 are

$$\lambda_{i_1}(a) - z^*, \ldots, \lambda_{i_{s-1}}(a) - z^* \quad \text{and} \quad z^* - \lambda_{i_{s+1}}(a), \ldots, z^* - \lambda_{i_{s-1}}(a),$$

respectively. Therefore, choosing different eigenvectors of a to represent the eigenspace corresponding to $\lambda_{i_j}(a)$ $(j = 1, \ldots, r)$ does not change v^* and w^*.

It follows from Lemma 3.4.1, that (by plugging $A(x)$ into the place of a in (3.4.35)) the problem (EV_k) can be formulated as the SDP

$$\begin{array}{lll} Min_{x,z,v,w} & kz + \langle I, v\rangle & \\ s.t. & v \in \mathcal{S}^n_+,\ w \in \mathcal{S}^n_+ & \quad (SDP_k) \\ & zI + v - w = A(x) & \end{array}$$

In the following we assume the mapping A to be fixed, and denote the set of optimal solutions of

- (3.4.35) (in the (z, v, w)-space) by $\Omega_k(a)$ for a given a.
- (EV_k) (in the x-space) by *Opt*(EV_k), and
- (SDP_k) (in the (x, z, v, w)-space) by *Opt*(SDP_k).

Theorem 3.4.1 *The following hold.*

(1) The point x^* is an extreme point of Opt(EV_k) if and only if $f_k \circ A$ is strictly convex at x^*.

(2) If nonempty, the set Opt(EV_k) has at least one extreme point.

(3) Let x^* be an extreme point of Opt(EV_k). If $m > k(n-k)$, then

$$\lambda_k(A(x^*)) = \lambda_{k+1}(A(x^*)). \qquad (3.4.37)$$

Sketch of Proof of (3) If x^* is an extreme point of *Opt(EV_k)*, then

$$F^* = \{x^*\} \times \Omega_k(A(x^*))$$

is a face of *Opt(SDP_k)*, therefore also of *Feas(SDP_k)*. Since the only possible degree of freedom in F^* is in choosing z^*, we have $\dim F^* \leq 1$. Now, let $(x^*, z^*, v^*, w^*) \in F^*$, $i = \operatorname{rank} v^*$, $j = \operatorname{rank} w^*$ and apply the Primal Faces Theorem with

$$X = \mathcal{R} \times \mathcal{R}^m \times \mathcal{S}^n \times \mathcal{S}^n, \quad Y = \mathcal{S}^n, \quad K = \mathcal{R} \times \mathcal{R}^m \times \mathcal{S}^n_+ \times \mathcal{S}^n_+$$

to conclude

$$1 + m + t(i) + t(j) \leq t(n) + 1. \qquad (3.4.38)$$

By (3.4.38) if $m > k(n-k)$, then $i = k$, $j = n-k$ cannot hold simultaneously. Since $(z^*, v^*, w^*) \in \Omega_k(A(x^*))$ was arbitrary, we conclude that (3.4.37) must hold. ■

3.4.2 *The geometry of a max-cut relaxation*

Consider the set

$$\mathcal{E}_n = \{x \in \mathcal{S}^n_+ \mid x_{ii} = 1 \, (i = 1, \ldots, n)\}.$$

Optimizing over $\mathcal{E}_n$ provides a strong relaxation of the maximum cut problem ([184], [285]); having this motivation, it was termed the *elliptope* and its geometry studied in [466] and [467].
In Theorem 3.4.2 we show that several results in these papers can be easily derived from our general framework. We consider $\mathcal{E}_n$ as a feasible set of a primal type SDP, ie. let $X = \mathcal{S}^n, Y = \mathcal{R}^n, K = K^* = \mathcal{S}^n_+$ and define A and A^* via $a^1, \ldots, a^n \in \mathcal{S}^n$, with a^i having 1 in the ith position on the main diagonal, and 0 everywhere else.

Theorem 3.4.2 *Let $\bar{x} \in \mathcal{E}_n$. Then*

(1) $\bar{x}$ is a nondegenerate point.

(2) $\dim \tan(\bar{x}, \mathcal{E}_n) = 0$ *if and only if* $\operatorname{rank} \bar{x} = 1$.

Proof Let

$$\mathcal{T} = \tan(\bar{x}, \mathcal{E}_n), \qquad F = \operatorname{face}(\bar{x}, \mathcal{S}^n_+), \qquad r = \operatorname{rank} \bar{x}.$$

(1): After exchanging rows and columns, we may assume that the elements of $\mathcal{R}(A^*)$ are of the form

$$\begin{pmatrix} I & 0 & 0 \\ 0 & -I & 0 \\ 0 & 0 & 0 \end{pmatrix}.$$

Suppose that such a matrix is in $\text{lin}\, F^\triangle$, with the size of the nonzero block equal to some $s \geq 0$. Then

$$\begin{pmatrix} I & 0 & 0 \\ 0 & -I & 0 \\ 0 & 0 & 0 \end{pmatrix} \in \text{lin}\, F^\triangle \Rightarrow \begin{pmatrix} I & 0 & 0 \\ 0 & I & 0 \\ 0 & 0 & 0 \end{pmatrix} \in F^\triangle \Rightarrow \left\langle \begin{pmatrix} I & 0 & 0 \\ 0 & I & 0 \\ 0 & 0 & 0 \end{pmatrix}, \bar{x} \right\rangle = 0.$$

Since $\bar{x} \succeq 0$, its lower $(n-s) \times (n-s)$ principal submatrix must be zero. But $\bar{x}_{ii} = 1$ $(i = 1, \ldots, n)$, hence $s = 0$, proving

$$\text{lin}\, F^\triangle \cap \mathcal{R}(A^*) \;=\; \{0\},$$

as required.
(2): From (2) in the Tangent Spaces Theorem,

$$\begin{aligned} \dim \mathcal{T} &= \dim F^{\triangle\perp} - \dim Y + \dim\, [\text{lin}\, F^\triangle \cap \mathcal{R}(A^*)] \\ &= (t(n) - t(n-r)) - n + 0 \\ &\geq 0, \end{aligned} \tag{3.4.39}$$

with equality holding iff $r = 0$. ∎

Remark 3.4.2 By the nondegeneracy of all points in $\mathcal{E}_n$ it also follows that for any $c \in \mathcal{S}^n$, the dual of

$$\begin{aligned} \min \quad & \langle c, x \rangle \\ st. \quad & x \in \mathcal{E}_n \end{aligned}$$

has a unique optimal solution, a result proved independently in [708] and [184]. Also, the formula for dim $\mathcal{T}$ given in (3.4.39) is equivalent to the formula for dim ncone $(\bar{x}, \mathcal{E}_n)$ in [467] since

$$\dim \text{ ncone}\,(\bar{x}, \mathcal{E}_n) \;=\; t(n) - \dim \mathcal{T}.$$

3.4.3 The embeddability of graphs

As one more example of semidefinite combinatorics, we give a simple proof of a theorem of Barvinok [76].

Definition 3.4.1 *Let $G = (V, E; w)$ be a graph, with $\{w_{ij} : (i,j) \in E\}$ a nonnegative weighting on the edges. We say that G is* realizable *in $\mathcal{R}^d$, if there exists vectors $x^1, \ldots, x^n \in \mathcal{R}^d$, such that*

$$||x^i - x^j|| = w_{ij} \;\; \forall (i,j) \in E.$$

Theorem 3.4.3 *Suppose that $G = (V, E; w)$ is realizable in some dimension. Then it is realizable in $\mathcal{R}^d$, with d satisfying*

$$t(d) \quad \leq \quad |E|.$$

Proof Define the matrices $m^{ij} \in S^n$ for $(i, j) \in E$ by

$$m^{ij}_{ii} = m^{ij}_{jj} = 1, \; m^{ij}_{ij} = m^{ij}_{ji} = -1,$$

and with all other components zero. Then G is realizable in $\mathcal{R}^d$, iff

$$\begin{array}{rll} \exists\, x^1, \ldots, x^n \in \mathcal{R}^d & s.t. \; \langle x^i, x^i \rangle - 2\langle x^i, x^j \rangle + \langle x^j, x^j \rangle = w^2_{ij} \; \forall (i,j) \in E & \Leftrightarrow \\ \exists\, x \in \mathcal{R}^{n \times d} & s.t. \; \langle m^{ij}, xx^{\mathrm{T}} \rangle = w^2_{ij} \;\; \forall (i,j) \in E & \Leftrightarrow \\ \exists\, y \in \mathcal{S}^n_+, \, \mathrm{rank}\, y = d & s.t. \; \langle m^{ij}, y \rangle = w^2_{ij} \;\; \forall (i,j) \in E. & \end{array}$$

Therefore, G is representable in *some* dimension, iff the SDP in the last line above has a feasible solution, and it is representable in dimension $\leq d^*$, iff it has a feasible solution of rank $\leq d^*$. However, the first of these statements implies the second with d^* satisfying

$$t(d^*) \quad \leq \quad |E|,$$

by taking a solution which is an extreme point of the feasible set.

3.5 TWO ALGORITHMIC ASPECTS

In this section we describe, how two algorithmic aspects of cone programming can be handled using our framework: transforming a feasible solution into one, which is an extreme point of the feasible set; and determining by how much one can perturb the objective function in a given direction, while keeping the current solution optimal.

3.5.1 Finding an extreme point solution

This section is devoted to the following question: Given a feasible solution $\bar{x}$ to the problem

$$\begin{array}{ll} Min & \langle c, x \rangle \\ s.t. & x \in K \\ & Ax = b \end{array} \qquad (P)$$

we want to find an extreme point feasible solution, called $\bar{\bar{x}}$ with no worse objective value. We claim that the algorithm below will do exactly this.

(1) Let $F = \mathrm{face}\,(\bar{x}, K)$.

(2) Find a nonzero $\Delta x \in \mathrm{lin}\, F \cap \mathcal{N}(A)$. If no such vector exists, set $\bar{\bar{x}} = \bar{x}$ and STOP.

(3) If $\langle c, \Delta x \rangle > 0$, set $\Delta x = -\Delta x$.

Determine $\alpha^* = \max\{\, \alpha \,|\, \bar{x} + \alpha \Delta x \in F^{\triangle} \,\}$.

If $\alpha^* = +\infty$, STOP; (P) is unbounded.

(4) Set $\bar{x} = \bar{x} + \alpha^* \Delta x$, and go to (1).

The current $\bar{x}$ is not an extreme point, iff a nonzero Δx can be found in Step (2), by (4) in the Primal Faces Theorem. Also, by elementary convex analysis, if α^* is found in Step (3), then

$$\text{face}\,(\bar{x} + \alpha^* \Delta x, K) \quad \subset \quad \text{face}\,(\bar{x}, K)$$

therefore the algorithm is correct, and finite.

3.5.2 Sensitivity Analysis

Consider the primal-dual pair of cone-lp's parametrized by the scalar $t \geq 0$.

$$(P_t) \quad \begin{array}{rl} Min & \langle c + t\Delta c, x \rangle \\ s.t. & x \in K \\ & Ax = b \end{array} \qquad\qquad \begin{array}{rl} Max & \langle b, y \rangle \\ s.t. & z \in K^* \\ & A^* y + z = c + t\Delta c \end{array} \quad (D_t)$$

Let $\bar{x}$ be an optimal solution of (P_0), and

$$t^* \quad = \quad \sup\{\, t \,|\, \bar{x} \text{ is an optimal solution of } (P_t) \,\}.$$

Theorem 3.5.1 *Suppose (P_0) satisfies the Slater condition, and $\bar{x}$ and $(\bar{y}, \bar{z})$ is an optimal solution-pair of (P_0) and (D_0). Then*

(1) t^* is the optimal value of

$$\begin{array}{lrcl} Max & & & t \\ st. & \bar{z} + t\Delta z & \in & F^{\triangle} \\ & \Delta z & \in & \text{lin}\, F^{\triangle} \\ & A^* \Delta y + \Delta z & = & \Delta c \end{array} \qquad (D_t')$$

(if the last two constraints of (D_t') are infeasible then the optimal value of (D_t') is understood to be 0).

(2) If $\bar{x}$ and $(\bar{y}, \bar{z})$ are strictly complementary, then $t^* > 0$.

(3) If $\bar{x}$ is nondegenerate, and there is $\Delta x \in \text{lin}\, F^{\triangle}$ s.t. $A^* \Delta y + \Delta z = \Delta c$, then

$$t^* \quad = \quad \max\{\, t \,|\, \bar{z} + t\Delta z \in F^{\triangle} \,\}$$

for some such fixed Δz.

Proof

(1): Since (P_0), equivalently (P_t) for an arbitrary $t \geq 0$ satisfies the Slater condition, (D_t) attains its optimal value. Therefore t^* is the optimal value of

$$\begin{array}{lrcl} Max & t & & \\ st. & z & \in & F^{\triangle} \\ & A^*y + z & = & c + t\Delta c \end{array} \qquad (D''_t)$$

Then

$$\begin{array}{rll} t^* > 0 & & \Leftrightarrow \\ \exists\, t > 0,\ (y(t), z(t)) & \text{feasible for } (D''_t) & \Leftrightarrow \\ \exists\, t > 0,\ (\Delta y, \Delta z) & \text{feasible for } (D'_t) & \end{array}$$

where the second equivalence follows by taking

$$(\Delta y, \Delta z) \;=\; \frac{1}{t}\left((y(t) - \bar{y}), (z(t) - \bar{z})\right)$$

and this also proves that (D'_t) and (D''_t) have the same optimal value.
(2): Straightforward.
(3): The primal optimal solution $\bar{x}$ is nondegenerate, iff

$$\operatorname{lin} F^{\triangle} \cap \mathcal{R}(A^*) \;=\; \{\,0\,\}.$$

In this case, the system

$$\begin{array}{rcl} \Delta z & \in & \operatorname{lin} F^{\triangle} \\ A^*\Delta y + \Delta z & = & \Delta c, \end{array}$$

which is part of (D'_t) has a unique solution. The claim follows. ■

3.6 LITERATURE

For Section 3.2 Several important papers on the facial structure of convex cones and their duals were written by Barker, see [68, 69, 70], and by Tam, see [762]. Nice cones, (although without this name) were introduced by Borwein and Wolkowicz [130, 131], and used in the theory of *facial reduction*, see also Chapter 7.
For Section 3.3 A general theorem stating that all faces of the intersection of two convex sets arise as the intersection of two corresponding faces, was proven by Dubins in [200]; parts of the Primal and Dual faces Theorems already follow from it. The bounds on the rank of extreme matrices in SDP's were proved by Pataki in [620], and in [623]. The existence of a solution with a small rank (a rank that satisfies the bound stated for an extreme point) was proved independently by Barvinok [76]. A recent interesting result obtained by him (see [77]) shows that for semidefinite programs, if one of the equality constraints is of the form $\langle I, x\rangle = 1$, then this bound can be improved by 1. A similar rankbound to the one in [623] was obtained independently in [191].

The faces of "spectrahedra" ie. of feasible sets of SDP's were characterized by Ramana and Goldman [654]. Nondegeneracy and strict complementarity for SDP were introduced and studied by Shapiro and Fan [715], Alizadeh, Haeberly and Overton [24], and for symmetric cones (ie. for the semidefinite, and second order cones) by Faybusovich [218]. For the second order cone, see also Alizadeh and Schmieta [26]. In [24] also the *genericity* of the property of strict complementarity was proved, ie. they showed that the instances of SDP which do not have a strictly complementary solution pair form a set of measure zero.
The general framework on the facial structure, nondegeneracy and strict complementarity for general cone programs was described by Pataki in [621], and in the dissertation [622]. Strict complementarity was also introduced independently by Luo, Sturm and Zhang [507].
For Section 3.4 The multiplicity of the critical eigenvalue in eigenvalue-optimization was studied in [623] by Pataki.
For Section 3.5 The algorithm to find an extreme point feasible solution of a cone-lp was given in [621] and [622] by Pataki. The method for sensitivity analysis is a generalization of the one given for SDP by Goldfarb and Scheinberg in [287]. The treatment given here is rather restrictive of course; it handles only a perturbation in a given direction, however, it is algorithmic, rather than purely structural. An extensive study on the structural properties of how the solution changes under perturbations is given by Bonnans and Shapiro [122]. For structural results on the sensitivity of a central solution of an SDP, see Sturm and Zhang [759]. A simple treatment on by how much the solution of an SDP can change, when the problem data is perturbed is described by Nayakkankuppam and Overton [563].

3.7 APPENDICES

3.7.1 A: The faces of the semidefinite cone

We will use some facts from linear algebra stated in

Proposition 3.7.1 *Suppose that x and y are in $\mathcal{S}^n_+$. Then*

(1)
$$\mathcal{R}(x+y) \;=\; \mathcal{R}(x)+\mathcal{R}(y).$$

(2) If $\mathcal{R}(y) \subseteq \mathcal{R}(x)$, then there exists $z \in \mathcal{S}^n_+$ such that
$$x \;\in\; (y,z).$$
In other words, the line-segment from y to x can be extended past x within $\mathcal{S}^n_+$.

(3) $\langle x,y\rangle = 0$ if and only if $\mathcal{R}(x) \perp \mathcal{R}(y)$.

To prove (3.2.6) in Example 3.2.2, it is enough to verify the PSD Faces Theorem below. Formula (3.2.7) for the conjugate face will then follow by using (3) in Proposition 3.7.1.

Theorem 3.7.1 (PSD Faces) *A set is a face of $\mathcal{S}_+^n$ if and only if it is of the form*

$$\{\, x \,|\, x \in \mathcal{S}_+^n,\ \mathcal{R}(x) \subseteq L \,\}$$

for some subspace L of $\mathcal{R}^n$.

Proof Denote the set in the statement of the theorem by $F(L)$.
(If) Proposition 3.7.1 (1) implies both the convexity of $F(L)$, and

$$\text{if } x + y \in F(L) \quad \text{then} \quad x \in F(L), \text{ and } y \in F(L)$$

(Only if) Let F be a face of $\mathcal{S}_+^n$. Define L as the subspace spanned by the range spaces of all matrices in F. We claim

$$F \;=\; F(L)$$

with the inclusion $\subseteq$ being obvious. To show the reverse, we first construct a matrix $\hat{x} \in F$ with $\mathcal{R}(\hat{x}) = L$. As there are matrices $x^1, \ldots, x^k$ in F such that

$$L \;=\; \sum_{i=1}^{k} \mathcal{R}(x^i) \;=\; \mathcal{R}(\sum_{i=1}^{k} x^i),$$

therefore

$$\hat{x} \;=\; \sum_{i=1}^{k} x^i$$

will do. Now, pick any $y \in F(L)$. By (2) in Proposition 3.7.1 there exists $z \in \mathcal{S}_+^n$ such that

$$\hat{x} \;\in\; (y, z)$$

Since F is a face of $\mathcal{S}_+^n$ containing $\hat{x}$, we conclude that y (and z) must be in F. ■

3.7.2 B: Proof of Lemma 3.3.1

Recall that $\phi(y) = c - A^* y$.
Proof of (1) We have

$$F \lhd Feas(D) \qquad \Leftrightarrow$$

$$(y^1, z^1), (y^2, z^2) \in Feas(D) \text{ and } \tfrac{1}{2}[(y^1, z^1) + (y^2, z^2)] \in F \tag{3.7.43}$$

$$\text{imply}$$

$$(y^1, z^1), (y^2, z^2) \;\in\; F \tag{3.7.44}$$

But, (3.7.43) is equivalent to

$$(y^1, \phi(y^1)), (y^2, \phi(y^2)) \in Feas(D) \;\text{ and }\; (\,\tfrac{1}{2}(y^1 + y^2), \phi(\tfrac{1}{2}(y^1 + y^2))\,) \;\in\; F \;\Leftrightarrow$$

$$y^1, y^2 \in Feas(D_y) \;\text{ and }\; \tfrac{1}{2}(y^1 + y^2) \in \text{Proj}_{\,y}(F) \tag{3.7.45}$$

and (3.7.44) to

$$\begin{aligned}(y^1, \phi(y^1)),\ (y^2, \phi(y^2)) \in F \ &\Leftrightarrow \\ y^1,\ y^2 \in \operatorname{Proj}_y(F)\end{aligned} \qquad (3.7.46)$$

Therefore,

$$F \lhd Feas(D) \quad \Leftrightarrow \quad (3.7.45) \text{ implies } (3.7.46) \quad \Leftrightarrow \quad \operatorname{Proj}_y(F) \lhd Feas(D_y).$$

This proves (1)(a) in Lemma 3.3.1. To see (1)(b), one only needs to note

$$F \;=\; \{\,(y, \phi(y)) \mid y \in \operatorname{Proj}_y(F)\,\}.$$

Proof of (2) By the proof of the Tangent Spaces Theorem

$$\operatorname{tcone}((\bar{y}, \bar{z}), Feas(D)) = \{(y, z) \mid A^* y + z = 0,\ z \in \operatorname{tcone}(\bar{z}, K^*)\}, \qquad (3.7.47)$$

therefore

$$\operatorname{Proj}_y[\operatorname{tcone}((\bar{y}, \bar{z}), Feas(D))] = \{y \mid -A^* y \in \operatorname{tcone}(c - A^*\bar{y}, K^*)\} \qquad (3.7.48)$$

A straightforward calculation shows

$$\begin{aligned}\operatorname{dir}(\bar{y}, Feas(D_y)) &= (-A^*)^{-1}[\operatorname{dir}(c - A^*\bar{y}, K^*)] \ \Rightarrow \\ \operatorname{Cl}\operatorname{dir}(\bar{y}, Feas(D_y)) &= (-A^*)^{-1}[\operatorname{Cl}\operatorname{dir}(c - A^*\bar{y}, K^*)] \ \Leftrightarrow \\ \operatorname{tcone}(\bar{y}, Feas(D_y)) &= (-A^*)^{-1}[\operatorname{tcone}(c - A^*\bar{y}, K^*)]\end{aligned} \qquad (3.7.49)$$

where the first implication follows by Theorem 6.7 in [679]. Putting (3.7.49) and (3.7.48) together yields

$$\begin{aligned}\operatorname{tcone}(\bar{y}, Feas(D_y)) &= \operatorname{Proj}_y[\operatorname{tcone}((\bar{y}, \bar{z}), Feas(D))] \ \Rightarrow \\ \tan(\bar{y}, Feas(D_y)) &= \operatorname{Proj}_y[\tan((\bar{y}, \bar{z}), Feas(D))]\end{aligned}$$

which proves (2)(a). Also, by (3.7.47)

$$\tan((\bar{y}, \bar{z}), Feas(D)) \;=\; \{\,(y, -A^* y) \mid -A^* y \in \tan(c - A^*\bar{y}, K^*)\,\},$$

which shows that projecting $\tan((\bar{y}, \bar{z}), Feas(D))$ onto the y-space preserves its dimension, proving (2)(b).

4 DUALITY AND OPTIMALITY CONDITIONS

Alexander Shapiro and Katya Scheinberg

The Lagrange multiplier method leads to several transformations which are important both theoretically and practically.

By means of thesee transformations new problems equivalent to a given problem can be so formulated that stationary conditions occur simultaneously in equivalent problems. In this way we are led to transformations of the problems which are important because of their symmetric character. **Moreover, for a given maximum problem with maximum M, we shall ofen be able to find an equivalent minimum problem with the same value M as minimum; this is a useful tool for bounding M from above and below.**

Courant and Hilbert, [168]
(taken from [452])

4.1 DUALITY, OPTIMALITY CONDITIONS, AND PERTURBATION ANALYSIS

Alexander Shapiro

4.1.1 *Introduction*

Consider the optimization problem

$$\operatorname*{Min}_{x \in C} f(x) \text{ subject to } G(x) \preceq 0, \tag{4.1.1}$$

where C is a convex closed cone in the Euclidean space $\mathbb{R}^n$, $f : \mathbb{R}^n \to \mathbb{R}$ and $G : \mathbb{R}^n \to \mathcal{Y}$ is a mapping from $\mathbb{R}^n$ into the space $\mathcal{Y} := \mathcal{S}^m$ of $m \times m$ symmetric matrices. We refer to the above problem as a nonlinear semidefinite programming problem. In particular, if $C = \mathbb{R}^n$, the objective function is linear, i.e. $f(x) := \sum_{i=1}^n b_i x_i$, and the constraint mapping is affine, i.e. $G(x) := A_0 + \sum_{i=1}^n x_i A_i$ where $A_0, A_1, ..., A_n \in \mathcal{S}^m$ are given matrices, problem (4.1.1) becomes a linear semidefinite programming problem

$$\operatorname*{Min}_{x \in \mathbb{R}^n} \sum_{i=1}^n b_i x_i \text{ subject to } A_0 + \sum_{i=1}^n x_i A_i \preceq 0. \tag{4.1.2}$$

In this article we discuss duality, optimality conditions and perturbation analysis of such nonlinear semidefinite programming problems.

Let us observe at this point that problem (4.1.1) can be formulated in the form

$$\operatorname*{Min}_{x \in C} f(x) \text{ subject to } G(x) \in K, \tag{4.1.3}$$

where $K := \mathcal{S}^m_-$ is the cone of negative semidefinite $m \times m$ symmetric matrices. That is, the feasible set of problem (4.1.1) can be defined by the "cone constraints" $\{x \in C : G(x) \in K\}$. Some of the results presented in this article can be formulated in the general framework of such "cone constrained" problems, while the others use a particular structure of the considered cone $\mathcal{S}^m_-$. Therefore we start our analysis by considering an optimization problem in the form (4.1.3) with $\mathcal{Y}$ being a finite dimensional vector space and $K \subset \mathcal{Y}$ being a convex closed cone. Note that if $C = \mathbb{R}^n$ and the cone K is given by the negative orthant $\mathbb{R}^p_- := \{y \in \mathbb{R}^p : y_i \leq 0,\ i = 1, ..., p\}$, then problem (4.1.3) becomes a standard (nonlinear) programming problem. As we shall see there are certain similarities between such nonlinear programming problems and semidefinite programming problems. There are also, however, some essential differences.

We assume that spaces $\mathbb{R}^n$ and $\mathcal{Y}$ are equipped with respective scalar products, denoted by "$\cdot$". In particular, in the Euclidean space $\mathbb{R}^n$ we use the standard scalar product $x \cdot z := x^T z$, and in the space $\mathcal{S}^m$ the scalar product $A \bullet B := \text{trace}(AB)$. With the cone K is associated its polar (negative dual) cone K^-,

$$K^- := \{y \in \mathcal{Y} : y \cdot w \leq 0,\ \forall\, w \in K\}.$$

Since the cone K is convex and closed we have the following duality relation $(K^-)^- = K$ (and similarly for the cone C). This is a classical result which can be easily derived from the separation theorem. It can be noted that the polar of the cone $\mathcal{S}^m_-$ is the cone $\mathcal{S}^m_+ := \{A \in \mathcal{S}^m : A \succeq 0\}$ of positive semidefinite matrices.

4.1.2 *Duality*

The Lagrangian function, associated with the problem (4.1.3), can be written in the form

$$L(x, \lambda) := f(x) + \lambda \cdot G(x), \quad (x, \lambda) \in \mathbb{R}^n \times \mathcal{Y}.$$

It follows from the duality relation $(K^-)^- = K$ that

$$\sup_{\lambda \in K^-} \lambda \cdot G(x) = \begin{cases} 0, & \textit{if } G(x) \in K, \\ +\infty, & \textit{otherwise.} \end{cases}$$

Therefore problem (4.1.3) can be also written in the form

$$\operatorname*{Min}_{x \in C} \left\{ \sup_{\lambda \in K^-} L(x, \lambda) \right\}. \tag{4.1.4}$$

By formally interchanging the "min" and "max" operators in (4.1.4) we obtain the following problem

$$\operatorname*{Max}_{\lambda \in K^-} \left\{ \inf_{x \in C} L(x, \lambda) \right\}. \tag{4.1.5}$$

We refer to (4.1.3) as the *primal* (P) and to (4.1.5) as its *dual* (D) problems, and denote their optimal values by val(P) and val(D) and their sets of optimal solutions by Sol(P) and Sol(D), respectively. In particular, the dual of the semidefinite problem (4.1.1) can be written in the form

$$\operatorname*{Max}_{\Omega \succeq 0} \left\{ \inf_{x \in C} L(x, \Omega) \right\}, \tag{4.1.6}$$

where $\Omega \in \mathcal{S}^m$ denotes the dual variable. Note that the optimal value of a minimization (maximization) problem with an empty feasible set is defined to be $+\infty$ $(-\infty)$.

In case the objective function is linear and the constraint mapping is affine the dual problem can be written explicitly. For example, in the case of the linear semidefinite programming problem (4.1.2) we have

$$\inf_{x \in \mathbb{R}^n} L(x, \Omega) = \begin{cases} \Omega \bullet A_0, & \text{if } b_i + \Omega \bullet A_i = 0, \ i = 1, ..., n, \\ -\infty, & \text{otherwise.} \end{cases}$$

Therefore the dual of (4.1.2) is

$$\operatorname*{Max}_{\Omega \succeq 0} \Omega \bullet A_0 \ \text{ subject to } \ \Omega \bullet A_i + b_i = 0, \ i = 1, ..., n. \tag{4.1.7}$$

Note that we can consider the above problem as a particular case of the cone constrained problem (4.1.3) by defining $f(\Omega) := \Omega \bullet A_0$, $C := \mathcal{S}_+^m$, $K := \{0\} \subset \mathbb{R}^n$ and $G(\Omega) := (\Omega \bullet A_1 + b_1, ..., \Omega \bullet A_n + b_n)$. Its dual then coincides with the primal problem (4.1.2). Therefore there is a complete symmetry between the dual pair (4.1.2) and (4.1.7), and which one is called primal and which is dual is somewhat arbitrary.

We say that $(\bar{x}, \bar{\lambda})$ is a *saddle point* of the Lagrangian $L(x, \lambda)$ if

$$\bar{x} = \arg\min_{x \in C} L(x, \bar{\lambda}) \ \ and \ \ \bar{\lambda} = \arg\max_{\lambda \in K^-} L(\bar{x}, \lambda). \tag{4.1.8}$$

Recall that the supremum of $\lambda \cdot G(\bar{x})$ over all $\lambda \in K^-$ equals 0 if $G(\bar{x}) \in K$ and is $+\infty$ otherwise. Therefore the second condition in (4.1.8) means that $G(\bar{x}) \in K$, $\bar{\lambda} \in K^-$ and the following, so-called *complementarity*, condition holds $\bar{\lambda} \cdot G(\bar{x}) = 0$. It follows that conditions (4.1.8) are equivalent to:

$$\bar{x} = \arg\min_{x \in C} L(x, \bar{\lambda}), \ \ \bar{\lambda} \cdot G(\bar{x}) = 0, \ \ G(\bar{x}) \in K, \ \ \bar{\lambda} \in K^-. \tag{4.1.9}$$

In particular, in the case of semidefinite programming problem (4.1.1) these conditions become

$$\bar{x} = \arg\min_{x \in C} L(x, \bar{\Omega}), \ \ \bar{\Omega} \bullet G(\bar{x}) = 0, \ \ G(\bar{x}) \preceq 0, \ \ \bar{\Omega} \succeq 0. \tag{4.1.10}$$

The following proposition is an easy consequence of the min-max representations of the primal and dual problems. It can be applied to the primal (4.1.1) and its dual (4.1.6) semidefinite programming problems in a straightforward way.

Proposition 4.1.1 *Let* (P) *and* (D) *be the primal and dual problems (4.1.3) and (4.1.5), respectively. Then* $\text{val}(D) \le \text{val}(P)$. *Moreover,* $\text{val}(P) = \text{val}(D)$ *and* $\bar{x}$ *and* $\bar{\lambda}$ *are optimal solutions of* (P) *and* (D), *respectively, if and only if* $(\bar{x}, \bar{\lambda})$ *is a saddle point of the Lagrangian* $L(x, \lambda)$.

Proof. For any $(x', \lambda') \in C \times K^-$ we have

$$\inf_{x \in C} L(x, \lambda') \le L(x', \lambda') \le \sup_{\lambda \in K^-} L(x', \lambda), \tag{4.1.11}$$

and hence

$$\sup_{\lambda \in K^-} \inf_{x \in C} L(x, \lambda) \le \inf_{x \in C} \sup_{\lambda \in K^-} L(x, \lambda). \tag{4.1.12}$$

It follows then from the min-max representations (4.1.4) and (4.1.5), of the primal and dual problems respectively, that $\text{val}(D) \le \text{val}(P)$.

Now if $(\bar{x}, \bar{\lambda})$ is a saddle point, then it follows from (4.1.8) that

$$\inf_{x \in C} L(x, \bar{\lambda}) = L(\bar{x}, \bar{\lambda}) = \sup_{\lambda \in K^-} L(\bar{x}, \lambda), \tag{4.1.13}$$

and hence

$$\sup_{\lambda \in K^-} \inf_{x \in C} L(x, \lambda) \ge L(\bar{x}, \bar{\lambda}) \ge \inf_{x \in C} \sup_{\lambda \in K^-} L(x, \lambda). \tag{4.1.14}$$

Inequalities (4.1.12) and (4.1.14) imply that $\text{val}(P) = \text{val}(D)$ and $\bar{x}$ and $\bar{\lambda}$ are optimal solutions of problems (P) and (D), respectively.

Conversely, if $\text{val}(P) = \text{val}(D)$ and $\bar{x}$ and $\bar{\lambda}$ are optimal solutions of problems (P) and (D), respectively, then (4.1.13) follows from (4.1.11), and hence $(\bar{x}, \bar{\lambda})$ is a saddle point. ■

By the above proposition the optimal value of the primal problem is always greater than or equal to the optimal value of the dual problem. The difference $\text{val}(P) - \text{val}(D)$ is called the *duality gap* between the primal and dual problems. It is said that there is no duality gap between the primal and dual problems, if $\text{val}(P) = \text{val}(D)$. It is well known that in the case of linear programming (i.e. if $f(x)$ is linear, $G(x)$ is affine, $C = \mathbb{R}^n$ and K is a convex polyhedral cone), there is no duality gap between the primal and dual problems, provided the feasible set of the primal or dual problem is nonempty. Also a linear programming problem always possesses an optimal solution provided its optimal value is finite. As the following examples show these properties do not always hold for linear semidefinite programming problems.

Example 4.1.1 Consider the following linear semidefinite programming problem

$$\text{Min } x_1 \text{ subject to } \begin{bmatrix} -x_1 & 1 \\ 1 & -x_2 \end{bmatrix} \preceq 0. \tag{4.1.15}$$

The feasible set of this problem is $\{(x_1, x_2) : x_1 \ge 0,\ x_2 \ge 0,\ x_1 x_2 \ge 1\}$, and hence its optimal value is 0 and this problem does not have an optimal solution. The dual of this problem is

$$\text{Max } 2\omega_{12} \text{ subject to } \begin{bmatrix} 1 & \omega_{12} \\ \omega_{21} & 0 \end{bmatrix} \succeq 0. \tag{4.1.16}$$

Its feasible set contains one point with $\omega_{12} = \omega_{21} = 0$, which is also its optimal solution, and hence its optimal value is 0. Therefore in this example there is no duality gap between the primal and dual problems, although the primal problem does not have an optimal solution.

Example 4.1.2 Consider the linear semidefinite programming problem

$$\text{Min } -x_2 \text{ subject to } \begin{bmatrix} x_2 - a & 0 & 0 \\ 0 & x_1 & x_2 \\ 0 & x_2 & 0 \end{bmatrix} \preceq 0, \tag{4.1.17}$$

where $a > 0$ is a given number. The dual of this problem is

$$\text{Max } -a\omega_{11} \text{ subject to } \Omega \succeq 0,\ \omega_{22} = 0,\ \omega_{11} + 2\omega_{23} = 1. \tag{4.1.18}$$

The feasible set of the primal problem is $\{(x_1, x_2) : x_1 \leq 0,\ x_2 = 0\}$ and hence its optimal value is 0. On the other hand any feasible Ω of the dual problem has $\omega_{11} = 1$, and hence the optimal value of the dual problem is $-a$. Therefore the duality gap in this example is a.

In case of linear semidefinite programming problem (4.1.2) and its dual (4.1.7), a pair $(\bar{x}, \bar{\Omega})$ is a saddle point iff $\bar{x}$ and $\bar{\Omega}$ are feasible points of (4.1.2) and (4.1.7), respectively, and the following complementarity condition

$$\bar{\Omega}\left(A_0 + \sum_{i=1}^{n} \bar{x}_i A_i\right) = 0 \tag{4.1.19}$$

holds. The above condition corresponds to the complementarity condition of (4.1.9), while feasibility of $\bar{\Omega}$ is equivalent to the first and forth conditions of (4.1.9) and feasibility of $\bar{x}$ is the third condition of (4.1.9). Note that since $\bar{\Omega} \succeq 0$ and $A_0 + \sum_{i=1}^{n} \bar{x}_i A_i \preceq 0$, the complementarity condition (4.1.19) is equivalent to $\bar{\Omega} \bullet (A_0 + \sum_{i=1}^{n} \bar{x}_i A_i) = 0$.

It is not always easy to verify existence of a saddle point, and moreover "no gap" property can hold even if the primal or/and dual problems do not have optimal solutions. We approach now the "no duality gap" problem from a somewhat different point of view. With the primal problem we associate the following parametric problem

$$\underset{x \in C}{\text{Min}}\, f(x) \text{ subject to } G(x) + y \in K, \tag{4.1.20}$$

depending on the parameter vector $y \in \mathcal{Y}$. We denote this problem by (P_y) and by $v(y)$ we denote its optimal value, i.e. $v(y) := \text{val}(P_y)$. Clearly problem (P_0) coincides with the primal problem (P) and $\text{val}(P) = v(0)$. The conjugate of the function $v(y)$ is defined as

$$v^*(y^*) := \sup_{y \in \mathcal{Y}} \{y^* \cdot y - v(y)\}.$$

We have then that

$$\begin{aligned} v^*(y^*) &= \sup\{y^* \cdot y - f(x) : (x, y) \in C \times \mathcal{Y},\ G(x) + y \in K\} \\ &= \sup\nolimits_{x \in C} \sup\nolimits_{y \in -G(x)+K} \{y^* \cdot y - f(x)\}. \end{aligned}$$

Now

$$\sup_{y \in -G(x)+K} y^* \cdot y = -y^* \cdot G(x) + \sup_{y \in K} y^* \cdot y,$$

and hence the above supremum equals $-y^* \cdot G(x)$ if $y^* \in K^-$ and is $+\infty$ otherwise.

We obtain that

$$-v^*(y^*) = \begin{cases} \inf_{x \in C} L(x, y^*), & \text{if } y^* \in K^-, \\ -\infty, & \text{otherwise.} \end{cases} \tag{4.1.21}$$

Therefore the dual problem (4.1.5) can be formulated as

$$\operatorname*{Max}_{\lambda \in \mathcal{Y}} \{-v^*(\lambda)\}. \tag{4.1.22}$$

Consequently we have that $\operatorname{val}(D) = v^{**}(0)$, where

$$v^{**}(y) := \sup_{y^* \in \mathcal{Y}} \{y^* \cdot y - v^*(y^*)\}$$

denotes the conjugate of the conjugate function $v^*(\cdot)$. The relations $\operatorname{val}(P) = v(0)$ and $\operatorname{val}(D) = v^{**}(0)$, between the primal and dual problems and the optimal value function $v(\cdot)$, allow to apply powerful tools of convex analysis.

In the subsequent analysis we deal with convex cases where the optimal value function $v(y)$ is convex. By the Fenchel-Moreau duality theorem we have that if $v(y)$ is *convex*, then

$$v^{**}(\cdot) = \begin{cases} \operatorname{lsc} v(\cdot), & \text{if } \operatorname{lsc} v(y) > -\infty \text{ for all } y \in \mathcal{Y}, \\ -\infty, & \text{otherwise,} \end{cases} \tag{4.1.23}$$

where

$$\operatorname{lsc} v(y) := \min\left\{ v(y), \liminf_{y' \to y} v(y') \right\}$$

denotes the lower semicontinuous hull of v, i.e. the largest lower semicontinuous function majorized by v.

The quantity

$$\operatorname{lsc} v(0) = \min\left\{ v(0), \liminf_{y \to 0} v(y) \right\}$$

is called the ***subvalue*** of the problem (P). The problem (P) is said to be ***subconsistent*** if its subvalue is less than $+\infty$, i.e. either if the feasible set of (P) is nonempty, and hence its optimal value is less than $+\infty$, or if by making "small" perturbations in the feasible set of (P) the corresponding optimal value function $v(y)$ becomes bounded from above. Note that if the lower semicontinuous hull of a convex function has value $-\infty$ in at least one point, then it can take only two possible values $+\infty$ or $-\infty$. Therefore if $\operatorname{lsc} v(0)$ is finite, then $v^{**}(0) = \operatorname{lsc} v(0)$. By the above discussion we obtain the following result.

Theorem 4.1.1 *Suppose that the optimal value function $v(y)$ is convex and that the primal problem (P) is subconsistent. Then*

$$\operatorname{val}(D) = \min\left\{\operatorname{val}(P), \liminf_{y\to 0} v(y)\right\}, \qquad (4.1.24)$$

and hence $\operatorname{val}(P) = \operatorname{val}(D)$ *if and only if* $v(y)$ *is lower semicontinuous at* $y = 0$.

We discuss now conditions ensuring convexity of $v(y)$. Let $Q \subset Y$ be a closed convex cone. We say that the mapping $G : \mathbb{R}^n \to Y$ is *convex* with respect to Q if for any $x_1, x_2 \in \mathbb{R}^n$ and $t \in [0,1]$ the following holds

$$tG(x_1) + (1-t)G(x_2) \succeq_Q G(tx_1 + (1-t)x_2). \qquad (4.1.25)$$

Here " $\succeq_Q$ " denotes the partial ordering associated with the cone Q, i.e. $a \succeq_Q b$ if $a - b \in Q$. We say that the problem (P) is convex, if the function $f(x)$ is convex and the mapping $G(x)$ is convex with respect to the cone $Q := -K$. Any affine mapping $G(x)$ is convex with respect to any cone Q, and hence any linear problem is convex.

Let us observe that if (P) is convex and $\lambda \in K^-$, then the function $L(\cdot, \lambda)$ is convex. Indeed, Condition (4.1.25) means that

$$tG(x_1) + (1-t)G(x_2) - G(tx_1 + (1-t)x_2) \in -K,$$

and hence for any $\lambda \in K^-$,

$$t\lambda \cdot G(x_1) + (1-t)\lambda \cdot G(x_2) - \lambda \cdot G(tx_1 + (1-t)x_2) \geq 0.$$

It follows that $\lambda \cdot G(\cdot)$ is convex, and since $f(x)$ is convex, we obtain that $L(\cdot, \lambda)$ is convex. Convexity of (P) also implies convexity of $v(\cdot)$.

Proposition 4.1.2 *Suppose that the primal problem (P) is convex. Then the optimal value function $v(y)$ is convex.*

Proof. Consider the following extended real valued function

$$\varphi(x,y) := \begin{cases} f(x), & \text{if } x \in C \text{ and } G(x) + y \in K, \\ +\infty, & \text{otherwise.} \end{cases}$$

Note that $v(y) = \inf_{x\in\mathbb{R}^n} \varphi(x,y)$. Let us show that the function $\varphi(\cdot,\cdot)$ is convex. Since $f(x)$ is convex, it suffices to show that the domain $\operatorname{dom}\varphi := \{(x,y) \in C \times \mathcal{Y} : \varphi(x,y) < +\infty\}$ of φ is a convex set. Consider $(x_1,y_1), (x_2,y_2) \in \operatorname{dom}\varphi$ and $t \in [0,1]$. By convexity of C, $tx_1 + (1-t)x_2 \in C$. By convexity of G, with respect to $-K$, we have that

$$k := G(tx_1 + (1-t)x_2) - [tG(x_1) + (1-t)G(x_2)] \in K.$$

Therefore, since K is a convex cone,

$$G(tx_1+(1-t)x_2)+ty_1+(1-t)y_2 = k+t[G(x_1)+y_1]+(1-t)[G(x_2)+y_2] \in K.$$

It follows that $t(x_1, y_1) + (1-t)(x_2, y_2) \in \operatorname{dom}\varphi$, and hence $\operatorname{dom}\varphi$ is convex.

Now, since φ is convex, we have for any $(x_1, y_1), (x_2, y_2) \in \mathbb{R}^n \times \mathcal{Y}$ and $t \in [0, 1]$,

$$t\varphi(x_1, y_1)+(1-t)\varphi(x_2, y_2) \geq \varphi(tx_1+(1-t)x_2, ty_1+(1-t)y_2) \geq v(ty_1+(1-t)y_2).$$

By minimizing the left hand side of the above inequality over x_1 and x_2, we obtain

$$tv(y_1) + (1-t)v(y_2) \geq v(ty_1 + (1-t)y_2),$$

which shows that $v(y)$ is convex. ■

In the case of semidefinite programming, i.e. when $K := \mathcal{S}^m_-$ and hence $Q := -K = \mathcal{S}^m_+$, convexity of $G(x)$ means that it is convex with respect to the Löwner partial order in the space $\mathcal{S}^m$. The inequality (4.1.25) is then equivalent to

$$tz^T G(x_1)z + (1-t)z^T G(x_2)z \geq z^T G(tx_1 + (1-t)x_2)z, \quad \forall\, z \in \Re^m. \quad (4.1.26)$$

This, in turn, means that the function $\psi(x) := z^T G(x)z$ is convex for any $z \in \Re^m$.

By theorem 4.1.1 we have that in the convex case, and if (P) is subconsistent, there is no duality gap between the primal and dual problems iff $v(y)$ is lower semicontinuous at $y = 0$. This is a topological condition which may be not easy to verify in a particular situation. We derive now more directly verifiable conditions ensuring the no duality gap property. Recall that a vector $z \in \mathcal{Y}$ is said to be a *subgradient* of $v(\cdot)$, at a point y such that $v(y)$ is finite, if

$$v(y') - v(y) \geq z \cdot (y' - y), \quad \forall\, y' \in \mathcal{Y}.$$

The set of all subgradients of $v(\cdot)$, at y, is called the *subdifferential* and denoted $\partial v(y)$. The function $v(\cdot)$ is said to be *subdifferentiable* at y if $v(y)$ is finite and the subdifferential $\partial v(y)$ is nonempty. Results of the following proposition are easy consequences of the definitions.

Proposition 4.1.3 *Suppose that the function $v(\cdot)$ has a finite value at a point $y \in \mathcal{Y}$. Then:*
(i) If $v(\cdot)$ is subdifferentiable at y, then $v(\cdot)$ is lower semicontinuous at y.
(ii)

$$\partial v^{**}(y) = \arg\max_{y^* \in Y}\{y^* \cdot y - v^*(y^*)\}. \quad (4.1.27)$$

(iii) If v is subdifferentiable at y, then $v^{**}(y) = v(y)$.
(iv) If $v^{**}(y) = v(y)$, then $\partial v^{**}(y) = \partial v(y)$.

Together with representation (4.1.22) of the dual problem, the above proposition implies the following results. Recall that $\operatorname{Sol}(D)$ denotes the set of optimal solutions of the dual problem (D).

Theorem 4.1.2 *Suppose that the primal problem (P) is convex. Then:*
(i) If $\text{val}(D)$ *is finite, then* $\text{Sol}(D) = \partial v^{**}(0)$.
(ii) If $v(y)$ is subdifferentiable at $y = 0$, then there is no duality gap between the primal and dual problems and $\text{Sol}(D) = \partial v(0)$.
(iii) If $\text{val}(P) = \text{val}(D)$ *and is finite, then the (possibly empty) set* $\text{Sol}(D)$, *of optimal solutions of the dual problem, coincides with $\partial v(0)$.*

Proof. Assertion (i) follows from representation (4.1.22) of the dual problem and formula (4.1.27) applied at $y = 0$.

If $v(y)$ is subdifferentiable at $y = 0$, then $v^{**}(0) = v(0)$ and $\partial v^{**}(y) = \partial v(y)$. Consequently, by theorem 4.1.1, we obtain that $\text{val}(P) = \text{val}(D)$. Together with (i) this proves assertion (ii).

If $\text{val}(P) = \text{val}(D)$ and is finite, then $v^{**}(y) = v(y)$ and hence $\partial v^{**}(y) = \partial v(y)$. Together with (i) this proves (iii). ■

Suppose now that $v(\cdot)$ is convex and that $v(y) < +\infty$ for all y in a neighborhood of zero. By convex analysis this implies that $v(y)$ is continuous at $y = 0$ (it still can happen that $v(y) = -\infty$ for all y in a neighborhood of zero). Moreover, if in addition $v(0)$ is finite, then $v(y)$ is subdifferentiable at $y = 0$ and $\partial v(0)$ is bounded. Note that $v(y) < +\infty$ iff the feasible set of (P_y) is nonempty, which means that there exists $x \in C$ such that $G(x) + y \in K$. Therefore the condition that $v(y) < +\infty$ for all y in a neighborhood of zero, can be written in the form

$$0 \in \text{int}\{G(C) - K\}, \tag{4.1.28}$$

where $G(C)$ denotes the set of points $G(x)$, $x \in C$, in the space $\mathcal{Y}$, and "int" stands for the interior of the corresponding set. We obtain the following result.

Theorem 4.1.3 *Suppose that the primal problem (P) is convex and that the regularity condition (4.1.28) holds. Then there is no duality gap between the primal and dual problems and, moreover, if their common optimal value is finite, then the set* $\text{Sol}(D)$, *of optimal solutions of the dual problem, is nonempty and bounded.*

It is interesting that in a sense a converse of the above result also holds. By theorem 4.1.5(i) we have that if the set of optimal solutions of the dual problem is nonempty (and hence $\text{val}(D)$ is finite) and bounded, then $\partial v^{**}(0)$ is nonempty and bounded. By convexity of $v(\cdot)$ this implies that $v(y)$ is continuous at $y = 0$, and hence (4.1.28) follows. Therefore we obtain the following result.

Theorem 4.1.4 *Suppose that the primal problem (P) is convex and that the dual problem has a nonempty and bounded set of optimal solutions. Then condition (4.1.28) holds and there is no duality gap between the primal and dual problems.*

We say that the *Slater condition* holds, for the primal problem, if there exists a point $\bar{x} \in C$ such that $G(\bar{x}) \in \text{int}(K)$. In the case of semidefinite programming, i.e. when $K = \mathcal{S}^m_-$, this means that $G(\bar{x}) \prec 0$, i.e. that the matrix $G(\bar{x})$ is negative definite. It is clear that the Slater condition implies the regularity condition (4.1.28). Converse of that is also true if the cone K has a nonempty interior, and hence in particular in the case of convex semidefinite programming problems.

Proposition 4.1.4 *Suppose that the mapping $G(x)$ is convex with respect to the cone $Q := -K$ and that K has a nonempty interior. Then (4.1.28) is equivalent to the Slater condition.*

Proof. It is clear that Slater condition implies (4.1.28). We proof now that the converse implication also holds. Suppose that (4.1.28) holds. Let x be a feasible point of (P), i.e. $x \in C$ and $G(x) \in K$. It is clear that existence of such a point follows from (4.1.28). Let $\bar{y}$ be an interior point of K, i.e. for some neighborhood $N \subset \mathcal{Y}$ of zero the inclusion $\bar{y} + N \subset K$ holds. It follows from (4.1.28) that for sufficiently small $\alpha > 0$, there is a point $x' \in C$ such that $\alpha(\bar{y} - G(x)) \in G(x') - K$. Consider $k_1 := G(x') - \alpha(\bar{y} - G(x)) \in K$ and let $\bar{x} := (x + x')/2$. By convexity of C, $\bar{x} \in C$, and by convexity of G we have

$$k_2 := -\tfrac{1}{2}G(x) - \tfrac{1}{2}G(x') + G(\bar{x}) \in K.$$

It follows that

$$G(\bar{x}) = \tfrac{1}{2}k_1 + k_2 + \tfrac{1}{2}(1-\alpha)G(x) + \tfrac{1}{2}\alpha\bar{y} = k + \tfrac{1}{2}\alpha\bar{y},$$

where $k := \frac{1}{2}k_1 + k_2 + \frac{1}{2}(1-\alpha)G(x) \in K$ for $\alpha < 1$. Consequently

$$G(\bar{x}) + \tfrac{1}{2}\alpha N = k + \tfrac{1}{2}\alpha(\bar{y} + N) \subset K,$$

and hence $G(\bar{x}) \in \text{int}(K)$, which completes the proof. ■

All the above results can be applied to convex semidefinite programming problems, and in particular to linear semidefinite programming problems. Consider, for instance, a linear semidefinite programming problem in the form (4.1.2) and its dual (4.1.7). If the matrix A_0 is negative definite, then the Slater condition holds (with $\bar{x} = 0$ for example) and hence in that case there is no duality gap, and moreover the dual problem has a nonempty and bounded set of optimal solutions. Suppose now that A_0 is positive definite and the optimal value of the dual problem is finite. Consider the set $\{\Omega \in \mathcal{S}^m_+ : \Omega \bullet A_0 = \text{val}(D)\}$. Since A_0 is positive definite this set is bounded and hence the set of optimal solutions of (D) is nonempty and bounded. We obtain then that the Slater condition, for the primal problem, holds and there is no duality gap between the primal and dual problems.

We can also apply such results to the dual linear semidefinite programming problem (4.1.7) as well. The regularity condition (4.1.28), for the problem (4.1.7), takes the form

$$0 \in \text{int}\{x \in \mathbb{R}^n : x_i = \Omega \bullet A_i + b_i,\ i = 1, ..., n,\ \Omega \succeq 0\}. \qquad (4.1.29)$$

If the above condition holds, then there is no duality gap between the primal and dual problems and the primal problem (4.1.2) has a nonempty and bounded set of optimal solutions. Conversely, if primal problem (4.1.2) has a nonempty and bounded set of optimal solutions, then condition (4.1.29) holds and there is no duality gap between the primal and dual problems. In particular, condition (4.1.29) holds if there exists a positive definite matrix Ω such that $\Omega \bullet A_i + b_i = 0$, $i = 1, \dots, n$.

4.1.3 *Optimality conditions*

In this section we discuss first and second order optimality conditions for the semidefinite programming problem (4.1.1). Again we adopt here an approach of considering a general "cone constrained" problem in the form (4.1.3), and then specifying the obtained results to the semidefinite programming setting. In order to simplify the presentation we assume in the remainder of this article that the cone C *coincides* with the space $\mathbb{R}^n$, that is $C = \mathbb{R}^n$. We also assume from now on that the function $f(x)$ and mapping $G(x)$ are sufficiently smooth, at least are continuously differentiable.

Suppose that the primal problem (P) is convex. Then, as we know, the function $L(\cdot, \lambda)$ is convex for any $\lambda \in K^-$, and hence $\bar{x}$ is a minimizer of $L(\cdot, \lambda)$, over $\mathbb{R}^n$, iff $\nabla_x L(\bar{x}, \lambda) = 0$. Therefore, for convex problems, conditions (4.1.9) can be written as

$$\nabla_x L(\bar{x}, \bar{\lambda}) = 0, \quad \bar{\lambda} \cdot G(\bar{x}) = 0, \quad G(\bar{x}) \in K, \quad \bar{\lambda} \in K^-. \tag{4.1.30}$$

In particular, for the semidefinite programming problem (4.1.1) the above conditions take the form

$$\nabla_x L(\bar{x}, \bar{\Omega}) = 0, \quad \bar{\Omega} \bullet G(\bar{x}) = 0, \quad G(\bar{x}) \preceq 0, \quad \bar{\Omega} \succeq 0. \tag{4.1.31}$$

Moreover, for the linear semidefinite programming problem (4.1.2) these conditions are reduced to feasibility of $A_0 + \sum_{i=1}^n \bar{x}_i A_i$ and $\bar{\Omega}$, considered as points of the respective primal (4.1.2) and dual (4.1.7) problems, and the complementarity condition (4.1.19).

Conditions (4.1.30) can be viewed as *first order* optimality conditions for the problem (4.1.3). A point $\bar{x} \in X$ is said to be a *stationary* point of the problem (P) if there exists $\bar{\lambda} \in Y$ such that conditions (4.1.30) hold. We refer to $\bar{\lambda}$, satisfying (4.1.30), as a *Lagrange multiplier* vector (in the case of semidefinite programming we refer to $\bar{\Omega}$ as a Lagrange multiplier matrix), and denote by $\Lambda(\bar{x})$ the set of all Lagrange multiplier vectors. In the convex case conditions (4.1.30) ensure that the point $(\bar{x}, \bar{\lambda})$ is a saddle point of the Lagrangian, and hence that $\bar{x}$ is an optimal solution of (P). That is, in the convex case conditions (4.1.30) are *sufficient* for optimality. However, even in the case of linear semidefinite programming, these conditions are not necessary and in order to ensure their necessity a *constraint qualification* is required.

Theorem 4.1.5 *Suppose that the primal problem (P) is convex and let $\bar{x}$ be its optimal solution. If the constraint qualification (4.1.28) holds, then the set*

$\Lambda(\bar{x})$, of Lagrange multiplier vectors, is nonempty and bounded and is the same for any optimal solution of (P). Conversely, if $\Lambda(\bar{x})$ is nonempty and bounded, then (4.1.28) holds.

Proof. If condition (4.1.28) holds, then by theorem 4.1.3 we have that there is no duality gap between the primal and dual problems, and moreover $\Lambda(\bar{x})$ coincides with the set of optimal solutions of the dual problem and is nonempty and bounded. The converse assertion follows from theorem 4.1.4. ∎

Recall that by proposition 4.1.4, in the convex case, constraint qualification (4.1.28) is equivalent to the Slater condition provided cone K has a nonempty interior. In particular, in the case of convex semidefinite programming we have that the set of Lagrange multiplier matrices is nonempty and bounded iff the Slater condition holds.

Let us consider now a possibly nonconvex optimization problem (P) of the form (4.1.3) (with $C = \mathbb{R}^n$), and let $\bar{x}$ be a feasible point of (P). We can derive then first order optimality conditions by considering the following linearization of (P) at $\bar{x}$:

$$\operatorname*{Min}_{h \in \mathbb{R}^n} Df(\bar{x})h \quad \text{subject to} \quad DG(\bar{x})h \in T_K(G(\bar{x})). \tag{4.1.32}$$

We denote here by $Df(\bar{x})$ the differential of $f(\cdot)$ at $\bar{x}$, i.e. $Df(\bar{x})h$ is a linear function of h given by $Df(\bar{x})h = h \cdot \nabla f(\bar{x})$, and similarly for the differential of the mapping $G(\cdot)$, $DG(\bar{x})h = \sum_{i=1}^n h_i G_i(\bar{x})$ where $G_i(\bar{x}) := \partial G(\bar{x})/\partial x_i$. By $T_K(y)$ we denote the tangent cone to K at a point $y \in K$, that is

$$T_K(y) := \{z \in \mathcal{Y} : \operatorname{dist}(y + tz, K) = o(t),\ t \geq 0\}.$$

It is also possible to linearize the constraint qualification (4.1.28), at the point $\bar{x}$ and for $C = \mathbb{R}^n$, as follows

$$0 \in \operatorname{int}\{G(\bar{x}) + DG(\bar{x})\mathbb{R}^n - K\}. \tag{4.1.33}$$

This constraint qualification was introduced by Robinson, and we refer to it as *Robinson constraint qualification.* Since the affine mapping $h \to G(\bar{x}) + DG(\bar{x})h$ is convex, with respect to any cone, we have by proposition 4.1.4 that if the cone K has a nonempty interior, then (4.1.33) is equivalent to existence of a vector $\bar{h} \in \mathbb{R}^n$ such that

$$G(\bar{x}) + DG(\bar{x})\bar{h} \in \operatorname{int}(K). \tag{4.1.34}$$

The above condition can be viewed as an extended Mangasarian-Fromovitz constraint qualification. Recall that if $K := \mathcal{S}_-^m$ is the cone of negative semidefinite matrices, then its interior is formed by negative definite matrices. Therefore in the case of semidefinite programming problem (4.1.1) the above constraint qualification takes the form: there exists a vector $\bar{h} \in \mathbb{R}^n$ such that

$$G(\bar{x}) + DG(\bar{x})\bar{h} \prec 0. \tag{4.1.35}$$

It is possible to show that if $\bar{x}$ is a locally optimal solution of (P) and Robinson constraint qualification holds, then $h = 0$ is an optimal solution of the linearized problem (4.1.32). Proof of that is based on the following stability result. Suppose that the constraint qualification (4.1.33) holds. Then there exists $\kappa > 0$ such that for for all x in a neighborhood of $\bar{x}$ one has

$$\operatorname{dist}(x, G^{-1}(K)) \leq \kappa \operatorname{dist}(G(x), K), \tag{4.1.36}$$

where $G^{-1}(K) := \{x : G(x) \in K\}$ is the feasible set of the problem (P). It follows from the Taylor expansion $G(\bar{x} + th) = G(\bar{x}) + tDG(\bar{x})h + o(t)$ and (4.1.36) that

$$T_{G^{-1}(K)}(\bar{x}) = \{h : DG(\bar{x})h \in T_K(G(\bar{x}))\},$$

i.e. the feasible set of the linearized problem (4.1.32) coincides with the tangent cone to the feasible set of the problem (P) at the point $\bar{x}$. By the linearization of the objective function $f(\cdot)$ at $\bar{x}$, we obtain then that it follows from local optimality of $\bar{x}$ that $Df(\bar{x})h$ should be nonnegative for any $h \in T_{G^{-1}(K)}(\bar{x})$, and hence $h = 0$ should be an optimal solution of (4.1.32).

Since the linearized problem (4.1.32) is convex, we can apply to it the corresponding first order optimality conditions, at $h = 0$. Note that the polar of the tangent cone $T_K(G(\bar{x}))$ is given by

$$[T_K(G(\bar{x}))]^- = \{\lambda \in K^- : \lambda \cdot G(\bar{x}) = 0\}.$$

Therefore the first order optimality conditions for the linearized problem are equivalent to conditions (4.1.30) for the problem (P) at the point $\bar{x}$. By theorem 4.1.30 we have then the following result.

Theorem 4.1.6 *Let $\bar{x}$ be a locally optimal solution of the primal problem (P), and suppose that Robinson constraint qualification (4.1.33) holds. Then the set $\Lambda(\bar{x})$, of Lagrange multiplier vectors, is nonempty and bounded. Conversely, if $\bar{x}$ is a feasible point of (P) and $\Lambda(\bar{x})$ is nonempty and bounded, then (4.1.33) holds.*

We discuss now *second order* optimality conditions. Since second order conditions are intimately related to a particular structure of the cone $\mathcal{S}^m_-$, we consider here only the case of semidefinite programming problems in the form (4.1.1), and with $C = \mathbb{R}^n$. In the remainder of this section we assume that $f(x)$ and $G(x)$ are twice continuously differentiable. In order to derive second order optimality conditions the following construction will be useful. Let $A \in \mathcal{S}^m_-$ be a matrix of rank $r < m$. Then there exists a neighborhood $\mathcal{N} \subset \mathcal{S}^m$ of A and a mapping $\Xi : \mathcal{N} \to \mathcal{S}^{m-r}$ such that: (i) $\Xi(\cdot)$ is twice continuously differentiable, (ii) $\Xi(A) = 0$, (iii) $\mathcal{S}^m_- \cap \mathcal{N} = \{X \in \mathcal{N} : \Xi(X) \in \mathcal{S}^{m-r}_-\}$, and (iv) $D\Xi(A)\mathcal{S}^m = \mathcal{S}^{m-r}$, i.e. the differential $D\Xi(A)$ maps the space $\mathcal{S}^m$ onto the space $\mathcal{S}^{m-r}$.

In order to construct such a mapping we proceed as follows. Denote by $\mathcal{E}(X)$ the eigenspace of $X \in \mathcal{S}^m$ corresponding to its $m - r$ largest eigenvalues, and let $P(X)$ be the orthogonal projection matrix onto $\mathcal{E}(X)$. Let E_0 be a (fixed) $m \times (m-r)$ matrix whose columns are orthonormal and span the space $\mathcal{E}(A)$. Consider $F(X) := P(X)E_0$ and let $U(X)$ be the $m \times (m-r)$ matrix whose columns are obtained by applying the Gram-Schmidt orthonormal procedure to the columns of $F(X)$. It is known that $P(X)$ is twice continuously differentiable (in fact even analytic) function of X in a sufficiently small neighborhood of A. Consequently $F(\cdot)$ and hence $U(\cdot)$ are twice continuously differentiable near A. Also by the above construction, $U(A) = E_0$, the column space of $U(X)$ coincides with $\mathcal{E}(X)$ and $U(X)^T U(X) = I_{m-r}$ for all X near A. Finally define $\Xi(X) := U(X)^T X U(X)$. It is straightforward to verify that this mapping Ξ satisfies the properties (i)-(iv). Note that since $D\Xi(A)$ is onto, Ξ maps a neighborhood of A *onto* a neighborhood of the null matrix in the space $\mathcal{S}^{m-r}$.

Now let $\bar{x}$ be a stationary point of the semidefinite problem (4.1.1), i.e. the corresponding set $\Lambda(\bar{x})$ of Lagrange multiplier matrices is nonempty. This, of course, implies that $\bar{x}$ is a feasible point of (4.1.1). Since it is assumed that $C = \mathbb{R}^n$, feasibility of $\bar{x}$ means that $G(\bar{x}) \in \mathcal{S}^m_-$. Let Ξ be a corresponding mapping, from a neighborhood of $G(\bar{x})$ into $\mathcal{S}^{m-r}$ (where r is the rank of $G(\bar{x})$), satisfying the corresponding properties (i)-(iv). Consider the composite mapping $\mathcal{G}(x) := \Xi(G(x))$, from a neighborhood of $\bar{x}$ into $\mathcal{S}^{m-r}$. Note that by the property (ii) of Ξ, we have that $\mathcal{G}(\bar{x}) = 0$. Since Ξ maps a neighborhood of $G(\bar{x})$ *onto* a neighborhood of the null matrix in the space $\mathcal{S}^{m-r}$, we have that for all x near $\bar{x}$ the feasible set of (4.1.1) can be defined by the constraint $\mathcal{G}(x) \preceq 0$. Therefore problem (4.1.1) is locally equivalent to the following, so called reduced, problem

$$\operatorname*{Min}_{x \in \mathbb{R}^n} f(x) \;\; \text{subject to} \;\; \mathcal{G}(x) \preceq 0. \tag{4.1.37}$$

It is relatively easy to write second order optimality conditions for the reduced problem (4.1.37) at the point $\bar{x}$. This is mainly because the "constraint cone" $\mathcal{S}^{m-r}_-$ coincides with its tangent cone at the point $\mathcal{G}(\bar{x}) = 0$.

Let

$$\mathcal{L}(x, \Psi) := f(x) + \Psi \bullet \mathcal{G}(x), \quad \Psi \in \mathcal{S}^{m-r},$$

be the Lagrangian of the problem (4.1.37). Since $D\Xi(\bar{y})$ is onto and (by the chain rule of differentiation) $D\mathcal{G}(\bar{x}) = D\Xi(\bar{y})DG(\bar{x})$, where $\bar{y} := G(\bar{x})$, we have that $\bar{x}$ is also a stationary point of the problem (4.1.37), and that Ψ is a Lagrange multiplier matrix of the problem (4.1.37) iff $\Omega := [D\Xi(\bar{y})]^* \Psi$ is a Lagrange multiplier matrix of the problem (4.1.1). We say that $h \in \mathbb{R}^n$ is a *critical direction* at the point $\bar{x}$, if $D\mathcal{G}(\bar{x})h \preceq 0$ and $Df(\bar{x})h = 0$. Note that since $\mathcal{G}(\bar{x}) = 0$, the tangent cone to $\mathcal{S}^{m-r}_-$ at $\mathcal{G}(\bar{x})$ coincides with $\mathcal{S}^{m-r}_-$, and hence $T_K(\bar{y}) = [D\Xi(\bar{y})]^{-1}\mathcal{S}^{m-r}_-$. Therefore the set of all critical directions can be written as

$$C(\bar{x}) = \{h \in \mathbb{R}^n : DG(\bar{x})h \in T_K(\bar{y}),\ Df(\bar{x})h = 0\}, \tag{4.1.38}$$

or, equivalently, for any $\bar{\Omega} \in \Lambda(\bar{x})$,

$$C(\bar{x}) = \{h \in \mathbb{R}^n : DG(\bar{x})h \in T_K(\bar{y}),\ \bar{\Omega} \bullet DG(\bar{x})h = 0\}. \tag{4.1.39}$$

Note that $C(\bar{x})$ is a closed convex cone, referred to as the *critical cone.* Note also that since $\Lambda(\bar{x})$ is nonempty, the optimal value of the linearized problem (4.1.32) is 0, and hence $C(\bar{x})$ coincides with the set of optimal solutions of the linearized problem (4.1.32). Cone $C(\bar{x})$ represents those directions for which the first order linearization of (P) does not provide an information about local optimality of $\bar{x}$.

Consider a critical direction $h \in C(\bar{x})$ and a curve $x(t) := \bar{x}+th+\frac{1}{2}t^2w+\varepsilon(t)$, $t \geq 0$, where the remainder $\varepsilon(t)$ is of order $o(t^2)$. Suppose that Robinson constraint qualification (4.1.33) holds. We have then that Robinson constraint qualification for the reduced problem (4.1.37) holds as well. It follows from the stability result (4.1.36) and the Taylor expansion

$$\mathcal{G}(x(t)) = \mathcal{G}(\bar{x}) + tD\mathcal{G}(\bar{x})h + \tfrac{1}{2}t^2\left[D\mathcal{G}(\bar{x})w + D^2\mathcal{G}(\bar{x})(h,h)\right] + o(t^2),$$

that $x(t)$ can be feasible, i.e. the remainder term $\varepsilon(t) = o(t^2)$ can be chosen in such a way that $\mathcal{G}(x(t)) \in S_-^{m-r}$ for $t > 0$ small enough, iff

$$\operatorname{dist}\left(D\mathcal{G}(\bar{x})h + t\left[D\mathcal{G}(\bar{x})w + D^2\mathcal{G}(\bar{x})(h,h)\right], S_-^{m-r}\right) = o(t).$$

That is, iff

$$D\mathcal{G}(\bar{x})w + D^2\mathcal{G}(\bar{x})(h,h) \in T_{S_-^{m-r}}(D\mathcal{G}(\bar{x})h). \tag{4.1.40}$$

Note that $D\mathcal{G}(\bar{x})h \in S_-^{m-r}$ since it is assumed that h is a critical direction.

We have that if $\bar{x}$ is a locally optimal solution of the problem (4.1.37) and $x(t)$ is feasible, then $f(x(t)) \geq f(\bar{x})$ for $t > 0$ small enough. By using the corresponding second order Taylor expansion of $f(x(t))$ and since $Df(\bar{x})h = 0$ we obtain that the optimization problem

$$\begin{array}{ll} \operatorname{Min}_{w\in\mathbb{R}^n} & Df(\bar{x})w + D^2f(\bar{x})(h,h) \\ \text{subject to} & D\mathcal{G}(\bar{x})w + D^2\mathcal{G}(\bar{x})(h,h) \in T_{S_-^{m-r}}(D\mathcal{G}(\bar{x})h), \end{array} \tag{4.1.41}$$

has a nonnegative optimal value. The above problem is a linear problem subject to "cone constraint". The dual of that problem consists in maximization of $h^T\nabla^2_{xx}\mathcal{L}(\bar{x},\Psi)h$ over $\Psi \in M(\bar{x})$, where $M(\bar{x})$ denotes the set of all Lagrange multiplier matrices of the problem (4.1.37). Therefore the following second order necessary conditions, for $\bar{x}$ to be a locally optimal solution of the problem (4.1.37), hold

$$\sup_{\Psi\in M(\bar{x})} h^T\nabla^2_{xx}\mathcal{L}(\bar{x},\Psi)h \geq 0, \quad \forall\, h \in C(\bar{x}). \tag{4.1.42}$$

The above second order necessary conditions can be formulated in terms of the original problem (4.1.1). By the chain rule we have

$$h^T\nabla^2_{xx}\mathcal{L}(\bar{x},\Psi)h = h^T\nabla^2_{xx}L(\bar{x},\Omega)h + \varsigma(\Omega,h), \tag{4.1.43}$$

where $\Omega := [D\Xi(\bar{y})]^*\Psi$ is a Lagrange multiplier matrix of the problem (4.1.1) and

$$\varsigma(\Omega,h) := \Omega \bullet D^2\Xi(\bar{y})(DG(\bar{x})h, DG(\bar{x})h). \tag{4.1.44}$$

The additional term $\varsigma(\Omega, h)$, which appears here, is a quadratic function of h and in a sense represents the curvature of the cone $\mathcal{S}^m_-$ at the point $\bar{y} := G(\bar{x})$. This term can be calculated explicitly by various techniques, and can be written as follows (unfortunately the involved derivations are not trivial and will be not given here)

$$\varsigma(\Omega, h) = h^T H(\bar{x}, \Omega) h, \tag{4.1.45}$$

where $H(\bar{x}, \Omega)$ is an $n \times n$ symmetric matrix with typical elements

$$[H(\bar{x}, \Omega)]_{ij} := -2\Omega \bullet \left(G_i(\bar{x}) [G(\bar{x})]^{\dagger} G_j(\bar{x}) \right), \quad i, j = 1, ..., n. \tag{4.1.46}$$

Here $G_i(\bar{x}) := \partial G(\bar{x})/\partial x_i$ is $m \times m$ matrix of partial derivatives and $[G(\bar{x})]^{\dagger}$ denotes the Moore-Penrose pseudo-inverse of $G(\bar{x})$, i.e. $[G(\bar{x})]^{\dagger} = \sum_{i=1}^{r} \alpha_i^{-1} e_i e_i^T$ where α_i are nonzero eigenvalues of $G(\bar{x})$ and e_i are corresponding orthonormal eigenvectors. Note that in the case of linear semidefinite problem (4.1.2), partial derivatives matrices $G_i(\bar{x})$ do not depend on $\bar{x}$, and $G_i(\bar{x}) = A_i$.

Together with (4.1.42) this implies the following second order necessary conditions.

Theorem 4.1.7 *Let $\bar{x}$ be a locally optimal solution of the semidefinite problem (4.1.1), and suppose that Robinson constraint qualification (4.1.33) holds. Then*

$$\sup_{\Omega \in \Lambda(\bar{x})} h^T \left(\nabla^2_{xx} L(\bar{x}, \Omega) + H(\bar{x}, \Omega) \right) h \geq 0, \quad \forall\, h \in C(\bar{x}). \tag{4.1.47}$$

The matrix $H(\bar{x}, \Omega)$ can be written in the following equivalent form

$$H(\bar{x}, \Omega) = -2 \left(\frac{\partial G(\bar{x})}{\partial x} \right)^T \left(\Omega \otimes [G(\bar{x})]^{\dagger} \right) \left(\frac{\partial G(\bar{x})}{\partial x} \right), \tag{4.1.48}$$

where $\partial G(\bar{x})/\partial x$ denotes the $m^2 \times n$ Jacobian matrix

$$\partial G(\bar{x})/\partial x := [\operatorname{vec} G_1(\bar{x}), ..., \operatorname{vec} G_n(\bar{x})],$$

and "$\otimes$" stands for the Kronecker product of matrices. Since $\Omega \succeq 0$ and $G(\bar{x}) \preceq 0$ we have that $\Omega \otimes [G(\bar{x})]^{\dagger}$ is a negative semidefinite matrix of rank rp, where $r = \operatorname{rank} G(\bar{x})$ and $p = \operatorname{rank} \Omega$. Therefore the matrix $H(\bar{x}, \Omega)$ is positive semidefinite of rank less than or equal to rp. It follows that the additional term $h^T H(\bar{x}, \Omega) h$ is always nonnegative. Of course if the semidefinite problem is linear, then $\nabla^2_{xx} L(\bar{x}, \Omega) = 0$. Nevertheless, even in the linear case the additional term can be strictly positive.

It is said that the *quadratic growth* condition holds, at a feasible point $\bar{x}$ of the problem (P), if there exists $c > 0$ such that for any feasible point x in a neighborhood of $\bar{x}$ the following inequality holds

$$f(x) \geq f(\bar{x}) + c\|x - \bar{x}\|^2. \tag{4.1.49}$$

Clearly this quadratic growth condition implies that $\bar{x}$ is a locally optimal solution of (P).

Theorem 4.1.8 *Let $\bar{x}$ be a stationary point of the semidefinite problem (4.1.1), and suppose that Robinson constraint qualification (4.1.33) holds. Then the quadratic growth condition (4.1.49) holds if and only if the following conditions are satisfied*

$$\sup_{\Omega \in \Lambda(\bar{x})} h^T \left(\nabla^2_{xx} L(\bar{x}, \Omega) + H(\bar{x}, \Omega) \right) h > 0, \quad \forall\, h \in C(\bar{x}) \setminus \{0\}. \qquad (4.1.50)$$

The above second order sufficient conditions can be proved in two steps. First, the corresponding second order sufficient conditions (without the additional term) can be derived for the reduced problem (4.1.37). That is, it is possible to show that second order conditions (4.1.42) become sufficient if the weak inequality sign is replaced by the strict inequality sign. Proof of that is based on the simple observation that for any $\Psi \succeq 0$, and in particular for any $\Psi \in M(\bar{x})$, and any feasible x, i.e. such that $\mathcal{G}(x) \preceq 0$, the inequality $f(x) \geq \mathcal{L}(x, \Psi)$ holds. Second, the obtained second order sufficient conditions for the reduced problem are translated into conditions (4.1.50) in exactly the same way as it was done for the corresponding second order necessary conditions.

It can be noted that the only difference between the second order necessary conditions (4.1.47) and the second order sufficient conditions (4.1.50) is the change of the weak inequality sign in (4.1.47) into the strict inequality in (4.1.50). That is, (4.1.47) and (4.1.50) give a pair of "no gap" second order optimality conditions for the semidefinite programming problem (4.1.1). As we mentioned earlier, even in the case of linear semidefinite programming problems the additional term can be positive and hence the corresponding quadratic growth condition can hold.

Note that it can happen that, for a stationary point $\bar{x}$, the critical cone $C(\bar{x})$ contains only one point 0. Of course, in that case the second order conditions (4.1.47) and (4.1.50) trivially hold. Since $\bar{x}$ is stationary, we have that $h = 0$ is an optimal solution of the linearized problem (4.1.32). Therefore condition $C(\bar{x}) = \{0\}$ implies that $Df(\bar{x})h > 0$ for any $h \neq 0$ such that $DG(\bar{x})h \in T_K(G(\bar{x}))$. This in turn implies that for some $\alpha > 0$ and all feasible x sufficiently close to $\bar{x}$, the inequality

$$f(x) \geq f(\bar{x}) + \alpha \|x - \bar{x}\| \qquad (4.1.51)$$

holds. That is, if $\bar{x}$ is a stationary point and $C(\bar{x}) = \{0\}$, then $\bar{x}$ is a *sharp* local minimizer of the problem (P).

Let us discuss now calculation of the critical cone $C(\bar{x})$. The tangent cone to $\mathcal{S}^m_-$ at $G(\bar{x})$ can be written as follows

$$T_{\mathcal{S}^m_-}(G(\bar{x})) = \{Z \in \mathcal{S}^m : E^T Z E \preceq 0\}, \qquad (4.1.52)$$

where $r = \text{rank}(G(\bar{x}))$ and E is an $m \times (m-r)$ matrix complement of $G(\bar{x})$, i.e. E is of rank $m - r$ and such that $G(\bar{x})E = 0$. By (4.1.39) we obtain then that for any $\Omega \in \Lambda(\bar{x})$,

$$C(\bar{x}) = \left\{ h \in \mathbb{R}^n : \sum_{i=1}^n h_i E^T G_i(\bar{x}) E \preceq 0,\ \sum_{i=1}^n h_i \Omega \bullet G_i(\bar{x}) = 0 \right\}. \qquad (4.1.53)$$

Let us observe that it follows from the first order optimality conditions (4.1.31) (i.e. since $\Omega \succeq 0$, $G(\bar{x}) \preceq 0$ and because of the complementarity condition $\Omega \bullet G(\bar{x}) = 0$) that for any $\Omega \in \Lambda(\bar{x})$, the inequality

$$\text{rank}(\Omega) + \text{rank}(G(\bar{x})) \leq m$$

holds. We say that the *strict complementarity* condition holds at $\bar{x}$ if there exists $\Omega \in \Lambda(\bar{x})$ such that

$$\text{rank}(\Omega) + \text{rank}(G(\bar{x})) = m. \qquad (4.1.54)$$

Consider a Lagrange multipliers matrix Ω. By the complementarity condition $\Omega \bullet G(\bar{x}) = 0$ we have that $\Omega = E_1 E_1^T$, where E_1 is an $m \times s$ matrix of rank $s = \text{rank}(\Omega)$ and such that $G(\bar{x})E_1 = 0$. Let E_2 be a matrix such that $E_2^T E_1 = 0$ and the matrix $E := [E_1, E_2]$ forms a complement of $G(\bar{x})$. (If the strict complementarity condition (4.1.54) holds, then $E = E_1$.) Then the critical cone can be written in the form

$$C(\bar{x}) = \left\{ h : \begin{array}{l} \sum_{i=1}^n h_i E_1^T G_i(\bar{x}) E_1 = 0,\ \sum_{i=1}^n h_i E_1^T G_i(\bar{x}) E_2 = 0, \\ \\ \sum_{i=1}^n h_i E_2^T G_i(\bar{x}) E_2 \preceq 0 \end{array} \right\}. \qquad (4.1.55)$$

In particular if the strict complementarity condition (4.1.54) holds, and hence $\Omega = EE^T$ for some complement E of $G(\bar{x})$, then

$$C(\bar{x}) = \left\{ h \in \mathbb{R}^n : \sum_{i=1}^n h_i E^T G_i(\bar{x}) E = 0 \right\}, \qquad (4.1.56)$$

and hence in that case $C(\bar{x})$ is a linear space.

Let us note that Robinson constraint qualification (4.1.33) is equivalent to the condition

$$DG(\bar{x})\mathbb{R}^n + T_K(G(\bar{x})) = \mathcal{S}^m. \qquad (4.1.57)$$

Consider the linear space

$$\mathcal{L}(G(\bar{x})) := \{ Z \in \mathcal{S}^m : E^T Z E = 0 \}, \qquad (4.1.58)$$

where E is a complement matrix of $G(\bar{x})$. (It is not difficult to see that this linear space does not depend on a particular choice of the complement matrix E.) This linear space represents the tangent space to the (smooth) manifold of matrices of rank $r = G(\bar{x})$ in the space $\mathcal{S}^m$ at the point $G(\bar{x})$. It follows from (4.1.52) that $\mathcal{L}(G(\bar{x})) \subset T_{\mathcal{S}^m_-}(G(\bar{x}))$. In fact $\mathcal{L}(G(\bar{x}))$ is the largest linear subspace of $T_{\mathcal{S}^m_-}(G(\bar{x}))$, i.e. it is the lineality space of the tangent cone $T_{\mathcal{S}^m_-}(G(\bar{x}))$. We have then that condition

$$DG(\bar{x})\mathbb{R}^n + \mathcal{L}(G(\bar{x})) = \mathcal{S}^m \tag{4.1.59}$$

is stronger than the corresponding Robinson constraint qualification, i.e. (4.1.59) implies (4.1.57). It is said that the *nondegeneracy condition* holds at $\bar{x}$ if condition (4.1.59) is satisfied. It is not difficult to show that the nondegeneracy condition (3.27) holds iff the n-dimensional vectors

$$f_{ij} := (e_i^T G_1(\bar{x})e_j, ..., e_i^T G_n(\bar{x})e_j), \quad 1 \le i \le j \le m - r,$$

are linearly independent. Here $e_1, ..., e_{m-r}$ are column vectors of the complement matrix E. Since there are $(m-r)(m-r+1)/2$ such vectors f_{ij}, the nondegeneracy condition can hold only if $n \ge (m-r)(m-r+1)/2$.

If the nondegeneracy condition holds, then $\Lambda(\bar{x})$ is a singleton. Moreover, if the strict complementarity condition holds, then $\Lambda(\bar{x})$ is a singleton iff the nondegeneracy condition is satisfied. If the nondegeneracy and strict complementarity conditions hold, then the critical cone $C(\bar{x})$ is defined by a system of $(m-r)(m-r+1)/2$ linearly independent linear equations and hence is a linear space of dimension $n - (m-r)(m-r+1)/2$.

4.1.4 Stability and sensitivity analysis

In this section we consider a parameterized semidefinite programming problem in the form

$$\operatorname*{Min}_{x\in\mathbb{R}^n} f(x,u) \text{ subject to } G(x,u) \preceq 0, \tag{4.1.60}$$

depending on the parameter vector u varying in a finite dimensional vector space $\mathcal{U}$. Here $f : \mathbb{R}^n \times \mathcal{U} \to \Re$, $G : \mathbb{R}^n \times \mathcal{U} \to \mathcal{S}^m$, and we assume that for a given value u_0 of the parameter vector, problem (4.1.60) coincides with the "unperturbed" problem (P), i.e. $f(\cdot, u_0) = f(\cdot)$ and $G(\cdot, u_0) = G(\cdot)$. We study continuity and differentiability properties of the optimal value $v(u)$ and an optimal solution $\hat{x}(u)$ of the parameterized problem (4.1.60) as functions of u near the point u_0. We assume throughout this section that the optimal value of the problem (P) is *finite*. Since the analysis is somewhat involved we only state some main results referring to the literature for proofs and extensions.

Let us consider first the parameterized problem (P_y), $y \in \mathcal{Y}$, in the form (4.1.20) and the corresponding optimal value function $v(y)$. We have then that if the problem (P) is convex and the regularity condition (4.1.28) holds, then $v(y)$ is continuous at $y = 0$ and $\partial v(0)$ is nonempty and bounded and coincides with the set of optimal solutions of the dual problem (D). By convex analysis we have that if a convex function $v(\cdot)$ is finite valued and continuous at a point $y \in \mathcal{Y}$, then for any $d \in \mathcal{Y}$,

$$v'(y,d) = \sup_{y^* \in \partial v(y)} y^* \cdot d, \tag{4.1.61}$$

where

$$v'(y,d) := \lim_{t \downarrow 0} \frac{v(y+td) - v(y)}{t}$$

denotes the directional derivative of $v(\cdot)$ at y in the direction d.

In particular, we can consider the following parameterization of the semidefinite problem (4.1.1)

$$\operatorname*{Min}_{x \in C} f(x) \ \text{ subject to } \ G(x) + Y \preceq 0, \tag{4.1.62}$$

parameterized by $Y \in \mathcal{S}^m$, and the corresponding optimal value function $v(Y)$. By the above discussion we have the following result.

Theorem 4.1.9 *Suppose that the semidefinite problem (4.1.1) is convex and that the Slater condition holds. Then the set* Sol(D), *of optimal solutions of the dual problem* (D), *is nonempty and bounded, the optimal value function* $v(Y)$ *of the problem (4.1.62) is convex, continuous at* $Y = 0$ *and*

$$v'(0,A) = \sup_{\Omega \in \mathrm{Sol}(D)} \Omega \bullet A. \tag{4.1.63}$$

In particular, we have that, under the assumptions of the above theorem, if the dual problem (D) has a unique optimal solution $\bar{\Omega}$, then the optimal value function $v(\cdot)$ is differentiable at $Y = 0$ and $\nabla v(0) = \bar{\Omega}$.

Let us consider now the parametric problem (P_u) in the general form (4.1.60). We denote by $L(x,\Omega,u)$ the Lagrangian associated with (P_u), that is

$$L(x,\Omega,u) := f(x) + \Omega \bullet G(x,u).$$

We assume that the functions $f(x,u)$ and $G(x,u)$ are sufficiently smooth, at least are continuously differentiable, and that the following, so-called *inf-compactness*, condition holds: there exist a number $\alpha > v(u_0)$ and a compact set $S \subset \mathbb{R}^n$ such that

$$\{x \in \mathbb{R}^n : f(x,u) \le \alpha,\ G(x,u) \preceq 0\} \subset S \tag{4.1.64}$$

for all u in a neighborhood of u_0. We have that under the above inf-compactness condition, for all u near u_0, the optimization is actually performed in the compact set S, and hence the set of optimal solutions of (P_u) is nonempty and bounded. In particular, it follows that the set Sol(P), of optimal solutions of the "unperturbed" problem (4.1.1), is nonempty and bounded.

In a sense the following result is an extension of formula (4.1.63).

Theorem 4.1.10 *Suppose that the semidefinite problem (4.1.1) is convex and that the Slater and inf-compactness conditions hold. Then the optimal value function $v(u)$ is directionally differentiable at u_0 and for any $d \in \mathcal{U}$,*

$$v'(u_0, d) = \inf_{x \in \mathrm{Sol}(P)} \sup_{\Omega \in \mathrm{Sol}(D)} d \cdot \nabla_u L(x, \Omega, u_0). \tag{4.1.65}$$

It can be noted that in the case of parametric problem (4.1.62), the corresponding Lagrangian $L(x, \Omega, Y)$ takes the form $L(x, \Omega) + \Omega \bullet Y$. Therefore its gradient, with respect to Y, coincides with Ω and does not depend on x. Consequently in that case formula (4.1.65) reduces to formula (4.1.63). Note, however, that formula (4.1.63) may hold even if the set $\mathrm{Sol}(P)$ is empty. Note also that in the formulation of the above theorem only the "unperturbed" problem is assumed to be convex while perturbed problems (P_u), $u \neq u_0$, can be nonconvex.

In a nonconvex case the analysis is more delicate and an analogue of formula (4.1.65) may not hold. Somewhat surprisingly, in nonconvex cases, the question of first order differentiability of the optimal value function may require a second order analysis of the considered optimization problem. Yet the following result holds.

Theorem 4.1.11 *Suppose that the inf-compactness condition holds and that for any $x \in \mathrm{Sol}(P)$ there exists a unique Lagrange multipliers matrix $\bar{\Omega}(x)$, i.e. $\Lambda(x) = \{\bar{\Omega}(x)\}$. Then the optimal value function is directionally differentiable at u_0 and*

$$v'(u_0, d) = \inf_{x \in \mathrm{Sol}(P)} d \cdot \nabla_u L(x, \bar{\Omega}(x), u_0). \tag{4.1.66}$$

Note that existence and uniqueness of the Lagrange multipliers matrix can be considered as a constraint qualification. Since then clearly $\Lambda(x)$ is nonempty and bounded, we have that this condition implies Robinson constraint qualification.

Let us discuss now continuity and differentiability properties of an optimal solution $\hat{x}(u)$ of the problem (P_u). We have that, under the inf-compactness condition, the distance from $\hat{x}(u)$ to the set $\mathrm{Sol}(P)$, of optimal solutions of the unperturbed problem, tends to zero as $u \to u_0$. In particular, if $\mathrm{Sol}(P) = \{\bar{x}\}$ is a singleton, i.e. (P) has unique optimal solution $\bar{x}$, then $\hat{x}(u) \to \bar{x}$ as $u \to u_0$. However, the rate at which $\hat{x}(u)$ converges to $\bar{x}$ can be slower than $O(||u - u_0||)$, i.e. $||\hat{x}(u) - \bar{x}||/||u - u_0||$ can tend to ∞ as $u \to u_0$, even if the quadratic growth condition (4.1.49) holds.

We assume in the remainder of this section that $\mathrm{Sol}(P) = \{\bar{x}\}$ is a singleton and that $f(x, u)$ and $G(x, u)$ are twice continuously differentiable. For $K := S^m_-$, a given direction $d \in \mathcal{U}$ and $t \geq 0$ consider the following linearization of the problem (P_{u_0+td}),

$$\operatorname*{Min}_{h \in \mathbb{R}^n} Df(\bar{x}, u_0)(h, d) \text{ subject to } DG(\bar{x}, u_0)(h, d) \in T_K(G(\bar{x}, u_0)), \tag{4.1.67}$$

where

$$Df(\bar{x}, u_0)(h, d) = D_x f(\bar{x}, u_0)h + D_u f(\bar{x}, u_0)d,$$

and similarly for $DG(\bar{x}, u_0)(h, d)$. This is a linear problem subject to "cone constraint", and its dual is given by

$$\operatorname*{Max}_{\Omega \in \Lambda(\bar{x})} d \cdot \nabla_u L(x, \Omega, u_0). \tag{4.1.68}$$

We refer to the above problems as (PL_d) and (DL_d), respectively. If Robinson constraint qualification holds, then there is no duality gap between problems (PL_d) and (DL_d), the set $\Lambda(\bar{x})$ is nonempty and bounded, and hence the set $\mathrm{Sol}(DL_d)$, of the problem (4.1.68), is also nonempty and bounded.

Consider the following, so-called strong form, of second order conditions

$$\sup_{\Omega \in \mathrm{Sol}(DL_d)} h^T \left(\nabla^2_{xx} L(\bar{x}, \Omega) + H(\bar{x}, \Omega)\right) h > 0, \quad \forall\, h \in C(\bar{x}) \setminus \{0\}. \tag{4.1.69}$$

Of course, if $\mathrm{Sol}(DL_d) = \Lambda(\bar{x})$, then the above second order conditions coincide with the second order conditions (4.1.50). In particular, this happens if $\Lambda(\bar{x})$ is a singleton or if $G(x, u)$ does not depend on u.

We can formulate now the basic sensitivity theorem for semidefinite programming problems.

Theorem 4.1.12 *Let $\hat{x}(t) := \hat{x}(u_0 + td)$ be an optimal solution of the problem (4.1.60), for $u = u_0 + td$ and $t \geq 0$, converging to $\bar{x}$ as $t \downarrow 0$. Suppose that Robinson constraint qualification holds at $\bar{x}$, that the strong second order sufficient conditions (4.1.69) are satisfied and that the set* $\mathrm{Sol}(PL_d)$*, of optimal solutions of the linearized problem (4.1.67), is nonempty. Then:*
(i) $\hat{x}(u)$ is Lipschitz stable at $\bar{x}$ in the direction d, i.e.

$$\|\hat{x}(t) - \bar{x}\| = O(t), \quad t \geq 0. \tag{4.1.70}$$

(ii) For $t \geq 0$, the optimal value function has the following second order expansion along the direction d,

$$v(u_0 + td) = v(u_0) + t\,\mathrm{val}(DL_d) + \tfrac{1}{2}t^2\,\mathrm{val}(Q_d) + o(t^2), \tag{4.1.71}$$

where $\mathrm{val}(DL_d)$ *is the optimal value of the problem (4.1.68) and* $\mathrm{val}(Q_d)$ *is the optimal value of the following min-max problem:*

$$\operatorname*{Min}_{h \in \mathrm{Sol}(PL_d)} \operatorname*{Max}_{\Omega \in \mathrm{Sol}(DL_d)} \left\{ D^2 L(\bar{x}, \Omega, u_0)((h, d), (h, d)) + \varsigma(\Omega, h, d) \right\}, \tag{4.1.72}$$

referred to as (Q_d), with $D^2 L(\bar{x}, \Omega, u_0)((h, d), (h, d))$ given by

$$h^T \nabla^2_{xx} L(\bar{x}, \Omega, u_0) h + 2h^T \nabla^2_{xu} L(\bar{x}, \Omega, u_0) d + d^T \nabla^2_{uu} L(\bar{x}, \Omega, u_0) d,$$

and

$$\varsigma(\Omega, h, d) := h^T H_{xx}(\Omega) h + 2h^T H_{xu}(\Omega) d + d^T H_{uu}(\Omega) d,$$

$$H_{xx}(\Omega) := -2\left(\frac{\partial G(\bar{x},u_0)}{\partial x}\right)^T \left(\Omega \otimes [G(\bar{x},u_0)]^\dagger\right)\left(\frac{\partial G(\bar{x},u_0)}{\partial x}\right),$$

$$H_{xu}(\Omega) := -2\left(\frac{\partial G(\bar{x},u_0)}{\partial x}\right)^T \left(\Omega \otimes [G(\bar{x},u_0)]^\dagger\right)\left(\frac{\partial G(\bar{x},u_0)}{\partial u}\right),$$

$$H_{uu}(\Omega) := -2\left(\frac{\partial G(\bar{x},u_0)}{\partial u}\right)^T \left(\Omega \otimes [G(\bar{x},u_0)]^\dagger\right)\left(\frac{\partial G(\bar{x},u_0)}{\partial u}\right).$$

(iii) Every accumulation point of $(\hat{x}(t)-\bar{x})/t$, as $t \downarrow 0$, is an optimal solution of the min-max problem (Q_d), given in (4.1.72). In particular, if (Q_d) has a unique optimal solution $\bar{h}$, then

$$\hat{x}(t) = \bar{x} + t\bar{h} + o(t), \quad t \geq 0. \tag{4.1.73}$$

Let us make the following remarks. If the constraint mapping $G(x,u)$ does not depend on u, then the set Sol(PL_d), of optimal solutions of the linearized problem, coincides with the critical cone $C(\bar{x})$. In any case $C(\bar{x})$ forms the recession cone of Sol(PL_d) provided Sol(PL_d) is nonempty. The above matrix $H_{xx}(\Omega)$ is exactly the same as the matrix $H(\bar{x},\Omega)$ defined in (4.1.46) and (4.1.48), and used in the second order optimality conditions. Strong second order conditions (4.1.69) ensure that the min-max problem (Q_d) has a finite optimal value and at least one optimal solution. It is possible to show that strong second order conditions (4.1.69) are "almost necessary" for the directional Lipschitzian stability of $\hat{x}(u)$. That is, if the left hand side of (4.1.69) is less than 0 for some $h \in C(\bar{x})$, then (4.1.70) cannot hold.

It follows from (4.1.71) that the directional derivative $v'(u_0, d)$ exists and is equal to the optimal value of the problem (4.1.68).

Existence of an optimal solution of the linearized problem (PL_d) is a *necessary* condition for the Lipschitzian stability (4.1.70) of an optimal solution. The following example shows that it can happen in semidefinite programming that the corresponding linearized problem does not possess an optimal solution.

Example 4.1.3 Consider the linear space $\mathcal{Y} := \mathcal{S}^2$, the cone $K := \mathcal{S}^2_+$, the mapping $G : \Re^2 \times \Re \to \mathcal{S}^2$ defined as $G(x_1,x_2,t) := \text{diag}(x_1,x_2) + tA$, where diag$(x_1,x_2)$ denotes the diagonal matrix with diagonal elements x_1 and x_2 and $A = (a_{ij})$ is the 2×2 symmetric matrix with zero diagonal elements and $a_{12} = a_{21} = 1$, and the parameterized problem

$$\operatorname*{Min}_{x\in\Re^2} x_1 + \tfrac{1}{2}x_1^2 + \tfrac{1}{2}x_2^2 \text{ subject to } G(x_1,x_2,t) \in \mathcal{S}^2_+. \tag{4.1.74}$$

It is not difficult to see that the feasible set of this problem is $\{x : x_1x_2 \geq t^2,\ x_1 \geq 0, x_2 \geq 0\}$, and that the unperturbed problem (for $t = 0$) has unique optimal solution $\bar{x} = (0,0)$ and the unique corresponding Lagrange matrix $\bar{\Omega} = \text{diag}(-1,0)$. Moreover, the above problem (4.1.74) is convex, the Slater condition holds and the strongest form of second order sufficient conditions is satisfied at the point $\bar{x}$. Yet it is not difficult to verify that the second coordinate of $\hat{x}(t)$ is of order $t^{2/3}$, as $t \downarrow 0$, and hence $\hat{x}(t)$ is not Lipschitz stable. The reason for such behavior of $\hat{x}(t)$ is that the corresponding linearized problem, at the direction $d = 1$,

$$\operatorname*{Min}_{x \in \Re^2} x_1 \text{ subject to } x_1 x_2 \geq 1,\ x_1 \geq 0,\ x_2 \geq 0, \tag{4.1.75}$$

does not possess an optimal solution.

Let Ω be an optimal solution of (DL_d), and let E_1 and E_2 be such matrices that $\Omega = E_1 E_1^T$, $E_2^T E_1 = 0$ and $E := [E_1, E_2]$ is a complement of $G(\bar{x}, u_0)$. Then similar to (4.1.55) we have

$$\operatorname{Sol}(PL_d) = \left\{ h : \begin{array}{l} \sum_i h_i E_1^T \dfrac{\partial G(\bar{x}, u_0)}{\partial x_i} E_1 + \sum_j d_j E_1^T \dfrac{\partial G(\bar{x}, u_0)}{\partial u_j} E_1 = 0 \\ \sum_i h_i E_1^T \dfrac{\partial G(\bar{x}, u_0)}{\partial x_i} E_2 + \sum_j d_j E_1^T \dfrac{\partial G(\bar{x}, u_0)}{\partial u_j} E_2 = 0 \\ \sum_i h_i E_2^T \dfrac{\partial G(\bar{x}, u_0)}{\partial x_i} E_2 + \sum_j d_j E_2^T \dfrac{\partial G(\bar{x}, u_0)}{\partial u_j} E_2 \preceq 0 \end{array} \right\}. \tag{4.1.76}$$

In particular, if the strict complementarity and nondegeneracy conditions hold, then $\Lambda(\bar{x}) = \{\bar{\Omega}\}$ is a singleton, $\operatorname{Sol}(DL_d) = \{\bar{\Omega}\}$ and

$$\operatorname{Sol}(PL_d) = \left\{ h : \sum_i h_i E^T \frac{\partial G(\bar{x}, u_0)}{\partial x_i} E + \sum_j d_j E^T \frac{\partial G(\bar{x}, u_0)}{\partial u_j} E = 0 \right\}. \tag{4.1.77}$$

In that case the corresponding problem (Q_d), defined in (4.1.72), becomes a problem of minimization of a quadratic function subject to linear constraints, and hence can be solved in a closed form. It follows then, under the second order sufficient conditions, that: (i) $\operatorname{val}(Q_d)$ is quadratic in d, and hence the optimal value function $v(u)$ is twice differentiable at u_0, and (ii) (Q_d) has a unique optimal solution $\bar{h} = \bar{h}(d)$ which is a linear function of d, and hence $v(u)$ is differentiable at u_0.

4.1.5 Notes

Lagrangian duality is a well developed concept in mathematical programming. Its origins go back to von Neumann's game theory. In the context of semidefinite programming particular examples of duality schemes were considered, for example, in [21, 710, 833]. Example 4.1.2, of a linear semidefinite program with a duality gap, is taken from [810]. The parametric approach to duality, by applying convex analysis to the parametric problem (4.1.20), was developed in Rockafellar [679, 680]. A proof of the Fenchel-Moreau duality theorem can be found in [679]. Convex semidefinite programming problems were discussed in [711].

First order necessary conditions of the form (4.1.30), for optimization problems subject to "cone constraints", were discussed in [456, 674, 874]. The constraint qualification (4.1.33) was introduced by Robinson in [676]. The stability result (4.1.37) is called metric regularity and is based on the Robinson-Ursescu stability theorem [675, 802]. Constraint qualifications (4.1.34) and (4.1.36) can be considered as extensions of the Mangasarian-Fromovitz [515] constraint qualification used in nonlinear programming. The result of theorem 4.1.6 is essentially due to Zowe and Kurcyusz [874].

The reduction approach to semidefinite programming (of considering the reduced problem (4.1.37)) is due to Bonnans and Shapiro [121]. A general formula for the additional term in second order optimality conditions under cone constraints is given in Cominetti [166], in a form of the support function of a certain second order tangent set, see also Kawasaki [407]. In the case of semidefinite programming this additional term is explicitly calculated and the second order necessary conditions (4.1.47) are given in Shapiro [711]. Sufficiency of second order conditions (4.1.50) is proved in [120]. The nondegeneracy condition in the form (4.1.59) was introduced in Shapiro and Fan [712] as a transversality condition (see also [25, 711] for a discussion of the nondegeneracy concept).

Theorem 4.1.10, giving the min-max formula (4.1.65) for the directional derivatives of the optimal value function in the convex case, is essentially due to Gol'shtein [291]. The result of theorem 4.1.11 is due to Levitin [476] and Lempio and Maurer [474]. Uniqueness of Lagrange multipliers in cone constrained, and in particular in semidefinite programming, problems is discussed in [712].

An extensive discussion of sensitivity type results of theorem 4.1.12 can be found in the review paper [122]. In the semidefinite programming first results of that type were obtained by an application of the Implicit Function Theorem in [711]. A proof of theorem 4.1.12 is given in [119]. Example 4.1.3 is taken from [122].

4.2 PARAMETRIC LINEAR SEMIDEFINITE PROGRAMMING

Katya Scheinberg

4.2.1 Optimality conditions

We will consider the case of linear semidefinite programming problems (we will refer to it simply as SDP). Such problems inherit certain properties of linear programming problems. We will exploit these properties to study the parametric behavior and sensitivity analysis of SDP problems. The sensitivity analysis in the previous part of the chapter studies SDP from nonlinear programming point of view. We begin this section by specifying the results of the previous section to the case of linear SDP. We will also depart from notation of the previous section.

To make the references to linear programming more apparent we henceforth will call the primal problem, what we called the dual (4.1.7) in the previous sections of this chapter. Thus, we return to standard notation:

$$(P) \qquad \begin{array}{ll} \min & C \bullet X \\ \text{s.t.} & A_i \bullet X = b_i, \quad i = 1, \ldots, m, \\ & X \succeq 0, \ X \in \mathcal{S}^n, \end{array}$$

where $C \in \mathcal{S}^n$, $A_i \in \mathcal{S}^n$, $i = 1, \ldots, m$, and $b \in \mathbb{R}^m$.

The problem dual to (P) is:

$$(D) \qquad \begin{array}{ll} \max & \sum_{i=1}^{m} y_i b_i \\ \text{s.t.} & \sum_{i=1}^{m} y_i A_i + Z = C, \\ & Z \succeq 0, \ Z \in \mathcal{S}^n. \end{array}$$

We will summarize some duality results that were derived above specializing them to the linear SDP case.

Assume that for the pair of primal-dual SDP problems the duality gap is zero at optimality and the optimal solutions exist. The first order optimality conditions then are

$$\begin{aligned} &X \bullet Z = 0 \\ &A_i \bullet X = b_i, \quad i = 1, \ldots, m, \\ &\sum_{i=1}^{m} y_i A_i + Z = C, \\ &X, Z \succeq 0, \ X, Z \in \mathcal{S}^n. \end{aligned}$$

For an optimal primal-dual solution (X, y, Z), it follows from $X \bullet Z = 0$ and $X, Z \succeq 0$ that $XZ = ZX = 0$. Thus X and Z share the same orthonormal base of eigenvectors; i.e., $X = Q\Lambda Q^T$ and $Z = Q\Omega Q^T$, where $QQ^T = I_n$ and Λ and Ω are diagonal matrices with the eigenvalues of X and Z, respectively, on their diagonals.

From complementary slackness, the common basis of eigenvectors of X and Z can be partitioned in the following way: $Q = [Q_P, Q_N, Q_D]$, $X = Q_P \Lambda_P Q_P^T$, $\Lambda_P \succ 0$ and $Z = Q_D \Omega_D Q_D^T$, $\Omega_D \succ 0$. If we let

$$\operatorname{rank}(X) = r, \ \operatorname{rank}(Z) = s,$$

then $r + s \leq n$. If $r + s = n$ and, hence, $Q_N = \emptyset$, then the primal-dual solution is strictly complementary.

The following definition of nondegeneracy was proposed in [24] and is the same as (4.1.59) when specialized to linear SDP.

Definition 4.2.1 *Let $\mathcal{M}_r = \{X \in \mathcal{S}^n : \operatorname{rank}(X) = r\}$ and, for a given $X \in \mathcal{M}_r$, let $\mathcal{T}_X$ be the tangent subspace to $\mathcal{M}_r$ at X; i.e.,*

$$\mathcal{T}_X = \left\{ Q \begin{bmatrix} U & V \\ V^T & 0 \end{bmatrix} Q^T : U \in \mathcal{S}^r, \ V \in \mathbb{R}^{r \times n-r} \right\},$$

where Q is the orthogonal matrix of the eigenvectors of X. Let $\mathcal{N} = \{X \in \mathcal{S}^n : A_i \bullet X = 0, \ i = 1, \ldots, m\}$.

We say that a solution X to problem (P) is primal nondegenerate if

$$\mathcal{T}_X + \mathcal{N} = \mathcal{S}^n.$$

Dual nondegeneracy is defined in a similar manner. We say that an SDP problem is nondegenerate if there is no degenerate optimal solution. We provide useful conditions equivalent to primal and dual nondegeneracy conditions due to [715] and [24]:

Lemma 4.2.1 *The primal problem (P) is nondegenerate if and only if the following matrices are linearly independent:*

$$\begin{bmatrix} Q_P^T A_i Q_P \\ Q_P^T A_i Q_{ND} \end{bmatrix} \quad i = 1, \ldots, m,$$

where $Q_{ND} = [Q_N, Q_D]$.

Lemma 4.2.2 *The dual problem (D) is nondegenerate if and only if the following matrices span $\mathcal{S}^{n-s}$:*

$$[Q_{PN}^T A_i Q_{PN}], \quad i = 1, \ldots, m,$$

where $Q_{PN} = [Q_P, Q_N]$.

Let us now study the *quadratic growth* condition and the equivalent second order sufficient optimality condition for linear SDP. We consider the form given earlier in the chapter, i.e., the quadratic growth and the second order sufficient optimality conditions for the *dual* solution. Recall that we say that the quadratic growth condition holds at a given feasible point $(\bar{y}, \bar{Z})$, whenever there exists $c > 0$ such that for any feasible (y, Z) the following special case of (4.1.49) holds

$$b^T y \geq b^T \bar{y} + c\|y - \bar{y}\|^2.$$

From Theorem 4.1.8 we know that if Slater's condition holds, then the quadratic growth condition is equivalent to the second order sufficient optimality condition. Let (y, Z) be an optimal dual solution and let $\mathcal{X}$ be the set of primal optimal solutions. Recall matrix H defined by (4.1.46),

$$[H(y, X)]_{ij} = -2X \bullet \left(G_i(y)[G(y)]^{\dagger} G_j(y)\right), \quad i, j = 1, ..., n.$$

Here $G_i(y) = \partial G(y)/\partial y_i$ and $[G(y)]^{\dagger}$ is the Moore-Penrose pseudo-inverse of $G(y)$. In the linear case $G(y) = C - \sum_{i=1}^m y_i A_i = Z$, then $G_i(y) = A_i$. Thus, in case of linear SDP

$$H_{ij}(y, X) = -2X \bullet [A_i Z^{\dagger} A_j].$$

In our notation the second order sufficient optimality condition (4.1.47) is

$$\sup_{X \in \mathcal{X}} h^T H(y, X) h < 0 \text{ for } h \in C(y) \setminus \{0\}$$

where $C(y)$ is the critical cone given by (4.1.55), which in our terms is written as

$$C(y) = \left\{ \begin{array}{ll} h \in \mathbb{R}^m: & \sum_i h_i Q_P^T A_i Q_P = 0, \quad \sum_i h_i Q_N^T A_i Q_P = 0, \\ & \sum_i h_i Q_N^T A_i Q_N \succeq 0. \end{array} \right.$$

Assume that the strict complementarity holds, then $Q_N = 0$ and $C(y) = \{h \in \mathbb{R}^m : \sum_i h_i Q_P^T A_i Q_P\} = 0$.

Notice that "$>$" in (4.1.47) and "$\preceq$" in (4.1.55) are replaced by "$<$' and "$\succeq$", respectively. This is due to the fact that in our notation the dual problem (D) is a maximization problem whereas problem (4.1.7) is a minimization problem.

From the decomposition of Z, $Z^{\dagger} = Q_D \Omega^{-1} Q_D^T$, where $\Omega \succ 0$. Thus,

$$H_{ij}(y, X) = 2Q_P \Lambda Q_P^T \bullet [A_i Q_D \Omega^{-1} Q_D^T A_j] = 2\Lambda \bullet [B_i \Omega B_j^T],$$

where $B_i = Q_P^T A_i Q_D$. Then

$$H(y, X) = -2B[\Lambda \otimes \Omega^{-1}]B^T$$

Clearly, from $\Lambda \succ 0$ and $\Omega \succ 0$, $\Lambda \otimes \Omega^{-1} \succ 0$.

Assume now that primal nondegeneracy holds, then, by Lemma 4.2.1, the system

$$\sum_{i=1}^{m} h_i Q_P^T A_i Q_P^T = 0, \ \sum_{i=1}^{m} h_i Q_P^T A_i Q_D = 0 \tag{4.2.78}$$

has only the trivial solution. Thus, for all nontrivial $h \in C(y)$ it holds that $B^T h \neq 0$. Then for all nontrivial $h \in C(\bar{x})$

$$\sup_{X \in \mathcal{X}} h^T H(y, X) h < 0. \tag{4.2.79}$$

Conversely, if the second order optimality condition holds, then for any $h \in C(y)$ $\exists X \in \mathcal{X}$ such that (4.2.79) holds. From results on duality and facial structure of linear SDP problems, there is $\bar{X} \in \operatorname{int} \mathcal{X}$, $\bar{X} = Q_P \Lambda Q_P^T$ such that for any $X \in \operatorname{int} \mathcal{X}$ $X = Q_P U Q_P^T$, $U \succeq 0$, $U \in \mathcal{S}^r$. Thus (4.2.79) can be written as

$$\sup_{U : Q_P U Q_P^T \in \mathcal{X}} \quad -h^T B[U \otimes \Omega^{-1}]B^T h < 0.$$

It follows that $Bh \neq 0$ for any $h \in C(y)$, hence, system (4.2.78) has only the trivial solution, thus by Lemma 4.2.1 primal nondegeneracy holds.

We conclude that, under strict complementarity, primal nondegeneracy is equivalent to the second order sufficient optimality condition and, thus, to the quadratic growth condition. Moreover, under strict complementarity, primal nondegeneracy is equivalent to uniqueness of the dual solution. Thus, we have the following result.

Theorem 4.2.1 *If Slater's condition and strict complementarity holds then an SDP problem has a unique optimal solution if and only if it satisfies the quadratic growth condition.*

Notice, that, since strict complementarity always holds in linear programming, in LP the above theorem reduces to the trivial fact that the quadratic growth condition holds if and only if the optimal solution is unique (Slater condition is not needed since strong duality holds for any pair of primal-dual feasible LP problems).

4.2.2 Parametric Objective Function

In this section we consider the following primal SDP problem in which the objective function coefficients depend linearly on a scalar parameter γ:

$$(P_\gamma) \qquad \begin{array}{ll} \min & (C+\gamma\bar{C}) \bullet X \\ \text{s.t.} & A_i \bullet X = b_i, \quad i=1,\ldots,m \\ & X \succeq 0, \ X \in \mathcal{S}^n. \end{array}$$

The dual problem then is

$$(D_\gamma) \qquad \begin{array}{ll} \max & \sum_{i=1}^{m} y_i b_i \\ \text{s.t.} & \sum_{i=1}^{m} y_i A_i + Z = C + \gamma\bar{C}, \\ & Z \succeq 0. \end{array}$$

This pair of parametric problems is equivalent to considering a directional linearization of the problem considered in the previous section:

$$\begin{array}{ll} \max & f(y) \\ \text{s.t.} & G(y) + Y \preceq 0, \end{array} \qquad (4.2.80)$$

where Y is a parameter. In our case $G(y) = \sum_{i=1}^{m} y_i A_i + C$ and $Y = \gamma\bar{C}$. Not only the properties of the problem (4.2.80) specialize to the case of (D_γ) and (P_γ) but also some properties of the latter are useful for studying the properties of (4.2.80).

The feasible region of the primal problem does not depend on γ. If the primal problem is infeasible, then (P_γ) is infeasible for all γ. Let us assume that the primal problem is strictly feasible; therefore, the Slater's condition holds for all primal problems (P_γ), which guarantees a zero duality gap at optimality. (If it is not strictly feasible, we consider the projection of the problem on the appropriate face of the cone of semidefinite matrices, such that the projected problem is strictly feasible).

Let Γ be the set of values of γ for which (P_γ) has a bounded solution. It is easy to show that Γ is a closed (possibly unbounded) interval. Hence, for any $\gamma \in \Gamma$ the dual problem is feasible (since the primal solution is bounded for $\gamma \in \Gamma$) and has a uniformly bounded solution (since the primal is feasible for all γ).

We do need the primal and the dual problems to have nonempty bounded optimal sets. This property is necessary to guarantee the convergence of an interior point method to a maximally complementary pair of primal-dual solutions, which is essential for our analysis. The dual problem has a nonempty bounded optimal set for all $\gamma \in \Gamma$ by Theorem 4.1.5, since the primal problem is strictly feasible.

Lemma 4.2.3 *The optimal set of the primal problem is either empty or unbounded for all $\gamma \in \Gamma$ or nonempty and bounded for any $\gamma \in \operatorname{int}\Gamma$.*

Proof. Let us assume that for some $\bar{\gamma} \in \operatorname{int}\Gamma$ (P_γ) has an empty or unbounded optimal face. Thus, there exists a nonzero direction $D \succeq 0$ such that $A_i \bullet D = 0$, $i = 1, \ldots, m$ and $(C + \bar{\gamma}\bar{C}) \bullet D = 0$. (This fact follows trivially in the case of an unbounded optimal set, and in the case of an empty optimal set it can be easily proved by choosing an infinite sequence of directions D_k, which converges to a direction D with desired properties.) Assume now that $\bar{C} \bullet D < 0$. Since $\bar{\gamma} \in \operatorname{int}\Gamma$, there exists an $\epsilon > 0$ such that $\bar{\gamma} + \epsilon \in \operatorname{int}\Gamma$ and $(C + (\bar{\gamma} + \epsilon)\bar{C}) \bullet D < 0$. Consequently, $(\mathrm{P}_{\bar{\gamma}+\epsilon})$ is unbounded, which contradicts the definition of Γ. Similarly we can dismiss the case $\bar{C} \bullet D > 0$. Thus $\bar{C} \bullet D = 0$ has to hold, which implies that $C \bullet D = 0$ and hence $C \bullet D + \gamma\bar{C} \bullet D = 0$ and (P_γ) has an empty or unbounded optimal face for any $\gamma \in \Gamma$. ∎

Assume that the dual Slater's condition does not hold. Let us consider the following procedure. By Lemma 4.2.3 there exists a feasible direction D, which is orthogonal to C and $\bar{C}$. We know, by an earlier assumption, that there exists a primal strictly feasible point, say X_0. We consider the following problem, obtained by adding a new constraint to problem (P_γ):

$$\begin{array}{ll} \min & (C + \gamma\bar{C}) \bullet X \\ \text{s.t.} & A_i \bullet X = b_i, \quad i = 1, \ldots, m \\ & D \bullet X = D \bullet X_0, \\ & X \succeq 0, \ X \in \mathcal{S}^n. \end{array}$$

It is clear that the feasible set of this problem is a subset of the feasible set of (P_γ), moreover the optimal set of this problem is a subset of the optimal set of (P_γ), provided the latter is not empty. The optimal value of the objective function is the same for both problems. This holds independently of the choice of γ.

The dual problem now is

$$\begin{array}{ll} \max & \displaystyle\sum_{i=1}^{m} y_i b_i + (D \bullet X_0)u \\ \text{s.t.} & \left[\displaystyle\sum_{i=1}^{m} y_i A_i + Du\right] + Z = C + \gamma\bar{C}, \\ & Z \succeq 0. \end{array}$$

The feasible and optimal sets of the new dual are supersets of, respectively, the feasible and the optimal set of (D_γ) and the optimal value of the objective is the same.

Now, notice that the primal Slater's condition still holds ($X_0 \succ 0$ is a feasible solution). If the dual Slater's condition holds then we are done. Otherwise, by

Lemma 4.2.3, there exists another feasible direction, say $\hat{D}$, which is orthogonal to C, $\bar{C}$ and D. By repeating the argument we obtain further modification of the primal-dual pair of problems (P_γ) and (D_γ). Since at every step the new direction is generated, which is orthogonal to all previously included directions, then the whole procedure has to terminate after a finite number of steps. When it terminates we obtain a pair of primal and dual problems, for which both primal and dual Slater's conditions are satisfied, the primal optimal set is s subset of the optimal set of (P_γ), the dual optimal set is a superset of the optimal set of (D_γ) and the optimal value of the objective equals the optimal value of (P_γ) and (D_γ).

This procedure allows us to study properties of the optimal value as a function of γ, while limitting our attention to the case when primal and dual problems are strictly feasible. Henceforth, we make the following assumption, which, by Theorem 4.1.5, guarantees that the optimal sets of (P_γ) and (D_γ) are nonempty and bounded for any $\gamma \in \operatorname{int}\Gamma$:

Assumption 4.2.1 *For all $\gamma \in \operatorname{int}\Gamma$, ($P_\gamma$) and ($D_\gamma$) are strictly feasible; i.e., the primal and dual Slater's conditions hold.*

Let F_p denote the feasible set of (P_γ). Let

$$C(\gamma) = C + \gamma\bar{C},$$

$$X(\gamma) \in \mathcal{X}(\gamma) = \{X_\gamma : X_\gamma = \operatorname{argmin}\{C(\gamma) \bullet X : X \in F_p\}\}$$

be some optimal solution (for example we can choose the analytic center of the optimal face) and

$$\phi(\gamma) = C(\gamma) \bullet X(\gamma).$$

Here we consider properties of $\phi(\gamma)$ and of the solutions as a function of γ. First let us specialize the results of the previous section to our case. The following results summarize Theorems 4.1.9-4.1.11.

Theorem 4.2.2 *For all $\gamma \in \operatorname{int}\Gamma$, $\phi(\gamma)$ is concave, continuous and has a subdifferential:*

$$\partial\phi(\gamma) = \{\bar{C} \bullet X | X \in \mathcal{X}(\gamma)\}.$$

Corollary 4.2.1 *If $\bar{C} \bullet X \equiv const$, $\forall X \in \mathcal{X}(\gamma)$ for given γ, then $\phi(\gamma)$ is continuously differentiable at γ. In particular, this holds when the primal optimal solution is unique.*

The next Theorem 4.2.3 is a corollary of Theorem 4.1.12.

Theorem 4.2.3 *Suppose $y(\gamma) \to y(\bar{\gamma})$ as $\gamma \to \bar{\gamma}$, $\gamma, \bar{\gamma} \in \operatorname{int}\Gamma$. Suppose also that $\bar{C} \bullet X \equiv const$, $\forall X \in \mathcal{X}(\gamma)$, and strict complementarity and primal nondegeneracy hold. Then*

1. *$(y(\gamma), Z(\gamma))$ is Lipschitz continuous at $(y(\bar{\gamma}), Z(\bar{\gamma}))$,*
2. *$\phi(\gamma)$ is twice differentiable.*

Under strict complementarity, primal nondegeneracy is equivalent to the uniqueness of the dual optimal solution. We, thus, proved Lipschitz continuity of the unique optimal solution of an SDP with the perturbed right-hand side. In [759] Sturm and Zhang address the question of Lipschitz continuity of the solution in the same framework without assuming uniqueness. They consider Lipschitz continuity of the central solution (i.e. solutions on the central path) of an SDP under perturbation of the right-hand side. They show the following result. Let $X(\gamma,\mu)$, $(y(\gamma,\mu), Z(\gamma,\mu))$ be feasible solutions of (P_γ) and (D_γ), respectively, satisfying $X(\gamma,\mu)Z(\gamma,\mu) = \mu I$ for $\mu > 0$. Then it is said that $X(\gamma,\mu)$ and $(y(\gamma,\mu), Z(\gamma,\mu))$ are on the primal-dual central path. It is known (see [288]) that the central path converges to the center of the optimal primal-dual face. The following is due to Sturm and Zhang.

Theorem 4.2.4 *Under strict complementarity, $(y(\gamma,\mu), Z(\gamma,\mu))$ is Lipschitz continuous for any fixed $\mu > 0$. As $\mu \to 0$ the Lipschitz constants converge to a finite limit, dependent on γ.*

Notice, that it does not follow from this result that the analytic center of the optimal face $(y(\gamma), Z(\gamma))$ is Lipschitz continuous, even though $(y(\gamma,\mu), Z(\gamma,\mu)) \to (y(\gamma), Z(\gamma))$ as $\mu \to 0$. The problem is that the limit of the Lipschitz constants may not be continuous in γ. It may even cease to exist in an arbitrarily small neighborhood of γ. In [759] there is an example which shows that strict complementarity may fail to hold in any neighborhood of a given $\gamma \in \operatorname{int}\Gamma$ even if it holds at γ. It is also shown that if strict complementarity fails to hold then Lipschitz continuity of the analytic center of the optimal face may fail to hold as well.

4.2.3 Optimal Partition

By $\mathbf{O}_P$ ($\mathbf{O}_D$) we denote the primal (dual) optimal face, i.e., the set of primal (dual) optimal solutions (possibly a single point). Suppose now that a partition $Q = [Q_P, Q_N, Q_D]$ corresponding to optimal solutions $\bar{X} \in \operatorname{ri}\mathbf{O}_P$ and $(\bar{y}, \bar{Z}) \in \operatorname{ri}\mathbf{O}_D$ is given. The following simple fact is shown in [288] and also follows from results in the previous Chapter 3. $X \in \mathbf{O}_P$ if and only if $X = Q_P U Q_P^T$, where U is a solution to

$$\begin{aligned} &A_i \bullet Q_P U Q_P^T = b_i, \qquad i = 1,\ldots,m, \\ &U \succeq 0,\ U \in \mathcal{S}^r. \end{aligned} \tag{4.2.81}$$

Similarly, $(y, Z) \in \mathbf{O}_D$ if and only if $Z = Q_D V Q_D^T$, where V is a solution to

$$\begin{aligned} &\sum_{i=1}^m y_i A_i + Q_D V Q_D^T = C, \\ &V \succeq 0,\ V \in \mathcal{S}^s. \end{aligned} \tag{4.2.82}$$

Clearly, this means that given a primal (dual) solution X $((y, Z))$ in the relative interior of the primal (dual) optimal face, one can completely describe this optimal face. We call such primal dual pairs of solutions, X and (y, Z), maximally complementary solutions, as in [96]. Moreover, the set of solutions to (4.2.81) and (4.2.82) is independent of the particular choice of the bases Q_P and Q_D as long as the subspaces spanned by these bases remain the same; that is as long as $\mathbf{R}_P \equiv \text{span}(Q_P)$ and $\mathbf{R}_D \equiv \text{span}(Q_D)$ stay fixed. Hence, $\mathbf{R}_P$ and $\mathbf{R}_D$ are invariant with respect to the choice of $\bar{X} \in \text{ri}\,\mathbf{O}_P$ and $(\bar{y}, \bar{Z}) \in \text{ri}\,\mathbf{O}_D$, and for every pair of primal-dual SDP problems there is a triple associated with it, $(\mathbf{R}_P, \mathbf{R}_N, \mathbf{R}_D)$, where $\mathbf{R}_P$ and $\mathbf{R}_D$ are as defined above and $\mathbf{R}_N = [\mathbf{R}_P \oplus \mathbf{R}_D]^\perp$. We refer to this triple as the *optimal partition.*

The optimal partition has been defined for parametric linear programming in [8] and [362]. For a standard form linear programming problem in which x denotes the primal variables and z denotes the dual slacks, the optimal partition is a partition of the indices $I = \{1, \ldots, n\}$ into two sets B and N so that $B = \{i \in I : \exists x \in \mathbf{O}_P,\ x_i > 0\}$ and $N = \{i \in I : \exists z \in \mathbf{O}_D,\ z_i > 0\}$. Clearly, from the properties of linear programming, $I = B \cup N$ and $B \cap N = \emptyset$. The concept of the optimal partition was also used in [315], [96] and [95] for quadratic programming problems and in [315] and [95] for linear complementarity problems. In this context the definition of the optimal partition is analogous to the linear programming case except that I is partitioned into *three* subsets of indices due to the possible absence of a strictly complementary solution. In the case of semidefinite programming we partition the space $\mathbb{R}^n$ rather than the set of indices I. Let us now give a formal definition of the optimal partition.

Definition 4.2.2 *Given a primal-dual pair of SDP's (P) and (D), the optimal partition for these problems is* $\pi = [\mathbf{R}_P, \mathbf{R}_N, \mathbf{R}_D]$ *if*

1. $\forall X \in \mathbf{O}_P$, $\text{span}(X) \subseteq \mathbf{R}_P$ *and* $\exists X \in \mathbf{O}_P$ *such that* $\text{span}(X) = \mathbf{R}_P$;
2. $\forall (y, Z) \in \mathbf{O}_D$, $\text{span}(Z) \subseteq \mathbf{R}_D$ *and* $\exists (y, Z) \in \mathbf{O}_D$ *such that* $\text{span}(Z) = \mathbf{R}_D$;
3. $\mathbf{R}_N = [\mathbf{R}_P \oplus \mathbf{R}_D]^\perp$.

Let us consider the following example.

For a given $\gamma \in \text{int}\,\Gamma$, let us denote the optimal partition corresponding to (P_γ) and (D_γ) by $\pi(\gamma) = [\mathbf{R}_P(\gamma), \mathbf{R}_N(\gamma), \mathbf{R}_D(\gamma)]$.

We will show how the optimal partition can be used in parametric and sensitivity analysis of SDP problems.

Suppose we have solved the problems (P_γ) and (D_γ) for some given $\bar{\gamma} \in \text{int}\,\Gamma$ and the optimal partition $\pi(\bar{\gamma}) = (\mathbf{R}_P(\bar{\gamma}), \mathbf{R}_N(\bar{\gamma}), \mathbf{R}_D(\bar{\gamma}))$ is available. Let Q_P be the matrix of an orthonormal spanning $\mathbf{R}_P(\bar{\gamma})$. We can compute the right and left derivatives of $\phi(\gamma)$ by the following lemma.

Lemma 4.2.4 *The solutions of the following problems produce the right and left derivatives of* $\phi(\gamma)$ *at* $\bar{\gamma}$:

$$d\phi/d\gamma_+ = \min\{\bar{C} \bullet Q_P U Q_P^T \mid A_i \bullet Q_P U Q_P^T = b_i,\ i = 1, \ldots, m,\ U \succeq 0\} \quad (4.2.83)$$

and

$$d\phi/d\gamma_- = \max\{\bar{C} \bullet Q_P U Q_P^T \,|\, A_i \bullet Q_P U Q_P^T = b_i,\ i = 1,\ldots,m,\ U \succeq 0\}. \tag{4.2.84}$$

The lemma follows immediately from Theorem 4.2.2.

Assume now that we are interested in determining the range of γ which we define as $\Gamma(\bar{\gamma}) = \{\gamma : \pi(\gamma) = \pi(\bar{\gamma})\}$. (The value of $\bar{\gamma}$ in this case is not necessarily the same as the one used in the previous lemma.)

In the next lemma we show that $\Gamma(\bar{\gamma})$ is either an open interval or the point $\{\bar{\gamma}\}$. In case $\Gamma(\bar{\gamma})$ is an interval we describe how to find its endpoints.

Lemma 4.2.5 *Let $\pi(\bar{\gamma}) = (\mathbf{R}_P(\bar{\gamma}), \mathbf{R}_N(\bar{\gamma}), \mathbf{R}_D(\bar{\gamma}))$ be the optimal partition corresponding to $\bar{\gamma}$, and let Q_P and Q_D be orthonormal bases of $\mathbf{R}_P(\bar{\gamma})$ and $\mathbf{R}_D(\bar{\gamma})$, respectively. The set of all values of γ, for which the optimal partition equals $\pi(\bar{\gamma})$, is either a single point $\{\bar{\gamma}\}$ or an open interval (a, b), where the endpoints a and b of the interval can be obtained by solving the following problems:*

$$a : \left\{\inf \gamma \,\Big|\, \sum_{i=1}^{n} A_i y_i + Q_D V Q_D^T = C + \gamma\bar{C},\ V \succ 0\right\} \tag{4.2.85}$$

$$b : \left\{\sup \gamma \,\Big|\, \sum_{i=1}^{n} A_i y_i + Q_D V Q_D^T = C + \gamma\bar{C},\ V \succ 0\right\}. \tag{4.2.86}$$

Proof. Let a and b be defined by (4.2.85) and (4.2.86). We need to show that for any $\gamma \in (a, b)$, $\pi(\gamma) = \pi(\bar{\gamma})$ and vice versa. From the definition of the optimal partition it follows that for any γ, such that $\pi(\gamma) = \pi(\bar{\gamma})$, there exists V satisfying

$$\sum_{i=1}^{n} A_i y_i + Q_D V Q_D^T = C + \gamma\bar{C},\ V \succ 0; \tag{4.2.87}$$

hence, it follows from (4.2.85) and (4.2.86) that $a \le \gamma \le b$.

Suppose now that there exists (γ, y, V) with $\gamma \neq \bar{\gamma}$ that satisfies (4.2.87). Consider $Z = Q_D V Q_D^T$. Z is feasible for (D_γ) from (4.2.87) and it is complementary to $X(\bar{\gamma})$ since $\mathcal{R}(Z) = \mathbf{R}_D(\bar{\gamma})$. Since $X(\bar{\gamma})$ is feasible for (P_γ) we conclude that $(X(\bar{\gamma}), y, Z)$ is an optimal primal-dual solution for γ. Thus, we can conclude that

$$\mathbf{R}_P(\gamma) \supseteq \mathbf{R}_P(\bar{\gamma})$$

and

$$\mathbf{R}_D(\gamma) \supseteq \mathbf{R}_D(\bar{\gamma})$$

(recall that $\pi(\gamma) = (\mathbf{R}_P(\gamma), \mathbf{R}_N(\gamma), \mathbf{R}_D(\gamma))$ is defined by a maximally complementary primal-dual solution). Let us now show that the partition $\pi(\gamma) = \pi(\bar{\gamma})$.

Assume that there is a primal solution to (P_γ) with a range larger than the range of $X(\bar{\gamma})$. This solution must be complementary to Z. Thus, it is complementary to $Z(\bar{\gamma})$ and, since it is feasible for $(P_{\bar{\gamma}})$, we obtain a contradiction to $X(\bar{\gamma})$ being a maximally complementary solution for $\bar{\gamma}$. Thus $\mathbf{R}_P(\gamma) = \mathbf{R}_P(\bar{\gamma})$.

Now, assume that there is a dual solution (y', Z') to (D_γ), such that the range of Z' is larger than the range of Z. Since both Z' and Z are feasible for (D_γ), we must have

$$\sum_{i=1}^{n} A_i[y_i' - y_i] + Z' - Z = 0. \tag{4.2.88}$$

If we multiply the above equation by an arbitrary $\epsilon > 0$ and add it to the feasibility equation for $(D_{\bar{\gamma}})$,

$$\sum_{i=1}^{n} A_i y_i(\bar{\gamma}) + Z(\bar{\gamma}) = C + \bar{\gamma}\bar{C},$$

we obtain

$$\sum_{i=1}^{n} A_i[y_i(\bar{\gamma}) + \epsilon(y_i' - y_i)] + Z(\bar{\gamma}) + \epsilon(Z' - Z) = C + \bar{\gamma}\bar{C}.$$

Since $\mathcal{R}(Z(\bar{\gamma})) = \mathcal{R}(Z) \subset \mathcal{R}(Z')$ and $Z, Z(\bar{\gamma}), Z' \succeq 0$, there exists a small enough ϵ, such that $\bar{Z} = Z(\bar{\gamma}) + \epsilon(Z' - Z) \succeq 0$ and $\mathcal{R}(\bar{Z}) \supset \mathcal{R}(Z(\bar{\gamma}))$. This contradicts the fact that $Z(\bar{\gamma})$ is a solution with the maximal range.

We have shown that any γ satisfies (4.2.87) if and only if the partition $\pi(\gamma)$ equals $\pi(\bar{\gamma})$. It is easy to see that the set of γ feasible for (4.2.87) is either a point or an open interval with endpoints a and b defined by (4.2.85) and (4.2.86). This completes the proof of the lemma. ■

Example. Let $n = 3$ and $m = 3$,

$$A_1 = \begin{bmatrix} 1 & 0 & 0 \\ 0 & 0 & 0 \\ 0 & 0 & 0 \end{bmatrix}, \quad A_2 = \begin{bmatrix} 0 & 0 & 1 \\ 0 & 1 & 0 \\ 1 & 0 & 0 \end{bmatrix}, \quad A_3 = \begin{bmatrix} 0 & 1 & 0 \\ 1 & 0 & 0 \\ 0 & 0 & 1 \end{bmatrix},$$

$$b^T = (1, 0, 0).$$

The feasible region of an SDP corresponding to the matrices A_1, A_2 and A_3 and the vector b is 3-dimensional (see Fig. 4.2) (this feasible region is taken from an example in [24]). Let us consider the objective function of the form $(C + \gamma\bar{C}) \bullet X$, where

$$C = \begin{bmatrix} -1 & -1 & -1 \\ -1 & 0 & 1 \\ -1 & 1 & 0 \end{bmatrix} \text{ and } \bar{C} = \begin{bmatrix} 1 & 1 & 1 \\ 1 & 1 & -1 \\ 1 & -1 & 1 \end{bmatrix}.$$

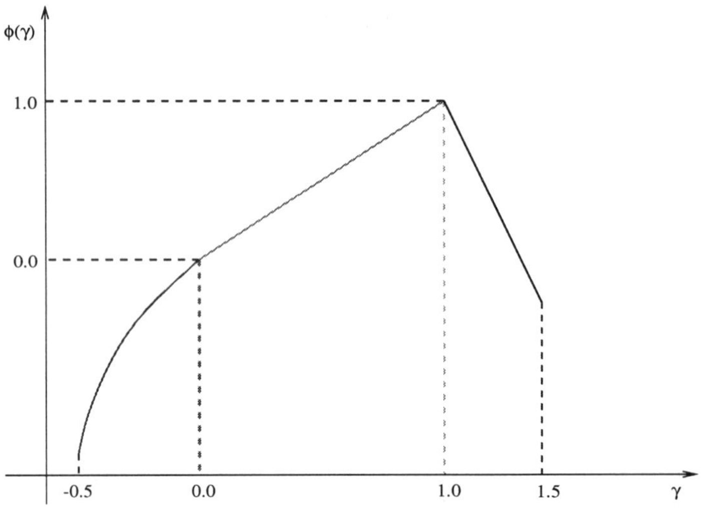

Figure 4.1 Shape of $\phi(\gamma)$

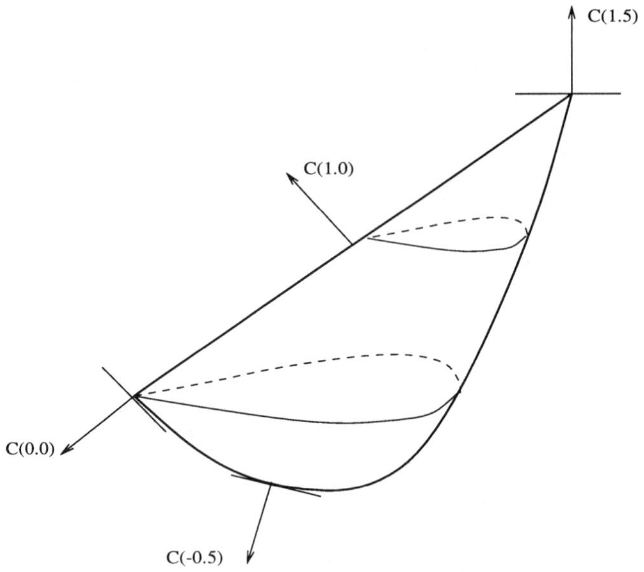

Figure 4.2 Examples of unique, multiple and not strictly complementary solutions

Let us vary γ in the interval $[-1.0, 2.4]$ (see Fig. 1).

1. In the interval $[-1, 0]$, $X(\gamma)$ and $\pi(\gamma)$ vary continuously, and $\phi(\gamma)$ is a concave nonlinear function.

2. For $\gamma = 0$,

$$X(0) = \begin{bmatrix} 1 & 0 & 0 \\ 0 & 0 & 0 \\ 0 & 0 & 0 \end{bmatrix}, \quad y(0) = (-1, -1, -1)^T, \quad Z(0) = \begin{bmatrix} 0 & 0 & 0 \\ 0 & 1 & 1 \\ 0 & 1 & 1 \end{bmatrix},$$

rank $(X(0)) = 1$, rank $(Z(0)) = 1$ and the primal-dual pair does not satisfy strict complementary slackness. In this case the optimal partition is

$$\mathbf{R}_P = \text{span}[(1, 0, 0)^T], \quad \mathbf{R}_D = \text{span}[(0, \tfrac{1}{\sqrt{2}}, \tfrac{1}{\sqrt{2}})^T],$$

$$\mathbf{R}_N = \text{span}[(0, \tfrac{1}{\sqrt{2}}, -\tfrac{1}{\sqrt{2}})^T].$$

3. In the interval $(0, 1)$, $X(\gamma)$ is constant and equals $X(0)$, the optimal partition

$$\mathbf{R}_P = \text{span}[(1, 0, 0)^T], \quad \mathbf{R}_D = \text{span}[(0, \tfrac{1}{\sqrt{2}}, \tfrac{1}{\sqrt{2}})^T, (0, \tfrac{1}{\sqrt{2}}, -\tfrac{1}{\sqrt{2}})^T],$$

$$\mathbf{R}_N = \emptyset$$

is constant and $\phi(\gamma)$ is linear. Notice that $\forall\, \gamma \in (0, 1)$, $X(0) = X(\gamma)$ but $\pi(0) \neq \pi(\gamma)$.

4. At $\gamma = 2$ the optimal solution is not unique and the optimal face is defined by

$$\mathcal{X}(2) = X : X = \theta \begin{bmatrix} 1 & 0 & 0 \\ 0 & 0 & 0 \\ 0 & 0 & 0 \end{bmatrix} + (1-\theta) \begin{bmatrix} 1 & -2 & -2 \\ -2 & 4 & 4 \\ -2 & 4 & 4 \end{bmatrix}, 0 \le \theta \le 1.$$

The dual solution is

$$y(2) = (1, 1, 1)^T, \quad Z(2) = \begin{bmatrix} 0 & 0 & 0 \\ 0 & 1 & -1 \\ 0 & -1 & 1 \end{bmatrix},$$

and rank $(X(2)) = 2$ and rank $(Z(2)) = 1$. The optimal partition $\pi(2) = (\mathbf{R}_P, \mathbf{R}_N, \mathbf{R}_D)$,

$$\mathbf{R}_P = \text{span}[(1, 0, 0)^T, (-\tfrac{1}{3}, \tfrac{2}{3}, \tfrac{2}{3})^T], \quad \mathbf{R}_D = \text{span}[(0, -\tfrac{1}{\sqrt{2}}, \tfrac{1}{\sqrt{2}})^T],$$
$$\mathbf{R}_N = \emptyset,$$

is different from $\pi(\gamma)$ $\forall\gamma \neq 2$. At $\gamma = 2$, $\phi(\gamma)$ has a break point. (See Fig. 4.1.) The right and left derivatives of $\phi(\gamma)$ at $\gamma = 2$ are

$$
\begin{aligned}
d\phi/d\gamma_+ &= \min\{\bar{C} \bullet X \mid X \in \mathcal{X}(2)\} \\
&= \begin{bmatrix} 1 & 1 & 1 \\ 1 & 1 & -1 \\ 1 & -1 & 1 \end{bmatrix} \bullet \begin{bmatrix} 1 & -2 & -2 \\ -2 & 4 & 4 \\ -2 & 4 & 4 \end{bmatrix} = -7
\end{aligned}
$$

and

$$
\begin{aligned}
d\phi/d\gamma_- &= \max\{\bar{C} \bullet X \mid X \in \mathcal{X}(2)\} \\
&= \begin{bmatrix} 1 & 1 & 1 \\ 1 & 1 & -1 \\ 1 & -1 & 1 \end{bmatrix} \bullet \begin{bmatrix} 1 & 0 & 0 \\ 0 & 0 & 0 \\ 0 & 0 & 0 \end{bmatrix} = 1.
\end{aligned}
$$

5. In the interval $[2, 2.4]$, $X(\gamma)$ is again constant and equals

$$
\begin{bmatrix} 1 & -2 & -2 \\ -2 & 4 & 4 \\ -2 & 4 & 4 \end{bmatrix},
$$

and $\phi(\gamma)$ is linear. The optimal partition is constant on $(2, 2.4]$,

$$
\begin{aligned}
\mathbf{R}_P &= \text{span}[(-\tfrac{1}{3}, \tfrac{2}{3}, \tfrac{2}{3})^T], \\
\mathbf{R}_D &= \text{span}[(0, -\tfrac{1}{\sqrt{2}}, \tfrac{1}{\sqrt{2}})^T, (\tfrac{4}{3\sqrt{2}}, \tfrac{1}{3\sqrt{2}}, \tfrac{1}{3\sqrt{2}})^T], \\
\mathbf{R}_N &= \emptyset.
\end{aligned}
$$

4.2.4 *Sensitivity Analysis*

We now study the problem of finding the interval of values of γ for which an optimal primal solution remains optimal. This problem is not exactly the same as (4.2.85)-(4.2.86) but, clearly, is closely related to it. Since we do not require the partition to stay the same, but merely want complementary slackness to hold, we allow the dual solution to have any range in $\mathbf{R}_N \oplus \mathbf{R}_D$. For brevity of notation let Q_{ND} denote now a matrix whose columns form an orthonormal basis of $\mathbf{R}_N \oplus \mathbf{R}_D$ and let $\text{rank}(Q_{ND}) = s$. The following problems then define the endpoints of the closed interval $[a, b]$:

$$
a = \min\left\{\gamma \middle| \sum_{i=1}^{n} A_i y_i + Q_{ND} V Q_{ND}^T = C + \gamma\bar{C},\ V \succeq 0,\ V \in \mathcal{S}^s\right\} \tag{4.2.89}
$$

$$
b = \max\left\{\gamma \middle| \sum_{i=1}^{n} A_i y_i + Q_{ND} V Q_{ND}^T = C + \gamma\bar{C},\ V \succeq 0,\ V \in \mathcal{S}^s\right\} \tag{4.2.90}
$$

In linear programming there is a finite set of such closed intervals, each of which corresponds to a particular optimal solution. In contrast, in semidefinite programming there can be whole regions of values of γ in which the optimal solution (and optimal partition) changes continuously. For example, the interval $-0.5 \leq \gamma \leq 0$ in Figure 4.1 is such a region. Each point in this region corresponds to the case where the interval $[a, b]$ collapses to a single point (i.e, $a = b$).

Problems (4.2.85) and (4.2.86) are semidefinite programming problems. Let us try to reduce them to simpler problems. Sensitivity analysis in linear programming under an assumption of nondegeneracy reduces to a minimum ratio test. However, as shown in [362], in the degenerate case the minimum ratio test produces incorrect results. One then has to solve a linear programming problem to obtain the correct range of γ. The situation is similar in our case.

Assume (without loss of generality) that $\gamma = 0 \in \operatorname{int}\Gamma$. We want to find $a \leq 0$ and $b \geq 0$, such that $X(\gamma) = X(0)$ for all $\gamma \in [a, b]$ Consider the problem of finding a. For b the analysis is analogous.

There are two possible cases:

Case 1: There does not exist $\alpha \in \mathbb{R}^m$ such that $\bar{C} = \sum \alpha_i A_i + Q_{ND} \Omega Q_{ND}^T$, $\Omega \in \mathcal{S}^s$. It is easy to see that $a = 0$, since a small perturbation of the original dual problem causes a change in the dual range.

Case 2: There exists $\alpha \in \mathbb{R}^m$ such that $\bar{C} = \sum \alpha_i A_i + Q_{ND} \Omega Q_{ND}^T$, $\Omega \in \mathcal{S}^s$. We can rewrite the constraints of our problem as $\sum_{i=1}^n A_i \bar{y}_i + Q_{ND} V Q_{ND}^T = C + \gamma Q_{ND} \Omega Q_{ND}^T$, $\bar{y} = y + \gamma\alpha$.

We consider Case 2. Let $Q = [Q_P, Q_{ND}]$, where Q_P and Q_{ND} are orthonormal bases of $\mathbf{R}_P$ and $\mathbf{R}_D \oplus \mathbf{R}_N$, respectively. By premultiplying the above system by Q^T and postmultiplying it by Q and considering the upper-left, upper-right (same as lower-left) and lower-right parts of the system separately, we obtain:

$$\sum \bar{y}_i Q_P^T A_i Q_P = Q_P^T C Q_P, \tag{4.2.91}$$

$$\sum \bar{y}_i Q_P^T A_i Q_{ND} = Q_P^T C Q_{ND}, \tag{4.2.92}$$

$$\sum \bar{y}_i Q_{ND}^T A_i Q_{ND} + V = Q_{ND}^T C Q_{ND} + \gamma\Omega. \tag{4.2.93}$$

Let $A(\bar{y}) = -\sum \bar{y}_i Q_{ND}^T A_i Q_{ND} + Q_{ND}^T C Q_{ND}$, and let $\bar{Y}$ be the set of solutions to equations (4.2.91) and (4.2.92). Then our problem reduces to

$$a = \min\{\gamma |\ A(\bar{y}) + \gamma\Omega \succeq 0,\ \bar{y} \in \bar{Y}\}. \tag{4.2.94}$$

Assume that the unperturbed problem is primal nondegenerate. By Lemma 4.2.1 the matrices $\begin{bmatrix} Q_P^T A_i Q_P \\ Q_P^T A_i Q_{ND} \end{bmatrix}$, $i = 1, \ldots, m$ are linearly independent. Hence, the equations (4.2.91) and (4.2.92) have a unique solution $\bar{y}$, so $A(\bar{y})$ does not depend on $\bar{y}$. $A(\bar{y}) = W_{ND}$, where $W_{ND} \succeq 0$ can be chosen to be the diagonal matrix of eigenvalues of the dual solution $Z(0)$ of the unperturbed problem.

If at $\bar{\gamma} = 0$ the solution is strictly complementary, then $W_{ND} \succ 0$. Hence, $W_{ND} + \gamma\Omega \succeq 0$ holds for some $\gamma < 0$ and (4.2.94) is equivalent to:

$$a = \min\{\gamma | \ I + \gamma(W_{ND}^{-1}\Omega) \succeq 0\}.$$

If $W_{ND}^{-1}\Omega \preceq 0$, then for any $\gamma < 0$, $W_{ND} + \gamma\Omega \succeq 0$, which mean that $\gamma_{\min} = -\infty$. If $(W_{ND}^{-1}\Omega)$ has at least one positive eigenvalue, then it is easy to see that

$$\gamma_{\min} = -1/\lambda_{\max}(W_{ND}^{-1}\Omega),$$

where $\lambda_{max}(W_{ND}^{-1}\Omega)$ is the largest (positive) eigenvalue of $W_{ND}^{-1}\Omega$. This computation is analogous to a minimum ratio test in linear programming.

Let us consider the case when the solution of the unperturbed problem is not strictly complementary; i.e., $W_{ND} \succeq 0$, but $W_{ND} \not\succ 0$. Recall that this can happen even assuming nondegeneracy. We can assume, without loss of generality, that $\text{null}(W_{ND}) \cap \text{null}(\Omega) = \emptyset$ (if this is not true we can reduce to an equivalent problem in lower dimension, for which it is true).

We use the following lemma on generalized eigenvalue problems from [292].

Lemma 4.2.6 *Suppose $A \in S^n$ and $B \in S^n$. If there exists a convex combination of A and B such that it is positive semidefinite and its nullspace is the intersection of the nullspace of A with the nullspace of B, then there exists a nonsingular X such that both $X^T AX$ and $X^T BX$ are diagonal.*

Three subcases are possible:

(i) The lemma holds for $A = W_{ND}$ and $B = \Omega$.

(ii) The lemma holds for $A = W_{ND}$ and $B = -\Omega$.

(iii) The lemma does not hold in either cases.

If (iii) holds then $[a, b] = \{0\}$; i.e., the solution changes for any $\gamma \neq 0$. This happens because Ω is indefinite on the nullspace of the dual solution W_{ND}, so by adding or subtracting small multiples of Ω we obtain a matrix with negative eigenvalues.

(i) and (ii) cannot hold simultaneously since then $\text{null}(W_{ND}) \subseteq \text{null}(\Omega)$ which was rules out earlier.

If (i) holds (and (ii) does not hold) then Ω is positive semidefinite on the nullspace of W_{ND} and we can only add positive multiples of Ω. So for small $\gamma > 0$, $W_{ND} + \gamma\Omega \succ 0$ and hence $a = 0$ and $b > 0$. Similarly, if (ii) holds (and (i) does not hold), $a < 0$ and $b = 0$.

To find a or b by the above lemma it suffices to find X, which diagonalizes A and B and then perform the minimum ratio test. A procedure of computing X is given in [292] for the case when one of the matrices is positive definite. Under assumption that $\text{null}(W_{ND}) \cap \text{null}(\Omega) = \emptyset$ and that, say, (ii) holds, there exists a convex combination of W_{ND} and $-\Omega$ which is positive definite, even if neither of them is. Thus we can consider a modified problem

$$a = \min\{\gamma|\ W_{ND} - \epsilon\Omega + [\gamma + \epsilon]\Omega \succeq 0\},$$

for some given $\epsilon > 0$, such that $W_{ND} - \epsilon\Omega \succ 0$. However, it may be difficult in practice to find such ϵ.

When the solution of the unperturbed problem is primal degenerate, $A(\bar{y})$ is a linear operator applied to $\bar{y}$. Thus we have an SDP problem to solve in order to find a and one to find b. Since $\bar{y}$ has to satisfy (4.2.91) and (4.2.92), it is reasonable to expect that the degrees of freedom of $\bar{y}$ is small. In linear programming in the case of primal degeneracy, one has to solve a linear programming problem (rather than performing a simple minimum-ratio test) to obtain the correct bounds on the parameter (e.g. see [362]).

Remark 1. Let us consider a dual problem with a parametric objective function vector $b + \gamma\bar{b}$ for some $\bar{b} \in \mathbb{R}^m$:

$$(D'_\gamma) \qquad \begin{array}{ll} \max & \sum_{i=1}^{m} y_i(b_i + \gamma\bar{b}_i) \\ \text{s.t.} & \sum_{i=1}^{m} y_i A_i + Z = C, \\ & Z \succeq 0. \end{array}$$

This is equivalent to considering the following primal problem with a parametric right-hand side:

$$(P'_\gamma) \qquad \begin{array}{ll} \min & C \bullet X \\ \text{s.t.} & A_i \bullet X = b_i + \gamma\bar{b}_i, \quad i = 1, \ldots, m \\ & X \succeq 0,\ X \in \mathcal{S}^n. \end{array}$$

Since (D'_γ) is an SDP problem with the parametric objective matrix, all the results of the paper apply to it. Suppose we have solved the pair of problems (D'_γ) and (P'_γ) for $\gamma = 0$. As before, assume that $X(0)$ and $Z(0)$ are a maximally complementary pair of primal-dual optimal solutions. We call the problem of finding the interval $[a, b]$ of γ on which $Z(0)$ stays optimal for (D'_γ) - dual sensitivity analysis. We want to reduce this problem to one of the form similar to (4.2.94).

By similar arguments to those used above we can show that a and b can be found by solving the problems

$$a = \min\{\gamma|\ A_i \bullet Q_{PN} U Q_{PN}^T = b_i + \gamma\bar{b}_i,\ i = 1, \ldots, m,\ U \succeq 0,\ U \in \mathcal{S}^r\},$$
$$b = \max\{\gamma|\ A_i \bullet Q_{PN} U Q_{PN}^T = b_i + \gamma\bar{b}_i,\ i = 1, \ldots, m,\ U \succeq 0,\ U \in \mathcal{S}^r\}.$$

Here Q_{PN} denotes a matrix whose columns form an orthonormal basis in $\mathbf{R}_P \oplus \mathbf{R}_N$ and $\text{rank}(Q_{PN}) = r$.

Let us consider the first problem (the second one is, clearly, similar). As before, we have two cases:

$$\text{Case 1: } \bar{b}_i = Q_{PN}^T A_i Q_{PN} \bullet V,\ i = 1, \ldots, m, \text{ for some } V \in \mathcal{S}^r.$$

$$\text{Case 2: } \nexists V \in \mathcal{S}^r \text{ such that } \bar{b}_i = Q_{PN}^T A_i Q_{PN} \bullet V,\ i = 1, \ldots, m.$$

In the second case, it is easy to see that $a = 0$. In the first case, if the dual solution $Z(0)$ is dual nondegenerate it follows by Lemma 4.2.2 that the matrices $Q_{PN}^T A_i Q_{PN}$ are linearly independent; thus the solution V to $\bar{b}_i = Q_{PN}^T A_i Q_{PN} \bullet V$ $i = 1, \ldots, m$ is unique. Then, to find a, one needs to solve the problem $\{\min \gamma : \Lambda_{PN} + \gamma V \succeq 0\}$. This can be done in the same way that problem (4.2.94) is solved in the case of primal nondegeneracy (see the discussion following (4.2.94)). In the case of dual degeneracy one has to solve an SDP problem, as in primal sensitivity analysis under primal degeneracy.

Remark 2. It is easy to see that if an SDP problem is reduced to an LP problem (by restricting all matrices to be diagonal) then our analysis reduces (under appropriate nondegeneracy assumptions) to standard linear programming sensitivity analysis; i.e., minimum ratio tests which guarantee nonnegativity of the reduced costs in the primal case and the nonnegativity of the primal basic variables in the dual case.

4.2.5 Conclusions

We have established a connection between properties of parametric linear programming problems and those of parametric SDP problems. Unfortunately, the nature of a parametric SDP is far more complicated, due to regions of nonlinearity of $\phi(\gamma)$. As far as we know there is no trivial way of describing $\phi(\gamma)$ completely. The results in Section 4 show that one can efficiently check if a given value of γ falls into an interval where $\phi(\gamma)$ is linear, and if this is the case, find the interval of linearity. Lemma 4.2.4 shows that one can also efficiently find the right and left derivatives of $\phi(\gamma)$.

An example of the application of sensitivity analysis in SDP within the context of combinatorial optimization can be found in [334]. There sensitivity analysis is used for fixing variables in a branch-and-bound framework which uses SDP relaxations. The approach suggested in [334] is similar to the approach for computing a sensitivity range involving the computation of the maximum eigenvalue of some matrix.

5 SELF-DUAL EMBEDDINGS

Etienne de Klerk, Tamás Terlaky, Kees Roos

5.1 INTRODUCTION

Most semidefinite programming algorithms found in the literature require strictly feasible starting points ($X^0 \succ 0, S^0 \succ 0$) for the primal and dual problems respectively. So-called 'big-M' methods (see e.g. [807]) are often employed in practice to obtain feasible starting points. For example, consider an SDP problem in the standard primal form:

$$(P): \ \ p^* := \inf_X \{\mathrm{Tr}(CX) \mid \mathrm{Tr}(A_i X) = b_i \ (i = 1, \dots, m), \ X \succeq 0\}.$$

Assume that a strictly feasible solution ($X^0 \succ 0$) of (P) is known, but no strictly primal feasible point is known for its Lagrangian dual (D) (which is in standard dual form):

$$(D): \quad d^* := \sup_{y,S} \left\{ b^T y \;\middle|\; \sum_{i=1}^m y_i A_i + S = C, \ S \succeq 0, \ y \in \mathbb{R}^m \right\}.$$

In order to apply primal–dual methods (where strictly feasible solutions to both (P) and (D) are required), one could solve the modified problem

$$\inf_X \{\mathrm{Tr}(CX) \mid \mathrm{Tr}(A_i X) = b_i \ (i = 1, \dots, m), \ \mathrm{Tr}(X) \le M, \ X \succeq 0\}. \qquad (5.1.1)$$

If M is 'large enough', then any optimal solution of (P) is an optimal solution of (5.1.1). The dual of problem (5.1.1) is

$$\sup_{y,S,\kappa \le 0} \left\{ b^T y + \kappa M \;\middle|\; \sum_{i=1}^m y_i A_i + S + \kappa I = C, \ S \succeq 0, \ y \in \mathbb{R}^m \right\}. \qquad (5.1.2)$$

One can now construct a strictly feasible starting point for problem (5.1.2) by choosing $-\kappa$ 'large enough', while no such starting point was available for (D). Note that the dual pair of problems (5.1.1) and (5.1.2) will both be solvable and the duality gap at the solutions will be zero. This follows form the fact that both problems are strictly feasible. In other words, problem (5.1.1) will have an optimal solution even if (P) does not.

An analogous strategy is available if only a strictly feasible dual solution is available. If neither a primal nor a dual feasible point is known, a similar strategy can again be employed by introducing two 'big-M' parameters (see [807] for details). The difficulty with these approaches is that no a priori choice for the 'big-M' parameters is available in general. For example, if κ cannot be driven to zero in problem (5.1.2), then one can only conclude that 'there is no optimal solution X^* of (P) with $\mathrm{Tr}(X^*) \leq$ M'. We therefore need an a priori bound on $\mathrm{Tr}(X^*)$ in order to give a certificate of the problem status of (P), while such information is not available in general.

In the LP case an elegant solution for the initialization problem is to embed the original problem in a skew–symmetric self–dual problem which has a known interior feasible solution on the central path. The solution of the embedding problem then yields the optimal solution to the original problem, or gives a certificate of either primal or dual infeasibility. In this way detailed information about the solution is obtained. The idea of self-dual embeddings for LP dates back to the 1950's and the work of Goldman and Tucker [290]. With the arrival of interior point methods, the embedding idea was revived to be used in infeasible start algorithms by Mizuno, Todd and Ye [860] (see also [359]).

Despite the desirable theoretical properties of self-dual embeddings, the idea did not receive immediate recognition in implementations, due to the fact that the embedding problem has a dense column in the coefficient matrix. This can lead to fill-in of Choleski factorizations during computation. In spite of this perception, Xu et al. [844] have made a successful implementation for LP using the embedding, and it has even been implemented as an option in the well-known commercial LP solver CPLEX-barrier and as the default option in the solver XPRESSMP. The common consensus now is that this strategy is a reliable way of detecting infeasibility, and promises to be competitive in practice [29] (see also [251] and [683]).

For semidefinite programming the homogeneous embedding idea was first developed by Potra and Sheng [642]. The embedding strategy was extended by De Klerk et al. in [418] and independently by Luo et al. [508] to obtain self-dual embedding problems with nonempty interiors. The resulting embedding problem has a known centered starting point, unlike the homogeneous embedding; it can therefore be solved using any feasible path-following interior point method. This is an advantage in the SDP case, where many possible primal-dual algorithms are available, while none has yet emerged as clear favourite.

A so-called *maximally complementary* solution (e.g. the limit of the central path) of the embedding problem yields one of the following alternatives about the original problem pair:

(I) an optimal solution with zero duality gap for the original problem is obtained;

(II) an improving ray is obtained for either the primal and/or dual problem (strong infeasibility is detected);

(III) a certificate is obtained that no optimal solution pair with zero duality gap exists and that neither the primal nor the dual problem has an improving ray. This can only happen if one or both of the primal and dual SDP problems fail to satisfy the Slater regularity condition.

Loosely speaking, the original primal and dual problems are solved if a complementary solution pair exists, or if one or both of the problems are strongly infeasible.

Unfortunately, some pathological duality effects can occur for SDP[1] which are absent from LP, for example:

- A positive duality gap at an optimal primal-dual solution pair;
- an arbitrarily small duality gap can be attained by feasible primal-dual pairs, but no optimal pair exists;
- an SDP problem may have a finite optimal value even though its (Lagrangian) dual is infeasible.

These situations cannot be identified with the embedding approach, unless an additional assumption is made: if the primal and dual problems are in so-called perfect duality, and one is solvable, then additional information can be obtained from the embedding, as was shown in [419]. This assumption holds if for example the primal is strictly feasible and the dual is feasible. Moreover, one can replace the problem to be solved by a larger problem which – together with its Lagrangian dual – satisfy these assumptions. In theory at least, we can therefore make the assumption without loss of generality.

5.2 PRELIMINARIES

Terminology

We say that a problem (P) and its Lagrangian dual (D) are in *standard form* if they are in the form

$$(P): \; p^* := \inf_X \{\mathrm{Tr}(CX) \mid \mathrm{Tr}(A_iX) = b_i \; (i = 1, \ldots, m), \; X \succeq 0\},$$

and

$$(D): \; d^* := \sup_{y,S} \left\{ b^Ty \;\middle|\; \sum_{i=1}^m y_iA_i + S = C, \; S \succeq 0, \; y \in \mathbb{R}^m \right\}.$$

[1] Examples of these effects will be given in Sections 5.2 and 5.8, and can also be found in [508] and [807].

The values p^* and d^* will be called the optimal values of (P) and (D), respectively. We use the convention that $p^* = -\infty$ if (P) is unbounded and $p^* = \infty$ if (P) is infeasible, with the analogous convention for (D).

We will assume that the matrices A_i are linearly independent. Under this assumption y is uniquely determined for a given dual feasible S.

If $p^* = d^*$ we say that (P) and (D) are in *perfect duality.* Note that if (P) and (D) are in perfect duality and one of the two is infeasible, then the other is unbounded.

The primal and dual feasible sets will be denoted by $\mathcal{P}$ and $\mathcal{D}$ respectively, and $\mathcal{P}^*$ and $\mathcal{D}^*$ will denote the respective optimal sets, *i.e.*

$$\mathcal{P}^* = \{X \in \mathcal{P} \,:\, \mathrm{Tr}(CX) = p^*\} \text{ and } \mathcal{D}^* = \left\{(S, y) \in \mathcal{D} \,:\, b^T y = d^*\right\}.$$

A problem (P) (resp. (D)) is called solvable if $\mathcal{P}^*$ (resp. $\mathcal{D}^*$) is nonempty.

Problems (P) and (D) satisfy the *weak duality* property: all feasible X, y, S satisfy

$$\mathrm{Tr}(CX) - b^T y = \mathrm{Tr}\left(\left(S - \sum_{i=1}^{m} y_i A_i\right) X\right) - \sum_{i=1}^{m} y_i \,\mathrm{Tr}(A_i X) = \mathrm{Tr}(SX) \geq 0,$$

where the inequality follows from $X \succeq 0$ and $S \succeq 0$. In other words, the *duality gap* is nonnegative for feasible solutions. Solutions (X, y, S) with zero duality gap

$$\mathrm{Tr}(CX) - b^T y = \mathrm{Tr}(SX) = 0 \tag{5.2.3}$$

are optimal. Optimal solutions $X \in \mathcal{P}^*$ and $S \in \mathcal{D}^*$ satisfying eq. (5.2.3) are called *complementary.* Since X and S are symmetric positive semi–definite matrices the complementarity of X and S is equivalent to $XS = 0$.

For LP, if either the primal or the dual problem has an optimal solution, then both have optimal solutions, and the duality gap at optimality is zero. This is the *strong duality* property. The SDP case is more subtle: One problem may be solvable and its dual infeasible, or the duality gap may be positive at optimality, etc. The existence of primal and dual optimal solutions is guaranteed if both (P) and (D) allow (strictly feasible) positive definite solutions, *i.e.* feasible $X \succ 0$ and $S \succ 0$. This is called the *Slater constraint qualification* (or Slater regularity condition), or strict feasibility condition.

Feasibility issues

To decide about possible infeasibility and unboundedness of the problems (P) and (D) we need the following definition.

Definition 5.2.1 (Primal and dual improving rays) *We say that the primal problem (P) has an improving ray if there is a symmetric matrix $\bar{X} \succeq 0$ such that $\mathrm{Tr}(A_i\bar{X}) = 0, \;\; \forall\, i$ and $\mathrm{Tr}(C\bar{X}) < 0$. Analogously, the dual problem (D) has an improving ray if there is a vector $\bar{y} \in \mathbb{R}^m$ such that $\bar{S} := -\sum_{i=1}^m \bar{y}_i A_i \succeq 0$ and $b^T\bar{y} > 0$.*

Primal improving rays cause infeasibility of the dual problem, and *vice versa.* Formally one has the following result.

Lemma 5.2.1 *If there is a dual improving ray $\bar{y}$ then (P) is infeasible. Similarly, a primal improving ray $\bar{X}$ implies infeasibility of (D).*

Proof.
Let a dual improving ray $\bar{y}$ be given. By assuming the existence of a primal feasible X one has

$$0 < b^T\bar{y} = \sum_{i=1}^{m} \operatorname{Tr}(A_i X)\bar{y}_i = -\operatorname{Tr}(X\bar{S}) \leq 0,$$

which is a contradiction. The proof in case of a primal improving ray proceeds similarly. ∎

Definition 5.2.2 (Strong infeasibility) *Problem (P) (resp. (D)) is called strongly infeasible if (D) (resp. (P)) has an improving ray.*

Every infeasible LP problem is strongly infeasible, but in the SDP case so-called *weak infeasibility* is also possible.

Definition 5.2.3 (Weak infeasibility) *Problem (P) is weakly infeasible if $\mathcal{P} = \emptyset$ and for each $\epsilon > 0$ exists an $X \succeq 0$ such that*

$$|\operatorname{Tr}(A_i X) - b_i| \leq \epsilon, \quad \forall i.$$

Similarly, problem (D) is called weakly infeasible if $\mathcal{D} = \emptyset$ and for every $\epsilon > 0$ exist $y \in \mathbb{R}^m$ and $S \succeq 0$ such that

$$\left\| \sum_{i=1}^{m} y_i A_i + S - C \right\| \leq \epsilon.$$

Example 5.2.1 *An example of weak infeasibility is given if (D) is defined by: find*

$$\sup y_1$$

subject to

$$y_1 \begin{bmatrix} 1 & 0 \\ 0 & 0 \end{bmatrix} \preceq \begin{bmatrix} 0 & 1 \\ 1 & 0 \end{bmatrix}$$

where we can construct an 'ϵ-infeasible solution' by setting

$$S = \begin{bmatrix} 1/\epsilon & 1 \\ 1 & \epsilon \end{bmatrix}, \; y_1 = -\frac{1}{\epsilon}.$$

∎

It has been shown (see *e.g.* [508]) that an infeasible SDP problem is either weakly infeasible or strongly infeasible.

5.3 THE EMBEDDING STRATEGY

In what follows, we make no assumptions about feasibility of (P) and (D).

Consider the following *homogeneous embedding* of (P) and (D):

$$\left.\begin{array}{llllll} & \mathrm{Tr}(A_iX) & -\tau b_i & & = 0 & \forall i \\ -\sum_{i=1}^m y_iA_i & & +\tau C & -S & = 0 & \\ b^Ty & -\mathrm{Tr}(CX) & & -\rho & = 0 & \\ y \in \mathbb{R}^m, & X \succeq 0, & \tau \geq 0, & S \succeq 0, \quad \rho \geq 0. & & \end{array}\right\} \tag{5.3.4}$$

A feasible solution to this system with $\tau > 0$ yields feasible solutions $\frac{1}{\tau}X$ and $\frac{1}{\tau}S$ to (P) and (D) respectively (by dividing the first two equations by τ). The last equation guarantees optimality by requiring a nonpositive duality gap. For this reason there is no strictly feasible solution to (5.3.4). The formulation (5.3.4) was first solved by Potra and Sheng [642] using an infeasible interior point method, and recently it has been incorporated in the SeDuMi software of Sturm [758].

Here we will describe an *extended self-dual embedding* [418], in order to have a strictly feasible, self–dual SDP problem with a known starting point on the central path. The advantage is that any feasible start path-following algorithm can be applied to such a problem. This is an important consideration in SDP, where many possible search directions and algorithms are available, with no clear method of choice at this time.

The strictly feasible embedding is obtained by extending the constraint set (5.3.4) and adding extra variables to obtain:

$$\min_{y,X,\tau,\vartheta,S,\rho,\nu} \vartheta\beta$$

subject to

$$\left.\begin{array}{llllllll} & \mathrm{Tr}(A_iX) & -\tau b_i & +\vartheta\bar{b}_i & & & = 0 & \forall i \\ -\sum_{i=1}^m y_iA_i & & +\tau C & -\vartheta\bar{C} & -S & & = 0 & \\ b^Ty & -\mathrm{Tr}(CX) & & +\vartheta\alpha & -\rho & & = 0 & \\ -\bar{b}^Ty & +\mathrm{Tr}(\bar{C}X) & -\tau\alpha & & & -\nu & = -\beta & \\ y \in \mathbb{R}^m, & X \succeq 0, & \tau \geq 0, & \vartheta \geq 0, & S \succeq 0, \quad \rho \geq 0, & \nu \geq 0 & & \end{array}\right\} \tag{5.3.5}$$

where

$$\begin{aligned} \bar{b}_i &:= b_i - \mathrm{Tr}(A_i) \\ \bar{C} &:= C - I \\ \alpha &:= 1 + \mathrm{Tr}(C) \\ \beta &:= n + 2. \end{aligned}$$

It is straightforward to verify that a feasible interior starting solution is given by

$$y^0 = 0, \; X^0 = S^0 = I, \; \vartheta^0 = \rho^0 = \tau^0 = \nu^0 = 1. \tag{5.3.6}$$

Note also that the solution with $\nu = \beta$ and all other variables zero is optimal, since the objective function is always nonnegative. In other words, $\vartheta = 0$ in any optimal solution. It is therefore a trivial matter to find an optimal solution, but we are only interested in *maximally complementary solutions.*

Definition 5.3.1 (Maximal complementarity) *An optimal solution* $(y^*, X^*, \tau^*, \vartheta^*, S^*, \rho^*, \nu^*)$ *of the embedding problem is called maximally complementary if* X^* *and* S^* *have largest possible rank and as many of the scalar variables as possible are strictly positive.*

Note that maximally complementary solutions exist due to the convexity of the optimal set.

The underlying idea is as follows: we wish to know whether an optimal solution to the embedding exists with $\tau^* > 0$. (Recall that this yields a complementary solution pair to (P) and (D).) A maximally complementary solution will yield such a solution, if it exists. In the next section it will become clear how maximally complementary solutions may be obtained as the limit of the *central path* of the embedding problem.

The dual of the embedding problem is essentially the same as the embedding problem itself, which explains the 'self-dual' terminology.

Theorem 5.3.1 *The embedding problem (5.3.5) is self-dual.*

Proof.

We will show here that the Lagrangian dual of problem (5.3.5) is equivalent to problem (5.3.5). The Lagrangian of problem (5.3.5) is given by

$$\begin{aligned} L(\xi, \Omega, \kappa, \gamma, y, X, \tau, \vartheta, S, \rho, \nu) \;=\;& \vartheta\beta - \sum_{i=1}^{m} \xi_i \left(\mathrm{Tr}(A_i X) - \tau b_i + \vartheta \bar{b}_i\right) \\ &+ \mathrm{Tr}\left(\Omega\left(\sum_{i=1}^{m} y_i A_i - \tau C + \vartheta \bar{C} + S\right)\right) \\ &- \kappa\left(b^T y - \mathrm{Tr}(CX) + \vartheta\alpha - \rho\right) \\ &- \gamma\left(-\bar{b}^T y + \mathrm{Tr}(\bar{C}X) - \tau\alpha - \nu + \beta\right), \end{aligned}$$

where the unrestricted Lagrange multipliers $\xi \in \mathbb{R}^m$, $\Omega \in \mathcal{S}^n$, $\kappa \in \mathbb{R}$, and $\gamma \in \mathbb{R}$ have been introduced. The dual problem now becomes

$$\max_{(\xi,\Omega,\kappa,\gamma)} \; \min_{(y,X,\tau,\vartheta,S,\rho,\nu)} \; L(\xi, \Omega, \kappa, \gamma, y, X, \tau, \vartheta, S, \rho, \nu) \tag{5.3.7}$$

where $y \in \mathbb{R}^m$, $X \succeq 0$, $\tau \geq 0$, $\rho \geq 0$, $\nu \geq 0$, and $S \succeq 0$.

One can regroup the variables in the Lagrangian to obtain

$$L(\xi, \Omega, \kappa, \gamma, y, X, \tau, \vartheta, S, \rho, \nu) \;=\; \sum_{i=1}^{m} y_i \left(\mathrm{Tr}(A_i \Omega) - \kappa b_i + \gamma \bar{b}_i\right)$$

$$+\operatorname{Tr}\left(X\left[-\sum_{i=1}^{m}\xi_i A_i+\kappa C-\gamma\bar{C}\right]\right)+\rho\kappa$$
$$+\tau\left(\sum_{i=1}^{m}\xi_i b_i-\operatorname{Tr}(C\Omega)+\gamma\alpha\right)+\nu\gamma$$
$$+\vartheta\left(-\sum_{i=1}^{m}\xi_i\bar{b}_i+\operatorname{Tr}(\bar{C}\Omega)-\kappa\alpha+\beta\right)+\operatorname{Tr}(S\Omega)-\beta\gamma.$$

This reformulation makes it clear that the inner minimization problem in (5.3.7) is only bounded from below if the following conditions hold:

$$\left.\begin{array}{rcl}\operatorname{Tr}(A_i\Omega)-b_i\kappa+\gamma\bar{b}_i &=& 0,\quad i=1,\ldots,m\\ -\sum_{i=1}^{m}\xi_i A_i+\kappa C-\gamma\bar{C} &\succeq& 0\\ \sum_{i=1}^{m}\xi_i b_i-\operatorname{Tr}(C\Omega)+\gamma\alpha &\geq& 0\\ -\sum_{i=1}^{m}\xi_i\bar{b}_i+\operatorname{Tr}(\bar{C}\Omega)-\kappa\alpha+\beta &\geq& 0\\ \kappa\geq 0,\ \gamma\geq 0,\ \Omega\succeq 0. && \end{array}\right\}\qquad(5.3.8)$$

Subject to these conditions, the inner minimization problem has minimum value

$$\min_{(y,X,\tau,\vartheta,S,\rho,\nu)} L(\xi,\Omega,\kappa,\gamma,y,X,\tau,\vartheta,S,\rho,\nu)=-\beta\gamma,$$

and the dual problem becomes

$$\max -\beta\gamma$$

subject to (5.3.8).

After changing to a minimization problem by inverting the sign of the objective, the dual becomes the embedding problem (5.3.5) where only the variables have been renamed and the slack variables have been omitted. ∎

The self-duality implies that the duality gap is equal to $2\vartheta\beta$. It is readily verified that

$$\vartheta\beta=\operatorname{Tr}(XS)+\tau\rho+\vartheta\nu. \qquad (5.3.9)$$

This shows that an optimal solution (where $\vartheta\beta=0$) must satisfy the complementarity conditions:

$$\left.\begin{array}{rcl}XS&=&0\\ \rho\tau&=&0\\ \vartheta\nu&=&0.\end{array}\right\}\qquad(5.3.10)$$

We can now use a maximally complementary solution of the embedding problem (5.3.5) to obtain information about the original problem pair (P) and (D). In particular, one can distinguish between the three possibilities as discussed in the Introduction, namely

(I) A primal–dual complementary solution pair (X^*, S^*) for $((P), (D))$ is obtained;

(II) A primal and/or dual improving ray is detected;

(III) A certificate is obtained that no complementary pair exists, and that neither (P) nor (D) has an improving ray.

Given a maximally complementary solution of the embedding problem, these cases are distinguished as follows:

Theorem 5.3.2 *Let* $(y^*,\ X^*,\ \tau^*,\ \vartheta^*,\ S^*,\ \rho^*,\ \nu^*)$ *be a maximally complementary solution to the self–dual embedding problem. Then:*

(i) if $\tau^* > 0$ *then case* (I) *holds;*

(ii) if $\tau^* = 0$ *and* $\rho^* > 0$ *then case* (II) *holds;*

(iii) if $\tau^* = \rho^* = 0$ *then case* (III) *holds.*

Proof.
Consider the two possibilities $\tau^* = 0$ and $\tau^* > 0$.

Case: $\tau^* > 0$
Here, $\frac{1}{\tau^*}X^*$ and $\frac{1}{\tau^*}S^*$ are maximally complementary and optimal for (P) and (D) respectively, *i.e.* case (I) holds.

Case: $\tau^* = 0$
In this case, one has $\tau = 0$ in any optimal solution of the embedding problem. This implies no complementary solution pair for (P) and (D) exists, because if such a pair exists we can construct an optimal solution of the embedding problem with $\tau > 0$ as follows: Let a complementary pair $(X_P, S_D) \in \mathcal{P}^* \times \mathcal{D}^*$ be given, and set $\vartheta^* = \rho^* = 0$, $X^* = \tau^* X_P$, and $S^* = \tau^* S_D$, where $\tau^* > 0$ will be specified presently. This choice of variables already satisfies the first three constraints of the embedding problem for any choice of $\tau^* > 0$ (see (5.3.5)). The fourth equation of the embedding problem can now be simplified to:

$$\begin{aligned} \nu^* &= n + 2 - \operatorname{Tr}(X^* + S^*) - \tau^* \\ &= n + 2 - \tau^* \left(\operatorname{Tr}(X_P + S_P) + 1\right). \end{aligned}$$

One can therefore choose any $\tau^* > 0$ which satisfies

$$\tau^* < \frac{n+2}{\operatorname{Tr}(X_P + S_P) + 1}, \tag{5.3.11}$$

to obtain a value $\nu^* \geq 0$. We have therefore constructed an optimal solution of the embedding problem with $\tau^* > 0$.

If $\tau^* = 0$ it also follows that $\mathrm{Tr}(A_i X^*) = 0$ for all i and $\sum_{i=1}^m y_i^* A_i \preceq 0$. Now we distinguish between two sub-cases: $\rho^* > 0$ and $\rho^* = 0$.

Sub-case: $\tau^* = 0$ and $\rho^* > 0$
Here one has $b^T y^* - \mathrm{Tr}(CX^*) > 0$, *i.e.* $b^T y^* > 0$ and/or $\mathrm{Tr}(CX^*) < 0$. In other words, there are primal and/or dual improving rays and case (II) applies. If $b^T y^* > 0$ then y^* is a dual improving ray. In this case (P) is infeasible by Lemma 5.2.1, and if (D) is feasible it is unbounded. If $\mathrm{Tr}(CX^*) < 0$ then there exists a primal improving ray. In this case (D) is (strictly) infeasible, and if (P) is feasible it is unbounded. If both $b^T y^* > 0$ and $\mathrm{Tr}(CX^*) < 0$ then both a primal and a dual improving ray exist and in this case both (P) and (D) are infeasible.

Conversely, one must show that if there exists a primal and/or dual improving ray, then any maximally complementary solution of the embedding problem must have $\rho^* > 0$ and $\tau^* = 0$. Given a primal improving ray $\bar{X} \succeq 0$, one can construct an optimal solution to the embedding by setting $X^* = \kappa \bar{X}$, where $\kappa > 0$ is a constant to be specified later, and further setting $\tau^* = 0$, $\vartheta^* = 0$ (which guarantees optimality), and $y^* = 0$, to obtain:

$$
\begin{aligned}
\rho^* &= -\kappa\,\mathrm{Tr}(C\bar{X}) > 0 \\
\kappa\,\mathrm{Tr}(A_i\bar{X}) = \mathrm{Tr}\left(A_i X^*\right) &= 0, \quad i = 1, \ldots, m \\
S^* &= 0 \\
\nu^* &= n + 2 + \kappa\,\mathrm{Tr}\left(C\bar{X} - \bar{X}\right).
\end{aligned}
$$

The first three equations show that ρ^*, X^* and S^* are feasible. It remains to prove that ν^* is nonnegative. This is ensured by choosing

$$
\kappa = \frac{-1}{\mathrm{Tr}\left(C\bar{X} - \bar{X}\right)} > 0,
$$

where the inequality follows from the definition of an improving ray. The proof for a dual improving ray proceeds analogously.

Sub-case: $\tau^* = \rho^* = 0$
Finally, if a maximally complementary solution is obtained with $\tau^* = \rho^* = 0$, then we again have that all optimal solutions yield $\rho = \tau = 0$, *i.e.* cases (I) and (II) cannot occur. This completes the proof. ∎

Three important questions now arise:

- How is the embedding problem actually solved?
- How does one decide if $\tau^* > 0$ and $\rho^* > 0$ in a maximally complementary solution, if only an ϵ-optimal solution of the embedding problem is available?
- What additional information can be obtained if case (III) holds?

These three questions will be addressed in turn in the following sections.

5.4 SOLVING THE EMBEDDING PROBLEM

The embedding problem can be solved by any central path following interior point method. But first we must formalize the notion of the central path of the embedding. To this end, one can relax the complementarity optimality conditions (5.3.10) of the embedding problem to

$$
\begin{aligned}
XS &= \mu I \\
\tau\rho &= \mu \\
\nu\vartheta &= \mu,
\end{aligned}
$$

If one defines new 'primal and dual variables' $\widetilde{X}, \widetilde{S}$ of dimension $\widetilde{n} := n + 2$ as follows:

$$
\widetilde{X} = \begin{bmatrix} X & & \\ & \tau & \\ & & \nu \end{bmatrix}, \quad \widetilde{S} = \begin{bmatrix} S & & \\ & \rho & \\ & & \vartheta \end{bmatrix},
$$

then the central path of the embedding problem can be defined as the (unique) solution of

$$
\widetilde{X}\widetilde{S} = \mu I, \quad \mu > 0,
$$

subject to (5.3.5), denoted by $(\widetilde{X}(\mu), \widetilde{S}(\mu))$ for each $\mu > 0$. We will sketch a constructive proof of the existence and uniqueness of the central path by showing that one can use interior point algorithms to converge to any 'μ-center', where $\mu \leq 1$ is fixed. Here we will utilize the fact that we know the μ-center for $\mu = 1$ (see (5.3.6)).

In the next section we therefore describe how interior point methods can be used to solve the embedding problem.

Search directions for the embedding problem

Let $(\Delta\widetilde{X}, \Delta\widetilde{S})$ denote any feasible direction for the embedding problem at the feasible point $(\widetilde{X}, \widetilde{S})$.

Using the primal and dual feasibility equations, it is straightforward to verify that $\operatorname{Tr}\left(\Delta\widetilde{X}\Delta\widetilde{S}\right) = 0$, *i.e.* the *orthogonality principle* holds.

The feasible directions of interior point methods can be computed from the following generic linear system:

$$
\left.
\begin{array}{lllllllll}
 & \operatorname{Tr}(A_i \Delta X) & -\Delta\tau b_i & +\Delta\vartheta \bar{b}_i & & & & = 0 & \forall i \\
-\sum_{i=1}^{m} \Delta y_i A_i & & +\Delta\tau C & -\Delta\vartheta \bar{C} & -\Delta S & & & = 0 & \\
b^T \Delta y & -\operatorname{Tr}(C\Delta X) & & +\Delta\vartheta\alpha & & -\Delta\rho & & = 0 & \\
-\bar{b}^T \Delta y & +\operatorname{Tr}(\bar{C}\Delta X) & -\Delta\tau\alpha & & & & -\Delta\nu & = 0 &
\end{array}
\right\} \tag{5.4.12}
$$

and

$$
\left.
\begin{aligned}
H_P(\Delta X S + X\Delta S) &= \mu I - H_P(XS), \\
\rho\Delta\tau + \tau\Delta\rho &= \mu - \tau\rho \\
\nu\Delta\vartheta + \vartheta\Delta\nu &= \mu - \vartheta\nu,
\end{aligned}
\right\} \tag{5.4.13}
$$

where H_P is the linear transformation given by

$$H_P(M) := \frac{1}{2}\left[PMP^{-1} + P^{-T}M^TP^T\right],$$

for any matrix M, and where the *scaling matrix* P determines the symmetrization strategy. The best choices of P from the literature are listed in Table 5.1.

P	Reference
$\left[X^{\frac{1}{2}}\left(X^{\frac{1}{2}}SX^{\frac{1}{2}}\right)^{-\frac{1}{2}}X^{\frac{1}{2}}\right]^{\frac{1}{2}}$	Nesterov and Todd [587];
$X^{-\frac{1}{2}}$	Monteiro [536], Kojima *et al.* [439];
$S^{\frac{1}{2}}$	Monteiro [536], Helmberg *et al.* [339], Kojima *et al.* [439];
I	Alizadeh, Haeberley and Overton [25];

Table 5.1 Choices for the scaling matrix P.

We will now prove (or derive sufficient conditions) for existence and uniqueness of the search directions corresponding each of the choices of P in Table 5.1. To this end, we will write the equations (5.4.12) and (5.4.13) as a single linear system and show that the coefficient matrix of this system is nonsingular. The approach used here is a straightforward extension of the analysis by Todd *et al.* in [782], where this result was proved for SDP problems in the standard form (P) and (D).

We will use the notation

$$\widetilde{A}_i := \begin{bmatrix} A_i & & \\ & -b_i & \\ & & \bar{b}_i \end{bmatrix} \quad (i = 1, \ldots, m),$$

and define $\widetilde{P}$ by replacing X by $\widetilde{X}$ and S by $\widetilde{S}$ in Table 5.1. We will rewrite (5.4.13) in terms of the following *symmetric Kronecker product notation*:

- $\operatorname{svec}(X) := [X_{11},\ \sqrt{2}X_{12},\ \ldots\ \sqrt{2}X_{1n},\ X_{22},\ \sqrt{2}X_{23},\ \ldots,\ X_{nn}]^T$;
- The *symmetric Kronecker product* $G \otimes_s K$ of $G, K \in \mathbb{R}^{n\times n}$ is implicitly defined via

$$(G \otimes_s K)\operatorname{svec}(H) := \frac{1}{2}\operatorname{svec}\left(KHG^T + GHK^T\right)\ (\forall\ H \in \mathcal{S}^n).$$

Using the symmetric Kronecker notation, we can combine (5.4.12) and (5.4.13) as

$$\begin{bmatrix} 0 & \widetilde{\mathcal{A}} & 0 \\ -\widetilde{\mathcal{A}}^T & \mathcal{S}_{kew} & I \\ 0 & E & F \end{bmatrix} \begin{bmatrix} \Delta y \\ \operatorname{svec}\left(\Delta\widetilde{X}\right) \\ \operatorname{svec}\left(\Delta\widetilde{S}\right) \end{bmatrix} = \begin{bmatrix} 0 \\ 0 \\ \operatorname{svec}\left(\mu I - H_{\widetilde{P}}(\widetilde{X}\widetilde{S})\right) \end{bmatrix} \tag{5.4.14}$$

where

$$\begin{aligned}
\widetilde{\mathcal{A}} &:= [\mathrm{svec}\,(\widetilde{A}_1)\ \ldots\ \mathrm{svec}\,(\widetilde{A}_m)]^T \\
\mathcal{S}_{kew} &:= \begin{bmatrix} 0 & \mathrm{svec}\,(C) & -\mathrm{svec}\,(\bar{C}) \\ -\mathrm{svec}\,(C)^T & 0 & \alpha \\ \mathrm{svec}\,(\bar{C})^T & -\alpha & 0 \end{bmatrix} \\
E &:= \widetilde{P} \otimes_s (\widetilde{P}^{-T}\widetilde{S}), \quad F := (\widetilde{P}\widetilde{X}) \otimes_s \widetilde{P}^{-T}.
\end{aligned}$$

The following lemma provides more information about the matrices E and F.

Lemma 5.4.1 (Todd *et al.* [782]) *Let $\widetilde{P}$ be invertible and $\widetilde{X}$ and $\widetilde{S}$ symmetric positive definite. Then the matrices E and F are invertible. If one also has $H_{\widetilde{P}}(\widetilde{X}\widetilde{S}) \succ 0$ then the symmetric part of $E^{-1}F$ is also positive definite.*

We are now in a position to prove a sufficient condition for uniqueness of the search direction.

Theorem 5.4.1 *The linear system (5.4.14) has a unique solution if $H_{\widetilde{P}}(\widetilde{X}\widetilde{S}) \succ 0$.*

Proof.
We consider the homogeneous system

$$\begin{bmatrix} 0 & \widetilde{\mathcal{A}} & 0 \\ -\widetilde{\mathcal{A}}^T & \mathcal{S}_{kew} & I \\ 0 & E & F \end{bmatrix} \begin{bmatrix} \Delta y \\ \mathrm{svec}\left(\Delta\widetilde{X}\right) \\ \mathrm{svec}\left(\Delta\widetilde{S}\right) \end{bmatrix} = \begin{bmatrix} 0 \\ 0 \\ 0 \end{bmatrix}, \tag{5.4.15}$$

and prove that it only has the zero vector as solution.

From (5.4.15) we have

$$\mathrm{svec}\left(\Delta\widetilde{S}\right) = \widetilde{\mathcal{A}}^T\Delta y - \mathcal{S}_{kew}\mathrm{svec}\left(\Delta\widetilde{X}\right)$$

and

$$\mathrm{svec}\left(\Delta\widetilde{S}\right) = -\left(F^{-1}E\right)\mathrm{svec}\left(\Delta\widetilde{X}\right). \tag{5.4.16}$$

Eliminating $\mathrm{svec}\left(\Delta\widetilde{S}\right)$ from the last two equations gives

$$\widetilde{\mathcal{A}}^T\Delta y - \mathcal{S}_{kew}\mathrm{svec}\left(\Delta\widetilde{X}\right) + \left(F^{-1}E\right)\mathrm{svec}\left(\Delta\widetilde{X}\right) = 0. \tag{5.4.17}$$

System (5.4.15) also implies

$$\widetilde{\mathcal{A}}\mathrm{svec}\left(\Delta\widetilde{X}\right) = 0. \tag{5.4.18}$$

From (5.4.17) we have

$$\begin{aligned}
\mathrm{svec}\left(\Delta\widetilde{X}\right)^T\widetilde{\mathcal{A}}^T\Delta y - \mathrm{svec}\left(\Delta\widetilde{X}\right)^T\mathcal{S}_{kew}\mathrm{svec}\left(\Delta\widetilde{X}\right)& \\
+\mathrm{svec}\left(\Delta\widetilde{X}\right)^T\left(F^{-1}E\right)\mathrm{svec}\left(\Delta\widetilde{X}\right) &= 0.
\end{aligned}$$

The first term on the left hand side is zero, by (5.4.18), and the second term is zero by the skew-symmetry of $\mathcal{S}_{kew}$. We therefore have

$$\text{svec}\left(\Delta\widetilde{X}\right)^T\left(F^{-1}E\right)\text{svec}\left(\Delta\widetilde{X}\right)=0,$$

which shows that $\Delta\widetilde{X}=0$, since EF^{-1} is assumed to be (non-symmetric) positive definite. It follows that $\Delta\widetilde{S}=0$ by (5.4.16). Furthermore, $\Delta y=0$ by (5.4.17), since $\widetilde{\mathcal{A}}$ has full rank (the matrices A_i $(i=1,\ldots,m)$ are linearly independent). ■

All that remains is to analyze the condition

$$H_{\widetilde{P}}(\widetilde{X}\widetilde{S})\succ 0 \tag{5.4.19}$$

in the theorem. For the first three choices of $\widetilde{P}$ in Table 5.1, condition (5.4.19) always holds. For $\widetilde{P}=I$ (the so-called AHO direction), (5.4.19) becomes the condition $\widetilde{X}\widetilde{S}+\widetilde{S}\widetilde{X}\succ 0$. It is shown in [546] and [437] that this latter condition holds if:

$$\left\{(\widetilde{X},\widetilde{S})\ :\ \left\|\frac{1}{2\mu}(\widetilde{X}\widetilde{S}+\widetilde{S}\widetilde{X})-I\right\|<1,\ \widetilde{X}\succ 0,\ \widetilde{S}\succ 0\right\},$$

for some $\mu>0$, i.e. if $(\widetilde{X},\widetilde{S})$ is in a (small) neighbourhood of the central path.

5.5 EXISTENCE OF THE CENTRAL PATH – A CONSTRUCTIVE PROOF

We now give a proof sketch of the existence of the central path of the embedding problem by using the Nesterov-Todd (NT) search direction as introduced in the previous section.

Recall that we know a feasible pair $(\widetilde{X},\widetilde{S})$ which satisfies $\widetilde{X}\widetilde{S}=\mu I$ if $\mu=1$. The idea is to reduce μ to some $\bar{\mu}\in(0,1)$ and then take steps along the NT direction which converge to a feasible solution $(\widetilde{X}(\bar{\mu}),\widetilde{S}(\bar{\mu}))$ of the embedding which satisfies $\widetilde{X}(\bar{\mu})\widetilde{S}(\bar{\mu})=\bar{\mu}I$.

The first step is to define a distance function, which measures the difference $\widetilde{X}\widetilde{S}-\mu I$ for a fixed value of $\mu\in(0,1)$ and for each feasible solution $(\widetilde{X},\widetilde{S})$:

$$\delta(\widetilde{X},\widetilde{S},\mu):=\frac{1}{2}\frac{1}{\sqrt{\mu}}\|D_V\|=\frac{1}{2}\left\|\sqrt{\mu}V^{-1}-\frac{1}{\sqrt{\mu}}V\right\|,$$

where $V^2:=\widetilde{X}^{\frac{1}{2}}\widetilde{S}\widetilde{X}^{\frac{1}{2}}$. This distance function was introduced by Jiang [368] (without the constant $\frac{1}{2}$).

Note that $\delta(\widetilde{X},\widetilde{S},\mu)\geq 0$ with equality if and only if $\widetilde{X}\widetilde{S}=\mu I$. We prove that δ can be reduced to zero by taking NT steps with respect to μ.

To analyse the progress along the NT steps, we will need the following lemmas from De Klerk *et al.* [420]:

Lemma 5.5.1 (Condition for a feasible full NT step) *If* $\delta := \delta(\widetilde{X}, \widetilde{S}, \mu) < 1$ *then the full NT step* $\left(\widetilde{X} + \Delta\widetilde{X}, \widetilde{S} + \Delta\widetilde{S}\right)$ *is strictly feasible. Moreover, the duality gap attains its target value:*

$$\operatorname{Tr}\left(\widetilde{X} + \Delta\widetilde{X}, \widetilde{S} + \Delta\widetilde{S}\right) = \widetilde{n}\mu.$$

Now let $\delta := \delta(\widetilde{X}, \widetilde{S}, \mu)$ and define $\delta^+ := \delta(\widetilde{X} + \Delta\widetilde{X}, \widetilde{S} + \Delta\widetilde{S}, \mu)$ (the value of the function δ after a full NT step with respect to μ).

Lemma 5.5.2 (Quadratic convergence) *After a full feasible NT step the distance measure satisfies*

$$\delta^+ \leq \frac{\delta^2}{\sqrt{2(1-\delta^2)}}.$$

In particular, if $\delta \leq \frac{1}{\sqrt{2}}$ *then* $\delta^+ \leq \delta^2$.

Loosely speaking, if the current product $\widetilde{X}\widetilde{S}$ is 'close to' μI, then the NT steps will converge quadratically to some feasible pair $(\widetilde{X}(\mu), \widetilde{S}(\mu))$ which satisfy $\widetilde{X}(\mu)\widetilde{S}(\mu) = \mu I$. The existence of the limit point $\left(\widetilde{X}(\mu)\widetilde{S}(\mu)\right)$ follows from the fact that all the NT steps with respect to μ yield feasible iterates $\left(\widetilde{X}, \widetilde{S}\right)$ satisfying $\operatorname{Tr}\left(\widetilde{X}\widetilde{S}\right) = \widetilde{n}\mu$. All the iterates therefore lie in the level set

$$\left\{\left(\widetilde{X}, \widetilde{S}\right) \in \widetilde{\mathcal{P}} \times \widetilde{\mathcal{D}} \mid \operatorname{Tr}\left(\widetilde{X}\widetilde{S}\right) = \widetilde{n}\mu\right\},$$

which is easily shown to be compact (cf. Lemma 5.6.1).

Given the pair $\left(\widetilde{X}(\mu), \widetilde{S}(\mu)\right)$, we can reduce μ so that we can again converge quadratically to the new μ-center.

Lemma 5.5.3 *Let* $\delta := \delta(\widetilde{X}, \widetilde{S}, \mu)$ *and let* $\operatorname{Tr}(\widetilde{X}\widetilde{S}) = \widetilde{n}\mu$. *If* $\mu^+ = (1-\theta)\mu$ *one has*

$$(\delta(\widetilde{X}, \widetilde{S}, \mu^+))^2 = \frac{\widetilde{n}\theta^2}{4(1-\theta)} + (1-\theta)\delta^2.$$

In particular, if $\delta(\widetilde{X}, \widetilde{S}, \mu) \leq \frac{1}{2}$, *and* $\theta = \frac{1}{2\sqrt{\widetilde{n}}}$, *then one has* $\delta(\widetilde{X}, \widetilde{S}, \mu^+) \leq \frac{1}{\sqrt{2}}$.

Thus we can compute all μ-centers for $\mu \in (0, 1]$. This proves the existence of the central path of the embedding problem for $\mu \leq 1$. The existence for $\mu > 1$ is proved similarly, by increasing μ at each step.

5.6 OBTAINING MAXIMALLY COMPLEMENTARY SOLUTIONS

Having established the existence of the central path of the embedding problem, the question remains how to obtain a maximally complementary solution. The key to the answer is that the limit of the central path (as $\mu \to 0$) is maximally complementary. This will be proved in this section, whereafter we will consider the practical question of how to obtain information about the limit of the central path from the ϵ-optimal output of a path-following interior point method.

In what follows we consider a fixed sequence $\{\mu^{(t)}\} \to 0$ with $\mu^{(t)} > 0$, $t = 1, \cdots$, and prove that there exists a subsequence of $\left\{\widetilde{X}(\mu^{(t)}), \widetilde{S}(\mu^{(t)})\right\}$ which converges to a maximally complementary solution.

The existence of limit points of the sequence is an easy consequence of the following lemma.

Lemma 5.6.1 *Given $\bar{\mu} > 0$, the set*

$$\left\{\left(\widetilde{X}(\mu), \widetilde{S}(\mu)\right) \;:\; 0 < \mu \leq \bar{\mu}\right\}$$

is bounded.

Proof.
Recall that $(\widetilde{X}^0, \widetilde{S}^0) = (I, I)$ is a strictly feasible solution of the embedding, and let $\left(\widetilde{X}(\mu), \widetilde{S}(\mu)\right)$ be a central solution corresponding to some $\mu > 0$. By orthogonality, one has

$$\operatorname{Tr}\left((\widetilde{X}(\mu) - \widetilde{X}^0)(\widetilde{S}(\mu) - \widetilde{S}^0)\right) = 0. \tag{5.6.20}$$

The centrality conditions imply $\operatorname{Tr}\left(\widetilde{X}(\mu)\widetilde{S}(\mu)\right) = \tilde{n}\mu$, which simplifies (5.6.20) to

$$\operatorname{Tr}(\widetilde{X}(\mu)\widetilde{S}^0) + \operatorname{Tr}(\widetilde{X}^0\widetilde{S}(\mu)) = \tilde{n}\mu + \operatorname{Tr}(\widetilde{X}^0\widetilde{S}^0). \tag{5.6.21}$$

Substituting $(\widetilde{X}^0, \widetilde{S}^0) = (I, I)$ one therefore has

$$\operatorname{Tr}\left(\widetilde{X}(\mu) + \widetilde{S}(\mu)\right) \leq \tilde{n}\mu + \tilde{n}.$$

Now using the fact that any positive semidefinite matrix $Z \succeq 0$ satisfies $||Z|| \leq \operatorname{Tr}(Z)$ for the Frobenius norm, one has

$$||\widetilde{X}(\mu) + \widetilde{S}(\mu)|| \leq (\mu + 1)\tilde{n},$$

which completes the proof. ■

Now let

$$\widetilde{X}(\mu^{(t)}) := Q(\mu^{(t)})\Lambda(\mu^{(t)})Q(\mu^{(t)})^T, \quad \widetilde{S}(\mu^{(t)}) := Q(\mu^{(t)})\Sigma(\mu^{(t)})Q(\mu^{(t)})^T$$

denote the spectral (eigenvector-eigenvalue) decompositions of $\widetilde{X}(\mu^{(t)})$ and $\widetilde{S}(\mu^{(t)})$. Lemma 5.6.1 implies that the eigenvalues of $\widetilde{X}(\mu^{(t)})$ and $\widetilde{S}(\mu^{(t)})$ are bounded. The matrices $Q(\mu^{(t)})$ are orthonormal for all t, and are therefore likewise restricted to a compact set. It follows that the sequence of triples $\left(Q(\mu^{(t)}), \Lambda(\mu^{(t)}), \Sigma(\mu^{(t)})\right)$ has an accumulation point, $(Q^*, \Lambda^*, \Sigma^*)$ say. Thus there exists a subsequence of $\{\mu^{(t)}\}$ (still denoted by $\{\mu^{(t)}\}$ for the sake of simplicity) such that

$$\lim_{t\to\infty} Q(\mu^{(t)}) = Q^*, \quad \lim_{t\to\infty} \Lambda(\mu^{(t)}) = \Lambda^*, \quad \lim_{t\to\infty} \Sigma(\mu^{(t)}) = \Sigma^*.$$

Note that $\Lambda(\mu^{(t)})\Sigma(\mu^{(t)}) = \mu I$. Thus, defining

$$\widetilde{X}^* := Q^*\Lambda^*Q^{*T} = \lim_{t\to\infty} \widetilde{X}(\mu^{(t)}), \quad \widetilde{S}^* := Q^*\Sigma^*Q^{*T} = \lim_{t\to\infty} \widetilde{S}(\mu^{(t)}), \quad (5.6.22)$$

we have $\Lambda^*\Sigma^* = 0$ and the pair $(\widetilde{X}^*, \widetilde{S}^*)$ is optimal.

Theorem 5.6.1 (Maximal complementarity) *The pair $(\widetilde{X}^*, \widetilde{S}^*)$ as defined in (5.6.22) is maximally complementary.*

Proof.
Let $(\widetilde{X}, \widetilde{S})$ be an arbitrary optimal pair. Applying the orthogonality property and $\mathrm{Tr}(\widetilde{X}\widetilde{S}) = 0$, $\mathrm{Tr}(\widetilde{X}(\mu^{(t)})\widetilde{S}(\mu^{(t)})) = \tilde{n}\mu^{(t)}$ we obtain

$$\mathrm{Tr}(\widetilde{X}(\mu^{(t)})\widetilde{S}) + \mathrm{Tr}(\widetilde{X}\widetilde{S}(\mu^{(t)})) = \tilde{n}\mu^{(t)}. \quad (5.6.23)$$

Since $\widetilde{X}(\mu^{(t)})\widetilde{S}(\mu^{(t)}) = \mu^{(t)}I$, dividing both sides by $\mu^{(t)}$ yields

$$\mathrm{Tr}(\widetilde{S}(\mu^{(t)})^{-1}\widetilde{S}) + \mathrm{Tr}(\widetilde{X}\widetilde{X}(\mu^{(t)})^{-1}) = \tilde{n} \quad (5.6.24)$$

for all t. This implies that

$$\mathrm{Tr}(\widetilde{X}\widetilde{X}(\mu^{(t)})^{-1}) \le \tilde{n}, \quad \mathrm{Tr}(\widetilde{S}(\mu^{(t)})^{-1}\widetilde{S}) \le \tilde{n}, \quad (5.6.25)$$

since both terms in the left hand side of (5.6.24) are nonnegative. We derive from this that $\widetilde{X}^*$ and $\widetilde{S}^*$ are maximally complementary. Below we give the derivation for $\widetilde{X}^*$; the derivation for $\widetilde{S}^*$ is similar and is therefore omitted.

Denoting the i-th column of the orthonormal (eigenvector) matrix $Q(\mu^{(t)})$ as $q_i(\mu^{(t)})$ and the i-th diagonal element of the (eigenvalue) matrix $\Lambda(\mu^{(t)})$ as $\lambda_i(\mu^{(t)})$ we have

$$\widetilde{X}(\mu^{(t)})^{-1} = Q(\mu^{(t)})\Lambda(\mu^{(t)})^{-1}Q(\mu^{(t)})^T = \sum_{i=1}^{\tilde{n}} \frac{1}{\lambda_i(\mu^{(t)})} q_i(\mu^{(t)})q_i(\mu^{(t)})^T. \quad (5.6.26)$$

Substituting (5.6.26) into the first inequality in (5.6.25) yields

$$\begin{aligned}\mathrm{Tr}\left(\widetilde{X}\widetilde{X}(\mu^{(t)})^{-1}\right) &= \sum_{i=1}^{\tilde{n}} \mathrm{Tr}\left(\frac{1}{\lambda_i(\mu^{(t)})}\widetilde{X}q_i(\mu^{(t)})q_i(\mu^{(t)})^T\right) \\ &= \sum_{i=1}^{\tilde{n}} \frac{q_i(\mu^{(t)})^T\widetilde{X}q_i(\mu^{(t)})}{\lambda_i(\mu^{(t)})} \le \tilde{n}.\end{aligned} \quad (5.6.27)$$

The last inequality implies

$$q_i(\mu^{(t)})^T \widetilde{X} q_i(\mu^{(t)}) \leq \widetilde{n}\lambda_i(\mu^{(t)}), i = 1, 2, \cdots, \widetilde{n}.$$

Letting t go to infinity we obtain

$$q_i^{*T} \widetilde{X} q_i^* \leq \widetilde{n}\lambda_i^*, i = 1, 2, \cdots, \widetilde{n},$$

where q_i^* denotes the i-th column of Q^* and λ_i^* the i-th diagonal element of Λ^*. Thus we have $q_i^{*T} \widetilde{X} q_i^* = 0$ whenever $\lambda_i^* = 0$. This implies

$$\widetilde{X} q_i^* = 0 \text{ if } \lambda_i^* = 0, \tag{5.6.28}$$

since $(q_i^*)^T \widetilde{X} q_i^* = \left\| \widetilde{X}^{\frac{1}{2}} q_i^* \right\|^2$, where $\widetilde{X}^{\frac{1}{2}}$ is the symmetric square root factor of $\widetilde{X}$. In other words, the row space of $\widetilde{X}$ is orthogonal to each column q_i^* of Q^* for which $\lambda_i^* = 0$. Hence the range space of $\widetilde{X}$ is a subspace of the space generated by the columns q_i^* of Q^* for which $\lambda_i^* > 0$. The latter space is simply $\mathbb{R}\left(\widetilde{X}^*\right)$. We conclude that $\mathbb{R}(\widetilde{X}) \subseteq \mathbb{R}\left(\widetilde{X}^*\right)$. This completes the proof, since $\widetilde{X}$ was an arbitrary optimal solution. ■

5.7 SEPARATING SMALL AND LARGE VARIABLES

A path following interior point method only yields an ϵ-optimal solution to the embedding problem. This solution may yield small values of ρ and τ, and to distinguish between cases (I) to (III) it is necessary to know if these values are zero in a maximally complementary solution. This is the most problematic aspect of the analysis at this time, and only partial solutions are given here. Two open problems are stated which would help resolve the current difficulties.

In what follows the set of feasible $\widetilde{X}$ for the embedding problems is denoted by $\widetilde{\mathcal{P}}$ and the optimal set by $\widetilde{\mathcal{P}}^*$. The sets $\widetilde{\mathcal{D}}$ and $\widetilde{\mathcal{D}}^*$ are defined similarly.

To separate 'small' and 'large' variables we need the following definition:

Definition 5.7.1 *The primal and dual condition numbers of the embedding are defined as*

$$\sigma_P := \sup_{\widetilde{X} \in \widetilde{\mathcal{P}}^*} f(\widetilde{X}), \quad \sigma_D := \sup_{\widetilde{S} \in \widetilde{\mathcal{D}}^*} f(\widetilde{S}).$$

where f is defined by

$$f(\widetilde{X}) := \begin{cases} \infty & \textit{if } \widetilde{X} = 0 \\ \min_{i:\lambda_i(\widetilde{X})>0} \lambda_i(\widetilde{X}) & \textit{otherwise.} \end{cases}$$

The condition number σ of the embedding is defined as $\sigma := \min\{\sigma_P, \sigma_D\}$.

Note that σ is positive and finite because the solution set of the self-dual embedding problem is bounded.

In linear programming a positive lower bound for σ can be given in terms of the problem data [683]. It is an open problem to give a similar bound in the semidefinite case.

Open problem 5.7.1 *Derive a lower bound for σ expressed as an explicit function of the data.*

If we have a centered solution to the embedding problem with centering parameter μ then we can use any knowledge of σ to decide the following:

Lemma 5.7.1 *For any positive μ one has:*

$$\begin{array}{llll} \tau(\mu) \geq \sigma/\tilde{n} & \text{and} & \rho(\mu) \leq \tilde{n}\mu/\sigma & \text{if } \tau^* > 0 \text{ and } \rho^* = 0 \\ \tau(\mu) \leq \tilde{n}\mu/\sigma & \text{and} & \rho(\mu) \geq \sigma/\tilde{n} & \text{if } \tau^* = 0 \text{ and } \rho^* > 0, \end{array}$$

where the superscript $*$ indicates a maximally complementary solution.

Proof.
Assume that ρ^* is positive in a maximally complementary solution. Let $\tilde{S}^* \in \tilde{\mathcal{D}}^*$ be such that ρ^* is as large as possible. By definition one therefore has $\rho^* \geq \sigma$. Recall that by (5.6.23) one has

$$\operatorname{Tr}\left(\tilde{X}(\mu)\tilde{S}^*\right) \leq \tilde{n}\mu,$$

which implies that the eigenvalues of $\tilde{X}(\mu)\tilde{S}^*$ satisfy

$$\lambda_i\left(\tilde{X}(\mu)\tilde{S}^*\right) \leq \tilde{n}\mu, \quad \forall\, i.$$

In particular

$$\tau(\mu)\rho^* \leq \tilde{n}\mu.$$

This shows that

$$\tau(\mu) \leq \frac{\tilde{n}\mu}{\rho^*} \leq \frac{\tilde{n}\mu}{\sigma}.$$

Since $\tau(\mu)\rho(\mu) = \mu$ one also has

$$\rho(\mu) \geq \frac{\sigma}{\tilde{n}}.$$

The case where $\tau^* > 0$ and $\rho^* = 0$ is proved in the same way. ■

The lemma shows that once the barrier parameter μ has been reduced to the point where $\mu \leq \left(\frac{\sigma}{\tilde{n}}\right)^2$, then it is known which of τ or ρ is positive in a maximally complementary solution, provided that one is indeed positive. The smaller the condition number, the more work will be needed in general to solve the embedding to sufficient accuracy.

The proof of Lemma 5.7.1 can easily be extended to the case where the ϵ-optimal solution is only approximately centered, where approximate centrality is defined by

$$\delta(\widetilde{X},\widetilde{S}) := \frac{\lambda_{\max}(\widetilde{X}\widetilde{S})}{\lambda_{\min}(\widetilde{X}\widetilde{S})} \leq \kappa,$$

for some parameter $\kappa > 1$. Formally one has the following result.

Lemma 5.7.2 *Let $(\widetilde{X},\widetilde{S})$ be a feasible solution of the embedding problem such that $\delta(\widetilde{X},\widetilde{S}) \leq \kappa$ for some $\kappa > 1$. One has the relations:*

$$\begin{array}{llll} \tau \geq \frac{\sigma}{\kappa n} & \text{and} & \rho \leq \frac{\mathrm{Tr}(\widetilde{X}\widetilde{S})}{\sigma} & \text{if } \tau^* > 0 \text{ and } \rho^* = 0 \\ \tau \leq \frac{\mathrm{Tr}(\widetilde{X}\widetilde{S})}{\sigma} & \text{and} & \rho \geq \frac{\sigma}{\kappa n} & \text{if } \tau^* = 0 \text{ and } \rho^* > 0 \end{array} \qquad (5.7.29)$$

where the superscript $*$ indicates a maximally complementary solution.

The condition number is related to the concept of *ill-posedness* (see [250, 252]) of the SDP problem to be solved. Consider the following example.

Example 5.7.1 *The problem in Example 5.2.1 is weakly infeasible, but it can be considered as ill-posed in the following sense: if we perturb the data slightly to obtain the problem*

$$\sup\left\{ y_1 \;\middle|\; y_1 \begin{bmatrix} 1 & 0 \\ 0 & \epsilon^2 \end{bmatrix} \preceq \begin{bmatrix} 0 & 1 \\ 1 & 0 \end{bmatrix} \right\} \qquad (5.7.30)$$

for some small $\epsilon > 0$, then the problem (5.7.30) and its dual are strictly feasible and therefore solvable. Problem (5.7.30) has the unique solution

$$S^* = \begin{bmatrix} 1/\epsilon & 1 \\ 1 & \epsilon \end{bmatrix}, \quad y_1^* = -\frac{1}{\epsilon},$$

and its dual has solution

$$X^* = \begin{bmatrix} \frac{1}{2} & -\frac{\epsilon}{2} \\ -\frac{\epsilon}{2} & \frac{1}{2\epsilon^2} \end{bmatrix}.$$

Assume now that we solve problem (5.7.30) via the embedding approach. It follows from inequality (5.3.11) that the optimal value of the embedding variable τ will satisfy

$$\tau^* = \frac{4}{1 + \frac{1}{\epsilon} + \epsilon + \frac{1}{2} + \frac{1}{2\epsilon^2}} < 8\epsilon^2.$$

This shows that the condition number σ will be $O(\epsilon^2)$. Thus, we will have to reduce μ to the point where $\mu < O\left(\epsilon^4\right)$ in order to correctly classify the problem status via the embedding approach. The required value of μ will typically be smaller than machine precision, if say $\epsilon \leq 10^{-4}$.

This illustrates the inherent numerical difficulty with problems like the weakly infeasible problem in Example 5.2.1. ■

In order to improve the results of Lemma 5.7.1 one must establish the rate at which $\tau(\mu)$ and $\rho(\mu)$ converges to zero in the case where both are zero in a maximally complementary solution. This is still an unresolved issue.

Open problem 5.7.2 *The spectrum of $\widetilde{X}(\mu) + \widetilde{S}(\mu)$ is a continuous function of μ defined on the central path.*

Establish an upper bound in terms of μ for the eigenvalues of $\widetilde{X}(\mu) + \widetilde{S}(\mu)$ which converge to zero as $\mu \to 0$.

Stoer and Wechs [754] considered the analogous problem in the case of *horizontal sufficient linear complementarity problems (LCP)*, and prove a bound of $O(\sqrt{\mu})$. Most recently, Illés et al. [353] have obtained explicit expressions for the constants appearing in this bound for LCP's with $P^*(\kappa)$ matrices. These explicit values are needed in order to obtain a lemma which improves on the analogy of Lemma 5.7.1, i.e. to obtain a value of μ which is small enough in order to distinguish between the three possible cases.

5.8 REMAINING DUALITY AND FEASIBILITY ISSUES

If $\rho^* = \tau^* = 0$ in a maximally complementary solution of the embedding problem (i.e. case (III) holds), then one of the following situations has occurred:

1) The problems (P) and (D) are solvable but have a positive duality gap;

2) either (P) or (D) (or both) are weakly infeasible;

3) both (P) and (D) are feasible, but one or both are unsolvable, e.g. $p^* \equiv \inf_{X \in \mathcal{P}^*} \mathrm{Tr}(CX)$ is finite but is not attained.

The case 2) was illustrated in Example 5.2.1. The remaining two cases occur in the following examples:

Example 5.8.1 *The following problem (adapted from [807]) is in the form (D): find*

$$\sup y_2$$

subject to

$$y_1 \begin{bmatrix} 0 & 0 & 0 \\ 0 & 1 & 0 \\ 0 & 0 & 0 \end{bmatrix} + y_2 \begin{bmatrix} 1 & 0 & 0 \\ 0 & 0 & 1 \\ 0 & 1 & 0 \end{bmatrix} \preceq \begin{bmatrix} 1 & 0 & 0 \\ 0 & 0 & 0 \\ 0 & 0 & 0 \end{bmatrix}.$$

This problem is solvable with optimal value $y_2^* = 0$ and the corresponding primal problem is solvable with optimal value 1. ■

Example 5.8.2 *Another difficulty is illustrated by the following problem (adapted from [807]): find*

$$\sup y_2$$

subject to

$$y_1 \begin{bmatrix} 1 & 0 \\ 0 & 0 \end{bmatrix} + y_2 \begin{bmatrix} 0 & 0 \\ 0 & 1 \end{bmatrix} \preceq \begin{bmatrix} 0 & 1 \\ 1 & 1 \end{bmatrix}.$$

This problem is not solvable but $\sup_{y \in \mathcal{D}} y_2 = 1$. *The corresponding primal problem is solvable with optimal value* 1. ■

The aim is therefore to see what further information can be obtained in the case $\tau^* = \rho^* = 0$. We will show that more information can be obtained if (P) and (D) are in perfect duality and if we moreover assume that $\mathcal{P}^*$ is nonempty if p^* is finite.

To this end, recall that along the central path of the embedding problem one has

$$\rho(\mu^{(t)})\tau(\mu^{(t)}) = \mu^{(t)} \text{ and } \vartheta(\mu^{(t)})\beta = \tilde{n}\mu^{(t)} \tag{5.8.31}$$

which shows that $\rho(\mu^{(t)}) \to \rho^* = 0$ implies

$$\vartheta(\mu^{(t)})/\tau(\mu^{(t)}) \to 0 \text{ as } t \to \infty. \tag{5.8.32}$$

This shows (by (5.3.5)) that:

$$\operatorname{Tr}\left(\frac{1}{\tau(\mu^{(t)})} A_i X(\mu^{(t)})\right) \to b_i, \quad \forall i \tag{5.8.33}$$

and

$$\sum_{i=1}^{m} \frac{y_i(\mu^{(t)})}{\tau(\mu^{(t)})} A_i + \frac{1}{\tau(\mu^{(t)})} S(\mu^{(t)}) \to C. \tag{5.8.34}$$

In other words if either or both of the sequences

$$\left\{\frac{1}{\tau(\mu^{(t)})} X(\mu^{(t)})\right\} \text{ and } \left\{\frac{1}{\tau(\mu^{(t)})} S(\mu^{(t)})\right\} \tag{5.8.35}$$

converge, the limit is feasible for (P) or (D) respectively. On the other hand, if (5.8.33) (resp. (5.8.34)) holds but (P) (resp. (D)) is infeasible, then (P) (resp. (D)) is weakly infeasible. If one also has

$$\frac{\rho(\mu^{(t)})}{\tau(\mu^{(t)})} \to 0 \text{ as } t \to \infty \tag{5.8.36}$$

then it also follows from (5.3.5) that

$$\frac{1}{\tau(\mu^{(t)})} b^T y(\mu^{(t)}) - \frac{1}{\tau(\mu^{(t)})} \operatorname{Tr}\left(C X(\mu^{(t)})\right) \to 0.$$

If this happens, at least one of the sequences in (5.8.35) diverges (or else an optimal pair with zero duality gap exists).

On the other hand, one always has $\vartheta(\mu^{(t)})/\rho(\mu^{(t)}) \to 0$ if $\tau(\mu^{(t)}) \to 0$, from (5.8.31). If it also holds that

$$\frac{\tau(\mu^{(t)})}{\rho(\mu^{(t)})} \to 0 \text{ as } t \to \infty \tag{5.8.37}$$

then

$$\frac{1}{\rho(\mu^{(t)})} b^T y(\mu^{(t)}) - \frac{1}{\rho(\mu^{(t)})} \operatorname{Tr}\left(CX(\mu^{(t)})\right) \to 1,$$

$$\operatorname{Tr}\left(\frac{1}{\rho(\mu^{(t)})} A_i X(\mu^{(t)})\right) \to 0, \quad \forall i \tag{5.8.38}$$

and

$$\sum_{i=1}^{m} \frac{y_i(\mu^{(t)})}{\tau(\mu^{(t)})} A_i + \frac{1}{\tau(\mu^{(t)})} S(\mu^{(t)}) \to 0. \tag{5.8.39}$$

A so-called *asymptotically improving ray* (or weakly improving ray) is thus detected for (P) and/or (D). It is shown in [508] that an asymptotically improving ray in (P) (resp. (D)) implies weak infeasibility in (D) resp. (P).

The problem is that none of these indicators gives a certificate of the status of a given problem. For example, there is no guarantee that (5.8.37) will hold if one (or both) of (P) and (D) have weakly improving rays. Luo et al. [508] derive similar detectors and show that these detectors yield no information in some cases. We therefore need the assumption of perfect duality together with primal or dual attainment. To fix our ideas, we assume the following.

Assumption 5.8.1 *Problems (P) and (D) are in perfect duality, and $\mathcal{P}^*$ is non-empty if d^* is finite.*

We will show in the next section that – in a well-defined sense – these assumptions can be made without loss of generality.

To proceed, we first show that (5.8.36) must hold if d^* is finite.

Lemma 5.8.1 *Assume that a given problem (D) has finite optimal value d^* and that (P) and (D) are in perfect duality. Then (5.8.36) holds for the embedding of (P) and (D).*

Proof.
Let $\epsilon^{(t)} := \vartheta(\mu^{(t)})/\tau(\mu^{(t)})$ and $(X, y, S) \in \mathcal{P} \times \mathcal{D}$. Note that $\epsilon^{(t)} \to 0$ as $t \to \infty$ by (5.8.32). For ease of notation we further define

$$X^{(t)} := \frac{1}{\tau(\mu^{(t)})} X(\mu^{(t)}), \quad S^{(t)} := \frac{1}{\tau(\mu^{(t)})} S(\mu^{(t)}), \quad y^{(t)} = \frac{1}{\tau(\mu^{(t)})} y(\mu^{(t)}).$$

In terms of this notation one has from (5.3.5):

$$\begin{aligned} \operatorname{Tr}\left(A_i X^{(t)}\right) + \epsilon^{(t)} \bar{b}_i &= b_i, \quad i = 1, \ldots, m \\ \sum_{i=1}^{m} (y^{(t)})_i A_i + S^{(t)} + \epsilon^{(t)} \bar{C} &= C. \end{aligned}$$

Using the feasibility of X and S and (5.6.20) it is easy to show that

$$\mathrm{Tr}\left(X^{(t)}S + S^{(t)}X\right) = \mathrm{Tr}(XS) - \epsilon^{(t)}\,\mathrm{Tr}(\bar{C}X) + \epsilon^{(t)}\bar{b}^T y + \left[\mathrm{Tr}(CX^{(t)}) - b^T y^{(t)}\right].$$

Substitution of $\bar{b}_i = b_i - \mathrm{Tr}(A_i)$ and $\bar{C} = C - I$, and using

$$\mathrm{Tr}(S) = \mathrm{Tr}\left(C - \sum_{i=1}^{m}(y)_i A_i\right)$$

yields

$$\begin{aligned}\mathrm{Tr}\left(X^{(t)}S + S^{(t)}X\right) &= (1+\epsilon^{(t)})\,\mathrm{Tr}(XS) - \epsilon^{(t)}\,\mathrm{Tr}(X+S) - \epsilon^{(t)}\,\mathrm{Tr}(C) \\ &\quad + \left[\mathrm{Tr}(CX^{(t)}) - b^T y^{(t)}\right] \qquad (5.8.40)\end{aligned}$$

If (5.8.36) does not hold, then there exists an $\bar{\epsilon} > 0$ such that

$$\left[\mathrm{Tr}(CX^{(\bar{t})}) - b^T y^{(\bar{t})}\right] < -\bar{\epsilon} \qquad (5.8.41)$$

for some $\bar{t}$ which can be chosen arbitrarily large.

Since X and S were arbitrary feasible solutions and perfect duality is assumed, we can assume that $\mathrm{Tr}(XS) < \bar{\epsilon}/2$. Choose $\bar{t}$ such that (5.8.41) holds and

$$\epsilon^{(\bar{t})}\,\mathrm{Tr}(XS) - \epsilon^{(\bar{t})}\,\mathrm{Tr}(X+S) - \epsilon^{(\bar{t})}\,\mathrm{Tr}(C) < \bar{\epsilon}/2.$$

The left hand side of (5.8.40) is always nonnegative, while the right hand side is negative for the above choice of $\bar{t}$. This contradiction shows that if a pair (X, S) exists with arbitrarily small duality gap, then (5.8.36) must hold. ■

The next question is how to obtain the value d^* if it is finite. The following lemma shows that this value can be obtained from a sequence of centered iterates of the embedding as a limit value.

Lemma 5.8.2 *If Assumption 5.8.1 holds for (P) and (D) and the optimal value of (D) is finite ($d^* < \infty$), then:*

$$d^* = \mathrm{Tr}\left(CX^*\right) = \lim_{t\to\infty}\frac{1}{\tau(\mu^{(t)})}b^T y(\mu^{(t)}) = \lim_{t\to\infty}\mathrm{Tr}\left(\frac{CX(\mu^{(t)})}{\tau(\mu^{(t)})}\right).$$

where X^* denotes an optimal solution of (P).

Proof.
Let X^* be any optimal solution of (P), which exists by Assumption 5.8.1. Using the 'superscript (t)' notation from the previous lemma, and the statement of the self–dual problem in (5.3.5), one can easily show that

$$\mathrm{Tr}\left(X^*S^{(t)}\right) = \mathrm{Tr}\left(CX^*\right) - b^T y^{(t)} - \epsilon^{(t)}\,\mathrm{Tr}\left(\bar{C}X^*\right)$$

or

$$\begin{aligned}\left|\operatorname{Tr}(CX^*) - b^T y^{(t)}\right| &= \left|\operatorname{Tr}\left(X^* S^{(t)}\right) - \epsilon^{(t)} \operatorname{Tr}(\bar{C}X^*)\right| \\ &\leq \left|\operatorname{Tr}\left(X^* S^{(t)}\right)\right| + \left|\epsilon^{(t)} \operatorname{Tr}(\bar{C}X^*)\right|.\end{aligned}$$

The second right hand side term converges to zero as $\epsilon^{(t)} \to 0$. The first right hand side term can be made arbitrarily small, as can easily be seen from (5.8.40). This completes the proof. ■

We now show how to detect infeasibility or unboundedness of (D).

Lemma 5.8.3 *If (P) and (D) are in perfect duality and (D) is unbounded or infeasible, then:*

$$\begin{aligned}\lim_{\mu\to 0} \operatorname{Tr}\left(\frac{CX(\mu)}{\tau(\mu)}\right) = \lim_{\mu\to 0} \operatorname{Tr}\left(\frac{b^T y(\mu)}{\tau(\mu)}\right) &= \infty \text{ if } (D) \text{ is unbounded;} \\ \lim_{\mu\to 0} \operatorname{Tr}\left(\frac{CX(\mu)}{\tau(\mu)}\right) = \lim_{\mu\to 0} \operatorname{Tr}\left(\frac{b^T y(\mu)}{\tau(\mu)}\right) &= -\infty \text{ if } (D) \text{ is infeasible.}\end{aligned}$$

Proof.
Recall from Assumption 5.8.1 that (D) is infeasible if and only if (P) is unbounded. Let us assume that (P) is unbounded, and let $K > 0$ be given. By the assumption, there exists a $X \in \mathcal{P}$ such that $\operatorname{Tr}(CX) \leq -K$. It is straightforward to derive the following relation from the statement of the self–dual problem (5.3.5):

$$\begin{aligned}\frac{1}{\tau(\mu^{(t)})}\operatorname{Tr}(CX(\mu^{(t)})) &= \operatorname{Tr}(CX) + \frac{\vartheta(\mu^{(t)})}{\tau(\mu^{(t)})}\alpha - \frac{\rho(\mu^{(t)})}{\tau(\mu^{(t)})} \\ &\quad - \frac{\vartheta(\mu^{(t)})}{\tau(\mu^{(t)})}\operatorname{Tr}(\bar{C}X) - \frac{1}{\tau(\mu^{(t)})}\operatorname{Tr}(S(\mu^{(t)})X) \\ &\leq -K + \frac{\vartheta(\mu^{(t)})}{\tau(\mu^{(t)})}\alpha - \frac{\rho(\mu^{(t)})}{\tau(\mu^{(t)})} - \frac{\vartheta(\mu^{(t)})}{\tau(\mu^{(t)})}\operatorname{Tr}(\bar{C}X),\end{aligned}$$

where we have discarded two nonpositive terms to obtain the inequality. Taking the limit as $t \to \infty$ yields

$$\lim_{t\to\infty} \frac{1}{\tau(\mu^{(t)})}\operatorname{Tr}(CX(\mu^{(t)})) \leq -K.$$

Since $K > 0$ was arbitrary, the second result follows. The case where (D) is unbounded is proved in a similar way. ■

In Table 5.2 the results of the lemmas are summarized.

Status of (D)	$\limsup_{\mu\to 0} \frac{\rho(\mu)}{\tau(\mu)}$	$\lim_{\mu\to 0} \mathrm{Tr}\left(\frac{CX(\mu)}{\tau(\mu)}\right)$	$\lim_{\mu\to 0} \frac{b^T y(\mu)}{\tau(\mu)}$
$d^* < \infty$	0	d^*	d^*
unbounded	$[0, \infty)$	∞	∞
infeasible	$[0, \infty)$	$-\infty$	$-\infty$

Table 5.2 Indicators of the status of problem (D) via the embedding of (D) and (P), if (P) and (D) satisfy Assumption 5.8.1.

5.9 EMBEDDING EXTENDED LAGRANGE-SLATER DUALS

The analysis of the previous section was based on Assumption 5.8.1. Here we show that this assumption (in theory) involves no loss of generality.

Assume now that the aim is to solve any given problem (D) in the standard dual form, like the problems in the examples. In other words, we wish to find the value

$$d^* \equiv \sup_{S,y\in\mathcal{D}} b^T y$$

if it is finite, or obtain certificate that (D) is infeasible, or alternatively, a certificate of unboundedness.

For the example problems the embedding of (D) and its Lagrangian dual (P) will be insufficient for this purpose. The solution discussed here is to solve a second embedding problem, using so-called *extended Lagrange-Slater duals*.

The so-called gap-free primal problem (P_{gf}) of (D) may be formulated instead of using the standard primal problem (P). The gapfree primal was first formulated by Ramana [653], and takes the form:

$$p^*_{gf} := \inf \mathrm{Tr}\,(C(U_0 + W_m))$$

subject to

$$\begin{aligned}
\mathrm{Tr}\,(A_i(U_0 + W_m)) &= b_i, \quad i = 1, \ldots, m \\
\mathrm{Tr}\,(C(U_i + W_{i-1})) &= 0, \quad i = 1, \ldots, m \\
\mathrm{Tr}\,(A_i(U_i + W_{i-1})) &= 0, \quad i = 1, \ldots, m \\
W_0 &= 0, \\
\begin{bmatrix} I & W_i^T \\ W_i & U_i \end{bmatrix} &\succeq 0, \quad i = 1, \ldots, m \\
U_0 &\succeq 0,
\end{aligned}$$

where the variables are $U_i \succeq 0$ and $W_i \in \mathbb{R}^{n\times n}$, $i = 0, \ldots, m$.

Note that the gap-free primal problem is easily cast in the standard primal form. Moreover, its size is polynomial in the size of (D). Unlike the Lagrangian dual (P) of (D), (P_{gf}) has the following desirable features:

- (Weak duality) If $(y, S) \in \mathcal{D}$ and (U_i, W_i) $(i = 0, \ldots, m)$ is feasible for (P_{gf}) then

$$b^T y \leq \mathrm{Tr}\,(C(U_0 + W_m)).$$

- (Dual boundedness) If (D) is feasible, its optimal value is finite if and only if (P_{gf}) is feasible.
- (Zero duality gap) The optimal value p^*_{gf} of (P_{gf}) equals the optimal value of (D) if and only if both (P_{gf}) and (D) are feasible.
- (Attainment) If the optimal value of (D) is finite, then it is attained by (P_{gf}).

The standard (Lagrangian) dual problem associated with (P_{gf}) is called the *corrected dual* (D_{cor}). The surprising result is that the pair (P_{gf}) and (D_{cor}) are now in perfect duality [655].

Moreover, a feasible solution to (D) can be extracted from a feasible solution to (D_{cor}). The only problem is that (D_{cor}) does not necessarily attain its supremum, even if (D) does.

A natural question is whether (D_{cor}) is strongly infeasible if (D) is only weakly infeasible. This would simplify matters greatly as strong infeasibility can be detected more easily. Unfortunately this is not the case — it is readily verified that the weakly infeasible problem (D) in Example 5.2.1 has a weakly infeasible corrected problem (D_{cor}).

The possible duality relations are listed in Table 5.3. The optimal value of D_{cor} is denoted by d^*_{cor}. The larger problems (P_{gf}) and (D_{cor}) therefore

Status of (D)	Status of (P_{gf})	Status of (D_{cor})
$d^* < \infty$	$p^*_{gf} = d^*$	$d^*_{cor} = d^*$
unbounded	infeasible	unbounded
infeasible	unbounded	infeasible

Table 5.3 Duality relations for a given problem (D), its gapfree dual (P_{gf}) and its corrected problem (D_{cor}).

satisfy Assumption 5.8.1, even though (D) and its Lagrangian dual don't. This means that we can use the results of the previous section in order to solve the embedding problem of (P_{gf}) and (D_{cor}). The solution of this embedding problem yields information about the status of (D) (see Table 5.3).

5.10 SUMMARY

Below we retrace the sequence of steps which (in theory) yield the status of a given problem:

$$(D): \quad d^* := \sup_{y,S} \left\{ b^T y \;\middle|\; \sum_{i=1}^{m} y_i A_i + S = C,\ S \succeq 0,\ y \in \mathbb{R}^m \right\}.$$

1. Solve the embedding of (D) and its Lagrangean dual (P) to obtain a maximally complementary solution $(y^*,\ X^*,\ \tau^*,\ \vartheta^*,\ S^*,\ \rho^*,\ \nu^*)$ of the embedding;

[i] If $\tau^* > 0$ then $\frac{1}{\tau^*}S^*$ is an optimal solution of (D); STOP

[ii] If $\rho^* > 0$ strong infeasibility of either (P) or (D) or both is detected; STOP

[ii] If $\tau^* = \rho^* = 0$ then go to step 2;

2. Solve the embedding of the Extended Lagrange-Slater dual of (D), namely (P_{gf}), and the Lagrangean dual of (P_{gf}) by following the central path of the embedding problem. Now one has

$$\lim_{\mu \to 0} \operatorname{Tr}\left(\frac{CX(\mu)}{\tau(\mu)}\right) = \begin{cases} \infty & \text{if } (D) \text{ is unbounded without an improving ray;} \\ -\infty & \text{if } (D) \text{ is weakly infeasible;} \\ d^* & \text{if } d^* \text{ is finite,} \end{cases}$$

where $X(\mu)$ and $\tau(\mu)$ refer to the values of the embedding variables on the central path.

6 ROBUSTNESS

Aharon Ben-Tal, Laurent El Ghaoui, Arkadi Nemirovski

6.1 INTRODUCTION

6.1.1 SDPs with uncertain data

We consider a semidefinite programming problem (SDP) of the form

$$\max\ b^T y \quad \text{subject to } F(y) = F_0 + \sum_{i=1}^{m} y_i F_i \succeq 0, \tag{6.1.1}$$

where $b \in \mathbf{R}^m$ is given, and F is an affine map from $y \in \mathbf{R}^m$ to $\mathcal{S}^n$.

In many practical applications, the "data" of the problem (the vector b and the coefficients matrices $F_0, \ldots, F_m$) is subject to possibly large uncertainties. Reasons for this include the following.

- *Uncertainty about the future.* The exact value of the data is unknown at the time the values of y should be determined and will be known only in the future. (In a risk management problem for example, the data may depend on future demands, market prices, etc, that are unknown when the decision has to be taken.)

- *Errors in the data.* The data originates from a process (measurements, computational process) that is very hard to perform error-free (errors in the measurement of material properties in a truss optimization problem, sensor errors in a control system, floating-point errors made during computation, etc).

- *Implementation errors.* The computed optimal solution y^* cannot be implemented exactly, which results in uncertainty about the feasibility of the implemented solution. For example, the coefficients of an optimal finite-impulse response (FIR) filter can often be implemented in 8 bits only. As it turns out, this can be modeled as uncertainties in the coefficients matrices F_i.

- *Approximating nonlinearity by uncertainty.* The mapping $F(y)$ is completely known, but (slightly) non linear. The optimizer chooses to approximate the nonlinearity by uncertainty in the data.

- *Infinite number of constraints.* The problem is posed with an infinite number of constraints, indexed on a scalar parameter ω (for example, the constraints express a property of a FIR filter at each frequency ω). It may be convenient to view this parameter as a perturbation, and regard the semi-infinite problem as a robustness problem.

Depending on the assumptions on the nature of uncertainty, several methods have been proposed in the Operations Research/Engineering literature. Stochastic programming works with random perturbations and probabilities of satisfaction of constraints, and thus requires correct estimates of the distribution of uncertainties; sensitivity analysis assumes the perturbations is infinitesimal, and can be used only as a "post-optimization" tool; the "Robust Mathematical Programming" approach recently proposed by Mulvey, Vanderbei and Zenios [556] is based on the (sometimes very restrictive) assumption that the uncertainty takes a finite number of values (corresponding to "worst-case scenarios"). Interval arithmetic is one of the methods that have been proposed to deal with uncertainty in numerical computations; references to this large field of study include the book by one of its founders, Moore [548], the more recent book by Hansen [327], and also the very extensive web site developed by Kosheler and Kreinovich [448].

The approach proposed here assumes (following the philosophy of interval calculus) that the data of the problem is only known to belong to some "uncertainty set" $\mathcal{U}$; in this sense, the perturbation to the nominal problem is deterministic, unknown-but-bounded. A *robust solution* of the problem is one which satisfies the perturbed constraints for every value of the data within the admissible region $\mathcal{U}$. The *robust counterpart* of the SDP is to minimize the worst-case value of the objective, among all robust solutions. This approach was introduced by the authors independently in [86, 84, 87] and [212, 210]; although apparently new in mathematical programming, the notion of robustness is quite classical in control theory (and practice).

Even for simple uncertainty sets $\mathcal{U}$, the resulting robust SDP is NP-hard. Our main result is to show how to compute, via SDP, *upper bounds* for the robust counterpart. Contrarily to the other approaches to uncertainty, the robust method provides (in polynomial time) *guarantees* (of, *e.g.*, feasibility), at the expense of possible conservatism. Note that there is no real conflict between other approaches to uncertainty and ours; for example, it is possible to solve via robust SDP a stochastic programming problem with unknown-but-bounded distribution of the random parameters.

6.1.2 *Problem definition*

To make our problem mathematically precise, we assume that the perturbed constraint is of the form

$$\mathbf{F}(y,\delta) \succeq 0,$$

where $y \in \mathbf{R}^m$ is the decision vector, δ is a "perturbation vector" that is only known to belong to some "perturbation set" $\mathcal{D} \subseteq \mathbf{R}^l$, and $\mathbf{F}$ is a mapping from $\mathbf{R}^m \times \mathcal{D}$ to $\mathcal{S}^n$. We assume that $\mathbf{F}(y,\delta)$ is affine in y for each δ, and rational in δ for each y; also we assume that $\mathcal{D}$ contains 0, and that $\mathbf{F}(y,0) = F(y)$ for every y. Without loss of generality, we assume that the objective vector b is independent of perturbation.

We consider the following problem, referred to as the *robust counterpart* to the "nominal" problem (6.1.1):

$$\max\ b^T y \text{ subject to } y \in \mathcal{X}_{\mathcal{D}} \tag{6.1.2}$$

where $\mathcal{X}_{\mathcal{D}}$ is the set of *robust feasible* solutions, that is,

$$\mathcal{X}_{\mathcal{D}} = \left\{ y \in \mathbf{R}^m \;\middle|\; \text{for every } \delta \in \mathcal{D},\ \mathbf{F}(y,\delta) \text{ is well-defined and } \mathbf{F}(y,\delta) \succeq 0 \right\}.$$

In this paper, we only consider *ellipsoidal uncertainty.* This means that the perturbation set $\mathcal{D}$ consists of block vectors, each block being subject to an Euclidean-norm bound. Precisely,

$$\mathcal{D} = \left\{ \delta \in \mathbf{R}^l \;\middle|\; \delta = \begin{bmatrix} \delta^1 \\ \vdots \\ \delta^N \end{bmatrix},\ \text{where } \delta^k \in \mathbf{R}^{n_k},\ \ \|\delta^k\|_2 \le \rho,\ \ k = 1,\dots,N \right\}, \tag{6.1.3}$$

where $\rho \geq 0$ is a given parameter that determines the "size" of the uncertainty, and the integers n_k denote the lengths of each block vector δ_k (we have of course $n_1 + \ldots + n_N = L$).

There are many motivations for considering the above framework. It can be used when the perturbation is a vector with each component bounded in magnitude, in which case each block vector δ^k is actually a scalar ($n_1 = \ldots = n_N = 1$). It also can be used when the perturbation is bounded in Euclidean norm (which is often the case when the bounds on the parameters are obtained from statistics, and a Gaussian distribution is assumed). In some applications, there is a mixture of Euclidean-norm and maximum-norm bounds. (For example, we might have some parameters of the form $\delta_1 = \rho\cos\theta$, $\delta_2 = \rho\sin\theta$, where both ρ and θ are uncertain.)

It turns out that already in the case of affine perturbations ($\mathbf{F}(\cdot,\delta)$ is affine in δ) the robust counterpart (6.1.2), generally speaking, is NP-hard. This is why we are interested not only in the robust counterpart itself, but also in its *approximations* – "computationally tractable" problems with the same objective as in (6.1.2) and feasible sets contained in the set $\mathcal{X}_{\mathcal{D}}$ of robust solutions. We will obtain an upper bound (approximation) on the robust counterpart in the form of an SDP. The size of this SDP is linear in both the length n_k of each block and the number N of blocks.

The paper is organized as follows. In Section 6.2, we consider the case when the perturbation vector affects the semidefinite constraint affinely, that is, the matrix $\mathbf{F}(y,\delta)$ is affine in δ; we provide not only an approximation of the robust counterpart, but also a result on the quality of the approximation, in an appropriately defined sense. Section 6.3 is devoted to the general case (the perturbation δ enters rationally in $\mathbf{F}(y,\delta)$). We provide interesting special cases (when the approximation is exact) in Section 6.4, while Section 6.5 describes several examples of application.

6.2 AFFINE PERTURBATIONS

In this section, we assume that the matrix function $\mathbf{F}(y,\delta)$ is given by

$$\mathbf{F}(y,\delta) = F^0(y) + \sum_{i=1}^{l} \delta_i F^i(y) \tag{6.2.4}$$

where each $F^i(y)$ is a symmetric matrix, affine in y.

We have the following result.

Theorem 6.2.1 *Consider uncertain semidefinite program with affine perturbation (6.2.4) and ellipsoidal uncertainty (6.1.3), and let $\nu_0 = 0$, $\nu_k = \sum_{s=1}^{k} n_s$. Then the semidefinite program*

$$\max\ b^T y$$

s.t.

$$(a)\quad \begin{bmatrix} S_k & \rho F_{\nu_{k-1}+1}(y) & \rho F_{\nu_{k-1}+2}(y) & \cdots & \rho F_{\nu_k}(y) \\ \rho F_{\nu_{k-1}+1}(y) & Q_k & & & \\ \rho F_{\nu_{k-1}+2}(y) & & Q_k & & \\ \vdots & & & \ddots & \vdots \\ \rho F_{\nu_k}(y) & & & & Q_k \end{bmatrix} \succeq 0,\ k = 1,2,\ldots,N; \tag{6.2.5}$$

$$(b)\quad \sum_{k=1}^{N} (S_k + Q_k) \preceq 2F_0(y);$$

in variables $y, S_1, \ldots, S_N, Q_1, \ldots, Q_N$ is an approximation of the robust counterpart (6.1.2), i.e., the projection of the feasible set of (6.2.5) on the space of y-variables is contained in the set of robust feasible solutions.

Proof. Let us fix a feasible solution $Y = (y, \{S_k\}, \{Q_k\})$ to (6.2.5), and let us set $F_i = F_i(y)$. We should prove that

(*) For every $\delta = \{\delta_i\}_{i=1}^{l}$ such that

$$\sum_{i=\nu_{k-1}+1}^{\nu_k} \delta_i^2 \le \rho^2,\ k = 1,\ldots,N, \tag{6.2.6}$$

one has

$$F_0 + \sum_i \delta_i F_i \succeq 0.$$

Since Y is feasible for (6.2.5), it follows that the matrices $F_0, S_1, \ldots, S_N, Q_1, \ldots, Q_N$ are positive semidefinite. By obvious regularization arguments, we may further assume these matrices to be positive definite. Finally, performing "scaling"

$$\begin{array}{rcl} S_k & \mapsto & F_0^{-1/2} S_k F_0^{-1/2}, \\ Q_k & \mapsto & F_0^{-1/2} Q_k F_0^{-1/2}, \\ F_i & \mapsto & F_0^{-1/2} F_i F_0^{-1/2}, \\ \Phi_0 & \mapsto & I, \end{array}$$

we reduce the situation to the one where $F_0 = I$, which we assume till the end of the proof.

Let I_k be the set of indices $\nu_{k-1}+1, \ldots, \nu_k$. Whenever δ satisfies (6.2.6) and $\xi \in \mathbf{R}^n$, n being the row size of F_i's, we have

$$\begin{array}{ll} \xi^T \left(I + \sum\limits_i \delta_i F_i \right) \xi & \\ \quad = \xi^T \xi + \sum\limits_k \left[Q_k^{1/2} \xi \right]^T \left[\sum\limits_{i \in I_k} \delta_i Q_k^{-1/2} F_i S_k^{-1/2} \xi_k \right] & \\ \quad [\xi_k = S_k^{1/2} \xi,\ k = 1, \ldots, N] & \\ \quad \geq \xi^T \xi - \sum\limits_k \|Q_k^{1/2} \xi\|_2 \left[\sum\limits_{i \in I_k} |\delta_i| \|Q_k^{-1/2} F_i S_k^{-1/2} \xi_k\|_2 \right] & \\ \quad \geq \xi^T \xi - \sum\limits_k \|Q_k^{1/2} \xi\|_2 \sqrt{\sum\limits_{i \in I_k} \rho^2 \|Q_k^{-1/2} F_i S_k^{-1/2} \xi_k\|_2^2} & \\ \quad [\text{we have used (6.2.6)}] & (6.2.7) \\ \quad = \xi^T \xi - \sum\limits_k \|Q_k^{1/2} \xi\|_2 \sqrt{\rho^2 \xi_k^T \left[\sum\limits_i \in I_k S_k^{-1/2} F_i Q_k^{-1} F_i S_k^{-1/2} \right] \xi_k} & \\ \quad \geq \xi^T \xi - \sum\limits_k \|Q_k^{1/2} \xi\|_2 \sqrt{\xi_k^T \xi_k} & \\ \quad [\text{we have used (6.2.5.}a)] & \\ \quad \geq \xi^T \xi - \sqrt{\sum\limits_k \|Q_k^{1/2} \xi\|_2^2} \sqrt{\sum\limits_k \xi_k^T \xi_k} & \\ \quad = \xi^T \xi - \sqrt{\xi^T \left[\sum\limits_k Q_k \right] \xi} \sqrt{\xi^T \left[\sum\limits_k S_k \right] \xi} & \end{array}$$

It remains to note that if $a = \xi^T \left[\sum_k Q_k\right] \xi$, $b = \xi^T \left[\sum_k S_k\right] \xi$, then $a + b \leq 2\xi^T \xi$ by (6.2.5.b) (recall that we are in the situation $F_0(y) = I$), so that $\sqrt{ab} \leq \xi^T \xi$. Thus, the concluding expression in (6.2.7) is nonnegative. ■

6.2.1 *Quality of approximation*

A general-type approximation of the robust counterpart (6.1.2) is an optimization problem

$$\max b^T y \quad \text{subject to } (y, z) \in \mathcal{Y} \tag{A}$$

(with variables y, z) such that the projection $\mathcal{X}(A)$ of its feasible set $\mathcal{Y}$ on the space of y-variables is contained in $\mathcal{X}_{\mathcal{D}}$, so that (A)-feasibility of (y, z) implies robust feasibility of y. For a particular approximation, a question of primary interest is how conservative the approximation is. A natural way to measure the "level of conservativeness" of an approximation (A) is as follows. Since (A) is an approximation of (6.1.2), we have $\mathcal{X}(A) \subset \mathcal{X}_{\mathcal{D}}$. Now let us increase the level of perturbations, i.e., let us replace the original set of perturbations $\mathcal{D}$ by its κ-enlargement $\kappa\mathcal{D}$, $\kappa \geq 1$. The set $\mathcal{X}_{\kappa\mathcal{D}}$ of robust feasible solutions associated with the enlarged set of perturbations shrinks as κ grows, and for large enough values of κ it may become a part of $\mathcal{X}(A)$. The lower bound of these "large enough" values of κ can be treated as the level of conservativeness $\lambda(A)$ of the approximation (A):

$$\lambda(A) = \inf\{\kappa \geq 1 : \mathcal{X}_{\kappa\mathcal{D}} \subset \mathcal{X}(A)\}.$$

Thus, we say that the level of conservativeness of an approximation (A) is $< \lambda$, if every y which is "rejected" by (A) (i.e., $y \notin \mathcal{X}(A)$) looses robust feasibility after the level ρ of perturbations in (6.1.3) is increased by factor λ.

The following theorem bounds the level of conservativeness of approximations we have derived so far:

Theorem 6.2.2 *Consider semidefinite program with affine perturbations* (6.2.4) *and ellipsoidal uncertainty* (6.1.3)*, the blocks δ^k of the perturbation vector being of dimensions n_k, $k = 1, \ldots, N$, and let $l = \sum_k n_k$ and n be the row size of $F(\cdot, \cdot)$. The level of conservativeness of the approximation* (6.2.5) *does not exceed* $\min\left[\sqrt{nN}; \sqrt{l}\right]$.

Proof. Of course, it suffices to consider the case of $\rho = 1$, which is assumed till the end of the proof.

Let $\mathcal{X}$ be the projection of the feasible set of (6.2.5) to the space of y-variables, and let $y \notin \mathcal{X}$. We should prove that $y \notin \mathcal{X}_{\lambda\mathcal{D}}$ at least in the following two cases:

(i.1): $\lambda > \sqrt{l}$; (i.2): $\lambda > \sqrt{nN}$.

To save notation, let us write F_i instead of $F_i(y)$. Note that we may assume that $F_0 \succeq 0$ – otherwise y is not robust feasible and there is nothing to prove. In fact we may assume even $F_0 \succ 0$, since from the structure of (6.2.5) it is clear that the relation $y \notin \mathcal{X}$, being valid for the original data $F_0, \ldots, F_l$, remains valid when we replace a positive semidefinite matrix F_0 with a close positive definite matrix. Note that this regularization may only increase the robust feasible set, so that it suffices to prove the statement in question in the case of $F_0 \succ 0$. Finally, the same scaling as in the proof of Theorem 6.2.1 allows to assume that $F_0 = I$. Let also I_k be the same index sets as in the proof of Theorem 6.2.1.

1^0. Consider the case of (i.1). Let us set

$$\begin{array}{rcl} Q_k & = & \frac{n_k}{l} I, \\ S_k & = & \frac{l}{n_k} \sum_{i \in I_k} F_i^2. \end{array}$$

The collection $(y, \{S_k, Q_k\})$ clearly satisfies (6.2.5.a), and therefore it must violate (6.2.5.b), since otherwise we would have $y \in \mathcal{X}$. Thus, there exists an n-dimensional vector ξ such that

$$\sum_{k=1}^{N} \frac{l}{n_k} \sum_{i \in I_k} \|F_i \xi\|_2^2 > \xi^T \xi. \tag{6.2.8}$$

Setting

$$p_k = \max_{i \in I_k} \|F_i \xi\|_2 = \|F_{i_k} \xi\|_2, \; i_k \in I_k,$$

we come to

$$\sum_{k=1}^{N} l p_k^2 > \xi^T \xi. \tag{6.2.9}$$

Now let $\delta \in \lambda \mathcal{D}$ be a random vector with independent coordinates distributed as follows: a coordinate δ_i with index $i \in I_k$ is zero, except the case of $i = i_k$, and the coordinate δ_{i_k} takes values $\pm \lambda$ with probabilities 1/2. The expected squared Euclidean norm of the random vector $\sum_i \delta_i F_i \xi$ clearly is equal to $\lambda^2 \sum_k p_k^2$; thus, $\lambda \mathcal{D}$ contains a perturbation δ such that

$$\| \left[\sum_{i=1}^{l} \delta_i F_i \right] \xi \|_2^2 \geq \lambda^2 \sum_k p_k^2 > \lambda^2 l^{-1} \xi^T \xi;$$

since $\lambda^2 l^{-1} > 1$ by (i.1), we conclude that the spectral norm of the matrix $\sum_i \delta_i F_i$ is > 1, whence either this matrix, or its negation is not $\preceq F_0 = I$. Thus, $y \notin \mathcal{X}_{\lambda \mathcal{D}}$, as claimed.

2^0. Now let (i.2) be the case. Let us set

$$\begin{array}{rcl} Q_k & = & N^{-1} I, \\ S_k & = & N \sum_{i \in I_k} F_i^2. \end{array}$$

By the same reasons as in 1^0, there exists an n-dimensional vector ξ such that

$$\sum_{k=1}^{N} N \sum_{i \in I_k} \|F_i \xi\|_2^2 > \xi^T \xi. \tag{6.2.10}$$

Denoting $e_1, \ldots, e_n$ the standard basic orths in $\mathbf{R}^n$, we conclude that there exists $p \in \{1, \ldots n\}$ such that

$$\sum_{k=1}^{N} N \sum_{i \in I_k} |e_p^T F_i \xi|^2 > \frac{1}{n} \xi^T \xi. \tag{6.2.11}$$

We clearly can choose $\delta \in \lambda \mathcal{D}$ in such a way that

$$\sum_{i \in I_k} \delta_i e_p^T F_i \xi = \lambda \sqrt{\sum_{i \in I_k} |e_p^T F_i \xi|^2}.$$

Setting

$$F = \sum_i \delta_i F_i,$$

we get

$$\begin{array}{rcl} e_p^T F \xi & = & \sum_k \sum_{i \in I_k} e_p^T \delta_i F_i \xi \\ & \geq & \lambda \sum_{k=1}^N \sqrt{\sum_{i \in I_k} |e_p^T F_i \xi|^2} \\ & \geq & \lambda \sqrt{\sum_{k=1}^N \sum_{i \in I_k} |e_p^T F_i \xi|^2} \\ & > & \frac{\lambda}{\sqrt{nN}} \|\xi\|_2 \quad \text{[see (6.2.11)]} \\ & > & \|\xi\|_2 \quad \text{[by (i.2)]} \end{array}$$

We conclude that the spectral norm of F is > 1, so that either F or $-F$ is not $\succeq F_0 = I$. Thus, $y \notin \mathcal{X}_{\lambda\mathcal{D}}$, as claimed. ■

6.3 RATIONAL DEPENDENCE

In this section, we seek to handle cases when the matrix-valued function $\mathbf{F}(y, \delta)$ is rational in δ for every y. There are many practical situations when the perturbation parameters enter rationally, and not affinely, in a perturbed SDP. One important example arises with the problem of checking *robust singularity* of a square (non symmetric) matrix $\mathbf{A}$, which depends affinely on parameters. One has to check if $\mathbf{A}^T\mathbf{A} \succ 0$ for every $\mathbf{A}$ in the affine uncertainty set; this is a matrix inequality condition in which the parameters enter quadratically.

We will introduce a versatile framework, called linear-fractional representations, for describing rational dependence, and devise approximate robust counterparts that are based on this linear-fractional representation. Our framework will of course cover the cases when the perturbation enters affinely in the matrix $\mathbf{F}(y, \delta)$, which is a case already covered by Theorem 6.2.1. At present we do not know if Theorem 6.2.1 always yields more accurate results than those described next, except in the case $N = 1$ (Euclidean-norm bounds), where both results are actually equivalent.

In this section, we take $\rho = 1$.

6.3.1 Linear-fractional representations

We assume that the function $\mathbf{F}$ is given by a "linear-fractional representation" (LFR):

$$\mathbf{F}(y, \Delta) = F(y) + L(y)\Delta(I - D\Delta)^{-1}R + R^T(I - \Delta^T D^T)^{-1}\Delta^T L(y)^T, \quad (a)$$
$$\Delta = \mathrm{diag}\left(\delta_1 I_{r_1}, \ldots, \delta_l I_{r_l}\right),$$

where $F(y)$ is defined in (6.1.1), $L(\cdot)$ is an affine mapping taking values in $\mathbf{R}^{n\times p}$, $R \in \mathbf{R}^{q\times n}$ and $D \in \mathbf{R}^{q\times p}$ are given matrices, and $r_1, \ldots, r_l$ are given integers. We assume that the above LFR is well-posed over $\mathcal{D}$, meaning that $\det(I - D\Delta)$ for every Δ of the form above, with $\delta \in \mathcal{D}$; we return to the issue of well-posedness later.

The above class of models seem quite specialized. In fact, these models can be used in a wide variety of situations. For example, in the case of affine dependence:

$$\mathbf{F}(\delta) = F^0(y) + \sum_{i=1}^{l} \delta_i F^i(y),$$

we can construct a linear-fractional representation, for example

$$L(y) = \frac{1}{\sqrt{2}} \begin{bmatrix} F^1(y) \\ \vdots \\ F^l(y) \end{bmatrix}^T, \quad R = \frac{1}{\sqrt{2}} \begin{bmatrix} I \\ \vdots \\ I \end{bmatrix}, \quad D = 0, \quad r_1 = \ldots = r_l = n. \tag{6.3.12}$$

Our framework also covers the case when the matrix $\mathbf{F}$ is rational in δ. The representation lemma [210], given below, illustrates this point.

Lemma 6.3.1 *For any rational matrix function* $\mathbf{M} : \mathbf{R}^l \to \mathbf{R}^{n \times c}$, *with no singularities at the origin, there exist nonnegative integers* $r_1, \ldots, r_l$, *and matrices* $M \in \mathbf{R}^{n \times c}$, $L \in \mathbf{R}^{n \times N}$, $R \in \mathbf{R}^{N \times c}$, $D \in \mathbf{R}^{N \times N}$, *with* $N = r_1 + \ldots + r_l$, *such that* $\mathbf{M}$ *has the following Linear-Fractional Representation (LFR): For all* δ *where* $\mathbf{M}$ *is defined,*

$$\mathbf{M}(\delta) = M + L\Delta\,(I - D\Delta)^{-1} R, \quad \textit{where } \Delta = \operatorname{diag}(\delta_1 I_{r_1}, \ldots, \delta_l I_{r_l}). \tag{6.3.13}$$

In the above construction, the sizes of the matrices involved are polynomial in the number l *of parameters.*

In the sequel, we denote by $\mathbf{\Delta}$ the set of matrices defined by

$$\mathbf{\Delta} = \{\Delta = \operatorname{diag}(\delta_1 I_{r_1}, \ldots, \delta_l I_{r_l}) \mid \delta \in \mathcal{D}\}. \tag{6.3.14}$$

The following developments are valid for a large class of matrix sets $\mathbf{\Delta}$.

6.3.2 Robustness analysis via Lagrange relaxations

The basic idea behind linear-fractional representation is to convert a a robustness condition such as

$$\xi^T \mathbf{F}(\Delta)\xi \geq 0 \text{ for every } \xi, \quad \Delta \in \mathbf{\Delta} \tag{6.3.15}$$

into a *quadratic* condition involving ξ and some additional variables p, q. Then, using Lagrange relaxation, we can obtain an SDP that yields a sufficient condition for robustness.

Using the LFR of $\mathbf{F}(\Delta)$, and assuming the latter is well-posed, we rewrite (6.3.15) as

$$\xi^T(F\xi + 2Lp) \geq 0, \text{for every } \xi, p, q, \Delta \text{ such that } q = R\xi + Dp, p = \Delta q, \Delta \in \mathbf{\Delta}. \tag{6.3.16}$$

where p, q are additional variables. In the above, the only non convex condition is $p = \Delta q$, $\Delta \in \mathbf{\Delta}$. It turns out that for the set $\mathcal{D}$ defined in (6.1.3) (as well as for many other sets), we can obtain a necessary and sufficient condition for $p = \Delta q$ for some $\Delta \in \mathbf{\Delta}$, in the form of a linear matrix inequality on the rank-one matrix zz^T, where

$$z = \begin{bmatrix} q \\ p \end{bmatrix} = \begin{bmatrix} R & D \\ 0 & I \end{bmatrix} \begin{bmatrix} \xi \\ p \end{bmatrix}.$$

set $\mathbf{\Delta}$	condition on p, q equivalent to $p = \Delta q$, $\Delta \in \mathbf{\Delta}$	$\Phi_{\mathbf{\Delta}}(Z)$, for $Z \in \mathcal{S}^{2N}$
$\{\Delta \mid \|\Delta\| \leq 1\}$	$q^Tq - p^Tp \geq 0$	$\mathrm{Tr}(Z_{22} - Z_{11})$
$\{sI \mid s \in \mathbf{C},\ \Re(s) \geq 0\}$	$pq^H + qp^H \succeq 0$	$Z_{12} + Z_{21}^H$
$\{\delta I \mid \delta \in \mathbf{R},\ \|\delta\| \leq 1\}$	$qq^T - pp^T \succeq 0$	$Z_{22} - Z_{11}$
$\{\mathrm{diag}(\delta_i I_{r_i})_{i=1}^l \mid \delta \in \mathbf{R}^l,\ \|\delta\|_\infty \leq 1\}$	$\mathrm{diag}(q_i^Tq_i - p_i^Tp_i)_{i=1}^l \succeq 0$	$\mathrm{diag}(Z_{22}(i) - Z_{11}(i))_{i=1}^l$
$\{\mathrm{diag}(\delta_i I_{r_i})_{i=1}^l \mid \delta \in \mathbf{R}^l,\ \|\delta\|_2 \leq 1\}$	$\mathrm{diag}(q_i^Tq_i)_{i=1}^l - pp^T \succeq 0$	$\mathrm{diag}(Z_{22}(i))_{i=1}^l - Z_{11}$

Table 6.1 Linear mappings $\Phi_{\mathbf{\Delta}}$ associated with various sets $\mathbf{\Delta} \subseteq \mathbf{R}^{N \times N}$. The notation Z_{kj}, $k, j \in \{1, 2\}$, refers to corresponding $N \times N$ blocks in matrix $Z \in \mathcal{S}^{2N}$, and $Z_{jj}(i)$ denotes the i-the diagonal block of Z_{jj}, of size $r_i \times r_i$.

Let us characterize this linear matrix inequality as

$$\Phi_{\mathbf{\Delta}}(zz^T) \succeq 0,$$

where $\Phi_{\mathbf{\Delta}}$ is a linear map from $\mathcal{S}^{2N}$ to $\mathcal{S}^N$ (recall N denotes the row size of matrix Δ, and is the size of vectors p, q). In table 6.1, we show the mappings $\Phi_{\mathbf{\Delta}}$ associated with various sets $\mathbf{\Delta}$.

Using this equivalent condition, we rewrite (6.3.16) as

$$\begin{array}{l} \mathrm{Tr} \begin{bmatrix} F & L \\ L^T & 0 \end{bmatrix} \begin{bmatrix} \xi \\ p \end{bmatrix} \begin{bmatrix} \xi \\ p \end{bmatrix}^T \geq 0, \text{ for every } \xi, p \text{ such that} \\ \Phi_{\mathbf{\Delta}}\left(\begin{bmatrix} R & D \\ 0 & I \end{bmatrix} \begin{bmatrix} \xi \\ p \end{bmatrix} \begin{bmatrix} \xi \\ p \end{bmatrix}^T \begin{bmatrix} R & D \\ 0 & I \end{bmatrix} \right) \succeq 0. \end{array} \tag{6.3.17}$$

The next step is to relax the non convex condition on p, ξ using Lagrange relaxation. Previous condition is true if there exist a positive semidefinite matrix S such that for every ξ, p, we have

$$\begin{bmatrix} \xi \\ p \end{bmatrix}^T \begin{bmatrix} F & L \\ L^T & 0 \end{bmatrix} \begin{bmatrix} \xi \\ p \end{bmatrix} \geq \operatorname{Tr} S \Phi_{\mathbf{\Delta}} \left(\begin{bmatrix} R & D \\ 0 & I \end{bmatrix} \begin{bmatrix} \xi \\ p \end{bmatrix} \begin{bmatrix} \xi \\ p \end{bmatrix}^T \begin{bmatrix} R & D \\ 0 & I \end{bmatrix} \right),$$

or, equivalently,

$$S \succeq 0, \quad \begin{bmatrix} F & L \\ L^T & 0 \end{bmatrix} \succeq \begin{bmatrix} R & D \\ 0 & I \end{bmatrix}^T \Phi_{\mathbf{\Delta}}^*(S) \begin{bmatrix} R & D \\ 0 & I \end{bmatrix}, \tag{6.3.18}$$

where $\Phi_{\mathbf{\Delta}}^*$ is the dual of $\Phi_{\mathbf{\Delta}}$.

Note that the dual of the above LMI condition amounts to enforce condition (6.3.17) not only on rank-one matrices, but for every positive semidefinite matrix.

Theorem 6.3.1 *149 Consider the uncertain semidefinite program with rational perturbation, described by the LFR (6.3.1), where the perturbation vector lies in an arbitrary set $\mathcal{D} \subseteq \mathbf{R}^l$. Let $\mathbf{\Delta}$ be defined by (6.3.14), and assume the LFR is well-posed over $\mathbf{\Delta}$. Assume that we can associate to this set a linear mapping $\Phi_{\mathbf{\Delta}}$ such that*

$$p = \Delta q \text{ for some } \Delta \in \mathbf{\Delta} \text{ if and only if } \Phi_{\mathbf{\Delta}} \left(\begin{bmatrix} q \\ p \end{bmatrix} \begin{bmatrix} q \\ p \end{bmatrix}^T \right) \succeq 0,$$

and denote by $\Phi_{\mathbf{\Delta}}^*$ the dual of this map.

Then the semidefinite program

$$\max\ b^T y \text{ subject to}$$
$$S \succeq 0, \quad \begin{bmatrix} F(y) & L(y) \\ L(y)^T & 0 \end{bmatrix} \succeq \begin{bmatrix} R & D \\ 0 & I \end{bmatrix}^T \Phi_{\mathbf{\Delta}}^*(S) \begin{bmatrix} R & D \\ 0 & I \end{bmatrix}$$

in variables y, S, is an approximation of the robust counterpart (6.1.2), i.e., the projection of the feasible set of (6.1.2) on the space of y-variables is contained in the set of robust feasible solutions.

It is now a simple matter to specialize the above result for the set $\mathcal{D}$ defined in (6.1.3). The condition on p, q that is equivalent to $p = \Delta q$, $\Delta \in \mathbf{\Delta}$ writes

$$\operatorname{diag}(q_i q_i^T)_{i \in I_k} - p^k (p^k)^T \succeq 0 \quad k = 1, \ldots, N, \tag{6.3.19}$$

where I_k be the set of indices $\nu_{k-1} + 1, \ldots, \nu_k$, with $\nu_0 = 0$, $\nu_k = \sum_{s=1}^k n_s$, and p^k is the vector with elements $(p_i)_{i \in I_k}$.

The following result is then a corollary of Theorem 6.3.1.

Corollary 6.3.1 *Consider the uncertain semidefinite program with rational perturbation, described by the LFR (6.3.1), where the perturbation vector lies* *in the set $\mathcal{D}$ defined in (6.1.3). Let $\mathbf{\Delta}$ be defined by (6.3.14), and assume the LFR is well-posed over $\mathbf{\Delta}$.*

Consider the semidefinite program

$$\max\ b^T y \text{ subject to } S \succeq 0,$$
$$\begin{bmatrix} F(y) & L(y) \\ L(y)^T & 0 \end{bmatrix} \succeq \begin{bmatrix} R & D \\ 0 & I \end{bmatrix}^T \begin{bmatrix} T & 0 \\ 0 & -S \end{bmatrix} \begin{bmatrix} R & D \\ 0 & I \end{bmatrix} \qquad (6.3.20)$$

where $S = \operatorname{diag}(S_1, \ldots, S_N)$, with each S_i of size $\sum_{i \in I_k} r_i$, and T is the block-diagonal matrix formed with the block-diagonal $r_i \times r_i$ blocks of S.

Then the above semidefinite program in variables y, S, is an approximation of the robust counterpart (6.1.2), i.e., the projection of the feasible set of (6.1.2) on the space of y-variables is contained in the set of robust feasible solutions.

Remark 6.3.1 *We note that the above condition, if it is strictly enforced, ensures well-posedness, meaning that $\det(I - D\Delta) \neq 0$ for every $\Delta \in \mathbf{\Delta}$. (To prove this, it suffices to apply the previous methodology to the matrix function $\mathbf{F}(\Delta) = (I - D\Delta)^T(I - D\Delta)$.)*

6.3.3 Comparison with earlier results

We do not have a general comparison theorem with the results of Section 6.2, in the case of affine dependence. However, we can prove that, when $\mathcal{D}$ represents Euclidean-norm bounds (that is, there is only one block: $N = 1$), then both results are equivalent.

Indeed, assume that $\mathbf{F}(\delta, y)$ has the form (6.2.4); a linear-fractional representation of this dependence is (6.3.1), with $F = F_0$, $D = 0$, and L, R given by (6.3.12). The linear matrix inequality (6.3.20) then involves a full matrix S of row size n^2, and writes (dropping the dependence on y, and exchanging L^T and R without loss of generality)

$$\begin{bmatrix} F_0 - \frac{1}{2}\sum_{i=1}^{l} F_i S_i F_i & \frac{1}{\sqrt{2}} I & \ldots & \frac{1}{\sqrt{2}} I \\ \frac{1}{\sqrt{2}} I & S_1 & * & * \\ \vdots & * & \ddots & * \\ \frac{1}{\sqrt{2}} I & * & * & S_l \end{bmatrix} \succeq 0,$$

where S_i are the $n \times n$ diagonal blocks of S, and the symbols $*$ refer to the other blocks of S. Using the elimination lemma [137], it is possible to get rid of these elements and rewrite the above as

$$\begin{bmatrix} F_0 - \frac{1}{2}\sum_{i=1} F_i S_i F_i & \frac{1}{\sqrt{2}} I \\ \frac{1}{\sqrt{2}} I & S_k \end{bmatrix} \succeq 0, \quad k = 1, \ldots, l.$$

Assuming (without loss of generality) that each S_k is positive definite, and setting

$$Q = 2F_0 - \sum_{i=1}^{l} F_i S_i F_i,$$

we get $Q \succeq S_k^{-1}$ for every k, and hence

$$2F_0 \succeq Q + \sum_{i=1}^{l} F_i Q^{-1} F_i,$$

which is precisely the result obtained from Theorem 6.2.1 , in the case $N = 1$.

6.4 SPECIAL CASES

In this section, we focus on several special cases when the above results yield "computationally tractable" *equivalent* forms of the robust counterpart rather than merely "tractable approximations" of it.

6.4.1 Linear programming with affine uncertainty

Linear programming can be treated as a very special case of Semidefinite Programming; here all considerations related to robust counterpart become especially simple. For self-contained derivation of the below results, see [84].

Consider an LP program in the form

$$\min c^T x \quad \text{subject to } Ax + b \geq 0 \tag{6.4.21}$$

"Simple" ellipsoidal uncertainty.. Let us start with the case when the data $[A, b] \in \mathbf{R}^{m\times(n+1)}$ in (6.4.21) are affinely parameterized by perturbation vector varying in an "elliptic cylinder" – the direct sum of an ellipsoid and a linear space:

$$[A; b] \in \mathcal{U} = \left\{ `[A^0; b^0] + \sum_{j=1}^{k} \xi_j [A^j; b^j] + \sum_{p=1}^{q} \zeta_p [C^p; d^p] : (\xi, \zeta) \in \mathcal{D} = \{\xi^T \xi \leq 1\} \right\}. \tag{6.4.22}$$

For this case the robust counterpart

$$\min c^T x \quad \text{subject to } Ax + b \geq 0 \quad \text{for all } [A; b] \in \mathcal{U} \tag{6.4.23}$$

of (6.4.21) is the program

$$\begin{array}{ll} & \min c^T x \\ \text{s.t.} & \\ & \begin{array}{ll} C_i^p x + d_i^p & = 0, \; p = 1, \ldots, q, \\ & i = 1, \ldots, m; \\ A_i^0 x + b_i^0 & \geq \sqrt{\sum_{j=1}^k (A_i^j x + b_i^j)^2}, \\ & i = 1, 2, \ldots, m; \end{array} \end{array} \tag{6.4.24}$$

here B_i denotes i-th row (treated as a row vector) of a matrix B.

Case of $\bigcap$-ellipsoidal uncertainty.. Now assume that the set $\mathcal{U}$ possible values of the data $[A;b]$ of uncertain LP problem (6.4.21) is intersection of finitely many sets of the form (6.4.22):

$$\begin{aligned}
[A;b] \in \mathcal{U} &= \bigcap_{s=0}^{t} \mathcal{U}_s, \\
\mathcal{U}_s &= \{[A^{s0};b^{s0}] + \sum_{j=1}^{k_s} \xi_j [A^{sj};b^{sj}] + \sum_{p=1}^{q_s} \zeta_p [C^{sp};d^{sp}] \mid \\
&\quad \xi^T\xi \le 1\}, \\
&\quad s = 0,1,\ldots,t.
\end{aligned} \tag{6.4.25}$$

Assume also that the set $\mathcal{U}$ is bounded and the ellipsoids $\mathcal{U}_s$ satisfy the following "Slater condition":

(*) *There exists $[A';b'] \in \mathcal{U}$ which, for every $s \le t$, can be represented* as

$$[A;b] = [A^{s0};b^{s0}] + \sum_{j=1}^{k_s} \xi_j^s [A^{sj};b^{sj}] + \sum_{p=1}^{q_s} \zeta_p^j [C^{sp};d^{sp}]$$

with $[\xi^s]^T\xi^s < 1$.

For this case the robust counterpart (6.4.23) of the uncertain LP program (6.4.21) is the program

$$\begin{aligned}
&\min\ c^T x \\
\text{s.t.}\ & \\
&C_i^{sp}\lambda^{is} + d_i^{sp}\mu^{is} = 0,\ 1 \le i \le m, 1 \le s \le t, 1 \le p \le q_s; \\
&C_i^{0p}x + d_i^{0p} = \sum_{s=1}^{t}(C_i^{0p}\lambda^{is} + d_i^{0p}\mu^{is}), \\
&i = 1,\ldots,m, p = 1,\ldots,q_0; \\
&\sum_{s=1}^{t}\{(A_i^{s0} - A_i^{00})\lambda^{is} + (b_i^{s0} - b_i^{00})\mu^{is}\} \\
&+A_i^{00}x + b_i^{00} \ge \left\| \begin{pmatrix} A_i^{01}x + b_i^{01} - \sum_{s=1}^{t}\{A_i^{01}\lambda^{is} + b_i^{01}\mu^{is}\} \\ A_i^{02}x + b_i^{02} - \sum_{s=1}^{t}\{A_i^{02}\lambda^{is} + b_i^{02}\mu^{is}\} \\ \cdots \\ A_i^{0k_0}x + b_i^{0k_0} - \sum_{s=1}^{t}\left\{A_i^{0k_0}\lambda^{is} + b_i^{0k_0}\mu^{is}\right\} \end{pmatrix} \right\|_2 \\
&+\sum_{s=1}^{t} \left\| \begin{pmatrix} A_i^{s1}\lambda^{is} + b_i^{01}\mu^{is} \\ A_i^{s2}\lambda^{is} + b_i^{02}\mu^{is} \\ \cdots \\ A_i^{sk_s}\lambda^{is} + b_i^{0k_s}\mu^{is} \end{pmatrix} \right\|_2, \\
&i = 1,\ldots,m,
\end{aligned}$$

in variables $x \in \mathbf{R}^n, \lambda^{is} \in \mathbf{R}^n, \mu^{is} \in \mathbf{R}, i = 1,\ldots,m, s = 1,\ldots,t$, see [84].

It is worthy of mentioning that the case of $\bigcap$-ellipsoidal uncertainty basically covers the case of affine perturbations with ellipsoidal uncertainty (6.2.4) – (6.1.3). Indeed, assume that the affine mapping (6.2.4) from the space of perturbation vectors to the space of data is an embedding. Since the set $\mathcal{D}$ given by (6.1.3) clearly is an intersection of elliptic cylinders, its image under the above embedding – i.e., the set of possible values of the perturbed data – is a $\bigcap$-ellipsoidal set. Note that this set clearly satisfies the "Slater condition" (*).

6.4.2 *Robust quadratic programming with affine uncertainty*

A (convex) quadratic problem is an optimization program of the form

$$\min c^T x \quad \text{subject to } -x^T[A^i]^T[A^i]x + 2[b^i]^T x + \gamma^i \geq 0, \ i = 1, \ldots, m; \quad (6.4.26)$$

such a problem can be easily reformulated as an SDP program with the data affinely depending on the data $\{A^i, b^i, \gamma^i\}_{i=1}^m$ of the original problem. Assume that the data (A^i, b^i, γ^i) of every quadratic constraint are uncertain and vary in respective ellipsoids:

$$(A^i, b^i, \gamma^i) \in \mathcal{U}_i = \left\{ (A^i, b^i, \gamma^i) = (A^{i0}, b^{i0}, \gamma^{i0}) + \sum_{j=1}^{k} \delta_j (A^{ij}, b^{ij}, \gamma^{ij}) \mid \delta^T \delta \leq 1 \right\}. \quad (6.4.27)$$

It turns out that the robust counterpart of uncertain conic quadratic program (6.4.26) – (6.4.27) is *equivalent* to the explicit semidefinite program as follows:

$$\min c^T x$$
$$\text{s.t.}$$
$$\left[\begin{array}{c|cccc|c}
\gamma^{i0} + 2x^T b^{i0} - \lambda^i & \frac{\gamma^{i1}}{2} + x^T b^{i1} & \frac{\gamma^{i2}}{2} + x^T b^{i2} & \cdots & \frac{\gamma^{ik}}{2} + x^T b^{ik} & [A^{i0}x]^T \\
\hline
\frac{\gamma^{i1}}{2} + x^T b^{i1} & \lambda^i & & & & [A^{i1}x]^T \\
\frac{\gamma^{i2}}{2} + x^T b^{i2} & & \lambda^i & & & [A^{i2}x]^T \\
\vdots & & & \ddots & & \cdots \\
\frac{\gamma^{ik}}{2} + x^T b^{ik} & & & & \lambda^i & [A^{ik}x]^T \\
\hline
A^{i0}x & A^{i1}x & A^{i2}x & \cdots & A^{ik}x & I_{l_i}
\end{array}\right] \succeq 0,$$
$$i = 1, 2, \ldots, m \quad (6.4.28)$$

with variables x and additional scalar variables $\lambda^1, \ldots, \lambda^m$, I_l being the unit $l \times l$ matrix. The result may be derived from Theorem 6.4.1 (for independent proof, see [87]).

6.4.3 *Robust conic quadratic programming*

A conic quadratic program is an optimization program of the form

$$\min c^T x \quad \text{subject to } \left\|A^i x + b^i\right\|_2 \leq [d^i]^T x + \gamma^i, \ i = 1, 2, \ldots, m\}. \quad (6.4.29)$$

such a problem can be easily reformulated as a semidefinite program with the data affinely depending on the data $\{A^i, b^i, d^i, \gamma^i\}_{i=1}^m$ of the original problem. Assume that the data of (6.4.29) are uncertain and that the uncertainty is of the following specific type:

- (I) The uncertainty is "constraint-wise": the data $(A^i, b^i, d^i, \gamma^i)$ of different conic quadratic constraint independently of each other run through respective uncertainty sets $\mathcal{U}_i$;
- (II) For every i, $\mathcal{U}_i$ is the direct product of two elliptic cylinders in the spaces of (A^i, b^i)- and (d^i, γ^i)-components of the data:

$$\begin{array}{rcl}
\mathcal{U}_i & = & \mathcal{V}_i \times \mathcal{W}_i, \\
\mathcal{V}_i & = & \{[A^i; b^i] = [A^{i0}; b^{i0}] + \sum\limits_{j=1}^{k_i} \xi_j [A^{ij}; b^{ij}] \\
& & + \sum\limits_{p=1}^{q_i} \zeta_p [E^{ip}; f^{ip}] \mid \xi^T \xi \leq 1\}, \\
\mathcal{W}_i & = & \{(d^i, \gamma^i) = (d^{i0}, \gamma^{i0}) + \sum\limits_{j=1}^{k'_i} \xi_j (d^{ij}, \gamma^{ij}) \\
& & + \sum\limits_{p=1}^{q'_i} \zeta_p (g^{ip}, h^{ip}) \mid \xi^T \xi \leq 1\}.
\end{array} \tag{6.4.30}$$

It turns out that the robust counterpart of uncertain conic quadratic program (6.4.29) – (6.4.30) is *equivalent* to the explicit semidefinite program as follows:

$$\begin{array}{l}
\min\ c^T x \\
\text{s.t.} \\
E^{ip} x + f^{ip} = 0,\ i = 1, \dots, m, p = 1, \dots, q_i; i = 1, \dots, m, p = 1, \dots, q'_i; \\
\begin{bmatrix}
[d^{i0}]^T x + \gamma^{i0} - \lambda^i & [d^{i1}]^T x + \gamma^{i1} & [d^{i2}]^T x + \gamma^{i2} & \cdots & [d^{ik'_i}]^T x + \gamma^{ik'_i} \\
[d^{i1}]^T x + \gamma^{i1} & [d^{i0}]^T x + \gamma^{i0} - \lambda^i & & & \\
[d^{i2}]^T x + \gamma^{i2} & & [d^{i0}]^T x + \gamma^{i0} - \lambda^i & & \\
\vdots & & & \ddots & \cdots \\
[d^{ik'_i}]^T x + \gamma^{ik'_i} & & & \cdots & [d^{i0}]^T x + \gamma^{i0} - \lambda^i
\end{bmatrix} \succeq 0, \\
i = 1, \dots, m; \\
\left[\begin{array}{c|cccc|c}
\lambda^i - \mu^i & & & \cdots & & [A^{i0} x + b^{i0}]^T \\
\hline
& \mu^i & & & & [A^{i1} x + b^{i1}]^T \\
& & \mu^i & & & [A^{i2} x + b^{i2}]^T \\
\vdots & & & \ddots & & \cdots \\
& & & & \mu^i & [A^{ik_i} x + b^{ik_i}]^T \\
\hline
A^{i0} x + b^{i0} & A^{i1} x + b^{i1} & A^{i2} x + b^{i2} & \cdots & A^{ik_i} x + b^{ik_i} & \lambda^i I_{l_i}
\end{array}\right] \succeq 0, \\
i = 1, \dots, m
\end{array} \tag{6.4.31}$$

with variables x and additional scalar variables λ^i, μ^i, $i = 1, \dots, m$.

The result again can be derived from Theorem 6.4.1 (for independent derivation, see [87]).

6.4.4 *Operator-norm bounds*

In several problems, the perturbed constraint writes as $\mathbf{F}(y, \Delta) \succeq 0$, where $\mathbf{F}(y, \Delta)$ is given in the linear-fractional form (6.3.1.a), and Δ is a matrix bounded in norm but otherwise arbitrary. We can handle this case with the Lagrange relaxation technique described in Section 6.3.2; the set $\boldsymbol{\Delta}$ has the form

$$\boldsymbol{\Delta} = \left\{ \Delta \in \mathbf{R}^{n_p \times n_q} \mid \|\Delta\| \leq 1 \right\}, \tag{6.4.32}$$

and the corresponding linear map $\Phi_{\boldsymbol{\Delta}}$ is given in table 6.1. It turns out that the resulting SDP approximation is exact in this case (this result is proved in [212]).

Theorem 6.4.1 *Consider the uncertain semidefinite program with rational perturbation, described by the LFR (6.3.1), where the perturbation matrix is bounded but otherwise arbitrary:* $\Delta \in \boldsymbol{\Delta}$, *where* $\boldsymbol{\Delta}$ *is defined in (6.4.32).*

Then the semidefinite program

$$\max\ b^T y \text{ subject to}$$
$$\tau \geq 0, \quad \begin{bmatrix} F(y) & L(y) \\ L(y)^T & 0 \end{bmatrix} \succeq \begin{bmatrix} R & D \\ 0 & I \end{bmatrix}^T \begin{bmatrix} \tau I_{n_q} & 0 \\ 0 & -\tau I_{n_p} \end{bmatrix} \begin{bmatrix} R & D \\ 0 & I \end{bmatrix}$$

in variables y, τ, *is equivalent to the robust counterpart (6.1.2).*

Remark 6.4.1 *We note that the above condition is strictly feasible if and only if the LFR is well-posed, meaning that* $\det(I - D\Delta) \neq 0$ *for every* $\Delta \in \mathbf{R}^{n_p \times n_q}$, $\|\Delta\| \leq 1$.

6.5 EXAMPLES

6.5.1 *A link with combinatorial optimization*

The method we have outlined is a way to solve a non convex optimization problem. It turns out that this method is similar in spirit to the one used in SDP relaxations for combinatorial optimization.

Consider the problem

$$\max_{\delta \in \mathbf{R}^l} \delta^T W \delta \text{ subject to } \delta_i^2 = 1, \quad i = 1, \ldots, l, \tag{6.5.33}$$

where W is a given symmetric matrix (of special structure, irrelevant here). The above problem is known in the combinatorial optimization literature as "the maximum cut" (MAX-CUT) problem [285], and is proven to be NP-hard.

This problem is a robust SDP problem. First we note that, without loss of generality, we may assume $W \succ 0$, and rewrite the problem as

$$\text{minimize } x \text{ subject to } \delta^T W_c \delta \leq x \text{ for every } \delta,\ \|\delta\|_\infty \leq 1.$$

Let us now apply Theorem 6.3.1, with

$$\mathbf{F}(y,\Delta) = \begin{bmatrix} x & \delta^T \\ \delta & W^{-1} \end{bmatrix}, \quad \Delta = \mathbf{diag}(\delta_1,\ldots,\delta_l).$$

The matrix $\mathbf{F}(y,\Delta)$ can be written in the LFR format as (6.3.1), with

$$F(y) = \begin{bmatrix} x & 0 \\ 0 & W^{-1} \end{bmatrix}, \quad R = \left[\begin{array}{c|c} 1 & \\ \vdots & 0 \\ 1 & \end{array}\right], \quad L = \begin{bmatrix} 0 \\ I \end{bmatrix}, \quad D = 0.$$

The perturbation set $\mathbf{\Delta}$ here is the set of diagonal $p \times p$ matrices. Theorem 6.3.1 shows that an upper bound on the MAX-CUT problem is given by the optimal value of the SDP

$$\min \operatorname{Tr} S \text{ subject to } W \preceq S, \; S \text{ diagonal.}$$

This upper bound is exactly the one obtained by Lovasz [496], and is dual (in the SDP sense) to the one obtained by Goemans and Williamson [285]. As shown in [285], the above relaxation is the most efficient currently available (in terms of closeness to the actual optimum, in a certain stochastic sense).

6.5.2 A link with Lyapunov theory in control

Semidefinite programming has many applications in control theory [137]. The basic idea is that using so-called quadratic Lyapunov functions, we may prove a number of interesting properties for uncertain dynamical systems; the search for quadratic Lyapunov functions $V(\xi) = \xi^T S \xi$ can be often written as an SDP in the matrix S.

Here we would like to show the link between classical Lyapunov theory and the Lagrange relaxations we used in Section 6.3.2. To illustrate this link, consider the problem of checking stability (convergence to zero of every trajectory) of the dynamical system

$$\dot{q} = Aq,$$

where $q \in \mathbf{R}^n$ is the state, and A is a (constant) square matrix. Taking the Laplace transform, we obtain that stability is equivalent to matrix A having no eigenvalues with positive real part:

$$(sI - A)^H (sI - A) \succ 0 \text{ for every } s, \;\; s + s^* \geq 0.$$

The above is a robustness analysis problem (the Laplace variable s is the uncertainty). Introduce $p = sq$, we have

$$p = sq \text{ for some } s, \;\; s + s^* \geq 0$$

if and only if $pq^H + qp^H \succeq 0$. Our problem is thus to check if

$$\|p - Aq\|^2 > 0 \text{ for every } (p,q) \neq (0,0), \;\; pq^H + qp^H \succeq 0.$$

Using Lagrange relaxation of the last matrix inequality constraint, we obtain a sufficient condition for stability: There exists a real matrix S such that

$$A^T S + SA \prec 0, \quad S \succ 0.$$

The above condition is the Lyapunov condition for stability that is well known in control (it turns out that this condition is necessary and sufficient if A is known and constant). If S satisfies this inequality, the quadratic function $V(\xi) = \xi^T S \xi$ can be interpreted as a Lypaunov function proving stability (that is, V decreases along every trajectory). The above is easily extended to the case when the matrix A is uncertain (see [137]).

6.5.3 Interval computations

A basic problem in interval computations is the following. We are given a function $\mathbf{f}$ from $\mathbf{R}^l$ to $\mathbf{R}^m$, and a set confidence $\mathcal{D}$ for $\delta \in \mathbf{R}^l$, in the form of a product of intervals. We seek to estimate intervals of confidence for the components of $x = \mathbf{f}(\delta)$ when δ ranges $\mathcal{D}$. Sometimes, $\mathbf{f}$ is given in implicit form, as in the interval linear algebra problem: here, we are given matrices $\mathbf{A} \in \mathbf{R}^{n \times n}$, $\mathbf{b} \in \mathbf{R}^n$ the elements of which are only known within intervals; in other words, $[\mathbf{A}\ \mathbf{b}]$ is only known to belong to an "interval matrix set" $\mathcal{U}$. we seek to compute *intervals of confidence* for the set of solutions, if any, to the equation $\mathbf{A}x = \mathbf{b}$.

Obtaining exact estimates for intervals of confidence for the elements of solutions x, even for the "linear interval algebra" problem, is already NP-hard [681, 682].

One classical approach to this problem resorts to interval calculus, where each one of the basic operations $(+, -, x, /)$ is replaced by an "interval counterpart", and standard (eg LU) linear algebra algorithms are adapted to this new "algebra". Many refinements of this basic idea have been proposed, but the algorithms based on this idea have in general exponential complexity.

Robust semidefinite programming can be used (at least as a subproblem in a global branch and bound method) for this problem, as follows. Assume we can describe $\mathbf{f}$ explicitly as a rational function of its arguments; from Lemma 6.3.1, we can construct (in polynomial time) a linear-fractional representation of $\mathbf{f}$, in the form

$$\mathbf{f}(\delta) = f + L\Delta\,(I - D\Delta)^{-1}\,r, \text{ where } \Delta = \operatorname{diag}(\delta_1 I_{r_1}, \ldots, \delta_l I_{r_l})\,.$$

Assume first that we seek an ellipsoid of confidence for the solution, in the form $\mathcal{E} = \{x \mid (x - x_0)(x - x_0)^T \preceq P\}$, where $x_0 \in \mathbf{R}^n$ and $P \succeq 0$ (our parametrization allows for degenerate, "flat", ellipsoids, to handle cases when some components of the solution are certain). We seek to minimize the "size" of $\mathcal{E}$ subject to $\mathbf{f}(\delta) \in \mathcal{E}$ for every $\delta \in \mathcal{D}$. Measuring the size of $\mathcal{E}$ by $\operatorname{Tr} P$ (other measures are possible, as seen below), we obtain the following equivalent formulation of the problem.

$$\min_{x_0, P} \operatorname{Tr} P \text{ subject to } \begin{bmatrix} P & (\mathbf{f}(\delta) - x_0) \\ (\mathbf{f}(\delta) - x_0)^T & 1 \end{bmatrix} \succeq 0 \text{ for every } \delta \in \mathcal{D}. \tag{6.5.34}$$

The above is obviously a robust semidefinite programming problem, for which an explicit SDP counterpart (approximation) can be devised, provided $\mathcal{D}$ takes the form of a (general) ellipsoidal set. (A typical set arising in interval calculus is a product of intervals $\Pi[\underline{\delta}_i\ \overline{\delta}_i]$, where $\underline{\delta}_i$, $\overline{\delta}_i$ are given.)

The above method finds ellipsoids of confidence, but it is also possible to find intervals of confidence for the components of $\mathbf{f}(\delta)$, by modifying the objective of the above robust SDP suitably (for example, if we minimize the $(1,1)$ component of the matrix variable P instead of its trace, we will obtain an interval of confidence for the first component of $\mathbf{f}(\delta)$, when δ ranges $\mathcal{D}$).

The resulting approximations have an interesting interpretation in the context of the "linear interval algebra problem" $\mathbf{A}x = \mathbf{b}$, where $[\mathbf{A}\ \mathbf{b}]$ is an uncertain matrix, subject to "unstructured perturbations". Assume

$$[\mathbf{A}\ \mathbf{b}] \in \mathcal{U} = \{[A + \Delta A\ b + \Delta b] \mid \|[\Delta A\ \Delta b]\| \le \rho\},$$

where $[A\ b] \in \mathbf{R}^{n \times (m+1)}$ and $\rho \ge 0$ are given. In this case, our results are exact, and yield a solution related to the notion of *total least squares* developed by Golub and Van Loan [292, 803]. Precisely, it can be shown that the center of the ellipsoid of confidence (corresponding to the variable x_0 in problem (6.5.34)) is of the form

$$x_0 = (A^T A - \rho^2 I)^{-1} A^T b$$

(We assume that $\sigma_{\min}([A\ b]) \ge \rho$, otherwise the ellipsoid of confidence is unbounded. Except in degenerate cases, this guarantees the existence of the inverse in the above.) When we let $\rho = \sigma_{\min}([A\ b])$, the ellipsoid of confidence can be shown to be reduced to the singleton $\mathcal{E} = \{x_0\}$, and x_0 is the "total least squares" solution to the problem $Ax = b$. (See [207] for details.)

As an example, consider the Vandermonde system

$$\begin{bmatrix} 1 & \mathbf{a}_1 & \mathbf{a}_1^2 \\ 1 & \mathbf{a}_1 & \mathbf{a}_1^2 \\ 1 & \mathbf{a}_1 & \mathbf{a}_1^2 \end{bmatrix} \begin{bmatrix} x_1 \\ x_2 \\ x_3 \end{bmatrix} = \begin{bmatrix} \mathbf{b}_1 \\ \mathbf{b}_2 \\ \mathbf{b}_3 \end{bmatrix},$$

where $\mathbf{a}$ $\mathbf{b}$ are interval vectors of $\mathbf{R}^3$.

In Figure 6.5.3, we show the box of confidence for the solution, computed by direct application of interval algebra; the right-hand side plots shows the ellipsoid of confidence obtained by robust semidefinite programming. We did not use elaborate algorithms to solve the problem via interval algebra, so the reader should not draw negative conclusions about it; rather, the instructive part is that the robust SDP method seems to behave well in this example.

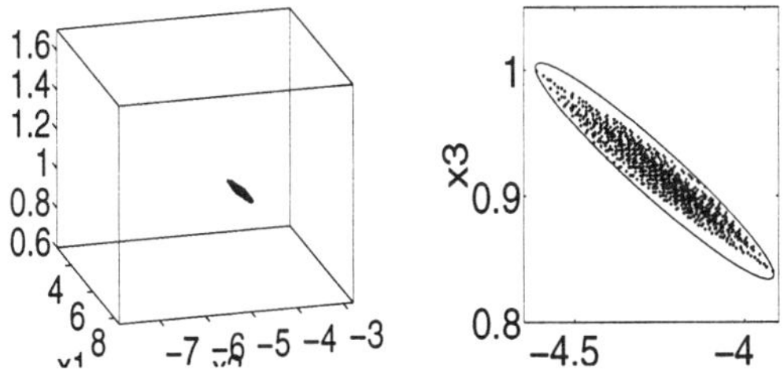

interval calculus *ellipsoid calculus*

Figure 6.1 Sets of confidence for an uncertain Vandermonde system.

6.5.4 *Worst-case simulation for uncertain dynamical systems*

The above ideas have been extended to do *worst-case simulation* for uncertain dynamical systems in [208, 208, 209]. Consider the uncertain, discrete-time dynamical system

$$x(k+1) = \mathbf{A}(k)x(k) + \mathbf{b}(k), \quad k = 0, 1, 2, \ldots$$

where the time-varying matrix $[\mathbf{A}(k)\ \mathbf{b}(k)]$ is only known to belong to a given set $\mathcal{U}$ (we assume $\mathcal{U}$ to be constant for simplicity only). We assume that the initial condition is only known to belong to a given ellipsoid $\mathcal{E}(0)$.

The idea developed in [209] is to compute recursively a sequence of ellipsoids of confidence $\mathcal{E}(k)$, for a given time horizon $k = 1, \ldots, N$. Each step is a robust semidefinite program of the type encountered in the previous section.

As an example, we consider a system described by an "interval AR" process

$$\xi(n+k) + \alpha_1(k)\xi(n+k-1) + \ldots + \alpha_n(k)\xi(k) = 0,$$

where each time-varying α_i is only known to belong to a given interval $[\alpha_i^- \ \alpha_i^+]$, where $\alpha_i^\pm = \alpha_i^{\text{nom}} \pm \rho$. Here, α^{nom} corresponds to the nominal system, and $\rho \geq 0$ measures the uncertainty size. Figure 6.5.4 illustrates typical results, on a system of order $n = 20$. We show the nominal time response and associated worst-case simulation bounds for the output vector $z(k) = (\xi(k) + \ldots + \xi(n+k-1))/n$, for three values of $\rho = 10^{-4}$, $2.5 10^{-4}$ and $5 10^{-4}$.

6.5.5 *Robust structural design*

A typical problem of (static) structural design is to specify a mechanical construction capable best of all withstand a given external load. As a concrete example of this type, consider the *Truss Topology Design* (TTD) problem (for more details, see [86]).

A *truss* is a construction comprised of thin elastic *bars* linked with each other at *nodes* – points from a given finite (planar or spatial) set. When subjected to a given *load* – a collection of external forces acting at some specific nodes

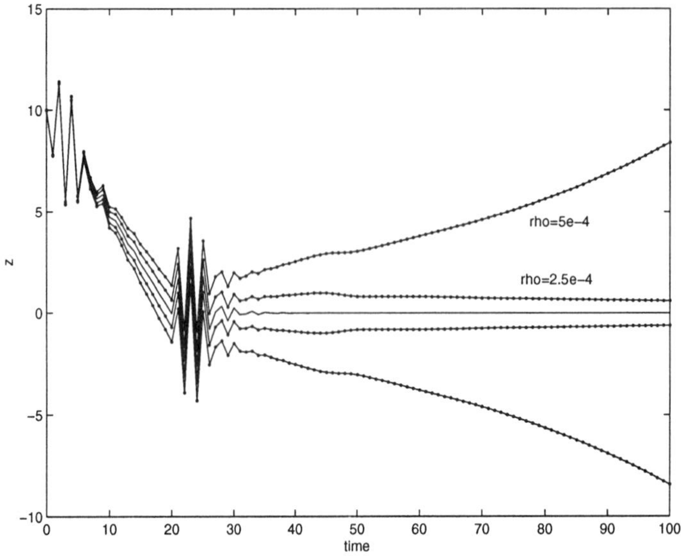

Figure 6.2 Worst-case simulation for an "interval AR" process of order 20 on a time horizon of $T = 100$ steps, for increasing values of the uncertainty size ρ.

– the construction is deformed, until the tensions caused by the distortion compensate the external load. The distorted truss capacitates certain potential energy, and this energy – the *compliance* – measures stiffness of the truss (its ability to withstand the load); the less is compliance, the more rigid is the truss.

In the usual TTD problem we are given the initial nodal set, the external "nominal" load and the total volume of the bars. The goal is to allocate this resource to the bars in order to minimize the compliance of the resulting truss. Mathematically the TTD problem can be modeled by the following semidefinite program:

$$\begin{array}{lrcl} & \min \tau & & \\ \text{s.t.} & & & \\ (a) & \begin{bmatrix} \tau & f^T \\ f & \sum_{i=1}^n t_i b_i b_i^T \end{bmatrix} & \succeq & 0, \\ (b) & t & \in & P \subset \mathbf{R}_+^n, \end{array} \tag{6.5.35}$$

with design variables $\tau \in \mathbf{R}$ and $t = (t_1, \ldots, t_n) \in \mathbf{R}^n$; t_i's are volumes of tentative bars. The data of the problem are

- vectors $b_i \in \mathbf{R}^m$; they are readily given by the geometry of the nodal set;
- vector $f \in \mathbf{R}^m$ representing the external load;
- a polytope P representing design restrictions like upper bound on the total bar volume, bounds on volumes of particular bars, etc.

In reality, the external load f should be treated as uncertain element of the data; the traditional approach to treat this uncertainty is to consider a number of "loads of interest" $f_1, \ldots, f_k$ and to optimize the worst-case, over this set of scenarios, compliance. Mathematically this approach is equivalent to replacing the LMI (6.5.35.a) (expressing the fact that τ is an upper bound on the compliance with respect to f) by k similar LMI's corresponding to $f = f_1, f = f_2, \ldots, f = f_k$. A disadvantage of the "scenario approach" is that it takes care just of a restricted number of "loads of interest" and ignores "occasional" loads, even small ones; as a result, there is a risk that the resulting construction will be crushed by a small "bad" load. An example of this type is depicted on Fig. 1. Fig. 1.a) shows a cantilever arm which withstands optimally the nominal load – the unit force f^* acting down at the most right node. The corresponding "nominal" optimal compliance is 1. It turns out, however, that the construction in question is highly instable: a small force f (10 times smaller than f^*) depicted by small arrow on Fig. 1.a) results in a compliance which is more than 3,000 times larger than the nominal one.

In order to improve design's stability, it makes sense to treat the load as uncertain element of the data varying through a "massive" uncertainty set rather than taking just a small number of "values of interest". From the mathematical viewpoint, it is convenient to deal with uncertainty set in the form of an ellipsoid centered at the origin:

$$f \in \mathcal{F} = \left\{ f = L\delta \mid \delta \in \mathcal{D} = \{\delta \in \mathbf{R}^k : \delta^T \delta \le 1\} \right\}. \tag{6.5.36}$$

Problem (6.5.35) with perturbation set given by (6.5.36) is a particular case of full matrix uncertainty (6.4.32); according to Theorem 6.4.1, the robust counterpart of (6.5.35) – (6.5.36) is *equivalent* to an explicit semidefinite program; this program can be finally converted to the form

$$\begin{array}{rl} & \min \ \tau \\ \text{s.t.} & \\ & \begin{bmatrix} \tau I & Q^T \\ Q & \sum_{i=1}^n t_i b_i b_i^T \end{bmatrix} \succeq 0, \\ & t \in P. \end{array} \tag{6.5.37}$$

(for details, see [86]).

To illustrate the potential of the outlined approach, let us come back to the above "cantilever arm" example. In this example a load is, mathematically, a collection of ten 2D vectors representing (planar) external forces acting at the ten non-fixed nodes of the cantilever arm; in other words, the data in our problem is a 20-dimensional vector. Let us pass from the nominal problem ("a singleton uncertainty set $\mathcal{F} = \{f\}$") to the problem with $\mathcal{F}$ being a "massive" ellipsoid, namely, the ellipsoid of the smallest volume containing the nominal load f and a 20-dimensional ball $B_{0.1}$ comprised of all 20-dimensional vectors ("occasional loads") of the Euclidean norm $\le 0.1\|f\|_2$.

Solving the robust counterpart (6.5.37) of the resulting uncertain SDP, we get the cantilever arm shown on Fig. 1.b).

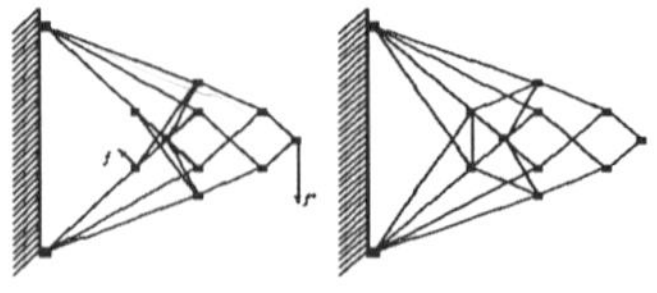

a) b)

Figure 1. Cantilever arm: nominal design (left) and robust design (right)

The compliances of the original and the new constructions with respect to the nominal load and their worst-case compliances with respect to the "occasional loads" from $B_{0.1}$ are as follows:

Design	Compliance w.r.t. f^*	Compliance w.r.t. $B_{0.1}$
nominal	1	> 3360
robust	1.0024	1.003

We see that in this example the robust counterpart approach improves dramatically the stability of the resulting construction, and that the improvement is in fact "costless" – the robust optimal solution is nearly optimal for the nominal problem as well.

6.6 CONCLUDING REMARKS

We have described a general methodology to handle deterministic uncertainty in semidefinite programming, which computes robust solutions via semidefinite programming. The method handles very general (nonlinear) uncertainty structures, and uses a special Lagrange relaxation (or, in a dual form, a "rank relaxation") to obtain the approximate robust counterpart in the form of an SDP. The method is actually an extension of techniques that are well-known in several (apparently unrelated) areas, such as rank relaxations in combinatorial optimization, or Lyapunov functions in control.

In the case of affine dependence, we can estimate the quality of the resulting approximation; in some other cases, the approximation is exact.

Further work should probably concentrate on reducing the level of conservativeness as much as possible, while keeping the size of the approximate robust counterpart reasonable.

Acknowledgements

This chapter has benefited from valuable comments and inputs from several people, including S. Niculescu, F. Oustry, G. Calafiore. The precious help of the Editors is gracefully acknowledged.

7 ERROR ANALYSIS

Zhiquan Luo and Jos Sturm

7.1 INTRODUCTION

We study a system of mixed linear, positive semidefinite (PSD) and second order cone (SOC) constraints:

$$\begin{cases} x \in b + \mathcal{A} \\ x \in \mathcal{K}, \end{cases} \tag{7.1.1}$$

where b is a given vector in $\Re^N$, $\mathcal{A}$ is a linear subspace of $\Re^N$, and $\mathcal{K} \subset \Re^N$ is a Cartesian product of second order cones and positive semidefinite cones. Thus,

$$\mathcal{K} = (\mathrm{SOC}(N_1) \times \cdots \times \mathrm{SOC}(N_p)) \times (\mathrm{PSD}(n_{p+1}) \times \cdots \times \mathrm{PSD}(n_q)), \tag{7.1.2}$$

where $\mathrm{SOC}(N_i)$ and $\mathrm{PSD}(n_j)$ stand for the second order cone in $\Re^{N_i}$, and the $n_j(n_j+1)/2$ dimensional cone of vectorized $n_j \times n_j$ (real symmetric) positive semidefinite matrices, respectively. We partition a vector $x \in \Re^N$ as $x = \left[\begin{array}{cccc} x_1^T, & x_2^T, & \ldots, & x_q^T \end{array}\right]^T$, according to the structure of $\mathcal{K}$; i.e.

$$x \in \mathcal{K} \Longleftrightarrow \begin{cases} x_i \in \mathrm{SOC}(N_i) \text{ for } i = 1, 2, \ldots, p \\ x_j \in \mathrm{PSD}(n_j) \text{ for } j = p+1, p+2, \ldots, q. \end{cases} \tag{7.1.3}$$

We use the following canonical form for a second order cone in $\Re^N$:

$$\mathrm{SOC}(N) = \{(x_1, x_2) \in \Re \times \Re^{N-1} \mid x_1 \geq \|x_2\|\}. \tag{7.1.4}$$

The positive semidefinite cone $\mathrm{PSD}(n)$ is represented in $\Re^{n(n+1)/2}$, by means of vectorization. Namely, we let $U_1, U_2, \cdots, U_{n(n+1)/2}$ denote an orthonormal set of $n \times n$ symmetric matrices, and

$$\text{mat}(x) := \sum_{i=1}^{n(n+1)/2} x_i U_i,$$

$$(\text{ vec}X)_i := \text{Trace}\,(U_i X) \text{ for } i = 1, 2, \ldots, n(n+1)/2.$$

The positive semidefinite cone in $\Re^{n(n+1)/2}$ is now defined as the cone of vectorized positive semidefinite matrices, i.e.

$$\text{PSD}(n) = \{x \in \Re^{n(n+1)/2} \mid \text{mat}(x) \text{ is positive semidefinite}\}.$$

Notice that $\text{PSD}(1) = \Re_+$, so that nonnegativity constraints are allowed in this setting.

The system (7.1.1) is in conic form, see e.g. [583, 756]. Systems of the form (7.1.1) can be solved numerically by recently developed interior point codes for mixed semidefinite and second order cone programming, such as SDPpack [22] and SeDuMi [758]. Such algorithms generate sequences of increasingly good approximate solutions, provided that the system is solvable. The quality of an approximate solution is typically measured by its *constraint violation* which is usually easily computable, nonnegative and is zero only when the approximate solution is a true solution. If the *constraint violation* is an upper bound on the distance from an approximate solution to the set of true solutions, then we can use the size of the constraint violation as a termination criterion for the iterative interior point algorithms. An inequality which bounds the distance to the solution set by the constraint violation is called an error bound.

In this chapter, we consider the solution set of the linear conic system (7.1.1) and show that the distance of a point to the solution set is at most $O(\epsilon^{2^{-d}})$, where ϵ is the amount of constraint violation defined as the sum of the distances to the cone $\mathcal{K}$ and the affine space b+$\mathcal{A}$. The nonnegative integer d is the so–called degree of singularity of the system. We also show that for pure second order cone programming, the degree of singularity is bounded above by the number of second order cone constraints. Our argument is based on the property of the regularization of the conic system (7.1.1). Our work generalizes many known error bound results, including Hoffman's global error bound [349] for linear inequality systems, Robinson's error bound for compact convex set with nonempty interior [678], Luo and Luo's error bound for convex quadratic systems [505], and Wang and Pang's error bound for convex quadratic systems [822].

The rest of the chapter is organized as follows. In Section 7.2, we introduce some basic facts about conic systems. In Section 7.3, the notion of a regularized backward error is introduced, and a Lipschitzian error bound in terms of this regularized backward error is established. It remains then to bound the regularized backward error in terms of the usual, computable backward error (i.e. the constraint violation). This is done in Section 7.4, yielding the general error bound result for mixed semidefinite and second order cone programming.

In Sections 7.5 and 7.6, the implications of this result are explored for infeasible systems and quadratic systems respectively.

Notation. The identity matrix is denoted by I, with columns $I = [e_1, \cdots, e_n]$, and we let $e := \sum_{i=1}^n e_i$ denote the all-one vector.

7.2 PRELIMINARIES

This preliminary section reviews the structure of the cones SOC(N) and PSD(n), focusing on spectral decomposition, faces, and Peirce decomposition.

Associated with a second order cone is a spectral decomposition for vectors $x \in \Re^N$. Namely, we can decompose $x = (x_1, x_2) \in \Re \times \Re^{N-1}$ as $x = F(x_1, x_2)\lambda(x_1, x_2)$, where

$$F(x_1, x_2) = \frac{1}{\sqrt{2}} \begin{bmatrix} 1 & 1 \\ -x_2/\|x_2\| & x_2/\|x_2\| \end{bmatrix}, \quad \lambda(x_1, x_2) = \frac{1}{\sqrt{2}} \begin{bmatrix} x_1 - \|x_2\| \\ x_1 + \|x_2\| \end{bmatrix}. \tag{7.2.5}$$

The $N \times 2$ matrix $F(x_1, x_2)$ is a Jordan frame, i.e. a complete system of idempotents [216]. The two scalars $\lambda_1(x_1, x_2) = (x_1 - \|x_2\|)/\sqrt{2}$ and $\lambda_2(x_1, x_2) = (x_1 + \|x_2\|)/\sqrt{2}$ are called the eigenvalues or spectral values of $x = (x_1, x_2)$. The rank of x is defined as the number of nonzero spectral values of x. A Jordan frame for SOC(N) is a matrix $F \in \Re^{N\times 2}$, such that

$$F^T F = I, \quad F\Re^2_+ = \operatorname{Img}(F) \cap \mathrm{SOC}(N), \tag{7.2.6}$$

i.e. the columns of F form an orthonormal basis of a 2–dimensional linear subspace of $\Re^N$, viz. Img (F), and the nonnegative orthant on this subspace, i.e. $F\Re^2_+$, is precisely the intersection of this subspace with the second order cone. The set of orthogonal transformations under which SOC(N) is invariant, is denoted by

$$\mathrm{OAut}(\mathrm{SOC}(N)) = \left\{ Q = \begin{bmatrix} 1 & 0^T \\ 0 & \tilde{Q} \end{bmatrix} \middle| \tilde{Q} \in \Re^{N-1\times N-1}, \tilde{Q}^T = \tilde{Q}^{-1} \right\}. \tag{7.2.7}$$

For any Jordan frame F, we can find a $Q \in$ OAut(SOC) such that $QF = E$, where E is a specially chosen Jordan frame,

$$E = \frac{1}{\sqrt{2}} \begin{bmatrix} 1 & 1 \\ -1 & 1 \\ 0 & 0 \end{bmatrix}. \tag{7.2.8}$$

This also means that SOC $= \cup\{QE\Re^2_+ \mid Q \in \mathrm{OAut(SOC)}\}$. The two columns of E are denoted by $E = \begin{bmatrix} E_1, & E_2 \end{bmatrix}$. Notice that $Fe = \sqrt{2}e_1$ for any Jordan frame.

The structure of the cone of $n \times n$ positive semidefinite matrices is quite similar. Namely, any $n \times n$ symmetric matrix can be diagonalized by an orthogonal matrix U, i.e. UXU^T is diagonal, and U can be chosen such that the diagonal entries of UXU^T are in nondecreasing order. Diagonalization is an orthogonal transformation, because

$$\text{Trace}\,((UXU^T)(UYU^T)) = \text{Trace}\,XY \text{ for any } X, Y.$$

This means that there is an $n(n+1)/2 \times n(n+1)/2$ orthogonal matrix Q, such that $\text{vec}(UXU^T) = Q\,\text{vec}(X)$. (In fact, $Q = U \otimes U$, where $\otimes$ denotes the symmetric Kronecker product [25].) Furthermore, $Q\text{PSD}(n) = \text{PSD}(n)$, because for any symmetric matrix Y, UYU^T has the same eigenvalues as Y.

Let $\text{OAut}(\text{PSD}(n))$ denote the set of orthogonal transformations under which $\text{PSD}(n)$ is invariant. It holds that for any $z \in \Re^N$, $N = n(n+1)/2$, there exists $Q \in \text{OAut}(\text{PSD}(n))$ such that $\text{mat}(Qz)$ is a diagonal matrix, for which the diagonal entries are in nondecreasing order. We let $\lambda_1(x), \lambda_2(x), \ldots, \lambda_n(x)$ denote the eigenvalues of the matrix $\text{mat}(x)$.

Since an $n \times n$ symmetric matrix has n eigenvalues, we say that the order of the cone $\text{PSD}(n)$ is n. Similarly, the order of a second order cone $\text{SOC}(N)$ is 2, since it has only 2 spectral values, irrespective of its dimension. Second order cones and semidefinite cones can sometimes be treated in a unified way, because they are both primitive symmetric cones [216]. Therefore, it is convenient to define

$$\mathcal{K}_i := \begin{cases} \text{SOC}(N_i) \text{ for } i = 1, 2, \ldots, p \\ \text{PSD}(n_i) \text{ for } i = p+1, p+2, \ldots, q, \end{cases} \tag{7.2.9}$$

see (7.1.2). We let $n(\mathcal{K}_i)$ denote the order of the cone $\mathcal{K}$, i.e. $n(\mathcal{K}_i) = n(\text{SOC}(N_i)) = 2$ for $i = 1, 2, \ldots, p$, and $n(\mathcal{K}_j) = n(\text{PSD}(n_j)) = n_j$ for $j = p+1, p+2, \ldots, q$. With respect to the cone $\mathcal{K} = \mathcal{K}_1 \times \cdots \times \mathcal{K}_q$, the vector of eigenvalues $\lambda(x)$ for $x \in \Re^N$ is

$$\lambda(x) = [\ \lambda(x_1)^T, \quad \cdots, \quad \lambda(x_q)^T\ ]^T.$$

Since $\lambda(x)$ has $\sum_{i=1}^q n(\mathcal{K}_i)$ components, we say that the order of $\mathcal{K}$ is

$$n(\mathcal{K}) = \sum_{i=1}^{q} n(\mathcal{K}_i). \tag{7.2.10}$$

The number of nonzero components of $\lambda(x)$ is called the rank of x, denoted as $\text{rank}\,(x)$. The smallest component of $\lambda(x)$ is denoted by $\lambda_{\min}(x)$. We let ι denote the unique vector in $\Re^N$ for which all spectral values are equal to one. Its components are $\text{vec}(I)$ for primitive cones of type $\text{PSD}(n)$, and $\sqrt{2}e_1$ for primitive cones of type $\text{SOC}(N)$.

7.2.1 Forward and backward error

Note that $x \in \mathcal{K}$ if and only if $\lambda_{\min}(x) \geq 0$. Given $x \in \Re^N$, we therefore say that x violates the constraint '$x \in \mathcal{K}$' by an amount of $[-\lambda_{\min}(x)]_+$, where the operator $[\cdot]_+$ yields the positive part. Similarly, we say that x violates the constraint '$x \in b + \mathcal{A}$' by an amount of $\text{dist}(x, b+\mathcal{A})$, where $\text{dist}(\cdot,\cdot)$ denotes the distance function (for a given norm). The total amount of constraint violation in x, i.e.

$$\text{dist}(x, b + \mathcal{A}) + [-\lambda_{\min}(x)]_+, \tag{7.2.11}$$

is called the *backward error* of x with respect to the system (7.1.1). The backward error indicates how much we should perturb the data of the problem, such that x is an exact solution to the perturbed problem. More importantly, the backward error is easily computable. Namely, it requires only a spectral factorization to determine $\lambda(x)$, and a projection to an affine subspace to determine $\text{dist}(x, b + \mathcal{A})$.

However, the backward error does not (immediately) tell us the distance from x to the solution set of the original system; this distance is called the *forward error* of x. The central issue in the study of error bounds is whether and how we can bound the forward error by the backward error.

7.2.2 Faces of the cone

A face of a self-dual cone $\mathcal{K}$ is of the form

$$\text{face}(\mathcal{K}, z) = \{x \in \mathcal{K} \mid z^T x = 0\}, \tag{7.2.12}$$

where z is a given vector in $\mathcal{K}$. In (7.1.1), we deal with a Cartesian product of second order cones and semidefinite cones, $\mathcal{K} = \mathcal{K}_1 \times \mathcal{K}_2 \times \cdots \times \mathcal{K}_q$, which is a (special case of a) self-dual cone. Partition the vector z according to this Cartesian product, as $z = \begin{bmatrix} z_1^T, & \ldots, & z_q^T \end{bmatrix}^T$. Since for any $x, z \in \mathcal{K}$, it holds that $x_i^T z_i \geq 0$ for all $i = 1, 2, \ldots, q$, it follows that

$$\text{face}(\mathcal{K}, z) = \text{face}(\mathcal{K}_1, z_1) \times \cdots \times \text{face}(\mathcal{K}_M, z_M). \tag{7.2.13}$$

Thus, a face of $\mathcal{K}$ is a Cartesian product of faces of second order cones and faces of semidefinite cones. Moreover, since $x^T z = (Qx)^T (Qz)$ for any orthogonal transformation Q, we have

$$Q\,\text{face}(\mathcal{K}_i, z) = \text{face}(\mathcal{K}_i, Qz) \text{ for any } Q \in \text{OAut}(\mathcal{K}_i), \tag{7.2.14}$$

where $i \in \{1, 2, \ldots, q\}$. Recall now that for the cone $\text{SOC}(N)$, we can find $Q \in \text{OAut}(\text{SOC}(N))$ such that $Qz = E\lambda$ for some $\lambda \in \Re_+^2$. Similarly, for $\text{PSD}(n)$ we can find $Q \in \text{OAut}(\text{PSD}(n))$ such that $Qz = \text{vec}(\Lambda)$ for some nonnegative diagonal matrix Λ.

Second order cone. To classify the faces of the second order cone, it suffices to consider $\text{face}(\text{SOC}(N), \lambda_1 E_1 + \lambda_2 E_2)$, with $0 \leq \lambda_1 \leq \lambda_2$; the structure of $\text{face}(\text{SOC}(N), z)$ then follows from (7.2.14). We distinguish three different cases, viz.

$$\begin{cases} \text{face}(\text{SOC}(N), 0) = \text{SOC}(N) \\ \text{face}(\text{SOC}(N), \lambda_2 E_2) = \{tE_1 \mid t \geq 0\}, \text{ for } \lambda_2 > 0 \\ \text{face}(\text{SOC}(N), \lambda_1 E_1 + \lambda_2 E_2) = \{0\}, \text{ for } \lambda_2 \geq \lambda_1 > 0. \end{cases} \tag{7.2.15}$$

The dual cone of $\text{face}(\text{SOC}(N), z)$, $z \in \text{SOC}(N)$, is defined as

$$\text{face}(\text{SOC}(N), z)^* := \{x \in \Re^N \mid x^T y \geq 0 \text{ for all } y \in \text{face}(\text{SOC}(N), z)\}.$$

We have

$$\begin{cases} \text{face}\,(\text{SOC}(N), 0)^* = \text{SOC}(N) \\ \text{face}\,(\text{SOC}(N), \lambda_2 E_2)^* = \{(x_1, x_2, x_3) | \, x_1 - x_2 \geq 0\}, \text{ for } \lambda_2 > 0 \\ \text{face}\,(\text{SOC}(N), \lambda_1 E_1 + \lambda_2 E_2)^* = \Re^N, \text{ for } \lambda_2 \geq \lambda_1 > 0, \end{cases} \quad (7.2.16)$$

where we partitioned x as $x = \begin{pmatrix} x_1, & x_2, & x_3 \end{pmatrix} \in \Re \times \Re \times \Re^{N-2}$.

In the above characterization of face $(\text{SOC}(N), E\lambda)$ and face $(\text{SOC}(N), E\lambda)^*$, there are only 3 linear subspaces involved: multiples of the column E_1, multiples of E_2, the kernel of E; this is known as the Peirce decomposition [216]. We will see later in this section that the Peirce decomposition allows for a unified characterization of the faces of $\text{SOC}(N)$ and $\text{PSD}(n)$.

As stipulated earlier, the structure of face $(\text{SOC}(N), z)$, $z \in \text{SOC}(N)$ follows by combining (7.2.14) and (7.2.15). In particular, we immediately have

$$\text{face}\,(\text{SOC}(N), 0) = \text{face}\,(\text{SOC}(N), 0)^* = \text{SOC}(N) \quad (7.2.17)$$

and

$$\text{face}\,(\text{SOC}(N), z) = \{0\} \text{ and } \text{face}\,(\text{SOC}(N), z)^* = \Re^N \text{ if rank}\,(z) = 2, \quad (7.2.18)$$

where we used that $\text{SOC}(N)$, $\Re^N$ and $\{0\}$ are invariant under orthogonal transformations from $\text{OAut}(\text{SOC}(N))$.

If rank $(z) = 1$ then z is on the boundary of $\text{SOC}(N)$ and hence $z_1 = \|z_2\|$. Let $Q \in \text{OAut}(\text{SOC}(N))$ and $t > 0$ be such that $z = tQE_2$, see (7.2.7). Then $tQE_1 = \begin{bmatrix} z_1, & -z_2^T \end{bmatrix}^T$. It now follows that if rank $(z) = 1$ then

$$\begin{cases} \text{face}\,(\text{SOC}(N), z) = \{x \in \Re^N \mid x_1 = tz_1, \, x_2 = -tz_2 \text{ for some } t \geq 0\}, \\ \text{face}\,(\text{SOC}(N), z)^* = \{x \in \Re^N \mid x_1 z_1 - x_2^T z_2 \geq 0\}. \end{cases} \quad (7.2.19)$$

For any $z \in \text{SOC}(N)$, we define an associated orthogonal decomposition for vectors $x \in \Re^N$:

$$x = x_{B(z)} + x_{N(z)} + x_{U(z)},$$

where the subscripts $B(z)$, $N(z)$, $U(z)$ signify the 'basic part', 'nonbasic part', and 'unrestricted part', respectively. The orthogonal decomposition is defined as follows:

- If rank $(z) = 0$, i.e. $z = 0$, then $x_{B(0)} = x$, $x_{N(0)} = 0$, $x_{U(0)} = 0$.
- If rank $(z) = 2$, i.e. $z \in \text{int}\,\text{SOC}(N)$, then $x_{B(z)} = 0$, $x_{N(z)} = x$, $x_{U(z)} = 0$.
- If rank $(z) = 1$, i.e. $z_1 = \|z_2\| > 0$, then $x_{B(z)} = (y^T x)y$, $x_{N(z)} = (z^T x/\|z\|^2)z$, $x_{U(z)} = x - x_{B(z)} - x_{N(z)}$, where $y := \begin{bmatrix} z_1, & -z_2 \end{bmatrix}^T / \|z\|$. Remark that $\begin{bmatrix} y, & z/\|z\| \end{bmatrix}$ is a Jordan frame for $\text{SOC}(N)$.

Notice that $z_{N(z)} = z$ for all $z \in \mathrm{SOC}(N)$.

With this notation, we can now write (7.2.17)–(7.2.19) more concisely, as

$$\left\{ \begin{array}{l} \mathrm{face}\,(\mathrm{SOC}(N), z) = \{x \mid x_{B(z)} \in \mathrm{SOC}(N),\ x_{U(z)} = 0,\ x_{N(z)} = 0\} \\ \mathrm{face}\,(\mathrm{SOC}(N), z)^* = \{x \mid x_{B(z)} \in \mathrm{SOC}(N),\ x_{U(z)} \text{ free},\ x_{N(z)} \text{ free}\}. \end{array} \right. \tag{7.2.20}$$

It is easily verified that if $x \in \mathrm{face}\,(\mathrm{SOC}(N), z)$ then $\mathrm{rank}\,(x + z) = \mathrm{rank}\,(x) + \mathrm{rank}\,(z) \leq 2$, and

$$x \in \text{ relint face}\,(\mathrm{SOC}(N), z) \iff \mathrm{rank}\,(x) + \mathrm{rank}\,(z) = 2, \tag{7.2.21}$$

where 'relint' stands for relative interior.

Positive semidefinite cone. The faces of $\mathrm{PSD}(n)$ can be treated in a similar fashion. First, we consider faces of the form $\mathrm{face}\,(\mathrm{PSD}(n),\ \mathrm{vec}(\Lambda))$, where Λ is a nonnegative diagonal matrix, with its diagonal entries in nondecreasing order. If Λ is nonzero but rank deficient then we can partition it as

$$\Lambda = \begin{bmatrix} 0 & 0 \\ 0 & \Lambda_2 \end{bmatrix} = 0 \oplus \Lambda_2, \tag{7.2.22}$$

where $\oplus$ denotes the direct sum of matrices. Partition $\mathrm{mat}(x)$ accordingly, as

$$\mathrm{mat}(x) = \begin{bmatrix} X_{11} & X_{12} \\ X_{12}^T & X_{22} \end{bmatrix}.$$

We let

$$\left\{ \begin{array}{rcl} x_{B(\,\mathrm{vec}(0\oplus\Lambda_2))} & = & \mathrm{vec}\left(\begin{bmatrix} X_{11} & 0 \\ 0 & 0 \end{bmatrix}\right) \\ x_{U(\,\mathrm{vec}(0\oplus\Lambda_2))} & = & \mathrm{vec}\left(\begin{bmatrix} 0 & X_{12} \\ X_{12}^T & 0 \end{bmatrix}\right) \\ x_{N(\,\mathrm{vec}(0\oplus\Lambda_2))} & = & \mathrm{vec}\left(\begin{bmatrix} 0 & 0 \\ 0 & X_{22} \end{bmatrix}\right). \end{array} \right. \tag{7.2.23}$$

For positive definite Λ, we let

$$x_{B(\,\mathrm{vec}(\Lambda))} = 0,\ x_{N(\,\mathrm{vec}(\Lambda))} = x,\ x_{U(\,\mathrm{vec}(\Lambda))} = 0.$$

Furthermore, we define

$$x_{B(0)} = x,\ x_{N(0)} = 0,\ x_{U(0)} = 0.$$

In analogy to (7.2.20), we easily derive from (7.2.12) that

$$\left\{ \begin{array}{l} \mathrm{face}\,(\mathrm{PSD}(n), z) = \{x \mid x_{B(z)} \in \mathrm{PSD}(n),\ x_{U(z)} = 0,\ x_{N(z)} = 0\} \\ \mathrm{face}\,(\mathrm{PSD}(n), z)^* = \{x \mid x_{B(z)} \in \mathrm{PSD}(n),\ x_{U(z)} \text{ free},\ x_{N(z)} \text{ free}\}, \end{array} \right. \tag{7.2.24}$$

where $z = \text{vec}(\Lambda)$, for some nonnegative diagonal matrix Λ. Using identity (7.2.14), we can extend relation (7.2.24) to any $z \in \text{PSD}(n)$, by letting

$$x_{B(z)} := Q(Q^T x)_{B(\text{vec}(\Lambda))},\ x_{N(z)} := Q(Q^T x)_{N(\text{vec}(\Lambda))},$$

and

$$x_{U(z)} := Q(Q^T x)_{U(\text{vec}(\Lambda))},$$

where $z = Q\ \text{vec}(\Lambda)$, $Q \in \text{OAut}(\text{PSD}(n))$ and Λ a nonnegative diagonal matrix. Notice that $z_{N(z)} = z$ for all $z \in \text{PSD}(n)$. Analogous to (7.2.21), we have for $x \in \text{face}\,(\text{PSD}(n), z)$ that $\text{rank}\,(x+z) = \text{rank}\,(x) + \text{rank}\,(z) \le n$, and

$$x \in \text{relint face}\,(\text{PSD}(n), z) \Longleftrightarrow \text{rank}\,(x) + \text{rank}\,(z) = n. \tag{7.2.25}$$

For further analysis on the faces of the cone $\text{PSD}(n)$, see Lewis [482].

General case. The extension to the Cartesian product $\mathcal{K} = \mathcal{K}_1 \times \cdots \times \mathcal{K}_q$ in (7.1.1) is straightforward, viz.

$$\begin{cases} \text{face}\,(\mathcal{K}, z) = \{x \mid x_{B(z)} \in \mathcal{K},\ x_{U(z)} = 0,\ x_{N(z)} = 0\} \\ \text{face}\,(\mathcal{K}, z)^* = \{x \mid x_{B(z)} \in \mathcal{K},\ x_{U(z)} \text{ free},\ x_{N(z)} \text{ free}\}, \end{cases} \tag{7.2.26}$$

and

$$\begin{cases} x \in \text{relint face}\,(\mathcal{K}, z) \Longrightarrow \text{rank}\,(x+z) = \text{rank}\,(x) + \text{rank}\,(z) = n(\mathcal{K}) \\ x \in \text{face}\,(\mathcal{K}, z) \Longrightarrow \text{rank}\,(x+z) = \text{rank}\,(x) + \text{rank}\,(z) \le n(\mathcal{K}) \\ x \in \text{relint face}\,(\mathcal{K}, z) \Longrightarrow y_{B(x)} = y_{N(z)} \text{ for all } y. \end{cases} \tag{7.2.27}$$

Using (7.2.21) and (7.2.25), we also have for $x \in \text{face}\,(\mathcal{K}, z)$ that

$$x \in \text{relint face}\,(\mathcal{K}, z) \Longleftrightarrow \text{rank}\,(x) + \text{rank}\,(z) = n(\mathcal{K}). \tag{7.2.28}$$

Remark from (7.2.26) that

$$\text{relint face}\,(\mathcal{K}, z)^* = \{x \mid x_{B(z)} \in \text{relint face}\,(\mathcal{K}, z)\}. \tag{7.2.29}$$

For future reference, we notice the obvious relation that

$$y \in \mathcal{K},\ z \in \mathcal{K} \Longrightarrow \text{face}\,(\mathcal{K}, y+z) \subseteq \text{face}\,(\mathcal{K}, y) \subseteq \mathcal{K}, \tag{7.2.30}$$

and its dual relation

$$y \in \mathcal{K},\ z \in \mathcal{K} \Longrightarrow \text{face}\,(\mathcal{K}, y+z)^* \supseteq \text{face}\,(\mathcal{K}, y)^* \supseteq \mathcal{K}. \tag{7.2.31}$$

Another useful relation is

$$x \in \mathcal{K} \Longrightarrow x_{B(y)} \in \mathcal{K},\ x_{N(y)} \in \mathcal{K} \quad \text{for any } y \in \mathcal{K}, \tag{7.2.32}$$

which follows directly from the definitions of $x_{B(y)}$ and $x_{N(y)}$. (Alternatively, this can be argued from the facts that $\mathcal{K} \subseteq \text{face}(\mathcal{K}, y)^*$ and $x_{N(y)} = x_{B(z)}$ for $z \in \text{relint face}(\mathcal{K}, y)$, see (7.2.28) and (7.2.31).)

Lemma 7.2.1 *Let $z \in \mathcal{K}$. Then*

$$\|x_{N(z)}\| = O(z^T x) \quad \textit{for all } x \in \mathcal{K}.$$

Proof. Suppose to the contrary that there is a sequence $x^{(1)}, x^{(2)}, \ldots$ such that

$$\lim_{i\to\infty} \frac{z^T x^{(i)}}{\|x^{(i)}_{N(z)}\|} = 0.$$

Since $z = z_{N(z)}$, we have $z^T x^{(i)} = z^T x^{(i)}_{N(z)}$. Therefore, it holds for any cluster point y of the sequence $\{x^{(i)}_{N(z)}/\|x^{(i)}_{N(z)}\|\}$ for $i \to \infty$ that $z^T y = 0$. However, since $x^{(i)} \in \mathcal{K}$, we have $x^{(i)}_{N(z)} \in \mathcal{K}$ and hence also $y \in \mathcal{K}$. We conclude that

$$y \in \text{face}(\mathcal{K}, z), \quad \|y_{N(z)}\| = 1,$$

which contradicts (7.2.26). ■

7.3 THE REGULARIZED BACKWARD ERROR

Let $\bar{\mathcal{A}}$ denote the smallest linear subspace containing $b + \mathcal{A}$, i.e.

$$\bar{\mathcal{A}} = \{x \in \Re^N \mid x + tb \in \mathcal{A} \quad \text{for some } t \in \Re\}. \tag{7.3.33}$$

We are naturally interested in the intersection of this linear subspace with the cone $\mathcal{K}$. For any self–dual cone $\mathcal{K}$, it holds that

$$\bar{\mathcal{A}} \cap \mathcal{K} = \{0\} \iff \bar{\mathcal{A}}^\perp \cap \text{int}\,\mathcal{K} \neq \emptyset; \tag{7.3.34}$$

the above characterization is a special case of a duality theorem for convex cones.

The general theorem states that given a linear subspace $\mathcal{L}$ and a convex cone $\Gamma \subseteq \Re^N$ with relative interior $\text{relint}\,\Gamma$ and dual cone $\Gamma^* := \{z \in \Re^N \mid z^T\Gamma \subseteq \Re_+\}$, it holds that

$$\mathcal{L} \cap \text{relint}\,\Gamma \neq \emptyset \iff \mathcal{L}^\perp \cap \Gamma^* \subseteq -\Gamma^*, \tag{7.3.35}$$

see Corollary 2 in Luo, Sturm and Zhang [508] and Corollary 2.2 in [756]. This result generalizes a classical duality theorem of Gordon and Stiemke for linear inequalities.

Let $x^* \in \text{relint}(\bar{\mathcal{A}} \cap \mathcal{K})$. For $z^* \in \text{relint face}(\mathcal{K}, x^*)$, we have $\text{rank}(x^*) + \text{rank}(z^*) = n(\mathcal{K})$, (and hence $x^* \in \text{relint face}(\mathcal{K}, z^*)$,) due to (7.2.28). As a straightforward application of (7.2.28), we have the following result:

Lemma 7.3.1 *Let* $x^* \in \text{relint}\,(\bar{\mathcal{A}} \cap \mathcal{K})$ *and* $z^* \in \text{relint face}\,(\mathcal{K}, x^*)$. *Then*

$$\text{relint}\,((b + \mathcal{A}) \cap \mathcal{K}) = (b + \mathcal{A}) \cap \text{ relint face}\,(\mathcal{K}, z^*).$$

Proof. Let $y \in \bar{\mathcal{A}} \cap \mathcal{K}$. By definition of x^*, there must exist $t > 0$ such that $x^* + \bar{t}y \in \mathcal{K}$ for all $|\bar{t}| \leq t$, which implies that

$$\bar{t}(z^*)^T y = (z^*)^T (x^* + \bar{t}y) \geq 0$$

for all $|\bar{t}| \leq t$. It follows that $y^T z^* = 0$, and so $y \in \text{ face}\,(\mathcal{K}, z^*)$. Hence

$$(b + \mathcal{A}) \cap \mathcal{K} = (b + \mathcal{A}) \cap \text{ face}\,(\mathcal{K}, z^*). \tag{7.3.36}$$

This trivially implies the inclusion $(b + \mathcal{A}) \cap \text{ relint face}\,(\mathcal{K}, z^*) \subseteq \text{relint}\,((b + \mathcal{A}) \cap \mathcal{K})$.

It remains to show for arbitrary $\xi \in \text{ relint}\,((b + \mathcal{A}) \cap \mathcal{K})$ that $\xi \in \text{ relint face}\,(\mathcal{K}, z^*)$. Let β be such that $x^* + \beta b \in \mathcal{A}$. Remark from (7.3.36) that $\xi \in \text{ face}\,(\mathcal{K}, z^*)$, and so $\xi = \xi_{B(z^*)}$. Furthermore, since $\xi \in \mathcal{K}$ and $x^* \in \mathcal{K}$ it holds that

$$\xi + t(\beta\xi + x^*) \in (b + \mathcal{A}) \cap \mathcal{K}, \tag{7.3.37}$$

for any $t \geq 0$ with $t\beta \geq -1$. As $\xi \in \text{ relint}\,((b + \mathcal{A}) \cap \mathcal{K})$, we must have (7.3.37) for t in a neighborhood of zero. Thus, if $\sigma^T \xi = 0$ for some $\sigma \in \mathcal{K}$, then also $\sigma^T x^* = 0$. This finally implies that $\xi \in \text{ relint face}\,(\mathcal{K}, z^*)$. ■

Due to the above result, face $(\mathcal{K}, z^*)$ is sometimes called the minimal cone [130] or the regularized cone [508] for the affine space $b + \mathcal{A}$. Notice also that face $(\mathcal{K}, z^*)$ is uniquely defined by $\bar{\mathcal{A}}$ and $\mathcal{K}$, whereas z^* is not.

Recalling (7.2.26), a natural definition of the backward error of $x(\epsilon)$ with respect to the regularized system

$$(b + \mathcal{A}) \cap \text{ face}\,(\mathcal{K}, z^*)$$

is the quantity

$$\text{dist}(x(\epsilon), b + \mathcal{A}) + [-\lambda_{\min}(x(\epsilon))]_+ + \|x_{U(z^*)}(\epsilon)\| + \|x_{N(z^*)}(\epsilon)\|. \tag{7.3.38}$$

The following lemma states, among others, that if $\{x(\epsilon) \mid \epsilon > 0\}$ is bounded, then the regularized backward error (7.3.38) is of the same order as the forward error $\text{dist}(x(\epsilon), (b + \mathcal{A}) \cap \mathcal{K})$.

Lemma 7.3.2 *Let* x^* *and* z^* *be as above. If* $\{x(\epsilon) \mid \epsilon > 0\}$ *is such that*

$$\text{dist}(x(\epsilon), b + \mathcal{A}) \leq \epsilon, \quad \lambda_{\min}(x(\epsilon)) \geq -\epsilon, \quad \|x_{U(z^*)}(\epsilon)\| + \|x_{N(z^*)}(\epsilon)\| \leq \epsilon, \tag{7.3.39}$$

for all $\epsilon > 0$, *then* $(b + \mathcal{A}) \cap \mathcal{K} \neq \emptyset$. *Moreover, there exists* $\{\delta(\epsilon) \in \Re \mid \epsilon > 0\}$, *such that*

$$\text{dist}((1 + \delta(\epsilon))x(\epsilon), (b + \mathcal{A}) \cap \mathcal{K}) = O(\epsilon), \quad |\delta(\epsilon)| = O(\epsilon),$$

for $0 < \epsilon \leq 1$.

Proof. As is well known [349] from Hoffman's error bound, the backward and forward error for a system of linear equations are of the same order. Therefore, the relations

$$\operatorname{dist}(x(\epsilon), b + \mathcal{A}) \leq \epsilon, \quad \|x_{U(z^*)}(\epsilon)\| \leq \epsilon, \quad \|x_{N(z^*)}(\epsilon)\| \leq \epsilon$$

imply that

$$\operatorname{dist}\left(x(\epsilon), \{x \in b + \mathcal{A} \mid x_{U(z^*)} = 0,\ x_{N(z^*)} = 0\}\right) = O(\epsilon).$$

This bound implies the existence of $\{y(\epsilon) \mid \epsilon > 0\}$, such that

$$x(\epsilon) + y(\epsilon) \in \{x \in b + \mathcal{A} \mid x_{U(z^*)} = 0,\ x_{N(z^*)} = 0\}, \quad \|y(\epsilon)\| = O(\epsilon). \quad (7.3.40)$$

Let

$$\alpha(\epsilon) := \frac{[-\lambda_{\min}(x(\epsilon))]_+ + \|\lambda(y(\epsilon))\|_\infty}{\min\{\lambda_i(x^*) \mid i \text{ for which } \lambda_i(x^*) > 0\}}, \qquad \text{if } x^* \neq 0,$$

and $\alpha(\epsilon) = 0$ for the case that $x^* = 0$. Notice that $\alpha(\epsilon) = O(\epsilon)$. Let

$$\hat{x}(\epsilon) := x(\epsilon) + y(\epsilon) + \alpha(\epsilon)x^*,$$

and notice $\hat{x}(\epsilon) = \hat{x}(\epsilon)_{B(z^*)}$, because of (7.3.40) and $x^* \in$ relint face $(\mathcal{K}, z^*)$. Therefore, it holds for any $u \in$ face $(\mathcal{K}, z^*)^*$ that

$$u^T\hat{x}(\epsilon) = u^T_{B(z^*)}\hat{x}(\epsilon).$$

Furthermore, since $u_{B(z^*)} \in \mathcal{K}$, we have

$$u^T_{B(z^*)}(x(\epsilon) + [-\lambda_{\min}(x(\epsilon))]_+\iota) \geq 0, \quad u^T_{B(z^*)}(y(\epsilon) + \|\lambda(y(\epsilon))\|_\infty\iota) \geq 0,$$

and since $x^* \in$ relint face $(\mathcal{K}, z^*)$,

$$(x^*)^T u_{B(z^*)} \geq \min\{\lambda_i(x^*) \mid i \text{ for which } \lambda_i(x^*) > 0\}\iota^T u_{B(z^*)}.$$

Combining the above relations, it follows that $u^T\hat{x}(\epsilon) \geq 0$ and hence

$$\hat{x}(\epsilon) = x(\epsilon) + y(\epsilon) + \alpha(\epsilon)x^* \in \text{face}\,(\mathcal{K}, z^*) \subseteq \mathcal{K}. \qquad (7.3.41)$$

Since $x^* \in \bar{\mathcal{A}}$, there must exist $t \in \Re$ such that $x^* - tb \in \mathcal{A}$. Let $\bar{\epsilon} > 0$ be such that $t\alpha(\epsilon) > -1$ for all $\epsilon \in (0, \bar{\epsilon}]$. Then, using also (7.3.41),

$$\frac{1}{1 + t\alpha(\epsilon)}(x(\epsilon) + y(\epsilon) + \alpha(\epsilon)x^*) \in (b + \mathcal{A}) \cap \mathcal{K} \text{ for } 0 < \epsilon \leq \bar{\epsilon}.$$

It follows from the above relation that

$$\operatorname{dist}\left(\frac{1}{1+t\alpha(\epsilon)}x(\epsilon),(b+\mathcal{A})\cap\mathcal{K}\right)=O(\epsilon),$$

for $0<\epsilon\leq\bar{\epsilon}$. ■

Under Slater's condition, i.e. if $(b+\mathcal{A})\cap \operatorname{int}\mathcal{K}\neq\emptyset$, we have $\operatorname{face}(\mathcal{K},z^*)=\mathcal{K}$. The regularized backward error and the usual (unregularized) backward error are then exactly the same. Thus, Lemma 7.3.2 generalizes Hoffman's error bound [349] for systems of linear inequalities and equations to mixed semidefinite and second order cone systems under Slater's condition. Notice in particular that no boundedness assumptions are made, i.e. the error bound holds globally over $\Re^N$. However, the lemma requires a scaling factor $1+\delta(\epsilon)$, which is not needed in case of linear inequalities and equations or convex quadratic inequalities [505]. This scaling factor is essential in the case of semidefinite or second order cone constraints. See e.g. Example 1 in Sturm [757] for the case of semidefinite constraints.

Some additional remarks are in order.

Remark 7.3.1 *Lemma 7.3.2 states that the mere existence of* $\{x(\epsilon)\mid\epsilon>0\}$ *satisfying* (7.3.39) *for all* $\epsilon>0$ *implies that* $(b+\mathcal{A})\cap\mathcal{K}\neq\emptyset$, *even though* $x(\epsilon)$ *is not necessarily bounded for* $\epsilon\downarrow 0$. *In the case of weak infeasibility, i.e. if*

$$\operatorname{dist}(b+\mathcal{A},\mathcal{K})=0,\quad (b+\mathcal{A})\cap\mathcal{K}=\emptyset,$$

we can therefore conclude that if the usual backward error of $x(\epsilon)$, *as defined in* (7.2.11), *is* $O(\epsilon)$ *then*

$$\liminf_{\epsilon\downarrow 0}\|x_{N(z^*)}(\epsilon)\|+\|x_{U(z^*)}(\epsilon)\|>0.$$

Remark 7.3.2 *If* $x^{(1)},x^{(2)},\ldots$ *is a bounded sequence with*

$$\operatorname{dist}(x^{(k)},b+\mathcal{A})\to 0 \text{ and } [-\lambda_{\min}(x^{(k)})]_+\to 0 \text{ for } k\to\infty,$$

then also $\|x^{(k)}_{U(z^*)}\|+\|x^{(k)}_{N(z^*)}\|\to 0$, *as follows from Lemma 7.3.1. Letting*

$$\epsilon_k:=\operatorname{dist}(x^{(k)},b+\mathcal{A})+[-\lambda_{\min}(x^{(k)})]_++\|x^{(k)}_{U(z^*)}\|+\|x^{(k)}_{N(z^*)}\|,$$

it follows from Lemma 7.3.2 and the boundedness of the sequence $\{x^{(k)}\mid k=1,2,\ldots\}$ *that*

$$\operatorname{dist}(x^{(k)},(b+\mathcal{A})\cap\mathcal{K})=O(\epsilon_k).$$

Remark 7.3.3 *The existence of an error bound for compact convex sets with Slater's condition is well known, see Robinson [678], Renegar [666] and Deng and Hu [187].*

We can also derive a Lipschitzian type error bound from Lemma 7.3.2 by imposing a regularity condition that is milder than Slater's condition.

Definition 7.3.1 *A pair (y, z) with $z \in \mathcal{K}$ is called regular if for any $k = 1, 2, \ldots, q$, it holds that*

$$(y_{B(z)})_k = 0 \text{ or } y_k \in \text{ relint face}\,(\mathcal{K}_k, z_k)^*.$$

Thus, if (y, z) is regular then for each component y_k, the 'basic' part (with respect to z) is either zero or of maximal rank. The connection between the above definition of regularity and the minimal cone, face $(\mathcal{K}, z^*)$, is given in the following lemma. Namely, it states that if y is a dual direction of maximal rank, i.e. $y \in$ relint $(\bar{\mathcal{A}}^\perp \cap \mathcal{K})$, and $(y, 0)$ is regular in the sense of Definition 7.3.1, then face $(\mathcal{K}, y)$ is the minimal cone for $\bar{\mathcal{A}}$. For the case that $(y, 0)$ is not regular, we will obtain a pair (y, z) which is regular and $0 \neq z \in \mathcal{K}$; the construction of such z is called regularization. Regularization is treated in Section 7.4.

Lemma 7.3.3 *Let $z \in \mathcal{K}$ and $y \in$ relint $(\bar{\mathcal{A}}^\perp \cap$ face $(\mathcal{K}, z)^*)$. If (y, z) is regular, then*

$$\bar{\mathcal{A}} \cap \text{ relint face}\,(\mathcal{K}, z + y_{B(z)}) \neq \emptyset.$$

Proof. Let $z' := z + y_{B(z)}$. We will show that $\bar{\mathcal{A}}^\perp \cap$ face $(\mathcal{K}, z')^* \subseteq -$ face $(\mathcal{K}, z')^*$. The lemma then follows as an application of (7.3.35).

Consider an arbitrary $y' \in \bar{\mathcal{A}}^\perp \cap$ face $(\mathcal{K}, z')^*$. As already stipulated, we want to show that $-y' \in$ face $(\mathcal{K}, z')^*$, or equivalently, $y'_{B(z')} = 0$, see (7.2.26). Consider $k \in \{1, 2, \ldots, q\}$. Since (y, z) is regular, it holds that either $(y_{B(z)})_k = 0$ or $y_k \in$ relint face $(\mathcal{K}_k, z_k)^*$.

1. If $(y_{B(z)})_k = 0$, we have $(y'_{B(z')})_k = (y'_{B(z)})_k$. We claim $(y'_{B(z)})_k = 0$. Suppose this is not the case so that $(y'_{B(z)})_k \neq 0$. Since $y \in$ relint $(\bar{\mathcal{A}}^\perp \cap$ face $(\mathcal{K}, z)^*)$ and $(y_{B(z)})_k = 0$, it follows that for any vector $\tilde{y}$ in $\bar{\mathcal{A}}^\perp \cap$ face $(\mathcal{K}, z)^*$, there must hold $(\tilde{y}_{B(z)})_k = 0$. This contradicts the fact that $y' \in \bar{\mathcal{A}}^\perp \cap$ face $(\mathcal{K}, z')^*$ and $(y'_{B(z)})_k = (y'_{B(z)})_k \neq 0$. Thus we have shown $(y'_{B(z)})_k = 0$.
2. If $y_k \in$ relint face $(\mathcal{K}_k, z_k)^*$ then $z'_k \in$ int $\mathcal{K}_k$. Therefore, $(y'_{B(z')})_k = (y'_k)_{B(z'_k)} = 0$ since z'_k is full rank (in the interior of the cone $\mathcal{K}_k$).

The above analysis shows that

$$y'_{B(z')} = \left((y'_1)_{B(z'_1)}, (y'_2)_{B(z'_2)}, \ldots, (y'_q)_{B(z'_q)}\right)^T = 0, \tag{7.3.42}$$

concluding the proof. ∎

We already mentioned that a Lipschitzian error bound holds under the condition that $(y, 0)$ is regular, with $y \in$ relint $(\bar{\mathcal{A}}^\perp \cap \mathcal{K})$. Remark that if Slater's condition holds then $y = 0$, see (7.3.35), and (0,0) trivially satisfies the definition of regularity. For $y \in \operatorname{relint}(\bar{\mathcal{A}}^\perp \cap \mathcal{K})$, we have from Lemma 7.2.1 that $\|x_{N(y)}(\epsilon)\| = O(\epsilon)$. Furthermore, if $(y, 0)$ is regular, then for all $k = 1, 2, \ldots, q$, y_k is either all-zero or full rank, and hence $x_{U(y)} = 0$ for any $x \in \Re^N$. So, there is no difference between the regularized backward error and the usual backward error in this case. For linear systems, we have $\mathcal{K} = \text{PSD}(1) \times \cdots \times \text{PSD}(1) = \Re^q_+$, and regularity of $(y, 0)$ always holds.

7.4 REGULARIZATION STEPS

In order to bound the regularized backward error (7.3.38) in terms of the original backward error (7.2.11), we use a sequence of regularization steps.

In Section 7.3, we worked with the minimal cone, i.e. $\operatorname{face}(\mathcal{K}, z^*)$. In this section, we will consider a shrinking set of faces of $\mathcal{K}$ that contain $\operatorname{face}(\mathcal{K}, z^*)$. In particular, starting from $z^0 = 0$, we will generate a sequence of dual vectors z^k with $0 \neq (z^{k+1} - z^k) \in \operatorname{face}(\mathcal{K}, z^k)$, until $\operatorname{face}(\mathcal{K}, z^{d+1}) = \operatorname{face}(\mathcal{K}, z^*)$. Notice from (7.2.30) that under such a construction,

$$\operatorname{face}(\mathcal{K}, z^*) = \operatorname{face}(\mathcal{K}, z^{d+1}) \subset \operatorname{face}(\mathcal{K}, z^d) \subset \cdots \subset \operatorname{face}(\mathcal{K}, z^0) = \mathcal{K}.$$

Lemma 7.4.1 *Let $\bar{\mathcal{A}}$ be a linear subspace of $\Re^N$ and let $z \in \mathcal{K}$. Suppose that $\{x(\epsilon) \mid 0 < \epsilon \leq 1\}$ is such that*

$$\operatorname{dist}(x(\epsilon), \bar{\mathcal{A}}) \leq \epsilon, \quad \|x_{U(z)}(\epsilon)\| + \|x_{N(z)}(\epsilon)\| \leq \epsilon, \quad \lambda_{\min}(x(\epsilon)) \geq -\epsilon,$$

for all $0 < \epsilon \leq 1$. Let

$$y \in \operatorname{relint}\left(\bar{\mathcal{A}}^\perp \cap \operatorname{face}(\mathcal{K}, z)^*\right)$$

and define $z' := z + y_{B(z)}$. Then $z' \in \mathcal{K}$, and

$$\|x_{N(z')}(\epsilon)\| = O(\epsilon).$$

Moreover, if (y, z) is regular, then

$$\|x_{U(z')}(\epsilon)\| \leq \|x_{U(z)}(\epsilon)\| = O(\epsilon).$$

and otherwise, i.e., if (y, z) is not regular, then

$$\|x_{U(z')}(\epsilon)\| = O(\sqrt{\epsilon \|x(\epsilon)\|}).$$

Proof. Since $\operatorname{dist}(x(\epsilon), \bar{\mathcal{A}}) \leq \epsilon$, there must exist $w(\epsilon)$, such that

$$x(\epsilon) + w(\epsilon) \in \bar{\mathcal{A}}, \quad \|w(\epsilon)\| \leq \epsilon, \tag{7.4.43}$$

for all $\epsilon > 0$. This implies that $y \perp (x(\epsilon) + w(\epsilon))$ because $y \in \bar{\mathcal{A}}^\perp$, and therefore

$$\begin{aligned} |y^T x_{B(z)}(\epsilon)| &= \left|y^T \left(x_{B(z)}(\epsilon) - x(\epsilon) - w(\epsilon)\right)\right| \\ &\leq \|y\| \left\|x_{U(z)}(\epsilon) + x_{N(z)}(\epsilon) + w(\epsilon)\right\|, \end{aligned}$$

where we used the Cauchy-Schwarz inequality. From the identity $z = z_{N(z)}$ we have $z_{B(z)} = 0$, so that also $z^T x_{B(z)}(\epsilon) = z^T_{B(z)} x_{B(z)}(\epsilon) = 0$. Hence

$$y^T x_{B(z)}(\epsilon) = (z+y)^T x_{B(z)}(\epsilon) = (z')^T x_{B(z)}(\epsilon).$$

Recall now that

$$\|x_{U(z)}(\epsilon)\| = O(\epsilon), \quad \|x_{N(z)}(\epsilon)\| = O(\epsilon), \quad \|w(\epsilon)\| = O(\epsilon),$$

so that we further obtain

$$|(z')^T x(\epsilon)| = O(\epsilon). \tag{7.4.44}$$

Applying Lemma 7.2.1, we obtain from (7.4.44) that

$$\|x_{N(z')}(\epsilon)\| = O(\epsilon). \tag{7.4.45}$$

It remains to bound $\|x_{U(z')}(\epsilon)\|$.

If (y,z) is regular, then for each $k = 1,2,\ldots,q$, either $(y_{B(z)})_k = 0$ or $y_k \in$ relint face$(\mathcal{K}_k, z_k)^*$. In the former case, $(x_{U(z')}(\epsilon))_k = (x_{U(z)}(\epsilon))_k$; in the latter case, $z'_k \in$ int$\,\mathcal{K}$ and so $(x_{U(z')}(\epsilon))_k = 0$. Therefore, we have

$$\|x_{U(z')}(\epsilon)\| \le \|x_{U(z)}(\epsilon)\| = O(\epsilon). \tag{7.4.46}$$

Suppose now that (y,z) is not regular, and partition $x(\epsilon)$ and z' as in (7.1.3). From (7.2.13), we have $(x_{U(z')}(\epsilon))_i = (x_i)_{U(z'_i)}(\epsilon)$, and so

$$\|x_{U(z')}(\epsilon)\|^2 = \sum_{i=1}^{q} \|(x_i)_{U(z'_i)}(\epsilon)\|^2.$$

Let $k \in \{1,2,\ldots,p\}$ and let $\xi(\epsilon) = x_k(\epsilon)$. If rank$\,(z'_k) \neq 1$ then $\xi_{U(z'_k)}(\epsilon) = 0$ by definition. Suppose now that rank$\,(z'_k) = 1$. By performing a basis rotation if necessary, we may assume that z'_k is a multiple of E_2. Partition $\xi(\epsilon)$ as $\xi(\epsilon) = [\ \xi_1(\epsilon),\ \ \xi_2(\epsilon),\ \ \xi_3(\epsilon)^T\ ]^T \in \Re \times \Re \times \Re^{N_k-2}$. Since $x(\epsilon) + \epsilon\iota \in \mathcal{K}$, we have

$$(\xi_1(\epsilon) + \sqrt{2}\epsilon)^2 \ge \xi_2(\epsilon)^2 + \|\xi_3(\epsilon)\|^2. \tag{7.4.47}$$

Recall from Section 7.2.2 that $\xi_{U(E_2)}(\epsilon) = \xi_3(\epsilon)$ and

$$\xi_{N(E_2)}(\epsilon) = (E_2^T \xi(\epsilon))E_2 = (\xi_1(\epsilon) + \xi_2(\epsilon))E_2/\sqrt{2}.$$

Using (7.4.45), it follows that $|\xi_1(\epsilon) + \xi_2(\epsilon)| = O(\epsilon)$. Together with (7.4.47), this yields

$$\begin{aligned}\|\xi_{U(E_2)}(\epsilon)\|^2 = \|\xi_3(\epsilon)\|^2 &\le \xi_1(\epsilon)^2 - \xi_2(\epsilon)^2 + 2\sqrt{2}\epsilon\xi_1(\epsilon) + 2\epsilon^2 \\ &= O(\epsilon(|\xi_1(\epsilon) + \xi_2(\epsilon)| + \xi_1(\epsilon) + \epsilon)) \\ &= O(\epsilon(\epsilon + \|\xi(\epsilon)\|)).\end{aligned}$$

This implies that

$$\|\xi_{U(z'_k)}(\epsilon)\|^2 = O(\epsilon\|\xi(\epsilon)\|).$$

Finally, consider $k \in \{p+1, p+2, \ldots, q\}$ and let $\xi(\epsilon) = x_k(\epsilon)$. Recall that $\xi_{U(z_k')}(\epsilon) = 0$ if z_k is either zero or full rank. Assume now that z_k is nonzero, but rank deficient. By applying a basis rotation if necessary, we may assume that $\mathrm{mat}(z_k) = 0 \oplus \Lambda_2$, for a certain positive diagonal matrix Λ_2, as in (7.2.22). We partition $\mathrm{mat}(\xi(\epsilon))$ accordingly, as

$$\mathrm{mat}(\xi(\epsilon)) = \begin{bmatrix} \Xi_{11}(\epsilon) & \Xi_{12}(\epsilon) \\ \Xi_{12}(\epsilon)^T & \Xi_{22}(\epsilon) \end{bmatrix};$$

$\xi_{B(z_k)}$, $\xi_{U(z_k)}$ and $\xi_{N(z_k)}$ are then defined as in (7.2.23). Since $\lambda_{\min}(x(\epsilon)) \geq -\epsilon$, we know that $\mathrm{mat}(\xi(\epsilon)) + 2\epsilon I$ is positive definite. The Schur complement of this matrix must therefore be positive definite as well, i.e.

$$(\Xi_{11}(\epsilon) + 2\epsilon I) - \Xi_{12}(\epsilon)(\Xi_{22}(\epsilon) + 2\epsilon I)^{-1}\Xi_{12}(\epsilon)^T \text{ is positive definite.}$$

However, all eigenvalues of $\Xi_{22}(\epsilon)$ are $O(\epsilon)$ in magnitude, see (7.4.45), and the eigenvalues of $(\Xi_{22}(\epsilon) + 2\epsilon I)^{-1}$ are therefore bounded below by $1/O(\epsilon)$. It thus follows that

$$||\Xi_{12}(\epsilon)||^2 = O(\epsilon ||\Xi_{11}(\epsilon) + 2\epsilon I||),$$

which implies that

$$||\xi_{U(z_k')}(\epsilon)||^2 = O(\epsilon ||\xi(\epsilon)||).$$

■

We create a sequence $\mathrm{face}(\mathcal{K}, z^0) \supseteq \mathrm{face}(\mathcal{K}, z^1) \supseteq \cdots$, as follows. Let

$$z^0 = 0,$$
$$\text{for } k = 0, 1, \ldots, \quad \begin{cases} y^k \in \mathrm{relint}\,(\bar{\mathcal{A}}^\perp \cap \mathrm{face}\,(\mathcal{K}, z^k)^*) \\ z^{k+1} = z^k + y^k_{B(z^k)}. \end{cases} \tag{7.4.48}$$

The level of singularity of a linear subspace $\bar{\mathcal{A}}$ with respect to the cone $\mathcal{K}$, denoted as $d(\bar{\mathcal{A}}, \mathcal{K})$, is by definition the number of iterative steps before (y^k, z^k) is regular, i.e.

$$d(\bar{\mathcal{A}}, \mathcal{K}) := \min\{k \in \{0, 1, 2, \ldots\} \mid (y^k, z^k) \text{ is regular}\}. \tag{7.4.49}$$

Theorem 7.4.1 *It holds that*

$$d(\bar{\mathcal{A}}, \mathcal{K}) \leq \min\left\{p + \sum_{i=p+1}^{q}(n_i - 1),\ \dim \bar{\mathcal{A}},\ \dim \bar{\mathcal{A}}^\perp\right\}.$$

Proof. For $z \in \mathcal{K}$, we let

$$\rho(z) := \sum_{i=1}^{q}[n(\mathcal{K}_i) - 1 - \mathrm{rank}\,(z_i)]_+.$$

In particular, $\rho(0) = p + \sum_{i=p+1}^{q}(n_i - 1)$. Based on Definition 7.3.1, we observe that

- If $y \in \text{face}(\mathcal{K}, z)^*$ is such that (y, z) is not regular, then $\rho(z + y_{B(z)}) \leq \rho(z) - 1$. Namely, if (y, z) is not regular, then there must exist i such that $0 < \text{rank}\,(y_{B(z)})_i < n(\mathcal{K}_i) - \text{rank}\,(z_i)$. This further implies that

$$\left[n(\mathcal{K}_i) - 1 - \text{rank}\,(z + y_{B(z)})_i\right]_+ \leq \left[n(\mathcal{K}_i) - 1 - \text{rank}\,(z_i)\right]_+ - 1.$$

- The above observation implies that if $\rho(z) = 0$ then (y, z) is regular for any $y \in \text{face}(\mathcal{K}, z)^*$.

Since (y^k, z^k) is not regular for $k = 0, 1, 2, \ldots, d(\bar{\mathcal{A}}, \mathcal{K}) - 1$, we have $0 \leq \rho(z^{d(\bar{\mathcal{A}},\mathcal{K})}) \leq \rho(z^{d(\bar{\mathcal{A}},\mathcal{K})-1}) - 1 \leq \cdots \leq \rho(0) - d(\bar{\mathcal{A}}, \mathcal{K})$. Therefore, we have

$$d(\bar{\mathcal{A}}, \mathcal{K}) \leq \rho(0) = p + \sum_{i=p+1}^{q}(n_i - 1). \tag{7.4.50}$$

Let $z \in \mathcal{K}$ and $y \in \bar{\mathcal{A}}^\perp \cap \text{face}(\mathcal{K}, z)^*$. If (y, z) is not regular, then $y_{B(z)} \neq 0$, and hence it is linearly independent of any vector $y' \neq 0$ with $y'_{B(z)} = 0$. Furthermore, it holds that $y_{B(z')} = 0$ for $z' = z + y_{B(z)}$, because $z' = z'_{N(z')}$ and $z = z_{N(z')}$. By induction, it follows that $y^0, y^1, \ldots, y^{d(\bar{\mathcal{A}},\mathcal{K})-1}$ are linearly independent vectors in $\bar{\mathcal{A}}^\perp$, and so

$$d(\bar{\mathcal{A}}, \mathcal{K}) \leq \dim \bar{\mathcal{A}}^\perp. \tag{7.4.51}$$

Let $z \in \mathcal{K}$ and $y \in \text{relint}\,(\bar{\mathcal{A}}^\perp \cap \text{face}(\mathcal{K}, z)^*)$, and $z' = z + y_{B(z)}$. Suppose that (y, z) is not regular, and let $y' \in \bar{\mathcal{A}}^\perp \cap \text{face}(\mathcal{K}, z')$. Since

$$\text{face}(\mathcal{K}, z') \subset \text{face}(\mathcal{K}, z) \subseteq \text{face}(\mathcal{K}, z)^*,$$

we have $y + y' \in \bar{\mathcal{A}}^\perp \cap \text{face}(\mathcal{K}, z)^*$. If $y' \neq 0$, we easily arrive at a contradiction to the fact that y is in the relative interior of $\bar{\mathcal{A}}^\perp \cap \text{face}(\mathcal{K}, z)^*$. This shows that

$$\bar{\mathcal{A}}^\perp \cap \text{face}(\mathcal{K}, z^k) = \{0\} \text{ for } k = 1, 2, \ldots, d(\bar{\mathcal{A}}, \mathcal{K}).$$

Applying (7.3.35), it follows that there exist

$$x^k \in \bar{\mathcal{A}} \cap \text{ relint face}(\mathcal{K}, z^k)^* \text{ for } k = 1, 2, \ldots, d(\bar{\mathcal{A}}, \mathcal{K}).$$

We have $\text{rank}\,(x^k_{B(z^k)}) = n - \text{rank}\,(z^k)$, where $\text{rank}\,(z^k) \geq \text{rank}\,(z^{k-1}) + 1$, for all $1 \leq k \leq d(\bar{\mathcal{A}}, \mathcal{K})$. This implies that $x^1, x^2, \ldots, x^d$ are linearly independent vectors in $\bar{\mathcal{A}}$, and hence

$$d(\bar{\mathcal{A}}, \mathcal{K}) \leq \dim \bar{\mathcal{A}}. \tag{7.4.52}$$

■

Lemma 7.4.2 *Suppose that* $\{x(\epsilon) \mid 0 < \epsilon \leq 1\}$ *is such that*

$$\operatorname{dist}(x(\epsilon), \bar{\mathcal{A}}) \leq \epsilon, \quad \lambda_{\min}(x(\epsilon)) \geq -\epsilon,$$

for all $0 < \epsilon \leq 1$. *Let*

$$z := z^{d(\bar{\mathcal{A}},\mathcal{K})+1}, \quad \gamma := 2^{-d(\bar{\mathcal{A}},\mathcal{K})}.$$

Then $\operatorname{face}(\mathcal{K}, z)$ *is the minimal cone for* $\bar{\mathcal{A}}$, *and*

$$\|x_{U(z)}(\epsilon)\| + \|x_{N(z)}(\epsilon)\| = O(\epsilon^{\gamma}\|x\|^{1-\gamma}).$$

Proof. By definition of $d(\bar{\mathcal{A}}, \mathcal{K})$, we know that $(y^{d(\bar{\mathcal{A}},\mathcal{K})}, z^{d(\bar{\mathcal{A}},\mathcal{K})})$ is regular, and hence $\operatorname{face}(\mathcal{K}, z)$ is the minimal cone for $\bar{\mathcal{A}}$, see Lemma 7.3.3.

Since $z^0 = 0$, we have $x_{U(z^0)}(\epsilon) = 0$ and $x_{N(z^0)}(\epsilon) = 0$. If $d(\bar{\mathcal{A}}, \mathcal{K}) = 0$, i.e. if (y^0, z^0) is regular, then the lemma follows immediately from Lemma 7.4.1. Otherwise, i.e. if (y^0, z^0) is not regular, we obtain from Lemma 7.4.1 that

$$\|x_{N(z^1)}(\epsilon)\| = O(\epsilon), \quad \|x_{U(z^1)}(\epsilon)\| = O(\sqrt{\epsilon\|x(\epsilon)\|}) \tag{7.4.53}$$

Let $k \in \{1, 2, \ldots, d(\bar{\mathcal{A}}, \mathcal{K}) - 1\}$, and suppose that

$$\|x_{N(z^k)}(\epsilon)\| = O(\epsilon^{2\phi}\|x(\epsilon)\|^{1-2\phi}), \quad \|x_{U(z^k)}(\epsilon)\| = O(\epsilon^{\phi}\|x(\epsilon)\|^{1-\phi}), \tag{7.4.54}$$

where $\phi := 2^{-k}$. Since (y^k, z^k) is not regular, we obtain from Lemma 7.4.1 that (7.4.54) also holds for $k + 1$. By induction, this proves that (7.4.54) holds for $k = d(\bar{\mathcal{A}}, \mathcal{K})$. For this particular k however, (y^k, z^k) is regular, and the lemma thus follows by applying Lemma 7.4.1 once more. ■

We now arrive at a general error bound result for the system (7.1.1).

Theorem 7.4.2 *Let* $\bar{\mathcal{A}} := \{x \in \Re^N \mid x + tb \in \mathcal{A} \text{ for some } t \in \Re\}$. *Suppose that* $\{x(\epsilon) \mid 0 < \epsilon \leq 1\}$ *is such that* $\|x(\epsilon)\|$ *is bounded, and*

$$\operatorname{dist}(x(\epsilon), \bar{\mathcal{A}}) \leq \epsilon, \quad \lambda_{\min}(x(\epsilon)) \geq -\epsilon,$$

for all $0 < \epsilon \leq 1$. *Then*

$$\operatorname{dist}(x(\epsilon), (b + \mathcal{A}) \cap \mathcal{K}) = O(\epsilon^{2^{-d(\bar{\mathcal{A}},\mathcal{K})}}).$$

Proof. This follows immediately by combining Lemma 7.3.2 and Lemma 7.4.2. ■

7.5 INFEASIBLE SYSTEMS

There is an extension of Farkas' lemma from linear inequalities to convex cones, which states that

$$\mathrm{dist}(b+\mathcal{A},\Gamma)>0 \Longleftrightarrow \exists y\in\mathcal{A}^{\perp}\cap\Gamma^{*} \text{ such that } b^{T}y<0. \tag{7.5.55}$$

where $\Gamma\subset\Re^{N}$ is a convex cone, and $\Gamma^{*}:=\{z\in\Re^{N}\mid z^{T}\Gamma\subseteq\Re_{+}\}$ is the associated dual cone. See e.g. Lemma 2.5 in [756].

Recall now from Lemma 7.3.2 that the system (7.1.1) is infeasible if and only if

$$\mathrm{dist}(b+\mathcal{A},\ \mathrm{face}\,(\mathcal{K},z^{*}))>0.$$

Applying (7.5.55), it follows that (7.1.1) is infeasible if and only if

$$\exists y\in\mathcal{A}^{\perp}\cap\ \mathrm{face}\,(\mathcal{K},z^{*})^{*} \text{ such that } b^{T}y<0. \tag{7.5.56}$$

We have the following characterization of infeasibility.

Theorem 7.5.1 *Let $\bar{\mathcal{A}}:=\{x\in\Re^{N}\mid x+tb\in\mathcal{A}$ for some $t\in\Re\}$. It holds that*

$$(b+\mathcal{A})\cap\mathcal{K}=\emptyset$$

if and only if

$$\exists y\in\mathcal{A}^{\perp}\cap\ \mathrm{face}\,(\mathcal{K},z^{d(\bar{\mathcal{A}},\mathcal{K})+1})^{*} \textit{ such that } b^{T}y<0.$$

Proof. We only need to observe the following equivalence:

$$\begin{aligned}(b+\mathcal{A})\cap\mathcal{K}=\emptyset &\Longleftrightarrow (b+\mathcal{A})\cap\ \mathrm{face}\,(\mathcal{K},z^{d(\mathcal{A},\mathcal{K})+1})^{*}=\emptyset\\ &\Longleftrightarrow \mathrm{dist}(b+\mathcal{A},\ \mathrm{face}\,(\mathcal{K},z^{d(\mathcal{A},\mathcal{K})+1})^{*})>0\\ &\Longleftrightarrow \exists y\in\mathcal{A}^{\perp}\cap\ \mathrm{face}\,(\mathcal{K},z^{d(\bar{\mathcal{A}},\mathcal{K})+1})^{*}\\ &\qquad \text{such that } b^{T}y<0.\end{aligned}$$

∎

It follows that for infeasible systems, there is a certificate of infeasibility, viz. the vector y defined in Theorem 7.5.1, along with the vectors $y^{0},y^{1},\ldots,y^{d(\bar{\mathcal{A}},\mathcal{K})}$, as defined in (7.4.48). Checking feasibility of such a certificate is straightforward, see (7.2.26). Furthermore, Theorem 7.4.1 gives an upper-bound on $d(\bar{\mathcal{A}},\mathcal{K})$. Therefore, (7.1.1) is an NP-decision problem, in the real number model. For the special case of linear matrix inequalities, this was already known from Ramana [653].

Remark that if $(y^{0},0)$ is regular, i.e. if $d(\bar{\mathcal{A}},\mathcal{K})=0$, then there exists $\delta>0$ such that $y^{0}+\delta y\in\mathcal{K}$, where y is specified in Theorem 7.5.1. Notice also that the requirement $y^{0}\in\bar{\mathcal{A}}^{\perp}$ is equivalent to $y^{0}\in\mathcal{A}^{\perp}$ and $b^{T}y^{0}=0$. Hence, letting $y':=y^{0}+\delta y$, we have

$$b^{T}(y')<0,\quad y'\in\mathcal{A}^{\perp}\cap\mathcal{K}. \tag{7.5.57}$$

Thus, for regular systems, Farkas dual solutions are of the form (7.5.57). Since in particular, linear systems are always regular, this yields the classical Farkas duality as a special case.

By taking a positive multiple of y', we can always obtain a vector y satisfying

$$b^T y \leq -1, \quad y \in \mathcal{A}^\perp \cap \mathcal{K}, \tag{7.5.58}$$

which is called a dual improving direction. For (possibly irregular) infeasible systems, it is known that there exist approximate dual improving directions, with arbitrarily small constraint violations. See e.g. Lemma 2.6 in [756]. The next theorem gives an upper bound for the minimal norm of such approximate dual improving directions in the case of infeasibility.

Theorem 7.5.2 *Let $\bar{\mathcal{A}} := \{x \in \Re^N \mid x + tb \in \mathcal{A}$ for some $t \in \Re\}$. If $(b+\mathcal{A}) \cap \mathcal{K} = \emptyset$ then there exist $\{y(\epsilon) \mid \epsilon > 0\}$ such that for all $0 < \epsilon \leq 1$, it holds that*

$$\operatorname{dist}(y(\epsilon), \mathcal{A}^\perp) = O(\epsilon), \quad [1 + b^T y(\epsilon)]_+ = O(\epsilon), \quad [-\lambda_{\min}(y(\epsilon))]_+ = O(\epsilon),$$

and

$$\|y(\epsilon)\| = O(\epsilon^{1-2^{d(\bar{\mathcal{A}},\mathcal{K})}}).$$

The solutions $y(\epsilon)$ in Theorem 7.5.2 can easily be constructed from y^0, $y^1, \ldots, y^{d(\bar{\mathcal{A}},\mathcal{K})}$ and the vector y as specified in Theorem 7.5.1. (The details are omitted.)

We have already seen that if $\bar{\mathcal{A}}$ is regular with respect to $\mathcal{K}$, then a dual improving direction exists if and only if (7.1.1) is infeasible, i.e. $(b+\mathcal{A}) \cap \mathcal{K} = \emptyset$. Irrespective of regularity, it is known that a dual improving direction exists if and only if $\operatorname{dist}(b + \mathcal{A}, \mathcal{K}) > 0$, which is known as strong infeasibility, see [508]. If a system is weakly infeasible, i.e. $(b + \mathcal{A}) \cap \mathcal{K} = \emptyset$ and $\operatorname{dist}(b + \mathcal{A}, \mathcal{K}) = 0$, there exist approximate solutions with arbitrarily small backward error. The following theorem gives a lower bound on the norm of approximate solutions to infeasible systems.

Theorem 7.5.3 *Let $\bar{\mathcal{A}} := \{x \in \Re^N \mid x + tb \in \mathcal{A}$ for some $t \in \Re\}$. Suppose that*

$$(b + \mathcal{A}) \cap \mathcal{K} = \emptyset.$$

If $\{x(\epsilon) \mid \epsilon > 0\}$ is such that

$$\operatorname{dist}(x(\epsilon), b + \mathcal{A}) \leq \epsilon \text{ and } \lambda_{\min}(x(\epsilon)) \geq -\epsilon \quad \text{for all } \epsilon > 0,$$

then $d(\bar{\mathcal{A}}, \mathcal{K}) \geq 1$. Moreover, for ϵ small enough, we have $x(\epsilon) \neq 0$ and

$$\frac{1}{\|x(\epsilon)\|} = O(\epsilon^{1/(2^{d(\bar{\mathcal{A}},\mathcal{K})}-1)}).$$

Proof. Let $\gamma := 2^{-d(\bar{\mathcal{A}},\mathcal{K})}$. Suppose to the contrary that

$$\epsilon_k^{\gamma/(\gamma-1)} \|x(\epsilon_k)\| \to 0$$

for some decreasing sequence $\{\epsilon_k\}$. Then we obtain from Lemma 7.3.2 and Lemma 7.4.2 that $(b + \mathcal{A}) \cap \mathcal{K} \neq \emptyset$, contradicting infeasibility. ∎

7.6 SYSTEMS OF QUADRATIC INEQUALITIES

In this section, the error bound for second order cones (Theorem 7.4.2) will be applied to systems of generalized convex quadratic inequalities. Such systems are of the form

$$\mathcal{S} := \{y \in \Re^l \mid g_i(y) \le 0 \text{ for } i = 1, 2, \ldots, m\}, \tag{7.6.59}$$

where each $g_i : \Re^l \to \Re$ is a convex or quasi-convex quadratic function. The backward error of an approximate solution y is defined as

$$\text{backward error } = \sum_{i=1}^{m} [g_i(y)]_+. \tag{7.6.60}$$

Let $i \in \{1, 2, \ldots, m\}$. Since $g_i(y)$ is quadratic, we have

$$g_i(y) = g_i(0) + (\nabla g_i(0))^T y + \frac{1}{2} y^T (\nabla^2 g_i) y \quad \text{for all } y \in \Re^l.$$

We will first consider the case that $g_i(\cdot)$ is convex for all $i = 1, 2, \ldots, m$.

7.6.1 *Convex quadratic systems*

Convex quadratic systems are known to have many attractive theoretical properties. We will give alternative proofs of several of these properties below, using Theorem 7.4.2. We first characterize the lower level set of a convex quadratic inequality in terms of the second order cone.

Lemma 7.6.1 *Let $g : \Re^l \to \Re$ be a convex quadratic function, and factorize its Hessian as $\nabla^2 g = LL^T$, where L is an $l \times r$ matrix, $r = \operatorname{rank}(\nabla^2 g)$. Let*

$$b := \begin{bmatrix} \frac{1}{2} - g(0) \\ \frac{1}{2} + g(0) \\ 0 \end{bmatrix}, \; A := \begin{bmatrix} -\nabla g(0), & \nabla g(0), & L \end{bmatrix}.$$

Denote the two eigenvalues of $b + A^T y$ with respect to the second order cone $\mathrm{SOC}(2+r)$ by $\lambda_1(b+A^T y)$ and $\lambda_2(b+A^T y)$, $\lambda_1(b+A^T y) \le \lambda_2(b+A^T y)$. Then $\lambda_2(b + A^T y) \ge 1/\sqrt{2}$ and

$$[g(y)]_+ = [-\lambda_1(b + A^T y)]_+ \lambda_2(b + A^T y) \ge [-\lambda_1(b + A^T y)]_+/\sqrt{2}.$$

This implies in particular that $g(y) \le 0$ if and only if $b + A^T y \in \mathrm{SOC}(2 + r)$.

Proof. Let $x := b + A^T y = \begin{bmatrix} x_1, & x_2, & x_3^T \end{bmatrix}^T$, with

$$x_1 = \frac{1}{2} - (g(0) + \nabla g(0)^T y), \quad x_2 = \frac{1}{2} + g(0) + \nabla g(0)^T y, \quad x_3 = L^T y.$$

From (7.2.5), it follows that

$$\lambda_1(x) = \left(x_1 - \sqrt{x_2^2 + \|x_3\|^2}\right)/\sqrt{2}, \quad \lambda_2(x) = \left(x_1 + \sqrt{x_2^2 + \|x_3\|^2}\right)/\sqrt{2}.$$

Since $x_1 = 1 - x_2$, we have

$$\sqrt{2}\lambda_2(x) \geq 1 - x_2 + |x_2| \geq 1. \tag{7.6.61}$$

Moreover,

$$\begin{aligned} \lambda_1(x)\lambda_2(x) &= \frac{x_1^2 - x_2^2 - \|x_3\|^2}{2} \\ &= -\left(g(0) + \nabla g(0)^T y + \frac{\|L^T y\|^2}{2}\right) \\ &= -g(y). \end{aligned} \tag{7.6.62}$$

The lemma follows immediately from (7.6.61)–(7.6.62). ■

We can now formulate the convex quadratic system (7.6.59) as a second order cone system of the form (7.1.1). To this end, we define

$$A = \begin{bmatrix} A_1, & A_2, & \ldots, & A_m \end{bmatrix}, b = \begin{bmatrix} b_1^T, & b_2^T, & \ldots, & b_m^T \end{bmatrix}^T, \tag{7.6.63}$$

with

$$A_i := \begin{bmatrix} -\nabla g_i(0), & \nabla g_i(0), & L_i \end{bmatrix}, \; L_i L_i^T = \nabla^2 g_i, \tag{7.6.64}$$

and

$$b_i = \begin{bmatrix} \frac{1}{2} - g_i(0), & \frac{1}{2} + g_i(0), & 0^T \end{bmatrix}^T, \tag{7.6.65}$$

for $i = 1, 2, \ldots, m$. Furthermore, we let

$$\mathcal{K} := \mathrm{SOC}(2 + r_1) \times \mathrm{SOC}(2 + r_2) \times \cdots \times \mathrm{SOC}(2 + r_m), \tag{7.6.66}$$

where r_i denotes the number of columns of L_i. It follows from Lemma 7.6.1 that

$$\mathcal{S} = \{y \in \Re^l \mid b + A^T y \in \mathcal{K}\}, \tag{7.6.67}$$

where $\mathcal{S}$ is the solution set of the quadratic system in (7.6.59).

Lemma 7.6.2 *Suppose that $\mathcal{K}$ can be decomposed as $\mathcal{K} = \mathcal{K}_1 \times \mathcal{K}_2$. Consider a conic system*

$$\{y \mid b_1 + A_1^T y \in \mathcal{K}_1, \, b_2 + A_2^T y \in \mathcal{K}_2\} = \{y \mid b + A^T y \in \mathcal{K}\},$$

where

$$b = \begin{bmatrix} b_1 \\ b_2 \end{bmatrix}, \; A = \begin{bmatrix} A_1, & A_2 \end{bmatrix}.$$

Let

$$\bar{\mathcal{A}}_2 := \{tb_2 + A_2^T y \mid t \in \Re, y \in \Re^l\}$$

and

$$\bar{\mathcal{A}} := \{tb + A^T y \mid t \in \Re, y \in \Re^l\}.$$

If there exists $t^* \in \Re$ and $y^* \in \Re^l$ such that $t^* b_1 + A_1^T y^* \in \operatorname{int} \mathcal{K}_1$ and $t^* b_2 + A_2^T y^* \in \mathcal{K}_2$ then

$$d(\bar{\mathcal{A}}, \mathcal{K}) = d(\bar{\mathcal{A}}_2, \mathcal{K}_2).$$

Proof. Consider an arbitrary dual direction $z \in \bar{\mathcal{A}}^\perp \cap \mathcal{K}$, i.e.,

$$z = [\ z_1^T,\ \ z_2^T\ ]^T$$

with

$$z_1 \in \mathcal{K}_1,\ z_2 \in \mathcal{K}_2,\ A_1 z_1 + A_2 z_2 = 0,\ b_1^T z_1 + b_2^T z_2 = 0.$$

Letting $x^* = t^* b + A^T y^*$, we have

$$0 = (x^*)^T z \geq (x_1^*)^T z_1,$$

where we used that $x_2^*, z_2 \in \mathcal{K}_2$. Since $x_1^* \in \operatorname{int} \mathcal{K}_1$, and $z_1 \in \mathcal{K}_1$, the above relation implies that $z_1 = 0$. We conclude that

$$(z_1, z_2) \in \bar{\mathcal{A}}^\perp \cap \mathcal{K} \iff z_1 = 0,\ z_2 \in \bar{\mathcal{A}}_2^\perp \cap \mathcal{K}_2.$$

It is now easy to see from (7.4.48)–(7.4.49) that

$$d(\bar{\mathcal{A}}, \mathcal{K}) = d(\bar{\mathcal{A}}_2, \mathcal{K}_2).$$

■

A nice property of convex quadratic inequalities is that a global error bound holds, i.e. no boundedness assumptions have to be made.

Theorem 7.6.1 *Let $g_i : \Re^l \to \Re$, $i = 1, 2, \ldots, m$ be convex quadratic functions. If $\{y(\epsilon) \mid 0 < \epsilon \leq 1\}$ is such that*

$$[g_i(y(\epsilon))]_+ \leq \epsilon \text{ for all } i = 1, 2, \ldots, m,$$

then

$$\operatorname{dist}(y(\epsilon), \mathcal{S}) = O(\epsilon^{2^{-d}}),$$

where d is the degree of singularity. More precisely, letting A, b and $\mathcal{K}$ as in (7.6.63)–(7.6.66), and

$$\bar{\mathcal{A}} := \{x \mid x = A^T y + tb \text{ for some } y \in \Re^l,\ t \in \Re\},$$

it holds that $d = d(\bar{\mathcal{A}}, \mathcal{K})$.

Proof. Consider an $l \times l$ orthogonal projection matrix P for which

$$g_i(Py) = g_i(y) \text{ for all } i = 1, 2, \ldots, m. \tag{7.6.68}$$

We claim

$$\operatorname{dist}(Py(\epsilon), \mathcal{S}) = \operatorname{dist}(y(\epsilon), \mathcal{S}).$$

Namely, consider an arbitrary $\epsilon > 0$ and suppose that $\text{dist}(Py(\epsilon), \mathcal{S}) = ||Py(\epsilon) - s||$ for some $s \in \mathcal{S}$. Let us write $y(\epsilon) = Py(\epsilon) + u$ for some u in the null space of P. By the property (7.6.68), we have

$$g_i(s+u) = g_i(Ps + Pu) = g_i(Ps) = g_i(s) \leq 0,$$

where the last step is due to $s \in \mathcal{S}$. This shows $s + u \in \mathcal{S}$. Thus

$$\begin{aligned} \text{dist}(y(\epsilon), \mathcal{S}) &\leq ||y(\epsilon) - (s+u)|| \\ &= ||(Py(\epsilon) + u) - (s+u)|| \\ &= ||Py(\epsilon) - s|| \\ &= \text{dist}(Py(\epsilon), \mathcal{S}). \end{aligned}$$

Similarly, we can show $\text{dist}(Py(\epsilon), \mathcal{S}) \leq \text{dist}(y(\epsilon), \mathcal{S})$, so we have

$$\text{dist}(Py(\epsilon), \mathcal{S}) = \text{dist}(y(\epsilon), \mathcal{S}).$$

We will prove Theorem 7.6.1 for quadratic system (7.6.59), where there exists P of rank at most k, such that (7.6.68) holds. We use induction on the rank of P, such that the lemma follows by setting $k = l$, $P = I$.

If the rank of P is zero, then $Py(\epsilon) = 0$ for all ϵ, and hence bounded. Notice from Lemma 7.6.1 that the backward error of $Py(\epsilon)$ with respect to the conic representation is $\sqrt{2}\epsilon$. The error bound thus follows as an application of Theorem 7.4.2.

Consider now the case that $\text{rank}\,(P) = k > 0$, and assume by induction that Theorem 7.6.1 holds for all convex quadratic systems that satisfy (7.6.68) for some $\widetilde{P}$ of rank at most $k-1$. If $\{Py(\epsilon) \mid 0 < \epsilon \leq 1\}$ is bounded, then

$$\text{dist}(Py(\epsilon), \mathcal{S}) = \text{dist}(y(\epsilon), \mathcal{S}) = O(\epsilon^{2^{-d}})$$

due to Theorem 7.4.2. Otherwise, there must exist a decreasing sequence $\epsilon_1, \epsilon_2, \cdots$ in $(0, 1)$ such that

$$\lim_{j \to \infty} ||Py(\epsilon_j)|| = \infty.$$

Let $\bar{y}$ be a cluster point of $\{Py(\epsilon_j)/||Py(\epsilon_j)|| | j = 1, 2, \ldots\}$. Define A, b and $\mathcal{K}$ as in (7.6.63)-(7.6.66). Since for all $i \in \{1, 2, \ldots, m\}$, we have $[g_i(Py(\epsilon))]_+ = O(\epsilon)$, it holds that $[-\lambda_{\min}(b + A^T Py(\epsilon))]_+ = O(\epsilon)$, see Lemma 7.6.1, and hence $A^T \bar{y} \in \mathcal{K}$. Notice now from (7.6.64) that

$$A_i^T \bar{y} = \left[\begin{array}{ccc} -\nabla g_i(0)^T \bar{y}, & \nabla g_i(0)^T \bar{y}, & L_i^T \bar{y} \end{array} \right].$$

Therefore, the fact that $A^T \bar{y} \in \mathcal{K}$ implies that

$$L_i^T \bar{y} = 0, \quad \nabla g_i(0)^T \bar{y} \leq 0, \quad \text{for all } i = 1, 2, \ldots, m. \tag{7.6.69}$$

Note that because $\bar{y}^T(\nabla^2 g_i)\bar{y} = ||L_i^T \bar{y}||^2 = 0$, we have

$$g_i(y + t\bar{y}) = g_i(y) + t\nabla g_i(0)^T \bar{y}, \quad \text{for all } y \in \Re^l,\ t \in \Re,\ i \in \{1, 2, \ldots, m\}. \tag{7.6.70}$$

Let

$$\mathcal{I} := \{i \in \{1,2,\ldots,m\} \mid \nabla g_i(0)^T \bar{y} = 0\},$$

and define

$$P' := (I - \bar{y}\bar{y}^T)P. \tag{7.6.71}$$

Remark that P' is a projection matrix of rank $k-1$, because $\|\bar{y}\| = 1$ and $\bar{y}$ is in the image of P. Furthermore, we have by definition of $\mathcal{I}$ and relation (7.6.69) that

$$b_i + A_i^T P' y = b_i + A_i^T P y = b_i + A_i^T y, \quad \text{for all } i \in \mathcal{I},\ y \in \Re^l,$$

and therefore

$$g_i(P'y) = g_i(y), \quad \text{for all } i \in \mathcal{I},\ y \in \Re^l.$$

By induction, there must exist $u(\epsilon) \in \Re^l$ such that

$$g_i(y(\epsilon) + u(\epsilon)) \le 0, \quad \text{for all } i \in \mathcal{I}, \quad \|u(\epsilon)\| = O\left(\epsilon^{2^{-\tilde{d}}}\right), \tag{7.6.72}$$

where $\tilde{d}$ is the level of singularity of the system that involves only the inequalities $g_i(y) \le 0$, $i \in \mathcal{I}$. Using (7.6.70) and the definitions of $\mathcal{I}$ and P', it further follows that

$$g_i(y(\epsilon) + u(\epsilon) + t\bar{y}) \le 0, \quad \text{for all } i \in \mathcal{I},\ t \in \Re. \tag{7.6.73}$$

Define

$$\bar{t} := \max\{[g_i(y(\epsilon) + u(\epsilon))]_+ / (-\nabla g_i(0)^T \bar{y}) \mid i \in \{1,2,\ldots,m\} \setminus \mathcal{I}\}.$$

We immediately get from (7.6.70) and (7.6.73) that

$$g_i(y(\epsilon) + u(\epsilon) + t\bar{y}) < 0, \quad \text{for all } i \in \{1,2,\ldots,m\} \setminus \mathcal{I},\ t \ge \bar{t}. \tag{7.6.74}$$

Due to (7.6.73) and (7.6.74), we may apply Lemma 7.6.2 to conclude that $\tilde{d} = d(\bar{\mathcal{A}}, \mathcal{K})$. Since $u(\epsilon) = O(\epsilon^\gamma)$, $\gamma := 2^{-\tilde{d}}$, we also have $\bar{t} = O(\epsilon^\gamma)$, concluding the proof. ∎

Our error bound for convex quadratic systems is a slight improvement of the error bound given by Wang and Pang [822], which in turn generalizes an error bound of Luo and Luo [505]. Wang and Pang have a different definition for the degree of singularity, namely the number of constraints for which a strict inequality $g_i(y) < 0$ cannot be satisfied. This is clearly an upper bound for our definition of singularity; the latter can be smaller, since one regularization step can regularize multiple quadratic inequalities at a time.

Since for quadratic systems, the error bound holds globally, such systems cannot be weakly infeasible. This implies, among others, that the optimum of a convex quadratically constrained convex quadratic program must always be attained. See also Luo and Zhang [510].

Notice also that the quadratic functions g_i are linear if and only if $r_i = 0$, in which case $\mathcal{K} = \text{SOC}(2)^m$. It is easily seen that SOC(2) is merely an orthogonal transformation of $\Re^2_+$, and hence the system is regular. We thus arrive at Hoffman's error bound as a special case.

7.6.2 Generalized convex quadratic systems

An advantage of the approach with second order cones is that it extends to quadratic functions $g_i(\cdot)$ that are quasi-convex on a properly chosen domain.

Lemma 7.6.3 *Let $g : \Re^l \to \Re$ be a quadratic function, for which the Hessian has only one negative eigenvalue, i.e.*

$$\nabla^2 g = LL^T - hh^T,$$

for some $l \times (r-1)$ matrix L, $r = \operatorname{rank}(\nabla^2 g)$, and a vector $h \in \Re^l$. Suppose further that there exist $\lambda \in \Re^l$ such that

$$(\nabla^2 g)\lambda = \nabla g(0), \quad g(\lambda) \geq 0.$$

Let

$$x := \begin{bmatrix} |h^T(y+\lambda)| \\ \sqrt{2g(\lambda)} \\ L^T(y+\lambda) \end{bmatrix}.$$

Denote the two eigenvalues of x with respect to the second order cone SOC$(1+r)$ *by $\lambda_1(x) \leq \lambda_2(x)$. Then*

$$[g(y)]_+ = [-\lambda_1(x)]_+ \lambda_2(x) \geq [-\lambda_1(x)]_+^2.$$

This implies in particular that $g(y) \leq 0$ if and only if $x \in$ SOC$(1+r)$.

Proof. Partition $x = \begin{bmatrix} x_1, & x_2, & x_3^T \end{bmatrix}^T$. From (7.2.5), it follows that

$$\lambda_1(x) = \left(x_1 - \sqrt{x_2^2 + \|x_3\|^2}\right)/\sqrt{2}, \quad \lambda_2(x) = \left(x_1 + \sqrt{x_2^2 + \|x_3\|^2}\right)/\sqrt{2},$$

and therefore

$$\begin{aligned} \lambda_1(x)\lambda_2(x) &= \frac{x_1^2 - x_2^2 - \|x_3\|^2}{2} \\ &= -\left(g(\lambda) + \frac{1}{2}(y+\lambda)^T(LL^T - hh^T)(y+\lambda)\right) \\ &= -g(y). \end{aligned}$$

Since the largest eigenvalue ($\lambda_2(x)$) is nonnegative, the lemma follows from the above relation. ■

The above second order cone characterization of merely quasi-convex quadratic functions involves the absolute value function $|h^T(y+\lambda)|$. Therefore, the set $\{y \mid g(y) \leq 0\}$ is the union of two convex regions. Remark also that if the quadratic formulation has a backward error $[g(y)]_+ \geq \epsilon$, then the conic formulation has a backward error $[-\lambda_{\min}(x)]_+ \leq \sqrt{\epsilon}$. Applying Theorem 7.4.2, we obtain an error bound for generalized convex quadratic systems. Unlike the convex case, we need to make a boundedness assumption.

Theorem 7.6.2 *Consider the quadratic system*

$$\mathcal{S} := \{y \in \Re^l \mid g_i(y) \leq 0, \quad \textit{for } i = 1, 2, \ldots, m\},$$

where for each $i = 1, 2, \ldots, m$, the function $g_i(\cdot)$ is quadratic, and either convex or merely quasi-convex. Let $\{y(\epsilon) \mid 0 < \epsilon \leq 1\}$ be such that $||y(\epsilon)||$ is bounded, and

$$\sum_{i=1}^{m} [g_i(y(\epsilon))]_+ \leq \epsilon.$$

Then

$$\text{dist}(y(\epsilon), \mathcal{S}) = O(\epsilon^{2^{-(d+1)}}),$$

where d is the degree of singularity of the system. In particular,

$$d \leq \min\{m, l+1\}.$$

II ALGORITHMS

The primary purpose of this book is to provide a unified body of theory and methods of transforming a constrained minimization problem into a sequence of unconstrained minimizations of an appropriate auxiliary function. The auxiliary functions considered are those that define "interior point" and "exterior point" methods which are characterized respectively according to whether the constraints are strictly satisfied by the minimizing sequence. Initial emphasis is on generality, and the central convergence theorems apply to the determination of local solutions of a nonconvex programming problem. Strong global and dual results follow for convex programming; a particularly important example is the fact that the exterior point methods do not require the Kuhn-Tucker constraint qualification [453] to ensure convergence or characterize optimality.

In addition to giving a rather comprehensive exposition of the rich theoretical foundation uncovered for this class of methods, we wish to emphasize the demonstrated practical applicability of their various realizations. This has been brought about largely by the adaptation and further development of effective computational algorithms for calculating an unconstrained minimum and by the development of special extrapolation techniques for accelerating convergence. Significant progress has also been made in the development of computational techniques that exploit the special structures characterizing large classes of problems. In addition to these efficiencies, which have been effected for the method proper, some exploration has been done in combining the present methods with other mathematical programming algorithms to obtain even more efficient composite algorithms.

Fiacco and McCormick 1968 [229] and second edition [230]

If I have seen further than others, it is by standing upon the shoulders of giants.

Newton, Isaac (1642-1727) b. Woolsthorpe, England

On how he made discoveries
By always thinking unto them. I keep the subject constantly before me and wait till the first dawnings open little by little into the full light.

Newton, Isaac (1642-1727) b. Woolsthorpe, England

8 SYMMETRIC CONES, POTENTIAL REDUCTION METHODS AND WORD-BY-WORD EXTENSIONS

Farid Alizadeh, Stefan Schmieta

8.1 INTRODUCTION

This is the first of three chapters in this book dealing with polynomial time complexity of interior point algorithms for semidefinite programming (SDP). As such it deals with, in a sense, the easiest class of algorithms for which polynomial time convergence can be established. More precisely, we present a "recipe" whereby polynomial time convergence proofs in linear programming (LP) can be extended "word-by-word" to analogous proofs in SDP. Our presentation closely follows [21].

We should mention some historical background. The program of "Word-by-Word" extension from LP to SDP was quite successful in a number of polynomial time algorithms, in particular, for those methods that were mostly primal oriented or mostly dual oriented. The recipe presented in [21] however, fails in primal-dual methods, for instance those put forward by Kojima et al in [431, 434] and Monteiro and Adler [538]. Such methods require somewhat more sophisticated techniques than their analogs in linear programming. This may seem disappointing at first, but as a byproduct we get not just one, but a number of primal-dual classes, each with its own nuance and characteristics. The subsequent two chapters present a detailed account of primal-dual algorithms.

In this chapter we focus on the situations where the "recipe" *does* work. More precisely, we examine certain LP potential reduction algorithms and show that they extend to SDP in a mechanical way. To be concrete we have chosen two particular algorithms from LP, both due to Ye [852, 853]. We demonstrate how the "recipe" may be used to extend these algorithms to semidefinite programming. It may prove useful if the reader has a copy of each of the two papers and matches the analysis presented in this chapter against their original counterparts.

Before proceeding to the extension, we should discuss the connection of semidefinite programming to the more general field of cone-LP optimization over *symmetric cones* (they are defined in § 8.3). First, let us provide some background and motivation. Shortly after [21], Nemirovskii and Scheinberg [569] presented a similar technique for cone-LP optimization problems over the *Lorenz* cone. We discuss the details later in this chapter, but the bottom line is that the "recipe" mentioned above worked for this class of cone-LP problems as well[1]. Cone-LP over the Lorenz cone is equivalent to the important problem of convex quadratically constrained quadratic programming (QCQP), and the Nemirovskii-Scheinberg paper revealed the very close connection between LP and QCQP in the context of interior point algorithms. Just like semidefinite programming, the Nemirovskii-Scheinberg extension does not work for all interior point algorithms. In particular, again as in SDP, it fails for the primal-dual methods. This brings us to the following two questions:

1. What is the largest class of optimization problems to which the "recipe" can be applied with both algorithms and polynomial-time analysis extending in a "word-by-word" fashion? In other words how far can we push the scheme of Alizadeh and Nemirovskii-Scheinberg?

2. Are the various more sophisticated techniques used for the primal-dual methods for SDP (and discussed in chapters 9 and 10) applicable to a more general setting than SDP?

In this chapter we study the first question and provide an answer. The second question is the subject of current research and is beyond the scope of the present chapter; the interested reader may wish to consult [697, 698].

Below we establish that the "recipe" for word by word extension of some interior point algorithms to SDP actually extends to all cone-LP problems over symmetric cones. Symmetric cones are intimately connected to *Euclidean Jordan algebras*; indeed we need the machinery of these algebras as a basic tool for carrying out our analysis.

The potential reduction methods studied in this chapter actually apply to all cone-LP problems for arbitrary convex cones. This was established by the seminal research of Nesterov and Nemirovskii [583]. However the extension to this level of generality requires the new concept of *self-concordant barrier functions* and the related notion of *normal barriers*. O. Güler showed that

[1]Nemirovskii and Scheinberg extended Karmarkar's original algorithm. But their methods would have worked just as well for other algorithms, say the two Ye methods discussed in this chapter.

this theory, when applied to symmetric cones, becomes especially simple and transparent. This is true because of the rich theory of these cones and their intimate connection to Euclidean Jordan algebras. Our approach in this chapter is in a sense in the opposite direction of Güler's. While he uses a top-down approach, specializing Nesterov-Nemirovskii theory to symmetric cones, we take a bottom-up approach and show that certain linear programming interior point methods and their analysis are actually applicable verbatim to symmetric cone optimization.

Simultaneously with Güler, Nesterov and Todd [587, 588] studied primal-dual interior point algorithms at the level of optimization problems over symmetric cones. Later Faybusovich [222, 224] also recognized the significance of Euclidean Jordan algebras in the context of cone LP over symmetric cones; in particular he used the machinery of these algebras to extend some primal-dual SDP methods to symmetric cone optimization. He also used Jordan algebraic techniques to extend the notions of degeneracy and strict complementarity from SDP to such optimization problems [223]. Tsuchyia in [791, 792] and Monteiro and Tsuchyia in [542] extended primal-dual methods to the cone LP over the Lorenz cone. Finally the authors of this chapter, in [697, 698], showed how to extend primal-dual methods from SDP to symmetric cone optimization in a "word-by-word" manner.

In §8.2 we briefly discuss general cone LP problems and review how symmetric cones fit this framework. In §8.3 we present the necessary background for Euclidean Jordan algebras and symmetric cones. Finally in §8.4 we proceed to outline the potential reduction interior point algorithms. We establish fundamental theorems about these methods which ultimately lead to polynomial time convergent algorithms. In the process we clarify the meaning of "polynomial time convergence" and contrast polynomiality results to linear programming. The key point here is that the analysis is a "word by word" extension of the corresponding LP algorithms. The bulk of this section is a generalization of results in [21] to optimization problems over symmetric cones.

Let us make a point about terminology:

> *For reasons outlined above and to become further clear in the following sections, in this chapter, and only in this chapter, the term "semidefinite programming" refers to the more general problem of cone-LP optimization over symmetric cones.*

A remark about notation

In this chapter we mostly work with column vectors, which are represented by boldface letters like $\mathbf{x}, \mathbf{y}$, etc. Row vectors are represented by using transpose: $\mathbf{x}^T, \mathbf{y}^T$. Also, $\mathbf{0}$ is the vector of all zeros, and $\mathbf{1}$ the vector of all ones. Finally individual entries of a vector $\mathbf{x}$ are written as x_j.

In many occasions we need to concatenate column vectors to construct new vectors. To save space we use the convention of many high level programming languages like MATLAB and use $(\cdot;\cdot)$ to join column vectors and $(\cdot,\cdot)$ to join row vectors: Thus, for instance:

$$(\mathbf{x};\mathbf{y}) = \begin{pmatrix} \mathbf{x} \\ \mathbf{y} \end{pmatrix} = (\mathbf{x}^T, \mathbf{y}^T)^T$$

8.2 SEMIDEFINITE PROGRAMMING: CONE-LP OVER SYMMETRIC CONES

We now define the notion of *cone-LP* problems. First let us recall some basic notions about convex cones.

Definition 8.2.1 *A cone $\mathcal{K} \subseteq \Re^n$ is called* proper *if it has nonempty interior in $\Re^n$ and is closed, pointed (i.e. $\mathcal{K} \cap (-\mathcal{K}) = \{\mathbf{0}\}$), and convex. The* dual *of $\mathcal{K}$ is the cone:*

$$\mathcal{K}^* \stackrel{\text{def}}{=} \{\mathbf{x} : \langle \mathbf{x}, \mathbf{z} \rangle \geq 0 \text{ for all } \mathbf{z} \in \mathcal{K}\}$$

where $\langle \cdot, \cdot \rangle$ is an appropriate inner product.

It is easy to show that if $\mathcal{K}$ is proper then so is $\mathcal{K}^*$.

Definition 8.2.2 *A proper cone $\mathcal{K}$ is* symmetric *if it is self-dual, i.e. $\mathcal{K} = \mathcal{K}^*$, and homogeneous, that is for any two points $\mathbf{x}, \mathbf{y} \in \operatorname{Int} \mathcal{K}$ (interior of $\mathcal{K}$) there exist a linear transformation L such that $L(\mathbf{x}) = \mathbf{y}$ and $L(\mathcal{K}) = \mathcal{K}$.*

Any convex optimization problem in $\Re^n$ can be put in the following pair of mutually dual "standard form cone-LP's":

$$\begin{array}{llcll} \min & \langle \mathbf{c}, \mathbf{x} \rangle & \qquad & \max & \mathbf{b}^T\mathbf{y} \\ \text{s.t.} & A\mathbf{x} = \mathbf{b} & & \text{s.t.} & A^T\mathbf{y} + \mathbf{s} = \mathbf{c} \\ & \mathbf{x} \in \mathcal{K} & & & \mathbf{s} \in \mathcal{K}^* \end{array} \tag{8.2.1}$$

where $\mathcal{K}$ is a proper cone, $\mathcal{K}^*$ its dual, and A is an $m \times n$ matrix which without loss of generality is assumed to be full rank. Somewhat arbitrarily, the minimization problem is called *the primal* while the maximization problem is called *the dual.*

It is well known that if either there is a primal-feasible $\mathbf{x} \in \operatorname{Int} \mathcal{K}$ or a dual-feasible $(\mathbf{y}, \mathbf{s})$ such that $\mathbf{s} \in \operatorname{Int} \mathcal{K}^*$ then the values of the solutions of the pair of dual problems coincide, see for example, [578].

What is less well-known is that at the optimum a general version of complementary slackness theorem holds. Given that the points $\mathbf{x}$ and $(\mathbf{y}, \mathbf{s})$ are optimal, respectively for the primal and the dual, it is easy to verify that $\mathbf{x}^T\mathbf{s} = 0$. However a stronger statement can be made.

Lemma 8.2.1 *Let $\mathbf{x}, \mathbf{s} \in \Re^n$ and consider the set*

$$C = \left\{(\mathbf{x};\mathbf{s}) : \mathbf{x} \in \mathcal{K}, \mathbf{s} \in \mathcal{K}^*, \text{ and } \mathbf{x}^T\mathbf{s} = 0\right\}.$$

Then the dimension of C is n.

For a simple proof due to Güler see [26].

The implication of this lemma is that there are a set of n, in a sense independent, equations $f_i(\mathbf{x}, \mathbf{s}) = 0$ that are satisfied at the optimum; they are the generalization of *complementary slackness relations* in linear programming. When we put these n equations along with the feasibility conditions $A\mathbf{x} = \mathbf{b}$, and $A^T\mathbf{y} + \mathbf{s} = \mathbf{c}$, we get a set of $2n + m$ equations with the same number of unknowns. In the absence of degeneracy and assuming that strict complementarity holds, this system of equations uniquely identifies the primal and dual optimal solutions. We do not go into details here; the interested reader is invited to consult [24], [223], [26], [624].

8.3 EUCLIDEAN JORDAN ALGEBRAS

In this section we present a minimal foundation of the theory of Euclidean Jordan algebras. This theory provides us with the basic toolbox for the analysis of primal-dual interior point methods. Our presentation mostly follows Faraut and Korányi [216].

8.3.1 Definitions and basic properties

Let $\mathfrak{J}$ be a n-dimensional vector space over real numbers with a multiplication "$\circ$" where the map $(\mathbf{x}, \mathbf{y}) \to \mathbf{x} \circ \mathbf{y}$ is bilinear, thus "$\circ$" distributes over addition:

$$\begin{aligned} &(\alpha\mathbf{x} + \beta\mathbf{y}) \circ \mathbf{z} = \alpha(\mathbf{x} \circ \mathbf{z}) + \beta(\mathbf{y} \circ \mathbf{z}) \\ &\mathbf{x} \circ (\alpha\mathbf{y} + \beta\mathbf{z}) = \alpha(\mathbf{x} \circ \mathbf{y}) + \beta(\mathbf{x} \circ \mathbf{z}) \qquad \text{for all} \quad \mathbf{x}, \mathbf{y}, \mathbf{z} \in \mathfrak{J} \quad \text{and} \quad \alpha, \beta \in \Re. \end{aligned}$$

Then $(\mathfrak{J}, \circ)$ is a Jordan algebra (represented simply by $\mathfrak{J}$ when $\circ$ is understood from the context) if for all $\mathbf{x}, \mathbf{y} \in \mathfrak{J}$

1. $\mathbf{x} \circ \mathbf{y} = \mathbf{y} \circ \mathbf{x}$,
2. $\mathbf{x} \circ (\mathbf{x}^2 \circ \mathbf{y}) = \mathbf{x}^2 \circ (\mathbf{x} \circ \mathbf{y})$ where $\mathbf{x}^2 = \mathbf{x} \circ \mathbf{x}$.

Note that Jordan algebras are not necessarily associative, that is $\mathbf{x} \circ (\mathbf{y} \circ \mathbf{z}) \neq (\mathbf{x} \circ \mathbf{y}) \circ \mathbf{z}$ in general. However they are power associative: $\mathbf{x} \circ (\mathbf{x} \circ \mathbf{x}) = (\mathbf{x} \circ \mathbf{x}) \circ \mathbf{x}$, for every $\mathbf{x} \in \mathfrak{J}$. Thus we may, without fear of ambiguity, write $\mathbf{x}^p$ for the product of p copies of $\mathbf{x}$, and $\mathbf{x}^{p+q} = \mathbf{x}^p \circ \mathbf{x}^q$ for all positive integers p and q.

A Jordan algebra $\mathfrak{J}$ is called *Euclidean* if there exists a symmetric, positive definite quadratic form Q on $\mathfrak{J}$ which is also associative that is

$$\text{for all } \mathbf{x}, \mathbf{y}, \mathbf{z} \in \mathfrak{J}, Q(\mathbf{x}, \mathbf{x}) > 0 \text{ when } \mathbf{x} \neq \mathbf{0}, \text{ and } Q(\mathbf{x} \circ \mathbf{y}, \mathbf{z}) = Q(\mathbf{x}, \mathbf{y} \circ \mathbf{z}).$$

A Jordan algebra has an identity if there exists a (necessarily unique) element $\mathbf{e}$ such that $\mathbf{x} \circ \mathbf{e} = \mathbf{e} \circ \mathbf{x} = \mathbf{x}$ for all $\mathbf{x} \in \mathfrak{J}$. It is known that all Euclidean Jordan algebras have an identity element.

Many of the results stated in this section actually hold true for the more general case of Jordan algebras, but this generality is not needed here and thus we restrict ourselves to Euclidean ones. Therefore, in the rest of this chapter, unless specifically stated otherwise, the term Jordan algebra refers to a Euclidean Jordan algebra.

Throughout most of this section we focus on the following two examples of Jordan algebras. They provide concrete examples of such algebras and also relate this theory to two of the most important optimization problems: SDP and QCQP. Towards the end of this section we state a fundamental theorem that characterizes all Euclidean Jordan algebras.

Example 8.3.1 (Jordan algebra of symmetric matrices) The set $\mathfrak{M}_n$ of real symmetric $n \times n$ matrices under the binary operation:

$$X \circ Y \stackrel{\text{def}}{=} \frac{XY + YX}{2}$$

forms a Jordan algebra with identity the ordinary identity matrix I. It is easy to see that "$\circ$" satisfies the axioms of Jordan algebra. Since the ordinary inner product $X \bullet Y = \operatorname{Trace} XY$ defines an associative and positive definite bilinear form, $\mathfrak{M}_n$ is Euclidean. This bilinear form can be expressed in matrix form by I_{n^2}, since $\operatorname{Trace} XY = \mathbf{vec}^T X I \mathbf{vec} Y$. If elements of $\mathfrak{M}_n$ are thought of as $n(n+1)/2$-vectors then the matrix of the quadratic form is an $n(n+1)/2 \times n(n+1)2$ diagonal matrix whose rows and columns are indexed by pair of numbers ij where $i \geq j$. The diagonal entries of this matrix are $\sqrt{2}$ at the position (ij, ij) if $i \neq j$, and 1 if $i = j$. ■

Example 8.3.2 (Jordan Algebra of quadratic forms) Take $\mathfrak{R}^{n+1}$ and assume that its vectors are indexed from zero. Also for each $\mathbf{x} \in \mathfrak{R}^{n+1}$ let $\bar{\mathbf{x}}$ be the n-vector consisting of entries 1 through n of $\mathbf{x}$. Thus, $\mathbf{x} = (x_0; \bar{\mathbf{x}})$. Define the following multiplication on $(n+1)$-vectors:

$$\mathbf{x} \circ \mathbf{y} \stackrel{\text{def}}{=} \begin{pmatrix} \mathbf{x}^T \mathbf{y} \\ x_0 \bar{\mathbf{y}} + y_0 \bar{\mathbf{x}} \end{pmatrix}$$

Then $\mathfrak{R}^{n+1}$ under this multiplication is a Euclidean Jordan algebra and we denote it by $\mathfrak{Q}_n$. Again it is straightforward to verify that "$\circ$" indeed satisfies axioms of Jordan algebra and $\mathbf{e} = (1; \mathbf{0})$ is the identity element. Also the bilinear form that maps $(\mathbf{x}, \mathbf{y}) \to \mathbf{x}^T \mathbf{y}$ is positive definite and associative. ■

Definition 8.3.1 *If $\mathfrak{J}$ is a Euclidean Jordan Algebra then its cone of squares is the set*

$$\mathcal{K}(\mathfrak{J}) \stackrel{\text{def}}{=} \{\mathbf{x}^2 : \mathbf{x} \in \mathfrak{J}\}.$$

The relevance of the theory of Euclidean Jordan algebras to $\mathcal{K}$-LP optimization stems from the following theorem.

Theorem 8.3.1 (Jordan algebraic characterization of symmetric cones) *A cone is symmetric iff it is the cone of squares of some Euclidean Jordan algebra.*

Example 8.3.3 (Cone of squares of $\mathfrak{M}_n$) For the algebra $\mathfrak{M}_n$ the cone of squares is the set of all symmetric matrices that are squares of other symmetric matrices. A moment's reflection reveals that this is but the cone of positive semidefinite matrices. ■

Example 8.3.4 (Cone of squares of $\mathfrak{Q}_n$) Let $\mathbf{x} \in \mathfrak{Q}_n$ and $\mathbf{y} = \mathbf{x}^2$. Then $y_0 = \mathbf{x}^T\mathbf{x} = ||\mathbf{x}||^2$, where $||\cdot||$ is the ordinary Euclidean norm. Also for $1 \leq i \leq n$ we have $y_i = 2x_0x_i$. The vector $\mathbf{y}$ can be expressed as:

$$\mathbf{y} = \begin{pmatrix} x_0^2 + ||\bar{\mathbf{x}}||^2 \\ 2x_0\bar{\mathbf{x}} \end{pmatrix}$$

Clearly , $y_0 \geq ||\bar{\mathbf{y}}|| \geq 0$. Conversely, if $\mathbf{y}$ is a vector with the property $y_0 \geq ||\bar{\mathbf{y}}||$ then we need to show that the system of equations:

$$\begin{aligned} x_0^2 + ||\bar{\mathbf{x}}||^2 &= y_0 \\ 2x_0\bar{\mathbf{x}} &= \bar{\mathbf{y}} \end{aligned}$$

has a solution in $\mathbf{x}$. But the second set of equations implies that $x_i = y_i/2x_0$ for $1 \leq i \leq n$. And thus the first equation results in the quartic:

$$4x_0^4 - 4y_0x_0^2 + ||\bar{\mathbf{y}}||^2 = 0$$

This equation has four solutions

$$x_0 = \pm\sqrt{\frac{y_0 \pm \sqrt{y_0^2 - ||\bar{\mathbf{y}}||^2}}{2}}$$

all of them real because $y_0 \geq ||\bar{\mathbf{y}}||$. Thus the cone of squares of $\mathfrak{Q}_n$ is given by

$$\mathcal{Q} = \{\mathbf{y} \in \Re^{n+1} : y_0 \geq ||\bar{\mathbf{y}}||\}$$

$\mathcal{Q}$ is known as the *Lorenz cone*[2]. ■

Since the binary operation "$\circ$" is bilinear, for every $\mathbf{x} \in \mathfrak{J}$ there exists a matrix $L(\mathbf{x})$ such that for every $\mathbf{y}$, $\mathbf{x} \circ \mathbf{y} = L(\mathbf{x})\mathbf{y}$. Since "$\circ$" is commutative $L(\mathbf{x})$ is symmetric.

Example 8.3.5 (L for $\mathfrak{M}_n$) For $\mathfrak{M}_n$ algebra one can express $L(X)$ matrix in two forms, depending on whether elements of $\mathfrak{M}_n$ are thought of as n^2-vectors or as $n(n+1)/2$-vectors. In the former case $L(X) = (X \otimes I + I \otimes X)/2$; the numerator is the *Kronecker sum* of X with itself. In the latter case $L(X) = X \otimes_s I$, the so-called symmetric Kronecker product of X and I, as defined in [25]. ■

Example 8.3.6 (L for $\mathfrak{Q}_n$) For $\mathfrak{Q}_n$ algebra the matrix $L(\mathbf{x})$ is defined as the *arrow-shaped* matrix

$$\mathbf{Arw}\,\mathbf{x} = \begin{pmatrix} x_0 & \bar{\mathbf{x}}^T \\ \bar{\mathbf{x}} & x_0 I_n \end{pmatrix}$$

[2] $\mathcal{Q}$ is known by various other names such as the ice cream cone, the second order cone, and the quadratic cone.

The transformation associated with this matrix is known as the *Lorenz transformation* in relativity theory literature. ∎

The matrix $L(\mathbf{x})$ defines a fundamental operator for Euclidean Jordan algebras. Another important matrix for these algebras is the *quadratic representation.* For each $\mathbf{x}, \mathbf{y} \in \mathfrak{J}$ define

$$\mathbf{Q}_{\mathbf{x},\mathbf{y}} \stackrel{\text{def}}{=} L(\mathbf{x})L(\mathbf{y}) + L(\mathbf{y})L(\mathbf{x}) - L(\mathbf{x} \circ \mathbf{y}) \qquad \mathbf{Q}_{\mathbf{x}} \stackrel{\text{def}}{=} 2L^2(\mathbf{x}) - L(\mathbf{x}^2).$$

$\mathbf{Q}_{\mathbf{x}}$ is *the quadratic representation* of $\mathbf{x}$. Clearly $\mathbf{Q}_{\mathbf{x},\mathbf{z}}\mathbf{y}$ and $\mathbf{Q}_{\mathbf{x}}\mathbf{y}$ are in $\mathfrak{J}$ for all $\mathbf{x}, \mathbf{y}, \mathbf{z} \in \mathfrak{J}$. The quadratic representation is an essential concept in the theory of Jordan algebras and plays an important role in our subsequent development. Indeed it is possible to start from a suitable axiomatization of the quadratic representation as starting point and define "$\circ$" and ultimately Jordan algebras from it, see for example [744] and [356].

Example 8.3.7 (The quadratic representation in $\mathfrak{M}_n$) The quadratic representation in $\mathfrak{M}_n$ is

$$\mathbf{Q}_{X,Y}(Z) = \frac{XZY + YZX}{2}, \text{ and in particular } \mathbf{Q}_X(Y) = XYX.$$

One can see that

$$\mathbf{vec}\,\mathbf{Q}_{X,Y}(Z) = \frac{1}{2}(Y \otimes X + X \otimes Y)\mathbf{vec}(Z), \quad \mathbf{vec}\,\mathbf{Q}_X(Z) = (X \otimes X)\mathbf{vec}(Z)$$

Thus in case of symmetric matrices we can succinctly express $\mathbf{Q}_X(\cdot)$ in terms of ordinary matrix multiplication, a concept external to the Jordan algebra of symmetric matrices. The definition of $\mathbf{Q}_X(\cdot)$ in general depends purely on Jordan algebraic operation of $L(\mathbf{x})$ and thus does not require existence of other multiplications outside of the algebra. ∎

Example 8.3.8 (Quadratic representation in $\mathfrak{Q}_n$) For $\mathfrak{Q}_n$ we have

$$\mathbf{Q}_{\mathbf{x},\mathbf{y}} = \begin{pmatrix} \mathbf{x}^T\mathbf{y} & x_0\bar{\mathbf{y}}^T + y_0\bar{\mathbf{x}}^T \\ x_0\bar{\mathbf{y}} + y_0\bar{\mathbf{x}} & (x_0y_0 - \bar{\mathbf{x}}^T\bar{\mathbf{y}})I + \bar{\mathbf{x}}\bar{\mathbf{y}}^T + \bar{\mathbf{y}}\bar{\mathbf{x}}^T \end{pmatrix}$$

and in particular

$$\mathbf{Q}_{\mathbf{x}} = \begin{pmatrix} ||\mathbf{x}||^2 & 2x_0\bar{\mathbf{x}}^T \\ 2x_0\bar{\mathbf{x}} & (x_0^2 - ||\bar{\mathbf{x}}||^2)I + 2\bar{\mathbf{x}}\bar{\mathbf{x}}^T \end{pmatrix}$$

∎

The following statements are taken from [216] and [356].

Lemma 8.3.1 (Properties of Q.) *Let* $\mathbf{x}$ *and* $\mathbf{y}$ *be elements of a Jordan algebra* $\mathfrak{J}$. *Then*

1. $\mathbf{Q}_{\mathbf{x},\mathbf{y}} = \frac{1}{2}\left(\mathbf{Q}_{\mathbf{x}+\mathbf{y}} - \mathbf{Q}_{\mathbf{x}} - \mathbf{Q}_{\mathbf{y}}\right)$
2. $\mathbf{Q}_{\mathbf{x}}\mathbf{x}^{-1} = \mathbf{x}$, $\mathbf{Q}_{\mathbf{x}}^{-1} = \mathbf{Q}_{\mathbf{x}^{-1}}$
3. $\mathbf{Q}_{\mathbf{x},\mathbf{x}^{-1}}\mathbf{Q}_{\mathbf{x}} = \mathbf{Q}_{\mathbf{x}}\mathbf{Q}_{\mathbf{x},\mathbf{x}^{-1}} = 2L(\mathbf{x})\mathbf{Q}_{e,\mathbf{x}} - \mathbf{Q}_{\mathbf{x}} = L(\mathbf{x}^2)$
4. $\mathbf{Q}_{\mathbf{Q}_{\mathbf{y}}\mathbf{x}} = \mathbf{Q}_{\mathbf{y}}\mathbf{Q}_{\mathbf{x}}\mathbf{Q}_{\mathbf{y}}$.

8.3.2 Eigenvalues, degree, rank and norms

Now consider any arbitrary power associative algebra with identity, that is a vector space with an additional binary operation of multiplication $(\mathbf{x}, \mathbf{y}) \to \mathbf{xy}$, that is bilinear in $\mathbf{x}$ and $\mathbf{y}$ and power associative. Then one can define for such an algebra concepts of minimal and characteristic polynomials, eigenvalue, trace, determinant, rank, etc. Since Jordan algebras are power associative, naturally this applies to them as well. We define these notions for Jordan algebras, but the reader should keep in mind that the definitions actually apply to all power associative algebras.

For $\mathbf{x} \in \mathfrak{J}$, $\mathfrak{J}$ a Jordan algebra, let r be the smallest integer such that the set $\{\mathbf{e}, \mathbf{x}, \mathbf{x}^2, \ldots, \mathbf{x}^r\}$ is linearly dependent. Then r is *degree* of $\mathbf{x}$ and we write $r = \deg(\mathbf{x})$; *rank* of $\mathfrak{J}$, $\mathrm{rk}\,(\mathfrak{J})$ is the largest $\deg(\mathbf{x})$ of any $\mathbf{x} \in \mathfrak{J}$. An element $\mathbf{x}$ is called *regular* in $\mathfrak{J}$ if its degree equals $\mathrm{rk}\,(\mathfrak{J})$.

For a regular element in a rank-r Jordan algebra $\mathfrak{J}$, since $\{\mathbf{e}, \mathbf{x}, \mathbf{x}^2, \ldots, \mathbf{x}^r\}$ is linearly dependent there are real numbers $a_1(\mathbf{x}), \ldots, a_r(\mathbf{x})$ such that

$$\mathbf{x}^r - a_1(\mathbf{x})\mathbf{x}^{r-1} + \cdots + (-1)^r a_r(\mathbf{x})\mathbf{e} = \mathbf{0}$$

It can be shown that each function $a_i(\mathbf{x})$ is in fact a homogeneous polynomial of degree i in entries x_j. The polynomial $\lambda^r - a_1(\mathbf{x})\lambda^{r-1} + \cdots + (-1)^r a_r(\mathbf{x})$ is the *characteristic polynomial* of $\mathbf{x}$. Its roots, $\lambda_1, \cdots, \lambda_r$ are the *eigenvalues* of $\mathbf{x}$. Notice that this definition applies to all $\mathbf{x} \in \mathfrak{J}$ and not just the regular elements.

Definition 8.3.2 *Let $\mathbf{x} \in \mathfrak{J}$ and $\lambda_1, \ldots, \lambda_r$ are the roots of its characteristic polynomial. Then*

1. $\mathrm{tr}\,(\mathbf{x}) \stackrel{\text{def}}{=} \lambda_1 + \cdots + \lambda_r = a_1(\mathbf{x})$ *is trace of* $\mathbf{x}$ *in* $\mathfrak{J}$*; trace is a linear function of* $\mathbf{x}$*;*

2. $\det(\mathbf{x}) \stackrel{\text{def}}{=} \lambda_1 \cdots \lambda_r = a_r(\mathbf{x})$ *is the determinant of* $\mathbf{x}$ *in* $\mathfrak{J}$.

3. *The multiplicity of the zero eigenvalue of* $\mathbf{x}$ *as a root of the characteristic polynomial is its* co-rank $\mathrm{cork}\,(\mathbf{x})$*; and* $\mathrm{rk}\,(\mathbf{x}) \stackrel{\text{def}}{=} r - \mathrm{cork}\,(\mathbf{x})$.

Example 8.3.9 (Characteristic polynomials of $\mathfrak{M}_n$) In $\mathfrak{M}_n$, the characteristic polynomial coincides with the usual definition, so do the notions of eigenvalues, trace, determinant, and rank. In addition we know that for symmetric matrices all eigenvalues are real numbers. We will shortly see that this is the property of all Euclidean Jordan algebras. ■

Example 8.3.10 (Characteristic polynomials of $\mathfrak{Q}_n$) Let $\mathbf{x} \in \Re^{n+1}$ be viewed as an element of $\mathfrak{Q}_n$. One can easily verify the identity

$$\mathbf{x}^2 - 2x_0\mathbf{x} + (x_0^2 - ||\bar{\mathbf{x}}||^2)\mathbf{e} = \mathbf{0}$$

(Recall that $\mathbf{e} = (1; \mathbf{0})$ is the identity element of $\mathfrak{Q}_n$.) Since this quadratic polynomial identity is true regardless of n it follows that always $\mathrm{rk}\,\mathfrak{Q}_n = 2$.

Furthermore:

$$\lambda_{1,2} = x_0 \pm ||\bar{\mathbf{x}}||$$
$$\text{tr}\,(\mathbf{x}) = 2x_0$$
$$\det(\mathbf{x}) = x_0^2 - ||\bar{\mathbf{x}}||^2$$

Again notice that eigenvalues are always real numbers. ■

Lemma 8.3.2 ([216]) *If* $\mathbf{x}, \mathbf{y} \in \mathfrak{J}$ *then*

$$\det(\mathbf{Q_y x}) = \det(\mathbf{y})^2 \det(\mathbf{x}) = \det(\mathbf{y}^2)\det(\mathbf{x}). \tag{8.3.2}$$

Together with the eigenvalues comes a decomposition of $\mathbf{x}$ into a linear combination of idempotents, known as *spectral decomposition.* An *idempotent* $\mathbf{c}$ is an element of $\mathfrak{J}$ where $\mathbf{c}^2 = \mathbf{c}$.

1. A *complete system of orthogonal idempotents* is a set $\{\mathbf{c}_1, \ldots, \mathbf{c}_k\}$ of nonzero idempotents where $\mathbf{c}_i \circ \mathbf{c}_j = \mathbf{0}$ for all $i \neq j$, and $\mathbf{c}_1 + \cdots + \mathbf{c}_k = \mathbf{e}$.

2. An idempotent is *primitive* if it is not sum of two other idempotents.

3. A complete system of orthogonal primitive idempotents is called a *Jordan frame.* In Jordan frames $k = r$, rank of $\mathfrak{J}$.

Theorem 8.3.2 (Spectral decomposition, version 1) *Let* $\mathfrak{J}$ *be a Euclidean Jordan algebra. Then for* $\mathbf{x} \in \mathfrak{J}$ *there exist unique real numbers* $\lambda_1, \ldots, \lambda_k$, *all distinct, and a unique complete system of orthogonal idempotents* $\mathbf{c}_1, \ldots, \mathbf{c}_k$ *such that*

$$\mathbf{x} = \lambda_1 \mathbf{c}_1 + \ldots + \lambda_k \mathbf{c}_k. \tag{8.3.3}$$

See [216].

Theorem 8.3.3 (Spectral decomposition, version 2) *Let* $\mathfrak{J}$ *be a Euclidean Jordan algebra with rank* r. *Then for* $\mathbf{x} \in \mathfrak{J}$ *there exists a Jordan frame* $\mathbf{c}_1, \ldots, \mathbf{c}_r$ *and real numbers* $\lambda_1, \ldots, \lambda_r$ *such that*

$$\mathbf{x} = \lambda_1 \mathbf{c}_1 + \ldots + \lambda_r \mathbf{c}_r \tag{8.3.4}$$

and the λ_i *are the eigenvalues of* $\mathbf{x}$.

Example 8.3.11 (Spectral decomposition in $\mathfrak{M}_n$) Let A be a symmetric matrix with eigenvalue decomposition $A = Q\Lambda Q^T$. Thus Q is an orthogonal matrix whose columns are $\mathbf{q}_i$, and Λ a diagonal matrix containing λ_i, the eigenvalues of A, in its diagonal. First notice that from each $\mathbf{q}_i$ we may construct an idempotent $\mathbf{c}_i = \mathbf{q}_i \mathbf{q}_i^T$. Furthermore since $\mathbf{q}_i$ are orthogonal any sum of $\mathbf{q}_{i_1}\mathbf{q}_{i_1}^T + \cdots + \mathbf{q}_{i_k}\mathbf{q}_{i_k}^T$ is also an idempotent. The set of rank-one matrices $\mathbf{q}_i \mathbf{q}_i^T$ form a Jordan frame. Thus the spectral decomposition of A can be written as

$$A = Q\Lambda Q^T = \sum_{i=1}^{n} \lambda_i \mathbf{q}_i \mathbf{q}_i^T$$

This gives the second version spectral decomposition. To get the first version suppose that a particular eigenvalue λ_i has multiplicity, say m with corresponding eigenvectors $\mathbf{q}_{i_1}, \ldots, \mathbf{q}_{i_m}$. Then, while $\mathbf{q}_{i_j}$ are not unique for $m > 1$, their sum $\mathbf{p}_i = \sum_j \mathbf{q}_{i_j}$ *is* unique. Thus if $\lambda_1, \ldots, \lambda_k$ are *distinct* eigenvalues of A then

$$A = \sum_{i=1}^{k} \lambda_i \mathbf{p}_i \mathbf{p}_i^T$$

gives the first version of eigenvalue decomposition with $\{\mathbf{p}_1\mathbf{p}_1^T, \ldots, \mathbf{p}_k\mathbf{p}_k^T\}$ the unique complete system of orthogonal idempotents. ■

Example 8.3.12 (Spectral decomposition in $\mathfrak{Q}_n$) For $\mathbf{x} \in \mathfrak{Q}_n$ the identity

$$\mathbf{x} = (x_0 + ||\bar{\mathbf{x}}||)\left(\frac{\mathbf{e}}{2} + \frac{\hat{\mathbf{x}}}{2||\hat{\mathbf{x}}||}\right) + (x_0 - ||\bar{\mathbf{x}}||)\left(\frac{\mathbf{e}}{2} - \frac{\hat{\mathbf{x}}}{2||\hat{\mathbf{x}}||}\right) \tag{8.3.5}$$

gives the eigenvalue decomposition of vector $\mathbf{x}$ (version 2). Here $\hat{\mathbf{x}} = (0; \bar{\mathbf{x}})$, and the pair of vectors $\frac{\mathbf{e}}{2} \pm \frac{\hat{\mathbf{x}}}{2||\bar{\mathbf{x}}||}$ form a Jordan frame. When the two eigenvalues are equal then $\mathbf{x}$ is either zero or a multiple of identity, and the unique system of orthogonal idempotents is $\{\mathbf{e}\}$. ■

Below, when we speak of spectral decomposition we always mean the second version, unless stated otherwise.

Now it is possible to extend the definition of any real or complex valued analytic function $f(\cdot)$ on real numbers to elements of Euclidean Jordan algebras through their eigenvalues. If the spectral decomposition of $\mathbf{x} = \lambda_1\mathbf{c}_1 + \cdots + \lambda_r\mathbf{c}_r$ then

$$f(\mathbf{x}) \stackrel{\text{def}}{=} f(\lambda_1)\mathbf{c}_1 + \cdots + f(\lambda_r)\mathbf{c}_r$$

We are in particular interested in

1. arbitrary powers of $\mathbf{x}$: $\mathbf{x}^t \stackrel{\text{def}}{=} \lambda_1^t\mathbf{c}_1 + \cdots + \lambda_r^t\mathbf{c}_r$, for any real (or even complex) t and nonzero λ_i (for $t > 0$, λ_i may be zero). For t a positive integer, this definition is consistent with earlier definition of powers of $\mathbf{x}$;

2. in particular inverse is defined as $\mathbf{x}^{-1} = \lambda_1^{-1}\mathbf{c}_1 + \cdots + \lambda_k^{-1}\mathbf{c}_k$ whenever all $\lambda_i \neq 0$ and undefined otherwise.

Remark 8.3.1 Unlike associative operations, for each $\mathbf{x} \in \mathfrak{J}$ there may be many vectors $\mathbf{y}$ where $\mathbf{y} \circ \mathbf{x} = \mathbf{e}$; only $\mathbf{x}^{-1}$ as defined above can be in a useful manner declared as *the* inverse (at least for our purposes). It should also be clear that $\mathbf{x}$ has an inverse if and only if it has no zero eigenvalue, that is $\text{rk}\,(\mathbf{x}) = \text{rk}\,(\mathfrak{J})$. Also an element $\mathbf{x}$ is invertible iff it is in the interior of the cone of squares of $\mathfrak{J}$, and thus all *rank-deficient* vectors, those with at least one zero eigenvalue, lie on the boundary of $\mathcal{K}(\mathfrak{J})$. ■

Definition 8.3.3 *Two elements* $\mathbf{x}, \mathbf{y} \in \mathfrak{J}$ operator commute, *or simply* commute *if there exist real numbers,* $\lambda_1, \ldots, \lambda_r$ *and* $\omega_1, \ldots, \omega_r$*, and a Jordan frame* $\{\mathbf{c}_1, \ldots, \mathbf{c}_r\}$*, such that*

$$\begin{aligned} \mathbf{x} &= \lambda_1 \mathbf{c}_1 + \cdots + \lambda_r \mathbf{c}_r \\ \mathbf{y} &= \omega_1 \mathbf{c}_1 + \cdots + \omega_r \mathbf{c}_r \end{aligned}$$

The term "commute" should not be confused with commutativity property of "$\circ$" operation. While all $\mathbf{x}, \mathbf{y} \in \mathfrak{J}$ commute with respect to $\circ$, only those that share a common Jordan frame in their spectral decomposition commute in the sense of Definition 8.3.3. We use the term *commute* because it can be shown that $\mathbf{x}$ and $\mathbf{y}$ commute iff $L(\mathbf{x})L(\mathbf{y}) = L(\mathbf{y})L(\mathbf{x})$ and equivalently $\mathbf{Q_x Q_y} = \mathbf{Q_y Q_x}$.

Clearly if $\mathbf{x}$ and $\mathbf{y}$ commute, then eigenvalues of $\mathbf{x} \circ \mathbf{y}$ are the pair-wise products of eigenvalues of $\mathbf{x}$ and $\mathbf{y}$, and $\mathbf{x} \circ \mathbf{y}$ also shares the common Jordan frame of $\mathbf{x}$ and $\mathbf{y}$. In particular, it follows that

$$\det(\mathbf{x} \circ \mathbf{y}) = \det(\mathbf{x}) \det \mathbf{y}. \tag{8.3.6}$$

Example 8.3.13 (commuting relationship in $\mathfrak{M}_n$) In symmetric matrices the commuting relationship as defined states that the two matrices X and Y share a common system of eigenvectors. This is equivalent to $XY = YX$. Thus, again, we can express the commuting relationship in terms of the concept of ordinary matrix multiplication which is outside of Jordan algebra. ■

Example 8.3.14 (Commuting relationship in $\mathfrak{Q}_n$) In the quadratic forms Jordan algebra if $\mathbf{x}$ and $\mathbf{y}$ commute then $\hat{\mathbf{x}} = \alpha \hat{\mathbf{y}}$ for some constant α. Thus, the orthogonal projections of commuting vectors onto the space of coordinates 1 through n are proportional. ■

We call $\mathbf{x} \in \mathfrak{J}$ *positive semidefinite* if all its eigenvalues are nonnegative; we call it *positive definite* if all its eigenvalues are positive. We write $\mathbf{x} \succcurlyeq \mathbf{0}$ (respectively $\mathbf{x} \succ \mathbf{0}$) if $\mathbf{x}$ is positive semidefinite (respectively positive definite.) It is clear that $\mathbf{x} \succcurlyeq \mathbf{0}$ is equivalent to $\mathbf{x} \in \mathcal{K}(\mathfrak{J})$ and $\mathbf{x} \succ \mathbf{0}$ to $\mathbf{x} \in \operatorname{Int} \mathcal{K}(\mathfrak{J})$, where $\mathcal{K}(\mathfrak{J})$ is the cone of squares of $\mathfrak{J}$.

In this chapter we use the terms "cone of squares of $\mathfrak{J}$", and the "positive semidefinite cone" interchangeably. Furthermore, as stated in the introduction, we use the term *semidefinite programming* to mean the same thing as cone-LP over some symmetric cones $\mathcal{K}$.

We may also define various norms on $\mathfrak{J}$ as functions of eigenvalues much the same way that unitarily invariant norms are defined on square matrices:

$$||\mathbf{x}||_{\mathrm{F}} \stackrel{\mathrm{def}}{=} \left(\sum \lambda_i^2\right)^{1/2} = \sqrt{\operatorname{tr}(\mathbf{x}^2)}, \qquad ||\mathbf{x}||_2 = \max_i |\lambda_i| \tag{8.3.7}$$

$||\cdot||_{\mathrm{F}}$ is the *Frobenius norm* and $||\cdot||_2$ the *spectral norm*. Observe that $||\mathbf{e}||_{\mathrm{F}} = \sqrt{r}$.

Since "$\circ$" is bilinear and trace is a symmetric positive definite quadratic form which is associative, $\mathrm{tr}\,(\mathbf{x}, \mathbf{y} \circ \mathbf{z}) = \mathrm{tr}\,(\mathbf{x} \circ \mathbf{y}, \mathbf{z})$, we define the inner product:

$$\langle \mathbf{x}, \mathbf{y} \rangle \stackrel{\text{def}}{=} \mathrm{tr}\,(\mathbf{x} \circ \mathbf{y}).$$

Example 8.3.15 (norms and scalar products for $\mathfrak{M}_n$) For symmetric matrices the concepts of Frobenius and spectral norms coincide with the familiar ones and the inner product of two such matrices X, Y is $\langle X, Y \rangle = X \bullet Y = \sum_{ij} X_{ij} Y_{ij}$. ■

Example 8.3.16 (norms and scalar products for $\mathfrak{Q}_n$) For $\mathbf{a} \in \mathfrak{Q}_n$, the Frobenius norm

$$||\mathbf{a}||_F = 2||\mathbf{a}||$$

Again we remind the reader that $||\cdot||$ refers to the ordinary Euclidean norm. The spectral norm is given by

$$||\mathbf{a}||_2 = \max\{|a_0 + ||\bar{\mathbf{a}}|||, |a_0 - ||\bar{\mathbf{a}}|||\}.$$

In particular for positive semidefinite $\mathbf{a}$, $||\mathbf{a}||_2 = a_0 + ||\bar{\mathbf{a}}||$. The scalar product is given by

$$\langle \mathbf{a}, \mathbf{b} \rangle = \mathrm{tr}\,(\mathbf{a} \circ \mathbf{b}) = 2\mathbf{a}^T \mathbf{b}$$

■

We need the following simple lemmas later:

Lemma 8.3.3 *For $\mathbf{a}, \mathbf{b} \in \mathfrak{J}$, with $\mathfrak{J}$ a Jordan algebra*

$$\mathrm{tr}\,(\mathbf{Q_a b}) = \mathrm{tr}\,\big((\mathbf{a} \circ \mathbf{b}) \circ \mathbf{a}\big).$$

Proof.

$$\begin{aligned}
\mathrm{tr}\,(\mathbf{Q_a b}) &= \mathrm{tr}\,\Big(\big(2L^2(\mathbf{a}) - L(\mathbf{a}^2)\big)\mathbf{b}\Big) = 2\mathrm{tr}\,\big(\mathbf{a} \circ (\mathbf{a} \circ \mathbf{b})\big) - \mathrm{tr}\,(\mathbf{a}^2 \circ \mathbf{b}) \\
&= 2\mathrm{tr}\,(\mathbf{a}^2 \circ \mathbf{b}) - \mathrm{tr}\,(\mathbf{a}^2 \circ \mathbf{b}) = \mathrm{tr}\,\big(\mathbf{a} \circ (\mathbf{a} \circ \mathbf{b})\big) \\
&= \mathrm{tr}\,\big((\mathbf{a} \circ \mathbf{b}) \circ \mathbf{a}\big)
\end{aligned}$$

Note that $\mathrm{tr}\,\big(\mathbf{a} \circ (\mathbf{a} \circ \mathbf{b})\big) = \mathrm{tr}\,(\mathbf{a}^2 \circ \mathbf{b})$ because of associativity of tr operation.

Lemma 8.3.4 *For $\mathbf{z} \in \mathfrak{J}$, with $\mathfrak{J}$ a Jordan algebra*

$$\mathbf{Q}_{\mathbf{x}^2} = \big(\mathbf{Q_x}\big)^2$$

Proof.
The proof follows from part 4 of Lemma 8.3.1, by noting that $\mathbf{Q_x e} = \mathbf{x}^2$. In fact one can more generally show that $\mathbf{Q}_{\mathbf{x}^t} = \big(\mathbf{Q_x}\big)^t$ for any real number t.

We now state another fundamental fact:

Theorem 8.3.4 *For all invertible $\mathbf{y} \in \mathfrak{J}$*

$$\mathbf{Q_y}\mathcal{K} = \mathcal{K}$$

See Faraut and Korányi [216]. Thus, the semidefinite cone remains invariant under $\mathbf{Q_y}$ operation for any invertible element $\mathbf{y} \in \mathfrak{J}$.

8.3.3 Simple Jordan algebras and decomposition theorem

So far we have dealt with only two concrete examples of Euclidean Jordan algebras: The real symmetric matrices under $X \circ Y = (XY + YX)/2$ operation and the $(n+1)$-vectors under $\mathbf{x} \circ \mathbf{y} = (\mathbf{x}^T \mathbf{y}; y_0 \bar{\mathbf{x}} + x_0 \bar{\mathbf{y}})$. There are a few other "atomic" Euclidean Jordan algebras. Before we discuss all of these algebras we need to remind the reader of the notion of direct sum of (Jordan) algebras and the concept of simple algebras.

Definition 8.3.4

1. *If* $(\mathfrak{J}_1, \circ_1)$ *and* $(\mathfrak{J}_2, \circ_2)$ *are two Jordan algebras then their* direct sum, *written as* $\mathfrak{J}_1 \oplus \mathfrak{J}_2$ *is the algebra* $(\mathfrak{J}_1 \times \mathfrak{J}_2, \circ)$ *which is itself a Jordan algebra if "$\circ$" is extended block component-wise:*

$$(\mathbf{x}_1; \mathbf{x}_2) \circ (\mathbf{y}_1; \mathbf{y}_2) \stackrel{\text{def}}{=} (\mathbf{x}_1 \circ_1 \mathbf{y}_1; \mathbf{x}_2 \circ_2 \mathbf{y}_2)$$

2. *If a Jordan algebra is isomorphic to direct sum of two other Jordan algebras, then it is called* reducible; *otherwise it is* irreducible *or* simple.

The following facts should be self-evident:

Theorem 8.3.5 *Let* $\mathfrak{J}_1 = (\mathfrak{J}_1, \circ_1)$ *and* $\mathfrak{J}_2 = (\mathfrak{J}_2, \circ_2)$ *be two Jordan algebras with identities* $\mathbf{e}_1$ *and* $\mathbf{e}_2$, *and* $\mathfrak{J}_1 \oplus \mathfrak{J}_2$ *their direct sum with identity* $(\mathbf{e}_1; \mathbf{e}_2)$. *Then*

1. *If* $\mathcal{K}_1$ *and* $\mathcal{K}_2$ *are each cone of squares of* $\mathfrak{J}_1$ *and* $\mathfrak{J}_2$, *respectively, then* $\mathcal{K}_1 \times \mathcal{K}_2$ *is the cone of squares of* $\mathfrak{J}_1 \oplus \mathfrak{J}_2$.

2. *If* $r_1 = \operatorname{rk}(\mathfrak{J}_1)$ *and* $r_2 = \operatorname{rk}(\mathfrak{J}_2)$, *then* $\operatorname{rk}(\mathfrak{J}_1 \oplus \mathfrak{J}_2) = r_1 + r_2$.

3. *If* p_1 *and* p_2 *are characteristic polynomials of* $\mathbf{x} \in \mathfrak{J}_1$ *and* $\mathbf{y} \in \mathfrak{J}_2$, *respectively, then* $p_1 p_2$ *is the characteristic polynomial of* $(\mathbf{x}; \mathbf{y}) \in \mathfrak{J}_1 \oplus \mathfrak{J}_2$. *If* $\Lambda = \{\lambda_1, \ldots, \lambda_{r_1}\}$ *is the (multi)set of eigenvalues of* $\mathbf{x}$, *and* $\Omega = \{\omega_1, \ldots, \omega_{r_2}\}$ *the (multi)set of eigenvalues of* $\mathbf{y}$, *then* $\Lambda \cup \Omega$ *is the (multi)set of eigenvalues of* $(\mathbf{x}; \mathbf{y})$; *and multiplicity of an eigenvalue* λ *in* $\mathfrak{J}_1 \oplus \mathfrak{J}_2$ *is the sum of its multiplicities in* $\mathfrak{J}_1$ *and* $\mathfrak{J}_2$. *Thus for any* $(\mathbf{x}; \mathbf{y}) \in \mathfrak{J}_1 \oplus \mathfrak{J}_2$, $\operatorname{tr}(\mathbf{x}; \mathbf{y}) = \operatorname{tr}(\mathbf{x}) + \operatorname{tr}(\mathbf{y})$, *and* $\det(\mathbf{x}; \mathbf{y}) = \det(\mathbf{x}) \det(\mathbf{y})$.

4. *If* $\{\mathbf{c}_1 \ldots, \mathbf{c}_{r_1}\}$ *is a Jordan frame in* $\mathfrak{J}_1$ *and* $\{\mathbf{d}_1, \ldots, \mathbf{d}_{r_2}\}$ *is a Jordan frame in* $\mathfrak{J}_2$, *then* $\{(\mathbf{c}_1; \mathbf{0}), \ldots, (\mathbf{c}_{r_1}; \mathbf{0}), (\mathbf{0}; \mathbf{d}_1), \ldots, (\mathbf{0}; \mathbf{d}_{r_2})\}$ *is a Jordan frame in* $\mathfrak{J}_1 \oplus \mathfrak{J}_2$. *Furthermore, if the spectral decomposition of* $\mathbf{x} \in \mathfrak{J}_1$ *and* $\mathbf{y} \in \mathfrak{J}_2$ *are, respectively,* $\mathbf{x} = \lambda_1 \mathbf{c}_1 + \cdots + \lambda_{r_1} \mathbf{c}_{r_1}$, *and* $\mathbf{y} = \omega_1 \mathbf{d}_1 + \cdots + \omega_{r_2} \mathbf{d}_{r_2}$, *then the spectral decomposition of* $(\mathbf{x}; \mathbf{y})$ *is*

$$(\mathbf{x}; \mathbf{y}) = \lambda_1 (\mathbf{c}_1; \mathbf{0}) + \cdots + \lambda_{r_1} (\mathbf{c}_{r_1}; \mathbf{0}) + \omega_1 (\mathbf{0}; \mathbf{d}_1) + \cdots + \omega_{r_2} (\mathbf{0}; \mathbf{d}_{r_2})$$

5. *For* $(\mathbf{x}; \mathbf{y}) \in \mathfrak{J}_1 \oplus \mathfrak{J}_2$

$$\|(\mathbf{x}; \mathbf{y})\|_{\mathrm{F}}^2 = \|\mathbf{x}\|_{\mathrm{F}}^2 + \|\mathbf{y}\|_{\mathrm{F}}^2 \qquad \|(\mathbf{x}; \mathbf{y})\|_2 = \max\{\|\mathbf{x}\|_2, \|\mathbf{y}\|_2\}$$

Example 8.3.17 (Direct sum of symmetric matrices) Let $A \in \mathfrak{M}_n$, and $B \in \mathfrak{M}_m$. The elements of the Jordan algebra $\mathfrak{M}_n \oplus \mathfrak{M}_m$ are pairs of symmetric matrices. However consider the direct sum of matrices as defined by:

$$A \oplus B = \begin{pmatrix} A & 0 \\ 0 & B \end{pmatrix}$$

The set of such matrices under $X \circ Y = (XY + YX)/2$ is a Jordan subalgebra of $\mathfrak{M}_{n+m}$ which is easily seen to be isomorphic to $\mathfrak{M}_n \oplus \mathfrak{M}_m$. $A \oplus B$ representation is less efficient than $(\mathbf{vec}A; \mathbf{vec}B)$ representation (mostly because of two redundant zero blocks), but is more convenient notationally. Obviously this example can be extended to direct sum of several matrices. ■

Example 8.3.18 (Linear Programming as symmetric cone LP) As a special case of the last example consider ordinary linear programming, that is cone-LP over the nonnegative orthant of $\Re^n$. The underlying algebra here is a direct sum of n copies of $\mathfrak{M}_1$. The cone of squares of $\mathfrak{M}_1$ is the half line $\mathbf{x} \geq 0$. The Jordan algebra $\mathfrak{M}_1$ is but the vector space of real numbers over real numbers, with "$\circ$" same as ordinary real number multiplication. This example shows that even when the underlying Jordan algebra is trivial and un-interesting, the induced cone-LP could be quite interesting with a rich structure. ■

Example 8.3.19 (Direct sum of Lorenz cones) Consider the optimization problem

$$\begin{array}{llll} \min & \mathbf{c}_1^T\mathbf{x}_1 + \cdots + \mathbf{c}_k^T\mathbf{x}_k & \min & \mathbf{b}^T\mathbf{y} \\ \text{s.t.} & A_1\mathbf{x}_1 + \cdots + A_k\mathbf{x}_k = \mathbf{b} & \text{s.t.} & A_i\mathbf{y} + \mathbf{s}_i = \mathbf{c}_i \quad i = 1, \ldots, k \\ & \mathbf{x}_i \in \mathcal{Q}_{n_i} & & \mathbf{s}_i \in \mathcal{Q}_{n_i} \end{array} \tag{8.3.8}$$

Where each $\mathbf{x}_i \in \Re^{n_i+1}$, and $\mathcal{Q}_{n_i}$ is the $(n_i + 1)$-dimensional Lorenz cone. This optimization problem is equivalent to quadratically constrained quadratic programming (QCQP) problem in that the problem

$$\begin{array}{ll} \min & \mathbf{x}^T C\mathbf{x} + \mathbf{c}^T\mathbf{x} \\ \text{s.t.} & \mathbf{x}^T A_i\mathbf{x} + \mathbf{a}_i^T\mathbf{x} \leq b_i \quad i = 1, \ldots, k \end{array}$$

can be reformulated into (8.3.8) if C and all A_i are symmetric positive semidefinite matrices. To see this notice that if $A = L^T L$ then $\mathbf{x}^T A\mathbf{x} + \mathbf{a}^T\mathbf{x} \leq b$ is equivalent to

$$\mathbf{x}^T L^T L\mathbf{x} + \left(\frac{\mathbf{a}^T\mathbf{x} - b + 1}{2}\right)^2 - \left(\frac{\mathbf{a}^T\mathbf{x} - b - 1}{2}\right)^2 \leq 0$$

which in turn says that

$$\left(\left(\frac{\mathbf{a}^T\mathbf{x} - b - 1}{2}\right); L\mathbf{x}; \left(\frac{\mathbf{a}^T\mathbf{x} - b + 1}{2}\right)\right) \in \mathcal{Q}.$$

Thus the cone-LP problem (8.3.8) is a semidefinite program whose cone of squares is the direct sum of k Lorenz cones. The Jordan algebra associated with (8.3.8), not surprisingly, is the direct sum of k quadratic form Jordan algebras. ■

In the next section we show that rank r of the underlying Jordan algebra solely determines the iteration complexity of the potential reduction interior point methods. Specifically, this complexity is proportional to $\sqrt{r}$ and the dimension of the vector space plays no role in it. For example in (8.3.8) it is k the number of blocks that determines rank and by implication iteration complexity. Thus if say, $k = 1$, no matter how big of a problem we have, the iteration complexity will be constant. (This is comforting because for $k = 1$ the QCQP problem is essentially a least squares problem and can be solved analytically.) These facts had already been proved by direct methods in [363] and [286] for the QCQP problem, and by Nemirovskii and Scheinberg [569] through word-by-word extension from LP. From our general development in this chapter we can derive the iteration complexity of QCQP problems transparently and as a simple corollary.

We are now ready to state the fundamental decomposition theorem of the Euclidean Jordan algebras.

Theorem 8.3.6 ([216]) *If $\mathfrak{J}$ is a simple Euclidean Jordan algebra then it must be one of the following:*

1. *The space $\Re^{n+1}$ with Jordan multiplication defined as $\mathbf{x} \circ \mathbf{y} = (\mathbf{x}^T \mathbf{y}; x_0\bar{\mathbf{y}} + s_0\bar{\mathbf{x}})$; this is the $\mathfrak{Q}_n$ algebra.*

2. *The space of real symmetric $n \times n$ matrices with $X \circ Y = (XY + YX)/2$; this is the $\mathfrak{M}_n$ algebra.*

3. *The space of complex Hermitian $n \times n$ matrices with $X \circ Y = (XY + YX)/2$ for Hermitian matrices X and Y; this is the $\mathfrak{C}_n$ algebra.*

4. *The space of Hermitian $n \times n$ matrices with quaternion entries and with $X \circ Y = (XY + YX)/2$ for quaternion Hermitian matrices X and Y; this is the $\mathfrak{H}_n$ algebra.*

5. *The space of 3×3 Hermitian matrices with octonion entries and multiplication $X \circ Y = (XY + YX)/2$. This algebra is called the* Albert algebra *and is represented by $\mathfrak{O}_3$.*

Some elaboration is in order.

Before we go into the details of these five classes, we note that the implication of the theorem is that all Euclidean Jordan algebras are either among these five, or are constructed by means of direct sum from any combination of them. Likewise cone-LP over symmetric cones is optimization over any of the five types of semidefinite cones associated with these algebras or a Cartesian product of a combination of them.

These five algebras can be grouped together into three classes. The first group is the algebra of real symmetric, complex hermitian and quaternion hermitian matrices. The 3×3 octonion Hermitian matrices and the quadratic forms algebra make up the other groups. Below we make some comments about each.

Group one: symmetric and Hermitian matrices. We have already extensively discussed the algebra of real symmetric matrices, and have noted that concepts such as eigenvalues, rank, determinant, norms, etc. coincide with the familiar notions from matrix theory. We could make the same statement about Hermitian complex and quaternion algebras. However, there is a small problem that has to be dispensed with. Our definition of Euclidean Jordan algebras requires that the underlying field of the linear space be real numbers. This is necessary if we want to use notions of norm and maximization/minimization properly[3]. Complex Hermitian matrices are ordinarily considered over the field of complex numbers and Quaternion Hermitian matrices over the division ring of quaternions; but we need them as algebras over reals. As it turns out, this is only a technicality. Note that the field of complex numbers is isomorphic to the field of 2×2 matrices of the form

$$\begin{pmatrix} x & -y \\ y & x \end{pmatrix}$$

under matrix addition and multiplication. Thus in every $n \times n$ complex hermitian matrix we can replace each entry $x + iy$ with the corresponding 2×2 matrix to get a $2n \times 2n$ real symmetric matrix. The resulting set of matrices under Jordan multiplication $(XY + YX)/2$ forms a Jordan subalgebra of $\mathfrak{M}_{2n}$; we call *this* algebra $\mathfrak{C}_n$. But this is not the end of story. A typical matrix X obtained in this way, when viewed as an element of $\mathfrak{M}_{2n}$ has rank $2n$, and $2n$ eigenvalues, etc. But $\mathfrak{C}_n$ is a rank n algebra and thus $\operatorname{rk}(X) = n$ in $\mathfrak{C}_n$. Furthermore, there are only n eigenvalues and a Jordan frame has n primitive idempotents. To see what is going on consider the eigenvalue decomposition of a matrix $A \in \mathfrak{C}_n$ represented in it original $n \times n$ complex Hermitian form. Then the eigenvalue decomposition is $A = U^*\Lambda U$ where U is a unitary complex $n \times n$ matrix and U^* its conjugate transpose, and Λ a necessarily real diagonal matrix. Just as in the case of real symmetric matrices we can write

$$A = U^*\Lambda U = \sum_{i=1}^{n} \lambda_i \mathbf{u}_i \mathbf{u}_i^*$$

where $\mathbf{u}_i$ the i'th column of U consists of complex numbers. Now if, following the construction above, we replace entries of $\mathbf{u}_i$ by representation of that entry as a real 2×2 matrix, then each $\mathbf{u}_i$ turns into a $2n \times 2$ real matrix. While

[3]Actually one can extend the notion of Euclidean Jordan algebras to those defined over complex numbers by a process known as complexification. This procedure is not relevant for our purposes however, and we won't consider it here; see [216] for further details.

$\mathbf{u}_i\mathbf{u}_i^*$ is not a primitive idempotent in $\mathfrak{M}_{2n}$, it *is* primitive in the subalgebra $\mathfrak{C}_n$. Thus $\{\mathbf{u}_1\mathbf{u}_1^*, \ldots, \mathbf{u}_n\mathbf{u}_n^*\}$ is a Jordan frame in $\mathfrak{C}_n$ and $\sum_i \lambda_i \mathbf{u}_i\mathbf{u}_i^*$ is the spectral decomposition (version 2) of A. This also implies that trace, determinant, norms, rank, etc, are the usual ones in the sense of $\mathfrak{C}_n$ (and not $\mathfrak{M}_{2n}$).

For Quaternions Hermitian matrices there is a similar construction Recall that each quaternion is a four dimensional number of the form $x + iy + ju + kv$ where x, y, u, and v are real and $i^2 = j^2 = k^2 = -1$, and $ij = k = -ji$, $jk = i = -kj$, and $ki = j = -ik$. It is easily verified that the division ring of quaternions is isomorphic to the set of 4×4 matrices of the form

$$\begin{pmatrix} x & -y & u & -v \\ y & x & v & u \\ -u & -v & x & y \\ v & -u & -y & x \end{pmatrix}$$

under ordinary matrix addition an multiplication. Thus an $n \times n$ Hermitian matrix with quaternion entries can be replaced by a $4n \times 4n$ real matrix by replacing the quaternions with the 4×4 matrix above. The set of such symmetric matrices is a subalgebra of $\mathfrak{M}_{4n}$; we call *this* algebra $\mathfrak{H}_n$. With an argument similar to the one we gave for complex Hermitian matrices we can argue that $\operatorname{rk}(\mathfrak{H}_n) = n$.

The three Jordan algebras $\mathfrak{M}_n$, $\mathfrak{C}_n$ and $\mathfrak{H}_n$ all have the property that their Jordan product is induced by ordinary matrix product which is by the way an associative operation. Indeed, one can show that any associative algebra induces a (not necessarily Euclidean) Jordan algebra by means of $(\mathbf{xy} + \mathbf{yx})/2$ operation, where $\mathbf{xy}$ is the associative product. Jordan subalgebras of such induced algebras are called *representable*; so are any Jordan algebras isomorphic to some representable one.

Group 2: The algebra of quadratic forms. The rank 2 algebra $\mathfrak{Q}_n$ was also studied extensively in the examples throughout this section. The only additional comment we should make about it is that this algebra is also representable. That is there is an associative algebra with a multiplication $\mathbf{xy}$ such that $\mathfrak{Q}_n$ is a subalgebra of the Jordan algebra induced by $(\mathbf{xy} + \mathbf{yx})/2$. This associative algebra is called the *Clifford algebra* and its vector space dimension is 2^n. We should mention that the relationship between $\mathfrak{Q}_n$ and its inducing associative algebra is a bit more complicated than those of the three matrix algebras in group one. In those matrix algebras we have a set and associative multiplication (matrix multiplication), and an *adjoint* operation (the transpose in $\mathfrak{M}_n$ and conjugate transpose in $\mathfrak{C}_n$ and $\mathfrak{H}_n$). The Euclidean Jordan algebra then is the set of *all* self-adjoint elements along with the Jordan multiplication $(XY + YX)/2$. Thus for instance for any element X in the associative algebra $X + X'$ and XX' are in the induced Jordan algebra (here "$'$" is the adjoint operator).

In the case of quadratic forms algebra $\mathfrak{Q}_n$ and Clifford algebras inducing them, there is also an adjoint operation, but $\mathfrak{Q}_n$ is a *proper subset* of self-adjoint elements of the Clifford algebra. Thus there are elements $\mathbf{x}$ in the

Clifford algebra where $\mathbf{x}+\mathbf{x}'$ or $\mathbf{x}\mathbf{x}'$ are not in $\mathfrak{Q}_n$. This issue does not affect our analysis of potential reduction interior point algorithms below. In fact we are not aware of any polynomiality proof where proper inclusion of $\mathfrak{Q}_n$ in self-adjoint elements of Clifford algebra has caused any problems. For further information and references on the Clifford algebras and their role in complexity analysis of some interior point methods see [698].

Group 3: The Exceptional Albert algebra. Octonions–also known as octavions and Cayley numbers–are 8-dimensional numbers that can be constructed from quaternions as follows. Let q_1 and q_2 be two quaternions and define a new element l with $l^2=-1$. Then octonions are the set of numbers of the form q_1+lq_2 with the properties:

$$q_1(q_2l)=(q_2q_1)l, \quad (q_1l)q_2=(q_1\overline{q_2})l, \quad \text{and } (q_1l)(q_2l)=-(\overline{q_2}q_1)$$

where for quaternions $\overline{a+ib+jc+kd}=a-ib-jd-kb$. Thus the product of two octonions is given by

$$(p_1+p_2l)(q_1+q_2l)=p_1q_1-\overline{q_2}p_2+(q_2p_1+p_2\overline{q_1})l$$

This product is not associative, thus if one constructs matrices out of octonions, then the matrix product is not associative either. Nevertheless, it turns out that the set of 3×3 Hermitian matrices of octonions, a 27-dimensional linear space, under the multiplication $X\circ Y=(XY+YX)/2$ is a Euclidean Jordan algebra known as the Albert algebra and written as $\mathfrak{O}_3$. The eigenvalues of matrices in $\mathfrak{O}_3$ are real, and $\text{rk}\,(\mathfrak{O}_3)=3$.

The main difference between this algebra and the other four is that $\mathfrak{O}_3$ is not representable (non-representable algebras are often called *exceptional*); that is there does not exist any associative operation "$*$" where "$\circ$" can be expressed as $(X*Y+Y*X)/2$. This issue does not affect the analysis of interior point algorithms considered in this chapter. However, it *does* affect other analyses in some primal-dual algorithms which use the properties of the underlying associative operation in a significant way; see [698] for more details.

8.3.4 Complementarity in semidefinite programming

In the remainder of this chapter we focus on the pair of standard primal and dual semidefinite programming problems

$$\begin{array}{llll} \min & \langle \mathbf{c},\mathbf{x}\rangle & \max & \mathbf{b}^T\mathbf{y} \\ \text{s.t.} & A\mathbf{x}=\mathbf{b} & \text{s.t.} & A^T\mathbf{y}+\mathbf{s}=\mathbf{c} \\ & \mathbf{x}\succcurlyeq \mathbf{0} & & \mathbf{s}\succcurlyeq \mathbf{0} \end{array} \tag{8.3.9}$$

where $\succcurlyeq$ is with respect to symmetric cone $\mathcal{K}=\mathcal{K}(\mathfrak{J})$ that is the cone of squares of a Euclidean Jordan algebra $\mathfrak{J}$. Let us set $r=\text{rk}\,(\mathfrak{J})$, otherwise make no assumptions on $\mathfrak{J}$, in particular, we do not require it to be simple. Furthermore, we remind the reader that the inner product $\langle \mathbf{x},\mathbf{s}\rangle$ is assumed to be induced by $\mathfrak{J}$: $\langle \mathbf{x}, \mathbf{s} \rangle = \operatorname{tr}(\mathbf{x} \circ \mathbf{s})$. A is taken to be a full rank $m \times n$ matrix and $A\mathbf{x}$ is understood to be the m-vector whose i'th entry is $\langle \mathbf{a}_i, \mathbf{x} \rangle$. Throughout this chapter we make the assumption that

1. The primal is feasible and furthermore there exists a feasible point $\mathbf{x} \succ \mathbf{0}$; such a point is referred to as *primal interior feasible.*

2. The dual problem is feasible and furthermore, there exists a feasible point $(\mathbf{y}, \mathbf{s})$ where $\mathbf{s} \succ \mathbf{0}$; such a point is referred to as *dual interior feasible.*

Lemma 8.3.5 (Complementary slackness in SDP) *Let $\mathcal{K} = \mathcal{K}(\mathfrak{J})$ be the cone of squares of the Euclidean Jordan algebra $\mathfrak{J}$. If $\mathbf{x}, \mathbf{s} \in \mathcal{K}$ and $\langle \mathbf{x}, \mathbf{s} \rangle = 0$ then $\mathbf{x} \circ \mathbf{s} = \mathbf{0}$.*

Proof.
Since $\mathbf{x}$ and $\mathbf{s}$ are positive semi-definite, their square roots exist and their multiplication matrices $L(\mathbf{x})$ and $L(\mathbf{s})$ are positive semi-definite. This is so because one can show that if $\mathbf{c} = \mathbf{c}^2$, then $L(\mathbf{c}) \succcurlyeq \mathbf{0}$ [216], and $L(\mathbf{x})$ being a nonnegative combination of positive semidefinite matrices, must itself be positive semidefinite. So,

$$0 = \langle \mathbf{x}, \mathbf{s} \rangle = \langle \mathbf{x}, (\mathbf{s}^{1/2})^2 \rangle = \langle L(\mathbf{x})\mathbf{s}^{1/2}, \mathbf{s}^{1/2} \rangle = \langle L^{1/2}(\mathbf{x})\mathbf{s}^{1/2}, L^{1/2}(\mathbf{x})\mathbf{s}^{1/2} \rangle \tag{8.3.10}$$

which means $L^{1/2}(\mathbf{x})\mathbf{s}^{1/2} = \mathbf{0}$ and therefore $L(\mathbf{x})\mathbf{s}^{1/2} = \mathbf{0}$. Now,

$$\langle \mathbf{x}, \mathbf{s}^2 \rangle = \langle \mathbf{x}, (\mathbf{s}^{1/2})^4 \rangle = \langle \mathbf{x} \circ \mathbf{s}^{1/2}, (\mathbf{s}^{1/2})^3 \rangle = \langle \mathbf{0}, (\mathbf{s}^{1/2})^3 \rangle = 0 \tag{8.3.11}$$

which implies $L(\mathbf{x})\mathbf{s} = \mathbf{x} \circ \mathbf{s} = \mathbf{0}$.

Thus, in semidefinite programming primal and dual feasibility and the complementarity condition $\mathbf{x} \circ \mathbf{x} = \mathbf{0}$ form a system of $2n+m$ equations in the same number of unknowns, which in the absence of various degeneracies completely determine the optimal solution.

8.4 POTENTIAL REDUCTION ALGORITHMS FOR SEMIDEFINITE PROGRAMMING

8.4.1 The logarithmic barrier function for symmetric cones

The central tool for all interior point methods is the *barrier function.* Barrier functions are convex functions defined on the relative interior of the feasible region of convex programs; they approach infinity as their argument gets close to the boundary of the feasible region. For a symmetric cone the interior points are positive definite vectors; so these points are all invertible, while the boundary of the cone consists of non-ivertible positive semidefinite vectors. In particular, at least one of the eigenvalues of every point on the boundary is zero.

Our development in this chapter starts with the *logarithmic barrier function*

$$-\ln \det \mathbf{x}.$$

For motivation one may think of the positive semidefinite cone expressed as $\{\mathbf{x} : \lambda_i(\mathbf{x}) \geq 0 \text{ for } i = 1, \ldots, r\}$. Then applying the standard logarithmic barrier on these constraints we obtain $\sum_{i=1}^{r} \ln \lambda_i(\mathbf{x}) = \ln \det \mathbf{x}$.

The classical barrier approach is to replace the following problem for the standard form primal $\mathcal{K}$-LP problem (8.3.9):

$$\begin{array}{ll} \min & f_\mu(\mathbf{x}) \stackrel{\text{def}}{=} \langle \mathbf{c}, \mathbf{x} \rangle - \mu \ln \det \mathbf{x} \\ \text{s.t.} & A\mathbf{x} = \mathbf{b} \end{array} \tag{8.4.12}$$

It can be shown that the gradient $\nabla_{\mathbf{x}} \ln \det \mathbf{x} = \mathbf{x}^{-T}$, (the transpose of the inverse of $\mathbf{x}$) and therefore the gradient of f_μ is given by:

$$\nabla_{\mathbf{x}} f_\mu = \mathbf{c}^T - \mu \mathbf{x}^{-T},$$

and the Lagrangian by

$$L(\mathbf{x}, \mathbf{y}) = \langle \mathbf{c}, \mathbf{x} \rangle - \mu \ln \det \mathbf{x} - \mathbf{y}^T (A\mathbf{x} - \mathbf{b}).$$

Let $\mathbf{x}_\mu$ be the optimal solution of (8.4.12). Then the optimality conditions imply that there exists an m-vector $\mathbf{y}_\mu$ such that

$$\nabla_{\mathbf{x}} L = \mathbf{c}^T - \mu \mathbf{x}_\mu^{-T} - \mathbf{y}_\mu^T A = \mathbf{0} \tag{8.4.13}$$

Now if we define $\mathbf{s}_\mu \stackrel{\text{def}}{=} \mathbf{c} - A^T \mathbf{y}_\mu$, then it is clear that $(\mathbf{y}_\mu, \mathbf{s}_\mu)$ is interior feasible for the dual problem. We call $\mathbf{x}_\mu$, $\mathbf{s}_\mu$ and $\mathbf{y}_\mu$, *μ-optimal.* The duality gap for the pair of primal and dual feasible points $\mathbf{x}_\mu$ and $(\mathbf{s}_\mu, \mathbf{y}_\mu)$ is

$$\langle \mathbf{c}, \mathbf{x} \rangle - \mathbf{b}^T \mathbf{y} = \langle \mathbf{x}, \mathbf{s} \rangle = \mu \text{tr}\,(\mathbf{e}) = \mu r$$

Therefore, as μ tends to zero, $\mathbf{x}_\mu$, $\mathbf{y}_\mu$ and $\mathbf{s}_\mu$ tend to the optimal solutions of the primal and dual problems. In fact, the identity $\mathbf{x}_\mu \circ \mathbf{s}_\mu = \mu \mathbf{e}$ may be thought of as a μ approximation to the complementary slackness relation.

As μ changes continuously, $\mathbf{x}_\mu$ traverses a smooth path through the interior of the feasible region of the primal $\mathcal{K}$-LP problem. This path induces another path in the interior of the feasible region in the dual problem, that is the path traversed by $(\mathbf{y}_\mu, \mathbf{s}_\mu)$. Such paths are often called, respectively, primal and dual *central paths.*

8.4.2 Potential functions

Now we define the concept of potential functions and study some of their properties that is needed later. Our development closely follows Ye's method for linear programming [853]. We should mention that Faybusovich in [224] also studies the potential functions for symmetric cones, but they are employed in primal-dual algorithms and may be viewed as generalizations of the subject matter of the next chapter.

Let $q > 0$ and $\underline{z}$ be given constants such that, if v^* is the optimal value of the objective function in the primal then $\underline{z} \leq v^*$. We define two classes of potential functions. Let $\mathbf{x}$ be interior primal feasible, $(\mathbf{y}, \mathbf{s})$ interior dual feasible, and $\mathbf{s} = \mathbf{c} - A^T\mathbf{y}$.

Define the *primal potential function* ϕ as:

$$\phi(\mathbf{x}, \underline{z}) = q \ln(\langle \mathbf{c}, \mathbf{x} \rangle - \underline{z}) - \ln \det \mathbf{x}, \tag{8.4.14}$$

and the *primal-dual potential function* ψ as:

$$\psi(\mathbf{x}, \mathbf{s}) = q \ln\langle \mathbf{x}, \mathbf{s} \rangle - \ln \det \mathbf{Q}_{\mathbf{x}^{1/2}} \mathbf{s}. \tag{8.4.15}$$

Observe that the primal-dual potential function is symmetric in $\mathbf{x}$ and $\mathbf{s}$ because by Lemma 8.3.2 $\ln \det \mathbf{Q}_{\mathbf{x}^{1/2}}\mathbf{s} = \ln \det \mathbf{Q}_{\mathbf{s}^{1/2}}\mathbf{x} = \ln \det \mathbf{x} + \ln \det \mathbf{s}$. Also if we set $\underline{z} = \mathbf{b}^T\mathbf{y}$ then,

$$\psi(\mathbf{x}, \mathbf{s}) = \phi(\mathbf{x}, \underline{z}) - \ln \det \mathbf{s}. \tag{8.4.16}$$

The purpose of potential functions is to measure the "accumulated progress" of the algorithm over many iterations. The main thrust of potential-reduction algorithms is to generate a sequence $\mathbf{x}_k$, $\mathbf{y}_k$ and $\mathbf{s}_k$ of primal and dual interior feasible points such that the value of $\psi(\mathbf{x}_k, \mathbf{s}_k)$ decreases by an absolute constant at each iteration. With an appropriate choice of q we are able to derive a bound of $\mathcal{O}(\sqrt{r}|\log \epsilon|)$ on the number of iterations that results in a duality gap below ϵ.

8.4.3 Potential reduction and polynomial time solvability

In this section we prove that any algorithm that generates a sequence of primal and dual interior feasible points $(\mathbf{x}_k, \mathbf{y}_k, \mathbf{s}_k)$ such that the value of the primal-dual potential function ψ is reduced by at least an absolute constant, say δ, from point k to point $k+1$ has *polynomial time convergence.* Then in the next section we show how to find an initial interior feasible point to start such an algorithm and in the subsequent sections we outline two algorithms that produce such a sequence of points.

Before we prove the main theorem we should explain what we mean by polynomial time convergence, as this concept is distinct from *polynomial time solvable.* The issue of polynomial time computability in the case of semidefinite programming (and indeed any nonlinear programming) is quite different from the situation in linear programming. Suppose that all input data in the SDP problem (8.3.9) are integral, that is the entries of A, $\mathbf{b}$ and $\mathbf{c}$ are integers. Furthermore assume that the largest integer appearing in this data has L bits. Then if this problem were only a linear program one would know that the optimal solution of the problems $(\mathbf{x}^*, \mathbf{y}^*, \mathbf{s}^*)$ consists of rational numbers of the form p/q where the number of bits in p and q is at most $2L$, see [304] for details. In a semidefinite program however the solution need not be a rational number (it is obviously an algebraic number though). Thus unlike linear programming, we cannot write down the solution in finite amount of space using the usual decimal or binary notation. Instead it is customary to have as part of input to the problem a tolerance level, that is a rational number, say ϵ, and ask for $\mathbf{x}, \mathbf{y}, \mathbf{s}$ interior feasible so that the size of duality gap $\langle \mathbf{x}, \mathbf{s} \rangle < \epsilon$. Furthermore, the number of iterations of the algorithm required to achieve such an ϵ-approximation has to be polynomial in $|\log \epsilon|$ and L. Replacement of an approximate in place of an exact solution in itself is not an impediment. Even in LP, mostly because of pragmatic reasons such as use of floating point arithmetic and other numerical reasons, the algorithms are not required to produce exact solutions, but only numerically satisfactory approximations.

The more serious problem in SDP as opposed to LP is that the magnitude of the solution is not necessarily bounded by a polynomial of 2^L as in linear programming. Consider the following optimization problem:

$$\begin{array}{ll} \min & x_n \\ \text{s.t.} & x_0 = 2 \\ & x_{i+1} \geq x_i^2 \quad \text{for } i = 0, \ldots n-1 \end{array}$$

First one can see that the optimal value of this optimization problem is $2^{2^{n-1}}$. It is easily seen that the example is actually a convex QCQP problem and thus a special kind of SDP. Second L here is a linear function of n, but the solution needs an exponential number of bits in n. It is also possible to create problems where the solution is *exponentially small*, that is entries have magnitudes of 2^{-2^n}. Khachyian and Porkolab [640] have shown that this is the worst that can happen in SDP problems; that is the magnitude of the solution is in the worst case only exponentially large or small.

To remedy this situation one may require that the input to the problem, in addition to $A, \mathbf{b}, \mathbf{c}$ and ϵ contain two other numbers R and r, and it be guaranteed that the feasible region of the primal-dual space contain a ball of radius r and be contained in a ball of radius R, the latter centered at the origin. Then an algorithm is called *polynomial time convergent* if its running time is polynomial in $L, \log \epsilon, \log r$ and $\log R$. In a sense, the role of these inner and outer balls is to allow a gracious exit if the solution is too small or too large. Large or small magnitude solutions are not peculiar to SDP and arise in most convex optimization problems.

Having these issues in mind we now present the main theorem regarding potential reduction algorithms:

Theorem 8.4.1 *Let* $(\mathbf{x}_0, \mathbf{y}_0, \mathbf{s}_0)$ *with* $\mathbf{s}_0 = \mathbf{c} - A^T\mathbf{y}_0$ *be a given initial interior feasible point for* (8.3.9). *Let also that* $q = r + \sqrt{r}$ *in the primal-dual potential function* ψ, *where* $r = \operatorname{rk}(\mathfrak{J})$ *the underlying Euclidean Jordan algebra. If* $\psi(\mathbf{x}_0, \mathbf{s}_0) \leq \mathcal{O}(\sqrt{r}E)$ *for some constant* E *and and if an algorithm generates a sequence of interior primal-dual points* $(\mathbf{x}_j, \mathbf{y}_j, \mathbf{s}_j)$ *such that* $\psi(\mathbf{x}_j, \mathbf{s}_j) \geq \psi(\mathbf{x}_{j+1}, \mathbf{s}_{j+1}) + \delta$ *for some fixed number* δ *then, after* $k = \mathcal{O}\big(\sqrt{r}(E + |\log \epsilon|)\big)$ *iterations, we have*

$$\langle \mathbf{c}, \mathbf{x}_k \rangle - \mathbf{b}^T \mathbf{y}_k < \epsilon$$

Proof.

Each iteration reduces the potential function by at least δ. Thus, if $\psi(\mathbf{x}_0, \mathbf{s}_0) < \mathcal{O}(\sqrt{r}E)$ then after $k = \mathcal{O}\big(\sqrt{r}(E + |\log \epsilon|)\big)$ iterations, for some constant c we have:

$$\begin{aligned}\psi(\mathbf{x}_k, \mathbf{s}_k) &< c\sqrt{r}E - c\sqrt{r}(E + |\log \epsilon|) \\ &\leq -c\sqrt{r}|\log(\epsilon)|.\end{aligned}$$

Therefore,

$$\begin{aligned}\sqrt{r}\ln\langle \mathbf{x}_k, \mathbf{s}_k\rangle &< -r \ln \operatorname{tr}\left(\mathbf{Q}_{\mathbf{x}_k^{1/2}} \mathbf{s}_k\right) + \ln\det \mathbf{Q}_{\mathbf{x}_k^{1/2}} \mathbf{s}_k - c\sqrt{r}|\log\epsilon| \\ &< -r\ln r - c\sqrt{r}|\ln\epsilon| \\ &< -c\sqrt{r}|\ln\epsilon|\end{aligned}$$

The first two inequalities come from noting Lemma 8.3.3 and applying the arithmetic-geometric inequality on the eigenvalues of $\mathbf{Q}_{\mathbf{x}_k^{1/2}}\mathbf{s}_k$. Thus, $\ln\langle \mathbf{x}_k, \mathbf{s}_k\rangle < c|\ln\epsilon|$, and since $\langle \mathbf{x}_k, \mathbf{s}_k\rangle = \langle \mathbf{c}, \mathbf{x}_k\rangle - \mathbf{b}^T\mathbf{y}_k$, the theorem follows.

In the next section we show that if we know an *a priori* upper bound on the magnitude of the optimal solution then we can always transform an SDP problem to one that possesses an initial interior feasible solution for both the primal and the dual. Moreover, the constant E for this initial feasible point is proportional to the number of bits in the bound.

8.4.4 *Feasibility and boundedness*

We now address the question of feasibility in semidefinite programming. Again the situation is quite different from linear programming. Given a semidefinite program, even a QCQP problem, there is no known polynomial time algorithm to decide whether it is feasible or not. To see the problem consider the set of constraints

$$\begin{gathered} x_1 \geq 2 \\ \begin{pmatrix} x_i & x_{i-1} \\ x_{i-1} & 1 \end{pmatrix} \succcurlyeq 0 \quad \text{for } i = 2, \ldots, n \end{gathered}$$

A moments reflection reveals that every feasible point has its coordinate $x_i \geq 2^{2^{i-1}}$. Now most natural feasibility checking algorithms in linear programming produce as a proof of feasibility a witness, a point that should satisfy the inequalities. In our small SDP feasibility problem any witness requires an exponential number of bits. Using extensions of Farkas lemma also requires producing witnesses with exponentially many bits. At present no indirect method of checking feasibility is known, and in fact the status of the complexity of this problem remains open, see [640] for further discussion of this problem.

As mentioned in the last section, the compromise is to require that an SDP be supplied with an *a priori* (lower and upper) bound on the magnitude of the solution. Then we can devise an algorithm that either produces a feasible point using polynomial time proportional to the number of bits in the upper or lower bound and other parameters of the problem or produces a certificate that indicates the original problem is infeasible or the supplied bounds are incorrect.

The trick is a familiar one: It is an extension of the so called *big M* method to SDP. Consider the following pair of primal and dual problems:

$$\begin{array}{ll} \min & \langle \mathbf{c}, \mathbf{x}\rangle + M x_0 \\ \text{s.t.} & A\mathbf{x} + (\mathbf{b} - A\mathbf{x}_0)x_0 = \mathbf{b} \\ & \langle A^T\mathbf{y}_0 + \mathbf{s}_0 - \mathbf{c}, \mathbf{x}\rangle + u = N \\ & \mathbf{x} \succcurlyeq \mathbf{0} \\ & x_0, u \geq 0 \end{array} \tag{8.4.17}$$

and its dual

$$\begin{array}{ll} \max & \mathbf{b}^T\mathbf{y} - N y_0 \\ \text{s.t.} & A^T\mathbf{y} + \mathbf{s} - (\mathbf{c} - A^T\mathbf{y}_0 - \mathbf{s}_0)y_0 = \mathbf{c} \\ & \left(\mathbf{b} - A\mathbf{x}_0\right)^T\mathbf{y} + v = M \\ & \mathbf{s} \succcurlyeq \mathbf{0} \\ & y_0, v \geq 0 \end{array} \tag{8.4.18}$$

where $\mathbf{x}_0$ and $\mathbf{s}_0$ are arbitrary positive definite vectors in $\mathfrak{J}$, and M and N are positive numbers. Clearly $\mathbf{x} = \mathbf{x}_0$, $x_0 = 1$ and $u = N - \langle A^T\mathbf{y}_0 + \mathbf{s}_0 - \mathbf{c}, \mathbf{x}_0\rangle$ are interior feasible for the primal (8.4.17) if N is sufficiently large to make u nonnegative. Similarly $\mathbf{s} = \mathbf{s}_0$, $\mathbf{y} = \mathbf{y}_0$, $y_0 = 1$ and $v = M - (\mathbf{b} - A\mathbf{x}_0)^T\mathbf{y}_0$ is interior feasible for the dual problem in (8.4.18) if M is sufficiently large enough to make v nonnegative. We may for instance set $\mathbf{x}_0 = \mathbf{s}_0 = \mathbf{e}$, $x_0 = y_0 = 1$, $\mathbf{y}_0 = \mathbf{0}$. Now if we have an *a priori* bound on the magnitude of the optimal solution, $\langle \mathbf{c}, \mathbf{x}^*\rangle = \mathbf{b}^T\mathbf{y}^* < R$ then we only need to choose M and N that are $\mathcal{O}(R + (m+n)2^L)$, where L is the number of bits in the largest integer in the input data $(A, \mathbf{b}, \mathbf{c})$. It is easy to see that if at the optimum the values of x_0 and y_0 are nonzero then the claim of $\mathbf{c}^T\mathbf{x}^* < R$ is false. If on the other hand $x_0 = y_0 = 0$ then the optimal $\mathbf{x}^*$ and $\mathbf{y}^*$ of (8.4.17) and (8.4.18) are also optimal for the original primal and dual (8.3.9). Furthermore, from the bounds on M and N given above, it can be easily verified that the value of the primal-dual potential function ψ at our chosen initial point is bounded by $\mathcal{O}\big(\sqrt{r}(L+\log R)\big)$. Therefore by Theorem 8.4.1 any algorithm that reduces potential function at each iteration brings the gap down to less than ϵ in $\mathcal{O}\big(\sqrt{r}(|\ln\epsilon| + L + \log R)\big)$ iterations.

All that remains to be done is to actually produce an algorithm that achieves this feat.

8.4.5 *Properties of Potential functions*

Before we delve into details of such algorithms and their analysis we need to derive some convenient relations. First, the gradient of the primal potential function is given by

$$\nabla_{\mathbf{x}}\phi(\mathbf{x}, \underline{z}) = \frac{q}{\langle \mathbf{c}, \mathbf{x}\rangle - \underline{z}}\mathbf{c}^T - \mathbf{x}^{-T}, \tag{8.4.19}$$

The following lemma is a direct generalization of a similar lemma that appears in the analysis of interior point linear programming methods (see for example, [853] and [404].)

Lemma 8.4.1 *Let* $\mathbf{x} \in \mathfrak{J}$. *If* $\mathbf{0} \prec \mathbf{x} \prec \mathbf{e}$, *then*

$$\ln \det \mathbf{x} \geq \operatorname{tr} \mathbf{x} - r - \frac{||(\mathbf{x}-\mathbf{e})||_F^2}{2[1 - ||(\mathbf{x}-\mathbf{e})||_2]}$$

Proof.
This is the direct extension of familiar result from literature on interior point methods in linear programming:

$$\sum_{j=1}^{r} \ln x_j \geq (\mathbf{1}^T \mathbf{x} - r) - \frac{||\mathbf{x}-\mathbf{1}||^2}{2(1 - ||\mathbf{x}-\mathbf{1}||||_\infty)} \tag{8.4.20}$$

whenever $\mathbf{x} \in \Re^r$ and $0 < x_j < 0$. The proof is almost identical to this LP version. Use the power series expansion

$$\begin{aligned} \ln x_j &= \ln\big(1 - (1-x_j)\big) = -(1-x_j) - \tfrac{(1-x_j)^2}{2} - \tfrac{(1-x_j)^3}{3} - \cdots \\ &\geq x_j - 1 - \tfrac{(1-x_j)^2}{2} - \tfrac{(1-x_j)^3}{2} - \cdots \\ &= x_j - 1 - \tfrac{(1-x_j)^2}{2\left(1-(1-x_j)\right)} \end{aligned}$$

Now plugging $\lambda_j(\mathbf{x})$ for x_j, summing over j on both sides of this inequality and replacing $||\mathbf{e}-\mathbf{x}||_2$ for $1 - \lambda_j(\mathbf{x})$ in all denominators proves the lemma.

As in linear programming, we need some form of scaling at each iteration of interior point algorithms. These scalings have typically the following properties:

1. They map the nonnegative orthant back to itself.
2. They are parameterized by a fixed point, say $\mathbf{x}_0 > \mathbf{0}$ which is mapped by the scaling to $\mathbf{1}$, the vector of all ones.
3. they do not affect many desirable properties of the linear program being analyzed or the appropriate potential functions associated with them.
4. They are typically either linear or projective scalings.

Below we describe a class of such scalings and their extension to the semidefinite cone.

8.4.6 Properties of Linear scalings

In LP a typical linear scaling with respect to a fixed point $\mathbf{x}_0$ is defined by

$$\widetilde{\mathbf{x}} \leftarrow (x_i/(\mathbf{x}_0)_i)_{i=1}^n = \big(\operatorname{Diag}(\mathbf{x}_0)\big)^{-1}\mathbf{x},$$

$\operatorname{Diag}(\mathbf{x}_0)$ a diagonal matrix constructed from $\mathbf{x}_0$.

the analogous transformation for SDP should map the fixed point $\mathbf{x}_0 \succ \mathbf{0}$ to the identity $\mathbf{e}$. At first it may seem natural to use the mapping $\widetilde{\mathbf{x}} \leftarrow \mathbf{x}_0^{-1} \circ \mathbf{x}$ for linear scaling. However it turns out that this is not quite appropriate. For instance $\mathcal{K}(\mathfrak{J})$, the cone of squares of $\mathfrak{J}$ in not invariant under this transformation. The correct choice is to use quadratic representation:

$$\tilde{\mathbf{x}} \leftarrow \mathbf{Q}_{\mathbf{x}_0^{-1/2}}\mathbf{x}$$

The inverse transformation is of course given by:

$$\underset{\sim}{\mathbf{x}} \stackrel{\text{def}}{=} \mathbf{Q}_{\mathbf{x}_0^{1/2}}\mathbf{x}$$

Remember from Theorem 8.3.4 that the semidefinite cone $\mathcal{K}$ is invariant under these linear scalings. In the transformed space the primal SDP problem is now given by:

$$\begin{array}{ll} \min & \langle \underset{\sim}{\mathbf{c}}, \tilde{\mathbf{x}} \rangle \\ \text{s.t.} & \underset{\sim}{A}\tilde{\mathbf{x}} = \mathbf{b} \\ & \tilde{\mathbf{x}} \succcurlyeq \mathbf{0} \end{array} \tag{8.4.21}$$

where $\underset{\sim}{A}$ is the matrix whose i'th row is $\underset{\sim}{\mathbf{a}_i}^T$. It is easily seen that

$$\underset{\sim}{A} = A\mathbf{Q}_{\mathbf{x}_0^{1/2}}$$

One effect of the linear scaling is that if $\mathbf{x}_0$ is an interior primal feasible point for the original SDP (8.3.9), then $\mathbf{e}$ is an interior primal feasible point for the scaled problem (8.4.21).

With these definitions in mind we may now show the following:

Corollary 8.4.1 *If* $\mathbf{x}_0 \succ \mathbf{0}$, $\mathbf{x}_1 \succ \mathbf{0}$, *and* $\left\| \mathbf{Q}_{\mathbf{x}_0^{-1/2}}(\mathbf{x}_1 - \mathbf{x}_0) \right\|_2 < 1$ *then*

$$\phi(\mathbf{x}_1, \underline{z}) - \phi(\mathbf{x}_0, \underline{z}) \leq \langle \nabla_{\mathbf{x}}^T \phi(\mathbf{x}_0, \underline{z}), (\mathbf{x}_1 - \mathbf{x}_0) \rangle + \frac{\left\| \mathbf{Q}_{\mathbf{x}_0^{-1/2}}(\mathbf{x}_1 - \mathbf{x}_0) \right\|_F^2}{2\left[1 - \left\| \mathbf{Q}_{\mathbf{x}_0^{-1/2}}(\mathbf{x}_1 - \mathbf{x}_0) \right\|_2\right]}$$

Proof.

Since the first term in the definition of ϕ is concave we have

$$\begin{aligned} \phi(\mathbf{x}_1, \underline{z}) - \phi(\mathbf{x}_0, \underline{z}) &\leq \langle \nabla_{\mathbf{x}}^T \phi(\mathbf{x}_0, \underline{z}), (\mathbf{x}_1 - \mathbf{x}_0) \rangle + \langle \mathbf{x}_0^{-1}, (\mathbf{x}_1 - \mathbf{x}_0) \rangle \\ &\qquad - \ln\det \mathbf{x} - \ln\det \mathbf{x}_0 \\ &\leq \langle \nabla_{\mathbf{x}}^T \phi(\mathbf{x}_0, \underline{z}), (\mathbf{x}_1 - \mathbf{x}_0) \rangle + \langle \mathbf{x}_0^{-1}, (\mathbf{x}_1 - \mathbf{x}_0) \rangle \\ &\qquad - \operatorname{tr}\left(\mathbf{Q}_{\mathbf{x}_0^{-1/2}}\mathbf{x}_1\right) + \frac{\left\| \mathbf{Q}_{\mathbf{x}_0^{-1/2}}(\mathbf{x}_1 - \mathbf{x}_0) \right\|_F^2}{2\left[1 - \left\| \mathbf{Q}_{\mathbf{x}_0^{-1/2}}(\mathbf{x}_1 - \mathbf{x}_0) \right\|_2\right]} \end{aligned}$$

Since $\langle \mathbf{x}_0^{-1}, (\mathbf{x}_1 - \mathbf{x}_0) \rangle - \operatorname{tr}\left(\mathbf{Q}_{\mathbf{x}_0^{-1/2}}\mathbf{x}_1\right) + r = 0$, the corollary follows.

8.4.7 *A potential reduction algorithm using linear scaling*

In this section we describe the first of two algorithms that with proper care decrease the primal-dual potential function at each iteration by at least a fixed amount, thus guaranteeing polynomial time convergence as was shown earlier. This method is based on linear transformation of the feasible region. This–and the projective algorithm to follow as well–are "word-byword" extensions of Ye's algorithm for linear programming, [853] and [852]. Consider the pair of primal and dual semidefinite programs (8.3.9).

Continuing our development from the previous section we assume that we have initial points $\mathbf{x}_0$ and $\mathbf{y}_0$, primal and dual feasible interior points, respectively and set $\mathbf{s}_0 = \mathbf{c} - A^T\mathbf{y}$. Therefore, $\mathbf{x}_0, \mathbf{s}_0 \succ \mathbf{0}$ Let $\underline{z} = \mathbf{b}^T\mathbf{y}_0$.

Consider the following restriction of the scaled primal problem (8.4.21):

$$\begin{array}{ll} \min & \langle \mathbf{Q}_{\mathbf{x}_0^{1/2}} \nabla_{\mathbf{x}}^T \phi(\mathbf{x}_0, \underline{z}), \widetilde{\mathbf{x}} - \mathbf{e}\rangle \\ \text{s.t.} & \underset{\sim}{A}(\widetilde{\mathbf{x}} - \mathbf{e}) = \mathbf{0} \\ & \|\widetilde{\mathbf{x}} - \mathbf{e}\|_{\mathrm{F}} \leq \beta \end{array} \tag{8.4.22}$$

where $\beta < 1$ is a fixed constant to be specified later. In this restriction we have replaced the feasible set of the original SDP–the intersection of an affine set $A\mathbf{x} = \mathbf{b}$ and the cone $\mathcal{K}$–with the intersection of that affine space with an ball, which gives us another, lower dimensional, ball. Furthermore, the new feasible set is entirely contained in the relative interior of the original feasible set, so its optimal points are interior feasible for the original. The restricted program (8.4.22) is now a least-squares problem and can be solved analytically: Starting from $\mathbf{e}$ we should take a step of length β in the direction of the orthogonal projection of $\mathbf{Q}_{\mathbf{x}_0^{1/2}} \nabla_{\mathbf{x}}^T \phi(\mathbf{x}_0, \underline{z})$. Define:

$$\mathbf{p}(\underline{z}) \stackrel{\text{def}}{=} \mathcal{P}_A\left(\mathbf{Q}_{\mathbf{x}_0^{1/2}} \nabla_{\mathbf{x}}^T \phi(\mathbf{x}_0, \underline{z})\right). \tag{8.4.23}$$

where

$$\mathcal{P}\underset{\sim}{A}(\mathbf{u}) = \left(I - \mathbf{Q}_{\mathbf{x}_0^{1/2}} A^T (A\mathbf{Q}_{\mathbf{x}_0} A^T)^{-1} A\mathbf{Q}_{\mathbf{x}_0^{1/2}}\right)\mathbf{u}$$

is the orthogonal projection of $\mathbf{u}$ onto the kernel of $\underset{\sim}{A}$ (remembering that $\mathbf{Q}^2_{\mathbf{x}^{1/2}} = \mathbf{Q}_{\mathbf{x}}$). Thus

$$\mathbf{x}' = \mathbf{e} - \beta \frac{\mathcal{P}\underset{\sim}{A}(\mathbf{Q}_{\mathbf{x}_0^{1/2}} \nabla_{\mathbf{x}}^T(\mathbf{x}_0, \underline{z}))}{\left\|\mathcal{P}\underset{\sim}{A}(\mathbf{Q}_{\mathbf{x}_0^{1/2}} \nabla_{\mathbf{x}}^T(\mathbf{x}_0, \underline{z}))\right\|_{\mathrm{F}}} = \mathbf{e} - \beta \frac{\mathbf{p}(\underline{z})}{\|\mathbf{p}(\underline{z})\|_{\mathrm{F}}}.$$

Scaling back to the original space we get an interior feasible solution for the primal problem.

$$\mathbf{x}(\underline{z}) = \mathbf{x}_0 - \beta \frac{\mathbf{Q}_{\mathbf{x}_0^{1/2}} \mathbf{p}(\underline{z})}{\|\mathbf{p}(\underline{z})\|_{\mathrm{F}}} \tag{8.4.24}$$

The new point $\mathbf{x}(\underline{z})$ serves as a candidate for the next iterate's primal solution.

Since $\mathcal{P}\underset{\sim}{A}$ is a projector it follows that:

$$\langle \nabla_{\mathbf{x}(\underline{z})} \phi(\mathbf{x}_0, \underline{z}), (\mathbf{x}(\underline{z}) - \mathbf{x}_0) \rangle = -\beta \|\mathbf{p}(\underline{z})\|_{\mathrm{F}} \tag{8.4.25}$$

Therefore, we may rewrite corollary (8.4.1) in the following form:

$$\phi(\mathbf{x}(\underline{z}), \underline{z}) - \phi(\mathbf{x}_0, \underline{z}) \leq -\beta \|\mathbf{p}(\underline{z})\|_{\mathrm{F}} + \frac{\beta^2}{2(1-\beta)}. \tag{8.4.26}$$

The second term comes from the fact that

$$\left\| \mathbf{Q}_{\mathbf{x}_0^{1/2}} (\mathbf{x}(\underline{z}) - \mathbf{x}_0) \right\|_2 \leq \left\| \mathbf{Q}_{\mathbf{x}_0^{1/2}} (\mathbf{x}(\underline{z}) - \mathbf{x}_0) \right\|_{\mathrm{F}} = \beta.$$

We now intend to obtain conditions under which ϕ decreases by a constant amount if we set $\mathbf{x}(\underline{z})$ according to (8.4.24). Then we show that this decrease translates into like decrease in ψ.

In order to bound reduction in ϕ we need to bound $\|\mathbf{p}(\underline{z})\|_{\mathrm{F}}$. Expanding it we get:

$$\mathbf{p}(\underline{z}) = \frac{q}{\langle \mathbf{c}, \mathbf{x}_0 \rangle - \underline{z}} \mathbf{Q}_{\mathbf{x}_0^{1/2}} \mathbf{s}(\underline{z}) - \mathbf{e} \tag{8.4.27}$$

where

$$\begin{aligned}
\mathbf{s}(\underline{z}) &\stackrel{\text{def}}{=} \mathbf{c} - A^T \left((A\mathbf{Q}_{\mathbf{x}_0} A^T)^{-1} A \mathbf{Q}_{\mathbf{x}_0^{1/2}} \left(\mathbf{Q}_{\mathbf{x}_0^{1/2}} \mathbf{c} - \tfrac{\langle \mathbf{c}, \mathbf{x}_0 \rangle - \underline{z}}{q} \mathbf{e} \right) \right) \\
&\stackrel{\text{def}}{=} \mathbf{c} - A^T \mathbf{y}(\underline{z}).
\end{aligned}$$

Having defined $\mathbf{y}(\underline{z})$ it is convenient to decompose it into sum of two other vectors:

$$\mathbf{y}_1 \stackrel{\text{def}}{=} (A\mathbf{Q}_{\mathbf{x}_0} A^T)^{-1} A \mathbf{Q}_{\mathbf{x}_0} \mathbf{c}, \tag{8.4.28}$$

$$\mathbf{y}_2 \stackrel{\text{def}}{=} (A\mathbf{Q}_{\mathbf{x}_0} A^T)^{-1} A \mathbf{Q}_{\mathbf{x}_0^{1/2}} \mathbf{e} = (A\mathbf{Q}_{\mathbf{x}_0} A^T)^{-1} \mathbf{b} \tag{8.4.29}$$

$$\mathbf{y}(\underline{z}) = \mathbf{y}_1 + \tfrac{\langle \mathbf{c}, \mathbf{x}_0 \rangle - \underline{z}}{q} \mathbf{y}_2. \tag{8.4.30}$$

Thus if $\mathbf{s}(\underline{z}) \succ \mathbf{0}$ then $(\mathbf{y}(\underline{z}), \mathbf{s}(\underline{z}))$ is dual interior feasible.

Here is what we are going to establish in the following lemmas. By construction $\mathbf{x}(\underline{z}) \succ \mathbf{0}$. First we show that if $(\mathbf{y}(\underline{z}), \mathbf{s}(\underline{z}))$ is not dual-feasible then the value of the primal potential function at the new point $\mathbf{x}(\underline{z})$ decrease by a constant amount. Second, if $\mathbf{s}(\underline{z}) \succ \mathbf{0}$ then we show that the value of the primal-dual potential function at the new dual feasible point $(\mathbf{y}(\underline{z}), \mathbf{s}(\underline{z}))$ decreases by a constant amount. Either way the primal-dual potential function decreases at least by a constant leading to polynomial-time convergent algorithm.

Lemma 8.4.2 *Define:*

$$\begin{aligned}
q &\stackrel{\text{def}}{=} r + \sqrt{r}, \\
\Delta_0 &\stackrel{\text{def}}{=} \tfrac{\langle \mathbf{x}_0, \mathbf{s}_0 \rangle}{r} = \tfrac{\langle \mathbf{c}, \mathbf{x}_0 \rangle - \underline{z}}{r}, \\
\Delta &\stackrel{\text{def}}{=} \tfrac{\langle \mathbf{x}_0, \mathbf{s}(\underline{z}) \rangle}{r}.
\end{aligned}$$

and let $0 < \alpha < 1$ *be a constant.*

i. *If* $\mathrm{s}(\underline{z}) \not\succeq \mathbf{0}$ *then*

$$\|\mathbf{p}(\underline{z})\|_{\mathrm{F}} \geq 1,$$

ii. *otherwise if* $\left\|\mathbf{Q}_{\mathbf{x}_0^{1/2}}\mathrm{s}(\underline{z}) - \Delta\mathbf{e}\right\|_{\mathrm{F}} \geq \alpha\Delta$ *then*

$$\|\mathbf{p}(\underline{z})\|_{\mathrm{F}} \geq \alpha\sqrt{\frac{r}{r+\alpha^2}},$$

iii. *otherwise if* $\Delta \geq (1 - \frac{\alpha}{2\sqrt{r}})\Delta_0$ *then*

$$\|\mathbf{p}(\underline{z})\|_{\mathrm{F}} \geq 1 - \alpha.$$

Proof.

For **i.** notice that if $\mathrm{s}(\underline{z})$ is not positive semidefinite then by Theorem 8.3.4 neither is $\mathbf{Q}_{\mathbf{x}_0^{1/2}}\mathrm{s}(\underline{z})$. Thus, $\mathbf{p}(\underline{z})$ must have an eigenvalue smaller than -1. This implies that $1 \leq \|\mathbf{p}(\underline{z})\|_2$. To show **ii** note that

$$\|\mathbf{p}(\underline{z})\|_{\mathrm{F}} = \left\|\frac{q}{r\Delta_0}\mathbf{Q}_{\mathbf{x}_0^{1/2}}\mathrm{s}(\underline{z}) - \frac{q\Delta}{r\Delta_0}\mathbf{e} + \frac{q\Delta}{r\Delta_0}\mathbf{e} - \mathbf{e}\right\|_{\mathrm{F}}$$

Since $\left(\frac{q}{r\Delta_0}\mathbf{Q}_{\mathbf{x}_0^{1/2}}\mathrm{s}(\underline{z}) - \frac{q\Delta}{r\Delta_0}\mathbf{e}\right) \perp \left(\frac{q\Delta}{r\Delta_0}\mathbf{e} - \mathbf{e}\right)$ we have:

$$\begin{aligned}\|\mathbf{p}(\underline{z})\|_{\mathrm{F}}^2 &= \left(\tfrac{q}{r\Delta_0}\right)^2 \left\|\mathbf{Q}_{\mathbf{x}_0^{1/2}}\mathrm{s}(\underline{z}) - \Delta\mathbf{e}\right\|_{\mathrm{F}}^2 + \left\|\tfrac{q\Delta}{r\Delta_0}\mathbf{e} - \mathbf{e}\right\|_{\mathrm{F}}^2 \\ &\geq \left(\tfrac{q\Delta}{r\Delta_0}\right)^2 \alpha^2 + \left(\tfrac{q\Delta}{r\Delta_0} - 1\right)^2 r \\ &\geq \alpha^2 \tfrac{r}{r+\alpha^2}\end{aligned} \tag{8.4.31}$$

For **iii** observe that

$$\frac{q\Delta}{r\Delta_0} \geq \left(1 + \frac{1}{\sqrt{r}}\right)\left(1 - \frac{\alpha}{2\sqrt{r}}\right) \geq 1.$$

Also (8.4.31) implies

$$\begin{aligned}\|\mathbf{p}(\underline{z})\|_{\mathrm{F}}^2 &\geq \left(\tfrac{q\Delta}{r\Delta_0} - 1\right)^2 r \\ &\geq \left[\left(1 + \tfrac{1}{\sqrt{r}}\right)\left(1 - \tfrac{\alpha}{2\sqrt{r}}\right) - 1\right]^2 r \\ &\geq \left(1 - \tfrac{\alpha}{2} - \tfrac{\alpha}{2\sqrt{r}}\right)^2 \\ &\geq (1-\alpha)^2.\end{aligned}$$

Lemma 8.4.3 *With the definitions in lemma* (8.4.2) *still valid, if neither of conditions* **i.**, **ii.** *or* **iii.** *apply then, by setting* $\mathbf{s}_1 \stackrel{\text{def}}{=} \mathbf{s}(\underline{z})$,

$$\psi(\mathbf{x}_0, \mathbf{s}_1) \le \psi(\mathbf{x}_0, \mathbf{s}_0) - \frac{\alpha}{2} + \frac{\alpha^2}{2(1-\alpha)}. \tag{8.4.32}$$

Proof.
We must have

$$\Delta < (1 - \frac{\alpha}{2\sqrt{r}})\Delta_0,$$

therefore,

$$\sqrt{r} \ln \frac{\langle \mathbf{x}_0, \mathbf{s}_1 \rangle}{\langle \mathbf{x}_0, \mathbf{s}_0 \rangle} = \sqrt{r} \ln \frac{\Delta}{\Delta_0} \le -\frac{\alpha}{2}. \tag{8.4.33}$$

Since

$$\left\| \mathbf{Q}_{\mathbf{x}_0^{1/2}} \mathbf{s}_1 - \Delta \mathbf{e} \right\|_F < \alpha \Delta.$$

we may apply lemma (8.4.1) to $\left(\mathbf{Q}_{\mathbf{x}_0^{1/2}} \mathbf{s}_1\right)/\Delta$ and get:

$$\begin{aligned}
r \ln \langle \mathbf{x}_0, \mathbf{s}_1 \rangle - \ln \det \mathbf{Q}_{\mathbf{x}_0^{1/2}} \mathbf{s}_1 \quad & = r \ln \frac{\langle \mathbf{x}_0, \mathbf{s}_1 \rangle}{\Delta} - \ln \det \frac{\mathbf{Q}_{\mathbf{x}_0^{1/2}} \mathbf{s}_1}{\Delta} \\
& = r \ln r - \ln \det \frac{\mathbf{Q}_{\mathbf{x}_0^{1/2}} \mathbf{s}_1}{\Delta} \\
& \le r \ln r + \frac{\left\| \frac{\mathbf{Q}_{\mathbf{x}_0^{1/2}} \mathbf{s}_1}{\Delta} - \mathbf{e} \right\|_F^2}{2\left[1 - \left\| \frac{\mathbf{Q}_{\mathbf{x}_0^{1/2}} \mathbf{s}_1}{\Delta} - \mathbf{e} \right\|_2\right]} \\
& \le r \ln \operatorname{tr}\left(\mathbf{Q}_{\mathbf{x}_0^{1/2}} \mathbf{s}_0\right) - \ln \det \mathbf{Q}_{\mathbf{x}_0^{1/2}} \mathbf{s}_0 + \frac{\alpha^2}{2(1-\alpha)}
\end{aligned}$$

The last inequality comes from the arithmetic-geometric inequality applied to the eigenvalues of $\mathbf{Q}_{\mathbf{x}_0^{1/2}} \mathbf{s}_0$. The lemma follows by adding the last inequality with inequality (8.4.33). Summing up lemmas 8.4.2 and 8.4.3 we get the potential reduction theorem:

Theorem 8.4.2 *Let* $\mathbf{x}_0$ *and* $\mathbf{y}_0$ *be any pair of primal and dual feasible interior points. If* $q = r + \sqrt{r}$, $\underline{z}_0 = \mathbf{b}^T \mathbf{y}_0$, $\mathbf{x}_1 = \mathbf{x}(\underline{z}_0)$ *given by* (8.4.24) *and* $\mathbf{s}_1 = \mathbf{s}(\underline{z}_0)$, *then there is an absolute constant* δ *such that either*

$$\psi(\mathbf{x}_1, \mathbf{s}_0) \le \psi(\mathbf{x}_0, \mathbf{s}_0) - \delta$$

or

$$\psi(\mathbf{x}_0, \mathbf{s}_1) \le \psi(\mathbf{x}_0, \mathbf{s}_0) - \delta$$

hold. Further, if we choose $\alpha = 0.43$ *and* $\beta = 0.3$ *then* $\delta > 0.005$.

Proof.

Let

$$\delta \stackrel{\text{def}}{=} \min\left(-\frac{\alpha}{2}+\frac{\alpha^2}{2(1-\alpha)},\ \beta\min\left[\alpha\sqrt{\frac{r}{r+\alpha^2}},\ 1-\alpha\right]+\frac{\beta^2}{2(1-\beta)}\right)$$

The second inequality comes from Lemma 8.4.3 For the first inequality notice that the primal potential function ϕ is reduced by Lemma 8.4.2 and (8.4.26); but in this case since $\mathbf{s}$ does not change, noting relation (8.4.16) we get the the same reduction δ in ψ. The constants given in the statement of the theorem are obviously consistent with the choices of α, β and δ.

The potential reduction theorem (8.4.2) now yields the linear scaling algorithm sketched in figure (8.1). This algorithm, not surprisingly, is a generalization of Ye's primal algorithm [853].

Input:

An n-vector $\mathbf{x}_0$, interior feasible for the primal in (8.3.9);
an m-vector $\mathbf{y}_0$ interior feasible for the dual problem;
a constant ϵ.

Output:

A primal feasible solution $\mathbf{x}$ and dual feasible solution $\mathbf{y}$ such that $\langle \mathbf{c}, \mathbf{x}\rangle - \mathbf{b}^T\mathbf{y} < \epsilon$.

Method:

1) Set $\underline{z}_0 = \mathbf{b}^T\mathbf{y}_0$.
2) Set $\mathbf{s}_0 \leftarrow \mathbf{c} - A^T\mathbf{y}_0$.
3) Set $k = 0$.
4) While $\langle \mathbf{c}, \mathbf{x}_k\rangle - \mathbf{b}^T\mathbf{y}_k \geq \epsilon$ do

begin

Set $\mathbf{s}(\underline{z}_k)$ according to (8.4.51)
and $\mathbf{p}(\underline{z}_k)$ according to (8.4.23).

If either **i.**, **ii.** or **iii.** in (8.4.2) apply then

a. Find $\beta^* \leftarrow \text{argmin}_{0\leq\beta\leq 1}\psi(\mathbf{x}_k - \beta\mathbf{Q}_{\mathbf{x}_k^{1/2}}\mathbf{p}(\underline{z}_k), \mathbf{s}_k)$, using a line search procedure.

b. Set $\mathbf{x}_{k+1} \leftarrow \mathbf{x}_k + \beta^*\mathbf{Q}_{\mathbf{x}_k^{1/2}}\mathbf{p}(\underline{z}_k)$.

c. Set $\mathbf{s}_{k+1} \leftarrow \mathbf{s}_k$, and $\underline{z}_{k+1} \leftarrow \underline{z}_k$.

Else

a. Find $\underline{z}^* \leftarrow \text{argmin}_{\underline{z}\leq \underline{z}_k}\psi(\mathbf{x}_k, \mathbf{s}(\underline{z}))$ by line search.

b. Set $\mathbf{s}_{k+1} \leftarrow \mathbf{s}(\underline{z}^*)$.

c. Set $\mathbf{x}_{k+1} \leftarrow \mathbf{x}_k$, and $\underline{z}_{k+1} \leftarrow \mathbf{b}^T\mathbf{y}(\underline{z}^*)$.

Set $k = k+1$.

end.

Figure 8.1 A potential reduction algorithm based on linear scaling .

As in Ye's linear programming algorithm, since the iteration complexity depends solely on the reduction in the potential function ψ, we may use line search to find the best possible β and $\underline{z}$.

8.4.8 Properties of projective scaling

In Karmarkar's original paper [404] projective rather than linear scalings are employed to facilitate the complexity analysis of interior point algorithms. In terms of worst case complexity analysis the best results have been obtained by using both linear and projective transformations. Using projective scalings adds some elegance to the theory, but it is not clear how much projective techniques improve on worst case or practical performance over linear versions, if at all. In any case, here we have included discussion of projective scaling methods and their extension to the SDP problem, in order to demonstrate that such techniques may also be generalized "word by word".

Let $\mathbf{x}_0 \succ \mathbf{0}$. Define the projective transformation $\mathcal{T} : \Re^n \to \Re^n \times \Re^p$, so that $(\widetilde{\mathbf{x}}; \hat{\mathbf{x}}) \stackrel{\text{def}}{=} \mathcal{T}(\mathbf{x})$, where

$$\widetilde{\mathbf{x}} \stackrel{\text{def}}{=} \frac{(r+p)\mathbf{Q}_{\mathbf{x}_0^{-1/2}}\mathbf{x}}{p+\langle \mathbf{x}_0^{-1}, \mathbf{x}\rangle} \tag{8.4.34}$$

$$\hat{\mathbf{x}} \stackrel{\text{def}}{=} \frac{r+p}{p+\langle \mathbf{x}_0^{-1}, \mathbf{x}\rangle}\mathbf{1} \tag{8.4.35}$$

where again, $\mathbf{1}$ is vector of all ones. The inverse transformation is given by:

$$\mathbf{x} = \mathcal{T}^{-1}(\widetilde{\mathbf{x}}, \hat{\mathbf{x}}) = \frac{\mathbf{Q}_{\mathbf{x}_0^{1/2}}\widetilde{\mathbf{x}}}{\sum \hat{x}_j / p} \tag{8.4.36}$$

Observe that

$$\langle \widetilde{\mathbf{x}}, \mathbf{e}\rangle + \mathbf{1}^T\hat{\mathbf{x}} = r + p, \qquad \mathcal{T}(\mathbf{x}_0) = (\mathbf{e}; \mathbf{1}).$$

Let $\underline{z} \leq v^*$, the value of the optimal primal objective function. Under $\mathcal{T}$, the primal SDP problem is transformed into the following SDP:

$$\begin{array}{ll} \min & \langle \underset{\sim}{\mathbf{c}}, \widetilde{\mathbf{x}}\rangle + \hat{\mathbf{c}}^T(\underline{z})\hat{\mathbf{x}} \\ \text{s.t.} & \underset{\sim}{A}\widetilde{\mathbf{x}} + \underset{\wedge}{A}\hat{\mathbf{x}} = \mathbf{0} \\ & \operatorname{tr}\widetilde{\mathbf{x}} + \mathbf{1}^T\hat{\mathbf{x}} = r + p \\ & \widetilde{\mathbf{x}} \succcurlyeq \mathbf{0}, \ \hat{\mathbf{x}} \geq \mathbf{0} \end{array} \tag{8.4.37}$$

where

$$\underset{\sim}{\mathbf{c}} \stackrel{\text{def}}{=} \mathbf{Q}_{\mathbf{x}_0^{1/2}}\mathbf{c} \tag{8.4.38}$$

$$\hat{\mathbf{c}}(\underline{z}) \stackrel{\text{def}}{=} \left(\frac{-\underline{z}}{p}\right)\mathbf{1} \tag{8.4.39}$$

$$\underset{\sim}{\mathbf{a}_i} \stackrel{\text{def}}{=} \mathbf{Q}_{\mathbf{x}_0^{1/2}}\mathbf{a}_i \tag{8.4.40}$$

$$\underset{\sim}{A} \stackrel{\text{def}}{=} A\mathbf{Q}_{\mathbf{x}_0^{1/2}} \tag{8.4.41}$$

$$\underset{\wedge}{A} \stackrel{\text{def}}{=} \left(\frac{-1}{p}\right)\mathbf{b}\mathbf{1}^T \tag{8.4.42}$$

Since the inequality $\hat{\mathbf{x}} \geq \mathbf{0}$ is a special kind of symmetric cone inequality, the transformed problem is an SDP and thus its primal potential function is:

$$\bar{\phi}(\widetilde{\mathbf{x}}, \hat{\mathbf{x}}, \underline{z}) = q\ln\left(\langle \underset{\sim}{\mathbf{c}}, \hat{\mathbf{x}}\rangle + \underset{\sim}{\mathbf{c}}(\underline{z})^T\mathbf{x}\right) - \ln\det\widetilde{\mathbf{x}} - \sum_{j=1}^{p}\ln\hat{x}_j \tag{8.4.43}$$

The following invariant property holds for the potential functions under projective transformations and is straightforward to prove:

Lemma 8.4.4 *If $\hat{\mathbf{x}}_1 = \cdots = \hat{\mathbf{x}}_p$, and $q = r + p$ then*

$$\phi(\mathbf{x}, \underset{\sim}{z}) - \phi(\mathbf{x}_0, \underset{\sim}{z}) = \overline{\phi}(\tilde{\mathbf{x}}, \hat{\mathbf{x}}, \underset{\sim}{z}) - \overline{\phi}(\mathbf{e}, \mathbf{1}, \underset{\sim}{z}). \qquad (8.4.44)$$

Expanding $\overline{\phi}$ and using Lemma 8.4.1 results in:

Corollary 8.4.2 *For $q = r + p$*

$$\begin{aligned} \overline{\phi}(\tilde{\mathbf{x}}, \hat{\mathbf{x}}, \underset{\sim}{z}) - \overline{\phi}(\mathbf{e}, \mathbf{1}, \underset{\sim}{z}) &\leq (r+p) \ln \left(\frac{\langle \underset{\sim}{\mathbf{c}}, \tilde{\mathbf{x}} \rangle + \hat{\mathbf{c}}^T (\underset{\sim}{z}) \hat{\mathbf{x}}}{\operatorname{tr} \underset{\sim}{\mathbf{c}} + \mathbf{1}^T \cosh (\underset{\sim}{z})} \right) \\ &- \frac{\|\tilde{\mathbf{x}} - \mathbf{e}\|_F^2 + \|\hat{\mathbf{x}} - \mathbf{1}\|^2}{2\left(1 - \|\tilde{\mathbf{x}} - \mathbf{e}\|_2 - \|\hat{\mathbf{x}} - \mathbf{1}\|_\infty\right)} \end{aligned} \qquad (8.4.45)$$

Again $\|\cdot\|$ refers to the ordinary Euclidean norm, and $\|\cdot\|_\infty$ is the max-norm of vectors; note that with respect to the Jordan algebra $\Re^p$ and Hadamard (component wise) multiplication of its vectors, $\|\cdot\|_F = \|\cdot\|$ and $\|\cdot\|_2 = \|\cdot\|_\infty$.

This corollary and Lemma 8.4.4 are used in the following sections to prove the reduction in the primal potential function.

8.4.9 Potential reduction with projective scaling

Like the linear scaling algorithm of the last section we replace the semidefinite constraint $\tilde{\mathbf{x}} \succcurlyeq \mathbf{0}, \hat{\mathbf{x}} \geq \mathbf{0}$ in the scaled problem (8.4.37) with ellipsoid constraints:

$$\begin{array}{ll} \min & \langle \underset{\sim}{\mathbf{c}}, \tilde{\mathbf{x}} \rangle + \hat{\mathbf{c}}^T \hat{\mathbf{x}} \\ \text{s.t.} & \tilde{A}\tilde{\mathbf{x}} + \hat{A}\hat{\mathbf{x}} = \mathbf{0} \\ & \operatorname{tr}(\tilde{\mathbf{x}}) + \mathbf{1}^T \hat{\mathbf{x}} = r + p \\ & \|\tilde{\mathbf{x}} - \mathbf{e}\|_F^2 + \|\hat{\mathbf{x}} - \mathbf{1}\|^2 \leq \beta^2 \end{array} \qquad (8.4.46)$$

where again $\beta < 1$ is given constant to be determined shortly. The solution of this least squares problem is a possible candidate for the new primal feasible interior point. Let

$$A' \stackrel{\text{def}}{=} \begin{pmatrix} \tilde{A} & \hat{A} \\ \mathbf{e}^T & \mathbf{1}^T \end{pmatrix} \qquad \mathbf{c}' \stackrel{\text{def}}{=} \begin{pmatrix} \underset{\sim}{\mathbf{c}} \\ \hat{\mathbf{c}} \end{pmatrix}$$

$$\mathbf{x}' \stackrel{\text{def}}{=} \begin{pmatrix} \tilde{\mathbf{x}} \\ \hat{\mathbf{x}} \end{pmatrix} \qquad \mathbf{e}' \stackrel{\text{def}}{=} \begin{pmatrix} \underset{\sim}{\mathbf{e}} \\ \mathbf{1} \end{pmatrix}$$

The solution is given by

$$\mathbf{x}' = \mathbf{e}' - \beta \frac{\mathbf{p}(\underset{\sim}{z})}{\|\mathbf{p}(\underset{\sim}{z})\|_F} \qquad (8.4.47)$$

with

$$\mathbf{p}(\underset{\sim}{z}) \stackrel{\text{def}}{=} \mathcal{P}A'(\mathbf{c}')$$

that is the orthogonal projection of $\mathbf{c}'$ onto the kernel of $\mathcal{A}'$.

After transforming back to the original space we get a new interior feasible point for the primal:

$$\mathbf{x}(\underline{z}) \stackrel{\text{def}}{=} \mathcal{T}^{-1}(\mathbf{x}') \tag{8.4.48}$$

Define:

$$\mathbf{y}(\underline{z}) \stackrel{\text{def}}{=} \left(A'A'^T\right)^{-1} A'\mathbf{c}' \tag{8.4.49}$$

Thus

$$\begin{aligned} \mathbf{y}(\underline{z}) &= ([\underset{\sim}{A}|\hat{A}][\underset{\sim}{A}|\hat{A}]^T)^{-1}[\underset{\sim}{A}|\hat{A}]\begin{pmatrix} \underset{\sim}{\mathbf{c}} \\ \hat{\mathbf{c}}(\underline{z}) \end{pmatrix} \\ &= \left(A\mathbf{Q}_{\mathbf{x}_0}A^T\mathbf{Q}_{\mathbf{x}^{1/2}} + (1/p)\mathbf{b}\mathbf{b}^T\right)^{-1}(A\mathbf{Q}_{\mathbf{x}_0}\mathbf{c} + (\underline{z}/p)\mathbf{b}) \end{aligned} \tag{8.4.50}$$

and

$$\mathbf{s}(\underline{z}) \stackrel{\text{def}}{=} \mathbf{c} - A^T\mathbf{y}(\underline{z}) \tag{8.4.51}$$

$\mathbf{s}(\underline{z})$ and $\mathbf{y}(\underline{z})$ are candidates for the new dual iterate. In terms of these quantities $\mathbf{p}(\underline{z})$ may be written as:

$$\mathbf{p}(\underline{z}) = \begin{pmatrix} \mathbf{Q}_{\mathbf{x}^{1/2}}\mathbf{s}(\underline{z}) \\ \frac{\mathbf{b}^T\mathbf{y}(\underline{z})-\underline{z}}{p}\mathbf{1} \end{pmatrix} - \frac{\langle \mathbf{c}, \mathbf{x}_0\rangle - \underline{z}}{r+p}\begin{pmatrix} \mathbf{e} \\ \mathbf{1} \end{pmatrix} \tag{8.4.52}$$

Now we show that either the new primal candidate $\mathbf{x}(\underline{z}_0)$, or the dual candidates $\mathbf{y}(\underline{z})$ and $\mathbf{s}(\underline{z})$ reduces the value of the primal-dual potential function ψ by a constant amount. First observe that since $\mathcal{P}A'$ is a projector, we have:

$$\langle \underset{\sim}{\mathbf{c}}, \tilde{\mathbf{x}} - \mathbf{e}\rangle + \hat{\mathbf{c}}^T(\underline{z})(\hat{\mathbf{x}} - \mathbf{1}) = -\beta\,\|\mathbf{p}(\underline{z})\|_{\mathrm{F}}\,.$$

Therefore, noting that $\ln(1+x) \le x$ for all nonnegative x, corollary (8.4.2) implies:

Corollary 8.4.3 *Let $q = r + p$ and $\mathbf{x}' = (\tilde{\mathbf{x}}; \hat{\mathbf{x}})$ be as in (8.4.47). Then*

$$\overline{\phi}(\tilde{\mathbf{x}}, \hat{\mathbf{x}}, \underline{z}) - \overline{\phi}(\mathbf{e}, \mathbf{1}, \underline{z}) \le -(r+p)\beta\frac{\|\mathbf{p}(\underline{z})\|_{\mathrm{F}}}{\mathbf{c}^T(\underline{z})\mathbf{1} + \operatorname{tr}\underset{\sim}{\mathbf{c}}} + \frac{\beta^2}{2(1-\beta)}$$

Let

$$\Delta_0 \stackrel{\text{def}}{=} \langle \mathbf{c}, \mathbf{x}_0\rangle - \underline{z}$$

and

$$\Delta_1 \stackrel{\text{def}}{=} \langle \mathbf{s}(\underline{z}), \mathbf{x}_0\rangle = \langle \mathbf{c}, \mathbf{x}_0\rangle - \mathbf{b}^T\mathbf{y}(\underline{z})$$

The following lemma is plays the same role as Lemma 8.4.2 in linear scaling algorithm.

Lemma 8.4.5 *If there is some real number α with $0 < \alpha < 1$, such that*

$$\|\mathbf{p}(\underline{z})\|_{\mathrm{F}} \le \alpha\frac{\Delta_0}{r+p}$$

then $\mathbf{s}(\underline{z}) \succcurlyeq \mathbf{0}$, *and* $\mathbf{b}^T\mathbf{y}(\underline{z}) > \underline{z}$. *Furthermore,*

$$\left\| \mathbf{Q}_{\mathbf{x}^{1/2}}\mathbf{s}(\underline{z}) - \frac{\Delta_0}{r}\mathbf{e} \right\|_{\mathrm{F}} \leq \frac{\Delta_0}{r}\alpha\sqrt{\frac{r+r^2/p}{r+r^2/p-\alpha^2}}, \tag{8.4.53}$$

and

$$\left| \frac{r+p}{r}\frac{\Delta_0}{\Delta_1} - 1 \right| \leq \frac{\alpha}{\sqrt{r+r^2/p}} \tag{8.4.54}$$

Proof.

Suppose $\mathbf{s}(\underline{z}) \not\succcurlyeq \mathbf{0}$. Then $\mathbf{Q}_{\mathbf{x}^{1/2}}\mathbf{s}(\underline{z})$ is not positive semidefinite and so some of its eigenvalues are less than zero. Therefore, from (8.4.52) we have

$$\|\mathbf{p}(\underline{z})\|_{\mathrm{F}} \geq \left\| \frac{\Delta_0}{r+p}\mathbf{e} - \left(\mathbf{Q}_{\mathbf{x}^{1/2}}\mathbf{s}(\underline{z})\right) \right\|_2 \geq \frac{\Delta_0}{r+p},$$

a contradiction. Also, If $\mathbf{b}^T\mathbf{y}(\underline{z}) > \underline{z}$ then from (8.4.52) we have

$$\|\mathbf{p}(\underline{z})\|_{\mathrm{F}} \geq \left\| \frac{\Delta_0}{r+p}\mathbf{1} - \frac{\mathbf{b}^T\mathbf{y}(\underline{z}) - \underline{z}}{p}\mathbf{1} \right\|_\infty \geq \frac{\Delta_0}{r+p},$$

which is again a contradiction. Now from (8.4.52) we have

$$\mathbf{p}(\underline{z}) = \begin{pmatrix} \left(\mathbf{Q}_{\mathbf{x}^{1/2}}\mathbf{s}(\underline{z}) - \frac{\Delta_1}{r}\mathbf{e}\right) - \left(\frac{\Delta_0}{r+p} - \frac{\Delta_1}{r}\right)\mathbf{e} \\ \left(\frac{\Delta_0 - \Delta_1}{p} - \frac{\Delta_0}{r+p}\right)\mathbf{1} \end{pmatrix}$$

Since $\langle \mathbf{e}, \mathbf{Q}_{\mathbf{x}^{1/2}}\mathbf{s}(\underline{z}) - (\Delta_1/r)\mathbf{e}\rangle = 0$, we have

$$\begin{aligned} \|\mathbf{p}(\underline{z})\|_{\mathrm{F}}^2 &= \left\|\mathbf{Q}_{\mathbf{x}^{1/2}}\mathbf{s}(\underline{z}) - \tfrac{\Delta_1}{r}\mathbf{e}\right\|_{\mathrm{F}}^2 + r\left(\tfrac{\Delta_0}{r+p} - \tfrac{\Delta_1}{r}\right)^2 + p\left(\tfrac{\Delta_0-\Delta_1}{p} - \tfrac{\Delta_0}{r+p}\right)^2 \\ &= \left\|\mathbf{Q}_{\mathbf{x}^{1/2}}\mathbf{s}(\underline{z}) - \tfrac{\Delta_1}{r}\mathbf{e}\right\|_{\mathrm{F}}^2 + \left(r + \tfrac{r^2}{p}\right)\left(\tfrac{\Delta_1}{p} - \tfrac{\Delta_0}{r+p}\right)^2. \end{aligned} \tag{8.4.55}$$

If (8.4.53) is false then from (8.4.55) we have

$$\begin{aligned} \|\mathbf{p}(\underline{z})\|_{\mathrm{F}}^2 \geq & \left(\tfrac{\Delta_1}{r}\right)^2\alpha^2\tfrac{r+r^2/p}{r+r^2/p-\alpha^2} + \left(r+\tfrac{r^2}{p}\right)\left(\tfrac{\Delta_1}{r} - \tfrac{\Delta_0}{r+p}\right)^2 \\ \geq & \qquad \alpha^2\left(\tfrac{\Delta_0}{r+p}\right)^2, \end{aligned}$$

again a contradiction. Finally from this last inequality it follows that

$$\left(r + \frac{r^2}{p}\right)\left(\frac{\Delta_1}{r} - \frac{\Delta_0}{r+p}\right)^2 \leq \alpha^2\left(\frac{\Delta_0}{r+p}\right)^2$$

from which (8.4.54) follows. Now we may prove the potential reduction theorem.

Theorem 8.4.3 *Let* $\mathbf{x}_0$ *be any interior feasible point for the primal problem* (8.3.9) *and* $\mathbf{y}_0$ *interior feasible for the dual. Let also,* $q \stackrel{\text{def}}{=} r + \lceil\sqrt{r}\rceil$, $\mathbf{s}_0 \stackrel{\text{def}}{=} \mathbf{c} - A^T\mathbf{y}_0$, $\underline{z}_0 \stackrel{\text{def}}{=} \mathbf{b}^T\mathbf{y}_0$, $\mathbf{x}_1 \stackrel{\text{def}}{=} \mathcal{T}^{-1}(\widetilde{\mathbf{x}}_1, \hat{\mathbf{x}}_1)$, *as in* (8.4.48), $\mathbf{y}_1 \stackrel{\text{def}}{=} \mathbf{y}(\underline{z}_0)$, *and* $\mathbf{s}_1 \stackrel{\text{def}}{=} \mathbf{s}(\underline{z}_0)$. *Then there exist an absolute constant* δ *such that either*

$$\psi(\mathbf{x}_1, \mathbf{s}_0) \leq \psi(\mathbf{x}_0, \mathbf{s}_0) - \delta,$$

or

$$\psi(\mathbf{x}_0, \mathbf{s}_1) \leq \psi(\mathbf{x}_0, \mathbf{s}_0) - \delta,$$

Furthermore, if we set $\alpha = 0.55$ *and* $\beta \stackrel{\text{def}}{=} 0.3$, *then* $\delta > 0.1$.

Proof.
If

$$||\mathbf{p}(\underline{z})||_{\mathrm{F}} \geq \alpha \frac{\Delta}{r+p}$$

then

$$\begin{aligned} \psi(\mathbf{x}_1, \mathbf{s}_0) - \psi(\mathbf{x}_0, \mathbf{s}_0) &= \phi(\mathbf{x}_1, \underline{z}_0) - \phi(\mathbf{x}_0, \underline{z}_0) \\ &= \overline{\phi}(\widetilde{\mathbf{x}}_1, \hat{\mathbf{x}}_1, \underline{z}_0) - \overline{\phi}(\mathbf{e}, 1, \underline{z}_0) \\ &\leq -\beta\alpha + \tfrac{\beta^2}{2(1-\beta)} \end{aligned}$$

(the last inequality is true by corollary 8.4.3. Otherwise, using the previous lemma, and also applying lemma (8.4.1) to $(r/\Delta_1)\mathbf{Q}_{\mathbf{x}_0^{1/2}}\mathbf{s}_1$, and setting $\gamma = \sqrt{\frac{r+r^2/p}{r+r^2/p-\alpha^2}}$ we have:

$$\begin{aligned} r\ln\langle \mathbf{x}_0, \mathbf{s}_1\rangle - \ln\det \mathbf{Q}_{\mathbf{x}_0^{1/2}}\mathbf{s}_1 &= r\ln\left(\tfrac{r\langle \mathbf{x}_0, \mathbf{s}_1\rangle}{\Delta_1}\right) - \ln\det \frac{r\mathbf{Q}_{\mathbf{x}_0^{1/2}}\mathbf{s}_1}{\Delta_1} \\ &= r\ln r - \ln\det \frac{r\mathbf{Q}_{\mathbf{x}_0^{1/2}}\mathbf{s}_1}{\Delta_1} \\ &\leq r\ln r + \frac{||r\mathbf{Q}_{\mathbf{x}^{1/2}}\mathbf{s}_1/\Delta_1 - \mathbf{e}||_{\mathrm{F}}^2}{2(1-||r\mathbf{Q}_{\mathbf{x}^{1/2}}\mathbf{s}_1/\Delta_1 - \mathbf{e}||_{\mathrm{F}})} \\ &\leq r\ln\operatorname{tr}\mathbf{Q}_{\mathbf{x}_0^{1/2}}\mathbf{s}_0 - \ln\det\mathbf{Q}_{\mathbf{x}_0^{1/2}}\mathbf{s}_0 + \tfrac{\gamma^2}{2(1-\gamma)}, \end{aligned}$$

where the last inequality again comes from applying the arithmetic-geometric mean inequality to the eigenvalues of $\mathbf{Q}_{\mathbf{x}_0^{1/2}}\mathbf{s}_0$. It is easily verified that the particular values provided for α, β and γ are consistent with the conditions imposed on them in the theorem.

Based on this result we present the projective version of the algorithm displayed in figure (8.2).

8.4.10 The Recipe

We have summarized the correspondence between proofs of polynomiality in linear programming interior point methods and analogous approaches for semidefinite programming. This correspondence is presented in Figure 8.1. Notice that

Input:

An $n \times n$ matrix $\mathbf{x}_0$, interior feasible for the primal in (8.3.9);
an m-vector $\mathbf{y}_0$ interior feasible for the dual problem;
a constant ϵ.

Output:

A primal feasible solution $\mathbf{x}$ and dual feasible solution $\mathbf{y}$ such that
$\langle \mathbf{c}, \mathbf{x} \rangle - \mathbf{b}^T \mathbf{y} < \epsilon$.

Method:

1) Set $\underline{z}_0 = \mathbf{b}^T \mathbf{y}_0$.
2) Set $\mathbf{s}_k := \mathbf{c} - A^T \mathbf{y}_0$.
3) Set $k = 0$.
4) While $\langle \mathbf{c}, \mathbf{x}_k \rangle - \mathbf{b}^T \mathbf{y}_k \geq \epsilon$ do

begin

Form $\mathbf{s}(\underline{z}_k)$ from (8.4.51) and $\mathbf{p}(\underline{z}_k)$ from (8.4.52).
If $\|\mathbf{p}(\underline{z}_k)\|_{\mathrm{F}} \geq \alpha(\langle \mathbf{c}, \mathbf{x}_k \rangle - \underline{z}_k)/(r+p)$ then

a.Find $\beta^* := \mathrm{argmin}_{0 \leq \beta \leq 1} \psi(\mathbf{x}_k - \beta \mathbf{Q}_{\mathbf{x}_0^{1/2}} \mathbf{p}(\underline{z}_k), \mathbf{s}_k)$,
using a line search procedure.

b. Set $(\tilde{\mathbf{x}}_{k+1}, \hat{\mathbf{x}}_{k+1}) = (\mathbf{e}, 1) - \beta^* \mathbf{p}(\underline{z})$,
and set $\mathbf{x}_{k+1} := \mathcal{T}^{-1}(\tilde{\mathbf{x}}_{k+1}, \hat{\mathbf{x}}_{k+1})$.

c. Set $\mathbf{s}_{k+1} := \mathbf{s}_k$, and $\underline{z}_{k+1} := \underline{z}_k$.

Else

a. Find $\underline{z}^* := \mathrm{argmin}_{\underline{z} \leq \underline{z}_k} \psi(\mathbf{x}_k, \mathbf{s}(\underline{z}))$ by a line search.

b. Set $\mathbf{s}_{k+1} = \mathbf{s}(\underline{z}^*)$.

c. Set $\mathbf{x}_{k+1} = \mathbf{x}_k$, and $\underline{z}_{k+1} = \mathbf{b}^T \mathbf{y}(\underline{z}^*)$.

Set $k = k + 1$.

end.

Figure 8.2 A potential reduction algorithm based on projective scaling.

the entries under the LP column are special case of those under SDP column. Given an appropriate interior point algorithm for linear programming we may construct, in a mechanical way, an algorithm for the SDP problem by replacing any references to the entries under the LP column, with the corresponding entries under the SDP column. Proofs of convergence or polynomial time complexity may also be extended mechanically in the same manner.

One can now ask the question: why is this "recipe" successful in the two algorithms presented in this chapter, but inadequate in certain other algorithms, most notably primal-dual methods such as those of Monteiro and Alder? One reason is that in the case of algorithms presented here the interaction between $\mathbf{x}$ and $\mathbf{s}$, or matrices, $L(\mathbf{x})$ and $L(\mathbf{s})$ is quite simple; in particular in no occasion we need to rely on the assumption that $\mathbf{x}$ and $\mathbf{s}$ operator commute. In case of linear programming, $L(\mathbf{x})$ and $L(\mathbf{s})$ are diagonal matrices, and thus

LP	SDP
unknown vector: $\mathbf{x}$	unknown Jordan algebra element: $\mathbf{x}$
inequality constraints: $\geq$	Löwner (SDP) constraints: $\succeq$
dual variable: $\mathbf{y}$	dual variable: $\mathbf{y}$
dual slack vector: $\mathbf{s}$	dual slack Jordan algebra element: $\mathbf{s}$
$\mathbf{1}$	$\mathbf{e}$
linear scaling: $\mathbf{x} \to \left(\mathrm{Diag}(\mathbf{x}_0)\right)^{-1}\mathbf{x}$	linear scaling: $\mathbf{x} \to \mathbf{Q}_{\mathbf{x}^{-1/2}}\mathbf{x}$
projective scaling: $\mathbf{x} \to \frac{c_1[\mathrm{Diag}(\mathbf{x}_0)]^{-1}\mathbf{x}}{c_2+\mathbf{1}^T[\mathrm{Diag}\,(\mathbf{x}_0)]^{-1}\mathbf{x}}$	projective scaling: $\mathbf{x} \to \frac{c_1\mathbf{Q}_{\mathbf{x}^{-1/2}}\mathbf{x}}{c_2+\mathrm{tr}\,\mathbf{Q}_{\mathbf{x}^{-1/2}}\mathbf{x}}$
barrier function: $\sum \ln x_i$	barrier function: $\ln \det \mathbf{x}$
norms: $\lVert\mathbf{x}\rVert$, $\lVert\mathbf{x}\rVert_\infty$, $\lVert\mathbf{x}\rVert_p$	norms: $\lVert\mathbf{x}\rVert_F$, $\lVert\mathbf{x}\rVert_2$, $\left(\sum \lvert\lambda_i(\mathbf{x})\rvert^p\right)^{1/p}$

Table 8.1 Correspondence between linear and semidefinite programming

all such matrices commute. In more general SDP's $\mathbf{x}$ and $\mathbf{s}$ do not in general operator commute and thus proofs that rely on this property cannot be extended in as straightforward a manner as prescribed by our recipe. For instance, in the work of Monteiro and Adler one key lemma uses the fact that $\left(L(\mathbf{x})L(\mathbf{s})\right)^{1/2} = L^{1/2}(\mathbf{x})L^{1/2}(\mathbf{s})$ for $\mathbf{x}$ and $\mathbf{s}$ in some neighborhood of the central path, ([538]). While this is true in LP, it is not generally true in SDP. Similar issues arise in all other approaches to primal-dual methods, including those relying on potential reduction methods. A side effect of this non-commutativity behavior is that we get not just one direct extension of primal-dual methods but in fact an infinite number of them. This comes from the fact that the relaxed complementarity relation $\mathbf{x} \circ \mathbf{s} = \mu\mathbf{e}$ can be written in many different equivalent forms. For instance we could write it in the form

$$\left(\mathbf{Q}_p\mathbf{x}\right) \circ \left(\mathbf{Q}_{p^{-1}}\mathbf{s}\right) = \mathbf{e}$$

for any arbitrary $\mathbf{p} \succ \mathbf{0}$, or

$$\mathbf{Q}_{\mathbf{x}^{1/2}}\mathbf{s} + \mathbf{Q}_{\mathbf{s}^{1/2}}\mathbf{x} = 2\mu\mathbf{e}.$$

Even thought these are mathematically equivalent, application of Newton's method to each results in a different set of directions. If $\mathbf{x}$ and $\mathbf{s}$ always operator commute (as in LP) then all these classes collapse to just one, that is all Newton directions will be the same too. But non-commutativity has made the analysis more challenging and interesting. This is also the reason behind existence of such a wide array of directions which have been studied in the past several years. The subsequent two chapters give a overview of these classes.

9 POTENTIAL REDUCTION AND PRIMAL-DUAL METHODS

Levent Tuncel

9.1 INTRODUCTION

We start with an abstract definition of *interior-point methods*. These are the methods of solving convex optimization problems by generating a sequence of points which lies in the relative interior of a convex set defined by the *difficult* constraints (see [797]). In this definition, we are thinking about a formulation of the underlying problem in which one has a maximal set of linear equality constraints and some convex set constraint. The fundamental idea of interior-point algorithms goes back at least to Frisch [257] and to Fiacco and McCormick [230]. However, almost all of the new interior-point algorithms have their foundations in Karmarkar's seminal work [404]. Karmarkar's work introduced many ingredients of the interior-point algorithms with polynomial iteration complexity that have been studied extensively since 1984. These algorithms possess polynomial bounds on the number of arithmetic operations required to solve a linear programming (LP) problem. In this chapter, we will be content with the bounds on the number of *iterations*. To date, it is not known whether even the SDP feasibility problem belongs to the class of decision problems solvable in polynomial time (assuming rational data and using the "bit model" on a Turing machine).

We consider the semidefinite programming(SDP) problems in the following primal (P) and dual (D) forms.

$$\begin{array}{llrcl} (P) & \min & C \bullet X & & \\ & & \mathcal{A}X & = & b, \\ & & X & \succeq & 0, \end{array}$$

$$\begin{array}{llrcl} (D) & \max & b^T y & & \\ & & \mathcal{A}^* y + S & = & C, \\ & & S & \succeq & 0, \end{array}$$

where $\mathcal{A}$ is a linear operator from $\mathcal{S}^n$ to $\mathbf{R}^m$, so that $b \in \mathbf{R}^m$. $\mathcal{A}^*$ denotes the adjoint of $\mathcal{A}$. Without loss of generality, we assume that $\mathcal{A}$ is surjective. If not, we can consider the representation $A_i \bullet X = b_i, i \in \{1, \ldots, m\}$, where $A_i \in \mathcal{S}^n$ for every i. $\mathcal{A}$ being surjective is equivalent to $A_1, A_2, \ldots, A_m$ being linearly independent. The latter can be assumed without loss of generality, since if they are linearly dependent, then either the system $A_i \bullet X = b_i, i \in \{1, \ldots, m\}$ has no solution, or there are some redundant equations which can be eliminated. In the first case, (P) is infeasible. In the second case, all redundant equations, and the corresponding A_i, b_i can be eliminated, to arrive at an equivalent problem satisfying the assumption. Under this assumption, for any solution, (y, S), of the equation $\sum_{i=1}^{m} y_i A_i + S = C$, the S part of the solution uniquely identifies the corresponding y. Sometimes, in interior-point algorithms, it is convenient to refer only to S when one mentions a feasible solution of (D).

Let K be a convex set. We denote by $\text{int}(K)$ and ∂K the interior and the boundary of K, respectively. We will include remarks after definitions and results to point out extensions (usually very easy) to more general settings. The generality will stem from the fact that we can replace the semidefiniteness of the matrix variable by a more general condition that the variable vector x lie in a convex cone and most of the results easily extend to the case when the underlying convex cone is symmetric, or homogeneous or simply the one with no specific property. Such approach will highlight the particular structure of the underlying cone helping us with what we aim to accomplish. So, the above setting of the primal-dual SDP pair can be embedded in the following more general setting of conic convex optimization problems:

$$\begin{array}{llrcl} (CP) & \inf & \langle c, x \rangle & & \\ & & \mathcal{A}(x) & = & b, \\ & & x & \in & K, \end{array}$$

where $\mathcal{A}$ is a surjective linear map and K is a pointed, closed, convex cone with non-empty interior. Sometimes it is convenient to think of K lying in a finite dimensional real vector space E (indeed, the properties are just the same as that of $\mathbf{R}^n$; however, following the approach of [583], we do not impose the Euclidean structure with a specific basis unless we need to do so). Moreover, $\langle \cdot, \cdot \rangle$ sometimes can denote an inner-product on E and sometimes a scalar product on E, E^* whichever is convenient to us in a given context (in the former usage, we identify E^* with E). Here, E^* denotes the dual of the vector space E, which is the set of all linear functionals on E. We define the dual of (CP) as

$$\begin{array}{llcccc} (CD) & \sup & \langle b, y\rangle & & & \\ & & \mathcal{A}^*(y) & + & s & = & c, \\ & & & & s & \in & K^*, \end{array}$$

where K^* is the dual of cone K with respect to $\langle\cdot,\cdot\rangle$, i.e.

$$K^* = \{s : \langle s, x\rangle \geq 0, \ \forall x \in K\}.$$

Let Y denote the finite dimensional real vector space y lives in. Then b lives in Y^*, the dual space to Y. Therefore, $\mathcal{A} : E \to Y^*$. We will refer to this setting as the conic convex optimization setting.

SDP problem fits into this general setting by letting

- $E := \mathcal{S}^n$,
- $K := \mathcal{S}^n_+$,
- $\langle x, s\rangle := X \bullet S$.

Under these definitions we have $K^* = K$. i.e., the cone of symmetric positive semidefinite matrices is self-dual under the trace inner-product. In addition to being self-dual, the cone $\mathcal{S}^n_+$ enjoys another symmetry property, in that it is *homogeneous*. That is, the set of nonsingular linear transformations keeping $\mathcal{S}^n_+$ the same (the *automorphism group* of $\mathcal{S}^n_+$) is rich enough to contain linear transformations which map any fixed interior point to any other fixed interior point of $\mathcal{S}^n_+$. Convex cones with both properties, i.e. homogeneous self-dual cones, are also called *symmetric*. (See [216] for the classification and many analytic properties of such cones.)

Let us define the notation for the feasible regions and their "interiors" as referred to in interior-point methods. $\mathcal{F}(P)$, $\mathcal{F}(D)$ denote the feasible regions of the (P) and (D) respectively. Intersection of the interior of the underlying cone constraint and the affine subspace constraint identifies the "interior" solutions of the underlying problem.

$$\mathcal{F}_+(P) := \{X \in \mathcal{S}^n : \mathcal{A}X = b, \ X \succ 0\}$$

and

$$\mathcal{F}_+(D) := \{(y, S) \in \mathbf{R}^m \oplus \mathcal{S}^n : \mathcal{A}^*y + S = C, \ S \succ 0\}.$$

Sometimes we will abuse the notation and use $S \in \mathcal{F}_+(D)$ etc., since the corresponding y is uniquely identified (recall that $\mathcal{A}$ is surjective). Occasionally, we will need to refer to the direct sum of $\mathcal{F}_+(P)$ and $\mathcal{F}_+(D)$, denoted by $\mathcal{F}_+$. The following *weak duality relation* is very well known.

Proposition 9.1.1 *Let $\bar{X} \in \mathcal{F}(P)$ and $(\bar{y}, \bar{S}) \in \mathcal{F}(D)$. Then*

$$C \bullet \bar{X} - b^T \bar{y} \;=\; \bar{X} \bullet \bar{S} \;\geq\; 0.$$

Proof.
We have

$$\begin{aligned} C \bullet \bar{X} - b^T \bar{y} &= (\mathcal{A}^* \bar{y} + \bar{S}) \bullet \bar{X} - (\mathcal{A}\bar{X})^T \bar{y} \\ &= \mathcal{A}^* \bar{y} \bullet \bar{X} - (\mathcal{A}\bar{X})^T \bar{y} + \bar{X} \bullet \bar{S} \\ &= \bar{X} \bullet \bar{S} \\ &\geq 0. \end{aligned}$$

The first equation above uses the fact that $\bar{X}$ and $(\bar{y}, \bar{S})$ satisfy the corresponding equations of (P) and (D). The second equation is evident. The third equation follows from the definition of the adjoint and the final equation from the fact that $\bar{X} \succeq 0$, $\bar{S} \succeq 0$. ∎

Remark 9.1.1 *This proposition and its proof trivially extend to conic convex optimization setting.*

We gave a detailed account of the above simple fact, since we will need some of the elementary identities of the proof in establishing an equivalent, symmetric, conic formulation for the primal-dual pair. In certain circumstances, the alternative formulation is more suitable.

Since the linear operator $\mathcal{A}$ is surjective, we can always find $\bar{X} \in \mathcal{S}^n$ such that

$$\mathcal{A}\bar{X} = b.$$

For the dual, because of the form we chose, we can always find $\bar{y} \in \mathbf{R}^m$, $\bar{S} \in \mathcal{S}^n$ such that

$$\mathcal{A}^* \bar{y} + \bar{S} = C.$$

Denoting

$$\mathcal{L} \;:=\; \{d \in \mathcal{S}^n \,:\, \mathcal{A}d = 0\},$$

we claim that (P) and (D) are *equivalent* to the following pair, in that the set of feasible as well as the optimal solution pairs (X, S) of the corresponding problems coincide.

$$\begin{array}{llll} (\tilde{P}) & \min & \bar{S} \bullet X & \\ & & X & \in \; (\mathcal{L} + \bar{X}), \\ & & X & \succeq \; 0, \end{array}$$

$$\begin{array}{llll} (\tilde{D}) & \min & \bar{X} \bullet S & \\ & & S & \in \; (\mathcal{L}^{\perp} + \bar{S}), \\ & & S & \succeq \; 0, \end{array}$$

here, $\mathcal{L}^{\perp}$ denotes the orthogonal complement of $\mathcal{L}$. All of our arguments related to the equivalence of $(P), (D)$ and $(\widetilde{P}), (\widetilde{D})$ go through in the conic convex optimization setting. Therefore we present the rest of the derivation in the general setting. We have

$$(C\widetilde{P}) \qquad \inf \ \langle \bar{s}, x \rangle$$

$$x \quad \in \quad (\mathcal{L} + \bar{x}) \cap K,$$

and

$$(C\widetilde{D}) \qquad \inf \ \langle s, \bar{x} \rangle$$

$$s \quad \in \quad (\mathcal{L}^{\perp} + \bar{s}) \cap K^*.$$

To establish the equivalence, first note that the feasible regions are preserved (in $(C\widetilde{D})$ we only refer to s). Recall the proof of the weak duality relation (Proposition 9.1.1). For every $x \in E$ satisfying $\mathcal{A}(x) = b$, and for every (y, s) satisfying $\mathcal{A}^*(y) + s = c$ we have

$$\langle c, x \rangle - \langle b, y \rangle \;=\; \langle x, s \rangle.$$

If we fix $(\bar{y}, \bar{s})$ such that

$$\mathcal{A}^*(\bar{y}) + \bar{s} = c$$

then for all $x \in E$ satisfying $\mathcal{A}(x) = b$ we have

$$\langle c, x \rangle \;=\; \langle x, \bar{s} \rangle + constant,$$

where the constant is $\langle b, \bar{y} \rangle$. Therefore, minimizing $\langle c, x \rangle$ subject to any set of constraints, containing the constraint $\mathcal{A}(x) = b$ is equivalent to minimizing $\langle x, \bar{s} \rangle$ subject to the same set of constraints. Similarly, we can establish the equivalence of the dual problems. We fix $\bar{x}$ such that $\mathcal{A}(\bar{x}) = b$ then for all (y, s) satisfying $\mathcal{A}^*(y) + s = c$, we have

$$-\langle b, y \rangle \;=\; \langle \bar{x}, s \rangle + constant,$$

where the constant is $-\langle c, \bar{x} \rangle$. Therefore, maximizing $\langle b, y \rangle$ subject to any set of constraints containing the constraint $\mathcal{A}^*(y) + s = c$ is equivalent to minimizing $\langle \bar{x}, s \rangle$ subject to the same set of constraints. For the properties of the pair $(C\widetilde{P}), (C\widetilde{D})$ see [583].

9.2 FUNDAMENTAL INGREDIENTS

Our development of the algorithms, starts with identifying a suitable barrier for the feasible solution sets. Then we define the primal-dual central path based on the barriers. Finally, we describe a functional proximity measure to evaluate the distance from the central path and the proximity measure will lead us to a potential function. The rest of the development is almost exclusively based on the primal-dual potential function.

Perhaps, *central path* is the most important concept in understanding the development and complexity analyses of interior-point methods. In the general setting of convex programming problems, one uses a self-concordant barrier for the underlying convex cone (or the convex set) to derive interior-point algorithms with polynomial iteration complexity (see [583] and [794]). A parameter of this barrier is one of the factors that determine the worst case iteration bound (in regards to the currently best theoretical analysis). For the cone of symmetric semidefinite matrices, the best (with respect to the smallest parameter value) barrier is

$$F(X) \;:=\; -\ln\det(X).$$

Even though this is the optimal barrier (with respect to the iteration complexity parameter, which is n in this case) for the underlying cone, in theory, better barriers for the feasible solution set exist, e.g., the universal barrier, when the dimension of the feasible solution set is much less than n, see [583]. However, to date, we do not know how to evaluate the values, the first and the second derivatives of such barriers efficiently.

Nesterov and Nemirovskii proved that if F is a self-concordant barrier for K then the Legendre-Fenchel conjugate of F, denoted F_* (slightly modified, due to our choice of the definition of dual cone) is a self-concordant barrier for K^* with the same parameter value.

$$F_*(s) \;:=\; \sup_{x\in int(K)} \left\{-\langle s, x\rangle - F(x)\right\}.$$

In our case (of SDP) we have

$$F(S) \;:=\; -\ln\det(S).$$

Note that $F_*(S)$ differs from $F(S)$ by a constant, n, which is irrelevant in potential reduction considerations. (This will become clear with the definition of the potential function and the underlying algorithms.) Now that we have barriers for the primal and the dual cones, we can define a pair of families of convex optimization problems parameterized by $\mu \geq 0$.

$$\begin{array}{llccc} (CP_\mu) & \inf & \langle c, x\rangle & + & \mu F(x) \\ & & \mathcal{A}(x) & = & b, \\ & & (x & \in & K), \end{array}$$

and

$$\begin{array}{llccccc} (CD_\mu) & \inf & -\langle b, y\rangle & + & \mu F_*(s) & & \\ & & \mathcal{A}^*(y) & + & s & = & c, \\ & & & & (s & \in & K^*). \end{array}$$

We wrote $x \in K$ and $s \in K^*$ in parentheses as the domains of F and F_* are $int(K)$ and $int(K^*)$ respectively.

Theorem 9.2.1 *Suppose $\mathcal{F}_+(CP) \neq \emptyset$, $\mathcal{F}_+(CD) \neq \emptyset$. Then (CP_μ) and (CD_μ) have unique optimal solutions $x(\mu)$, $(y(\mu), s(\mu))$, for each $\mu > 0$.*

Definition 9.2.1 $\{(x(\mu), y(\mu), s(\mu)) : \mu > 0\}$ *is called the* primal-dual central path *for the pair* $(CP), (CD)$.

Sometimes we refer only to $(x(\mu), s(\mu))$.

Theorem 9.2.2 *Suppose* $\mathcal{F}_+(CP) \neq \emptyset$, $\mathcal{F}_+(CD) \neq \emptyset$. *Then for every* $\mu > 0$, *the unique solution* $x(\mu)$ *of* (CP_μ) *and the unique solution* $s(\mu)$ *of* (CD_μ) *make up the unique solution of the following system*

$$\mathcal{A}(x) = b, \quad x \in int(K) \tag{9.2.1}$$
$$\mathcal{A}^*(y) + s = c, \quad s \in int(K^*) \tag{9.2.2}$$
$$s = -\mu F'(x). \tag{9.2.3}$$

We have established the central path. However, it is very difficult to ensure, in a practical, efficient algorithm, that all iterates lie on the central path. Therefore, we need a measure of proximity. We define

$$\psi(x, s) := \vartheta \ln\left(\frac{\langle s, x\rangle}{\vartheta}\right) + F(x) + F_*(s) + \vartheta.$$

The next proposition shows that for every pair of interior solutions (x, s), the proximity measure is non-negative and it is equal to zero if and only if the point (x, s) lies on the central path.

Proposition 9.2.1 *[583] Let* F *be a* $\vartheta-$*normal barrier for* K. *Then*

$$\vartheta \ln\left(\frac{\langle s, x\rangle}{\vartheta}\right) + F(x) + F_*(s) \geq -\vartheta,$$

for all $x \in int(K)$, $s \in int(K^*)$. *Moreover, the inequality above is held as equality iff*

$$s = -F'(tx),$$

for some $t > 0$.

Proof.
Note that by the definition of Legendre-Fenchel conjugate we have

$$F_*(s) = \sup_{x \in int(K)} \{-\langle s, x\rangle - F(x)\}.$$

This implies

$$F_*(s) + F(x) + \langle s, x\rangle \geq 0 \ \ \forall\, x \in int(K), \ \forall\, s \in int(K^*).$$

If we plug in $s := -F'(x)$ in the left hand side, we obtain

$$F_*(-F'(x)) + F(x) + \langle -F'(x), x\rangle = 0.$$

Since the function $\langle s, x\rangle + F(x)$ is a strictly convex function of x, such x (satisfying $s := -F'(x)$ for a fixed s) is the unique x achieving the supremum. Thus, we have equality above if and only if $s := -F'(x)$. Since K and K^* are cones, we also have for every $\lambda_1, \lambda_2 > 0$,

$$F_*(\lambda_1 s) + F(\lambda_2 x) + \lambda_1\lambda_2\langle s, x\rangle \ \geq\ 0 \ \ \forall\, x \in int(K),\ \forall\, s \in int(K^*).$$

Defining $t := \lambda_1\lambda_2$, and using the ϑ−logarithmic homogeneity of F and F^*, we arrive at for all $t > 0$, $x \in int(K), s \in int(K^*)$,

$$F(x) + F_*(s) + t\langle x, s\rangle - \vartheta \ln t \ \geq\ 0.$$

Moreover, we have equality above iff $s = -F'(tx)$. For a fixed pair x, s, the LHS is a convex function of t, the value of t minimizing the LHS is

$$t^* \ =\ \frac{\vartheta}{\langle s, x\rangle}.$$

Therefore,

$$\vartheta \ln\left(\frac{\langle s, x\rangle}{\vartheta}\right) + F(x) + F_*(s) \ \geq\ -\vartheta,$$

for all $x \in int(K)$, $s \in int(K^*)$. Moreover, the inequality above is held as equality iff

$$s = -F'(tx),$$

for some $t > 0$. ■

Now, let us consider SDP as a special case of the above general conic setting. $K := \mathcal{S}^n_+$, $F(x) := -\ln\det X$, $\vartheta = n$.

$$-F'(x) \ =\ -X^{-1},$$

$$\langle x, s\rangle := X \bullet S,$$

$(X(\mu), y(\mu), S(\mu))$ is on the central path iff

$$\begin{aligned} \mathcal{A}X &= b, \ \ X \succeq 0 \\ \mathcal{A}^* y + S &= C, \ \ S \succeq 0 \\ S &= \mu X^{-1}. \end{aligned}$$

The dual barrier is

$$\begin{aligned} F_*(S) &= \sup\{-X \bullet S + \ln\det(X)\} \\ &= -\ln\det(S) - n. \end{aligned}$$

We know that the supremum above is attained by X satisfying

$$-F'(X) \ =\ S.$$

Therefore we have, $X = S^{-1}$ and

$$F_*(S) \;=\; -n - \ln\det(S).$$

Now, let us compute the measure of centrality. First, note that $\mu = X \bullet S/n$.

$$\psi(x,s) \;=\; \vartheta \ln\left(\frac{\langle s, x\rangle}{\vartheta}\right) + F(x) + F_*(s) + \vartheta$$

is equal to

$$n\ln(\mu) - \sum_{j=1}^{n} \ln(\lambda_j) \;=\; \ln\left(\frac{\mu^n}{\prod_{j=1}^{n} \lambda_j}\right),$$

where we denoted by λ_j the eigenvalues of $S^{1/2}XS^{1/2}$. So, in this case, Proposition 9.2.1 reduces to the arithmetic-geometric mean inequality applied to the eigenvalues of $S^{1/2}XS^{1/2}$.

9.3 WHAT ARE THE USES OF A POTENTIAL FUNCTION ?

- Potential functions measure how "good" a given point is. A potential function designed for feasible start algorithms measures the distance to the central path and the objective function value and returns a single value for the feasible point based on these two attributes.

- Potential functions can provide information about "good" search directions. e.g., if the potential function is continuously differentiable, we can consider the steepest descent direction (suitably projected to the null space of the equality constraints to maintain feasibility) or an approximation of it as the search direction.

- Once the search direction is computed, the potential function value along the given search direction becomes a function of only the step size (or the step sizes, if different step sizes in primal and dual are to be used). So, we can perform a line search (or a plane search) to determine the best step size(s) to reduce the potential function value as much as possible.

- Potential functions can provide ease and new insights in complexity analyses. A general approach first establishes a large first order decrease in the potential function. Such decrease is usually ensured by the choice of the steepest-descent direction for the potential function as the search direction. Next, the second and higher order effects must be estimated. As we illustrate shortly, this is one of the places where the self-concordance of the underlying barrier function plays a key role.

In this chapter, we adapt the potential functions that are the most studied ones. However, there are many other possibilities e.g., generalizations of those in [793]. We do not consider here the infeasible-start algorithms which require the usage of an additional condition (which ensures that infeasibilities do not decrease much faster than the complementarity gap) or a third part in the potential function so that the potential function accounts not only for complementarity and proximity to the central path but the also for the decrease in infeasibilities relative to the decrease in complementarity. For SDP extensions of related potential reduction algorithms of Mizuno, Kojima and Todd [533] see Kojima, Shindoh and Hara [439]. It is also possible to extend the algorithms of Seifi and Tuncel [702].

Let us define a primal-dual potential function

$$\phi_q(x,s) = q\ln(\langle x,s\rangle) + \psi(x,s),$$

for $q \geq 1$. This potential function is a generalization of Tanabe-Todd-Ye potential function (see [770], [783]) to the general conic setting, as used in [583].

Next we present the main steps of a primal-dual potential reduction algorithm for SDP.

Input-Instance($C, A_1, \ldots, A_m, b, X^0, y^0, S^0, \epsilon$);
iteration counter $k := 0$;
WHILE $X^k \bullet S^k > \epsilon$ DO
  find a search direction (d_x, d_y, d_s);
  find a step size $\alpha > 0$ such that
    $\phi_q(X^k + \alpha d_x, S^k + \alpha d_s) \leq \phi_q(X^k, S^k) - \delta$;
  set $(X^{k+1}, y^{k+1}, S^{k+1}) := (X^k, y^k, S^k) + \alpha(d_x, d_y, d_s)$;
  set $k := k + 1$;
 END{WHILE};
OUTPUT(X^k, y^k, S^k)

Main steps of a primal-dual potential reduction algorithm

In addition to the problem data $(\mathcal{A}, b, C)$, we also provide the algorithm with an initial point (X^0, y^0, S^0) and a (small) positive constant ϵ. X^0 and S^0 must lie in the interior of the respective cones $(X^0 \succ 0,\ S^0 \succ 0)$. In this chapter, we only discuss feasible-start algorithms. Therefore, we must have

$$\mathcal{A}X^0 = b, \quad \text{and} \quad \mathcal{A}^* y^0 + S^0 = C.$$

$\epsilon > 0$ represents the desired quality of the final solution in terms of its duality gap.

In each iteration, the algorithm computes a search direction. In many variants of the potential reduction method, one chooses a steepest descent direction for the potential function, subject to satisfying all the equality constraints (that is, we must have $\mathcal{A}d_x = 0$ and $\mathcal{A}^* d_y + d_s = 0$). In some other variants of the potential reduction method, one chooses the search direction from a description of the central path in terms of a system of linear and nonlinear equations and inequalities. We present examples of these two variants on SDP problems, in the next two sections.

After the search direction is determined, the algorithm chooses a step size $\alpha > 0$. Here, $\delta > 0$ is an absolute constant, independent of the input (e.g., 1 or (1/4)). We will prove in the next theorem that decreasing the potential function value by at least δ in every iteration ensures the convergence of the algorithm. Note that since the potential function includes barrier terms for both X and S, the facts that $X^k \succ 0$, $S^k \succ 0$ and the potential function value decreases, imply $X^{k+1} \succ 0$, $S^{k+1} \succ 0$. (Unless, $\langle X + \alpha d_x, S + \alpha d_s \rangle$ is tending to zero faster than $(X^k, S^k) \to \partial(\mathcal{S}_+^n \oplus \mathcal{S}_+^n)$; but, in this case we get optimality of the next iterate!) In this chapter, we are applying the same step size α to both the primal and the dual iterates; however, in implementation it might be beneficial to allow different step sizes for the primal and the dual iterates.

As we mentioned, while designing and/or implementing interior-point methods, potential functions can be utilized in various ways:

(i) in choosing the search direction

(ii) in choosing the step size

(iii) in the complexity analysis.

There is some value in having all of the three uses in the same algorithm. It is possible to adapt other strategies when suitable. E.g., see [793], [571], [798].

The following result is very well-known and standard in the analyses of interior-point algorithms.

Theorem 9.3.1 *Let $\epsilon \in (0,1)$ and $(x^0, y^0, s^0) \in \mathcal{F}_+$ such that*

$$\psi(x^0, s^0) \;\le\; q \ln\left(\frac{1}{\epsilon}\right)$$

be given. If we reduce the value of the potential function ϕ_q by at least an absolute constant in every iteration, then in $O(q\ln(1/\epsilon))$ iterations we have $(x^k, y^k, s^k) \in \mathcal{F}_+$ such that

$$\langle x^k, s^k \rangle \;\le\; \epsilon \langle x^0, s^0 \rangle.$$

Proof.
As above, let $\delta > 0$ denote the absolute constant by which the potential function is decreased in every iteration. First, note that since the potential value does not increase, all iterates lie in $\mathcal{F}_+$. After k iterations we have

$$\phi_q(x^k, s^k) - \phi_q(x^0, s^0) \;=\; q \ln\left(\frac{\langle x^k, s^k \rangle}{\langle x^0, s^0 \rangle}\right) + \psi(x^k, s^k) - \psi(x^0, s^0) \;\le\; -k\delta.$$

Next, noting that $\psi(x^k, s^k) \ge 0$ by Proposition 9.2.1, we obtain

$$q \ln\left(\frac{\langle x^k, s^k \rangle}{\langle x^0, s^0 \rangle}\right) \;\le\; -k\delta + \psi(x^0, s^0).$$

Since $\psi(x^0, s^0) \leq q \ln(1/\epsilon)$, we conclude

$$\ln\left(\frac{\langle x^k, s^k\rangle}{\langle x^0, s^0\rangle}\right) \leq \frac{-k\delta + q\ln(1/\epsilon)}{q}.$$

If

$$k \geq \frac{2q\ln(1/\epsilon)}{\delta}$$

then we have

$$\langle x^k, s^k\rangle \leq \epsilon\langle x^0, s^0\rangle,$$

as desired. ■

The above theorem shows that our problem is reduced to finding a search direction and a step size such that ϕ_q is decreased by a constant. First we mention an important property of self-concordant barriers. For $x \in \text{int}(K)$, we define

$$||v||_x := \langle v, F''(x)v\rangle^{1/2},$$

where $F''(x)$ denotes the Hessian of F at x. All the other norms represent the norm induced by the trace inner-product, unless specified otherwise.

Theorem 9.3.2 *([583]) Let F be a ϑ-normal barrier for K, also let $x \in$ int(K). Then for every $z \in$ int(K) such that $\alpha := ||x - z||_x < 1$, we have*

$$(1-\alpha)^2||h||_x^2 \leq ||h||_z^2 \leq \frac{1}{(1-\alpha)^2}||h||_x^2,$$

for all $h \in E$.

Proof.
See [583], pp. 13-15. ■

The following basic result is a generalization of Karmarkar's related lemma on $-\sum_{j=1}^n \ln(x_j)$. The generalization of Karmarkar's result to semidefinite programming (using the self-concordant barrier $-\ln\det(X)$) was noticed independently by Alizadeh [21]. Below, we cite the general result due to Nesterov and Nemirovskii [583].

Proposition 9.3.1 *Let F be a $\vartheta-$normal barrier for K and also let $d \in E$ such that $||d||_x \leq 1$. Then*

$$F(x) + \langle F'(x), d\rangle \leq F(x+d) \leq F(x) + \langle F'(x), d\rangle + \frac{||d||_x^2}{2(1-||d||_x)^2}.$$

Proof.
The LHS inequality follows by the convexity of F. For the RHS, we use Taylor's Theorem: "Let $f : \Xi \subseteq E \to \mathbf{R}$ be a C^r (r times continuously differentiable)

function. Suppose that x, $(x+d)$, and the line segment joining x and $(x+d)$ lie in Ξ. Then there exists $\xi \in [x,(x+d)]$ such that

$$f(x+d) = f(x) + \sum_{k=1}^{r-1} \frac{1}{k!} D^k f(x)[d,d,\ldots,d] + \frac{1}{r!} D^r f(\xi)[d,d,\ldots,d]."$$

We use Taylor's Theorem for $r := 2$ to obtain, "there exists $z \in int(K)$ such that

$$F(x+d) = F(x) + \langle F^{'}(x), d\rangle + \frac{1}{2}\langle F^{''}(z)d, d\rangle."$$

By Theorem 9.3.2, using $||d||_x \leq 1$, we have

$$(1-||d||_x)^2 \langle F''(x)d,d\rangle \leq \langle F^{''}(z)d,d\rangle \leq \frac{1}{(1-||d||_x)^2}\langle F''(x)d,d\rangle.$$

■

Using such results, Nesterov and Nemirovskii [583], Alizadeh [21] (via a generalization of Ye's algorithm [853]), Vandenberghe and Boyd [809] (via a generalization of the work of Gonzaga and Todd [293]) established polynomial iteration bounds for extensions of various interior-point algorithms.

The main focus of this paper is on the symmetric primal-dual algorithms. These algorithms are completely symmetric in the way they treat the primal and the dual problems and they are expected to take long steps in both the primal and the dual spaces. Practical versions of such algorithms have been very successful in computation. Such algorithms possess very nice theoretical properties as well. Because of these two reasons and the increasing popularity and demand for solving semidefinite programming problems, many researchers started studying primal-dual algorithms for SDP as extensions of the corresponding algorithms for LP and LCP. Some of the initial claims that "everything essential in the LP case goes through for SDP" were not well-founded. Nevertheless, it is true that many of the essential ingredients for primal-dual algorithms for LP do go through for SDP, some do not. During the last five years many primal-dual algorithms for SDP have been proposed. In this chapter, we focus on primal-dual potential reduction algorithms only. We must note however, that there are very close connections between the path-following and potential reduction algorithms.

About the same time, in 1994, two research reports were released, generalizing symmetric, polynomial-time, primal-dual potential reduction algorithms and the underlying iteration-complexity proof of Kojima, Mizuno and Yoshise [433]. One by Kojima, Shindoh and Hara [439], the other by Nesterov Todd [587]. The first paper above considers the setting of $K = \mathcal{S}_+^n$ however, it also addresses the monotone Linear Complementarity Problems (LCP). The first paper yields a continuum of search directions (later called KSH family) to be used in the potential reduction (or in a path-following) scheme. The results of the second paper yield a unique direction (later called Nesterov-Todd direction) and the algorithms there, apply to conic convex optimization problems when K is a symmetric cone. When $K = \mathcal{S}^n_+$, Nesterov-Todd direction is a member of the KSH family (Kojima, Shida and Shindoh [436]).

Now we motivate and very briefly describe the above approaches.

9.4 KOJIMA-SHINDOH-HARA APPROACH

Recall the general definition of the central path. In case of SDP, the nonlinear equations (9.2.3) become

$$S = \mu X^{-1}.$$

Utilizing the properties of the algebra of $n \times n$ matrices over real numbers, there are many equivalent ways of expressing this system of equations. One that has been considered by many is

$$XS = \mu I.$$

With a superficial understanding of the LP case, one may be tempted to use a *linearization* following the substitutions $X \leftarrow (X + d_x)$ and $S \leftarrow (S + d_s)$. However, the linear system of equations obtained from this procedure may not always have solutions (d_x, d_s) in $\mathcal{S}^n \oplus \mathcal{S}^n$, see [439]. Kojima-Shindoh-Hara proposed the following modification of the nonlinear equations. Let $\widetilde{\mathcal{S}}^n$ denote the set of $n \times n$ skew-symmetric matrices. Let $X, S \in \mathcal{S}^n_{++}$, $\gamma \in \mathbf{R}_+$ be given. Recall that $\mu := (X \bullet S)/n$.

Definition 9.4.1 *The following system is called the* KSH equation

$$\begin{aligned} X(d_s + \widetilde{d}_s) + (d_x + \widetilde{d}_x)S &= \gamma\mu I - XS \\ d_x, d_s \in \mathcal{S}^n, &\qquad \widetilde{d}_x, \widetilde{d}_s \in \widetilde{\mathcal{S}}^n. \end{aligned}$$

A very important condition is enforced on the auxiliary variables $\widetilde{d}_x, \widetilde{d}_s$ in that they must lie in an $n(n-1)/2$ dimensional monotone linear subspace of $\widetilde{\mathcal{S}}^n \oplus \widetilde{\mathcal{S}}^n$. Kojima, Shindoh and Hara show that the search direction defined above is unique (for every choice of the $n(n-1)/2$–dimensional monotone subspace of $\widetilde{\mathcal{S}}^n \oplus \widetilde{\mathcal{S}}^n$). They develop primal-dual potential reduction and path following algorithms based on such search directions and prove that such algorithms possess polynomial iteration bounds. Next, we give different characterizations of the monotonicity condition.

Theorem 9.4.1 *Let $d_x, d_s, \widetilde{d}_x, \widetilde{d}_s$ satisfy the KSH equation. Then*

$$\begin{aligned} \langle \widetilde{d}_x, \widetilde{d}_s \rangle \geq 0 \quad &\textit{if and only if} \quad \langle \hat{d}_x, \hat{d}_x \rangle \geq \langle \bar{d}_x, \bar{d}_x \rangle \\ &\textit{if and only if} \quad \langle \hat{d}_s, \hat{d}_s \rangle \geq \langle \hat{d}_s, \bar{d}_s \rangle, \end{aligned}$$

where

$$\begin{aligned} \bar{d}_x &:= X^{-1/2} d_x S^{1/2}, \qquad & \bar{d}_s &:= S^{-1/2} d_s X^{1/2} \\ \hat{d}_x &:= X^{-1/2} \widetilde{d}_x S^{1/2}, \qquad & \hat{d}_s &:= S^{-1/2} \widetilde{d}_s X^{1/2}. \end{aligned}$$

Proof.
Every solution of the KSH equation satisfies

$$X(d_s + \widetilde{d}_s) + (d_x + \widetilde{d}_x)S = \gamma\mu I - XS. \tag{9.4.4}$$

Using the facts that $\widetilde{d}_x, \widetilde{d}_s \in \widetilde{\mathcal{S}}^n$ and $X, S, S^{-1}, d_x, d_s \in \mathcal{S}^n$, we conclude that Equation (9.4.4) is equivalent to

$$X(d_s + \widetilde{d}_s)S^{-1} + (d_x + \widetilde{d}_x) = \gamma\mu S^{-1} - X. \tag{9.4.5}$$

Using the symmetry or skew-symmetry of the underlying matrices, we see that (9.4.5) is equivalent to the following

$$S^{-1}(d_s - \widetilde{d}_s)X + (d_x - \widetilde{d}_x) = \gamma\mu S^{-1} - X. \tag{9.4.6}$$

Subtracting (9.4.6) from (9.4.5) we obtain

$$\widetilde{d}_x = \frac{1}{2}\left[S^{-1}(d_s - \widetilde{d}_s)X - X(d_s + \widetilde{d}_s)S^{-1}\right]. \tag{9.4.7}$$

Similarly, multiplying Equation (9.4.4) from left by X^{-1}, transposing both sides and subtracting one equation from the other, we also obtain

$$\widetilde{d}_s = \frac{1}{2}\left[S(d_x - \widetilde{d}_x)X^{-1} - X^{-1}(d_x + \widetilde{d}_x)S\right]. \tag{9.4.8}$$

(This also follows from Equation (9.4.7) and the fact that Definition 9.4.1 is completely symmetric in the way it treats X-space and S-space.) Using (9.4.7) and (9.4.8) we have

$$\begin{aligned}
\langle \widetilde{d}_x, \widetilde{d}_s\rangle &= \frac{1}{2}\Big[\text{Trace}\,(\widetilde{d}_x S\widetilde{d}_x X^{-1}) + \text{Trace}\,(\widetilde{d}_x X^{-1} d_x S) + \text{Trace}\,(\widetilde{d}_x X^{-1}\widetilde{d}_x S) \\
&\qquad -\text{Trace}\,(\widetilde{d}_x S d_x X^{-1})\Big] \\
&= \frac{1}{2}\Big[\text{Trace}\,(\widetilde{d}_x X^{-1}\widetilde{d}_x S) + \text{Trace}\,(X^{-1} d_x S\widetilde{d}_x) + \text{Trace}\,(\widetilde{d}_x X^{-1}\widetilde{d}_x S) \\
&\qquad -\text{Trace}\,(\widetilde{d}_x S d_x X^{-1})\Big] \\
&= \frac{1}{2}\Big[2\text{Trace}\,(\widetilde{d}_x X^{-1}\widetilde{d}_x S) + \text{Trace}\,(\widetilde{d}_x^T S d_x X^{-1}) - \text{Trace}\,(\widetilde{d}_x S d_x X^{-1})\Big] \\
&= \frac{1}{2}\Big[2\text{Trace}\,(\widetilde{d}_x X^{-1}\widetilde{d}_x S) - 2\text{Trace}\,(\widetilde{d}_x S d_x X^{-1})\Big] \\
&= \langle \hat{d}_x, \hat{d}_x\rangle - \langle \hat{d}_x, \bar{d}_x\rangle.
\end{aligned}$$

Therefore,

$$\langle \widetilde{d}_x, \widetilde{d}_s\rangle \geq 0 \quad \text{if and only if} \quad \langle \hat{d}_x, \hat{d}_x\rangle \geq \langle \hat{d}_x, \bar{d}_x\rangle.$$

Proof of the other characterization can be obtained through a similar computation. It also follows from the underlying symmetry of Definition 9.4.1. ■

One of the most popular search directions from KSH family coincides with the search direction proposed (independent of the work of KSH) by Helmberg, Rendl, Vanderbei and Wolkowicz [339]. This search direction was later rediscovered by Monteiro [536] through yet another approach (which is useful for some other purposes—see the chapter on Path Following Methods) and is known as the H../K../M direction.

9.5 NESTEROV-TODD APPROACH

In this section, we follow the motivation in [780] to introduce the Nesterov-Todd direction. As in Karmarkar's algorithm [404], we can try the steepest descent direction for ϕ_q. So, we consider for some $u \in int(K)$,

$$\begin{array}{lrcl} \min & \langle \nabla_x \phi, d_x \rangle & + & \frac{1}{2}\langle F''(u) d_x, d_x \rangle \\ & \mathcal{A}(d_x) & = & 0. \end{array}$$

Note that here the usage of $F''(u)$ is in place of using the constraint

$$||d_x||_u^2 \leq 1.$$

Similarly, we can write the steepest descent direction for the dual. For some $v \in int(K^*)$, consider

$$\begin{array}{lrcl} \min & \langle \nabla_s \phi, d_s \rangle & + & \frac{1}{2}\langle F_*''(v) d_s, d_s \rangle \\ & \mathcal{A}^*(d_y) + (d_s) & = & 0. \end{array}$$

Both problems have strictly convex (quadratic) objective functions. The unique solution of the primal steepest descent direction problem is given by the unique solution of the following system of equations:

$$\begin{array}{rcl} \mathcal{A}(d_x) & = & 0 \\ \mathcal{A}^*(d_y) + d_s & = & 0 \\ F''(u) d_x + d_s & = & -\nabla_x \phi. \end{array}$$

The unique solution of the following system gives the steepest descent direction for the dual variable.

$$\begin{array}{rcl} \mathcal{A}(d_x) & = & 0 \\ \mathcal{A}^*(d_y) + d_s & = & 0 \\ F_*''(v) d_s + d_x & = & -\nabla_s \phi. \end{array}$$

Choosing $u = x$ or $v = s$ leads to generalizations of Ye's algorithm [853]. In the worst case, the dual updates (that is, if we choose the first system above with $u = x$, and base our decision of which iterate to update on the size of d_x

with respect to the local norm defined by the barrier function and point x, then the update of s^k to s^{k+1}) can be *short-step*, almost as bad as the theoretical guarantees. As a result, this algorithm has not been very successful in practice. But, its simplicity and elegance improved our understanding of interior-point methods and lead to generalizations. This remark brings us into the main topic of this chapter: Symmetric primal-dual algorithms. As we mentioned before, these algorithms are completely symmetric in the way they treat the primal and the dual problems and they are expected to take long steps in both the primal and the dual spaces. Polynomial time symmetric primal-dual algorithms for LP have been proposed by Kojima, Mizuno and Yoshise [431] and by Monteiro and Adler [538]. Currently, the most successful implementations of interior-point algorithms for LP are based on symmetric primal-dual algorithms. A very simple and elegant analysis of a symmetric primal-dual potential reduction algorithm for Linear Complementarity Problems (with positive semidefinite coefficient matrix) was given by Kojima, Mizuno and Yoshise [433].

Recall the linear systems of equations yielding the search direction. Primal steepest-descent direction is given as the unique solution of

$$\begin{aligned} \mathcal{A}(d_x) &= 0 \\ \mathcal{A}^*(d_y) + d_s &= 0 \\ F''(u)d_x + d_s &= -\nabla_x \phi. \end{aligned}$$

On the other hand, the dual steepest-descent direction was given as the unique solution of the same first two equations and the third one being different:

$$F_*''(v)d_s + d_x = -\nabla_s \phi.$$

We consider the question, "Given x, s interior points in the corresponding cones, K, K^*, is it possible to choose $u \in int(K)$ and $v \in int(K^*)$ such that the unique solutions of the above two systems coincide ?"

Such choice for u, v would make the underlying algorithm symmetric with respect to switching the roles of the primal and the dual problems. Let us consider a sufficient condition for a "Yes" answer to the above question. One sufficient condition is obtained by making the third equations equivalent (i.e. choose u, v such that the third equations have the same solution set). Let us look at the LP case first. (Only for this discussion, we follow the standard notation of LP, in that, X denotes diag(x) etc.) We have as the third equations

$$U^{-2}d_x + d_s = X^{-1}e - \frac{n+q}{x^T s}s$$

and

$$d_x + V^{-2}d_s = S^{-1}e - \frac{n+q}{x^T s}x.$$

These equations are equivalent to (multiply both sides by $X^{1/2}S^{-1/2}$ and $X^{-1/2}S^{1/2}$) respectively

$$X^{1/2}S^{-1/2}U^{-2}d_x + X^{1/2}S^{-1/2}d_s = X^{-1/2}S^{-1/2}e - \frac{n+q}{x^T s}X^{1/2}S^{1/2}e$$

and

$$X^{-1/2}S^{1/2}d_x + X^{-1/2}S^{1/2}V^{-2}d_s = X^{-1/2}S^{-1/2}e - \frac{n+q}{x^T s}X^{1/2}S^{1/2}e.$$

Now that the RHS of both are the same, we see that it is sufficient to have,

$$U = X^{1/2}S^{-1/2},$$

and

$$V = X^{-1/2}S^{1/2}.$$

Note that it turns out $U = V^{-1}$. This leads to the primal-dual search directions used in LP. The usual (and original) way of obtaining these search directions is by considering a Newton type direction to follow the central path (as expressed by Megiddo [528]) see, for instance, [430].

It is easy to see that (by a similar computation, left for the reader), in general, it suffices to have

- $F_*''(v) = \left[F''(u)\right]^{-1}$,
- $F''(u)x = s$,
- $F''(u)F_*'(s) = F'(x)$.

These conditions motivate the following definition of a special barrier function.

9.5.1 Self-scaled Barriers and Long Steps

Definition 9.5.1 *([587]) Let K be a pointed convex cone with non-empty interior and F be a $\vartheta-$normal barrier for K. Then F is a $\vartheta-$*self-scaled barrier *for K if $\forall\ x, w \in int(K)$*
(a) $F''(w)x \in int(K^*)$
and
(b) $F_*(F''(w)x) = F(x) - 2F(w) - \vartheta$.

F_* is a $\vartheta-$self-scaled barrier for K^* if and only if F is a $\vartheta-$self-scaled barrier for K. We mention here an equivalent definition of self-scaled barriers.

Theorem 9.5.1 *[795] Let K be a pointed convex cone with non-empty interior and F be a $\vartheta-$normal barrier for K. Then F is a $\vartheta-$*self-scaled barrier *for K if and only if*
$\forall\ x, w \in int(K)$ there exists unique $z \in int(K)$ such that
(a) $-F'(z) = F''(w)x$

and
(b) $F(w) = \frac{1}{2}[F(x) + F(z)]$.

As Nesterov-Todd prove, one very distinguishing property of self-scaled barriers (versus the more general self-concordant barriers) is that the Hessian at any interior point x of K, can be used to estimate the Hessian at other interior points by the same functions as in Theorem 9.3.2 even if those interior points are **outside** the unit ellipsoid centered at x, and defined with respect to the local norm given by $F''(x)$. Given $x \in int(K)$ and $d \in E$, they define

$$\sigma_x(d) := \frac{1}{\sup\{\alpha : (x - \alpha d) \in K\}}.$$

Then they prove the following important properties

- The function $g(z) := \langle -F'(x), z\rangle$ is convex for every $z \in K$.
- For all $x \in int(K)$, $d \in E$ and $\alpha \in \left[0, \frac{1}{\sigma_x(d)}\right)$,

$$\frac{1}{(1+\alpha\sigma_x(d))^2}F''(x) \preceq F''(x-\alpha d) \preceq \frac{1}{(1-\alpha\sigma_x(d))^2}F''(x).$$

The second property above ensures the possibility of long steps. Some similar approaches are possible for cones that are not necessarily symmetric. (See [311], [566].)

Nesterov and Todd [587], [588] generalize various interior-point algorithms and their (polynomial iteration) complexity analyses, for LP and LCP to conic convex optimization problems, where the difficult constraints are described by symmetric cones. In the next section, we discuss some issues related to one of their algorithms: The joint scaling primal-dual interior-point method.

9.6 SCALING, NOTIONS OF PRIMAL-DUAL SYMMETRY AND SCALE INVARIANCE

Since Karmarkar's polynomial-time interior-point algorithm, the set of linear transformations keeping the *difficult* constraints invariant, loosely referred to as *scalings*, played a key role in polynomial time interior-point algorithms and their extensions. In the case of LP, it was well understood that by *scalings* one meant the set of $n \times n$, positive definite, diagonal matrices. For SDP problems, Todd, Toh and Tütüncü [782], and others, considered linear transformations T of the form $T(X) := PXP^T$ for some nonsingular $P \in \mathcal{M}^n$. Clearly, any such transformation is an automorphism of $\mathcal{S}^n_+$. Using results of Lim [485] and Waterhouse [823], Güler [312] pointed out that any automorphism of $\mathcal{S}^n_+$ can be written as $P \cdot P^T$ for some nonsingular $P \in \mathcal{M}^n$. Therefore, we have

Theorem 9.6.1 *The automorphism group of the cone $\mathcal{S}^n_+$ over the real or complex numbers consists of linear transformations $T : \mathcal{S}^n \to \mathcal{S}^n$ of the form*

$$T(X) = PXP^T,$$

where P is a non-singular $n \times n$ matrix.

There are many closely related results in the algebra literature. We pick the following result (see Theorem 1 in Marcus and Moyls [517]).

Theorem 9.6.2 *A nonsingular linear transformation T on $\mathbf{R}^{n\times n}$ preserves rank iff T is either*

$$T(X) = UXV$$

or

$$T(X) = UX^TV$$

for some non-singular matrices U and V.

One of the key properties established by the proof of Theorem 9.6.1 is the fact that the action of the automorphism T on rank-1 matrices, can be identified by the action of a nonsingular linear transformation L on $\mathbf{R}^n$. That is, if T takes $e_ie_i^T$ (e_i denotes the ith unit vector in $\mathbf{R}^n$), to $f_if_i^T$, then defining $L : \mathbf{R}^n \to \mathbf{R}^n$

$$L(e_i) := f_i,$$

for each i, we can identify the action of T on $\mathcal{S}^n$ as follows:

$$T(e_ie_j^T + e_je_i^T) \;=\; L(e_i)[L(e_j)]^T + L(e_j)[L(e_i)]^T,$$

for all pairs i, j. Using these ideas we have another corollary of Theorem 9.6.2.

Corollary 9.6.1 *Every automorphism $\bar{T}$ of the cone $\mathcal{S}^n_+$ can be extended to a nonsingular linear transformation T on n by n matrices, preserving rank.*

Let's explain why the above Corollary holds. Recall, $\tilde{\mathcal{S}}^n$ denotes the set of $n \times n$ skew-symmetric matrices. Then

$$\mathbf{R}^{n\times n} \;=\; \mathcal{S}^n \oplus \tilde{\mathcal{S}}^n.$$

Let $\bar{T}$ in the automorphism group of $\mathcal{S}^n_+$ be given. Then clearly, $\bar{T}$ is a non-singular linear transformation on $\mathcal{S}^n$ preserving rank. To extend $\bar{T}$ to a linear transformation T on the whole space we will define the action of the extension on the orthogonal complement, $\tilde{\mathcal{S}}^n$ of $\mathcal{S}^n$. For a given $M \in \mathbf{R}^{n\times n}$, we write

$$M = \bar{M} + \widetilde{M}$$

(the orthogonal decomposition of M into symmetric and skew-symmetric components) and define

$$T(M) \;:=\; \bar{T}(\bar{M}) + \tilde{T}(\widetilde{M}).$$

Given $\bar{T}$ we describe the corresponding $\tilde{T}$ as follows: For each i,

$$f_if_i^T \;:=\; \bar{T}(e_ie_i^T),$$

so that

$$\tilde{T}(e_i e_j^T - e_j e_i^T) := f_i f_j^T - f_j f_i^T.$$

We can obtain from the proof of Theorem 9.6.1 that

$$\bar{T}(e_i e_j^T + e_j e_i^T) = f_i f_j^T + f_j f_i^T.$$

This gives

$$T(e_i e_j^T) = f_i f_j^T.$$

Now, it is clear that $\tilde{T}$ preserves rank two matrices (note that the rank of a skew symmetric matrix is always even) and that the extension T is nonsingular. It is easy to check that T preserves rank.

Actually, the rank preserving extension of $\bar{T}$ is always unique. Note that any automorphism of $\mathcal{S}^n_{++}$ must preserve the rank, since the Carathéodory number of a point in the cone is invariant under the automorphism group (see Güler and Tuncel [314]). Also note that as we showed in Corollary 9.6.1, any automorphism of $\mathcal{S}^n_{++}$ can be extended to a nonsingular linear transformation on the space $\mathbf{R}^{n\times n}$ preserving rank. Now, since we also want T to preserve symmetric positive definite matrices, we must have

$$UV^{-T}Y \in \mathcal{S}^n_{++}, \quad \text{for all } Y \in \mathcal{S}^n_{++},$$

we used the automorphism $V^{-T} \cdot V^{-1}$ of $\mathcal{S}^n_{++}$. First, plugging in $Y := I$, we see that $UV^{-T} \in \mathcal{S}^n_{++}$. Secondly, we have

$$(UV^{-T})Y = YV^{-1}U^T = Y(UV^{-T}), \quad \text{for all } Y \in \mathcal{S}^n_{++}.$$

That is, (UV^{-T}) commutes with every symmetric positive definite matrix. Therefore, $UV^{-T} = \alpha I$ for some scalar $\alpha > 0$, and without loss of generality, $T(X) = UXU^T$. This shows that the scale-invariance notion used in the SDP literature coincides with the invariance under the automorphism group for the cone $\mathcal{S}^n_+$. If we apply this scaling to the primal iterate X, then the corresponding dual scaling is the inverse of the adjoint of $U \cdot U^T$ which is $U^{-T} \cdot U^{-1}$.

Results of the remainder of this section are based on [795]. Recall the symmetric conic formulations

$$(P) \qquad \inf \ \langle c, x\rangle \qquad x \in (\mathcal{L} + b) \cap K,$$

$$(D) \qquad \inf \ \langle s, b\rangle \qquad s \in (\mathcal{L}^\perp + c) \cap K^*,$$

where we abused the notation and wrote b for $\bar{x}$ and c for $\bar{s}$. We start with a set of desired properties: *symmetry under duality* and *scale-invariance*. These desired properties are often discussed in articles studying primal-dual interior-point methods, however, not at the level of formality or generality that we present. Our definition of scale-invariance is also slightly different than the usual understanding of invariance under any non-singular linear transformation of the underlying space. We are rather interested in the linear transformations which preserve the feasible region defined by the *difficult* constraints (linear or non-linear **inequalities**) including nontrivial set inclusion constraints. We describe a way of achieving both properties via an abstraction of the so-called *v−space* construction. This construction was presented by Kojima et al. [430] for LCP. A generalization to semidefinite programming was proposed by Sturm and Zhang [760].

A *primal-dual algorithm (PDA)* takes as an input $(\mathcal{L}, b, c, x^0, s^0; K, K^*)$. Here, we do not describe in detail how each element should be formatted (indeed, it is easy to imagine examples: e.g. $\mathcal{L}$ can be described by a basis, K can be described by a barrier function oracle, given a description of K and the inner-product, we may not require any explicit description of K^*). Only assumption in our general approach is that the interiors of K and K^* are non-empty and $x^0 \in int(K)$ and $s^0 \in int(K^*)$ are given. (Note that x^0 and s^0 are not necessarily feasible.)

When constructing an algorithm we have many criteria. First, the algorithm must be well-defined, and it must generate approximately optimal solutions if they exist. Secondly, in interior-point methods, we also expect that the algorithm require only polynomially many iterations in n and the desired accuracy, in the worst case. Here, each iteration may involve approximately solving, $O(1)$ linear systems of equations of dimension at most $n \times n$. All other work per iteration must be of lower order of complexity. In addition to the above, we may have some other criteria that we would like the algorithm to satisfy. For instance, in case that (P) is infeasible, we may require the algorithm to converge to a point in K that is the closest to the affine space $(\mathcal{L} + b)$ or generate a certificate of *approximate infeasibility* of (P). Here, we will discuss two other criteria: primal-dual symmetry and scale-invariance. These two criteria have been sought in interior-point algorithms in many different forms. Now, let us consider equivalent representations of (P) and (D). Let $T \in Aut(K)$ (automorphism group of K). Then, we can *scale* (P) by T and have an equivalent scaled problem:

$$(\bar{P}) \qquad \inf \ \langle T^{-*}(c), T(x) \rangle$$

$$T(x) \quad \in \quad (T(\mathcal{L}) + T(b)) \cap K,$$

we used the fact that $T(K) = K$. Indeed, the given inner-product forces the scaling of c to be T^{-*} (this is the inverse of the adjoint of T). The dual of $(\bar{P})$ is given by:

$$(\bar{D}) \qquad \inf \ \langle T^{-*}(s), T(b)\rangle$$

$$T^{-*}(s) \quad \in \quad (T^{-*}(\mathcal{L}^{\perp}) + T^{-*}(c)) \cap K^{*}.$$

We used the well-known fact that $T \in Aut(K)$ if and only if $T^{-*} \in Aut(K^*)$. (Note that

$$\begin{aligned} T \in Aut(K) \quad & \text{if and only if} \quad K^* = \{s : \langle s, T(x)\rangle \geq 0, \forall x \in K\} \\ & \text{if and only if} \quad K^* = \{s : \langle T^*(s), x\rangle \geq 0, \forall x \in K\} \\ & \text{if and only if} \quad T^{-*}(K^*) = K^*.) \end{aligned}$$

We consider the pair of problems $(\bar{P}), (\bar{D})$ equivalent to the pair $(P), (D)$. This is our notion of *scale invariance* which we make more precise in the next definition.

Definition 9.6.1 *([795]) A primal dual algorithm is called* symmetric *if for all inputs*
$(\mathcal{L}, b, c, x^0, s^0; K, K^*)$ it satisfies the following properties:
(I) Symmetry under duality:

$$\{x^k, s^k\} = PDA(\mathcal{L}, b, c, x^0, s^0; K, K^*)$$

if and only if

$$\{s^k, x^k\} = PDA(\mathcal{L}^{\perp}, c, b, s^0, x^0; K^*, K).$$

That is, the algorithm is symmetric with respect to switching the roles of (P) and (D).
(II) Scale invariance:
For every $T \in Aut(K)$,

$$\{x^k, s^k\} = PDA(\mathcal{L}, b, c, x^0, s^0; K, K^*)$$

if and only if

$$\{T(x^k), T^{-*}(s^k)\} = PDA(T(\mathcal{L}), T(b), T^{-*}(c), T(x^0), T^{-*}(s^0); K, K^*).$$

Since (I) and (II) can be applied in any combination, the corresponding dual statement to (II) must hold as well.

Condition (I) (symmetry under duality) actually enforces consistency of the algorithm under the switching of the roles of the primal and the dual problem. Combined with some other criteria, it can be more effective e.g. among all symmetric primal-dual algorithms we would prefer the one with the smallest iteration bound in the worst case. Note that a symmetric primal-dual algorithm does not have to work equally hard at both problems (P) and (D). For instance, given a method and a fixed pair of problems (P) and (D), one of these problems might be *easier* for the method at hand. In this case, the definition can be used to test if the method *knows* duality. If it does, and if it can recognize the easier problem quickly, then it would solve the easier problem regardless of which one of (P) or (D) is identified as the primal problem in the input.

9.6.1 An Abstraction of the $v-$space Approach

One way of achieving this symmetry (especially in primal-dual interior point algorithms) is to describe a subset G of $Aut(K)$ such that for each $(x, s) \in int(K) \oplus int(K^*)$, the equation

$$T(x) \quad = \quad T^{-*}(s), \quad T \in G \tag{9.6.9}$$

has a unique solution. Then, in each iteration, a primal-dual algorithm can compute the unique solution above, apply the transformation T to the primal problem and T^{-*} to the dual problem and after the application of a Newton type method, the algorithm can translate the new iterate back to the *original scaling* and continue.

Let us formalize this in the following definition.

Definition 9.6.2 *([795]) A conic convex optimization problem of the form* (P) *is said to* admit a symmetric $v-$space construction *if there exists* $G \subseteq Aut(K)$ *such that for each* $(x, s) \in int(K) \oplus int(K^*)$*, the equation*

$$T(x) \quad = \quad T^{-*}(s), \quad T \in G$$

has a unique, self-adjoint solution T.

We now mention that for each pair $(x, s) \in int(K) \oplus int(K^*)$ the existence of a self-adjoint linear transformation T solving (9.6.9), implies that K is a homogeneous self-dual cone. Combining this with the powerful results of Nesterov and Todd [587], we arrive at the following theorem.

Theorem 9.6.3 *([795]) A conic convex optimization problem of the form* (P) *admits a symmetric* $v-$*space construction if and only if* K *is a homogeneous self-dual cone.*

Let F be a $\vartheta-$self-scaled barrier for K. The algorithm of [587] computes at iteration k (the current iterate is (x^k, s^k)), the unique scaling point $w \in int(K)$ such that

$$F''(w)x^k \quad = \quad s^k. \tag{9.6.10}$$

Since $F''(w)$ is positive definite, it has a unique symmetric positive definite square-root. We define

$$T := \left(F''(w)\right)^{\frac{1}{2}}.$$

Thus, defining

$$G \quad := \quad \bigcup_{w \in int(K)} F''(w),$$

suffices to show that (9.6.9) is solvable. (It is known that $\forall w \in int(K)$, $\exists \bar{w} \in int(K)$ such that $F''(\bar{w}) = \left[F''(w)\right]^{1/2}$; see, e.g., Rothaus [687].) It is easy to see that the rest of the computations of the Nesterov-Todd algorithm can be done in the *scaled space* with the current iterate

$$(v^k, v^k) := \left(T(x^k), T^{-1}(s^k)\right).$$

One consequence of the above theorem is the following. If we would like to design a symmetric primal-dual interior-point algorithm for convex programming problems in the conic form, one way of achieving this is via a symmetric $v-$space construction. Our result shows that the class of problems for which such a joint scaling exists is exactly the class of problems which can be solved by the primal-dual algorithm of Nesterov and Todd.

Our view of the $v-$space approach leads us to the next section. In the next section, we do not insist that T lie in the automorphism group of $\mathcal{S}_+^n$. Instead, we consider a set of (minimal) axioms that allows extensions of primal-dual interior-point algorithms for LP and LCP to SDP and more general convex optimization problems.

9.7 A POTENTIAL REDUCTION FRAMEWORK

In this section, we study the primal-dual potential-reduction algorithm of Kojima, Mizuno and Yoshise [433]. We illustrate a set of minimal assumptions guaranteeing an extension of the results to SDP and some related problems. As one of the consequences of our results, we obtain a short complexity analysis for the primal-dual joint-scaling algorithm of Nesterov-Todd. Our proof relies on an elementary extension of a central lemma of [433] which also had a very short proof.

Given $X, S \in \mathcal{S}_{++}^n$, we define a set of linear transformations $\mathcal{T}(X,S)$.

Definition 9.7.1 *([313]) A linear transformation T on the space of $n \times n$ symmetric matrices is in $\mathcal{T}(X,S)$ if*

(a) T is symmetric ($T^* = T$) and nonsingular,

(b) $T(S) = T^{-1}(X) =: V$,

(c) $T(X^{-1}) = T^{-1}(S^{-1}) = V^{-1}$,

(d) $\lambda := \lambda_{\min}(V) = \left(\lambda_{\min}(X^{1/2}SX^{1/2})\right)^{1/2}$.

(e) Any one of the following is sufficient:

(e.1)

$$\langle \bar{d}_x, TF''(X)T(\bar{d}_x)\rangle \le \frac{\eta}{\lambda^2} \text{ and } \langle \bar{d}_s, T^{-1}F''(S)T^{-1}(\bar{d}_s)\rangle \le \frac{\eta}{\lambda^2},$$

where $\eta > 0$ is an absolute constant.

(e.2)

$$\langle \bar{d}_x, TF''(X)T(\bar{d}_s)\rangle \ge 0, \text{ and } \langle \bar{d}_x, T^{-1}F''(S)T^{-1}(\bar{d}_s)\rangle \ge 0.$$

(e.3) *Operator equation:* $T^{-1}(F''(S)(T^{-1}(\cdot))) = T(F''(X)(T(\cdot)))$ *and* T *is positive definite.*

Remark 9.7.1 *It will become clear in the analysis that condition (d) can be relaxed to the requirement that*

$$\lambda_{\min}(V) = \xi \left(\lambda_{\min}(X^{1/2}SX^{1/2})\right)^{1/2},$$

where $\xi > 0$ *is an absolute constant. We will prove that (e.2) and (e.3) are special cases of (e.1). As we prove in Proposition 9.7.2, condition (e.3) makes the underlying transformation* T *unique and the search direction determined by such* T *is the Nesterov-Todd direction. We note that all these conditions can be generalized to symmetric cones. Instead of* $\lambda_{\min}$ *use the minimum generalized eigenvalue function and for* X^{-1} *use the duality mapping* $-F'(X)$. *It is also worth noticing that the condition (e.1) is similar to the conditions of Theorem 9.4.1 describing the KSH family of directions.*

Remark 9.7.2 *Note that the set of linear transformations* $\mathcal{T}(X,S)$ *can vary significantly from one iteration to the next. For instance, let us suppose* $X = S = I$. *Then choosing* $V = I$, *we see that for a linear transformation* T *to satisfy conditions (b), (c) and (d), it suffices to have*

$$T(I) = I.$$

This is a single linear equation on T.

Remark 9.7.3 *Using the conditions (a)-(d), Kojima and Tuncel [441] modified slightly the Nesterov-Todd search directions and obtained complexity analyses for path following algorithms based on these search directions. These algorithms achieve the current best iteration bounds and, they improve both the primal and the dual objective values strictly in every iteration. Moreover the analyses of [441] is not much more complicated than the analyses of related algorithms from the LP case.*

For any $T \in \mathcal{T}(X,S)$, we can define a search direction as follows. First we define $\bar{\mathcal{A}} := \mathcal{A}(T(\cdot))$ $W := V^{-1} - \frac{n+\sqrt{n}}{\langle x,s\rangle}V$ and $U := W/||W||$. Since

$$\langle V, V\rangle = \langle T(V), T^{-1}(V)\rangle = \langle X, S\rangle$$

and $n+\sqrt{n} > n$, we have $||V^{-1} - \frac{n+\sqrt{n}}{\langle x,s\rangle}V|| > 0$, thus U is well-defined. Let $(\bar{d}_x, \bar{d}_y, \bar{d}_s)$ denote the unique solution of the following system of linear equations:

$$\bar{\mathcal{A}}\bar{d}_x = 0 \quad (9.7.11)$$

$$\bar{\mathcal{A}}^*\bar{d}_y + \bar{d}_s = 0. \quad (9.7.12)$$

$$\bar{d}_x + \bar{d}_s = U. \quad (9.7.13)$$

Finally our search directions are defined as

$$d_x := T(\bar{d}_x), \;\; d_s := T^{-1}(\bar{d}_s).$$

We described how to find the search direction. Once we have the search direction (d_x, d_s), we find a value of the step size $\alpha > 0$ which minimizes the potential function ϕ_q along the given search direction. Clearly, the above algorithm is well-defined. Now, we analyze its computational complexity in terms of the number iterations required to attain a desired (small, but positive) duality gap.

Proposition 9.7.1 *Let positive definite matrices X and S be given. Then for any $T \in \mathcal{T}(X, S)$, we have*

$$\langle U, T^{-1}F''(S)T^{-1}(U)\rangle \;=\; \langle U, TF''(X)T(U)\rangle \;\leq\; \frac{1}{\lambda^2}.$$

Proof.
For the first equation, we simply compute

$$\langle U, T^{-1}F''(S)T^{-1}(U)\rangle \;=\; \frac{\left\|X^{-1/2}S^{-1}X^{-1/2} - \frac{n+\sqrt{n}}{n\mu}I\right\|^2}{||W||^2},$$

and

$$\langle U, TF''(X)T(U)\rangle \;=\; \frac{\left\|X^{-1/2}S^{-1}X^{-1/2} - \frac{n+\sqrt{n}}{n\mu}I\right\|^2}{||W||^2}.$$

To prove the inequality, we note that

$$\begin{aligned}
&\left\|X^{-1/2}S^{-1}X^{-1/2} - \frac{n+\sqrt{n}}{n\mu}I\right\|^2 \\
=\;& \text{Trace}\left[X^{-1}S^{-1}X^{-1}S^{-1} - 2\frac{n+\sqrt{n}}{n\mu}X^{-1}S^{-1} + \left(\frac{n+\sqrt{n}}{n\mu}\right)^2 I\right] \\
=\;& \text{Trace}\left[(X^{-1}S^{-1})\left(X^{-1}S^{-1} - 2\frac{n+\sqrt{n}}{n\mu}I + \left(\frac{n+\sqrt{n}}{n\mu}\right)^2 SX\right)\right].
\end{aligned}$$

Now, recall that $T^{-1}(V^{-1}) = X^{-1}$ and $T(V^{-1}) = S^{-1}$ and by condition (e), $\lambda = \left(\lambda_{\min}(X^{1/2}SX^{1/2})\right)^{1/2}$. Therefore, we have

$$\begin{aligned}
&\left\|X^{-1/2}S^{-1}X^{-1/2} - \frac{n+\sqrt{n}}{n\mu}I\right\|^2 \\
\leq\;& \frac{1}{\lambda^2}\text{Trace}\left[V^{-2} - 2\frac{n+\sqrt{n}}{n\mu}I + \left(\frac{n+\sqrt{n}}{n\mu}\right)^2 V^2\right] \\
=\;& \frac{1}{\lambda^2}||W||^2.
\end{aligned}$$

Now, we conclude

$$\langle U, TF''(X)T(U)\rangle \leq \frac{1}{\lambda^2}.$$

■

Proposition 9.7.2 *Each of the conditions (e.2) and (e.3) implies condition (e.1) with $\eta = 1$. Moreover, the set of all $T \in \mathcal{T}(X,S)$ satisfying condition (e.3) is a singleton for every pair of positive definite matrices X, S and the underlying algorithm is the primal-dual potential reduction algorithm of Nesterov and Todd.*

Proof.
Consider Equation (9.7.13):

$$\bar{d}_x + \bar{d}_s = U$$

we have

$$\left[F''(X)\right]^{1/2} T(\bar{d}_x) + \left[F''(X)\right]^{1/2} T(\bar{d}_s) = \left[F''(X)\right]^{1/2} T(U).$$

Now, taking the norm of both sides, we obtain

$$\begin{aligned}\langle \bar{d}_x, TF''(X)T(\bar{d}_x)\rangle &= \langle U, TF''(X)T(U)\rangle - \langle \bar{d}_s, TF''(X)T(\bar{d}_s)\rangle \\ &\quad -2\langle \bar{d}_x, TF''(X)T(\bar{d}_s)\rangle.\end{aligned}$$

Similarly we obtain

$$\begin{aligned}\langle \bar{d}_s, T^{-1}F''(S)T^{-1}(\bar{d}_s)\rangle &= \langle U, T^{-1}F''(S)T^{-1}(U)\rangle - \langle \bar{d}_x, T^{-1}F''(S)T^{-1}(\bar{d}_x)\rangle \\ &\quad -2\langle \bar{d}_s, T^{-1}F''(S)T^{-1}(\bar{d}_x)\rangle.\end{aligned}$$

It is clear from these two identities that if (f.2) holds, then we have

$$\langle \bar{d}_x, TF''(X)T(\bar{d}_x)\rangle \leq \langle U, TF''(X)T(U)\rangle$$

and

$$\langle \bar{d}_s, T^{-1}F''(S)T^{-1}(\bar{d}_s)\rangle \leq \langle U, T^{-1}F''(S)T^{-1}(U)\rangle.$$

Using Proposition 9.7.1, we see that condition (e.2) implies condition (e.1) with $\eta = 1$. For the second part of the proposition, we use Theorem 3.2 of [587] and conclude that for each positive definite pair (X, S) there exists unique positive definite R such that

$$T^2 = \left[F''(R)\right]^{-1}.$$

This implies, since we insist that T be positive definite, T is unique. The rest follows by employing Corollary 4.1 of [587]. ■

The following lemma is a generalization of Lemma 2.5 of Kojima, Mizuno and Yoshise [433]. The main steps of the proof below are almost identical to theirs.

Lemma 9.7.1 *([313]) Given $X, S \in \mathcal{S}^n_{++}$, and $T \in \mathcal{T}(X,S)$, let $V := T(S)$ and W as above. Then*

$$\|W\| \geq \frac{\sqrt{3}}{2\lambda}.$$

Proof.
Let $\mu := \frac{\langle x,s \rangle}{n}$. Recall that $\langle V, V\rangle = \langle X, S\rangle = n\mu$. Since $\langle V, V^{-1} - \frac{1}{\mu}V\rangle = 0$, we have

$$\begin{aligned} \|W\|^2 &= \|V^{-1} - \frac{1}{\mu}V\|^2 + \frac{1}{n\mu^2}\|V\|^2 \\ &\geq \left(\frac{1}{\lambda} - \frac{\lambda}{\mu}\right)^2 + \frac{1}{\mu} \\ &= \frac{1}{\lambda^2}\left[\left(\frac{\lambda^2}{\mu} - \frac{1}{2}\right)^2 + \frac{3}{4}\right], \end{aligned}$$

where the inequality above follows by utilizing the eigenvalue decompositions of V and V^{-1} since, V is symmetric and positive definite. Thus we conclude that

$$\|W\| \geq \frac{\sqrt{3}}{2\lambda}.$$

∎

Remark 9.7.4 *The inequality in the proof of the above lemma extends to all symmetric cones, via the usage of the generalized eigenvalues of V. Therefore, the lemma extends Theorem 5.2 of [587] in the way that it applies to all $T \in \mathcal{T}(X,S)$.*

Theorem 9.7.1 *([313]) Let feasible interior-points X^0, S^0 and a positive constant ϵ such that*

$$\psi(X^0, S^0) \leq \sqrt{n}\ln(1/\epsilon)$$

be given. Then the interior-point algorithm described above using at each iteration k, any $T \in \mathcal{T}(X^k, S^k)$, will generate in $O(\sqrt{n}\ln(1/\epsilon))$ iterations feasible interior points X, S such that

$$\langle X, S\rangle \leq \epsilon\langle X^0, S^0\rangle.$$

Proof.
We prove that in each iteration the value of the potential function is decreased by at least an absolute constant. Then the conclusion of the theorem follows from the standard arguments. Let us denote, as in the path-following section,

$$X(\alpha) := X + \alpha d_x,$$

$$S(\alpha) := S + \alpha d_s.$$

Then

$$\begin{aligned}\phi(X(\alpha), S(\alpha)) - \phi(X, S) &= (n+\sqrt{n})\ln\left(\frac{\langle X, S\rangle + \alpha(\langle X, d_s\rangle + \langle S, d_x\rangle)}{\langle X, S\rangle}\right) \\ &\quad + F(X+\alpha d_x) - F(X) + F(S+\alpha d_s) - F(S),\end{aligned}$$

we used the fact that $\langle d_x, d_s\rangle = \langle T^{-1}(d_x), T(d_s)\rangle = \langle \bar{d}_x, \bar{d}_s\rangle = 0$. Using Proposition 9.3.1 we arrive at the following relation.

$$\phi(X(\alpha), S(\alpha)) - \phi(X, S) \le \alpha(\langle d_x, p_x\rangle + \langle d_s, p_s\rangle) + \alpha^2\beta,$$

where

$$\begin{aligned}p_x &:= F'(X) + \frac{n+\sqrt{n}}{\langle X, S\rangle}S = -X^{-1} + \frac{n+\sqrt{n}}{\langle X, S\rangle}S, \\ p_s &:= F'(S) + \frac{n+\sqrt{n}}{\langle X, S\rangle}X = -S^{-1} + \frac{n+\sqrt{n}}{\langle X, S\rangle}X, \\ \beta &:= \frac{\|d_x\|_x^2}{2(1-\alpha\|d_x\|_x)^2} + \frac{\|d_s\|_s^2}{2(1-\alpha\|d_s\|_s)^2}.\end{aligned}$$

Note that

$$T(p_x) = T^{-1}(p_s) = -V^{-1} + \frac{n+\sqrt{n}}{\langle X, S\rangle}V = -W.$$

Thus,

$$\begin{aligned}\langle d_x, p_x\rangle + \langle d_s, p_s\rangle &= \langle T^{-1}(d_x), T(p_x)\rangle + \langle T(d_s), T^{-1}(p_s)\rangle \\ &= \langle \bar{d}_x + \bar{d}_s, -W\rangle \\ &= -\|W\|.\end{aligned}$$

The last equation follows from the definition of U and that $\bar{d}_x + \bar{d}_s = U$. From property (e) of the linear transformations in $\mathcal{T}(X, S)$, it follows that

$$\langle \bar{d}_x, TF''(X)T(\bar{d}_x)\rangle \le \frac{\eta}{\lambda^2} \text{ and } \langle \bar{d}_s, T^{-1}F''(S)T^{-1}(\bar{d}_s)\rangle \le \frac{\eta}{\lambda^2}.$$

Hence we arrive at

$$\phi(X(\alpha), S(\alpha)) - \phi(X, S) \le -\alpha\|W\| + \alpha^2\frac{\eta/\lambda^2}{(1-\alpha\sqrt{\eta}/\lambda)^2}.$$

Using Lemma 9.7.1 and choosing a suitable value for α, e.g., $\alpha = \frac{1}{3\lambda}$ if $\eta \le 1$ and $\alpha = \frac{1}{4\eta\lambda}$ if $\eta > 1$, we see that the value of ϕ is decreased by at least an absolute constant. Using Theorem 9.3.1, with $q := \sqrt{n}$, we conclude that in $O(\sqrt{n}\ln(1/\epsilon))$ iterations, we have

$$\langle X^k, S^k\rangle \le \epsilon\langle X^0, S^0\rangle,$$

as desired. This completes the proof. ■

Throughout most of the chapter, we paid special attention to SDP. However, most of the results, with a careful choice of substitutions, carry over to symmetric cones. Some of our results and formulations can be considered even more generally (e.g., when K is only homogeneous).

Acknowledgement: This paper was written during my visit to Tokyo Institute of Technology on a sabbatical leave from the University of Waterloo. I thank Masakazu Kojima and T.I.T. for their support and hospitality.

10 PATH-FOLLOWING METHODS

Renato Monteiro, Michael Todd

Follow the yellow brick road!

L. Frank Baum, The Wizard of Oz

10.1 INTRODUCTION

In this chapter we study interior-point primal-dual path-following algorithms for solving the semidefinite programming (SDP) problem. In contrast to linear programming, there are several ways one can define the Newton-type search directions used by these algorithms. We discuss several ways in which this can done by motivating and introducing several search directions and families of directions. Polynomial convergence results for short- and long-step path-following algorithms using the Monteiro and Zhang family of directions are derived in detail; similar results are only surveyed, without proofs, for the Kojima, Shindoh and Hara family and the Monteiro and Tsuchiya family.

Semidefinite programming (SDP) has a wide range of applications in several areas of engineering and science including structural optimization, statistics, probability, matrix and eigenvalue optimization problems, control theory, pattern recognition etc. (See Part III, Chapters 12 through 20, of this handbook for a description of some of these applications.) SDP has also been useful in combinatorial and discrete optimization, where several discrete formulations of NP-hard problems have provably tight SDP relaxations (see Chapter 12 of this handbook). Interior-point methods for the solution of SDP problems were first proposed by Alizadeh [21] and Nesterov and Nemirovskii [577, 579, 578]. These first methods were primal-only (or dual-only) type algorithms. Alizadeh [21] directly consider algorithms for the SDP problem and shows that most, if not all, primal-only (or dual-only) algorithms for linear programming can be extended to SDP in a mechanical way. On the other hand, Nesterov and Nemirovskii [577, 579, 578] presented a deep and unified theory of interior-point methods for solving the more general cone programming problem using the notion of a self-concordant barrier; see their book [583] for a comprehensive treatment of this subject. In particular, they show that geometric programming, convex quadratic programs with convex quadratic constraints, approximation in L_p-norm, and semidefinite programs can be "solved in polynomial time."

The first algorithms for SDP which are extensions of the well-known primal-dual LP algorithms, such as the long-step path-following algorithm of Kojima, Mizuno and Yoshise [432] and Tanabe [769, 770], the short-step path-following algorithm of Kojima, Mizuno and Yoshise [431] and Monteiro and Adler [538, 539], and the predictor-corrector algorithm of Mizuno, Todd and Ye [534], were introduced approximately five years after the advent of the primal-only methods. Part of the reason for this delay was that the natural SDP analogue of the Newton direction used in the primal-dual LP algorithms was not evident. In fact, several Newton-type SDP directions have been proposed which generalize the one for LP. The first three and most popular of them are: i) the Alizadeh, Haeberly and Overton (AHO) direction proposed in [23], ii) a direction independently proposed by Helmberg, Rendl, Vanderbei and Wolkowicz [339] and Kojima, Shindoh and Hara [439], and later rediscovered by Monteiro [536], which we refer to as the HRVW/KSH/M direction and, iii) the Nesterov and Todd (NT) direction introduced in [587, 588].

Application of the Newton method to the central path equation $XS = \sigma\mu I$ results in an equation of the form

$$X\Delta S + \Delta X S = \sigma\mu I - XS, \tag{10.1.1}$$

which in general yields non-symmetric directions. The AHO direction corresponds to the symmetric equation obtained by directly symmetrizing both sides of (10.1.1). Another way of symmetrizing (10.1.1) is to first apply a similarity transformation $P(\cdot)P^{-1}$ to both sides of (10.1.1) and then symmetrize it. Such an approach was first introduced by Monteiro [536] for the cases $P = X^{-1/2}$ and $P = S^{1/2}$. The resulting directions were found to be equivalent to two special directions of the KSH family of directions introduced earlier by Kojima, Shindoh and Hara [439] using a different approach. The second direction (with $P = S^{1/2}$), which is the HRVW/KSH/M direction, was also proposed by Helmberg, Rendl, Vanderbei and Wolkowicz [339] independently of [439]. (For simplicity, we refer to the first direction with $P = X^{-1/2}$ as the dual HRVW/KSH/M direction. We use the term HRVW/KSH/M directions to refer to both of them.) To unify the NT direction and the HRVW/KSH/M directions, Zhang [867] formally introduced the above scaling and symmetrization scheme for a general nonsingular scaling matrix P, which leads to a class of search directions parametrized by P, usually referred to as the Monteiro and Zhang (MZ) family. Subsequently, Todd, Toh and Tütüncü [782] and Kojima, Shida and Shindoh [436] showed that the NT direction is a member of the MZ family and the KSH family, respectively. In contrast, it is known that the AHO direction does not belong to the KSH family.

Unified convergence analyses for the MZ family have been given by Monteiro and Zhang [547] and Monteiro [537]. In the paper [547], iteration-complexity bounds are derived for the long-step primal-dual path-following method based on a subclass of the MZ family of search directions, which contains the NT and HRVW/KSH/M directions but not the AHO direction. In particular, it is shown that the corresponding algorithms based on the NT and the HRVW/KSH/M directions perform $\mathcal{O}(nL)$ and $\mathcal{O}(n^{3/2}L)$ iterations, respectively, to reduce the duality gap by a factor of at least $2^{-\mathcal{O}(L)}$. (The $\mathcal{O}(n^{3/2}L)$ iteration-complexity bound for the HRVW/KSH/M directions was in fact obtained earlier by Monteiro [536].) More recently, Monteiro [537] proves the polynomiality of the short-step primal-dual path-following algorithm and the Mizuno-Todd-Ye predictor-corrector-type algorithm based on any member of the MZ family, thus obtaining as a by-product the important result that Frobenius-norm-type algorithms based on the AHO direction are polynomially convergent.

Unified analysis for the KSH family of directions are provided in Kojima, Shindoh and Hara [439] and Monteiro and Tsuchiya [544]. The paper [439] introduces the KSH family and establishes: 1) the polynomiality of the short-step path-following (feasible) method based on the two KSH/HRVW/M directions (both members of the KSH family); and 2) the polynomiality of a potential reduction (feasible and infeasible) algorithm based on *any* direction of the KSH family. Using techniques developed in Monteiro [537], the paper [544] extends the result 1) above to any direction of the KSH family. It also proves polynomial convergence of a Mizuno-Todd-Ye predictor-corrector-type algorithm based on the whole KSH family.

Primal-dual algorithms based on the Newton direction for the central path equation $X^{1/2}SX^{1/2} - \mu I = 0$, and more generally for the scaled central path equation

$$(PXP^T)^{1/2}(P^{-T}SP^{-1})(PXP^T)^{1/2} - \mu I = 0, \qquad (10.1.2)$$

where P is a nonsingular matrix, were introduced and studied in Monteiro and Tsuchiya [543]. The resulting family of Newton directions associated with (10.1.2) as P changes is usually referred to as the MT family. It is shown in [543] that the short-step and the semilong-step path-following algorithms based on the whole MT family have iteration-complexity bounds $\mathcal{O}(\sqrt{n}L)$ and $\mathcal{O}(nL)$, respectively, and that the long-step path-following algorithm based on a subclass of the MT family has iteration-complexity bound $\mathcal{O}(n^{3/2}L)$ to reduce the duality by a factor of $2^{-\mathcal{O}(L)}$.

Tseng [790] also introduced another family of search directions and analyzed the polynomial convergence of path-following primal-dual algorithms based on it. As for the KSH family, both the MT and the Tseng families contain the HRVW/KSH/M directions and the NT direction but not the AHO direction. Kruk et al. [450] introduced a Gauss-Newton direction based on the least-squares solution to an over-determined linear (Newton) system rather than the exact solution to a square system. Todd [780] investigated the properties of twenty primal-dual directions in terms of scale invariance, primal-dual symmetry and whether they are always well-defined. Monteiro and Zanjácomo [545] discussed the computational work involved in computing six primal-dual directions, including the AHO, the NT and the HRVW/KSH/M directions which had their computational complexities substantially improved. A general and unified convergence analysis of short-step methods was presented in Kojima, Shida and Shindoh [438] by considering a family of inexact Newton search directions, which includes all the families of directions mentioned above. Burer and Monteiro [146] on the other hand presented a unified convergence treatment of long-step primal-dual path-following methods by considering scaled Newton directions with respect to central path equations of the form $\Phi(X,S) - \mu I = 0$, where Φ is a map from $\mathcal{S}^n_+ \times \mathcal{S}^n_+$ into $\mathcal{S}^n$ satisfying some suitable properties.

The organization of this chapter is as follows. In Section 10.2, we define and show the existence and uniqueness of the primal-dual central path for a standard form SDP problem and its dual. We also define the main neighborhoods of this path which are used to describe primal-dual path-following algorithms in Section 10.4. Section 10.3 motivates and introduces the main Newton-type directions used in primal-dual interior-point algorithms for SDP. Section 10.4 studies primal-dual path-following (feasible only) algorithms for solving the SDP problem. It consists of four subsections. The first three subsections are dedicated to studying algorithms based on the MZ family. The first subsection describes a general primal-dual (PD) framework based on the MZ family and a scaling scheme that is used in later subsections on the analysis of some specific methods in the PD framework. The second subsection analyzes the short-step path-following method and a Mizuno-Todd-Ye-type predictor-corrector method. The long-step path-following method based on a commutative class of the MZ family is studied in the third subsection. The last subsection introduces the MT and KSH families and surveys the main convergence results for them.

10.2 THE CENTRAL PATH

We consider the SDP given in the following standard form:

$$
\begin{array}{llrcll}
(P) & \min_X & C \bullet X & & & \\
 & & A_i \bullet X & = & b_i, & i = 1, \ldots, m, \\
 & & X & \succeq & 0, &
\end{array}
\qquad (10.2.3)
$$

where each $A_i \in \mathcal{S}^n$, $b \in I\!R^m$, and $C \in \mathcal{S}^n$ are given, and $X \in \mathcal{S}^n$. Here $\mathcal{S}^n$ denotes the space of $n \times n$ symmetric matrices, and $X \succeq 0$ indicates that X is positive semidefinite. We also write $\mathcal{P}$ for $\{X \in \mathcal{S}^n : X \succeq 0\}$, and $\mathcal{P}_+$ for its interior, the set of positive definite matrices in $\mathcal{S}^n$. For $X \in \mathcal{P}$, $X^{1/2}$ denotes its positive semidefinite square root. If $X \in \mathcal{P}_+$, we write $X^{-1/2}$ for the inverse of $X^{1/2}$, or equivalently the positive semidefinite square root of X^{-1}. We use $A \bullet B$ to denote the inner product $\mathrm{Trace}\, A^T B$ of two $m \times n$ matrices; if U and V are in $\mathcal{S}^n$, $U \bullet V = \mathrm{Trace}\, U^T V = \mathrm{Trace}\, UV = \sum_{i,j} u_{ij} v_{ij} = \mathrm{Trace}\, VU = V \bullet U$. We write $\mathcal{S}^n_\perp$ for the space of $n \times n$ skew-symmetric matrices; this is the orthogonal complement, with respect to the inner product above, of $\mathcal{S}^n$ in $\Re^n$. We assume that the set $\{A_i\}$ is linearly independent. The dual problem associated with

(P) is:

$$(D) \qquad \max_{y,S} \quad b^T y \qquad \begin{array}{rcl} \sum_{i=1}^m y_i A_i + S & = & C, \\ S & \succeq & 0, \end{array} \qquad (10.2.4)$$

where $y \in \mathbb{R}^m$ and $S \in \mathcal{S}^n$. We write $F(P)$ and $F(D)$ for the sets of feasible solutions to (P) and (D) respectively, and correspondingly $F^0(P)$ and $F^0(D)$ for the sets of strictly feasible solutions to (P) and (D) respectively; here "strictly" means that X or S is required to be positive definite. Hence

$$F(P) := \{X \in \mathcal{P} : A_i \bullet X = b_i, \, i = 1, \cdots, m\}, \qquad (10.2.5)$$

$$F^0(P) := \{X \in F(P) : X \in \mathcal{P}_+\}, \qquad (10.2.6)$$

$$F(D) := \{(y, S) \in \mathbb{R}^m \times \mathcal{P} : \sum_{i=1}^m y_i A_i + S = C\}, \qquad (10.2.7)$$

$$F^0(D) := \{(y, S) \in F(D) : S \in \mathcal{P}_+\}. \qquad (10.2.8)$$

The aim of this section is to motivate and define the *central path* for (P) and (D), and then prove its existence and uniqueness.

The first easy result is weak duality:

Proposition 10.2.1 *If X and (y, S) are feasible in (P) and (D) respectively, then*

$$C \bullet X - b^T y = X \bullet S \geq 0. \qquad (10.2.9)$$

Proof.

For $X \in F(P)$ and $(y, s) \in F(D)$, we have

$$C \bullet X - b^T y = (\sum_{i=1}^m y_i A_i + S) \bullet X - \sum_{i=1}^m (A_i \bullet X) y_i = X \bullet S,$$

and since X and S and thus $X^{1/2} S X^{1/2}$ are positive semidefinite,

$$X \bullet S = \operatorname{Trace} XS = \operatorname{Trace} X^{1/2} S X^{1/2} \geq 0. \qquad (10.2.10)$$

■

(Thus $X \bullet S$ is the "excess" of the primal objective function value $C \bullet X$ over the dual value $b^T y$.)

Corollary 10.2.1 *If X and (y, S) are feasible in (P) and (D) respectively, with $X \bullet S = 0$, or equivalently $XS = 0$, then X and (y, S) are optimal in their respective problems.*

Proof.

It is only necessary to prove the stated equivalence. It is clear that $XS = 0$ implies $X \bullet S = 0$. For the converse, note that the latter implies (using (10.2.10)) that the positive semidefinite matrix $X^{1/2}SX^{1/2}$ has zero trace, and hence is the zero matrix. Now by diagonalizing X we easily see that $XS = 0$ also. ■

Unfortunately, strong duality need not hold for semidefinite programming, but it does hold if either (P) or (D) has a strictly feasible solution. For now, we shall assume that both these problems have strictly feasible solutions, i.e., that $F^0(P)$ and $F^0(D)$ are nonempty. In this case, strong duality holds, and both problems have bounded nonempty optimal solution sets. In fact, we shall prove this as a byproduct of our analysis of the central path.

From the corollary above, it is clear that the conditions below (together with X and S belonging to $\mathcal{P}$) are sufficient for X and (y, S) to be optimal solutions:

$$\begin{array}{rcccl} & \sum_{i=1}^m y_i A_i & + & S & = & C, \\ A_i \bullet X & & & & = & b_i, \quad \text{for } i = 1, \ldots, m, \\ & XS & & & = & 0. \end{array} \tag{10.2.11}$$

Let us simplify the notation slightly: we define the operator $\mathcal{A} : \mathcal{S}^n \to I\!R^m$ by

$$(\mathcal{A}X)_i := A_i \bullet X, \quad i = 1, ..., m.$$

Then the adjoint of this operator is $\mathcal{A}^* : I\!R^m \to \mathcal{S}^n$ satisfying

$$\mathcal{A}^* y = \sum_{i=1}^m y_i A_i.$$

Using this notation, we can rewrite the equations above as

$$\begin{array}{rcccl} & \mathcal{A}^* y & + & S & = & C, \\ \mathcal{A}X & & & & = & b, \\ & XS & & & = & 0. \end{array} \tag{10.2.12}$$

In both (10.2.11) and (10.2.12) we could replace the last equation equivalently by $X^{1/2}SX^{1/2} = 0$, or by $S^{1/2}XS^{1/2} = 0$.

The central path is defined as the set of solutions $(X, y, S) = (X(\nu), y(\nu), S(\nu)) \in \mathcal{P} \times I\!R^m \times \mathcal{P}$ to

$$\begin{array}{rcccl} & \mathcal{A}^* y & + & S & = & C, \\ \mathcal{A}X & & & & = & b, \\ & XS & & & = & \nu I, \end{array} \tag{10.2.13}$$

for all $\nu > 0$. Clearly any solution to these equations gives strictly feasible solutions to both (P) and (D), since the last condition implies that X and S are nonsingular, hence positive definite. We shall show that the existence of strictly feasible solutions to both problems is also sufficient for the existence and uniqueness of solutions to (10.2.13) for every positive ν.

The key to this proof is the analysis of a certain barrier problem associated with (P). For this, we define the following barrier function for the cone of positive semidefinite matrices $\mathcal{P}$:

$$f(X) := -\ln\det X \tag{10.2.14}$$

(by convention, we call this a barrier function for $\mathcal{P}$, even though it is defined only for points in $\mathcal{P}_+$; it clearly tends to $+\infty$ as X in $\mathcal{P}_+$ converges to a point on the boundary of $\mathcal{P}$). We need to deal with the derivatives of f. The first derivative of f at $X \in \mathcal{P}_+$ is $f'(X) = -X^{-1}$, in the usual sense that

$$f(X+H) = f(X) + [-X^{-1}] \bullet H + o(\|H\|).$$

For the second derivative, we introduce the useful notation $P \odot Q$ for $n \times n$ matrices P and Q (usually P and Q are symmetric). This is an operator defined from $\mathcal{S}^n$ to $\mathcal{S}^n$ by

$$(P \odot Q)U := \frac{1}{2}(PUQ^T + QUP^T). \tag{10.2.15}$$

Then it is not too hard to show that the second derivative of f is $f''(X) = X^{-1} \odot X^{-1}$, in the usual sense that

$$f(X+H) = f(X) + f'(X) \bullet H + \frac{1}{2}[(X^{-1} \odot X^{-1})H] \bullet H + o(\|H\|^2).$$

Note that $f''(X)$ is a self-adjoint and positive definite operator:

$$f''(X)U \bullet V = [X^{-1}UX^{-1}] \bullet V = \operatorname{Trace} X^{-1}UX^{-1}V = \operatorname{Trace} X^{-1}VX^{-1}U = f''(X)V \bullet U$$

and

$$f''(X)U \bullet U = \operatorname{Trace} X^{-1}UX^{-1}U = \operatorname{Trace}(X^{-1/2}UX^{-1/2})^2 = \|X^{-1/2}UX^{-1/2}\|_F^2,$$

where $\|\cdot\|_F$ denotes the Frobenius norm, the norm induced by the inner product we are using; this quantity is nonnegative, and positive unless $X^{-1/2}UX^{-1/2}$ (and hence U) is zero.

Since f'' is positive definite everywhere, f is strictly convex. Now we consider the primal barrier problem:

$$\begin{array}{llrcl} (PBP) & \min_X & C \bullet X + \nu f(X) & & \\ & & \mathcal{A}X & = & b, \\ & & X & \in & \mathcal{P}_+, \end{array}$$

for positive ν. Note that the KKT (or Lagrange) conditions for this problem are necessary and sufficient, since the objective function is convex and the constraints linear (apart from the open set constraint). These optimality conditions can be written as

$$\mathcal{A}X = b, \qquad C - \nu X^{-1} - \mathcal{A}^* y = 0,$$

for $X \in \mathcal{P}_+$, which can alternatively be expressed as (10.2.13) by setting $S := \nu X^{-1} \in \mathcal{P}_+$.

Let us note that (10.2.13) also gives the optimality conditions for the dual barrier problem:

$$
\begin{array}{lllcll}
(DBP) & \max_{y,S} & b^T y & - & \nu f(S) & \\
 & & \mathcal{A}^* y & + & S & = \; C, \\
 & & & & S & \in \; \mathcal{P}_+.
\end{array}
$$

We are now prepared to prove existence and uniqueness of the central path:

Theorem 10.2.1 *Suppose that both (P) and (D) have strictly feasible solutions. Then, for every positive ν, there is a unique solution $(X(\nu), y(\nu), S(\nu))$ in $\mathcal{P}_+ \times I\!R^m \times \mathcal{P}_+$ to the central path equations (10.2.13).*

Proof.
We start with uniqueness, which is easy. Any solution to (10.2.13) gives an X which solves (PBP), by the sufficiency of the optimality conditions. Hence X is unique by the strict convexity of the objective function of (PBP). Then the last equation shows that S is also unique, and finally y is unique because of the first equation and our assumption that the A_i's are linearly independent. (The uniqueness of S also follows from the strict convexity of the objective function of (DBP) in S and the remark above the theorem.)

To prove existence, we show that (PBP) has an optimal solution, which gives the result by the necessity of the optimality conditions. For this, we want to reduce (PBP) to the minimization of a continuous function over a nonempty compact set and use the Weierstrass theorem. Unfortunately, at present the feasible region is possibly unbounded and certainly relatively open, and if we take its closure, the objective function is not even defined on the boundary.

Let $\hat{X}$ be strictly feasible in (P) and $(\hat{y}, \hat{S})$ be strictly feasible in (D). Then we can replace the objective function of (PBP) by $\hat{S} \bullet X + \nu f(X)$, since by (10.2.9) it differs from the original objective function by the constant $b^T \hat{y}$. Moreover, $\hat{X}$ is feasible in (PBP), and we can add without loss of generality the constraint

$$\hat{S} \bullet X + \nu f(X) \leq \hat{S} \bullet \hat{X} + \nu f(\hat{X}).$$

We aim to show that

$$\mathcal{U} := \{X \in \mathcal{P}_+ : \mathcal{A}X = b, \quad \hat{S} \bullet X + \nu f(X) \leq \hat{S} \bullet \hat{X} + \nu f(\hat{X})\}$$

is compact (it clearly contains $\hat{X}$), and that $\hat{S} \bullet X + \nu f(X)$ is continuous on $\mathcal{U}$. We do this by bounding the eigenvalues $\lambda_j(X)$ of X from above and below on $\mathcal{U}$. Let $\sigma > 0$ denote the smallest eigenvalue of $\hat{S}$, and note that for $X \in \mathcal{U}$,

$$\sum_j (\sigma \lambda_j(X) - \nu \ln \lambda_j(X)) = \sigma I \bullet X + \nu f(X) \leq \hat{S} \bullet \hat{X} + \nu f(\hat{X}).$$

The left-hand side above can be written as $\sum_j \phi(\lambda_j(X))$, where $\phi(\lambda)$ is defined as $\sigma\lambda - \nu \ln \lambda$ for positive λ. But note that ϕ is strictly convex, attains its minimum, and tends to ∞ as λ tends to either 0 or ∞. It follows that $\phi(\lambda_j(X)))$ is bounded above on $\mathcal{U}$, and hence that each eigenvalue $\lambda_j(X)$ lies between some positive $\underline{\lambda}$ and some finite $\bar{\lambda}$. Hence $\mathcal{U}$ is a subset of a compact set of positive definite matrices. Since it is defined by linear equations and an inequality on a function that is continuous on such sets, $\mathcal{U}$ is itself compact, and since $\mathcal{U}$ contains only positive definite matrices, the objective function $\hat{S} \bullet X + \nu f(X)$ is continuous on it. This completes the proof. ∎

So far we have shown the existence and uniqueness of points on the central path, but we have not justified calling it a path. This will follow if we show that the equations defining it are differentiable, with a derivative that is square and nonsingular at points on the path. Unfortunately, while the equations of (10.2.13) are certainly differentiable, the derivative is not even square since the left-hand side maps $(X, y, S) \in \mathcal{P}_+ \times I\!R^m \times \mathcal{P}_+ \subseteq \mathcal{S}^n \times I\!R^m \times \mathcal{S}^n$ to a point in $\mathcal{S}^n \times I\!R^m \times I\!R^{n\times n}$; XS is usually not symmetric even if X and S are. We therefore need to change the equations defining the central path. There are many possible approaches, which as we shall see lead to different search directions for our algorithms, but for now we choose a simple one: we replace $XS = \nu I$ by $-\nu X^{-1} + S = 0$. As in our discussion of the barrier function f, the function $X \to -\nu X^{-1}$ is differentiable, with derivative $\nu(X^{-1} \odot X^{-1})$. So the central path is defined by the equations

$$\Phi_P(X, y, S) := \begin{pmatrix} & \mathcal{A}^* y & + & S \\ \mathcal{A}X & & & \\ -\nu X^{-1} & & + & S \end{pmatrix} = \begin{pmatrix} C \\ b \\ 0 \end{pmatrix}, \qquad (10.2.16)$$

whose derivative is

$$\Phi'_P(X, y, S) := \begin{pmatrix} 0 & \mathcal{A}^* & \mathcal{I} \\ \mathcal{A} & 0 & 0 \\ \nu(X^{-1} \odot X^{-1}) & 0 & \mathcal{I} \end{pmatrix}, \qquad (10.2.17)$$

where $\mathcal{I}$ denotes the identity operator on $\mathcal{S}^n$. We have been rather loose in writing this in matrix form, since the blocks are operators rather than matrices, but the meaning is clear. We want to show that this derivative is nonsingular, and for this it suffices to prove that its null-space is trivial. Since similar equations will occur frequently, let us derive this from a more general result.

Theorem 10.2.2 *Suppose the operators $\mathcal{E}$ and $\mathcal{F}$ map $\mathcal{S}^n$ to itself, and that $\mathcal{E}$ is nonsingular and $\mathcal{E}^{-1}\mathcal{F}$ is positive definite. Then the solution to*

$$\begin{array}{rcrcl} & \mathcal{A}^*\Delta y & + & \Delta S & = & R_d, \\ \mathcal{A}\,\Delta X & & & & = & r_p, \\ \mathcal{E}\,\Delta X & & + & \mathcal{F}\,\Delta S & = & R_{EF} \end{array} \qquad (10.2.18)$$

is uniquely given by

$$\begin{array}{rcl} \Delta y & = & (\mathcal{A}\mathcal{E}^{-1}\mathcal{F}\mathcal{A}^*)^{-1}(r_p - \mathcal{A}\mathcal{E}^{-1}(R_{EF} - \mathcal{F}R_d)), \\ \Delta S & = & R_d - \mathcal{A}^*\Delta y, \\ \Delta X & = & \mathcal{E}^{-1}(R_{EF} - \mathcal{F}\Delta S). \end{array} \qquad (10.2.19)$$

Proof.
The formulae for ΔS and ΔX follow directly from the first and third equations. Now substituting for ΔS in the formula for ΔX, and inserting this in the second equation, we obtain after some manipulation

$$(\mathcal{A}\mathcal{E}^{-1}\mathcal{F}\mathcal{A}^*)\Delta y = r_p - \mathcal{A}\mathcal{E}^{-1}(R_{EF} - \mathcal{F}R_d).$$

Since $\mathcal{E}^{-1}\mathcal{F}$ is positive definite and the A_i's are linearly independent, the $m \times m$ matrix on the left is positive definite and hence nonsingular. This verifies that Δy is uniquely determined as given, and then so are ΔS and ΔX. Moreover, these values solve the equations. ∎

In our case, $\mathcal{F}$ is the identity, while $\mathcal{E}$ is $\nu(X^{-1} \odot X^{-1})$ with inverse $\nu^{-1}(X \odot X)$. This is easily seen to be positive definite, in the same way we showed that $f''(X)$ was. Hence the theorem applies, and so the derivative of the function Φ_P is nonsingular on the central path (and throughout $\mathcal{P}_+ \times I\!R^m \times \mathcal{P}_+$); thus the central path is indeed a differentiable path.

By taking the trace of the last equation of (10.2.13), we obtain the last part of the following theorem, which summarizes what we have observed:

Theorem 10.2.3 *Assume that both (P) and (D) have strictly feasible solutions. Then the set of solutions to (10.2.13) for all positive ν forms a nonempty differentiable path, called the central path. If $(X(\nu), y(\nu), S(\nu))$ solve these equations for a particular positive ν, then $X(\nu)$ is a strictly feasible solution to (P) and $(y(\nu), S(\nu))$ a strictly feasible solution to (D), with duality gap*

$$C \bullet X(\nu) - b^T y(\nu) = X(\nu) \bullet S(\nu) = n\nu. \qquad (10.2.20)$$

∎

We claimed above that we could use the central path to prove strong duality. Indeed, we have:

Theorem 10.2.4 *The existence of strictly feasible solutions to (P) and (D) implies that both have bounded nonempty optimal solution sets, with zero duality gap.*

Proof.
The last part follows from the existence of the central path, since by (10.2.20) the duality gap associated to $X(\nu)$ and $(y(\nu), S(\nu))$ is $n\nu$, and this approaches zero as ν tends to zero. (In fact, the central path approaches optimal solutions to the primal and dual problems as ν decreases to zero [506, 288], but we shall not prove this here.)

To show that (P) has a bounded nonempty set of optimal solutions, we proceed as in the proof of Theorem 10.2.1. Clearly, the set of optimal solutions is unchanged if we change the objective function of (P) to $\hat{S} \bullet X$ and add the constraint $\hat{S} \bullet X \leq \hat{S} \bullet \hat{X}$. But this latter constraint (for $X \in \mathcal{P}$) implies that all the eigenvalues of X are bounded by $(\hat{S} \bullet \hat{X})/\sigma$, where again $\sigma > 0$ denotes the smallest eigenvalue of $\hat{S}$. This shows that all optimal solutions of (P) (if any) lie in a compact set of feasible solutions; but the minimum of the continuous function $\hat{S} \bullet X$ over this compact set (containing $\hat{X}$) is attained, and so the set of optimal solutions is nonempty and bounded. The proof that the set of optimal dual solutions is bounded and nonempty is similar: we start by noting that the objective of maximizing $b^T y$ can be replaced by that of minimizing $\hat{X} \bullet S$ using (10.2.9). ■

Given that we start with strictly feasible solutions to (P) and (D), it is simple to maintain equality in the first two equations of (10.2.13), but satisfying the last nonlinear equation exactly is hard. Hence *path-following methods* require that the iterates satisfy these last equations in some approximate sense. The requirement can be stated in terms of certain *neighborhoods* of the central path, defined either in terms of norms or semi-norms of $X^{1/2}SX^{1/2} - \nu I$ or in terms of the eigenvalues of this matrix. In fact, we will replace ν by $\mu := \mu(X, S) := (X \bullet S)/n$ (cf. (10.2.20)). We then define the following centrality measures for $X, S \in \mathcal{P}$:

$$
\begin{array}{rcll}
d_F(X,S) & := & \|X^{1/2}SX^{1/2} - \mu I\|_F & = & [\sum_{i=1}^n (\lambda_i(XS) - \mu)^2]^{1/2}, \\
d_\infty(X,S) & := & \|X^{1/2}SX^{1/2} - \mu I\| & = & \max_{1 \leq i \leq n} |\lambda_i(XS) - \mu|, \\
d_{-\infty}(X,S) & := & \|X^{1/2}SX^{1/2} - \mu I\|_{-\infty} & := & \max_{1 \leq i \leq n} (\mu - \lambda_i(XS)).
\end{array}
\tag{10.2.21}
$$

Here $\|\cdot\|$ for a matrix denotes the operator norm with respect to the Euclidean norm on vectors. The last equation basically defines $\|M\|_{-\infty}$ for symmetric M as the negative of the smallest eigenvalue of M; it is easy to check that this is a semi-norm on $\mathcal{S}^n$. The equations hold because $X^{1/2}SX^{1/2}$ and XS, being similar, have the same eigenvalues. It is easy to see that these measures are zero, for strictly feasible iterates, only for points on the central path. Also, for $X, S \in \mathcal{P}$,

$$
d_{-\infty}(X,S) \leq d_\infty(X,S) \leq d_F(X,S),
$$

and $d_{-\infty}(X,S) < \mu(X,S)$ holds if and only if X and S are positive definite.

For a given $0 < \gamma < 1$, we then define the neighborhoods

$$
\begin{array}{rcl}
\mathcal{N}_F(\gamma) & := & \{(X,y,S) \in F^0(P) \times F^0(D) : d_F(X,S) \leq \gamma\mu(X,S)\}, \\
\mathcal{N}_\infty(\gamma) & := & \{(X,y,S) \in F^0(P) \times F^0(D) : d_\infty(X,S) \leq \gamma\mu(X,S)\}, \\
\mathcal{N}_{-\infty}(\gamma) & := & \{(X,y,S) \in F^0(P) \times F^0(D) : d_{-\infty}(X,S) \leq \gamma\mu(X,S)\}.
\end{array}
$$

The first and last are usually called the *narrow* and *wide* neighborhoods of the central path.

10.3 SEARCH DIRECTIONS

At each iteration, a path-following algorithm has available an iterate (X, y, S) in one of the neighborhoods above, and obtains another such iterate by taking a step in some search direction $(\Delta X, \Delta y, \Delta S)$. Usually this direction is chosen so that $(X + \Delta X, y + \Delta y, S + \Delta S)$ should be a good approximation to $(X(\sigma\mu), y(\sigma\mu), Z(\sigma\mu))$, with $\mu := \mu(X, S)$ and $\sigma \in [0, 1]$, and then the step size is chosen so that the next iterate lies in the appropriate neighborhood. Since the central path is defined by a system of nonlinear equations, the direction is frequently taken as the Newton step for some such system of equations, but choosing the appropriate set is not straightforward.

For simplicity, we are assuming that all our iterates are feasible; see Section 5 for infeasible-interior-point path-following methods. Thus our generic feasible-interior-point path-following method can be described as follows:

Generic feasible-interior-point path-following method:

Choose $L > 1$, $0 < \gamma < 1$, and an associated neighborhood $\mathcal{N}(\gamma)$.
Let $(X^0, y^0, S^0) \in \mathcal{N}(\gamma)$ be given and set $\mu_0 := (X^0 \bullet S^0)/n$.
Repeat until $\mu_k \leq 2^{-L}\mu_0$, **do**

(1) Choose a direction $(\Delta X^k, \Delta y^k, \Delta S^k)$.
(2) Set $(X^{k+1}, y^{k+1}, S^{k+1}) := (X^k, y^k, S^k) + \alpha_k(\Delta X^k, \Delta y^k, \Delta S^k)$ for some $\alpha_k > 0$ such that $(X^{k+1}, y^{k+1}, S^{k+1}) \in \mathcal{N}(\gamma)$;
(4) Set $\mu_{k+1} := (X^{k+1} \bullet S^{k+1})/n$ and increment k by 1.

End do

End

In this section, we are interested in the choice of the search direction. This would appear to be easy: merely take a Newton step for the central path equations (10.2.13) for $\nu := \sigma\mu$. However, as we observed above (10.2.16), these equations map a point in $\mathcal{S}^n \times I\!R^m \times \mathcal{S}^n$ to a point in $\mathcal{S}^n \times I\!R^m \times I\!R^{n\times n}$, and thus a Newton step is not defined. Instead we can use the equations (10.2.16), and the Newton step for these defines the so-called *primal* direction for SDP. Indeed, it is easy to see that the resulting ΔX is exactly the Newton step for the primal barrier problem (PBP) from the current iterate X. Moreover, Theorem 10.2.2 and the discussion below it implies that this Newton step is well-defined for all $(X, y, S) \in \mathcal{N}(\gamma)$.

Such a primal Newton barrier method was proposed by Nesterov and Nemirovskii [583] as an extension of the corresponding method for linear programming. However, it does not seem to be as attractive, either theoretically or practically, as the primal-dual methods (where ΔX, Δy, and ΔS all depend on both the primal iterate X and the dual iterate (y, S)) to be described below. Theoretical results have only been obtained for short-step methods using the narrow neighborhood $\mathcal{N}_F(\gamma)$, and practically, these methods do not seem to allow as aggressive step sizes as primal-dual methods. The reason is not too hard to see. These primal methods are based on taking Newton steps for systems of equations involving the inverse map $X \rightarrow X^{-1}$, and this map is not as well-behaved (particularly as the solution is approached, or after a long step) as the map $(X, S) \to XS$ that appears in (10.2.13). Thus we would like to modify the equations (10.2.13) to allow a Newton step to be taken without losing this nice behavior.

Because we are assuming our current iterates are feasible, the search directions we consider below can all be obtained as solutions of (10.2.18) with r_p and R_d both zero, and suitable $\mathcal{E}$, $\mathcal{F}$, and R_{EF}. (Other directions require a slightly more complicated 4 by 4 block system; see [780].) Computing the corresponding directions follows the scheme of (10.2.19), but the details vary by method, and we refer to the original papers for elaboration.

The first method we consider is due to Alizadeh, Haeberly, and Overton [23], and is motivated by symmetrizing the last equation of (10.2.13) directly; thus we replace $XS = \nu I$ by $XS + SX = \nu I$. Then the equations do indeed map $\mathcal{S}^n \times I\!R^m \times \mathcal{S}^n$ to itself. Linearizing this gives the following equation for the direction (in addition to the feasibility equations):

$$\frac{1}{2}(\Delta X\, S + S\, \Delta X + X\, \Delta S + \Delta S\, X) = \nu I - \frac{1}{2}(XS + SX).$$

Thus the resulting Newton step $(\Delta X, \Delta y, \Delta S)$ solves (10.2.18) for

$$\mathcal{E} = S \odot I, \quad \mathcal{F} = X \odot I, \quad R_{EF} = \nu I - \frac{1}{2}(XS + SX).$$

We call this the AHO direction. Note that the last equation is symmetric between the primal and dual: it is unchanged if we interchange X and S and correspondingly ΔX and ΔS. We call directions obtained from such systems *primal-dual symmetric*, and if a direction is not primal-dual symmetric, we call the direction resulting from such an interchange its *dual* direction.

The AHO direction appears to be very natural, and indeed it does perform very well in practice, obtaining highly accurate solutions in a small number of iterations in almost all cases. However, it does have some drawbacks. From a theoretical point of view, it does not have the attractive property of scale invariance (see Todd, Toh, and Tütüncü [782] and [780]); polynomial-time convergence results seem much harder to obtain; and the search directions may not be well-defined if the iterates do not lie in the neighborhood $\mathcal{N}_\infty(1/2)$ (see [782] and Monteiro and Zanjácomo [546]). From a practical point of view, obtaining $\mathcal{E}^{-1}U$ for a symmetric matrix U as needed to solve the system requires the solution of a Lyapunov system; while this is not too onerous, the $m \times m$ matrix $\mathcal{A}\mathcal{E}^{-1}\mathcal{F}\mathcal{A}^*$ that arises in the solution is not in general symmetric, and this leads to operation counts (and times) that are up to twice as high per iteration as for other methods.

The next method we describe, instead of symmetrizing the last equation, allows non-symmetric search directions to make the dimensions match. Thus we initially allow ΔX to be an arbitrary matrix (note that the dual feasibility equations imply that ΔS must be symmetric), so that the central path equations map a point in $I\!R^{n\times n} \times I\!R^m \times \mathcal{S}^n$ to one in $\mathcal{S}^n \times I\!R^m \times I\!R^{n\times n}$. Then the resulting value for ΔX is symmetrized. Thus the first linearization gives (the feasibility equations and)

$$\Delta X\, S + X\, \Delta S = \nu I - XS,$$

or equivalently

$$\Delta X + X\, \Delta S\, S^{-1} = \nu S^{-1} - X.$$

To symmetrize ΔX it suffices to symmetrize the second term, since the others are already symmetric. Thus the resulting direction solves (10.2.18) for

$$\mathcal{E} = I \odot I, \quad \mathcal{F} = X \odot S^{-1}, \quad R_{EF} = \nu S^{-1} - X.$$

This direction was obtained, by roughly the argument above, independently by Helmberg, Rendl, Vanderbei, and Wolkowicz [339] and Kojima, Shindoh, and Hara [439]. Kojima et al. also described its dual direction with

$$\mathcal{E} = S \odot X^{-1}, \quad \mathcal{F} = I \odot I, \quad R_{EF} = \nu X^{-1} - S.$$

Later, Monteiro [536] gave another derivation of these directions. Each can be viewed in a way very similar to the AHO approach as the Newton step for a symmetrized system of equations, but before symmetrizing XS, a similarity transformation is applied. It is useful to introduce the following two operators $\mathcal{S}_P(\cdot)$ and $\mathcal{A}_P(\cdot)$ mapping $\Re^n$ into $\mathcal{S}^n$ and into the space $\mathcal{S}^n_\perp$ of $n \times n$ skew-symmetric matrices, defined for every $U \in \Re^n$ by

$$\mathcal{S}_P(U) = \frac{1}{2}\left[PUP^{-1} + (PUP^{-1})^T\right],$$
$$\mathcal{A}_P(U) = \frac{1}{2}\left[PUP^{-1} - (PUP^{-1})^T\right],$$

where P is a given nonsingular matrix. (If P is the identity, these operators are the usual projection operators into these two spaces.) Thus the directions above are the Newton steps for the system of equations (10.2.13) where the last equation is replaced by

$$\mathcal{S}_P(XS) = \nu I, \tag{10.3.22}$$

where $P = S^{1/2}$ for the first and $P = X^{-1/2}$ for the second direction. Because of these three contributions, the first direction is often called the HRVW/KSH/M direction. Zhang [867] then generalized Monteiro's work and analyzed the resulting directions for arbitrary nonsingular P; the resulting set of directions is called the Monteiro-Zhang family. Clearly this family also contains the AHO direction, corresponding to $P = I$.

Now let us consider the so-called Nesterov-Todd direction, which is the specialization to semidefinite programming of a symmetric primal-dual direction for a certain class of convex programming problems studied in [587, 588]. We will derive this direction as a way to make the primal direction primal-dual symmetric. Recall that the primal direction is defined by the feasibility equations and

$$\nu X^{-1}\, \Delta X\, X^{-1} + \Delta S = \nu X^{-1} - S.$$

This last equation can be written equivalently as

$$\Delta X + (\nu^{-1/2}X)\,\Delta S\,(\nu^{-1/2}X) = X - \nu^{-1}XSX.$$

On the other hand, its dual direction, which is the Newton step for the dual barrier problem, solves the feasibility equations and

$$\Delta X + (\nu^{1/2}S^{-1})\,\Delta S\,(\nu^{1/2}S^{-1}) = \nu S^{-1} - X,$$

which coincides with the equation displayed above if the current iterate is exactly the point on the central path corresponding to the parameter ν. We will therefore replace the terms $\nu^{-1/2}X$ and $\nu^{1/2}S^{-1}$ by a matrix that is in some respects half way between them (or halfway between X and S^{-1}).

Suppose we substitute for these two expressions $W \in \mathcal{P}_+$, so that the first equation above becomes

$$W^{-1}\,\Delta X\,W^{-1} + \Delta S = \nu X^{-1} - S$$

and the third

$$\Delta X + W\,\Delta S\,W = \nu S^{-1} - X. \tag{10.3.23}$$

These two equations are equivalent as long as $WSW = X$, which implies that $WX^{-1}W = S^{-1}$ also. By pre- and post-multiplying by $S^{1/2}$, we find that

$$W = W_{nt} := S^{-1/2}(S^{1/2}XS^{1/2})^{1/2}S^{-1/2}; \tag{10.3.24}$$

this matrix is called the *metric geometric mean* of X and S^{-1} [32]; it is similarly the metric geometric mean of $\nu^{-1/2}X$ and $\nu^{1/2}S^{-1}$. In interior-point methods, it is called the *scaling* matrix corresponding to the primal and dual iterates X and S.

The direction defined by the feasibility equations and (10.3.23), or equivalently by (10.2.18) for

$$\mathcal{E} = I \odot I, \quad \mathcal{F} = W \odot W, \quad R_{EF} = \nu S^{-1} - X,$$

where W is the scaling matrix defined by (10.3.24), is called the Nesterov-Todd (NT) direction. Let us note that the resulting ΔX also solves the quadratic approximation to the primal barrier problem

$$\min\{(C + \nu f'(X)) \bullet \Delta X + \frac{1}{2}[f''(W)(\Delta X)] \bullet \Delta X : \mathcal{A}\Delta X = 0\},$$

which differs from the Newton step in that the quadratic term $[f''(W)(\Delta X)] \bullet \Delta X/2$ replaces the expected $\nu[f''(X)(\Delta X)] \bullet \Delta X/2$. Similarly, the resulting $(\Delta y, \Delta S)$ also solves the quadratic approximation to the dual barrier problem

$$\max\{b^T\Delta y - \nu f'(S) \bullet \Delta S - \frac{1}{2}[(f''(W))^{-1}(\Delta S)] \bullet \Delta S : \mathcal{A}^*\,\Delta y + \Delta S = 0\},$$

which is also a slight modification of the quadratic problem defining the Newton step. Todd, Toh, and Tütüncü [782] also showed that the NT direction is a member of the Monteiro-Zhang family, corresponding to $P = W^{-1/2}$ in (10.3.22).

The AHO, HRVW/KSH/M, and NT directions are those most used in practice, as well as being the ones most analyzed theoretically. Before ending this section, however, we briefly mention several other directions which have been addressed in the literature. First, in addition to the HRVW/KSH/M direction and its dual, Kojima, Shindoh, and Hara [439] described a whole family of directions (including also the NT direction, but not the AHO direction). Next, Monteiro and Tsuchiya [543] introduced and analyzed a family of directions called the Monteiro-Tsuchiya family, basically defined by taking a Newton step for the central path equations (10.2.13) where the final equation is replaced by

$$X^{1/2}SX^{1/2} = \nu I.$$

A whole family arises if this approach is applied to a scaled problem, involving a nonsingular matrix P as in the Monteiro-Zhang family. We refer to the original paper and to [780] for more details. Tseng [790] introduced yet another family of search directions. While the Tseng and Monteiro-Tsuchiya families differ from the Kojima-Shindoh-Hara family, and from each other, all contain the HRVW/KSH/M direction, its dual, and the NT direction, but exclude the AHO direction. Finally, Gu [308] and Toh [784] have recently proposed two other members of the Monteiro-Zhang family which seem promising. Kojima, Shida, and Shindoh [438] describe most of these directions and families and show the relationships between them; Todd [780] describes twenty particular search directions and analyzes their theoretical properties.

10.4 PRIMAL-DUAL PATH-FOLLOWING METHODS

In this section we discuss the main convergence results that have been obtained for feasible path-following algorithms based on three families of search directions, namely: the MZ, MT, and KSH families. Due to space constraints, it is impossible to discuss the three families in detail. For this reason, we have decided to treat the MZ family in detail and to survey the main results for the other two families. This section is divided into four subsections. The first three subsections are dedicated to studying algorithms based on the MZ family. The first subsection describes a general primal-dual (PD) framework based on the MZ family and a scaling scheme that is used in later subsections on the analysis of some specific methods in the PD framework. The second subsection analyzes the short-step path-following method and a Mizuno-Todd-Ye-type predictor-corrector method. The long-step path-following method based on a commutative class of the MZ family is studied in the third subsection. The last subsection introduces the MT and KSH families and surveys the main convergence results for them.

10.4.1 The MZ primal-dual framework and a scaling procedure

Given $(X, y, S) \in \mathcal{P}_+ \times I\!R^m \times \mathcal{P}_+$ and a nonsingular matrix $P \in I\!R^{n\times n}$, consider the linear system of equations

$$\begin{array}{rcl} \mathcal{A}^*\Delta y \quad +\Delta S & = & C - S - \mathcal{A}^* y, \\ \mathcal{A}\Delta X \qquad\qquad & = & b - \mathcal{A}X, \\ \mathcal{S}_P(\Delta X S \quad + \quad X\Delta S) & = & \sigma\mu I - \mathcal{S}_P(XS), \end{array} \tag{10.4.25a}$$

where $(\Delta X, \Delta y, \Delta S) \in \mathcal{S}^n \times I\!R^m \times \mathcal{S}^n$, $\sigma \in [0, 1]$ is the centering parameter and $\mu = \mu(X, S) := (X \bullet S)/n$ is the normalized duality gap corresponding to (X, y, S). This is the Newton system for the feasibility equations and the central path equations in the form $\mathcal{S}_P(XS) = \sigma\mu I$. Under certain conditions on (X, y, S) which will be studied later on, it can be shown that system (10.4.25) has a unique solution, denoted by $(\Delta X_P, \Delta y_P, \Delta S_P)$. As P varies over the set of nonsingular $n \times n$-matrices, the set of resulting directions gives rise to the MZ family of directions with parameter σ at the point (X, y, S).

The following simple result shows that there is no loss of generality in restricting our attention to those scaling matrices P that are in $\mathcal{P}_+$, since they yield all the possible directions that can be generated by system (10.4.25) as P varies over the set of nonsingular matrices.

Proposition 10.4.1 *The set of solutions to system (10.4.25) remains invariant as long as the matrix $V = P^T P$ does not change.*

Proof.
Note that (10.4.25a) is equivalent to the equation

$$V(X\Delta S + \Delta XS) + (\Delta SX + S\Delta X)V = V(\sigma\mu I - XS) + (\sigma\mu I - SX)V$$

obtained by multiplying (10.4.25a) on the left by P^T and on the right by P. Since this last equation depends on V only, the result follows. ■

Hence, for fixed $V \in \mathcal{P}_+$, there is no loss of generality in considering only the matrix $V^{1/2}$ among all those scaling matrices P such that $P^T P = V$, since their corresponding system (10.4.25) all have the same solution set. Thus, we assume from now on that $P \in \mathcal{P}_+$.

We now state the generic path-following algorithm which will be analyzed in detail in this section.

Framework MZ-PD:

Let $L > 1$ and $(X^0, y^0, S^0) \in \mathcal{F}^0(P) \times \mathcal{F}^0(D)$ be given.
Set $\mu_0 := (X^0 \bullet S^0)/n$.
Repeat until $\mu_k \leq 2^{-L}\mu_0$, **do**

(1) Choose a matrix $P^k \in \mathcal{P}_+$ and a centrality parameter $\sigma_k \in [0,1]$;

(2) Compute the solution $(\Delta X^k, \Delta y^k, \Delta S^k)$ of system (10.4.25) with $P = P^k$, $\mu = \mu_k$, $\sigma = \sigma_k$ and $(X, y, S) = (X^k, y^k, S^k)$;

(3) Set $(X^{k+1}, y^{k+1}, S^{k+1}) := (X^k, y^k, S^k) + \alpha_k(\Delta X^k, \Delta y^k, \Delta S^k)$ for some $\alpha_k > 0$ such that $(X^{k+1}, S^{k+1}) \in \mathcal{P}_+ \times \mathcal{P}_+$;

(4) Set $\mu_{k+1} := (X^{k+1} \bullet S^{k+1})/n$ and increment k by 1.

End do

End

Specialized versions of Framework MZ-PD will be studied in Subsections 10.4.2 and 10.4.3. In Subsection 10.4.2 we study algorithms based on the neighborhood $\mathcal{N}_F(\cdot)$, namely the short-step method and the predictor-corrector method. In Subsection 10.4.3 we study the long-step method which is based on the larger neighborhood $\mathcal{N}_{-\infty}(\cdot)$. In the analysis of these algorithms, we will make use of an important scaling procedure that will simplify the proofs of the main results. This scaling procedure essentially allows us to assume that $P^k = I$ by viewing each iteration in the original space as an equivalent one in the scaled space. In other works, in the scaled space the iterations reduce to steps along the AHO direction.

We now introduce the aforementioned scaling procedure. Consider the following change of variables

$$\widetilde{X} := PXP, \qquad (\widetilde{y}, \widetilde{S}) := (y,\, P^{-1}SP^{-1}). \tag{10.4.26}$$

Letting

$$\widetilde{C} := P^{-1}CP^{-1}, \qquad (\widetilde{A}_i, \widetilde{b}_i) := (P^{-1}A_iP^{-1},\, b_i), \quad \text{for } i = 1, \ldots, m,$$

we easily see that problems (P) and (D) are equivalent to the pair of problems

$$(\widetilde{P}) \qquad \min\left\{\widetilde{C} \bullet \widetilde{X} : \widetilde{\mathcal{A}}\widetilde{X} = \widetilde{b},\, \widetilde{X} \succeq 0\right\},$$

$$(\widetilde{D}) \qquad \max\left\{\widetilde{b}^T\widetilde{y} : \widetilde{\mathcal{A}}^*\widetilde{y} + \widetilde{S} = \widetilde{C},\, \widetilde{S} \succeq 0\right\},$$

where the operator $\widetilde{\mathcal{A}} : \mathcal{S}^n \to I\!R^m$ is defined as $(\widetilde{\mathcal{A}}X)_i = \widetilde{A}_i \bullet X$ for every $X \in \mathcal{S}^n$ and $i = 1, \ldots, m$. It can be easily verified that if (X, y, S) and $(\widetilde{X}, \widetilde{y}, \widetilde{S})$ in $\mathcal{P}_+ \times I\!R^m \times \mathcal{P}_+$ are related according to (10.4.26), then $(\widetilde{X}, \widetilde{y}, \widetilde{S})$ is feasible in $(\widetilde{P})$ and $(\widetilde{D})$ if and only if (X, y, S) is feasible in (P) and (D), and $\mu = \widetilde{\mu}$, $d_F(\widetilde{X}, \widetilde{S}) = d_F(X, S)$, $d_\infty(\widetilde{X}, \widetilde{S}) = d_\infty(X, S)$, $d_{-\infty}(\widetilde{X}, \widetilde{S}) = d_{-\infty}(X, S)$, where $\widetilde{\mu} := (\widetilde{X} \bullet$

$\widetilde{S})/n$. If we let $\widetilde{N}_F(\gamma)$, $\widetilde{N}_\infty(\gamma)$, $\widetilde{N}_{-\infty}(\gamma)$ denote the neighborhoods associated with the pair of problems $(\widetilde{P})$ and $(\widetilde{D})$, the above observation immediately implies that

$$(\widetilde{X},\widetilde{y},\widetilde{S}) \in \widetilde{N}_F(\gamma) \iff (X,y,S) \in \mathcal{N}_F(\gamma), \tag{10.4.27a}$$

$$(\widetilde{X},\widetilde{y},\widetilde{S}) \in \widetilde{N}_\infty(\gamma) \iff (X,y,S) \in \mathcal{N}_\infty(\gamma), \tag{10.4.27b}$$

$$(\widetilde{X},\widetilde{y},\widetilde{S}) \in \widetilde{N}_{-\infty}(\gamma) \iff (X,y,S) \in \mathcal{N}_{-\infty}(\gamma). \tag{10.4.27c}$$

Moreover, if $(\widetilde{X}_\nu,\widetilde{y}_\nu,\widetilde{S}_\nu)$ denotes the point on the central path with parameter $\nu > 0$ for the pair of problems $(\widetilde{P})$ and $(\widetilde{D})$, then $(\widetilde{X}_\nu,\widetilde{y}_\nu,\widetilde{S}_\nu) = (PX_\nu P, y_\nu, P^{-1}S_\nu P^{-1})$.

A solution $(\Delta X, \Delta y, \Delta S)$ of the MZ system (10.4.25) corresponds in the scaled space to the scaled direction $(\widetilde{\Delta X}, \widetilde{\Delta y}, \widetilde{\Delta S})$ defined as

$$\widetilde{\Delta X} := P\Delta X P, \quad \widetilde{\Delta S} := P^{-1}\Delta S P^{-1}, \quad \widetilde{\Delta y} = \Delta y. \tag{10.4.28}$$

The direction $(\widetilde{\Delta X}, \widetilde{\Delta y}, \widetilde{\Delta S})$ is easily seen to be a solution of the AHO system:

$$\begin{array}{rcl} \widetilde{\mathcal{A}}^*\widetilde{\Delta y} + \widetilde{\Delta S} &=& \widetilde{C} - \widetilde{S} - \widetilde{\mathcal{A}}^*\widetilde{y}, \\ \widetilde{\mathcal{A}}\widetilde{\Delta X} &=& \widetilde{b} - \widetilde{\mathcal{A}}\widetilde{X}, \\ \mathcal{S}_I(\widetilde{\Delta X}\widetilde{S} + \widetilde{X}\widetilde{\Delta S}) &=& \mathcal{S}_I(\sigma\widetilde{\mu}I - \widetilde{X}\widetilde{S}). \end{array}$$

In other words, $(\widetilde{\Delta X}, \widetilde{\Delta y}, \widetilde{\Delta S})$ is the AHO direction (with parameter σ) at the point $(\widetilde{X},\widetilde{y},\widetilde{S})$. Hence, an iteration of Framework MZ-PD in the (X,y,S)-space corresponds to a step along the AHO direction in the $(\widetilde{X},\widetilde{y},\widetilde{S})$-space.

We use the following notation throughout this section. For every $\alpha \in I\!R$, let

$$X(\alpha) := X + \alpha\Delta X, \quad y(\alpha) := y + \alpha\Delta y, \quad S(\alpha) := S + \alpha\Delta S, \tag{10.4.30}$$

$$\widetilde{X}(\alpha) := \widetilde{X} + \alpha\widetilde{\Delta X}, \quad \widetilde{y}(\alpha) := \widetilde{y} + \alpha\widetilde{\Delta y}, \quad \widetilde{S}(\alpha) := \widetilde{S} + \alpha\widetilde{\Delta S}, \tag{10.4.31}$$

$$\mu(\alpha) := \frac{X(\alpha) \bullet S(\alpha)}{n}, \quad \widetilde{\mu}(\alpha) := \frac{\widetilde{X}(\alpha) \bullet \widetilde{S}(\alpha)}{n}. \tag{10.4.32}$$

Clearly, we have $\widetilde{X}(\alpha) = PX(\alpha)P$, $\widetilde{S}(\alpha) = P^{-1}S(\alpha)P^{-1}$, $\widetilde{y}(\alpha) = y(\alpha)$ and $\mu(\alpha) = \widetilde{\mu}(\alpha)$ for all $\alpha \in I\!R$.

Finally, we make some comments about the NT direction in regards to the above scaling scheme. Recall that the solution $(\Delta X_P, \Delta y_P, \Delta S_P)$ of (10.4.25) is the NT direction when $P = P_{nt} := W_{nt}^{-1/2}$, where W_{nt} is the matrix defined in (10.3.24). The defining property of the matrix W_{nt} is that it is the unique solution $W \in \mathcal{P}_+$ of the equation $WSW = X$ (see the discussion following (10.3.23)). This implies that P_{nt} is the unique solution $P \in \mathcal{P}_+$ of the equation $PXP = P^{-1}SP^{-1}$. Hence,

$$P = P_{nt} = W_{nt}^{-1/2} \Longrightarrow \widetilde{X} = \widetilde{S}. \tag{10.4.33}$$

10.4.2 Short-step and predictor-corrector algorithms

In this subsection we study the convergence analysis of two primal-dual feasible algorithms based on the MZ family and the $\mathcal{N}_F(\cdot)$ central path neighborhood, namely: i) the short-step path-following method, and ii) a Mizuno-Todd-Ye-type predictor-corrector method. The analysis of this section is based on Monteiro [537] and Monteiro and Zanjácomo [546]. However, our development differs from these two papers in a few aspects. For example, we prove that the directions of the MZ family are well-defined over a larger neighborhood of the central path. Moreover, some of the bounds obtained here are sharper and our convergence analysis is slightly simpler than that in [537].

We adopt the style of stating the main results first while postponing their proofs to the end of the subsection. The following result guarantees that the direction $(\Delta X_P, \Delta y_P, \Delta S_P)$ is well-defined for any (X, y, S) close to the central path.

Proposition 10.4.2 *If $(X, y, S) \in \mathcal{P}_+ \times \mathbb{R}^m \times \mathcal{P}_+$ is such that*

$$||S^{1/2}XS^{1/2} - \nu I|| < \frac{\nu}{\sqrt{2}}$$

for some scalar $\nu > 0$, then system (10.4.25) has a unique solution for any nonsingular matrix P. In particular, $(\Delta X_P, \Delta y_P, \Delta S_P)$ is well-defined for any $(X, y, S) \in \mathcal{P}_+ \times \mathbb{R}^m \times \mathcal{P}_+$ such that $d_\infty(X, S) < \mu/\sqrt{2}$.

Monteiro and Zanjácomo [546] established that $(\Delta X_P, \Delta y_P, \Delta S_P)$ is well-defined in a smaller neighborhood of the central path, namely that consisting of points $(X, y, S) \in \mathcal{P}_+ \times \mathbb{R}^m \times \mathcal{P}_+$ such that $||S^{1/2}XS^{1/2} - \nu I|| < \nu/2$ for some $\nu > 0$. Our proof of the above result, which will be given after Lemma 10.4.3, is quite different than the one given in [546] and is more related to the flow of ideas used in this subsection.

In view of the above result, it is important to restrict (at least in theory) the iterates of Framework MZ-PD to neighborhoods of the central path that are contained in the region of points specified in Proposition 10.4.2. The methods studied in this section all do that.

We now describe the short-step and predictor-corrector methods. We also state their corresponding convergence results, postponing their proofs until the end of this subsection. Since the two methods are specializations of Framework MZ-PD, we only need to describe how the sequences $\{\sigma_k\}$ and $\{\alpha_k\}$ are chosen.

Short-step method: *for all $k \geq 0$, let $\alpha_k = 1$ and $\sigma_k := 1 - \delta/\sqrt{n}$, where $\delta > 0$ is a constant which is specified in Theorem 10.4.1 below.*

The following convergence result for the short-step method based on the MZ family is due to Monteiro [537] (with slightly different conditions on γ and δ).

Theorem 10.4.1 *Let $\gamma \in (0, 1/\sqrt{2})$ and $\delta \in (0, 1)$ be constants satisfying*

$$\frac{2(\gamma^2 + \delta^2)}{(1 - \sqrt{2}\gamma)^2}\left(1 - \frac{\delta}{\sqrt{n}}\right)^{-1} \leq \gamma,$$

and assume that $(X^0, y^0, S^0) \in \mathcal{N}_F(\gamma)$. *Then, every iterate* (X^k, y^k, S^k) *generated by the short-step method is in the neighborhood* $\mathcal{N}_F(\gamma)$ *and satisfies* $X^k \bullet S^k = (1 - \delta/\sqrt{n})^k (X^0 \bullet S^0)$. *Moreover, the short-step method terminates in at most* $\mathcal{O}(\sqrt{n}L)$ *iterations.*

We next give the description of the predictor-corrector method.

Predictor-corrector method: *Let* $\tau \in (0, 1/15]$ *be given. For every* $k \geq 0$ *even, let* $\sigma_k = 0$ *and* α_k *be the largest* $\alpha > 0$ *such that* $(X^k, y^k, S^k) + \widetilde{\alpha}(\Delta X^k, \Delta y^k, \Delta S^k) \in \mathcal{N}_F(2\tau)$ *for all* $\widetilde{\alpha} \in [0, \alpha]$; *for every* $k \geq 0$ *odd, let* $\sigma_k = 1$ *and* $\alpha_k = 1$.

The following convergence result for the predictor-corrector method based on the MZ family is given in Monteiro [537].

Theorem 10.4.2 *Assume that* $\tau \in (0, 1/15]$. *Then, the predictor-corrector algorithm satisfies the following statements:*

a) for every $k \geq 0$ *even,* $(X^k, y^k, S^k) \in \mathcal{N}_F(\tau)$, $\alpha_k = 1/\mathcal{O}(\sqrt{n})$ *and* $X^{k+1} \bullet S^{k+1} = (1 - \alpha_k)(X^k \bullet S^k)$;

b) for every $k \geq 0$ *odd,* $(X^k, y^k, S^k) \in \mathcal{N}_F(2\tau)$ *and* $X^{k+1} \bullet S^{k+1} = X^k \bullet S^k$.

As a consequence, the algorithm terminates in at most $\mathcal{O}(\sqrt{n}L)$ *iterations.*

In the remaining part of this subsection, we concentrate on giving the proofs of Proposition 10.4.2, Theorem 10.4.1 and Theorem 10.4.2. The following two lemmas are mainly technical.

Lemma 10.4.1 *For every* $E \in I\!R^{n \times n}$, *we have:*

a) $\operatorname{Trace} E^2 = \|\mathcal{S}_I(E)\|_F^2 - \|\mathcal{A}_I(E)\|_F^2$ *and* $\|E\|_F^2 = \|\mathcal{S}_I(E)\|_F^2 + \|\mathcal{A}_I(E)\|_F^2$;

b) for every pair of nonsingular matrices $P, Q \in I\!R^{n \times n}$, $\|\mathcal{S}_P(E)\|_F^2 - \|\mathcal{A}_P(E)\|_F^2 = \|\mathcal{S}_Q(E)\|_F^2 - \|\mathcal{A}_Q(E)\|_F^2$.

Proof.
The proof of a) is straightforward and is based on the following identities: $S \bullet A = \operatorname{Trace} AS = \operatorname{Trace} SA = 0$, $\|S\|_F^2 = \operatorname{Trace} S^2$ and $\|A\|_F^2 = -\operatorname{Trace} A^2$ for every $S \in \mathcal{S}^n$ and $A \in \mathcal{S}^n_\perp$.

Statement b) is an immediate consequence of the identity

$$\begin{aligned} \|\mathcal{S}_P(E)\|_F^2 - \|\mathcal{A}_P(E)\|_F^2 &= \|\mathcal{S}_I(PEP^{-1})\|_F^2 - \|\mathcal{A}_I(PEP^{-1})\|_F^2 \\ &= \operatorname{Trace} (PEP^{-1})^2 \\ &= \operatorname{Trace} PE^2P^{-1} = \operatorname{Trace} E^2. \end{aligned}$$

■

Lemma 10.4.2 *Assume that $E \in I\!R^{n\times n}$ is such that $\mathcal{S}_P(E) = 0$ for some nonsingular matrix $P \in I\!R^{n\times n}$. Then:*

a) $\|\mathcal{S}_I(E)\|_F \le \|E\|_F/\sqrt{2} \le \|\mathcal{A}_I(E)\|_F$;

b) if $E = R + T$ *where* $R \in \mathcal{S}^n$ *and* $T \in I\!R^{n\times n}$ *then* $\|R\|_F \le \sqrt{2}\|T\|_F$.

Proof.
Using Lemma 10.4.1(b) with $Q = I$ and the assumption that $\mathcal{S}_P(E) = 0$, we easily see that $\|\mathcal{S}_I(E)\|_F \le \|\mathcal{A}_I(E)\|_F$. It is now easy to see that this inequality together with the second identity of Lemma 10.4.1(a) imply a).

To prove b), first observe that $\mathcal{S}_I(E) = R + \mathcal{S}_I(T)$ and $\mathcal{A}_I(E) = \mathcal{A}_I(T)$. These two equations together with a) and the second identity of Lemma 10.4.1(a) imply that

$$\begin{aligned}\|R\|_F &\le \|R + \mathcal{S}_I(T)\|_F + \|\mathcal{S}_I(T)\|_F = \|\mathcal{S}_I(E)\|_F + \|\mathcal{S}_I(T)\|_F \\ &\le \|\mathcal{A}_I(E)\|_F + \|\mathcal{S}_I(T)\|_F = \|\mathcal{A}_I(T)\|_F + \|\mathcal{S}_I(T)\|_F \\ &\le \sqrt{2}\left(\|\mathcal{A}_I(T)\|_F^2 + \|\mathcal{S}_I(T)\|_F^2\right)^{1/2} = \sqrt{2}\,\|T\|_F.\end{aligned}$$

■

Using the above lemmas, it is now possible to establish that the direction $(\Delta X_P, \Delta y_P, \Delta S_P)$ is well-defined for points (X, y, S) lying close to the central path. The important technical result that allows us to do this is stated next.

Lemma 10.4.3 *Let $(\widetilde{X}, \widetilde{y}, \widetilde{S}) \in \mathcal{P}_+ \times I\!R^m \times \mathcal{P}_+$ be a point such that $\|\widetilde{S}^{1/2}\widetilde{X}\widetilde{S}^{1/2} - \nu I\| \le \gamma\nu$ for some $\gamma \in (0, 1/\sqrt{2})$ and $\nu > 0$. Assume that $H \in I\!R^{n\times n}$ and $(U, V) \in \mathcal{S}^n \times \mathcal{S}^n$ satisfy*

$$\mathcal{S}_I(\widetilde{X}V + U\widetilde{S}) = \mathcal{S}_I(H), \quad U \bullet V \ge 0. \qquad (10.4.34)$$

Then:

$$\max\{\delta_u, \delta_v\} \le \frac{1}{1-\sqrt{2}\gamma}\left(\sqrt{2}\,\|\mathcal{A}_{\widetilde{X}^{-1/2}}H\|_F + \|\mathcal{S}_{\widetilde{X}^{-1/2}}H\|_F\right) \qquad (10.4.35)$$

$$\|\mathcal{S}_I(E)\|_F \le \|\mathcal{A}_I(E)\|_F \le \gamma\delta_u + \|\mathcal{A}_{\widetilde{X}^{-1/2}}H\|_F, \qquad (10.4.36)$$

where

$$\delta_u := \nu\|\widetilde{X}^{-1/2}U\widetilde{X}^{-1/2}\|_F, \quad \delta_v := \|\widetilde{X}^{1/2}V\widetilde{X}^{1/2}\|_F, \qquad (10.4.37)$$

$$E := \widetilde{X}^{-1/2}(U\widetilde{S} + \widetilde{X}V - H)\widetilde{X}^{1/2}. \qquad (10.4.38)$$

Proof.
Using (10.4.38), we easily see that $E = R + T$, where

$$R := \nu\widetilde{X}^{-1/2}U\widetilde{X}^{-1/2} + \widetilde{X}^{1/2}V\widetilde{X}^{1/2} - \mathcal{S}_{\widetilde{X}^{-1/2}}(H), \qquad (10.4.39)$$

$$T := \widetilde{X}^{-1/2}U\widetilde{X}^{-1/2}(\widetilde{X}^{1/2}\widetilde{S}\widetilde{X}^{1/2} - \nu I) - \mathcal{A}_{\widetilde{X}^{-1/2}}(H). \qquad (10.4.40)$$

Using the norm inequality $||PQ||_F \le ||P||_F||Q||$, the assumption that $||\widetilde{S}^{1/2}\widetilde{X}\widetilde{S}^{1/2} - \nu I|| \le \gamma\nu$ and relations (10.4.40) and (10.4.35), we obtain

$$\begin{aligned} ||T||_F &\le ||\widetilde{X}^{-1/2}U\widetilde{X}^{-1/2}||_F||\widetilde{X}^{1/2}\widetilde{S}\widetilde{X}^{1/2} - \nu I|| + ||\mathcal{A}_{\widetilde{X}^{-1/2}}(H)||_F \\ &\le \gamma\delta_u + ||\mathcal{A}_{\widetilde{X}^{-1/2}}(H)||_F. \end{aligned} \tag{10.4.41}$$

The first relation of system (10.4.34) implies that $\mathcal{S}_{\widetilde{X}^{1/2}}(E) = 0$, and since $R \in \mathcal{S}^n$, it follows from Lemma 10.4.2(b) that $||R||_F \le \sqrt{2}||T||_F$. Moreover, the last relation of system (10.4.34) implies that $0 \le U \bullet V = (\widetilde{X}^{-1/2}U\widetilde{X}^{-1/2}) \bullet (\widetilde{X}^{1/2}V\widetilde{X}^{1/2})$. Using these two last conclusions and expressions (10.4.39), (10.4.40) and (10.4.41), we obtain

$$\begin{aligned} \max\{\delta_u, \delta_v\} &\le (\delta_u^2 + \delta_v^2)^{1/2} \le ||\nu\widetilde{X}^{-1/2}U\widetilde{X}^{-1/2} + \widetilde{X}^{1/2}V\widetilde{X}^{1/2}||_F \\ &= ||R + \mathcal{S}_{\widetilde{X}^{-1/2}}(H)||_F \le ||R||_F + ||\mathcal{S}_{\widetilde{X}^{-1/2}}(H)||_F \\ &\le \sqrt{2}||T||_F + ||\mathcal{S}_{\widetilde{X}^{-1/2}}(H)||_F \\ &\le \sqrt{2}\left(\gamma\delta_u + ||\mathcal{A}_{\widetilde{X}^{-1/2}}(H)||_F\right) + ||\mathcal{S}_{\widetilde{X}^{-1/2}}(H)||_F \\ &\le \sqrt{2}\gamma\max\{\delta_u, \delta_v\} + \sqrt{2}||\mathcal{A}_{\widetilde{X}^{-1/2}}(H)||_F + ||\mathcal{S}_{\widetilde{X}^{-1/2}}(H)||_F, \end{aligned}$$

from which (10.4.35) easily follows.

The fact that $\mathcal{S}_{\widetilde{X}^{1/2}}(E) = 0$ and Lemma 10.4.2(a) imply the first inequality in (10.4.36). The second inequality in (10.4.36) follows from (10.4.41) as follows:

$$||\mathcal{A}_I(E)||_F = ||\mathcal{A}_I(T)||_F \le ||T||_F \le \gamma\delta_u + ||\mathcal{A}_{\widetilde{X}^{-1/2}}(H)||_F. \tag{10.4.42}$$

■

Using the above result, we can now give a proof of Proposition 10.4.2.

Proof of Proposition 10.4.2: Consider system (10.4.29) with $\widetilde{X}$ and $\widetilde{S}$ defined by (10.4.26) and note that $||\widetilde{S}^{1/2}\widetilde{X}\widetilde{S}^{1/2} - \nu I|| = ||S^{1/2}XS^{1/2} - \nu I|| < \nu/\sqrt{2}$. We know that (10.4.25) has a unique solution if and only if (10.4.29) does. To show that (10.4.29) has a unique solution, let (U, w, V) be a solution of the homogeneous system associated with (10.4.29). Then, $\mathcal{S}_I(\widetilde{X}V + U\widetilde{S}) = 0$ and $U \bullet V = 0$. By Lemma 10.4.3 with $H = 0$, we conclude that $U = V = 0$. Since the matrices A_i's, and hence the $\widetilde{A}_i$'s, are linearly independent and $\widetilde{\mathcal{A}}^*w = -V = 0$, we must have $w = 0$. We have thus shown that $(U, w, V) = (0, 0, 0)$. This implies that system (10.4.29), and hence (10.4.25), has a unique solution. ■

We now establish a number of technical lemmas that will be used in the proof of Theorems 10.4.1 and 10.4.2.

Lemma 10.4.4 *Let* $(\widetilde{X}, \widetilde{y}, \widetilde{S}) \in \mathcal{F}^0(\widetilde{P}) \times \mathcal{F}^0(\widetilde{D})$ *and* $(\widetilde{\Delta X}, \widetilde{\Delta y}, \widetilde{\Delta S})$ *be a solution of (10.4.29) for some* $\sigma \in I\!R$. *Then, for every* $\alpha \in I\!R$, *we have*

$$\widetilde{\mu}(\alpha) = (1-\alpha+\sigma\alpha)\widetilde{\mu}, \tag{10.4.43}$$

$$\mathcal{S}_{\widetilde{X}^{-1/2}}\left(\widetilde{X}(\alpha)\widetilde{S}(\alpha) - \widetilde{\mu}(\alpha)I\right) = (1-\alpha)\left(\widetilde{X}^{1/2}\widetilde{S}\widetilde{X}^{1/2} - \widetilde{\mu}I\right) + \alpha\,\mathcal{S}_I(E_x) + \alpha^2\mathcal{S}_{\widetilde{X}^{-1/2}}(\widetilde{\Delta X}\widetilde{\Delta S}), \tag{10.4.44}$$

where

$$E_x := \widetilde{X}^{-1/2}(\widetilde{X}\widetilde{\Delta S} + \widetilde{\Delta X}\widetilde{S} + \widetilde{X}\widetilde{S} - \sigma\widetilde{\mu}I)\widetilde{X}^{1/2}. \tag{10.4.45}$$

Proof.
Using (10.4.29), (10.4.32) and the identity $\operatorname{Trace}\mathcal{S}_I(M) = \operatorname{Trace} M$ for $M \in I\!R^{n\times n}$, we obtain

$$\begin{aligned}
\widetilde{X} \bullet \widetilde{\Delta S} + \widetilde{S} \bullet \widetilde{\Delta X} &= \operatorname{Trace}(\widetilde{X}\widetilde{\Delta S} + \widetilde{\Delta X}\widetilde{S}) = \operatorname{Trace}\mathcal{S}_I(\widetilde{X}\widetilde{\Delta S} + \widetilde{\Delta X}\widetilde{S}) \\
&= \operatorname{Trace}\mathcal{S}_I(\sigma\widetilde{\mu}I - \widetilde{X}\widetilde{S}) = \operatorname{Trace}(\sigma\widetilde{\mu}I - \widetilde{X}\widetilde{S}) \\
&= \sigma\widetilde{\mu}n - (\widetilde{X} \bullet \widetilde{S}) = (\sigma - 1)(\widetilde{X} \bullet \widetilde{S}).
\end{aligned}$$

Using (10.4.29) and (10.4.29) and the fact that $\widetilde{\mathcal{A}}\widetilde{X} = \widetilde{b}$ and $\widetilde{\mathcal{A}}^*\widetilde{y} + \widetilde{S} = \widetilde{C}$, we easily see that $\widetilde{\Delta X} \bullet \widetilde{\Delta S} = 0$. These two equalities together with (10.4.31) then imply

$$\begin{aligned}
\widetilde{X}(\alpha) \bullet \widetilde{S}(\alpha) &= (\widetilde{X} + \alpha\widetilde{\Delta X}) \bullet (\widetilde{S} + \alpha\widetilde{\Delta S}) \\
&= \widetilde{X} \bullet \widetilde{S} + \alpha(\widetilde{X} \bullet \widetilde{\Delta S} + \widetilde{S} \bullet \widetilde{\Delta X}) + \alpha^2\widetilde{\Delta X} \bullet \widetilde{\Delta S} \\
&= \widetilde{X} \bullet \widetilde{S} + \alpha(\sigma - 1)\widetilde{X} \bullet \widetilde{S} = (1 - \alpha + \alpha\sigma)\widetilde{X} \bullet \widetilde{S}.
\end{aligned}$$

Dividing this expression by n and noting (10.4.32), we obtain (10.4.43). Using (10.4.31), (10.4.43) and (10.4.45), we conclude that for every $\alpha \in I\!R$,

$$\begin{aligned}
\widetilde{X}(\alpha)\widetilde{S}(\alpha) - \widetilde{\mu}(\alpha)I &= (\widetilde{X} + \alpha\widetilde{\Delta X})(\widetilde{S} + \alpha\widetilde{\Delta S}) - \widetilde{\mu}(\alpha)I \\
&= \widetilde{X}\widetilde{S} + \alpha(\widetilde{\Delta X}\widetilde{S} + \widetilde{X}\widetilde{\Delta S}) + \alpha^2\widetilde{\Delta X}\widetilde{\Delta S} - (1 - \alpha + \sigma\alpha)\widetilde{\mu}I \\
&= (1-\alpha)(\widetilde{X}\widetilde{S} - \widetilde{\mu}I) + \alpha(\widetilde{\Delta X}\widetilde{S} + \widetilde{X}\widetilde{\Delta S} + \widetilde{X}\widetilde{S} - \sigma\widetilde{\mu}I) + \alpha^2\widetilde{\Delta X}\widetilde{\Delta S} \\
&= (1-\alpha)(\widetilde{X}\widetilde{S} - \widetilde{\mu}I) + \alpha\widetilde{X}^{1/2}E_x\widetilde{X}^{-1/2} + \alpha^2\widetilde{\Delta X}\widetilde{\Delta S}.
\end{aligned}$$

Applying the operator $\mathcal{S}_{\widetilde{X}^{-1/2}}(\cdot)$ to both sides of this expression, we obtain (10.4.44). ∎

Lemma 10.4.5 *Suppose that* $(\widetilde{X}, \widetilde{y}, \widetilde{S}) \in \widetilde{\mathcal{N}}_F(\gamma)$ *for some* $\gamma \in I\!R_+$. *Then, for any* $\sigma \in I\!R$, *we have*

$$\|\widetilde{X}^{1/2}\widetilde{S}\widetilde{X}^{1/2} - \sigma\widetilde{\mu}I\|_F \le \left[\gamma^2 + (1-\sigma)^2 n\right]^{1/2}\widetilde{\mu} \tag{10.4.46}$$

Proof.
This result follows by the Pythagorian theorem and the fact that $\widetilde{X}^{1/2}\widetilde{S}\widetilde{X}^{1/2} - \widetilde{\mu}\sigma I$ and $(\sigma - 1)\widetilde{\mu}I$ are orthogonal to one another. ■

Lemma 10.4.6 *Assume that $(\widetilde{X}, \widetilde{y}, \widetilde{S}) \in \widetilde{\mathcal{N}}_F(\gamma)$ for some $\gamma \in (0, 1/\sqrt{2})$. Then, for every $\alpha \in [0,1]$, we have*

$$\left\| \mathcal{S}_{\widetilde{X}^{-1/2}} \left(\widetilde{X}(\alpha)\widetilde{S}(\alpha) - \widetilde{\mu}(\alpha)I \right) \right\|_F \leq \left[(1-\alpha)\gamma + \alpha\gamma\Theta + \alpha^2\Theta^2 \right] \widetilde{\mu}, \qquad (10.4.47)$$

where

$$\Theta := \frac{\left(\gamma^2 + (1-\sigma)^2 n\right)^{1/2}}{1 - \sqrt{2}\gamma}.$$

Proof.
Let $\alpha \in [0,1]$ be given and define $\delta_x := \widetilde{\mu}\|\widetilde{X}^{-1/2}\widetilde{\Delta X}\widetilde{X}^{-1/2}\|_F$ and $\delta_s := \|\widetilde{X}^{1/2}\widetilde{\Delta S}\widetilde{X}^{1/2}\|_F$. Applying Lemma 10.4.3 with $H = \sigma\widetilde{\mu}I - \widetilde{X}\widetilde{S}$ and $\nu = \widetilde{\mu}$, and noting that $\mathcal{A}_{\widetilde{X}^{-1/2}}(H) = 0$, we conclude from (10.4.35), (10.4.36) and Lemma 10.4.5 that

$$\begin{aligned} \max\{\delta_x, \delta_s\} &\leq \frac{1}{1-\sqrt{2}\gamma} \|\widetilde{X}^{1/2}\widetilde{S}\widetilde{X}^{1/2} - \widetilde{\mu}\sigma I\|_F \leq \Theta\widetilde{\mu}, \\ \|\mathcal{S}_I(E_x)\|_F &\leq \gamma\delta_x \leq \gamma\Theta\widetilde{\mu}, \end{aligned}$$

where E_x is given by (10.4.45). Using Lemma 10.4.4, the triangle inequality for norms, the fact that $\alpha \in [0,1]$ and $(\widetilde{X}, \widetilde{y}, \widetilde{S}) \in \widetilde{\mathcal{N}}_F(\gamma)$ and the above two inequalities, we obtain

$$\begin{aligned} &\left\| \mathcal{S}_{\widetilde{X}^{-1/2}} \left(\widetilde{X}(\alpha)\widetilde{S}(\alpha) - \widetilde{\mu}(\alpha)I \right) \right\|_F \\ &\leq (1-\alpha)\|\widetilde{X}^{1/2}\widetilde{S}\widetilde{X}^{1/2} - \widetilde{\mu}I\|_F + \alpha\|\mathcal{S}_I(E_x)\|_F + \alpha^2\|\mathcal{S}_{\widetilde{X}^{-1/2}}(\widetilde{\Delta X}\widetilde{\Delta S})\|_F \\ &\leq (1-\alpha)\gamma\widetilde{\mu} + \alpha\gamma\Theta\widetilde{\mu} + \alpha^2\frac{\delta_x\delta_s}{\widetilde{\mu}} \leq \left[(1-\alpha)\gamma + \alpha\gamma\Theta + \alpha^2\Theta^2\right]\widetilde{\mu}. \end{aligned}$$

■

Lemma 10.4.7 *Suppose that $(\hat{X}, \hat{S}) \in \mathcal{P}_+ \times \mathcal{P}_+$ and $Q \in I\!R^{n\times n}$ is a nonsingular matrix. Then, for every $\nu \in I\!R$, we have*

$$\|\hat{X}^{1/2}\hat{S}\hat{X}^{1/2} - \nu I\|_F \leq \|\mathcal{S}_Q(\hat{X}\hat{S} - \nu I)\|_F. \qquad (10.4.48)$$

Proof.

Using Lemma 10.4.1(b) with $P = \widetilde{X}^{-1/2}$ and $E = \widetilde{X}\widetilde{S} - \nu I$, and noting that $\mathcal{A}_P(E) = 0$, we obtain

$$||\mathcal{S}_P(E)||_F^2 \leq ||\mathcal{S}_Q(E)||_F^2 - ||\mathcal{A}_Q(E)||_F^2 \leq ||\mathcal{S}_Q(E)||_F^2,$$

which is equivalent to (10.4.48). ∎

Lemma 10.4.8 *Let $V, Q \in I\!R^{n\times n}$ be given and assume that Q is nonsingular. If $\lambda_{\min}[\mathcal{S}_Q(V)] > 0$ then V is nonsingular. In particular, if $||\mathcal{S}_Q(V) - I|| < 1$ then V is nonsingular.*

Proof.
Define $W := QVQ^{-1}/2$. The assumption $\lambda_{\min}[\mathcal{S}_Q(V)] > 0$ implies $W + W^T \succ 0$. If $u \in I\!R$ is such that $Wu = 0$, then we have $u^T(W + W^T)u = 2u^TWu = 0$, from which it follows that $u = 0$, due to the fact that $W + W^T \succ 0$. This implies that W, and hence V, is nonsingular. ∎

The proof of Theorem 10.4.1 follows as one of the consequences of the result stated below.

Theorem 10.4.3 *Let $\gamma \in (0, 1/\sqrt{2})$ and $\delta \in [0, n^{1/2})$ be constants satisfying*

$$\Gamma := \frac{2(\gamma^2 + \delta^2)}{(1 - \sqrt{2}\gamma)^2}\left(1 - \frac{\delta}{\sqrt{n}}\right)^{-1} < 1. \qquad (10.4.49)$$

Suppose that $(X, y, S) \in \mathcal{N}_F(\gamma)$ and let $(\Delta X, \Delta y, \Delta S)$ denote the solution of system (10.4.25) with $\sigma := 1 - \delta/\sqrt{n}$ and P a nonsingular matrix. Then,

(a) $(X(1), y(1), S(1)) := (X + \Delta X, y + \Delta y, S + \Delta S) \in \mathcal{N}_F(\Gamma)$;

(b) $X(1) \bullet S(1) = (1 - \delta/\sqrt{n})(X \bullet S)$.

Proof.
The assumption of the theorem implies that $(\widetilde{X}, \widetilde{y}, \widetilde{S}) \in \widetilde{\mathcal{N}}_F(\gamma)$ and that $(\widetilde{\Delta X}, \widetilde{\Delta y}, \widetilde{\Delta S})$ is a solution of (10.4.29). It follows from Lemma 10.4.6, the definition of σ, the fact that $\gamma \leq \Theta$ and relation (10.4.49) that for every $\alpha \in [0, 1]$,

$$\begin{aligned}
&\left\|\mathcal{S}_{\widetilde{X}^{-1/2}}\left(\widetilde{X}(\alpha)\widetilde{S}(\alpha) - \widetilde{\mu}(\alpha)I\right)\right\|_F \\
&\leq [(1-\alpha)\gamma + 2\alpha\Theta^2]\,\widetilde{\mu} = \left\{(1-\alpha)\gamma + 2\alpha\frac{(\gamma^2+\delta^2)}{(1-\sqrt{2}\gamma)^2}\right\}\widetilde{\mu} \\
&= \left\{(1-\alpha)\gamma + \alpha\left(1 - \frac{\delta}{\sqrt{n}}\right)\Gamma\right\}\widetilde{\mu} = \{(1-\alpha)\gamma + \alpha\sigma\Gamma\}\widetilde{\mu}, \qquad (10.4.50)
\end{aligned}$$

and hence, in view of (10.4.43) and (10.4.49), we have

$$\left\|\mathcal{S}_{\widetilde{X}^{-1/2}}\left(\frac{\widetilde{X}(\alpha)\widetilde{S}(\alpha)}{\widetilde{\mu}(\alpha)}\right) - I\right\|_F \leq \frac{[(1-\alpha)\gamma + \alpha\sigma\Gamma]}{1 - \alpha + \sigma\alpha} \leq \max\{\gamma, \Gamma\} < 1.$$

By Lemma 10.4.8, this relation implies that the matrix $\widetilde{X}(\alpha)\widetilde{S}(\alpha)/\widetilde{\mu}(\alpha)$, and hence each one of the matrices $\widetilde{X}(\alpha)$ and $\widetilde{S}(\alpha)$, is nonsingular for every $\alpha \in [0,1]$. Using this conclusion, the fact that $(\widetilde{X},\widetilde{S}) \in \mathcal{P}_+ \times \mathcal{P}_+$, and a trivial continuity argument, we see that $(\widetilde{X}(\alpha),\widetilde{S}(\alpha)) \in \mathcal{P}_+ \times \mathcal{P}_+$ for every $\alpha \in [0,1]$. Using the fact that $\widetilde{\mathcal{A}}\widetilde{X} = \widetilde{b}$, $\widetilde{\mathcal{A}}\widetilde{\Delta X} = 0$, $\widetilde{\mathcal{A}}^*\widetilde{y} + \widetilde{S} = \widetilde{C}$ and $\widetilde{\mathcal{A}}^*\widetilde{\Delta y} + \widetilde{\Delta S} = 0$, we easily see that $\widetilde{\mathcal{A}}\widetilde{X}(\alpha) = \widetilde{b}$ and $\widetilde{\mathcal{A}}^*\widetilde{y}(\alpha) + \widetilde{S}(\alpha) = \widetilde{C}$ for every α. We have thus shown that $(\widetilde{X}(\alpha),\widetilde{y}(\alpha),\widetilde{S}(\alpha)) \in \mathcal{F}^0(\widetilde{P}) \times \mathcal{F}^0(\widetilde{D})$, for every $\alpha \in [0,1]$. Using Lemma 10.4.7 with $(\hat{X},\hat{S}) = (\widetilde{X}(1),\widetilde{S}(1))$, $Q = \widetilde{X}^{-1/2}$ and $\nu = \widetilde{\mu}(1)$, and relations (10.4.43) with $\alpha = 1$ and (10.4.50) with $\alpha = 1$, we obtain

$$d_F(\widetilde{X}(1),\widetilde{S}(1)) \le \left\| \mathcal{S}_{\widetilde{X}^{-1/2}}[\widetilde{X}(1)\widetilde{S}(1) - \widetilde{\mu}(1)I] \right\|_F \le \Gamma\widetilde{\mu}(1).$$

We have thus shown that $(\widetilde{X}(1),\widetilde{y}(1),\widetilde{S}(1)) \in \widetilde{\mathcal{N}}_F(\Gamma)$. In view of (10.4.27a), we conclude that (a) holds. Statement (b) is an immediate consequence of relations (10.4.32) and (10.4.43) with $\alpha = 1$, the definition of σ and the identity $\mu(1) = \widetilde{\mu}(1)$. ∎

The proof of Theorem 10.4.1 follows immediately from Theorem 10.4.3 by using a straightforward induction argument and the fact that its assumption implies that $\Gamma \le \gamma$. We now state two results which immediately imply Theorem 10.4.2 in a similar way.

Theorem 10.4.4 *Suppose that $(X,y,S) \in \mathcal{N}_F(\tau)$ for some $\tau \in (0,1/2)$ and let $(\Delta X, \Delta y, \Delta S)$ denote the solution of system (10.4.25) with $\sigma = 0$ and P a nonsingular matrix. Let $\bar{\alpha}$ denote the unique positive root of the quadratic $p(\alpha) = \Theta^2\alpha^2 + (\tau\Theta + \tau)\alpha - \tau$, where $\Theta := (\tau^2 + n)^{1/2}/(1 - \sqrt{2}\,\tau)$. Then, for any $\alpha \in [0,\bar{\alpha}]$, we have:*

(a) $(X(\alpha),y(\alpha),S(\alpha)) \in \mathcal{N}_F(2\tau)$;

(b) $X(\alpha) \bullet S(\alpha) = (1-\alpha)(X \bullet S)$.

Moreover, $\bar{\alpha} = 1/\mathcal{O}(n^{1/2})$.

Proof.
The assumption of the theorem implies that $(\widetilde{X},\widetilde{y},\widetilde{S}) \in \widetilde{\mathcal{N}}_F(\tau)$ and that $(\widetilde{\Delta X},\widetilde{\Delta y},\widetilde{\Delta S})$ is a solution of (10.4.29). Using Lemma 10.4.6 with $\gamma = \tau$ and $\sigma = 0$, the fact that $p(\alpha) \le 0$ for $\alpha \in [0,\bar{\alpha}]$ and relations (10.4.43) with $\sigma = 0$, we obtain

$$\begin{aligned}\left\| \widetilde{X}^{-1/2}[\widetilde{X}(\alpha)\widetilde{S}(\alpha) - \widetilde{\mu}(\alpha)I]\widetilde{X}^{1/2} \right\|_F &\le \left\{(1-\alpha)\tau + \alpha\tau\Theta + \alpha^2\Theta^2\right\}\widetilde{\mu} \\ &= 2\tau\widetilde{\mu}(\alpha) + p(\alpha)\widetilde{\mu} \le 2\tau\widetilde{\mu}(\alpha).\end{aligned}$$

An argument similar to that used in Theorem 10.4.3 together with the fact that $2\tau < 1$ can be used to show that (a) and (b) hold. The last assertion of the theorem follows by a straightforward verification. ∎

Theorem 10.4.5 *Suppose that $(X, y, S) \in \mathcal{N}_F(2\tau)$ for some $\tau \in (0, 1/15]$ and let $(\Delta X, \Delta y, \Delta S)$ denote the solution of system (10.4.29) with $\sigma = 1$. Then,*

$$\begin{aligned} (X, y, S) + (\Delta X, \Delta y, \Delta S) &\in \mathcal{N}_F(\tau), \\ (X + \Delta X) \bullet (S + \Delta S) &= X \bullet S. \end{aligned}$$

Proof.
The result follows immediately from Theorem 10.4.3 with $\delta = 0$ and $\gamma = 2\tau$ and $\Gamma = \tau$, and the fact that Γ given by (10.4.49) satisfies $\Gamma \leq \tau$ when $\tau \in (0, 1/15]$. ■

10.4.3 Long-step method

In this subsection we provide the convergence analysis of the long-step method based on a subclass of the MZ family of directions. The methods of the previous subsection have the best iteration-complexity bounds although in practice they are not effective due to the restrictive condition that the iterates have to remain in a relatively small central path neighborhood $\mathcal{N}_F(\gamma)$, and hence generate small steps. A practical advantage of the long-step path-following method is that it is based on the larger central path neighborhood $\mathcal{N}_{-\infty}(\cdot)$, which allows for more freedom in the choice of the next iterate and hence larger steps (hence, the name long-step method). However, the long-step method has worse theoretical iteration-complexity bounds than the methods of the previous subsection. The main result of this subsection establishes that the long-step method based on a subclass of the MZ family of directions has an iteration-bound of $\mathcal{O}(\sqrt{\kappa_\infty}\, nL)$ where $\kappa_\infty \geq 1$ is the quantity defined in (10.4.53) below. Hence, only when $\kappa_\infty = 1$ does this bound equal that of the long-step method for solving LP problems. The latter condition is shown to hold if the NT direction is used at every iteration.

Our first goal is to give a result that guarantees that the direction $(\Delta X_P, \Delta y_P, \Delta S_P$ computed at the point (X, y, S) is well-defined when P belongs to a certain subclass (to be defined later) of $\mathcal{P}_+$ which depends on the pair (X, S). Since the particular long-step method analyzed in this subsection always chooses the scaling matrix P within this subclass, the search directions are guaranteed to be well-defined everywhere in $\mathcal{F}^0(P) \times \mathcal{F}^0(D)$.

We first need a few preliminary technical results about the operator $\Xi_A : \mathcal{S}^n \to \mathcal{S}^n$ defined as $\Xi_A := I \odot A = A \odot I$, where $A \in \mathcal{S}^n$. By (10.2.15), we have $\Xi_A(U) = (AU + UA)/2$ for all $U \in \mathcal{S}^n$.

Lemma 10.4.9 *Let $A \in \mathcal{S}^n$ be given. Then:*

a) Ξ_A is symmetric (or self-adjoint), that is $\Xi_A(U) \bullet V = U \bullet \Xi_A(V)$ for all $U, V \in \mathcal{S}^n$;

b) if $A \in \mathcal{P}_+$ then Ξ_A is positive definite, that is $\Xi_A(U) \bullet U > 0$ for all $0 \neq U \in \mathcal{S}^n$; in particular, Ξ_A is invertible.

Proof.
A simple direct verification. ∎

Recall that a symmetric operator $\Xi : \mathcal{S}^n \to \mathcal{S}^n$ has a basis of eigenvectors and that all its eigenvalues are real. In what follows, we denote its smallest and largest eigenvalues by $\lambda_{\min}(\Xi)$ and $\lambda_{\max}(\Xi)$, respectively. Moreover, when Ξ is positive definite, we denote its spectral condition number by $\text{cond}\,(\Xi) := \lambda_{\max}(\Xi)/\lambda_{\min}(\Xi)$. Recall that $\lambda_{\min}(\Xi) = \min\{V \bullet \Xi(V) : V \in \mathcal{S}^n, \|V\|_F = 1\}$.

Lemma 10.4.10 *Let $A, B \in \mathcal{S}^n$ be given. Then:*

a) if A and B commute then so do the operators Ξ_A and Ξ_B;

b) if $A, B \in \mathcal{P}_+$ then

$$\lambda_{\min}[\Xi_A \Xi_B^{-1} + \Xi_B^{-1}\Xi_A] > \lambda_{\min}[\mathcal{S}_I(AB^{-1})]; \qquad (10.4.51)$$

c) if $A, B \in \mathcal{P}_+$ commute then $\text{cond}\,(\Xi_A\Xi_B^{-1}) < 4\,\text{cond}\,(AB^{-1})$.

Proof.
It is easy to verify a) directly. To show b), it suffices to show that

$$2V \bullet \Xi_A\Xi_B^{-1}(V) > \lambda_{\min}[\mathcal{S}_I(AB^{-1})]$$

for $V \in \mathcal{S}^n$ such that $\|V\|_F = 1$. Indeed, letting $U := \Xi_B^{-1}(V)$, using Lemma 10.4.9(a) and noting that $BU + UB = 2V$ and $U \neq 0$, we obtain

$$\begin{aligned}
2\,V \bullet \Xi_A\Xi_B^{-1}(V) &= 2\,\Xi_A(V) \bullet \Xi_B^{-1}(V) \;=\; (AV + VA) \bullet U \\
&= 2\,\text{Trace}\,(VAU) \;=\; \text{Trace}\,[(BU + UB)AU] \\
&= \|A^{1/2}UB^{1/2}\|_F^2 + \frac{1}{2}\text{Trace}\,\left[UB(B^{-1}A + AB^{-1})BU\right] \\
&> \text{Trace}\,\left[\mathcal{S}_I(AB^{-1})BU^2B\right] \\
&= \text{Trace}\,\left[(BU^2B)^{1/2}\mathcal{S}_I(AB^{-1})(BU^2B)^{1/2}\right] \\
&\geq \lambda_{\min}[\mathcal{S}_I(AB^{-1})]\,\|(BU^2B)^{1/2}\|_F^2 \\
&= \lambda_{\min}[\mathcal{S}_I(AB^{-1})]\,\|BU\|_F^2 \;\geq\; \lambda_{\min}[\mathcal{S}_I(AB^{-1})],
\end{aligned}$$

where in the last line we used the facts that $(BU^2B)^{1/2} = BUQ$ for some orthogonal matrix Q and $\|BU\|_F \geq \|(BU + UB)/2\|_F = \|V\|_F = 1$.

For the proof of c), assume that $A, B \in \mathcal{P}_+$ commute. By a), b), and Lemma 10.4.9(b), we have

$$0 < \lambda_{\min}[AB^{-1}] < 2\lambda_{\min}[\Xi_A\Xi_B^{-1}], \quad 0 < \lambda_{\min}[BA^{-1}] < 2\lambda_{\min}[\Xi_B\Xi_A^{-1}].$$

Hence, we obtain

$$\begin{aligned}
\left[\text{cond}\,(\Xi_A\Xi_B^{-1})\right]^{-1} &= \lambda_{\min}[\Xi_A\Xi_B^{-1}]\,\lambda_{\min}[\Xi_B\Xi_A^{-1}] \\
&> \frac{1}{4}\lambda_{\min}[AB^{-1}]\,\lambda_{\min}[BA^{-1}] \;=\; \left[4\,\text{cond}\,(AB^{-1})\right]^{-1}.
\end{aligned}$$

■

The following result due to Todd, Toh and Tütüncü (see Theorem 3.1 of [782]) gives a sufficient condition for system (10.4.25) to have a unique solution. It is a generalization of a result of Shida, Shindoh and Kojima [729] corresponding to the case in which $P = I$.

Proposition 10.4.3 *Let $X, S, P \in \mathcal{P}_+$ be given. Then, a sufficient condition for system (10.4.25) to have a unique solution is that $\Xi_{\widetilde{X}}\Xi_{\widetilde{S}} + \Xi_{\widetilde{S}}\Xi_{\widetilde{X}} \succ 0$. Moreover, the latter condition holds if $\mathcal{S}_I(\widetilde{X}\widetilde{S}) = \mathcal{S}_P(XS) \succeq 0$.*

Proof.
To show the first assertion, it is sufficient to prove that system (10.4.29) has a unique solution. Assume that (U, w, V) is a solution of the homogeneous system associated with (10.4.29). We will show that $(U, w, V) = (0, 0, 0)$ from which the result follows. Indeed, we have $\mathcal{S}_I(U\widetilde{S} + \widetilde{X}V) = 0$, $\widetilde{\mathcal{A}}^* w + V = 0$, $\widetilde{\mathcal{A}}U = 0$. The first equation implies that $\Xi_{\widetilde{S}}(U) + \Xi_{\widetilde{X}}(V) = 0$ and the last two imply that $U \bullet V = 0$. Hence, letting $\Gamma := \Xi_{\widetilde{X}}\Xi_{\widetilde{S}} + \Xi_{\widetilde{S}}\Xi_{\widetilde{X}}$ and $\widetilde{U} := \Xi_{\widetilde{X}}^{-1}(U)$, we obtain

$$0 = U \bullet V = -U \bullet \Xi_{\widetilde{X}}^{-1}\Xi_{\widetilde{S}}(U) = -\widetilde{U} \bullet \Xi_{\widetilde{S}}\Xi_{\widetilde{X}}(\widetilde{U}) = -\frac{1}{2}\widetilde{U} \bullet \Gamma(\widetilde{U}).$$

Since $\Gamma \succ 0$, the last relation implies that $\widetilde{U} = 0$, and hence that $U = 0$. In a similar way, we can show that $V = 0$. Using the linear independence of the $\mathcal{A}_i$'s, and hence of the $\widetilde{\mathcal{A}}_i$'s, we can now easily see that $w = 0$.

The last assertion follows from

$$\mathcal{S}_I(\widetilde{X}\widetilde{S}) \succeq 0 \Leftrightarrow \mathcal{S}_I(\widetilde{X}^{-1}\widetilde{S}) \succeq 0 \Rightarrow \Xi_{\widetilde{X}}^{-1}\Xi_{\widetilde{S}} + \Xi_{\widetilde{S}}\Xi_{\widetilde{X}}^{-1} \succ 0 \Leftrightarrow \Xi_{\widetilde{X}}\Xi_{\widetilde{S}} + \Xi_{\widetilde{S}}\Xi_{\widetilde{X}} \succ 0,$$

where the implication is due to Lemma 10.4.10(b). ■

In the development of the long-step algorithm, we further restrict P to the following class of scaling matrices:

$$\mathcal{P}(X, S) := \{P \in \mathcal{P}_+ : P^2XS = SXP^2\} = \{P \in \mathcal{P}_+ : \widetilde{X}\widetilde{S} = \widetilde{S}\widetilde{X}\}. \quad (10.4.52)$$

Since the scaling matrices in $\mathcal{P}(X, S)$ make $\widetilde{X}$ and $\widetilde{S}$ commute, we will refer to $\mathcal{P}(X, S)$ as the *commutative class* of scaling matrices at (X, S). Clearly, by Proposition 10.4.3, $(\Delta X_P, \Delta y_P, \Delta S_P)$ is well-defined whenever $P \in \mathcal{P}(X, S)$ (indeed, if $\widetilde{X}$ and $\widetilde{S}$ commute, then $\mathcal{S}_I(\widetilde{X}\widetilde{S}) = \widetilde{X}\widetilde{S} = \widetilde{X}^{1/2}\widetilde{S}\widetilde{X}^{1/2}$, which is positive definite).

Observe that the scaling matrices $P = W_{nt}^{1/2}$, $P = S^{1/2}$ and $P = X^{-1/2}$, corresponding to the NT direction, the HRVW/KSH/M direction and the dual HRVW/KSH/M direction, are contained in $\mathcal{P}(X, S)$, and hence our analysis applies to these directions. However, $P = I$ does not belong to $\mathcal{P}(X, S)$, except for the unlikely case when X and S commute; hence, our analysis does not apply to the AHO direction.

We now give the description of the long-step method.

Long-step method: *Let $\gamma, \sigma \in (0,1)$ be given. For every $k \geq 0$, assume that $P^k \in \mathcal{P}(X^k, S^k)$ and let $\sigma_k = \sigma$ and α_k be the largest $\alpha > 0$ such that $(X^k, y^k, S^k) + \tilde{\alpha}(\Delta X^k, \Delta y^k, \Delta S^k) \in \mathcal{N}_{-\infty}(\gamma)$ for all $\tilde{\alpha} \in [0, \alpha]$.*

The next result, which is the main result of this subsection, gives an iteration-complexity bound for the long-step algorithm in terms of a certain quantity κ_∞ defined as

$$\kappa_\infty := \sup\left\{ \text{cond}\left[(\widetilde{X}^k)^{-1}\widetilde{S}^k\right] \; : \; k = 0, 1, 2, \ldots\right\}. \tag{10.4.53}$$

Clearly, $\kappa_\infty \geq 1$.

Theorem 10.4.6 *Assume that $\kappa_\infty < \infty$. Then the sequence $\{\mu_k\}$ generated by the long-step algorithm satisfies*

$$\mu_{k+1} = (1 - (1 - \sigma)\alpha_k)\mu_k,$$

where

$$\alpha_k \geq \min\left(1, \frac{\sigma\gamma}{1 - 2\sigma + \sigma^2/(1-\gamma)} \frac{1}{\sqrt{\kappa_\infty}\, n}\right).$$

Consequently, the long-step algorithm terminates in at most $O(\sqrt{\kappa_\infty}\, nL)$ iterations.

We postpone the proof of Theorem 10.4.6 to the end of this subsection after developing some necessary technical lemmas. We now specialize Theorem 10.4.6 to the three special choices of the sequence $\{P^k\}$ that lead to the NT direction, the HRVW/KSH/M direction and the dual HRVW/KSH/M direction.

Theorem 10.4.7 *The long-step algorithms based on the NT direction, the HRVW/KSH/M direction and the dual HRVW/KSH/M direction have iteration-complexity bounds equal to $\mathcal{O}(nL)$, $\mathcal{O}(n^{3/2}L)$ and $\mathcal{O}(n^{3/2}L)$, respectively.*

Proof.
The sequence of scaling matrices $\{P^k\}$ for the long-step algorithm based on the NT direction is given by $P^k = (W^k_{nt})^{1/2}$ for all k. In view of (10.4.33), this implies that $\widetilde{X}^k = \widetilde{S}^k$ for all k, and hence $\kappa_\infty = 1$ from (10.4.53). For the long-step algorithm based on the HRVW/KSH/M direction, we have $P^k = (S^k)^{1/2}$ for all k. In this case, we have $\widetilde{X}^k = (S^k)^{1/2}X^k(S^k)^{1/2}$ and $\widetilde{S}^k = I$. Therefore, in view of (10.4.53), we have $\kappa_\infty = \sup_k\{\text{cond}\,((X^k)^{1/2}S^k(X^k)^{1/2})\} \leq n/(1-\gamma)$, where the last inequality follows from the fact that $\lambda_{\max}[(X^k)^{1/2}S^k(X^k)^{1/2}] = \lambda_{\max}[X^kS^k] \leq X^k \bullet S^k = n\mu_k$ and $\lambda_{\min}[(X^k)^{1/2}S^k(X^k)^{1/2}] = \lambda_{\min}[X^kS^k] \geq (1-\gamma)\mu_k$ for all k. The argument for the dual HRVW/KSH/M direction is similar to that for the HRVW/KSH/M direction. ■

We note that the iteration-complexity bounds obtained in Theorem 10.4.7 for the HRVW/KSH/M and the dual HRVW/KSH/M directions were first obtained by Monteiro [536] within a more specialized context. We also note that, in the context of the long-step algorithm, Theorem 10.4.7 guarantees that the Nesterov-Todd direction achieves the best possible iteration-complexity bound of $O(nL)$ that can provably be derived from our analysis. This is due to the fact that the condition number κ_∞ achieves its lowest possible value, namely $\kappa_\infty = 1$, for the sequence $\{P^k\}$ corresponding to the NT direction.

We now develop a number of technical results that will be used in the proof of Theorem 10.4.6.

Lemma 10.4.11 *Let $(\widetilde{X}, \widetilde{y}, \widetilde{S}) \in \mathcal{P}_+ \times \mathbb{R}^m \times \mathcal{P}_+$ and $(\widetilde{\Delta X}, \widetilde{\Delta y}, \widetilde{\Delta S})$ satisfy (10.4.29). Then*

$$\mathcal{S}_I(\widetilde{X}(\alpha)\widetilde{S}(\alpha)) = (1-\alpha)\mathcal{S}_I(\widetilde{X}\widetilde{S}) + \alpha\sigma\widetilde{\mu}I + \alpha^2\mathcal{S}_I(\widetilde{\Delta X}\widetilde{\Delta S}). \qquad (10.4.54)$$

Proof.
Using (10.4.31), (10.4.29) and the fact that $\mathcal{S}_I(\cdot)$ is a linear operator, we obtain

$$\begin{aligned}
\mathcal{S}_I\left(\widetilde{X}(\alpha)\widetilde{S}(\alpha)\right) &= \mathcal{S}_I\left(\widetilde{X}\widetilde{S} + \alpha(\widetilde{X}\widetilde{\Delta S} + \widetilde{S}\widetilde{\Delta X}) + \alpha^2\widetilde{\Delta X}\widetilde{\Delta S}\right) \\
&= \mathcal{S}_I(\widetilde{X}\widetilde{S}) + \alpha\mathcal{S}_I(\widetilde{X}\widetilde{\Delta S} + \widetilde{S}\widetilde{\Delta X}) + \alpha^2\mathcal{S}_I(\widetilde{\Delta X}\widetilde{\Delta S}) \\
&= \mathcal{S}_I(\widetilde{X}\widetilde{S}) + \alpha\mathcal{S}_I(\sigma\widetilde{\mu}I - \widetilde{X}\widetilde{S}) + \alpha^2\mathcal{S}_I(\widetilde{\Delta X}\widetilde{\Delta S}),
\end{aligned}$$

from which (10.4.54) follows. ■

Lemma 10.4.12 *Suppose that $(\widetilde{X}, \widetilde{y}, \widetilde{S}) \in \widetilde{\mathcal{N}}_{-\infty}(\gamma)$ for some constant $\gamma \in (0,1)$ with $\widetilde{X}\widetilde{S} = \widetilde{S}\widetilde{X}$ and that $(\widetilde{\Delta X}, \widetilde{\Delta y}, \widetilde{\Delta S}) \in \mathcal{S}^n \times \mathbb{R}^m \times \mathcal{S}^n$ is a solution of system (10.4.29) for some $\sigma \in (0,1]$. Then, $(\widetilde{X}(\alpha), \widetilde{y}(\alpha), \widetilde{S}(\alpha)) \in \widetilde{\mathcal{N}}_{-\infty}(\gamma)$ for every $\alpha \in [0, \widetilde{\alpha}]$, where*

$$\widetilde{\alpha} := \min\left(1, \frac{\sigma\gamma\widetilde{\mu}}{\|\mathcal{S}_I(\widetilde{\Delta X}\widetilde{\Delta S})\|}\right). \qquad (10.4.55)$$

Proof.
Using the fact that $(\widetilde{X}, \widetilde{y}, \widetilde{S}) \in \mathcal{F}^0(\widetilde{P}) \times \mathcal{F}^0(\widetilde{D})$, (10.4.29) and (10.4.29), we easily see that $\widetilde{\mathcal{A}}\widetilde{X}(\alpha) = \widetilde{b}$ and $\widetilde{\mathcal{A}}^*\widetilde{y}(\alpha) + \widetilde{S}(\alpha) = \widetilde{C}$ for every $\alpha \in \mathbb{R}$. Using (10.4.54), (10.4.43), the fact that $\lambda_{\min}[\cdot]$ is a homogeneous concave function on the space of symmetric matrices and the assumption that $(\widetilde{X}, \widetilde{y}, \widetilde{S}) \in \widetilde{\mathcal{N}}_{-\infty}(\gamma)$,

we obtain for every $\alpha \in [0, \widetilde{\alpha}]$ that

$$\begin{aligned}
\lambda_{\min}\left[\mathcal{S}_I\left(\widetilde{X}(\alpha)\widetilde{S}(\alpha)\right)\right] &\geq (1-\alpha)\lambda_{\min}[\mathcal{S}_I(\widetilde{X}\widetilde{S})] + \alpha\sigma\widetilde{\mu} + \alpha^2\lambda_{\min}[\mathcal{S}_I(\widetilde{\Delta X}\widetilde{\Delta S})] \\
&\geq (1-\gamma)(1-\alpha)\widetilde{\mu} + \alpha\sigma\widetilde{\mu} - \alpha^2\|\mathcal{S}_I(\widetilde{\Delta X}\widetilde{\Delta S})\| \\
&= (1-\gamma)\widetilde{\mu}(\alpha) + \gamma\alpha\sigma\widetilde{\mu} - \alpha^2\|\mathcal{S}_I(\widetilde{\Delta X}\widetilde{\Delta S})\| \\
&\geq (1-\gamma)\widetilde{\mu}(\alpha) \; > \; 0.
\end{aligned}$$

By Lemma 10.4.8, this implies that the matrix $\widetilde{X}(\alpha)\widetilde{S}(\alpha)$, and hence each factor $\widetilde{X}(\alpha)$ and $\widetilde{S}(\alpha)$, is nonsingular for every $\alpha \in [0, \widetilde{\alpha}]$. Since $\widetilde{X}(0)$ and $\widetilde{S}(0)$ are in $\mathcal{P}_+$, a continuity argument implies that $\widetilde{X}(\alpha)$ and $\widetilde{S}(\alpha)$ are also in $\mathcal{P}_+$ for $\alpha \in [0, \widetilde{\alpha}]$. Using the fact that the real part of the spectrum of a real matrix is contained between the largest and the smallest eigenvalues of its Hermitian part (see p. 187 of [352], for example), we conclude that for $\alpha \in [0, \widetilde{\alpha}]$,

$$\lambda_{\min}\left[\widetilde{X}(\alpha)\widetilde{S}(\alpha)\right] \geq \lambda_{\min}\left[\mathcal{S}_I\left(\widetilde{X}(\alpha)\widetilde{S}(\alpha)\right)\right] \geq (1-\gamma)\widetilde{\mu}(\alpha).$$

We have thus shown that $(\widetilde{X}(\alpha), \widetilde{y}(\alpha), \widetilde{S}(\alpha)) \in \widetilde{\mathcal{N}}_{-\infty}(\gamma)$ for every $\alpha \in [0, \widetilde{\alpha}]$. ∎

Lemma 10.4.13 *Let $(\widetilde{X}, \widetilde{y}, \widetilde{S}) \in \widetilde{\mathcal{N}}_{-\infty}(\gamma)$ satisfy $\widetilde{X}\widetilde{S} = \widetilde{S}\widetilde{X}$ and let $G = \widetilde{X}\widetilde{S}^{-1}$. Then, the unique solution $(\widetilde{\Delta X}, \widetilde{\Delta y}, \widetilde{\Delta S})$ of (10.4.29) satisfies*

$$\|\widetilde{\Delta X}\|_F\|\widetilde{\Delta S}\|_F \leq n\widetilde{\mu}\sqrt{\operatorname{cond}(G)}\left(1 - 2\sigma + \frac{\sigma^2}{1-\gamma}\right). \qquad (10.4.56)$$

Proof.
Let $H := \sigma\widetilde{\mu}I - \mathcal{S}_I(\widetilde{X}\widetilde{S})$ and observe that $\Xi_{\widetilde{X}}^{-1}(H) = \sigma\widetilde{\mu}\widetilde{X}^{-1} - \widetilde{S}$ and $\Xi_{\widetilde{S}}^{-1}(H) = \sigma\widetilde{\mu}\widetilde{S}^{-1} - \widetilde{X}$. Denoting the eigenvalues of $\widetilde{X}\widetilde{S}$ by $\lambda_1 \leq \cdots \leq \lambda_n$ and noting that $\lambda_i \geq (1-\gamma)\widetilde{\mu}$ for all i due to the assumption that $(\widetilde{X}, \widetilde{y}, \widetilde{S}) \in \widetilde{\mathcal{N}}_{-\infty}(\gamma)$, we then obtain

$$\begin{aligned}
\Xi_{\widetilde{X}}^{-1}(H) \bullet \Xi_{\widetilde{S}}^{-1}(H) &= (\sigma\widetilde{\mu}\widetilde{X}^{-1} - \widetilde{S}) \bullet (\sigma\widetilde{\mu}\widetilde{S}^{-1} - \widetilde{X}) \\
&= (\widetilde{X} \bullet \widetilde{S}) - 2\sigma n\widetilde{\mu} + \sigma^2\widetilde{\mu}^2(\widetilde{X}^{-1} \bullet \widetilde{S}^{-1}) \\
&= n\widetilde{\mu} - 2\sigma n\widetilde{\mu} + \sigma^2\widetilde{\mu}^2\sum_{i=1}^{n}\frac{1}{\lambda_i} \\
&\leq n\widetilde{\mu}\left(1 - 2\sigma + \frac{\sigma^2}{1-\gamma}\right). \qquad (10.4.57)
\end{aligned}$$

Observe that (10.4.29) is equivalent to

$$\Xi_{\widetilde{S}}(\widetilde{\Delta X}) + \Xi_{\widetilde{X}}(\widetilde{\Delta S}) = H. \qquad (10.4.58)$$

Letting $\mathcal{G} := \Xi_{\tilde{X}}^{-1}\Xi_{\tilde{S}}$ and using (10.4.58), the identity $\widetilde{\Delta X} \bullet \widetilde{\Delta S} = 0$ and the fact that the operators $\Xi_{\tilde{X}}$, $\Xi_{\tilde{S}}$ and their inverses all commute, we obtain

$$\begin{aligned}
\Xi_{\tilde{X}}^{-1}(H) \bullet \Xi_{\tilde{S}}^{-1}(H) &= \left(\Xi_{\tilde{X}}^{-1}\Xi_{\tilde{S}}(\widetilde{\Delta X}) + \widetilde{\Delta S}\right) \bullet \left(\widetilde{\Delta X} + \Xi_{\tilde{S}}^{-1}\Xi_{\tilde{X}}(\widetilde{\Delta S})\right) \\
&= \widetilde{\Delta X} \bullet \mathcal{G}(\widetilde{\Delta X}) + \widetilde{\Delta S} \bullet \mathcal{G}^{-1}(\widetilde{\Delta S}) \\
&\geq \lambda_{\min}[\mathcal{G}]\, \|\widetilde{\Delta X}\|_F^2 + \lambda_{\min}[\mathcal{G}^{-1}]\, \|\widetilde{\Delta S}\|_F^2 \\
&\geq 2\sqrt{\lambda_{\min}[\mathcal{G}]\lambda_{\min}[\mathcal{G}^{-1}]}\, \|\widetilde{\Delta X}\|_F \|\widetilde{\Delta S}\|_F \\
&= 2\left(\mathrm{cond}\,[\mathcal{G}]\right)^{-1/2}\, \|\widetilde{\Delta X}\|_F \|\widetilde{\Delta S}\|_F,
\end{aligned}$$

where the last inequality follows from the relation $\beta_1^2 + \beta_2^2 \geq 2\beta_1\beta_2$ for $\beta_1, \beta_2 \in \mathbb{R}$. The result now follows by combining the last inequality with inequality (10.4.57) and using the fact that $\mathrm{cond}\,(\mathcal{G}) < 4\mathrm{cond}\,(G)$, due to Lemma 10.4.10(c). ■

Theorem 10.4.8 *Suppose that $(X, y, S) \in \mathcal{N}_{-\infty}(\gamma)$ for some $\gamma > 0$ and let $(\Delta X, \Delta y, \Delta S)$ denote the unique solution of system (10.4.25), where $\sigma \in (0,1)$ and matrix $P \in \mathcal{P}(X, S)$. Let*

$$\hat{\alpha} := \min\left\{1, \frac{\sigma\gamma}{n\sqrt{\mathrm{cond}\, G}}\left(1 - 2\sigma + \frac{\sigma^2}{1-\gamma}\right)^{-1}\right\},$$

where $G := \tilde{X}\tilde{S}^{-1} = PXP^2S^{-1}P$. Then, $(X(\alpha), y(\alpha), S(\alpha)) \in \mathcal{N}_{-\infty}(\gamma)$ and $\mu(\alpha) = (1 - \alpha + \sigma\alpha)\mu$ for any $\alpha \in [0, \hat{\alpha}]$.

Proof.
The assumption of the theorem implies that $(\tilde{X}, \tilde{y}, \tilde{S}) \in \tilde{\mathcal{N}}_{-\infty}(\gamma)$ and that $(\widetilde{\Delta X}, \widetilde{\Delta y}, \widetilde{\Delta S})$ is a solution of (10.4.29). Hence by Lemma 10.4.12 it follows that $(\tilde{X}(\alpha), \tilde{y}(\alpha), \tilde{S}(\alpha)) \in \tilde{\mathcal{N}}_{-\infty}(\gamma)$ for any $\alpha \in [0, \tilde{\alpha}]$, where $\tilde{\alpha}$ is given by (10.4.55). By Lemma 10.4.13 and the fact that

$$\|S_I(\widetilde{\Delta X}\widetilde{\Delta S})\| \leq \|\widetilde{\Delta X}\widetilde{\Delta S}\| \leq \|\widetilde{\Delta X}\widetilde{\Delta S}\|_F \leq \|\widetilde{\Delta X}\|_F \|\widetilde{\Delta S}\|_F,$$

we easily see that $\tilde{\alpha} \geq \hat{\alpha}$. Due to (10.4.27c), we conclude that $(X(\alpha), y(\alpha), S(\alpha)) \in \mathcal{N}_{-\infty}(\gamma)$ for all $\alpha \in [0, \hat{\alpha}]$. ■

Finally, we note that Theorem 10.4.6 follows as an immediate consequence of Theorem 10.4.8, the definition of κ_∞ given in (10.4.53) and definition of the sequence of step-sizes $\{\alpha_k\}$.

10.4.4 Convergence results for other families of directions

In this subsection we describe two other families of search directions and survey the main convergence results that have been obtained for feasible primal-dual algorithms based on them.

Monteiro and Tsuchiya family. Here we discuss the MT family of search directions which was introduced in Monteiro and Tsuchiya [543]. All the results stated below for the MT family are due to Monteiro and Tsuchiya [543] and proofs can be found there.

Recall that each direction $(\Delta X_P, \Delta y_P, \Delta S_P)$ within the MZ family can be transformed into the AHO direction in a certain scaled space obtained by a scaling transformation that depends on P (see (10.4.26)). This property of the MZ family could in fact be used to describe it. More precisely, given a scaling matrix $P \in \mathcal{P}_+$, a point $(X, y, S) \in \mathcal{P}_+ \times I\!R^m \times \mathcal{P}_+$ and a parameter σ, the corresponding MZ direction $(\Delta X, \Delta y, \Delta S)$ can be obtained by performing the following steps:

MZ construction:

i) map (X, y, S) into the point $(\widetilde{X}, \widetilde{y}, \widetilde{S}) \in \mathcal{P}_+ \times I\!R^m \times \mathcal{P}_+$ in the scaled space computed by using (10.4.26);

ii) compute the Newton direction $(\widetilde{\Delta X}, \widetilde{\Delta y}, \widetilde{\Delta S})$ for the system $\mathcal{S}_I(\widetilde{X}\widetilde{S}) - \sigma\widetilde{\mu}I = 0$, $\widetilde{\mathcal{A}}\widetilde{X} - \widetilde{b} = 0$ and $\widetilde{\mathcal{A}}^*\widetilde{y} + \widetilde{S} - \widetilde{C} = 0$; and

iii) map the direction $(\widetilde{\Delta X}, \widetilde{\Delta y}, \widetilde{\Delta S})$ back into the original space using (10.4.28) to obtain $(\Delta X, \Delta y, \Delta S)$.

A similar type of construction can be used to define a general direction of the MT family. The only difference with respect to the above construction is that in ii) the equation $\mathcal{S}_I(\widetilde{X}\widetilde{S}) = \sigma\widetilde{\mu}I$ is replaced by the alternative central path equation $\widetilde{X}^{1/2}\widetilde{S}\widetilde{X}^{1/2} - \sigma\widetilde{\mu}I = 0$. (In [543], arbitrary nonsingular scaling matrices P can be chosen instead of just symmetric positive definite ones as here; however, as with the MZ family, it is not hard to show the direction depends only on PP^T, so without loss of generality we restrict attention to $\mathcal{P}_+$.) To define the procedure more precisely, given a scaling matrix $P \in \mathcal{P}_+$, a point $(X, y, S) \in \mathcal{P}_+ \times I\!R^m \times \mathcal{P}_+$ and a parameter σ, the corresponding MT direction $(\Delta X, \Delta y, \Delta S)$ can be obtained by performing the following steps:

MT construction:

i') same as i) as above;

ii') compute the Newton direction $(\widetilde{\Delta X}, \widetilde{\Delta y}, \widetilde{\Delta S})$ for the system $\widetilde{X}^{1/2}\widetilde{S}\widetilde{X}^{1/2} - \sigma\widetilde{\mu}I = 0$, $\widetilde{\mathcal{A}}\widetilde{X} - \widetilde{b} = 0$ and $\widetilde{\mathcal{A}}^*\widetilde{y} + \widetilde{S} - \widetilde{C} = 0$; and

iii') same as iii) above.

We now discuss how to compute the Newton direction at step ii'). Recall that if $A \in \mathcal{P}_+$ then the operator $\Xi_A := A \odot I = I \odot A$ is positive definite, and hence has an inverse. The following result expresses the directional derivative of the square root function $X \to X^{1/2}$ in terms of the inverse of $\Xi_{X^{1/2}}$.

Lemma 10.4.14 *Let $\theta : \mathcal{P}_+ \to \mathcal{P}_+$ denote the square root function $\theta(X) = X^{1/2}$. Then, θ is an analytic function, and*

$$\theta'(X)H = \frac{1}{2}\left(\Xi_{X^{1/2}}\right)^{-1} H, \quad \textit{for every } X \in \mathcal{P}_+ \textit{ and } H \in \mathcal{S}^n,$$

where $\theta'(X)$ is the derivative of θ at X and $\theta'(X)H$ is the linear map $\theta'(X)$ evaluated at H.

Using Lemma 10.4.14, it is now easy to see that the Newton direction $(\widetilde{\Delta X}, \widetilde{\Delta y}, \widetilde{\Delta S})$ for the system in ii') is the solution of the following system of linear equations:

$$\begin{aligned} \tilde{\mathcal{A}}^* \widetilde{\Delta y} + \widetilde{\Delta S} &= \tilde{C} - \tilde{\mathcal{A}}^* \tilde{y} - \tilde{S}, && (10.4.59a)\\ \tilde{\mathcal{A}} \widetilde{\Delta X} &= b - \tilde{\mathcal{A}}\tilde{X}, && (10.4.59b)\\ \mathcal{S}_I\left(\Xi^{-1}_{\tilde{X}^{1/2}}(\widetilde{\Delta X})\tilde{S}\tilde{X}^{1/2}\right) + \tilde{X}^{1/2}\widetilde{\Delta S}\tilde{X}^{1/2} &= \sigma\tilde{\mu} I - \tilde{X}^{1/2}\tilde{S}\tilde{X}^{1/2}. && (10.4.59c) \end{aligned}$$

As with step ii) of the MZ construction, step ii') of the MT construction is not always well-defined in that the coefficient matrix of system (10.4.59) may be singular. The following result gives two conditions which guarantee that this coefficient matrix is nonsingular.

Proposition 10.4.4 *Let $(X, y, S) \in \mathcal{P}_+ \times I\!R^m \times \mathcal{P}_+$, $\sigma \in I\!R$ and $P \in \mathcal{P}_+$ be given and assume that one of the following conditions hold:*

a) $d_\infty(X, S) = d_\infty(\tilde{X}, \tilde{S}) < \mu/\sqrt{2}$;

b) $\tilde{X}^{1/2}\tilde{S} + \tilde{S}\tilde{X}^{1/2} = (PXP)^{1/2}(P^{-1}SP^{-1}) + (P^{-1}SP^{-1})(PXP)^{1/2} \succ 0$.

Then, the corresponding direction obtained from the MT construction is well-defined in the sense that system (10.4.59) has exactly one solution.

The algorithms considered in this subsection are all special cases of the following framework.

Framework MT-PD:

Same as Framework MZ-PD but with step (2) replaced by

(2') Compute the MT direction $(\Delta X^k, \Delta y^k, \Delta S^k)$ according to the MT construction with $P = P^k$, $\sigma = \sigma_k$ and $(X, y, S) = (X^k, y^k, S^k)$.

End

Condition a) of Proposition 10.4.4 gives the foundation for the first two algorithms that we consider, namely: the short-step method and the semilong-step method based on the MT family of directions. These two methods, while not imposing any condition on the choice of scaling matrix P, require that the iterates be generated inside the region of points specified in Proposition 10.4.4(a), namely: $\mathcal{N}_F(\gamma)$ and $\mathcal{N}_\infty(\gamma)$, respectively, where $\gamma \in (0, 1/\sqrt{2})$.

The main results regarding the polynomial convergence of the short-step method and the semilong-step method based on the MT family are stated in the next two theorems. Their proofs can be found in Monteiro and Tsuchiya [543].

Theorem 10.4.9 (Polynomiality of the MT Short-Step Path-Following Algorithm) *Suppose that $\gamma \in (0, 1/\sqrt{2})$ and $\delta \in (0, 1)$ are constants satisfying*

$$\frac{40\,(\gamma^2 + \delta^2)}{(1 - \sqrt{2}\gamma)^2} \leq \left(1 - \frac{\delta}{\sqrt{n}}\right)\gamma.$$

Assume that in Framework MT-PD, $(X^0, y^0, S^0) \in \mathcal{N}_F(\gamma)$, $\sigma_k = 1 - \delta/\sqrt{n}$ *and* $\alpha_k = 1$ *for every* $k \geq 0$. *Then, every iterate* (X^k, y^k, S^k) *generated by this short-step path-following algorithm is in the neighborhood* $\mathcal{N}_F(\gamma)$ *and satisfies* $X^k \bullet S^k = (1 - \delta/\sqrt{n})^k (X^0 \bullet S^0)$. *Moreover, the method terminates in at most* $\mathcal{O}(\sqrt{n}L)$ *iterations.*

Theorem 10.4.10 (Polynomiality of the MT Semilong-Step Path-Following Algorithm) *Let constants* $\gamma \in (0, 1/\sqrt{2})$ *and* $\bar{\sigma} \in (0, 1)$ *be given. Assume that in Framework MT-PD,* $(X^0, y^0, S^0) \in \mathcal{N}_\infty(\gamma)$ *and the sequences* $\{\sigma_k\}$ *and* $\{\alpha_k\}$ *are chosen according to*

$$\begin{aligned} \sigma_k &:= \bar{\sigma}, \\ \alpha_k &:= \max\left\{\alpha \in [0,1] : (X^k, y^k, S^k) + \alpha(\Delta X^k, \Delta y^k, \Delta S^k) \in \mathcal{N}_\infty(\gamma)\right\}, \end{aligned}$$

Then, the sequence of iterates $\{(X^k, y^k, S^k)\} \subset \mathcal{N}_\infty(\gamma)$ *generated by this semilong-step path-following algorithm satisfies* $X^k \bullet S^k \leq (1 - \bar{\eta})^k (X^0 \bullet S^0)$ *for all* $k \geq 0$, *where*

$$\bar{\eta} := \frac{\bar{\sigma}(1-\bar{\sigma})\gamma(1-\sqrt{2}\gamma)^2}{40n\left[\gamma^2 + (1-\bar{\sigma})^2\right]}.$$

Moreover, if the quantity $\max\{\gamma^{-1}, (1-\sqrt{2}\gamma)^{-1}, \bar{\sigma}^{-1}, (1-\bar{\sigma})^{-1}\}$ *is independent of* n, *then the method terminates in at most* $\mathcal{O}(nL)$ *iterations.*

Note that the choices of the sequences $\{\alpha_k\}$ and $\{\sigma_k\}$ in the semilong-step path-following method described in Theorem 10.4.10 are similar to those used in the long-step method except that it is based on the smaller neighborhood $\mathcal{N}_\infty(\gamma)$. It is unknown whether a corresponding semilong-step path-following method with the above complexity can be obtained for the whole MZ family.

Condition b) of Proposition 10.4.4 forms the foundation for the long-step path-following algorithm which we now discuss. Given $(X, S) \in \mathcal{P}_+ \times \mathcal{P}_+$, denote by $\mathcal{P}_0(X, S)$ the class of scaling matrices satisfying condition (b) of Proposition 10.4.4. We refer to the family of directions corresponding to these scaling matrices as the MT* family. By restricting P to the class $\mathcal{P}_0(X, S)$, we obtain the following polynomial convergence result for the long-step path-following method.

Theorem 10.4.11 (Polynomiality of the MT* Long-Step Path-Following Algorithm) *Let constants* $\gamma, \bar{\sigma} \in (0, 1)$ *be given. Assume that in Framework MT-PD,* $(X^0, y^0, S^0) \in \mathcal{N}_{-\infty}(\gamma)$, $P^k \in \mathcal{P}_0(X^k, S^k)$ *for all* k, *and the sequences* $\{\sigma_k\}$ *and* $\{\alpha_k\}$ *are chosen according to*

$$\begin{aligned} \sigma_k &= \bar{\sigma}, \\ \alpha_k &= \max\left\{\alpha \in [0,1] : (X^k, y^k, S^k) + \alpha(\Delta X^k, \Delta y^k, \Delta S^k) \in \mathcal{N}_{-\infty}(\gamma)\right\}. \end{aligned}$$

Then, the sequence of iterates $\{(X^k, y^k, S^k)\} \subset \mathcal{N}_{-\infty}(\gamma)$ *generated by this long-step path-following algorithm satisfies* $X^k \bullet S^k \leq (1 - \bar{\eta})^k (X^0 \bullet S^0)$ *for all* $k \geq 0$, *where*

$$\bar{\eta} := \frac{\bar{\sigma}(1-\bar{\sigma})\gamma(1-\gamma)^{1/2}}{30n^{3/2}[1 - 2\bar{\sigma} + \bar{\sigma}^2/(1-\gamma)]}.$$

Moreover, if the quantity $\max\{\gamma^{-1}, (1-\gamma)^{-1}, \bar{\sigma}^{-1}, (1-\bar{\sigma})^{-1}\}$ *is independent of* n*, then the method terminates in at most* $\mathcal{O}(n^{3/2}L)$ *iterations.*

We end this subsection by giving a few remarks. It is possible to derive a symmetric MT family based on the alternative central path equation $\widetilde{S}^{1/2}\widetilde{X}\widetilde{S}^{1/2} - \sigma\widetilde{\mu}I = 0$, obtained from the one used in this subsection by interchanging the role of $\widetilde{X}$ and $\widetilde{S}$. It is easy to see that the symmetric MT family obtained by applying the Newton method to this equation plus the feasibility equations (in the scaled space) has similar properties to the one studied in this subsection and that it contains the NT direction and the HRVW/KSH/M direction.

We refer to the MT direction corresponding to $P = I$ as the $X^{1/2}SX^{1/2}$-Newton direction. Like the AHO direction, this is a pure Newton direction in the sense that it is a Newton direction for a system of the form $\Phi(X, S) = \sigma\mu I$ where the map $\Phi(\cdot,\cdot)$ is independent of the current iterate or any parameter such as the scaling matrix P. On the basis of the results obtained so far, the $X^{1/2}SX^{1/2}$-Newton direction has clear theoretical advantages over the AHO direction in the sense that polynomial convergence of the semilong-step path-following algorithm is only known for the former direction. So far this is the only pure Newton path-following algorithm which is polynomially convergent and is based on a wide neighborhood of the central path.

The MT* family also has theoretical advantages over the MZ* family based on the results so far. While for the MZ* family the iteration-complexity bound depends on a certain condition number associated with the sequence $\{P^k\}$ of scaling matrices, the corresponding bound for the MT* family does not depend on this sequence.

Finally, Monteiro and Zanjácomo [545] have reported promising computational results for algorithms based on the $X^{1/2}SX^{1/2}$-Newton direction and two other pure Newton directions based on the alternative central path equations:

$$\begin{aligned} S^{1/2}XS^{1/2} &= \sigma\mu I, \\ L_S^T X L_S &= \sigma\mu I, \end{aligned}$$

respectively, where L_S denotes the Cholesky lower triangular factor of S, that is $S = L_S L_S^T$ with L_S lower triangular.

KSH family. Here we analyze the KSH family of search directions which was introduced in Kojima, Shindoh and Hara [439].

In contrast to the MT and MZ families, whose directions are parametrized by a nonsingular scaling matrix P (in addition to a centrality parameter σ), the directions of the KSH family are parametrized by subspaces $H \subset \mathcal{S}^n_\perp \times \mathcal{S}^n_\perp$ having the property that $\dim(H) = n(n-1)/2$ and H is monotone, that is $U \bullet V \geq 0$ for all $(U, V) \in H$. In what follows, we denote this set of subspaces by $\mathcal{H}$. Given $(X, y, S) \in \mathcal{P}_+ \times I\!R^m \times \mathcal{P}_+$ and $H \in \mathcal{H}$, consider the following Newton-type system of linear equations

$$\mathcal{A}\Delta X = b - \mathcal{A}X, \quad \mathcal{A}^*\Delta y + \Delta S = C - \mathcal{A}^*y - S, \qquad (10.4.60a)$$

$$X(\Delta S + \Delta S') + (\Delta X + \Delta X')S = \sigma\mu I - XS, \qquad (10.4.60b)$$
$$(\Delta X, \Delta S) \in \mathcal{S}^n \times \mathcal{S}^n, \quad (\Delta X', \Delta S') \in H. \qquad (10.4.60c)$$

It has been shown in Theorem 4.2 of Kojima et al. [439] that, for every $H \in \mathcal{H}$, system (10.4.60) always has a unique solution $(\Delta X, \Delta y, \Delta S, \Delta X', \Delta S')$ (see also Theorem 2.2 of Monteiro and Tsuchiya [544] for a simpler proof). The component $(\Delta X, \Delta y, \Delta S)$ of this solution is then the KSH search direction at (X, y, S) corresponding to H and σ.

Kojima [428] (see Lemma 2.1 of Monteiro [536]) has shown that the KSH search directions corresponding to the subspaces $H = \mathcal{S}^n_\perp \times \{0\}$ and $H = \{0\} \times \mathcal{S}^n_\perp$ are equal to the MZ directions with $P = S^{1/2}$ and $P = X^{-1/2}$, respectively (that is, the HRVW/KSH/M direction and its dual direction, respectively). Also, it is shown in Kojima, Shida and Shindoh [436] that the NT direction is a member of the KSH family (the corresponding H is not so simple to describe). On the other hand, the AHO direction is not in the KSH family since, as opposed to the directions of the KSH family, the AHO direction does not necessarily exist at every point $(X, y, S) \in \mathcal{P}_+ \times I\!R^m \times \mathcal{P}_+$ (see the discussion before Theorem 3.2 of Todd et al. [782]).

The HRVW/KSH/M direction can be equivalently obtained by first solving the Newton system

$$\mathcal{A}^* \widehat{\Delta y} + \widehat{\Delta S} = C - \mathcal{A}^* y - S, \quad \mathcal{A}\widehat{\Delta X} = b - \mathcal{A}X, \qquad (10.4.61)$$
$$X\widehat{\Delta S} + \widehat{\Delta X}S = \sigma\mu I - XS, \qquad (10.4.62)$$

with the restriction that $(\widehat{\Delta X}, \widehat{\Delta y}, \widehat{\Delta S}) \in I\!R^{n\times n} \times I\!R^m \times \mathcal{S}^n$. It is easily seen that this system has a unique solution and that $(\Delta X, \Delta y, \Delta S) := ((\widehat{\Delta X} + \widehat{\Delta X}^T)/2, \widehat{\Delta y}, \widehat{\Delta S})$ is the direction of the KSH family corresponding to $H = \mathcal{S}^n_\perp \times \{0\}$, and hence the HRVW/KSH/M direction according to the discussion in the previous paragraph.

We are now ready to state a general framework based on the KSH family.

Framework KSH-PD:

Same as Framework MZ-PD but with steps (1) and (2) replaced by

(1') Choose a linear subspace $H_k \in \mathcal{H}$ and a centrality parameter $\sigma_k \in [0, 1]$;

(2') Compute the KSK direction $(\Delta X^k, \Delta y^k, \Delta S^k)$ at (X^k, y^k, S^k) corresponding to $H = H_k$ and $\sigma = \sigma_k$.

End

Polynomial convergence results have been obtained for several variants of the above framework. By taking $H_k = \mathcal{S}^n_\perp \times \{0\}$ or $H_k = \{0\} \times \mathcal{S}^n_\perp$ for all k, that is by selecting the search directions to be either the HRVW/KSH/M direction or its dual direction at every iteration, Kojima et al. [439] establish that the short-step path-following method converges in $\mathcal{O}(\sqrt{n}\,L)$ iterations. (An alternative derivation of this result was also given in Monteiro [536] by viewing these directions as members of the MZ family.) This result was then generalized by Monteiro and Tsuchiya [544], who establish the same iteration-complexity bound for a short-step path-following method and a predictor-corrector method with no restriction on the subspaces H_k's. Kojima et al. [439] have also established the polynomial convergence of a potential-reduction algorithm with no restriction on the subspaces H_k's.

Finally, we observe that it is an open question whether it is possible to establish the polynomial convergence of path-following algorithms based on neighborhoods other than the Frobenius neighborhood $\mathcal{N}_F(\cdot)$ and whose search directions are from the KSH family with no restriction on the H_k's.

11 BUNDLE METHODS TO MINIMIZE THE MAXIMUM EIGENVALUE FUNCTION

Christoph Helmberg, Francois Oustry

11.1 INTRODUCTION

In the last ten years the study of interior point methods dominated algorithmic research in semidefinite programming. Only recently interest in nonsmooth optimization methods revived again, the impetus coming from two different directions. On the one hand alternative possibilities were sought to solve structured large scale semidefinite programs which were not amenable to current interior point codes [338], on the other hand new developments in the second order theory of nonsmooth convex optimization suggested the specialization of these theoretic techniques to semidefinite programming [597, 598]. We present these new methods under the common framework of bundle methods and survey the underlying theory as well as some implementational aspects. In order to illustrate the efficiency and potential of the algorithms we also present numerical results.

In this chapter we consider a class of semidefinite programs with a specific structure:

$$\min \lambda_{\max}(C - \mathcal{A}^T y) + b^T y, \tag{11.1.1}$$

where $C \in \mathcal{S}^n$ and $b \in \mathbb{R}^m$ are constants and $\mathcal{A}^T$ is a linear operator from $\mathbb{R}^m$ to $\mathcal{S}^n$.

Constant trace semidefinite programs. Eigenvalue optimization problems of the form (11.1.1) are equivalent to the dual of semidefinite programs whose primal feasible set has constant trace one and contains a strictly feasible solution, *i.e.*, $\text{Trace}\, X = 1$ for all $X \in \{X \succ 0 : \mathcal{A}X = b\} \neq \emptyset$. In order to explain this equivalence we add the redundant constraint $\text{Trace}\,(X) = \langle I, X\rangle = 1$ to the primal program,

$$
\text{(P)} \quad \begin{array}{ll} \text{maximize} & \langle C, X\rangle \\ \text{subject to} & \langle I, X\rangle = 1 \\ & \mathcal{A}X = b \\ & X \succeq 0 \end{array} \qquad \text{(D)} \quad \begin{array}{ll} \text{minimize} & y_0 + b^T y \\ \text{subject to} & Z = y_0 I + \mathcal{A}^T y - C \\ & Z \succeq 0. \end{array}
$$

By assumption, the primal is strictly feasible. The dual has a strictly feasible solution (even without $y_0 I$, since I is in the range of $\mathcal{A}^T$ by assumption). Therefore the duality gap is zero and the optimum value is attained for both problems. Let $\bar{X}$ and $\bar{Z}$ with appropriate $(\bar{y}_0, \bar{y})$ be a pair of primal and dual optimal solutions. Then the optimality condition implies $\bar{X}\bar{Z} = 0$. Thus, $\bar{X}$ and $\bar{Z}$ are simultaneously diagonalizable, *i.e.*, there is an orthonormal matrix $P \in \mathbb{R}^{n\times n}$ such that $\bar{X} = P\Lambda_{\bar{X}}P^T$ and $\bar{Z} = P\Lambda_{\bar{Z}}P^T$ where $\Lambda_{\bar{X}}$, $\Lambda_{\bar{Z}}$ are diagonal matrices with nonnegative entries. It follows that $\Lambda_{\bar{X}}\Lambda_{\bar{Z}} = 0$. Since $\bar{X} \neq 0$ this implies

$$
0 = \lambda_{\min}(\Lambda_{\bar{Z}}) = \lambda_{\min}(\bar{y}_0 I + \mathcal{A}^T\bar{y} - C) = \bar{y}_0 + \lambda_{\min}(\mathcal{A}^T\bar{y} - C).
$$

Thus

$$
\bar{y}_0 = -\lambda_{\min}(\mathcal{A}^T\bar{y} - C) = \lambda_{\max}(C - \mathcal{A}^T\bar{y})
$$

and therefore (D) is equivalent to

$$
\text{minimize}\, \lambda_{\max}(C - \mathcal{A}^T y) + b^T y.
$$

Observe, that any nontrivial primal program with bounded feasible set could be brought into form (P) by projection and scaling. However, in connection with bundle methods these operations are useful only if they also preserve structure, which is not the case in general.

Graph problems. Semidefinite programs satisfying this constant trace property arise, *e.g.*, in some combinatorial applications, in particular in semidefinite relaxations of quadratic 0-1 programming and graph coloring/graph partitioning. These relaxations have been around for some time [496, 304, 733, 184, 634], they became widely known when Goemans and Williamson and subsequently others proved their good approximation properties [285, 393, 256]. §§ 11.9.1 presents numerical results for the semidefinite relaxation of max-cut for graphs with up to 10000 nodes and the Lovász ϑ-function for graphs with up to 10000 edges.

Control theory. Another important field of application where (11.1.1) appears as such is the problem of analyzing the stability of Linear Differential Inclusions (LDIs) using quadratic Lyapunov functions. Numerous references can be found in [137, Chapter 6, Notes and References]. In §§ 11.9.2 and §§ 11.9.3, we present numerical results for various polytopic LDIs.

Outline. In §11.2 we review some basic properties and formulations of the maximum eigenvalue and its subdifferential. The algorithmic framework common to all our algorithms is explained in §11.3, it is a bundle framework open to various realizations of the cutting plane model as well as the regularization term. The proximal bundle method of §11.4 uses a polyhedral cutting plane model and is a general method for nonsmooth convex optimization that has repeatedly been used for eigenvalue optimization problems. The spectral bundle method presented in §11.5 specializes the proximal bundle method to eigenvalue optimization problems in employing a semidefinite and thus nonpolyhedral cutting plane model; it is particularly well suited for large scale problems because of its aggregation possibilities. For reasonably sized matrices but huge m it may be worth to consider the mixed polyhedral-semidefinite method of §11.6; it determines ϵ-subgradients of high quality by computing the projection onto the δ_ϵ-subdifferential of the current matrix explicitly and uses these to form a polyhedral cutting plane model. This method is also well suited for combination with the second order method explained in §11.7. Finally, in §11.8 we discuss various topics of implementational importance such as large scale eigenvalue computation, setting up the semidefinite cutting plane model, and computing the projection explicitly. We also present numerical results for the spectral bundle method, the mixed polyhedral-semidefinite method, and the second order method.

11.2 THE MAXIMUM EIGENVALUE FUNCTION

Convexity is the crucial property enjoyed by the maximum eigenvalue function: $\lambda_{\max}$ is a max-function over a compact set. Indeed, from *Rayleigh's variational formulation* we have, for all $Z \in \mathcal{S}^n$,

$$\lambda_{\max}(Z) = \max_{p \in \mathbb{R}^m,\, \|p\|=1} p^T Z p .$$

Setting,

$$\mathcal{W} := \{W \in \mathcal{S}^n_+ : \text{Trace}\, W = 1\}\,, \tag{11.2.2}$$

we obtain, what can be referred to as Ky Fan's variational formulation of $\lambda_{\max}$ [604]:

$$\lambda_{\max}(Z) = \max_{W \in \mathcal{W}} \langle Z, W \rangle\ . \tag{11.2.3}$$

In other words, $\lambda_{\max}$ is the support function of $\mathcal{W}$. As we will see in § 3, models of this convex function will be easily obtained by taking inner approximations of $\mathcal{W}$.

From (11.2.3), we can derive all first-order properties of $\lambda_{\max}$. In particular, the subdifferential of $\lambda_{\max}$ will be the face of $\mathcal{W}$ exposed by Z. This face can be explicitly described in terms of eigenvectors of Z associated with $\lambda_{\max}$. Explicit descriptions of ∂f were first given in [241, 600]. The terminology of support functions and exposed faces is emphasized in [604, 344] for the sum of the k

largest eigenvalues function. Various formulations can be found.

$$\begin{aligned}\partial\lambda_{\max}(Z) &= \{W \in \mathcal{W} : \langle W, Z\rangle = \lambda_{\max}(Z)\}\\ &= \operatorname{Conv}\{pp^T : p \text{ is a normalized eigenvector to } \lambda_{\max}(Z)\}\\ &= \{PVP^T : V \in \mathcal{S}^r_+, \operatorname{Trace} V = 1\},\end{aligned} \tag{11.2.4}$$

where $P \in I\!R^{n\times r}$ with $P^T P = I_r$ is an orthonormal matrix whose columns span the eigenspace of $\lambda_{\max}(Z)$. From (11.2.4), the expression of the subdifferential of the composite convex function

$$I\!R^m \ni y \mapsto f(y) := \lambda_{\max}(C - \mathcal{A}^T y) + b^T y\,,$$

can be derived easily. Applying a classical chain rule [346, Theorem XI.3.2.1], we obtain indeed

$$\partial f(y) = \{b - \mathcal{A}W : W \in \partial\lambda_{\max}(C - \mathcal{A}^T y)\}\,. \tag{11.2.5}$$

Another first-order object of importance for algorithmical perspectives is the ϵ-subdifferential of a convex function [346, Definition XI.1.1.1]: for $\epsilon \geq 0$ and $y \in I\!R^m$,

$$\partial_\epsilon f(y) := \{s \in I\!R^m : f(z) - f(y) \geq s^T(z - y),\ \forall z \in I\!R^m\}\,.$$

Using formulation (11.2.3), it is shown in [344] that for $\epsilon \geq 0$ and $Z \in \mathcal{S}^n$

$$\partial_\epsilon\lambda_{\max}(Z) = \mathcal{W} \cap \{W \in \mathcal{S}^n : \langle W, Z\rangle \geq \lambda_{\max}(Z) - \epsilon\}\,. \tag{11.2.6}$$

Applying again a chain rule [346, Theorem XI.3.2.1], we deduce the expression of the ϵ-subdifferential of our composite function f at $y \in I\!R^m$:

$$\partial_\epsilon f(y) = \{b - \mathcal{A}W : W \in \partial_\epsilon\lambda_{\max}(C - \mathcal{A}^T y)\}\,. \tag{11.2.7}$$

11.3 GENERAL SCHEME

All algorithms of this chapter are various realizations of the following scheme. At iteration k we denote by y_k the current iterate and construct, either from information collected in previous iterations or from local information,

- a *model* φ_k of our convex function f:

$$\begin{aligned}\varphi_k(y) &:= \max_{W\in\mathcal{W}_k} \langle W, C - \mathcal{A}^T y + (b^T y)\, I_n\rangle\,,\\ &= (b^T y) + \max_{W\in\mathcal{W}_k} \langle W, C - \mathcal{A}^T y\rangle\,,\end{aligned} \tag{11.3.8}$$

  where the (convex) set $\mathcal{W}_k$ is an approximation of $\mathcal{W}$ satisfying

$$\{p_k p_k^T\} \subset \mathcal{W}_k \subset \mathcal{W}\,, \tag{11.3.9}$$

and p_k is a normalized 'approximate' eigenvector to $\lambda_{\max}(C - \mathcal{A}^T y_k)$: $p_k^T(C - \mathcal{A}^T y_k)p_k \geq \lambda_{\max}(C - \mathcal{A}^T y_k) - \epsilon$ for some $\epsilon > 0$, *i.e.*, according to (11.2.6), $p_k p_k^T \in \partial_\epsilon \lambda_{\max}(C - \mathcal{A}^T y_k)$. This, together with inclusions (11.3.9), ensure that the model satisfies

$$\begin{aligned} \varphi_k(y) &\leq f(y) \quad \text{for all } y \in I\!R^m , \\ \text{and } \varphi_k(y_k) &\geq f(y_k) - \epsilon . \end{aligned} \tag{11.3.10}$$

- a *norming* on $I\!R^m$ of the form $\|\cdot\|_k := (\cdot)^T M_k(\cdot)$, where $M_k \succ 0$ is a positive definite matrix.

A new candidate y_{k+1} is determined as the minimizer of the model φ_k augmented with a penalty term to control the distance, induced by this norm, of the new candidate to a so called *stability center* x_k,

$$y_{k+1} = \operatorname*{argmin}_{y \in I\!R^m} \varphi_k(y) + \frac{\mu_k}{2} \|y - x_k\|_k^2$$

The *weight* $\mu_k > 0$ may be regarded as a Lagrange multiplier for the trust region constraint $\|y - x_k\|_k^2 \leq R^2$. The x_k form a subsequence of selected iterates of y_k. Whenever a candidate y_{k+1} exhibits 'sufficient decrease' , x_{k+1} will be set to y_{k+1} (a *serious step*), otherwise the old stability center is kept, $x_{k+1} = x_k$ (a *null step*).

Algorithm 11.3.1 (General Bundle Method)
Input: *An initial point* $x_0 \in I\!R^m$, *a* $\delta > 0$ *for termination, an improvement parameter* $m_L \in (0, \frac{1}{2})$, *a weight* $\mu_0 > 0$, *a minimal weight* $\mu_{\min}$, *a positive definite matrix* M_0.

1. *(Initialization)* $k = 0$, *compute* $f(x_0)$ *and some first-order information, at least* p_0 *a normalized eigenvector to* $\lambda_{\max}(C - \mathcal{A}^T x_0)$, *set* $\mathcal{W}_0$ *satisfying* (11.3.9).

2. *(Candidate finding) Compute* y_{k+1} *solution of*

$$\min_{y \in I\!R^m} \varphi_k(y) + \frac{\mu_k}{2} \|y - x_k\|_k^2 , \tag{11.3.11}$$

   where φ_k *is defined by* (11.3.8).

3. *(Evaluation) Determine* $f(y_{k+1})$ *and some first order information, at least* p_{k+1} *a normalized (*ϵ*-approximate) eigenvector to* $\lambda_{\max}(C - \mathcal{A}^T y_{k+1})$.

4. *(Model Updating) Choose a set* $\mathcal{W}_{k+1}$ *satisfying* (11.3.9).

5. *(Termination) If* $(f(x_k) - \varphi_k(y_{k+1})) \leq \delta(|f(x_k)| + 1)$ *then* **stop**.

6. *(Serious step) If*

$$f(x_k) - f(y_{k+1}) \geq m_L(f(x_k) - \varphi_k(y_{k+1})) \tag{11.3.12}$$

then set $x_{k+1} = y_{k+1}$, *select* $\mu_{k+1} \in [\mu_{\min}, \mu_k]$, *choose the metric* M_{k+1} *and continue with Step 8. Otherwise continue with Step 7.*

7. *(Null step) Set* $x_{k+1} = x_k$, *select* $\mu_{k+1} \in [\mu_k, 10\mu_k]$.

8. *Increase* k *by 1 and go to Step 2.*

All the first-order algorithms described in this chapter will be obtained by solving, for various choices of $\mathcal{W}_k$, the dual of (11.3.11).

Theorem 11.3.1 *For the general model* φ_k *of* (11.3.8), *the solution* y_{k+1} *of subproblem* (11.3.11) *is given by*

$$y_{k+1} = x_k - \frac{1}{\mu_k} M_k^*(b - \mathcal{A}\overline{W}_k), \tag{11.3.13}$$

where $M_k^* := M_k^{-1}$ *and* $\overline{W}_k$ *is the solution of the dual problem*

$$\begin{array}{ll} \text{minimize} & \frac{1}{2\mu_k}\|b - \mathcal{A}W\|_{k,*}^2 - (b - \mathcal{A}W)^T x_k - \langle C, W\rangle \\ \text{subject to} & W \in \mathcal{W}_k\,, \end{array} \tag{11.3.14}$$

and $\|\cdot\|_{k,*} := (\cdot)^T M_k^*(\cdot)$ *is the dual norm at iteration* k.

Proof.
Since $\mathcal{W}_k$ is convex and bounded, it follows from Corollary 37.3.2 of [679], that

$$\min_{y \in I\!R^m} \varphi_k(y) + \frac{\mu_k}{2}\|y - x_k\|_k^2 = \max_{W \in \mathcal{W}_k} \min_{y \in I\!R^m} b^T y + \langle W, C - \mathcal{A}^T y\rangle + \frac{\mu_k}{2}\|y - x_k\|_k^2. \tag{11.3.15}$$

The inner minimization with respect to y is unconstrained, first order optimality conditions require

$$y = x_k - \frac{1}{\mu_k} M_k^*(b - \mathcal{A}W).$$

Substituting this into the right hand side of (11.3.15) gives rise to the dual problem. ■

In practice we do not need to store information in the space of matrices: we rather manage at iteration k a *bundle* of augmented vectors

$$B_k := \left\{ \begin{bmatrix} \langle C, W\rangle \\ b - \mathcal{A}W \end{bmatrix} : W \in \mathcal{W}_k \right\} \subset I\!R^{m+1}. \tag{11.3.16}$$

and (11.3.14) takes the form

$$\begin{array}{ll} \text{minimize} & \frac{1}{2\mu_k}\|g\|_{k,*}^2 - \begin{bmatrix} 1 \\ x_k \end{bmatrix}^T \begin{bmatrix} e \\ g \end{bmatrix} \\ \text{subject to} & \begin{bmatrix} e \\ g \end{bmatrix} \in B_k\,. \end{array} \tag{11.3.17}$$

An *invariant feature* of all these bundle methods is that the solution of (11.3.17) provides us with an ϵ_k-subgradient of f at x_k.

Proposition 11.3.1 *Let* $\begin{bmatrix} \bar{e}_k \\ \bar{g}_k \end{bmatrix}$ *be a solution of* (11.3.17). *Then* $\bar{g}_k$ *is an* ϵ_k-*subgradient of* f *at* x_k *where* ϵ_k *is given by*

$$\epsilon_k := f(x_k) - \bar{e}_k - \bar{g}_k^T x_k \geq 0 . \tag{11.3.18}$$

Furthermore setting the decreasing of the model

$$\delta_k := f(x_k) - \varphi_k(y_{k+1}) - \frac{\mu_k}{2} ||y_{k+1} - x_k||_k^2 , \tag{11.3.19}$$

we have the relation

$$\delta_k = \epsilon_k + \frac{1}{2\mu_k} ||\bar{g}_k||_{k,*}^2 . \tag{11.3.20}$$

Proof.
According to (11.3.16), $\begin{bmatrix} \bar{e}_k \\ \bar{g}_k \end{bmatrix} = \begin{bmatrix} \langle C, \overline{W}_k \rangle \\ b - \mathcal{A}\overline{W}_k \end{bmatrix}$ where $\overline{W}_k$ is a solution of (11.3.14). On the other hand, (11.2.6) implies that $\overline{W}_k \in \partial_{\epsilon_k} \lambda_{\max}(C - \mathcal{A}^T x_k)$ for

$$\begin{aligned} 0 \leq \epsilon_k &= \lambda_{\max}(C - \mathcal{A}^T x_k) - \langle \overline{W}_k, C - \mathcal{A}^T x_k \rangle \\ &= (\lambda_{\max}(C - \mathcal{A}^T x_k) + b^T x_k) - \langle \overline{W}_k, C \rangle - (b - \mathcal{A}\overline{W}_k)^T x_k \\ &= f(x_k) - \bar{e}_k - \bar{g}_k^T x_k ; \end{aligned}$$

chain rule (11.2.7) gives us $\bar{g}_k \in \partial_{\epsilon_k} f(x_k)$.
Furthermore, starting from (11.3.19) and (11.3.15), we have

$$\begin{aligned} \delta_k &= f(x_k) - \bar{e}_k - \bar{g}_k^T y_{k+1} - \tfrac{\mu_k}{2} ||y_{k+1} - x_k||_k^2 \\ &= f(x_k) - \bar{e}_k - \bar{g}_k^T x_k + \tfrac{1}{2\mu_k} ||\bar{g}_k||_{k,*}^2 && \text{[using (11.3.13)]} \\ &= \epsilon_k + \tfrac{1}{2\mu_k} ||\bar{g}_k||_{k,*}^2 && \text{[from (11.3.18)]} \end{aligned} .$$

This completes the proof. ■

In what follows we denote by **Algorithm 11.3.1*** the 'dual version' of Algorithm 11.3.1, where we solve the dual subproblem (11.3.17) instead of (11.3.11) and y_{k+1} is updated by (11.3.13). Note that in this dual variant, we choose a dual metric M_k^* instead of choosing M_k and computing its inverse.

11.4 THE PROXIMAL BUNDLE METHOD

The proximal bundle method, also referred to as *cutting planes with stabilization by penalty*, is a classical approach for minimizing general nonsmooth convex functions f where f is given by a routine that returns for any point

$y \in I\!R^m$ the function value and some arbitrary subgradient of the function at this point. Detailed descriptions of such algorithms and their implementation can be found in [414, 700, 346].

For the eigenvalue optimization problem the underlying model in iteration k is the cutting plane model

$$\check{f}_k(y) := \max_{i=1,\ldots,k} f(y_i) + g_i^T(y - y_i), \tag{11.4.21}$$

where the subgradients $g_i \in \partial f(y_i)$ are of the form $g_i = b - \mathcal{A} p_i p_i^T$ and p_i is a normalized eigenvector to $\lambda_{\max}(C - \mathcal{A}^T y_i)$ for $i = 1, \ldots, k$. To formulate this algorithm within the framework depicted in the previous section set

$$\begin{aligned} \mathcal{W}_k &:= \operatorname{Conv}\{p_i p_i^T : i = 1, \ldots, k\} \\ \epsilon &:= 0 \\ M_k^* &:= I. \end{aligned}$$

Indeed, model (11.3.8) coincides with the cutting plane model:

$$\begin{aligned} \varphi_k(y) &= (b^T y) + \max_{W \in \mathcal{W}_k} \langle W, C - \mathcal{A}^T y \rangle \\ &= \max_{i=1,\ldots,k} \langle p_i p_i^T, C - \mathcal{A}^T y \rangle + b^T y \\ &= \max_{i=1,\ldots,k} \langle p_i p_i^T, C - \mathcal{A}^T y_i \rangle + b^T y_i + (b - \mathcal{A} p_i p_i^T)^T (y - y_i) \\ &= \check{f}_k(y). \end{aligned}$$

Therefore, in our context, the classical proximal bundle method considers only polyhedral approximations of $\mathcal{W}$. In practice it is usually impossible to include the subgradients of all previous iterates in the model. Without endangering convergence, the number of subgradients can be controlled by introducing an aggregate subgradient [413], [346, § XV.3].

Here, the aggregate subgradient of iteration k corresponds to $\overline{W}_k$ of Theorem 11.3.1. The decisive two conditions for the convergence analysis of the proximal bundle method are

$$\overline{W}_k \in \mathcal{W}_{k+1} \text{ and } p_{k+1} p_{k+1}^T \in \mathcal{W}_{k+1} \text{ with } p_{k+1} \in \partial f(y_{k+1}). \tag{11.4.22}$$

These conditions ensure that after a null step the value of (11.3.11) has to increase. Indeed, if we assume $x_{k+1} = x_k$, then for all $y \in I\!R^m$

$$\begin{aligned} &\varphi_{k+1}(y) + \frac{\mu_{k+1}}{2} \|y - x_{k+1}\|^2 = \\ &= \max_{W \in \mathcal{W}_{k+1}} \langle W, C - \mathcal{A}^T y \rangle + b^T y + \frac{\mu_{k+1}}{2} \|y - x_k\|^2 \\ &\geq \langle \overline{W}_k, C - \mathcal{A}^T y \rangle + b^T y + \frac{\mu_k}{2} \|y - x_k\|^2 \\ &\qquad\qquad [\text{since } \overline{W}_k \in \mathcal{W}_{k+1} \text{ and } \mu_k \leq \mu_{k+1}] \\ &= \langle \overline{W}_k, C \rangle + \langle b - \mathcal{A}\overline{W}_k, y \rangle + \frac{\mu_k}{2} \left\| y - y_{k+1} - \frac{1}{\mu_k}(b - \mathcal{A}\overline{W}_k) \right\|^2 \end{aligned}$$

[using (11.3.13)]

$$= \varphi_k(y_{k+1}) + \frac{\mu_k}{2}\|y_{k+1} - x_k\|^2 + \frac{\mu_k}{2}\|y - y_{k+1}\|^2.$$

[using (11.3.15)]

On the other hand

$$\begin{aligned} \varphi_{k+1}(y_{k+1}) &= f(y_{k+1}) && [\text{since } p_{k+1}p_{k+1}^T \in \mathcal{W}_{k+1}] \\ &> (1-m_L)f(x_k) + m_L\varphi_k(y_{k+1}) && [\text{since (11.3.12) is violated}], \end{aligned}$$

so the value of (11.3.11) has to increase after a null step. This also means that δ_{k+1} of (11.3.19) has to decrease after a null step.

These considerations may serve to illustrate the idea of the algorithm, but for a proof of convergence much more work is needed. We refrain from giving a detailed proof and cite the main result only.

Theorem 11.4.1 (Kiwiel[414]) *Let x^k denote the sequence of points generated by Algorithm 11.3.1 with $\delta = 0$, $M_k = I_m$, and the update step satisfying (11.4.22). Either the x^k converge to a minimizer of f or there is no minimizer and $\|x^k\| \to \infty$. In both cases $f(x^k) \downarrow \inf f$.*

The choice of the weight μ has a strong influence on the performance of the algorithm. So far the dynamic updating strategies suggested in [414] and [700] are considered to be the most robust rules. [700] includes some numerical results for eigenvalue optimization problems of moderate sizes.

It should be clear that for practical implementations the minimal requirements of (11.4.22) are too extreme, they would yield a cutting plane model of poor quality. In order to obtain a reasonable cutting plane model one tries to keep all "important" cutting planes and aggregates the least important cutting planes only, where "important" is usually associated with the size of the dual variables.

11.5 THE SPECTRAL BUNDLE METHOD

The spectral bundle method of [338] specializes the proximal bundle method to eigenvalue optimization problems in that it uses a semidefinite cutting plane model exploiting the semidefinite structure of the convex set $\mathcal{W}$.

Inspired by representation (11.2.4) of the subdifferential we collect the eigenspace information of the vectors p_i in an orthonormal matrix $P_k \in \mathbb{R}^{n \times r_k}$, $P_k^T P_k = I_{r_k}$, with the property that the columns of P_k span the p_i, *i.e.*,

$$\text{for } i \in \{1, \ldots k\}\ \exists \pi \in \mathbb{R}^{r_k} \text{ with } \|\pi_i\| = 1 \text{ so that } p_i = P_k \pi_i \qquad (11.5.23)$$

and thus

$$p_i p_i^T = P_k \pi_i \pi_i^T P_k^T. \qquad (11.5.24)$$

This P_k is used to define a nonpolyhedral cutting plane model via the set

$$\mathcal{W}_k = \text{Conv}\left\{P_k V P_k^T : V \in \mathcal{S}_+^{r_k}, \text{Trace}(V) = 1\right\}. \qquad (11.5.25)$$

The corresponding model function is best represented as

$$\varphi_k(y) = \lambda_{\max}(P_k^T(C - \mathcal{A}^T y)P_k) + b^T y,$$

Thus, the maximal eigenvalue is taken with respect to the projected matrix, or alternatively, the maximal eigenvalue is approximated by means of vectors in the subspace spanned by P_k.

Since $\pi_i\pi_i^T \in \mathcal{S}_+^{r_k}$ and Trace $(\pi_i\pi_i^T) = 1$, (11.5.24) implies that

$$\text{Conv}\,\{p_i p_i^T : i = 1, \ldots, k\} \subset \mathcal{W}_k,$$

the nonpolyhedral cutting plane model of the spectral bundle method improves on the polyhedral cutting plane model of the proximal bundle method without requiring more information. Why is this possible? For a general convex function, a subgradient $g \in \partial f(x)$ does not allow to construct a hyperplane supporting f without evaluating the function in x. For eigenvalue optimization problems (11.1.1), however, the supporting hyperplane corresponding to a subgradient generated by some $W \in \mathcal{W}$ is easily obtained via (11.2.3) as in (11.3.16). In this case, the purpose of the bundle method is not so much the construction of cutting planes, but rather the identification of a good subspace for approximating the problem.

It is easy to check that maintaining (11.5.23) and (11.5.25) ensures (11.4.22), therefore this update ensures convergence by Theorem 11.4.1. Unfortunately, each iteration of the algorithm will in general result in the addition of a new column to P_k, since (11.5.23) has to be satisfied for the new eigenvector. Eventually P_k will span the full space ($r_k = n$) and there would be no advantage in comparison to solving the problem directly with interior point methods.

As in the proximal bundle method, we can use aggregation to keep the number r_k of columns of P_k small. To this end we introduce an aggregation matrix $\widetilde{W}_k \in \mathcal{S}_+^n$ with Trace $(\widetilde{W}_k) = 1$,

$$\mathcal{W}_k = \text{Conv}\left\{P_k V P_k^T + \alpha \widetilde{W}_k : V \in \mathcal{S}_+^{r_k}, \alpha \geq 0, \text{Trace}\,(V) + \alpha = 1\right\}. \tag{11.5.26}$$

It is important to note that P_k itself offers ample opportunity for aggregation, as well. In order to illustrate these possibilities we will discuss three aggregation schemes: An update rule for $r_k = 1$ (P_k will hold the new eigenvector only), an implementable update with aggregation in both $\widetilde{W}_k$ and P_k, and finally a version without $\widetilde{W}_k$.

Consider (11.3.14) for $\mathcal{W}_k$ as in (11.5.26),

$$\begin{array}{ll} \text{minimize} & \frac{1}{2\mu_k}\|b - \mathcal{A}W\|^2 - (b - \mathcal{A}W)^T x_k - \langle C, W\rangle \\ \text{subject to} & W = P_k V P_k^T + \alpha \widetilde{W}_k \\ & \text{Trace}\,(V) + \alpha = 1 \\ & V \in \mathcal{S}_+^{r_k}, \alpha \geq 0. \end{array} \tag{11.5.27}$$

Let $\overline{W}_k = P_k V_k P_k^T + \alpha_k \widetilde{W}_k$ denote an optimal solution and y_{k+1} the iterate determined by (11.3.13). Then, in the extreme case, we set

$$\widetilde{W}_{k+1} = \overline{W}_k \quad \text{and} \quad P_{k+1} = p_{k+1}, \tag{11.5.28}$$

where p_{k+1} is an eigenvector to $\lambda_{\max}(C - \mathcal{A}^T y_{k+1})$. This choice satisfies (11.4.22) and convergence of the method is guaranteed even for $r_k = 1$.

Proposition 11.5.1 *If* **Algorithm 11.3.1*** *is run for $\delta = 0$, $M_k^* = I_m$, and $\mathcal{W}_k$ of (11.5.26) with update step (11.5.28) then (11.4.22) is satisfied in all iterations and Theorem 11.4.1 applies.*

In practical algorithms the number of columns provided in P_{k+1} will be selected so that computation time for solving (11.5.27) is in balance with function evaluation time. The available columns of P_{k+1} should span the subspace (of the space spanned by P_k) that is most important for the optimization process. Some information in this respect is hidden in V_k. Let $V_k = Q\Lambda Q^T$ be an eigenvalue decomposition of V_k with $Q^T Q = I_{r_k}$ and $\Lambda \in \mathcal{S}^{r_k}$ the diagonal matrix of eigenvalues. Then, by

$$\overline{W}_k = P_k Q \Lambda Q^T P_k^T + \alpha_k \widetilde{W}_k,$$

the columns of $P_k Q$ corresponding to the largest eigenvalues in Λ carry the most important information of $\overline{W}_k$. In order to extract this information we partition the columns of Q into two parts, Q_1 and Q_2, and correspondingly Λ into Λ_1 and Λ_2. Q_1 contains the collection of eigenvectors belonging to the largest eigenvalues Λ_1 of V_k. These are the vectors that we want to keep for P^{k+1}. The columns of Q_2 carry the eigenvectors of the remaining eigenvalues Λ_2 of V_k. These will be aggregated in $\widetilde{W}_{k+1}$. For this partition the update rule is

$$\widetilde{W}_{k+1} = \frac{P_k Q_2 \Lambda_2 Q_2^T P_k^T + \alpha_k \widetilde{W}_k}{\operatorname{Trace} \Lambda_2 + \alpha_k} \quad \text{and} \quad P_{k+1} = \operatorname{orth}\{P_k Q_1, p_{k+1}\}, \tag{11.5.29}$$

where orth refers to taking an orthonormal basis of the columns of $P_k Q_1$ and p_{k+1}. It is not difficult to show that this choice ensures (11.4.22).

Proposition 11.5.2 *If* **Algorithm 11.3.1*** *is run for $\delta = 0$, $M_k^* = I_m$, and $\mathcal{W}_k$ of (11.5.26) with update step (11.5.29) then (11.4.22) is satisfied in all iterations and Theorem 11.4.1 applies.*

The aggregation hidden in the updates of P adds significantly to the strength of the spectral bundle method. The effect of the aggregation in P is best visible if r is chosen large enough so that the usual aggregate $\widetilde{W}$ is not needed at all. To determine an upper bound for such an r, observe that for all optimal solutions $\overline{W}$ of (11.5.27) (assume $\widetilde{W}_k = 0$) the vector $\bar{b} = \mathcal{A}\overline{W}$ is the same because of the uniqueness of the solution to (11.3.11). Thus, any $\overline{W}$ arises from a solution $\overline{V}$ of the semidefinite program

$$\begin{array}{ll} \text{minimize} & -\langle C - \mathcal{A}^T x_k, P_k V P_k^T \rangle \\ \text{subject to} & \mathcal{A}(P_k V P_k^T) = \bar{b} \\ & \operatorname{Trace}(V) = 1 \\ & V \succeq 0. \end{array}$$

This program has $m+1$ constraints, therefore it has a solution $\overline{V}$ with rank $\bar{r}$ bounded by $\binom{\bar{r}}{2} \leq m+1$ and, of course, by n (see [623]). Choose r to be the minimum of n and the largest $\bar{r}$ satisfying this bound plus one (for the new eigenvector). Then, by computing a rank minimal solution V_k, all information about $\overline{W}_k$ can be stored in $P \in I\!R^{n\times r}$ and there is no need for the aggregate $\widetilde{W}$. Note, that in general the solution $\overline{W}$ will be unique, hence it does not make sense to care about minimal rank solutions in implementations.

We will see in §11.8.3 that the information which is actually stored in each iteration is not $\widetilde{W}_k$ itself, but

- the aggregate augmented subgradient:

$$\begin{bmatrix} \bar{e}_k \\ \bar{g}_k \end{bmatrix} := \begin{bmatrix} \left\langle C, \widetilde{W}_k \right\rangle \\ b - \mathcal{A}\widetilde{W}_k \end{bmatrix} \in I\!R^{m+1},$$

- and an $n \times r_k$ normal matrix P_k.

Information from the past can be found in $\begin{bmatrix} \bar{e}_k \\ \bar{g}_k \end{bmatrix}$ as well as in P_k. In practice we may store more than one vector from the past to reduce the size of the semidefinite part in (11.3.17) or add, in each iteration, several vectors to P_{k+1} to enhance the local quality of the model. The mixed polyhedral-semidefinite method presented in § 11.6 can be seen as a variant of the spectral bundle method where the semidefinite part is reduced to a minimal size and contains only local information whereas global information is recovered through a polyhedral model.

11.6 THE MIXED POLYHEDRAL-SEMIDEFINITE METHOD

We propose here to modify the update of $\mathcal{W}_k$, *i.e.*, the *oracle* that computes the first-order information, by introducing a parameter $\epsilon > 0$. This parameter is going to play a bivalent role:

(*i*) ϵ will control the final precision we want on the solution.

(*ii*) ϵ is a *viscosity parameter* which, at the point y_k, *locally stabilizes* the first-order (see this section) and second-order (see the following section) objects.

Going along the lines of [178], we introduce the following enlargement of the subdifferential of $\lambda_{\max}$ at $Z_k := (C - \mathcal{A}^T y_k) \in \mathcal{S}^n$:

$$\delta_\epsilon \lambda_{\max}(Z_k) = \{P_k V P_k^T : V \in \mathcal{S}_+^{r_k}, \operatorname{Trace} V = 1\}, \qquad (11.6.30)$$

where r_k is the number of eigenvalues in $]\lambda_{\max}(Z_k) - \epsilon, \lambda_{\max}(Z_k)]$ and P_k is an $n \times r_k$ matrix whose columns form an orthonormal basis of the space spanned by the r_k first eigenvectors. By analogy with (11.2.5) and (11.2.7), a natural enlargement of $\partial f(y_k)$ is then

$$\delta_\epsilon f(y_k) := \{b - \mathcal{A}W : W \in \delta_\epsilon \lambda_{\max}(C - \mathcal{A}^T y_k)\}. \qquad (11.6.31)$$

In [598], it is proved that

$$\partial f(y_k) \subset \delta_\epsilon f(y_k) \subset \partial_\epsilon f(y_k),$$

and a quantitative analysis of the (Hausdorff) distance between the convex (compact) sets $\delta_\epsilon f(y_k)$ and $\partial_\epsilon f(y_k)$ is proposed. One consequence of this analysis is to provide us with a 'good' ϵ-subgradient at y_k: $g_k = b - \mathcal{A} W_k$, the unique solution of the local quadratic-semidefinite program

$$\begin{array}{ll} \text{minimize} & \|g\|_{k,*}^2 \\ \text{subject to} & g \in \delta_\epsilon f(y_k). \end{array} \tag{11.6.32}$$

From here we propose to partition the information managed by the bundle method into a *local semidefinite model* and a *global polyhedral model.*

The local semidefinite model. At the local stage a *rich oracle* evaluates $f(y_k)$ and produces a pair (g_k, W_k) such that

$$\begin{cases} W_k \in \delta_\epsilon \lambda_{\max}(C - \mathcal{A}^T y_k) \\ g_k = b - \mathcal{A} W_k \text{ is the unique solution of (11.6.32)} \end{cases} \tag{11.6.33}$$

We explain in §§ 11.8.4 how to implement efficiently this rich oracle.

The global polyhedral model. In this paragraph we show that, even if the rich oracle (11.6.33) produces ϵ-subgradients instead of exact subgradients, the management of the global polyhedral cutting plane model can be done exactly as for the proximal bundle methods of § 11.4. The main difference will come in the convergence analysis: we will not be allowed to let $\delta \to 0$ as in Theorem 11.4.1; we will obtain finite convergence together with approximate optimality conditions.

We consider approximations of $\mathcal{W}$ of the form

$$\mathcal{W}_k := \text{Conv}\{W_1, \ldots, W_k\},$$

where the W_i's are ϵ-subgradients produced by the rich oracle (11.6.33) at the respective points y_i's. The model of (11.3.8) is the following approximate cutting plane function

$$\begin{array}{rcl} \varphi_k(y) & := & \max_{i=1,\ldots,k} \; b^T y + \langle W_i, C - \mathcal{A}^T y \rangle \\ & = & \max_{i=1,\ldots,k} \; e_i + g_i^T y, \end{array}$$

where the augmented vectors $\begin{bmatrix} e_i \\ g_i \end{bmatrix}$'s are the vertices of the set B_k (11.3.16):

$$\begin{bmatrix} e_i \\ g_i \end{bmatrix} := \begin{bmatrix} \langle W_i, C \rangle \\ b - \mathcal{A} W_i \end{bmatrix}, \quad i = 1, \ldots, k.$$

The (dual) subproblem (11.3.17) , as for the classical proximal bundle method, is a quadratic program over the simplex
$\Delta_k := \{\alpha \in \mathbb{R}^k : \sum_{i=1}^k \alpha_i = 1,\ \alpha_i \geq 0,\ i = 1,\dots,k\}$:

$$\begin{array}{ll} \text{minimize} & \frac{1}{2\mu_k} \left\| \sum_{l=1}^k \alpha_i g_i \right\|_{k,*}^2 - \sum_{l=1}^k \alpha_i e_i - x_k^T (\sum_{l=1}^k \alpha_i g_i) \\ \text{subject to} & \alpha \in \Delta_k . \end{array} \tag{11.6.34}$$

Let $[\bar{\alpha}_1, \dots, \bar{\alpha}_k]^T$ denote an optimal solution of (11.6.34) and set

$$\begin{bmatrix} \bar{e}_k \\ \bar{g}_k \end{bmatrix} := \sum_{i=1}^k \bar{\alpha}_i \begin{bmatrix} e_i \\ g_i \end{bmatrix} . \tag{11.6.35}$$

Despite the fact that the g_i's are not exact subgradients, the aggregate subgradient $\bar{g}_k$ can still be interpreted as an ϵ_k-subgradient thanks to Proposition 11.3.1. As for the proximal bundle method, when the size of the polytope B_k is too large, some vertices are deleted and replaced by $\begin{bmatrix} \bar{e}_k \\ \bar{g}_k \end{bmatrix}$.

To control proximity to optimality, we evaluate the decreasing of the model δ_k of (11.3.19). Then the *(termination)* and *(serious)* steps of **Algorithm 11.3.1*** are modified accordingly

5′ . *(Termination) If $\delta_k \leq \delta$ then stop.*

6′ . *(Serious step) if*

$$f(x_k) - f(y_{k+1}) \geq m_L \delta_k$$

then set $x_{k+1} = y_{k+1}$, select $\mu_{k+1} \in [\mu_{min}, \mu_k]$, choose the dual metric M^*_{k+1} and continue with Step 8. Otherwise continue with Step 7.

Global convergence of this mixed polyhedral-sdp bundle method can be analyzed just as in [346, §XV.3.2]. Assuming that $\epsilon > 0$, $\delta > 0$ and controlling the sequence with a condition number κ such that $\{\frac{\lambda_{\min}(M_k^*)}{2\mu_k}\} \geq \frac{1}{\kappa}$, the algorithm stops at some iteration k; then using Proposition 11.3.1, the last iterate satisfies the inequality

$$f(y) \geq f(x_k) - ||\bar{g}_k||\,||y - x_k|| - \epsilon_k \quad \text{for all } y \in \mathbb{R}^m ,$$

and $\delta_k = \epsilon_k + \frac{1}{2\mu_k} ||\bar{g}_k||_{k,*}^2 \leq \delta$. This implies that the following approximate optimality condition holds

$$f(y) \geq f(x_k) - \sqrt{\kappa\delta}||y - x_k|| - \delta \quad \text{for all } y \in \mathbb{R}^m .$$

11.7 A SECOND-ORDER PROXIMAL BUNDLE METHOD

In this section we show that second-order information can be used to improve the asymptotic properties of the first order schemes presented in the previous sections: we present a second-order bundle method which converges globally and which enjoys asymptotically a quadratic rate of convergence.

11.7.1 *Second-order development of f*

As pointed out first by R. Fletcher [241] and as developed further by M. Overton through several works [254, 599, 600, 605], the maximum eigenvalue function enjoys a very specific structure: in a neighborhood of a point $Z \in \mathcal{S}^n$ belonging to the set

$$\mathcal{M}_r := \{X \in \mathcal{S}^n : \lambda_{\max}(X) = \ldots = \lambda_r(X) > \lambda_{r+1}(X)\},$$

the function $\lambda_{\max}$ is "smooth on $\mathcal{M}_r$". More explicitly the set $\mathcal{M}_r$ is a submanifold of $\mathcal{S}^n$ (see [597] for a formal proof) and there exists a neighborhood Ω of Z in $\mathcal{S}^n$ such that the function $\mathcal{M}_r \cap \Omega \ni X \mapsto \lambda_{\max}(X)$ is C^∞ (for a definition of C^∞ functions on a manifold see [2, Definition 3.2.5]). To obtain a similar result for f, some precautions must be taken: the intersection of $\mathcal{M}_r$ with the image of the affine mapping $\mathbb{R}^m \ni y \mapsto F(y) := C - \mathcal{A}^T y + (b^T y)I_n$ may have some singularities. To avoid these situations, a *transversality* assumption can be made: this is the case in [605, Theorem 3], though not stated as such, or in [715, 711]. Denoting by $\mathcal{R}$ the range of $\mathbb{R}^m \ni y \mapsto (b^T y)I_n - \mathcal{A}^T y$ and by $T_{\mathcal{M}_r}(Z)$ the tangent space to $\mathcal{M}_r$ at $Z := C - \mathcal{A}^T x + (b^T x)I_n$, the transversality condition can be written $\mathcal{R} + T_{\mathcal{M}_r}(Z) = \mathcal{S}^n$. In practice, this condition is often too conservative, a weaker one consists in requiring that the image of F *intersects cleanly* [2, Corollary 3.5.13]: the dimension of the subspace $\mathcal{R} \cap T_{\mathcal{M}_r}(X)$ must be constant when X varies locally in the intersection of $\mathcal{M}_r$ with the image of F.

Under transversality or clean intersection conditions, we obtain that the set $F^{-1}\mathcal{M}_r$ is a submanifold of $\mathbb{R}^m$ and the function f is locally C^∞ on $F^{-1}\mathcal{M}_r$. We say that such a point is *regular*. We are now in a position to write the second-order development of f along $F^{-1}\mathcal{M}_r$ at x.

Theorem 11.7.1 ([597, Theorem 5.10, Corollary 5.11]) *Let $x \in F^{-1}\mathcal{M}_r$ be* regular. *Take g in the relative interior of $\partial f(x)$. Then for all $d \in \mathbb{R}^m$ such that $F(x+d) \in \mathcal{M}_r$, we have*

$$f(x+d) = f(x) + g^T d + \frac{1}{2} d^T H(x,g) d + o(\|d\|^2), \qquad (11.7.36)$$

where $H(x,g) \in \mathcal{S}^m$ is given by

$$H(x,g)_{i,j} = 2 \operatorname{Tr} A_i W A_j (f(x)I_n - F(x))^\dagger \qquad i,j = 1,\ldots,m,$$

and $W \in \partial\lambda_{\max}(C - \mathcal{A}^T x)$ with $g = b - \mathcal{A}W$.

11.7.2 *Quadratic step*

At iteration k, assume that $y_k = x_k$ (*i.e.*, the previous step was a serious step). The rich oracle (11.6.33) has provided us with an approximate multiplicity r_k (depending on ϵ fixed at the beginning of the algorithm), the matrix of local ϵ-maximal eigenvectors P_k, the local ϵ-subgradient g_k and a corresponding W_k for which $g_k = b - \mathcal{A}W_k$. The quadratic step we consider here consists in minimizing the second-order approximation (11.7.36) of f at x_k under the first-order approximation of the constraint $F(x_k + d) \in \mathcal{M}_{r_k}$. This latter approximation is shown in [598, §5.3] to be equivalent to the constraint $F(x_k + d) \in T_{\mathcal{M}_{r_k}}(\hat{F}_k)$ where $\hat{F}_k$ is the projection of $F(x_k)$ onto $\mathcal{M}_{r_k}$. This in turn can be written $[D\phi]_k\, d = L_k$, where

$$\begin{cases} \mathbb{R}^m \ni d \mapsto [D\phi]_k\, d := P_k^T(-\mathcal{A}^T d)P_k - \frac{\operatorname{Trace}(P_k^T(-\mathcal{A}^T d)P_k)}{r_k} I_{r_k} \\ \text{and } L_k := \left(\frac{1}{r_k}\sum_{i=1}^{r_k} \lambda_i(F(x_k)) I_{r_k} - \operatorname{diag}(\lambda_1(F(x_k)), \ldots, \lambda_{r_k}(F(x_k))\right). \end{cases} \tag{11.7.37}$$

We therefore obtain the following quadratic step:

$$\begin{array}{ll} \text{minimize} & g_k^T d + \frac{1}{2} d^T H_k d \\ \text{subject to} & [D\phi]_k\, d = L_k\,, \end{array} \tag{11.7.38}$$

where the matrix $H_k \in \mathcal{S}^m$ is given by

$$[H_k]_{i,j} = 2 \operatorname{Tr} A_i W_k A_j (f(x_k) I_n - F(x_k))^\dagger \qquad i, j = 1, \ldots, m,$$

or, using the Kronecker product as in [711]:

$$H_k = 2 \left[\ \mathbf{vec} A_1 \cdots \mathbf{vec} A_m\ \right]^T (W_k \otimes [f(x_k) I_n - F(x_k)]^\dagger) \left[\ \mathbf{vec} A_1 \cdots \mathbf{vec} A_m\ \right].$$

Remark 11.7.1 *The structure of H_k is similar to that of the system matrix $[M]_{ij} = \operatorname{Tr} A_i X A_j Z^{-1}$ arising in primal-dual interior point methods. Indeed, by (11.6.33) W_k may be interpreted as the best approximation of a primal feasible solution X of the semidefinite program (P) over $\{P_k V P_k^T : \operatorname{Tr} V = 1, V \succeq 0\}$. On the other hand, $Z = f(x) I_n - F(x) = \lambda_{\max}(C - \mathcal{A}^t y) I + \mathcal{A}^T y - C \succeq 0$ is exactly the dual slack matrix appearing in (D) with $y_0 = \lambda_{\max}(C - \mathcal{A}^t y)$ being the best possible choice for the specific y. The eigenvalue decomposition of $Z = Q_k \Lambda_Z Q_k^T$, with diagonal $\Lambda_Z \succ 0$ and $Q_k^T Q_k = I_{n-r_k}$, $Q_k^T P_k = 0$, yields $Z^\dagger = Q_k \Lambda_Z^{-1} Q_k^T$. So the major difference between H_k and M is that the eigenvalues of Z converging to zero do not appear in $Z^\dagger$ of H_k while they dominate Z^{-1} in M. The reason behind this difference is that in the quadratic step (11.7.38) the influence of $\mathcal{A}^T d$ within the maximal eigenspace, governed by $P_k^T \mathcal{A}^T d P_k$, is already controlled by the constraint $[D\phi]_k\, d = L_k$ whereas in interior point methods the barrier term has to control $P_k^T \mathcal{A}^T d P_k$.*

Decomposition of the space. Let us decompose the space of variables in $\mathbb{R}^m = \mathcal{U}_k \oplus \mathcal{V}_k$ where $\mathcal{U}_k$ is the kernel of $[D\phi]_k$ and $\mathcal{V}_k := \mathcal{U}_k^\perp$, *i.e.*, $\mathcal{V}_k$ is the range of $[D\phi]_k^T$. Then we can give two different interpretations of the subspaces $\mathcal{U}_k$ and $\mathcal{V}_k$. Note that no regularity assumption is needed here.

Proposition 11.7.1 ([598, Theorem 5.5, Propositions 5.2, 5.6])
GEOMETRY *Recalling definition* (11.7.37), *the projection of $F(x_k)$ onto M_{r_k} is given by $\hat{F}_k = F(x_k) + P_k L_k P_k^T$ and we have:*

$$\mathcal{U}_k = (\mathcal{A}^T)^{-1} T_{\mathcal{M}_{r_k}}(\hat{F}_k) \quad \text{and} \quad \mathcal{V}_k = \mathcal{A} N_{\mathcal{M}_{r_k}}(\hat{F}_k),$$

where $N_{\mathcal{M}_{r_k}}(\hat{F}_k)$ *denotes the normal space to* $\mathcal{M}_{r_k}$ *at* $\hat{F}_k$.
CONVEX ANALYSIS $\mathcal{U}_k$ *is the largest subspace where the support function of* $\delta_\epsilon f(x_k)$ *is linear;* $\mathcal{V}_k$ *is the subspace parallel to the affine hull of* $\delta_\epsilon f(x_k)$.

Normal Step. Under transversality condition the operator $[D\phi]_k$ is onto [598, Remark 3.2] (we could also assume that the range of $[D\phi]_k$ is locally a fixed subspace when the point $\hat{F}_k$ varies on the submanifold $\mathcal{M}_{r_k}$; this would be a particular case of the clean intersection condition); $[D\phi]_k$ has therefore a right-inverse $[D\phi]_k^- := [D\phi]_k^T \circ ([D\phi]_k \circ [D\phi]_k^T)^{-1}$. We therefore define the *normal step* as

$$v_k := [D\phi]_k^- L_k \,. \tag{11.7.39}$$

We have $v_k \in \mathcal{V}_k$ and $[D\phi]_k \, v_k = L_k$.

Tangent step. By definition of $\mathcal{U}_k$ and v_k, the set of directions d satisfying the linear equation (11.7.38) is $v_k + \mathcal{U}_k$. Then, setting $u = d - v_k$, (11.7.38) amounts to

$$\text{minimize}_{u \in \mathcal{U}_k} \; (g_k + H_k \, v_k)^T u + \frac{1}{2} u^T H_k u \,, \tag{11.7.40}$$

which has a unique solution

$$u_k := -Z_k [Z_k^T H_k Z_k]^{-1} Z_k^T (g_k + H_k \, v_k) \,, \tag{11.7.41}$$

when H_k is positive definite on $\mathcal{U}_k$, *i.e.,* introducing the $n \times \dim \mathcal{U}_k$ matrix Z_k whose columns form an orthonormal basis of $\mathcal{U}_k$, when $Z_k^T H_k Z_k$ is positive definite.

The unique solution of (11.7.38) is then $d_k = u_k + v_k$.

11.7.3 The dual metric

To the decomposition of the space $I\!R^m = \mathcal{U}_k \oplus \mathcal{V}_k$, corresponds a natural decomposition of the dual metric $M_{k+1}^* = [M_{k+1}^T]_{\mathcal{U}_k} \oplus [M_{k+1}^*]_{\mathcal{V}_k}$.
Equation (11.7.41) indicates the choice of

$$[M_{k+1}^T]_{\mathcal{U}_k} := Z_k [Z_k^T H_k Z_k]^{-1} Z_k^T \,. \tag{11.7.42}$$

In other words, the $\mathcal{U}_k$-part of the dual metric is forced by Newton's method.

More flexibility is offered for the $\mathcal{V}_k$-part of the dual metric. Here we describe $[M_k^*]_{\mathcal{V}_k}$ as a pre-scaling at x_k so that a very efficient resolution of (11.6.32) will be available. The update of the $\mathcal{V}_k$-part we present here, is therefore anterior to the the call of the rich-oracle (11.6.33). After a null step the norm of the global polyhedral model will not be changed, nonetheless we may use the local norm to compute the solution of (11.6.32).

Proposition 11.7.2 *By choosing*

$$[M_k^*]_{\mathcal{V}_k} := [D\phi]_k^T ([D\phi]_k \circ [D\phi]_k^T)^{-2} [D\phi]_k \,, \tag{11.7.43}$$

the quadratic-semidefinite problem (11.6.32) *amounts to solving the projection problem in* $\mathcal{S}^{r_k}$

$$\begin{array}{ll} \text{minimize} & \|Z - B_k\|^2 \\ \text{subject to} & Z \in \mathcal{S}^{r_k} \\ & \operatorname{Trace} Z = 1,\ Z \succeq 0\,, \end{array} \tag{11.7.44}$$

where $B_k := -[D\phi]_k [M_k^*]_{\mathcal{V}_k} g_c$ *and* g_c *is the 'center' of* $\delta_\epsilon f(x_k)$*:* $g_c := b - \frac{1}{r_k} \mathcal{A} P_k P_k^T$.

Proof.
Reformulate (11.6.32) starting with

$$\delta_\epsilon f(x_k) = \{[D\phi]_k^T Z + g_c : Z \in \mathcal{S}^{r_k},\ Z \succeq 0,\ \operatorname{Trace} Z = 1\}\,,$$

and use (11.7.43) to complete the proof. ∎

The projection problem (11.7.44) has a very specific structure: Z varies in a transversal section of the cone of $r \times r$ semidefinite matrices; this will enable us in §11.8.4 to solve the projection problem exactly.

11.7.4 The second-order proximal bundle method

We are now in a position to insert the steps (11.7.39) and (11.7.41) in the mixed polyhedral-semidefinite proximal bundle method of § 11.6, *i.e.*, **Algorithm 11.3.1*** with steps 5′ and 6′.

Algorithm 11.7.1 Second-order proximal bundle method
Input: *An initial point* $x_0 \in \mathbb{R}^m$, *a* $\delta > 0$ *for termination, the viscosity parameter* $\epsilon > 0$, *an improvement parameter* $m_L \in (0, \frac{1}{2})$, *a weight* $u_0 > 0$, *a minimal weight* $u_{\min}$.

- *STEP 0.* $k = 0$, *compute* $f(x_0)$, $[M_0^*]_{\mathcal{V}_0}$ *of* (11.7.43) *and* g_0 *solution of* (11.6.32).

- *STEP 1. If* $y_k = x_k$ *goto Step 1.1 else goto Step 1.2.*

  - *STEP 1.1 Compute* $d_k := v_k + u_k$ *where* v_k *and* u_k *are respectively given by* (11.7.39) *and* (11.7.41). *Set* $[M_k^*]_{\mathcal{U}_k}$ *from* (11.7.42) *and*

$$\begin{aligned} y_{k+1} &:= x_k + \frac{1}{\mu_k} d_k \\ \delta_k &:= g_k^T (y_{k+1} - x_k) + \tfrac{1}{2}\mu_k \|y_{k+1} - x_k\|_k^2\,. \end{aligned}$$

  - *STEP 1.2 Compute* $\bar{\alpha}$ *solving* (11.6.34) *and set*

$$\begin{aligned} \bar{g}_k &:= \textstyle\sum_{i=1}^k \bar{\alpha}_i g_i \\ y_{k+1} &:= x_k - \frac{1}{\mu_k} M_k^*\, \bar{g}_k \\ \delta_k &:= \epsilon_k + \frac{1}{2\mu_k} \|\bar{g}_k\|_{k,*}^2\,, \end{aligned}$$

  and ϵ_k *is defined by* (11.3.18).

- *STEP 2. If $\delta_k \leq \delta$ stop.*
- *STEP 3. Call the rich oracle (11.6.33). Compute a solution $g_{k+1} \in \delta_\epsilon f(y_{k+1})$ of (11.6.32). If*

$$f(y_{k+1}) \leq f(x_k) - m\delta_k$$

set $x_{k+1} = y_{k+1}$, select $\mu_{k+1} \in [\mu_{min}, \mu_k]$ and set $[M_{k+1}^*]_{\mathcal{V}_{k+1}}$ from (11.7.43). Otherwise set $x_{k+1} = x_k$ (null-step).
Replace k by $k+1$ and loop to Step 1.

As far as the bundling mechanism is concerned, this algorithm is formally the same as the mixed polyhedral-semidefinite method of § 11.6. Its global convergence is therefore established likewise, via appropriate safeguards on the various metrics used such as $\{\frac{\lambda_{\min}(M_k^*)}{2\mu_k}\} \geq \frac{1}{\kappa}$. Its local convergence can also be analyzed through existing works: when $\mu_k = 1$ is accepted for the descent-test, the algorithm becomes the second-order bundle method of [598] and therefore enjoys the same asymptotic properties.

11.8 IMPLEMENTATIONS

11.8.1 Computing the eigenvalues

Assume that the structure of a symmetric matrix $A \in S^n$ (the matrix we have in mind is $Z = C - \mathcal{A}^T y$) is such that the product of the matrix with a vector can be computed quickly. The current state of the art approach for computing a few extremal eigenvalues of such a matrix is the Lanczos method. It is an iterative method that constructs for some normalized starting vector q_1 incrementally a tridiagonal symmetric matrix $T_i = Q_i^T A Q_i \in S^i$ where the columns of $Q_i \in \mathbb{R}^{n \times i}$ form an orthonormal basis of the Krylov subspace

$$\begin{aligned}\mathcal{K}(A, q_1, i) &= \text{span}\left\{q_1, Aq_1, \ldots, A^{i-1}q_1\right\} \\ &= \text{span}\left\{q_1, P\Lambda P^T q_1, \ldots, P\Lambda^{i-1}P^T q_1\right\}.\end{aligned}$$

Here, $P\Lambda P^T = A$ is an eigenvalue decomposition of A with $P^T P = I_n$ and $\Lambda = \text{diag}(\lambda_1, \ldots, \lambda_n)$. Each iteration involves only one matrix vector multiplication with A. An eigenvector v to the maximal eigenvalue of T_i gives rise to a polynomial $p(\cdot)$ of degree $i-1$ (via $Q_i v = p(A)q_1$) that maximizes the Rayleigh quotient

$$\frac{v^T T_i v}{v^T v} = \frac{q_1^T p(A) A p(A) q_1}{q_1^T p(A)^2 q_1} = \frac{q_1^T P p(\Lambda)^2 \Lambda P^T q_1}{q_1^T P p(\Lambda)^2 P^T q_1} = \frac{\sum_{j=1}^n \lambda_j p(\lambda_j)^2 (P_{\cdot,j}^T q_1)^2}{\sum_{j=1}^n p(\lambda_j)^2 (P_{\cdot,j}^T q_1)^2}.$$

Thus, if the maximal eigenvalue of A is well separated and the component of q_1 pointing into the direction of the corresponding eigenvector is large then a polynomial of small degree, *i.e.*, a few matrix vector multiplications will suffice to produce reasonable approximations to $\lambda_{\max}(A)$ and its eigenvector (see [292] for a quick introduction or [689] for a detailed description of Lanczos methods).

In theory, the method stops with $\lambda_{\max}(A)$ after at most n iterations. In practice, however, the number of columns of Q_i has to be kept small for reasons of computational efficiency. This is usually achieved by restarting the process after a certain number of iterations. Various strategies for restarting are available, *e.g.*, one might use the vector $Q_i v$, where v is the eigenvector to $\lambda_{\max}(T_i)$, as a new starting vector [689] or employ the implicit restarting approach of [743]. The restriction of the number of columns of Q together with numerical difficulties due to generic cancellations appearing in the Lanczos algorithm may lead to iteration numbers much larger than n if the maximal eigenvalue is not well separated.

In the first few iterations of the bundle algorithms the maximal eigenvalue is in general well separated and the maximal eigenvector of the previous iterate is a reasonable starting vector. However, as the algorithm proceeds, more and more eigenvalues cluster around $\lambda_{\max}(A)$. This clustering is a generic property of optimal solutions in semidefinite programming, see [623]. Thus, it gets more and more difficult to compute good approximations via the Lanczos method. In consequence the eigenvalue routine often turns out to be the bottleneck of the algorithm.

The situation can be improved somewhat by stopping early in the case of null steps (this modification was suggested by K.C. Kiwiel and will be elaborated in [335], also for bound-constrained problems [336]). Essentially, the Lanczos method produces successively better lower estimates

$$\lambda_{\max}(T_i) \leq \lambda_{\max}(A) = \lambda_{\max}(C - \mathcal{A}^T y)$$

for the maximal eigenvalue and therefore for the true function value $\lambda_{\max}(C - \mathcal{A}^T y) + b^T y$. If the current estimate $\lambda_{\max}(T_i)$ is already good enough to prove that the current y will result in a null step, then the corresponding (normalized) eigenvector v of $\lambda_{\max}(T_i)$ gives rise to a cutting plane

$$\begin{bmatrix} e \\ g \end{bmatrix} = \begin{pmatrix} \langle C, Q_i v v Q_i^T \rangle \\ b - \mathcal{A} Q_i v v Q_i^T \end{pmatrix}$$

that improves the cutting plane model sufficiently. This may lead to considerable speed-up because in iterative methods eigenvalues converge significantly faster than eigenvectors [183]. The true eigenvectors of A are only needed if we have to prove that $\lambda_{\max}(T_i)$ is indeed close to an eigenvalue of A.

The Lanczos method offers further useful information. Typically, all of the large eigenvalues of T_i are close to large eigenvalues of A. Thus, in addition to the eigenvector to $\lambda_{\max}(T_i)$, the eigenvectors to the other large eigenvalues of T_i can also be used to construct good cutting planes for the model at no extra cost.

11.8.2 Structure of the mapping

In control the variable y is often a symmetric matrix itself, say $P \in \mathcal{S}^n$ and:

$$\begin{array}{rcl} \mathcal{S}^n & \to & \oplus_{s=1}^L \mathcal{S}^n \\ P & \mapsto & \mathcal{A}P := \oplus_{s=1}^L PA_i + A_i^T P , \end{array}$$

where the A_i are given matrices in $\mathbb{R}^{n\times n}$. Assuming that the A_i are dense, it is crucial to integrate this structure in our oracle rather than decomposing the variable P in the standard basis of $\mathcal{S}^n$.

In combinatorial applications C is typically sparse and the A_i are either sparse or have low rank structure, *e.g.*, $A_i = vv^T$ with $v \in \mathbb{R}^n$ sparse or dense.

11.8.3 *Solving the quadratic semidefinite program*

In this section we show how to solve (11.3.14) for

$$\mathcal{W}_k = \left\{PVP^T + \alpha\widetilde{W} : V \in \mathcal{S}^r_+, \alpha \geq 0, \text{Trace}\,(V) + \alpha = 1\right\}$$

and diagonal $M^* = D^* = \text{diag}(d_1, \ldots, d_m) \succ 0$. For simplicity the set $\mathcal{W}_k$ is formed by a semidefinite and one linear variable, the algorithm is easily extended to several linear variables. In this setting (11.3.14) reads (we drop all iteration indices)

$$\text{(QSDP)}\quad \begin{array}{ll} \text{minimize} & \left\langle b - \mathcal{A}W, \frac{1}{2\mu}D^*(b - \mathcal{A}W) - x\right\rangle - \langle C, W\rangle \\ \text{subject to} & W = PVP^T + \alpha\widetilde{W} \\ & \text{Trace}\,(V) + \alpha = 1 \\ & V \in \mathcal{S}^r_+, \alpha \geq 0. \end{array}$$

The process of solving (QSDP) can be decomposed into two steps. First we compute the cost coefficients of V and α, then we solve the reduced size (QSDP) with interior point methods. After some technical linear algebra the reduced problem reads (constants are neglected)

$$\begin{array}{ll} \min & \frac{1}{2}\binom{\text{svec}\,(V)}{\alpha}^T \begin{pmatrix} Q_{11} & q_{12} \\ q_{12}^T & q_{22} \end{pmatrix} \binom{\text{svec}\,(V)}{\alpha} + \binom{c_1}{c_2}^T \binom{\text{svec}\,(V)}{\alpha} \\ & \text{Trace}\, V + \alpha = 1 \\ & V \succeq 0, \alpha \geq 0, \end{array}$$

where

$$\begin{aligned} Q_{11} &= \frac{1}{\mu}\sum_{i=1}^{m} d_i \text{svec}\,(P^T A_i P)\text{svec}\,(P^T A_i P)^T \\ q_{12} &= \frac{1}{\mu}\text{svec}\,(P^T \mathcal{A}^T (D^* \mathcal{A}\widetilde{W})P) \\ q_{22} &= \frac{1}{\mu}\left\langle \mathcal{A}\widetilde{W}, D^*\mathcal{A}\widetilde{W}\right\rangle \\ c_1 &= -\text{svec}\,(P^T(\mathcal{A}^T(\tfrac{1}{\mu}D^* b - x) + C)P) \\ c_2 &= -\left\langle \tfrac{1}{\mu}D^* b - x, \mathcal{A}\widetilde{W}\right\rangle - \left\langle C, \widetilde{W}\right\rangle. \end{aligned}$$

If $P \in \mathbb{R}^{n\times r}$ then $P^T C P$ and $P^T A_i P \in \mathcal{S}^r$. These matrices have to be determined only once for each (QSDP). Since, by assumption, matrix times vector is an efficient operation, they are rather cheap to compute if r is small. Also note, that there is no need to store the matrix $\widetilde{W}$. Only the terms $\langle C, \widetilde{W}\rangle$ and $\mathcal{A}\widetilde{W}$ appear in the formulas above and these can be updated without $\widetilde{W}$. The critical operation is the accumulation of $Q_{11} \in \mathcal{S}^{\binom{r+1}{2}}$ which requires $O(m\binom{r+1}{2}^2)$ flops. In particular, if $m \gg \binom{r+1}{2}$ the computation of the cost coefficients is more expensive than solving the reduced (QSDP).

The development of an interior point code for the reduced (QSDP) is rather straight forward. Both primal and dual programs have strictly feasible solutions that can be used as starting points for a feasible primal dual method. The main work per iteration is the inversion of a symmetric positive definite matrix of order $\binom{r+1}{2}$. Thus the amount of work spent in the interior point code is roughly $O(r\binom{r+1}{2}^3)$.

11.8.4 The rich oracle

In the mixed polyhedral-semidefinite method of § 11.6, we have encountered a quadratic program over a transversal section of the cone of $r_k \times r_k$ positive semidefinite matrices (11.6.32). Several methods were proposed so far to solve this subproblem: one can use an interior-point method as in §§ 11.8.3 or a support-black box method as in [598]. Here we show that, via a reformulation of the problem with the metric $[M_k^*]_{V_k}$ of §§ 11.7.3, the solution is in fact obtained explicitly.

Lemma 11.8.1 *Let $B \in \mathcal{S}^r$. The solution of the projection problem*

$$\begin{array}{ll} \text{minimize} & \langle Z, Z\rangle - 2\langle Z, B\rangle \\ \text{subject to} & Z \in \mathcal{S}^r,\ Z \succeq 0, \end{array} \qquad (11.8.45)$$

is $\bar{Z} := B^+$, the matrix obtained by taking the eigenvalue decomposition of B and replacing all negative eigenvalues by 0.

Proof.
It is a strictly convex problem; it is sufficient to verify that B^+ satisfies the first-order optimality condition of the nonlinear SDP. These conditions are (see Chapter 4 of this book): there exists $\Omega \in \mathcal{S}_+^r$ such that

$$B = Z - \Omega \text{ and } Z\Omega = 0\,.$$

This is exactly the decomposition of B into its positive and negative parts. ■

Lemma 11.8.1 establishes that the projection of a symmetric matrix onto the cone of positive semidefinite matrices can be obtained via its eigenvalue decomposition. It is a well-known fact in statistics. Proposition 11.8.1 shows that it is still true when $\mathcal{S}_+^r$ is intersected with $\{Z \in \mathcal{S}^r : \text{Trace}\, Z = 1\}$.

Proposition 11.8.1 *The unique solution of* (11.7.44) *is* $\bar{Z}_k := [B_k + \bar{\beta} I_{r_k}]^+$, *where* $\bar{\beta}$ *is the solution of the equation*

$$\sum_{i=1}^{r_k} (\beta + \lambda_i(B_k))^+ = 1\,,$$

i.e.,

$$\bar{\beta} = \begin{cases} \frac{1-\sum_{l=1}^{i} \lambda_l(B_k)}{i} & \text{if there exits } i \in \{1,\ldots,r_k-1\} \\ \text{such that} & \sum_{l=1}^{i} (\lambda_l(B_k) - \lambda_i(B_k)) \le 1 \le \sum_{l=1}^{i+1} (\lambda_l(B_k) - \lambda_{i+1}(B_k)), \\ \frac{1-\operatorname{Trace} B_k}{r_k} & \text{otherwise.} \end{cases} \tag{11.8.46}$$

Proof.
Let us dualize the constraint Trace $Z = 1$ in (11.7.44): the corresponding Lagrangian is $L(Z,\beta) = \langle Z, Z\rangle - 2\langle Z, B_k\rangle + 2\beta(1 - \operatorname{Trace} Z)$. The dual function is thereby $q(\beta) := \inf_{Z \in S_+^{r_k}} L(Z,\beta)$. Applying Lemma 11.8.1, with $B := B_k + \beta I_{r_k}$ we obtain, for all $\beta \in I\!R$

$$q(\beta) = -\left\|[B_k + \beta I_{r_k}]^+\right\|^2 + 2\beta\,.$$

Now, $\|[B_k + \beta I_{r_k}]^+\|^2 = \sum_{i=1}^{r_k} ([\lambda_i(B_k) + \beta]^+)^2$ is a differentiable function of β. Then q is also differentiable and $q'(\beta) = -2\sum_{i=1}^{r_k} [\lambda_i(B_k) + \beta]^+ + 2$ is a continuous and nonincreasing piecewise linear function:

$$\frac{1}{2}\,q'(\beta) = \begin{cases} 1 & \text{if } \beta \in]-\infty, -\lambda_1(B_k)] \\ 1 - i\beta - \sum_{l=1}^{i} \lambda_l(B_k) & \text{if } \beta \in]-\lambda_i(B_k), -\lambda_{i+1}(B_k)], \\ & \text{for } i = 1,\ldots,r_k - 1 \\ 1 - r_k\beta - \operatorname{Trace} B_k & \text{if } \beta \in]-\lambda_n(B_k), +\infty[\,. \end{cases}$$

The optimality condition $q'(\beta) = 0$, has clearly a unique solution $\bar{\beta}$ defined by (11.8.46) which is the unique solution of the (convex) dual problem $\max_{\beta \subset I\!R} q(\beta)$. Since the primal problem (11.7.44) is (strongly) convex, there is no duality gap and $\bar{Z}_k = B_k + \bar{\beta} I_{r_k}$ is the unique primal solution. ■

11.9 NUMERICAL RESULTS

In this section we solve semidefinite programs from combinatorial optimization or control theory for the three bundle methods presented in this paper.

11.9.1 The spectral bundle method

We first describe some implementational aspects of the spectral bundle method.

For computing the maximal eigenvalue of $A = C - \mathcal{A}^T y_{k+1}$ we use the Lanczos method on $p(A)$ with p being a Chebyshev polynomial of variable degree (usually between 20 and 50). We stop the eigenvalue computation as soon as the lower bound on the maximal eigenvalue guarantees a null step. Otherwise the maximal eigenvalue is computed to the relative precision

$$\min\left\{10^{-3}, 10^{-1}\frac{m_L(f(x_k) - \varphi_k(y_{k+1}))}{|\lambda_{\max}^k| + 1}\right\},$$

where $\lambda_{\max}^k$ denotes the result of the last eigenvalue computation which is a lower bound on the maximal eigenvalue of the last matrix. This choice ensures that the objective value is computed to a reasonable relative precision with respect to the gap between model value and last objective value. In order to ensure that we obtain, in case of a serious step, an upper bound on $\lambda_{\max}(A)$ we add the (positive) relative error to the result.

As starting vector we use a slightly perturbed vector $P_k v$ where v is an eigenvector to the maximal eigenvalue of $P_k^T A P_k$. We restart the method for the first time after 15 Lanczos iterations and subsequently after 20 Lanczos iterations with the eigenvector to the largest eigenvalue of T_i. With this choice we hope to terminate quickly for poor candidates. We do not employ a block Lanczos method any longer because for larger matrix dimensions the eigenvalues corresponding to the block of starting vectors tended to emerge before the component into the direction of an eigenvector to the maximal eigenvalue was strong enough.

To form P_{k+1} we keep n_K (n_K will be specified below) eigenvectors corresponding to large eigenvalues of V_k, and add n_A Lanczos vectors corresponding to the largest eigenvalues of T_i. The inclusion of several Lanczos vectors, say 8 or 10, added considerably to the quality of the model. Intuitively, the fact that single vector Lanczos methods cannot generate multiple eigenvectors to eigenvalues with large multiplicity helps that less attractive regions will not be probed.

We have changed the stopping criterion for the bundle algorithm slightly, we stop when

$$\mu_k(f(x_k) - \varphi_k(y_{k+1})) \leq \delta(|f(x_k)| + 1).$$

Since

$$f(x_k) - \varphi_k(y_{k+1}) = \lambda_{\max}(C - \mathcal{A}^T x_k) - \langle C - \mathcal{A}^T x_k, W_k\rangle + \frac{1}{\mu_k}\left\|\mathcal{A}(W^k) - b\right\|^2$$

this ensures greater independence of the stopping criterion with respect to the current choice of μ_k while taking both, the gap and the norm of the subgradient, into account. In our computations we will use $\delta = 10^{-5}$. In comparing the numerical results to the exact solutions this stopping criterion proved very reliable in practice.

The current realization of the eigenvalue computation favors small μ. Not only does the eigenvalue computation stop quickly for poor candidates. A larger trust region also helps to avoid early clustering of eigenvalues. Therefore our initial choice for μ is

$$\mu_0 = \frac{\|\mathcal{A}(p_0 p_0^T) - b\|}{\sqrt{m}}.$$

Intuitively speaking, in (11.3.13) each coordinate of y is allowed to differ by one unit with respect to x. We decrease μ according to the same rules outlined in [414]. In addition we decrease μ if the last few serious steps moved into roughly the same direction and exhibited comparable decrease in both model and function value. We do not increase μ unless several serious steps that moved in 'opposite' directions are followed by ten null steps.

In our numerical experiments we concentrate on the quadratic $\{1, -1\}$ optimization problem

$$\text{maximize } x^T Cx \text{ subject to } x \in \{1, -1\}^n, \tag{11.9.47}$$

where $C \in \mathcal{S}^n$ is the cost matrix. Observe, that $x^T Cx = \langle C, xx^T \rangle$ with $xx^T \succeq 0$ and $\text{diag}(xx^T) = e$ for all $x \in \{1, -1\}^n$, e denoting the vector of all ones. The canonical semidefinite relaxation of (11.9.47) associates a matrix $X \in \mathcal{S}^n$ with xx^T and requires the same two constraints to hold for this matrix,

$$\text{maximize } \langle C, X \rangle \text{ subject to } \text{diag}(X) = e, X \succeq 0. \tag{11.9.48}$$

This is a semidefinite program enjoying the constant trace property, so its dual can be formulated as an eigenvalue optimization problem of the form (11.1.1).

If A is the (weighted) adjacency matrix of a graph then for $C = \frac{1}{4}(\text{Diag}(Ae) - A)$ (11.9.48) is the basic semidefinite relaxation of the max-cut problem. Most of our examples for the spectral bundle method will be of this type with A generated by the machine independent graph generator **rudy** written by G. Rinaldi.

The second set of examples corresponds to the Lovász ϑ-function for bounding the size of the largest stable set in a graph. For implementational convenience we use its formulation as a constrained version of (11.9.48) (see, *e.g.*, [334]). For a graph with n nodes and m edges this yields a matrix $C \in \mathcal{S}^{n+1}$ and $n + 1 + m$ linear constraints.

All examples were computed on a SUN, Ultra 10 with a 299 MHz SUNW, UltraSPARC-IIi CPU which is about twice as fast as the machine used in [338]. Computation times will be given in the format hh:mm:ss (hours:minutes: seconds) where the time measured is the user time. The maximum amount of memory our code used on the examples below is about 50 MByte, but there is still room for improvement.

Tables 11.1 and 11.2 give the results for max-cut examples G_1 to G_{42} specified in [338]. In these examples we set $n_K = 20$ and $n_A = 8$. The first column of the table refers to the name of the instance, n and m give the size of the matrix and the number of constraints, respectively. Column f displays the function value at termination. *time* is the computation time. *serious* gives the number of serious steps, *iter* displays the total number of iterations, including serious and null steps. $\|g\|$ is the norm of the subgradient corresponding to the last W_k.

	n = m	f	time	serious	iter	$\|\|g\|\|$
G_1	800	12083.24	1:11	11	16	0.15
G_2	800	12089.49	1:33	17	20	0.17
G_3	800	12084.40	1:09	10	16	0.14
G_4	800	12111.49	1:17	12	16	0.14
G_5	800	12100.17	1:11	15	19	0.14
G_6	800	2656.21	1:22	16	21	0.06
G_7	800	2489.27	1:14	15	20	0.06
G_8	800	2506.94	1:08	12	16	0.06
G_9	800	2528.74	1:10	13	19	0.08
G_{10}	800	2485.07	1:08	11	17	0.07
G_{11}	800	629.17	3:46	32	200	0.04
G_{12}	800	623.88	2:22	30	110	0.05
G_{13}	800	647.14	2:01	28	99	0.05
G_{14}	800	3191.58	1:50	23	37	0.08
G_{15}	800	3171.58	2:16	23	43	0.10
G_{16}	800	3175.03	1:12	20	25	0.20
G_{17}	800	3171.34	1:54	23	44	0.08
G_{18}	800	1166.01	54	20	26	0.08
G_{19}	800	1082.02	49	20	25	0.09
G_{20}	800	1111.40	43	20	26	0.10
G_{21}	800	1104.29	1:14	22	38	0.04

Table 11.1 Max-Cut, $n_K = 20$, $n_A = 8$, $\delta = 10^{-5}$.

The results of the tables show considerable improvements, in particular for the bad examples of [338]. The solution of G_{15} required about 42 minutes in [338], now, on a machine twice as fast, it takes only 4 minutes and the result is more accurate. Since the number of iterations is roughly the same, we locate the main improvement in the new realization of the eigenvalue computation.

For all examples of Table 11.3 and Table 11.5 we stop the algorithm after the first serious step occurring after one hour of computation time and use $n_K = 8$ and $n_A = 10$. The reason for this choice of n_K and n_A is that for large m the computation of the coefficients of (QSDP) becomes increasingly expensive and soon outgrows the cost of the eigenvalue computation. In particular, it is no longer efficient to choose n_K large enough so that it exceeds the multiplicity of the maximal eigenvalue in the optimal solution. Therefore it is important to control the eigenspace that is close to but does *not* belong to the optimal eigenspace of $\lambda_{\max}$. Indeed, if we select $n_A = 1$ and n_K not large enough to cover a bit more than the optimal eigenspace, then the model function tends to produce candidates that avoid the promising regions. Intuitively, the model is rather tight for 'good' candidates and linear on less promising subspaces. Therefore 'good' candidates will be found only if, after time, the aggregate subgradient balances the linear part nicely.

Table 11.3 displays results for max-cut examples G_{55} to G_{67} generated by S. Benson using the graph generator rudy. Since these have not yet appeared

	n = m	f	time	serious	iter	$\|g\|$
G_{22}	2000	14136.03	8:03	17	31	0.18
G_{23}	2000	14145.60	8:03	18	38	0.17
G_{24}	2000	14140.92	7:14	16	30	0.18
G_{25}	2000	14144.28	6:49	18	27	0.22
G_{26}	2000	14132.95	5:54	17	25	0.25
G_{27}	2000	4141.68	8:31	18	35	0.09
G_{28}	2000	4100.82	12:15	19	35	0.11
G_{29}	2000	4208.91	7:54	19	34	0.11
G_{30}	2000	4215.40	7:35	20	36	0.09
G_{31}	2000	4116.70	7:17	21	31	0.14
G_{32}	2000	1567.65	18:09	37	278	0.08
G_{33}	2000	1544.32	15:17	41	239	0.07
G_{34}	2000	1546.70	11:41	37	195	0.10
G_{35}	2000	8014.78	13:43	30	68	0.16
G_{36}	2000	8006.02	18:02	31	63	0.19
G_{37}	2000	8018.67	14:20	30	61	0.21
G_{38}	2000	8015.01	12:50	29	63	0.15
G_{39}	2000	2877.66	9:59	31	69	0.08
G_{40}	2000	2864.80	10:30	30	59	0.08
G_{41}	2000	2865.23	9:37	32	64	0.08
G_{42}	2000	2946.26	9:47	32	59	0.07

Table 11.2 Max-Cut, $n_K = 20$, $n_A = 8$, $\delta = 10^{-5}$.

	n = m	f	time	serious	iter	$\|g\|$
G_{55}	5000	12870.47	1:11:27	28	464	0.45
G_{56}	5000	4761.37	1:18:27	21	577	1.30
G_{57}	5000	3885.01	1:03:30	29	525	0.46
G_{58}	5000	20137.84	1:05:53	39	219	0.42
G_{59}	5000	7314.38	1:10:05	36	257	0.24
G_{60}	7000	15226.06	1:05:18	19	344	0.78
G_{61}	7000	6832.61	1:01:57	19	334	0.58
G_{62}	7000	5601.25	1:08:06	29	380	0.78
G_{63}	7000	28247.77	1:08:00	35	168	0.78
G_{64}	7000	10470.75	1:00:11	36	153	0.54
G_{65}	8000	6216.27	1:05:52	28	334	1.05
G_{66}	9000	6968.97	1:01:04	27	274	1.32
G_{67}	10000	7746.76	1:04:36	25	263	1.58

Table 11.3 Max-Cut, $n_K = 8$, $n_A = 10$, $\delta = 10^{-5}$, time limit one hour.

in the literature we list the corresponding arguments for **rudy** in Table 11.4. Although none of the examples could be solved to the required precision within the time limit of one hour, we are quite confident that the solutions differ by about five units at most. This confidence is based on the fact that for well

G_{55}	`-rnd_graph 5000 0.12 50001`
G_{56}	`-rnd_graph 5000 0.1 50001 -random 0 1 50001 -times 2 -plus -1`
G_{57}	`-toroidal_grid_2D 50 100 -random 0 1 50001 -times 2 -plus -1`
G_{58}	`-planar 5000 99 50001 -planar 5000 99 50002 +`
G_{59}	`-planar 5000 99 50001 -planar 5000 99 50002 + -random 0 1 50001 -times 2 -plus -1`
G_{60}	`-rnd_graph 7000 0.07 70001`
G_{61}	`-rnd_graph 7000 0.07 70001 -random 0 1 70001 -times 2 -plus -1`
G_{62}	`-toroidal_grid_2D 70 100 -random 0 1 70001 -times 2 -plus -1`
G_{63}	`-planar 7000 99 70001 -planar 7000 99 70002 +`
G_{64}	`-planar 7000 99 70001 -planar 7000 99 70002 + -random 0 1 70001 -times 2 -plus -1`
G_{65}	`-toroidal_grid_2D 80 100 -random 0 1 100001 -times 2 -plus -1`
G_{66}	`-toroidal_grid_2D 90 100 -random 0 1 100001 -times 2 -plus -1`
G_{67}	`-toroidal_grid_2D 100 100 -random 0 1 100001 -times 2 -plus -1`

Table 11.4 Arguments for generating the graphs G_{55} to G_{67} by the graph generator rudy as communicated by S. Benson.

scaled problems and a small value of μ, φ behaves like a lower bound. For the examples of Table 11.3, $\varphi_k(y_k)$ seems to converge from below to the same value as the $f(x_k)$ from above.

	n	m	f	time	serious	iter	$\|g\|$
G_{43}	1001	10991	281.50	1:09:32	43	371	0.31
G_{44}	1001	10991	281.99	1:00:41	65	312	0.46
G_{45}	1001	10991	281.31	1:06:45	65	357	0.33
G_{46}	1001	10991	280.78	1:07:46	47	366	0.29
G_{47}	1001	10991	282.86	1:03:21	46	341	0.35
G_{48}	3001	9001	1500.01	5:30	32	52	0.08
G_{49}	3001	9001	1500.02	6:21	35	60	0.10
G_{50}	3001	9001	1496.98	7:26	39	66	0.08
G_{51}	1001	6910	352.57	1:09:37	41	643	0.27
G_{52}	1001	6917	350.50	1:10:17	46	635	0.23
G_{53}	1001	6915	351.96	1:01:44	45	578	0.27
G_{54}	1001	6917	345.06	1:10:52	47	664	0.29

Table 11.5 ϑ-function, $n_K = 8$, $n_A = 10$, $\delta = 10^{-5}$, time limit one hour.

Finally, Table 11.5 refers to the same examples of the Lovász ϑ-function used in [338]. Again the results are considerably better. In particular, problems G_{48} to G_{50} could be solved to the desired precision within a few minutes. For all other examples, within one hour we obtain significantly smaller upper bounds on the optimum than we obtained in [338] within four hours. This time, the reason is not only the improvement of the code, but also a special scaling of the additional node that enters in the transformation of the ϑ-function to its constrained quadratic $\{1, -1\}$ programming formulation. More advanced scaling techniques could be of relevance for these problems. The gap between model value and function value as well as the gradient suggest, that there are still a few units to be gained for the remaining examples, but probably not much more. In particular, feasible solutions for G_{51} to G_{54} prove that the gap is at most 10 units for these four examples.

11.9.2 The mixed polyhedral-semidefinite bundle method

In this paragraph, we apply the mixed polyhedral-semidefinite bundle method of § 11.6 on the problem of analyzing the stability of a polytopic linear differential inclusion [137, Chapter 5] of the form:

$$\dot{x} = A(t)\,x, \quad A(t) \in \operatorname{Conv}\{A_1, \ldots, A_L\}, \tag{11.9.49}$$

where the A_i's are given stable $n \times n$ real matrices. Then we define the structured mapping:

$$\begin{array}{lcl} \mathcal{S}^n & \to & \oplus_{i=1}^L \mathcal{S}^n \\ P & \mapsto & \mathcal{F} \cdot P := \oplus_{i=1}^L PA_i + A_i^T P \,; \end{array}$$

a sufficient condition for (11.9.49) to be quadratically stable is

$$\exists P > 0 : \;\; \lambda_{\max}(\mathcal{F} \cdot P) < 0 \,. \tag{11.9.50}$$

Then an equivalent condition for (11.9.50) to hold is to require that the value of the following eigenvalue optimization problem:

$$\begin{array}{l} \inf \;\; \lambda_{\max}(\mathcal{F} \cdot P) \\ P > 0 \\ \operatorname{Tr} P = n \,, \end{array} \tag{11.9.51}$$

is negative. Actually, using a generalization of Lyapunov's theorem [352, Theorem 2.4.10], we can drop the constraint $P > 0$; the equality constraint can also be removed by eliminating one variable. Yet for the implementation of the rich oracle (11.6.33) we do keep, as announced in § 11.8.2, the matricial structure of the decision variable P instead of decomposing P onto a basis of $\mathcal{S}^n$; the minimized function has the form

$$f(P) := \lambda_{\max}(\mathcal{F} \cdot E + \mathcal{F} \cdot \widetilde{P}) \,,$$

where E is the symmetric matrix with zeros everywhere except $E_{nn} = n$ and $\widetilde{P}$ equal P everywhere except $\widetilde{P}_{nn} := -\sum_{i=1}^{n-1} P_{ii}$. The adjoint of $P \mapsto \mathcal{F} \cdot \widetilde{P}$ has a similar matricial structure. This formulation eases the resolution of (11.9.51) when n and $m = \frac{n(n+1)}{2} - 1$ are large numbers.

In Table 11.6, we take $L = 2$ and we generate stable random matrices A_1 and A_2 of various sizes. The most difficult situations are when the value of (11.9.51) is nonnegative. We report for such situations the final value of the function, the norm of the corresponding aggregate subgradient, the number of total iterations (descent or null steps) and the total cputime. The mixed polyhedral-semidefinite method is run with the viscosity parameter $\epsilon = 10^{-3}$ and the stopping tolerance $\delta = 10^{-2}$.

n	m	f^*	$\|\bar{g}\|$	# iter	cputime
500	125249	10^{-2}	10^{-2}	80	7 h
1000	500499	$5.1\,10^{-2}$	$6.3\,10^{-2}$	47	20 h

Table 11.6 PLDI, $n \geq 500$

11.9.3 The second-order proximal bundle method

In this paragraph we focus our attention on a small instance of (11.9.51): $n = 5$ and the matrices A_1 and A_2 are given by

$$A1 = \begin{bmatrix} -0.30514981684203 & -0.12717167399150 & -0.37080266389640 & -0.57112026933673 \\ -0.22727849390700 & -0.24318911206550 & -0.36426275722518 & -0.02224065695781 \\ -0.36873022242947 & -0.25327247833998 & -0.68348855951185 & -0.74119819525553 \\ -0.42560381552175 & -0.27639287914406 & -0.57459988147874 & -0.55530006009433 \end{bmatrix}$$

$$A2 = \begin{bmatrix} -0.70166259253876 & -0.05779033486378 & -0.10942688756431 & -0.48637313513570 \\ 0.10902133192757 & -0.14730314658672 & 0.01058304534831 & -0.25766505826345 \\ -0.26311036762919 & -0.10429623489082 & -0.04975331091086 & -0.10833871463463 \\ -0.56854215687151 & -0.10666376499017 & -0.30404398482014 & -0.60107828929368 \end{bmatrix}$$

Although A_1 and A_2 are both stable, their respective spectral abscissa are $\alpha(A_1) = -2.93e - 05$ and $\alpha(A_2) = -3.25e - 06$, there is no quadratic Lyapunov function proving the stability of (11.9.49). Indeed, using the second-order proximal bundle method 11.7.1, we solve (11.9.51) to machine precision: at the final point P^* generated by the algorithm we have $\bar{g} \in \partial f(P^*)$ with $\|\bar{g}\| \leq 2e - 15$. The final value is $f^* := f(P^*) = 0.01149788039455$ with multiplicity 4 to the machine precision; the next eigenvalue is $\lambda_5(\mathcal{F} \cdot E + \mathcal{F} \cdot \widetilde{P^*}) = -0.00670989012524$. The corresponding solution is positive definite,

$$P^* = \begin{bmatrix} 1.40104588451064 & 0.47711672234709 & 0.47017792285932 & 1.14391695175544 \\ 0.47711672234709 & 0.23379477434643 & 0.30129912349098 & 0.44304589342816 \\ 0.47017792285932 & 0.30129912349098 & 0.87145527971679 & 0.82344103313835 \\ 1.14391695175544 & 0.44304589342816 & 0.82344103313835 & 1.49370406142614 \end{bmatrix}.$$

Figure11.9.3 shows us the acceleration produced by the introduction of second-order information. In particular we observe the asymptotic superlinear convergence. The mixed polyhedral-sdp bundle method and Algorithm 11.7.1 are run with $\epsilon = 10^{-8}$.

Acknowledgements. The first author thanks K.C. Kiwiel and D.C. Sorensen for very helpful discussions and suggestions that led to considerable improvements in the spectral bundle code. The second author thanks C. Lemaréchal and M. Overton for their numerous constructive advices.

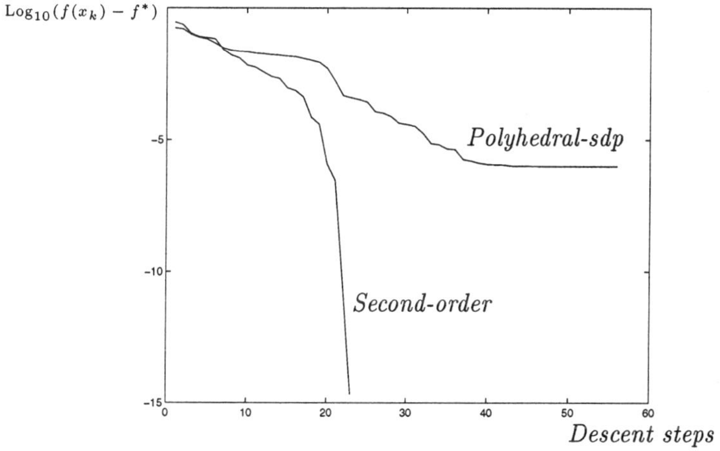

Figure 11.1 Acceleration with second-order information

III APPLICATIONS and EXTENSIONS

Mathematical ideas originate in emperics, although the genealogy is sometimes long and obscure. But, once they are so conceived, the subject begins to live a peculiar life of its own and is better compared to a creative one, governed by almost entirely aesthetical motivations, than to anything else and, in particular, to an empirical science. There is, however, a further point which, I believe, needs stressing. As a mathematical discipline travels far from its empirical source, or still more, if it is a second and third generation only indirectly inspired by ideas coming from 'reality', it is beset with very grave dangers. It becomes more and more purely aestheticising, more and more purely l'art pour l'art. This need not be bad, if the field is surrounded by correlated subjects, which still have closer empirical connections, or if the discipline is under the influence of men with an exceptionally well-developed taste. But there is grave danger that the subject will develop along the line of least resistance, that the stream, so far from its source, will separate into a multitude of insignificant branches, and that the discipline will become a disorganised mass of details and complexities. In other words, at a great distance from its empirical source, or after much 'abstract' inbreeding, a mathematical subject is in danger of degeneration.

Professor John von Neumann in the first paper of his collected works

12 COMBINATORIAL OPTIMIZATION

Michel Goemans, Franz Rendl

12.1 FROM COMBINATORIAL OPTIMIZATION TO SDP

In this section we investigate various ways to derive semidefinite relaxations of combinatorial optimization problems. We start out with a generic way to obtain an SDP relaxation for problems in binary variables. The key step is to linearize quadratic functions in the original vector $x \in \mathbb{R}^n$ through a new $n \times n$ matrix X, see also Chapter 13.

We also look at some specific combinatorial optimization problems where the special structure of the problem yields SDP models, different from the generic one.

12.1.1 Quadratic problems in binary variables as SDP

To see how SDP relaxations of combinatorial optimization problems can be derived, we start with quadratic problems in $-1, 1$ variables.

$$z_{mc} := \max \quad x^T L x \text{ such that } x \in \{-1, 1\}^n.$$

Here L is a given (symmetric) matrix. This problem can be used to model the *Maximum Cut problem* defined as follows: Given a graph $G = (V, E)$ with a weight w_e for each edge $e \in E$, find a partition of V into S and $V \setminus S$ such that the edges across the partition, or in the *cut* $\delta(S) = \{(i,j) \in E : |\{i,j\} \cap S| = 1\}$, have maximum total weight. This problem can be formulated as a quadratic problem in $-1, 1$ variables by letting L be (up to a multiplicative factor of

1/4) the (weighted) Laplacian of the graph, i.e. $L_{ij} = -w_{ij}$ for $(i,j) \in E$, $L_{ii} = \sum_{j:(i,j)\in E} w_{ij}$, and $L_{ij} = 0$ otherwise.

To obtain a tractable relaxation of this NP-hard problem, we note that

$$x^T Lx = L \bullet (xx^T),$$

and introduce a new variable X for xx^T, i.e. we linearize. We have thus lifted a problem in a vector space of dimension n to the space of symmetric matrices of size $n \times n$. If we require now that $\mathrm{diag}(X) = e, X \succeq 0$ and $\mathrm{rank}\,(X) = 1$, then X is necessarily to be of the form $X = xx^T$ for some $-1, 1$ vector x. Unfortunately, the rank constraint is not convex and, by dropping it, we get the following relaxation:

$$z_{sdp} = \max \quad L \bullet X \text{ such that } \mathrm{diag}(X) = e, X \succeq 0. \tag{12.1.1}$$

If the objective also contains a linear term $c^T x$, the easiest way to handle it is to introduce a new variable $x_0 \in \{-1, 1\}$, and define X'_{0i} to represent x_i and X'_{ij} to represent $x_i x_j$. We can thus express the augmented matrix X' as

$$X' = \begin{pmatrix} 1 & x^T \\ x & X \end{pmatrix}$$

where the first block corresponds to the index 0. Observing that

$$\begin{pmatrix} 1 & 0 \\ -d & I \end{pmatrix} \begin{pmatrix} 1 & d^T \\ d & X \end{pmatrix} \begin{pmatrix} 1 & -d^T \\ 0 & I \end{pmatrix} = \begin{pmatrix} 1 & 0 \\ 0 & X - dd^T \end{pmatrix},$$

we obtain that $X' \succeq 0$ is equivalent to $X - xx^T \succeq 0$. In other words, we have replaced the (non-convex) equality $X' = xx^T$ by a convex inequality $X - xx^T \succeq 0$.

Let us now consider problems in 0, 1 variables. Linearizing as before leads again to $X = xx^T$, but now we require $\mathrm{diag}(X) = x$ since $x_i^2 = x_i$ for $x_i \in \{0, 1\}$. The constraint $X - \mathrm{diag}(X)\mathrm{diag}(X)^T = 0$ is not convex, but we relax it to

$$X - \mathrm{diag}(X)\mathrm{diag}(X)^T \succeq 0.$$

As above, we can equivalently rewrite this constraint as

$$\begin{pmatrix} 1 & \mathrm{diag}(X)^T \\ \mathrm{diag}(X) & X \end{pmatrix} \succeq 0.$$

The corresponding relaxation is now:

$$z_{sdp} = \max \quad L \bullet X \text{ such that } X - \mathrm{diag}(X)\mathrm{diag}(X)^T \succeq 0.$$

Of course, a quadratic problem in 0, 1 variables can easily be transformed (via a linear transformation) into a quadratic problem in $-1, 1$ variables and vice versa, provided we allow for linear terms. It is a simple (but slightly tedious) matter to derive that, in the process, the corresponding semidefinite programming relaxations are transformed into one another, see for instance [337].

Another equivalent way to derive these relaxations in the above cases is through the use of Lagrangean relaxation. Consider

$$z_{mc} := \max \quad x^T Lx \text{ such that } x \in \{-1,1\}^n,$$

and suppose we write the constraints as $x_i^2 = 1$ for $i = 1, \cdots, n$. Dualizing these quadratic constraints with multipliers μ_i for $i = 1, \cdots, n$, we obtain the Lagrangean:

$$\begin{aligned} L(\mu) &:= \max \quad x^T Lx - \sum_i \mu_i x_i^2 + \sum_i \mu_i \\ &= \max \quad x^T (L - \mathrm{Diag}\,(\mu))x + \sum_i \mu_i. \end{aligned}$$

Since x is now unconstrained, this Lagrangean has a finite value if and only if $L - \mathrm{Diag}\,(\mu) \preceq 0$, and thus we obtain the following Lagrangean dual:

$$\min_\mu L(\mu) = \min \sum_i \mu_i : \mathrm{Diag}\,(\mu) \succeq L.$$

This is precisely the dual semidefinite program to (12.1.1). See Lemaréchal and Oustry [473] for a detailed account of this use of Lagrangean relaxation.

Yet another way to derive z_{sdp} for the quadratic problem in $-1, 1$ variables is to relax the constraint $x \in \{-1,1\}^n$ to $x^T x = n$. We then get

$$\begin{aligned} &\max \quad x^T Lx \text{ such that } x^T x = n \\ &= \max_{x \neq 0} n \frac{x^T Lx}{x^T x} \\ &= n\lambda_{max}(L), \end{aligned}$$

where $\lambda_{max}(L)$ denotes the largest eigenvalue of L, by the Rayleigh-Ritz theorem, see e.g. Section 2.3.3. Noticing that $x^T \mathrm{Diag}\,(\eta)x$ is constant over $x \in \{-1,1\}^n$, we can improve this bound to:

$$\min_{\eta : e^T \eta = 0} n\lambda_{max}(L + \mathrm{Diag}\,(\eta)),$$

which can be seen again to be z_{sdp}.

12.1.2 *Modeling linear inequalities*

Lifting linear inequalities, say

$$a^T x \leq \alpha, \tag{12.1.2}$$

into the space of symmetric matrices is less straightforward. Let us assume that we have $x_i \in \{0,1\}$. One obvious way to lift (12.1.2) would be to ask that

$$\mathrm{Diag}\,(a) \bullet X \leq \alpha.$$

Formally, this amounts to using $x_i = x_i^2$ and replacing (12.1.2) by the quadratic inequality $\sum_i a_i x_i^2 \leq \alpha$. Let us call this *diagonal lifting.*

In [340] the following idea is exploited. Consider constraints of the form

$$|a^T x| \leq \alpha. \tag{12.1.3}$$

(Constraint (12.1.2) is of this form if $a \geq 0$.) Then (12.1.3) can be squared to give

$$(aa^T) \bullet X \leq \alpha^2.$$

We call this *squared lifting.* When $a \geq 0$ (and hence $a^T x \geq 0$ for all x), Bauvin (personal communication, 1999) proposes an extension of this idea by replacing (12.1.3) with

$$(aa^T) \bullet X \leq \alpha(a^T \operatorname{diag}(X)),$$

and imposing $X \succeq \operatorname{diag}(X)\operatorname{diag}(X)^T$. Let us call this *extended squared lifting.*

Lovász and Schrijver [497] investigate in great detail how linear inequalities can be turned into quadratic ones for lifting. They propose the following basic procedure. Take (12.1.2) and multiply it by x_j and $(1-x_j)$, leading to quadratic inequalities in x, hence linear in X once we identify $x_i x_j$ with X_{ij}. This is a very powerful idea, but one has to keep in mind that we obtain $2n$ lifted inequalities for each original inequality. Let us call this *Lovász-Schrijver lifting.*

It is not hard to show that diagonal lifting is dominated by squared lifting, in turn dominated by extended square lifting, and this is in turn dominated by the Lovász-Schrijver lifting, see e.g. [340].

Example 12.1.1 *The following example from [340] shows that the difference between these liftings can be quite drastic. Consider*

$$\max \quad x^T(J-I)x \text{ such that } \sum_i x_i \leq k, \; x_i \in \{0,1\}.$$

(Here J is the matrix of all ones and I is the identity matrix.) The optimal solution to the diagonal lifting turns out to be $(n-1)k$, while squared lifting yields $k^2 - \frac{k}{n}k$. Finally, both the extended squared lifting and the Lovász-Schrijver lifting gives the exact optimum, which is $k^2 - k$.

In summary, there are many different ways of lifting linear inequalities, and in practice there is a trade-off between the tightness of the relaxation and the computational cost for solving it; this area is still not very well understood and more work needs to be done.

12.2 SPECIFIC COMBINATORIAL OPTIMIZATION PROBLEMS

In this section we look in some detail at several graph optimization problems, which have SDP relaxations that are derived through the specific structure of the problem, and differ from the generic relaxation, introduced in the previous section.

We will use the following notation. If A is a symmetric matrix, we define its *Laplacian* $L(A)$ to be the matrix $L(A) := D(A) - A$, where $D(A)$ is the diagonal matrix having the row sums of A on the main diagonal, $D(A) := \mathrm{Diag}(Ae)$. The eigenvalues of a symmetric matrix A will usually be numbered nondecreasingly, $\lambda_1(A) \leq \ldots \leq \lambda_n(A)$.

12.2.1 Equipartition

The *k-Equipartition problem (EQP)* is defined as follows. A graph G on $n = m \cdot k$ vertices is given through its (weighted) adjacency matrix A. The goal is to find a partition $S = (S_1, \ldots, S_k)$ of $V(G)$ into k sets $S_1, \ldots, S_k$ of equal cardinality, i.e. $|S_j| = m$, such that the total weight of edges joining different sets S_j is minimized. We associate to S the $n \times k$ *partition matrix* $Y = (y_{ij})$ with $y_{ij} = 1$ if $i \in S_j$ and $y_{ij} = 0$ otherwise. In other words, column j of Y is the indicator vector of S_j.

For such matrices, we note that $\mathrm{Trace}\, Y^T DY = \mathrm{Trace}\, D$, if D is diagonal. On the other hand $y_j^T A y_j$ is twice the weight of edges with both ends in S_j, hence $\mathrm{Trace}\, Y^T AY = 2w_{uncut}$, the weight of all edges not cut by the partition defined by Y. The total weight of edges cut by Y is therefore

$$w_{cut} = \tfrac{1}{2}\mathrm{Trace}\, Y^T LY.$$

To obtain a relaxation, the cost function $\mathrm{Trace}\, Y^T LY = \mathrm{Trace}\, L(YY^T)$ is linearized replacing YY^T by a new variable X. This leads to the investigation of the set

$$F_k := \mathrm{Conv}\,\{YY^T : Y \text{ partition matrix}\}.$$

(Note that in this linearization we replace a sum of quadratic terms, namely $\sum_r y_{ir} y_{jr}$ by a new variable x_{ij}. Hence X is $n \times n$ even though there are $n \cdot k$ original binary variables in Y. A direct linearization of the binary variables would have ended up with a matrix of order $n \cdot k$. This later linearization is investigated in the thesis of Zhao, see [869] and also [839].) The following properties of matrices $X \in F_k$ are immediate:

1. $Xe = me$ (because $Y^T e_n = me_k, Ye_k = e_n$)
2. $\mathrm{diag}(X) = e$ (because each row of Y has exactly one entry equal 1)
3. $X \geq 0$, (because $Y \geq 0$)
4. $X \succeq 0$ (because $YY^T \succeq 0$)
5. $mI - X \succeq 0$ (because $Y^T Y = mI_k$ and YY^T has the same nonzero eigenvalues as $Y^T Y$).

Depending on which subsets of these constraints are selected, we can derive most of the relaxations commonly in use for EQP. Linear relaxations are investigated mostly for $k = 2$, and use additional combinatorial cutting planes, see [160, 144]. Surprisingly enough, the purely linear relaxation was not the first model investigated. Donath and Hoffman [197] use the following well-known characterization for the sum of the k smallest eigenvalues of a symmetric matrix A,

$$\sum_{j=1}^{k} \lambda_j(A) = \min_{Y^TY=I_k} \operatorname{Trace} Y^T AY.$$

They propose the following relaxation,

$$\begin{aligned} w_{cut} &\geq \max_{u^Te=0} \min_{Y^TY=I_k} \operatorname{Trace} \tfrac{1}{2} Y^T(L + \operatorname{Diag}(u))Y \\ &= \max_{u^Te=0} \tfrac{1}{2} \sum_j \lambda_j(L + \operatorname{Diag}(u)) \\ &=: w_{DH}. \end{aligned}$$

It is a simple exercise using duality for SDP, see [21], to rewrite this lower bound as

$$w_{DH} = \min \tfrac{1}{2} L \bullet X \text{ such that } mI \succeq X \succeq 0, \operatorname{diag}(X) = e,$$

corresponding to the constraints 2., 4. and 5 for F_k. Rendl and Wolkowicz [660] improved this model to get

$$\begin{aligned} w_{cut} &\geq \max_{u^Te=0} \min_{Y^TY=I_k, Ye=me, Y^Te=e} \tfrac{1}{2} \operatorname{Trace} Y^T(L + \operatorname{Diag}(u))Y \\ &=: w_{RW}. \end{aligned}$$

Using duality again, it can be shown, see [396] that

$$w_{RW} = \min \tfrac{1}{2} L \bullet X \text{ such that } \operatorname{diag}(X) = e, Xe = me, \ mI \succeq X \succeq 0,$$

and thus constraints 1. are also taken into consideration.

The bounds w_{DH} and w_{RW}, written as SDP, have two independent semidefiniteness constraints. This is of no theoretical importance, but it increases the computational effort in practical computations. These bounds often give quite good approximations. In [213], computational experience is given for real world problems with $k = 2$ and n up to 15000, showing that the bound w_{RW} yields tight approximations of the true integer optimum with a gap in the range of a few percentage points.

The following SDP relaxation for k-equipartition, introduced in [396], has only one semidefiniteness constraint:

$$\min \tfrac{1}{2} L \bullet X \text{ such that } \operatorname{diag}(X) = e, Xe = me, X \geq 0, X \succeq 0.$$

Since $X \geq 0, Xe = me$ implies by the Perron-Frobenius theorem that X has no eigenvalue larger than m, it is concluded in [396] that this model dominates the eigenvalue relaxations w_{DH} and w_{RW}; it thus corresponds to combining all constraints 1.-5. discussed previously. Practical experience based on this relaxation for problems arising in telecommunication are reported in [489]. It turns out that this model approximates the equipartition problem quite nicely, even for large values of k.

Finally it should be mentioned that the purely linear relaxation of k-EQP, using only $\text{diag}(X) = e$, $Xe = me$, $X \geq 0$ gives surprisingly good results for problems where k is large, say $k \approx \sqrt{n}$ or larger. Computational details are given in section 12.3.

If $A \geq 0$, then $L \succeq 0$, hence $\text{Trace}\, Y^T LY$ is convex. Therefore one might also think of minimizing $\frac{1}{2}\text{Trace}\, Y^T LY$ over the convex hull of partition matrices Y, i.e. over $\sum_j y_{ij} = 1$, $\sum_i y_{ij} = m$, $y \geq 0$. Unfortunately, $\hat{y_{ij}} = \frac{1}{k}$, the average of all partition matrices gives $\text{Trace}\, \hat{Y}^T L\hat{Y} = 0$, so the (convex) quadratic programming relaxation is too weak.

12.2.2 Stable sets and the θ function

Given a graph $G = (V, E)$, a *stable* set is a subset of pairwise non-adjacent vertices. The problem of finding a stable set of maximum cardinality (or of maximum weight if the vertices are weighted) is a basic (NP-hard) combinatorial optimization problem. Let $\alpha(G)$ denote the size of a maximum stable set in G. In a seminal paper [496] (see also Grötschel et al. [304]), Lovász proposed an upper bound on $\alpha(G)$ known as the theta function $\vartheta(G)$. The theta function can be expressed in many equivalent ways, as an eigenvalue bound, as a semidefinite program, or in terms of orthonormal representations.

As an eigenvalue bound, $\vartheta(G)$ can be derived as follows. Consider $P = \{A \in S_n : a_{ij} = 1 \text{ if } (i,j) \notin E \text{ (or } i = j)\}$. If there exists a stable set of size k, the corresponding principal submatrix of any $A \in P$ will be J_k, the all ones matrix of size k. By interlacing of eigenvalues for symmetric matrices, we derive that $\lambda_{max}(A) \geq \lambda_{max}(J_k) = k$ for any $A \in P$, where $\lambda_{max}(\cdot)$ denotes the largest eigenvalue. As a result, $\min_{A \in P} \lambda_{max}(A)$ is an upper bound on $\alpha(G)$, and this is one of the equivalent formulations of Lovász's theta function.

This naturally leads to a semidefinite program since $\lambda_{max}(A) = \min\{t : tI - A \succeq 0\}$. We also observe that $A \in P$ is equivalent to $A - J$ being generated by E_{ij} for $(i,j) \in E$, where all entries of E_{ij} are zero except for (i,j) and (j,i). Thus, we can write

$$\vartheta(G) = \min\{t : tI + \sum_{(i,j) \in E} x_{ij} E_{ij} \succeq J\}.$$

By strong duality, we can also write:

$$\vartheta(G) = \max J \bullet Y \text{ such that } Y_{ij} = 0 \text{ for } (i,j) \in E, I \bullet Y = 1, Y \succeq 0.$$

Lovász's first definition of $\vartheta(G)$ was in terms of orthonormal representations. An *orthonormal representation* of G is a system $v_1, \cdots, v_n$ of unit vectors in R^n such that v_i and v_j are orthogonal (i.e. $v_i^T v_j = 0$) whenever i and j are not adjacent. The value of the orthonormal representation is $z = \min_{c:||c||=1} \max_{i \in V} \frac{1}{(c^T v_i)^2}$. This is an upper bound on $\alpha(G)$, since $||c||^2 \geq \sum_{i \in S} (c^T v_i)^2 \geq |S|/z$ for any stable set S. Taking the minimum value over all orthonormal representations of G, one derives another expression for $\vartheta(G)$. This result can be restated in a slightly different form. If x denotes the incidence vector of a stable set then we have that

$$\sum_i (c^T v_i)^2 x_i \leq 1. \tag{12.2.4}$$

In other words, the *orthonormal representation constraints* (12.2.4) are valid inequalities for $STAB(G)$, the convex hull of incidence vectors of stable sets of G. Grötschel et al. [303] show that if we let $TH(G) = \{x : x \text{ satisfies (12.2.4) and } x \geq 0\}$, then $\vartheta(G) = \max\{\sum_i x_i : x \in TH(G)\}$. Yet more formulations of ϑ are known.

Perfect graphs. A graph G is called *perfect* if, for every induced subgraph G', its chromatic number is equal to the size of the largest clique in G'. It is not known if the recognition problem of deciding whether a graph is perfect is in P or NP-complete. However, the theta function gives some important characterizations (but not a "good" or NP∩co-NP characterization) of perfect graphs.

Theorem 12.2.1 (Grötschel et al. [303]) *The following are equivalent:*

- *G is perfect,*
- $TH(G) = \{x \geq 0 : \sum_{i \in C} x_i \leq 1$ *for all cliques* $C\}$
- *$TH(G)$ is polyhedral.*

Moreover, even though recognizing perfect graphs is still open, one can find the largest stable set in a perfect graph in polynomial time by computing the theta function using semidefinite programming (Grötschel et al. [302, 304]); similarly one can solve the weighted problem, or find the chromatic number or the largest clique.

Although $\vartheta(G) = \alpha(G)$ for perfect graphs, $\vartheta(G)$ can provide a fairly poor upper bound on $\alpha(G)$ for general graphs. Feige [226] has shown the existence of graphs for which $\vartheta(G)/\alpha(G) \geq \Omega(n^{1-\epsilon})$ for any $\epsilon > 0$. See [281] for further details and additional references on the quality of $\vartheta(G)$.

12.2.3 Traveling salesman problem

An interesting SDP model for the Traveling Salesman Problem (TSP) was recently introduced by Cvetković et al [179]. The key idea of this approach is to use the concept of *algebraic connectivity* to enforce connectivity. The algebraic connectivity $\alpha(A)$ of a graph given by its adjacency matrix A was introduced by Fiedler [232], and is defined to be the second-smallest (Laplacian) eigenvalue of $L(A)$, $\alpha(A) := \lambda_2(L(A))$. Fiedler [232] shows that $\alpha(A) := \lambda_2(L(A)$, behaves in many ways quite similar to the edge-connectivity $\mu(A)$ of the graph with (weighted) adjacency matrix A:

$$\mu(A) := \min\left\{ \sum_{i\in S, j\notin S} a_{ij} : 0 < |S| < |V(G)|, S \subset V(G) \right\}.$$

We recall that e is an eigenvector of $L(A)$, corresponding to the eigenvalue 0. Since $L(A) \succeq 0$ if $A \geq 0$, $\alpha(A)$ can be expressed as

$$\alpha(A) = \min\{x^T L(A)x : x^T x = 1, x^T e = 0\}.$$

From this characterization it is clear that $\alpha(A)$ is concave on the set of symmetric nonnegative matrices.

Now let C_n be the cycle on n vertices. Its Laplacian eigenvalues are $2(1 - \cos(\frac{2k\pi}{n}))$ for $k = 0, \cdots, n-1$, see [232]. Therefore

$$\alpha(C_n) = 2\left(1 - \cos\frac{2\pi}{n}\right) =: h_n. \tag{12.2.5}$$

In [179] it is therefore concluded that if X is in the convex hull of all Hamiltonian cycles through n vertices, then

$$\alpha(X) \geq h_n.$$

This is turned into a semidefiniteness condition on the Laplacian $L(X)$. Since $\lambda_1(L(X)) = 0$, with eigenvector e, we get that the eigenvalues of $L(X)+\beta J$ are βn (for the eigenvector e), and $\lambda_2(L(X)), \ldots, \lambda_n(L(X))$. If $\beta n \geq h_n$, which for example holds for $\beta = 1$ and $n \geq 3$, then it follows that $\lambda_1(L(X) + \beta J) \geq h_n$. Conversely, if all eigenvalues of $L(X) + \beta J$ are at least as large as h_n, then $\alpha(X) \geq h_n$. Hence we have shown that a nonnegative symmetric matrix X of order $n \geq 3$ has

$$\alpha(X) \geq h_n \text{ if and only if } L(X) + J - h_n I \succeq 0.$$

This is used in [179] to characterize Hamiltonian cycles as follows.

Lemma 12.2.1 *[179] The following statements are equivalent for all $n \geq 3$.*

1. *H is the adjacency matrix of a Hamiltonian cycle of length n.*
2. *H is a symmetric 0-1 matrix satisfying (1) $He = 2e$, (2) $\operatorname{diag}(H) = 0$, (3) $L(H) + J - h_n I \succeq 0$.*

Proof.
We just argued that (1) implies (2). Conversely, if a 0-1 matrix H satisfies (2), then it is the adjacency matrix of a 2-regular graph by the first two equations in (2). The last condition shows that the resulting graph is connected.

This leads to the following integer semidefinite program to model TSP, proposed for the first time in [179]:

$$\min \tfrac{1}{2} C \bullet X \text{ such that}$$

$\text{diag}(X) = 0, Xe = 2e, L(X) + J - h_n I \succeq 0$, and X is a symmetric 0-1 matrix. Relaxing integrality of X to $0 \leq X \leq J$ yields a semidefinite relaxation with simple inequality constraints:

$$\min \tfrac{1}{2} C \bullet X \text{ such that}$$
$$\text{diag}(X) = 0, Xe = 2e, L(X) + J - h_n I \succeq 0, \text{ and } 0 \leq X \leq J.$$

How tight is this relaxation in comparison to purely linear relaxations? It should be noted first that $L(X) + J \succeq 0$ for any nonnegative symmetric matrix X, hence

$$\lambda_1(L(X) + J - h_n I) \geq -h_n.$$

Since $h_n \searrow 0$ as $n \to \infty$, one might think that the semidefiniteness condition on X should get weaker as n is increased. We will in fact show more, namely that the subtour elimination model dominates this relaxation for all $n \geq 3$.

The *subtour elimination* relaxation is a commonly used linear relaxation for TSP. In our matrix notation, this problem reads

$$\min \tfrac{1}{2} C \bullet X, \text{ such that } \text{diag}(X) = 0, Xe = 2e, \mu(X) \geq 2, 0 \leq X \leq J.$$

The constraint $\mu(X) \geq 2$ expands into $O(2^n)$ linear inequalities, $\sum_{i \in S, j \notin S} x_{ij} \geq 2$ for all nontrivial sets S of vertices. To compare these relaxations we need to understand how $\alpha(X)$ and $\mu(X)$ relate to each other. The main tool is the following result proved by Fiedler [231].

Theorem 12.2.2 *[231] Let A be a symmetric stochastic $n \times n$ matrix with edge connectivity $\mu = \mu(A)$ and eigenvalues*

$$1 = \lambda_n(A) \geq \ldots \geq \lambda_1(A).$$

Then

$$1 - \lambda_{n-1}(A) \geq \max \left\{ 2(1 - \cos \frac{\pi}{n})\mu,\ 1 - 2(1 - \mu)\cos \frac{\pi}{n} - (2\mu - 1) cos \frac{2\pi}{n} \right\}.$$

Using this result, we can now show that the subtour elimination model is at least as strong as the SDP relaxation, modeling $\alpha(X) \geq h_n$.

Theorem 12.2.3 *Let X be a symmetric $n \times n$ matrix. Suppose further that* $\text{diag}(X) = 0, Xe = 2e, 0 \leq X \leq J, \mu(X) = 2$. *Then $\alpha(X) \geq h_n$.*

Proof.
(The proof of this result is contained in [232] in the case of unweighted graphs.) The matrix $Y := \frac{1}{2} X$ is stochastic with $\mu = \frac{1}{2}\mu(X) = 1$. Theorem 12.2.2 says that

$$1 - \lambda_{n-1}(Y) \geq 1 - 2(1 - \mu)\cos \frac{\pi}{n} - (2\mu - 1)\cos \frac{2\pi}{n} = 1 - \cos \frac{2\pi}{n} = \tfrac{1}{2} h_n.$$

Now $L(Y) = I - Y$ and thus $L(X) = 2(I - Y)$, implying that

$$\alpha(X) = \lambda_2(L(X)) = 2(1 - \lambda_{n-1}(Y)) \geq h_n.$$

12.2.4 Quadratic assignment problem

The *Quadratic Assignment Problem (QAP)* is one of the most difficult combinatorial optimization problems. It is usually formulated in the following way. For given symmetric matrices A, B and arbitrary matrix C, all of order n, find a permutation matrix $X = (x_{ij})$ minimizing

$$\operatorname{Trace}((AXB + C)X^T).$$

We refer to [153] for a comprehensive summary on QAP. A direct attempt to model this quadratic function using SDP leads immediately to matrices of order n^2. On the other hand, the Hoffman-Wielandt inequality provides the following bound on the quadratic part of the cost function.

Theorem 12.2.4 *Let A and B be symmetric matrices of order n, with spectral decompositions $A = PDP^T, B = QEQ^T, D = \operatorname{diag}(\lambda_1(A), \ldots, \lambda_n(A))$ and $E = \operatorname{diag}(\lambda_n(B), \ldots, \lambda_1(B))$. Then*

$$\min\{\operatorname{Trace}(AXBX^T) : X^T X = I\} = \sum_i \lambda_i(A)\lambda_{n+1-i}(B). \tag{12.2.6}$$

The minimum is attained for $X = PQ^T$.

It should be noted that P and Q diagonalize the respective matrices in such a way that the scalar product of the eigenvalues of A and B is minimized.

Since permutation matrices are orthogonal, this result can be used to bound the quadratic part of the cost function of QAP.

It was recently shown by Anstreicher and Wolkowicz [41] that in fact the minimization in (12.2.6) can equivalently be expressed as an SDP.

Theorem 12.2.5 *[41]*

$$\begin{aligned} &\min\{\operatorname{Trace}(AXBX^T) : X^T X = I\} \\ &= \max \ \{\operatorname{Trace} S + \operatorname{Trace} T : (B \otimes A) - (I \otimes S) - (T \otimes I) \succeq 0\} \end{aligned} \tag{12.2.7}$$

The minimization problem in (12.2.6) can be solved quite efficiently, by computing the eigenvalue decompositions of A and B. The SDP on the right hand side however involves matrices of order n^2. The linear term from the original QAP can be included in the right hand side model as follows:

$$\max \ \operatorname{Trace} S + \operatorname{Trace} T \text{ such that } (B \otimes A) + \operatorname{Diag}(C) - (I \otimes S) - (T \otimes I) \succeq 0.$$

It is not known whether this model can be solved directly in the original space of $n \times n$ matrices. It is possible, though, to use the left hand side model in Theorem 12.2.5 to warm-start it. Further refinements and variations of this model along with computational results are contained in [870].

12.3 COMPUTATIONAL ASPECTS

It is well known that SDP can be solved in polynomial time to some fixed precision. In practice, interior point methods have turned out most efficient for SDP whenever dense linear algebra is feasible.

SDP relaxations of combinatorial optimization problems often have the following property. Aside from the semidefiniteness constraint, such as $X \succeq 0$, there are some generic (easy) linear equalities, such as $\mathrm{diag}(X) = e$ in the case of the maximum cut problem, or $Xe = me$ for the equipartition problem EQP. Then there is a large set (not necessarily polynomial in size) of combinatorially defined inequalities, such as the *triangle inequalities* for the maximum cut problem (in this case, there are $O(n^3)$ inequalities).

The computational effort to solve SDP grows significantly with the number of constraints. Therefore it is crucial to identify among all possible (combinatorial) constraints those which are most likely to be active at the optimum. This is usually done in an iterative way. Some of the most violated inequalities are included, and the new SDP is resolved. Then inactive constraints are dropped and the process of adding constraints is iterated. In general it is computationally infeasible to iterate this process until all combinatorial constraints are satisfied both because of computer memory and computation time limitations. Therefore one is interested in finding the most important constraints quickly.

To give a flavor of how this works, we refer to [338] for computational results in the case of the Max-Cut problem, and to [396] for general graph bisection. In Tables 12.1 and 12.2 we summarize some more recent computational results on real world data for EQP. The data are taken from [489] and represent a subnetwork from the french telecommunications network. The number of vertices for this problem is $n = 100$, with edge density about 95%. We consider partitioning the vertices into k subsets of equal cardinality, where $k \in \{5, 10, 20\}$. Our goal is to solve the relaxation

$$\min \tfrac{1}{2} L \bullet X \text{ such that } \mathrm{diag}(X) = e, Xe = me, X \geq 0, X \succeq 0.$$

However, as just explained, it is computationally prohibitive to solve this SDP directly, because the sign constraints $X \geq 0$ amount to 4950 inequalities. To get at least a good approximation to this relaxation, we can proceed as follows.

SDP based approach.. Solve the SDP

$$\min \tfrac{1}{2} L \bullet X \text{ such that } \mathrm{diag}(X) = e, Xe = me, X \succeq 0,$$

and iteratively add only a few of the most violated sign constraints $x_{ij} \geq 0$. Table 12.1 contains some results using this approach. We note for instance that in the case of partitioning into $k = 10$ subsets, the initial lower bound of 275.51 is improved to a value of 317.48 by adding a total of 268 sign constraints to the initial SDP. The largest violation of nonnegativity is initially -0.8, and is reduced to -0.45. It is also worth pointing out that the computational effort grows quickly with the number of sign constraints. The computation times are minutes on a PC (Pentium 2, 233 Mhz). We use the matrix X from the relaxation to generate an integer solution obtained through rounding and some post-processing. The value of the best feasible solution found is given in the column labeled 'cut'. The computational behavior shown in Table 12.1 is quite typical for SDP relaxations of combinatorial optimization problems. Most of the improvement is obtained in the first few iterations. In later stages of adding cutting planes, the computational effort outweighs the small marginal improvement. This is also the reason why we did not go on with the iterations.

LP-based approach.. Looking at the results of the SDP-based approach, we note that as k gets larger the sign constraints seem to get increasingly important. This suggests to use linear programming as a starting point, and add 'SDP cuts'. Thus we consider solving the LP

$$\min \tfrac{1}{2} L \bullet X \text{ such that } \operatorname{diag}(X) = e, Xe = me, 0 \leq x_{ij} \leq 1.$$

(The upper bound constraints $x_{ij} \leq 1$ are not included in the SDP model, because $X \succeq 0, \operatorname{diag}(X) = e$ implies them.) If the resulting X from this LP is not positive semidefinite, then there exist vectors a such that $a^T X a < 0$. Thus we may add SDP cuts of the form $a^T X a \geq 0$ for appropriately chosen vectors a. We can do a little better however, by adding triangle constraints of the form

$$x_{ik} - x_{ij} - x_{jk} \geq -1$$

which essentially state that if i is in the same set as j, and j is in the same set as k, then i and k must also be in the same set. In Table 12.2 we summarize some results using this approach. We also indicate the smallest eigenvalue $\lambda_{\min}(X)$ of X. The qualitative behavior seems to be complementary to the SDP-based approach. Here the results for large value of k, $k = 20$ are by far superior to the SDP approach.

Further details and also computational experience with larger instances are contained in [489].

12.4 SDPS REDUCING TO EIGENVALUE BOUNDS

In this section we investigate simplifications of SDP in which they reduce to linear programs involving the eigenvalues of the input matrices, when these input matrices satisfy a special property. The results of this section are based on [283]. We start out with the following dual pair of problems.

$$(P) \quad \min A_0 \bullet X \text{ such that } A_i \bullet X = b_i, \; i = 1, \ldots, m, \quad X \succeq 0,$$

and

$$(D) \quad \max \; b^T y \text{ such that } A_0 - \sum_{i \geq 1} y_i A_i =: Z \succeq 0.$$

Let us suppose that all the input matrices commute, i.e.

$$A_i A_j = A_j A_i \quad \forall i, j \in \{0, 1, \ldots, m\}. \tag{12.4.8}$$

k	cut	SDP-bound	# sign constr.	largest viol.	time
5	283.16	244.90	0	-.60	0.13
		259.98	101	-.44	0.76
		264.33	176	-.39	1.74
		267.21	229	-.24	2.86
10	364.07	275.51	0	-.80	0.14
		296.61	95	-.72	0.84
		308.57	177	-.55	1.81
		317.48	268	-.45	3.07
20	467.22	290.81	0	-.90	0.17
		315.87	96	-.83	0.96
		329.21	165	-.77	2.07
		341.47	256	-.67	3.38

Table 12.1 SDP approximation for EQP by iteratively adding sign constraints. The number and largest violation of sign constraints are displayed together with computation time and the lower bound.

k	cut	LP-bound	$\lambda_{\min}(X)$	time
5	283.16	118.53	-5.69	-
		254.30	-3.08	10
10	364.07	273.02	-3.76	-
		346.89	-1.79	5.3
20	467.22	429.98	-2.19	-
		454.95	-1.05	1.2

Table 12.2 LP-based approximation for EQP. Improvement using triangle inequalities.

Then there exists an orthogonal matrix P, such that

$$P^T A_i P = D_i = \mathrm{Diag}\,(d_i).$$

In other words, the columns of P form a set of eigenvectors common to all matrices A_i. Since the dual slack matrix is a linear combination of the A_i, it also commutes with all A_i. Therefore $Z \succeq 0$ if and only if $\sum_i y_i d_i \leq d_0$. If the eigenvalue vectors $d_1, \ldots, d_m$ are collected in the matrix $D = (d_1, \ldots, d_m)$, we can rewrite (D) as an ordinary linear program (LP-D):

$$\text{(LP-D)} \quad \max \; b^T y \text{ such that } \; Dy \leq d_0.$$

If we assume in addition that both (P) and (D) are feasible, then (LP-D) has a finite optimum, and hence also its dual, which reads

$$\text{(LP-P)} \quad \min d_0^T x \text{ such that } D^T x = b, x \geq 0.$$

The optimal solution x of this LP can be used to define the matrix $X = P\text{Diag}(x)P^T$. By construction, this matrix is feasible for the SDP (P). It has the same objective function value as (D), hence this solution is optimal. In summary, we have proved the following theorem.

Theorem 12.4.1 *[283] Suppose that the matrices defining (P) and (D) commute, and that both problems are feasible. Then there exist optimal solutions X and y, Z to both problems that can be obtained through the linear programs (LP-P) and (LP-D). Strong duality holds without further constraint qualification.*

Example 12.4.1 *As an application of this result we derive a well known lower bound for the* graph bisection problem, *or 2-EQP. This problem asks to partition the vertices of a graph into two sets of equal cardinality so as to minimize the total weight of edges cut by the partition. Boppana [123] shows that a lower bound to the equicut problem is given by $\frac{n}{4}\lambda_2(L)$, if L denotes the weighted Laplacian. Modeling bisections by $(1,-1)$ vectors x, the equicut problem can be written as*

$$\min \frac{1}{4} x^T Lx \text{ such that } x^T e = 0, x \in \{-1,1\}^n.$$

The following relaxation leads to an SDP with all data matrices commuting. (Here xx^T is replaced by X. $e^T x = e^T xx^T e = 0$ is turned into $J \bullet X = 0$.)

$$\min \frac{1}{4} L \bullet X \text{ such that } I \bullet X = n, J \bullet X = 0, X \succeq 0.$$

(Note that $LJ = 0$, hence L and J commute.) The matrix P diagonalizing L, J and I is determined by the vector parallel to e and the nontrivial eigenvectors of L. The solution of (LP-D) can be written down explicitly for this problem. The resulting vectors of eigenvalues are

$$d_0 = \frac{1}{4}\begin{pmatrix} 0 \\ \lambda_2(L) \\ \vdots \\ \lambda_n(L) \end{pmatrix}, d_1 = \begin{pmatrix} 1 \\ \vdots \\ 1 \end{pmatrix}, d_2 = \begin{pmatrix} n \\ 0 \\ \vdots \\ 0 \end{pmatrix}.$$

The optimal dual solution is therefore $y_1 = \frac{1}{4}\lambda_2(L)$ and $y_2 = -\frac{y_1}{n}$. Therefore the optimal objective function value is $\frac{n}{4}\lambda_2(L)$, in agreement with the Boppana bound.

In general, the input matrices arising from SDP relaxations of combinatorial optimization problems do not commute. However, in some cases, if the underlying graph has special properties then some of the inequalities of the relaxation can be aggregated without weakening it and in such a way that the resulting input matrices commute. This is for example the case for many combinatorial optimization problems defined on association schemes, see [283] and references therein for details.

12.5 APPROXIMATION RESULTS THROUGH SDP

One of the advantages of semidefinite programming is that sometimes it leads to much tighter relaxations than what can be easily expressed with purely linear constraints. This is for example the case for the stable set problem, as we saw in Section 12.2.2. In the worst-case though, we saw that $\vartheta(G)$ can be arbitrarily far away from $\alpha(G)$. In this section, we consider combinatorial optimization problems where semidefinite programming relaxations are always within a constant factor of the optimum, and where these relaxations can be used to efficiently (i.e., in polynomial-time) find a feasible solution of the problem of value within a certain factor, say α, of the optimum value. Such an algorithm is referred to as an approximation algorithm, or as an α-approximation algorithm.

The simplest and most illustrative example where semidefinite programming helps in the design of approximation algorithms is the maximum cut problem, as was discovered by Goemans and Williamson [285]. Let us assume that we are given a graph $G = (V, E)$ with nonnegative edge weights w_e for $e \in E$ and we would like to find a cut of maximum weight. It is easy to obtain a 0.5-approximation algorithm but improving on this appeared quite challenging. First of all, a classical linear programming relaxation of the problem (involving cycle constraints, and based on the fact that any cycle intersects a cut in an even number of edges) can be arbitrarily close to twice the optimum value [636]. Thus if we compare the solution obtained by any algorithm to this linear programming relaxation we cannot expect a performance guarantee α better than 0.5. Furthermore, recently, Hastad [329], based on a long series of developments in the area of probabilistically checkable proofs, has shown that it is NP-hard to approximate the maximum cut problem within $16/17 + \epsilon = 0.94117\cdots$ for any $\epsilon > 0$. This sets a limit to how close we can hope to be able to approximate the problem.

Goemans and Williamson [285] propose the following randomized approximation algorithm based on semidefinite programming with a performance guarantee of 0.87856. First solve (within a small tolerance ϵ) the SDP relaxation derived in (12.1.1):

$$z_{sdp} = \max \sum_{(i,j)\in E} w_{ij} \frac{1 - X_{ij}}{2} \text{ such that } \operatorname{diag}(X) = e, X \succeq 0.$$

Then using a Cholevsky decomposition, express X_{ij} as $v_i^T v_j$ for some vectors $v_i \in \mathbb{R}^n$ where $n = |V|$. Since $\operatorname{diag}(X) = e$, the vectors are of unit norm. Let r be a vector uniformly selected on the unit sphere $\{x \in \mathbb{R}^n : ||x|| = 1\}$. The hyperplane orthogonal to r separates the vectors into two sets, and this is the (random) cut output: $S = \{i : r^T v_i > 0\}$. This is referred to as *the random hyperplane technique.*

The analysis is elementary. The probability that a given edge (i, j) is in the cut can be seen to be equal to the angle between the vectors v_i and v_j normalized so that the probability is 1 when the angle is π. As a result, the expected value of the cut produced has a nice closed form formula:

$$E[w(\delta(S)] = \sum_{(i,j)\in E} w_{ij} \frac{\arccos(v_i^T v_j)}{\pi}.$$

If the weights are nonnegative then, comparing the corresponding terms of $E[w(\delta(S))]$ and z_{sdp}, we derive that $E[w(\delta(S))] \geq 0.87856 z_{sdp}$ since $\frac{\arccos(x)}{\pi} \geq 0.87856\frac{1-x}{2}$ for $-1 \leq x \leq 1$.

Karloff [402] has shown that the ratio $E[w(\delta(S))]/z_{sdp}$ can be arbitrarily close to z_{sdp}. Goemans and Williamson [285] and Zwick [877] have shown that improved bounds can be given as a function of $\rho = z_{sdp}/\sum_{(i,j)\in E} w_{ij}$, unless ρ is $0.84458\cdots$. Furthermore, for any value of ρ, this improved analysis of the random hyperplane technique is also tight, as shown by Alon and Sudakov [28] and Alon et al. (see [877]). We should point out though that the exact worst-case value of OPT/z_{sdp} is unknown where OPT denotes the optimum maximum cut value, although this worst-case lies between $0.87856\cdots$ and $0.88445\cdots$ where the latter corresponds to an unweighted 5-cycle.

When we express the objective function of z_{sdp} as $L \bullet X$ (as it would be more natural for quadratic problems), the expected value of the solution obtained by the random hyperplane technique can be conveniently written as $\frac{2}{\pi}L \bullet \arcsin[X]$ where $\arcsin[X]$ denotes the matrix whose entries are $\arcsin(X_{ij})$. A performance guarantee of β then means that $\frac{2}{\pi}L \bullet \arcsin[X] \geq \beta L \bullet X$ or $L \bullet (\frac{2}{\pi}\arcsin[X] - \beta X) \geq 0$. If we assume that L belongs to a cone K, we can bound β by imposing that $\frac{2}{\pi}\arcsin[X] - \beta X \in K^*$ whenever $K \succeq 0, \mathrm{diag}(X) = e$. When the cone K is the positive semidefinite cone, β can be chosen to be $\frac{2}{\pi}$ as shown by Nesterov [572].

The random hyperplane technique has been applied to several problems that fit fairly easily as quadratic problems, including the maximum directed cut problem [285, 227], the maximum k-cut problem and bisection problems [256, 858], maximum satisfiability problems (see [285, 227, 403, 877, 46]), graph coloring [393], machine scheduling problems [740], and more general quadratic problems (see Section 13.2 for details).

Although the random hyperplane technique is the only currently known method for turning the solution of a semidefinite program into an integer solution, a few variants and extensions have been proposed.

It is sometimes useful to apply the random hyperplane technique, not directly on the optimum $X \succeq 0$ given by a positive semidefinite relaxation, but rather on a new matrix $X' \succeq 0$. Two modifications have been used in different contexts. First, one could let X' be a convex combination of X and the identity matrix (or another positive semidefinite matrix). Observe that the random hyperplane technique applied to the identity matrix places every element i on either side of

S independently, and thus a convex combination of X and the identity allows to make the choices of where to place each element "more independently" of each other. This was exploited in [574, 740, 858, 877] for example.

In other cases, one variable in the original quadratic formulation, say x_1 plays a special role; this variable represents "true" for maximum satisfiability problems [285, 227], one specific side of the cut for the maximum dicut problem [285, 227], or one specific machine in the case of scheduling problems on 2 machines [740]. If the variable plays a more important role in the objective function than other variables, it is then useful to bias the random hyperplane technique in the direction of the corresponding vector v_1. For example, one can map each vector v_i to w_i where $w_1 = v_1$ and $w_i = (1 - \lambda(v_1^T v_i))v_i + \lambda(v_1^T v_i)v_1$, for some function $\lambda(x)$. This has been used successfully, see [227, 876, 740] for example.

The random hyperplane technique can also be used to obtain an ordering. So far, there has been only one such application, by Chor and Sudan [161] for the betweenness problem. In this case, we are given a ground set V and triplets $(x_i, x_j, x_k) \in V \times V \times V$ and we would like to find an ordering which maximizes the number of given triplets (x_i, x_j, x_k) for which x_j is between x_i and x_k. In the case of satisfiable instances, they set up an appropriate semidefinite program, solve it, derive the vectors v_i as before and then find an ordering by sorting the $r^T v_i$ for a random vector r on the unit sphere.

13 SEMIDEFINITE PROGRAMMING RELAXATIONS OF NONCONVEX QUADRATIC OPTIMIZATION

Yuri Nesterov, Henry Wolkowicz, Yinyu Ye

13.1 INTRODUCTION

Quadratically constrained quadratic programs, denoted Q^2P, are an important modelling tool, e.g.: for hard combinatorial optimization problems, Chapter 12; and SQP methods in nonlinear programming, Chapter 20. These problems are too hard to solve in general. Therefore, relaxations such as the Lagrangian relaxation are used. The dual of the Lagrangian relaxation is the SDP relaxation. Thus SDP has enabled us to efficiently solve the Lagrangian relaxation and find good approximate solutions for these hard, possibly nonconvex, Q^2P. This area has generated a lot of research recently. This has resulted in many strong and elegant theorems that describe the strength/performance of the bounds obtained from solving relaxations of these Q^2P.

For the simple Q^2P case of one quadratic constraint (the trust region subproblem) strong duality holds, even though both the objective function and constraint may be nonconvex, i.e. there is a zero duality gap and the dual is attained. In addition, necessary and sufficient (strengthened) second order optimality conditions and efficient algorithms exist. However, these nice duality results already fail for the two trust region subproblem (CDT problem).

Surprisingly, there are other classes of nonconvex Q^2P where strong duality holds. This includes the special cases of orthogonality type constraints.

Throughout this chapter we emphasize the theme (or open problem) that *Lagrangian duality is best*, i.e. in every case that we have a good (tractable) bound we show that it is equivalent to that obtained from the Lagrangian relaxation of an appropriate problem. Moreover, we include several results on the strength of these bounds. These results follow the pioneering paper [285] and study the theme that a solution of an indefinite quadratic maximization problem with some linear constraints on the squared variables can be approximated with a constant relative accuracy.

In parts 13.2 and 13.3, we present several complexity results on the quality of the SDP relaxations. We present a convex conic relaxation for a problem of maximizing an indefinite quadratic form over a set of convex constraints on the squared variables. We show that for all these problems we get at least $\frac{12}{37}$-relative accuracy of the approximation. In the second part of the paper we derive the conic relaxation by another approach based on the second order optimality conditions. We show that for l_p-balls, $p \geq 2$, intersected by a linear subspace, it is possible to guarantee $(1-\frac{2}{p})$-relative accuracy of the solution. As a consequence, we prove $(1-\frac{1}{e\ln n})$-relative accuracy of the conic relaxation for an indefinite quadratic maximization problem over an n-dimensional unit box with homogeneous linear equality constraints. We discuss the implications of the results for the discussion around the question $P = NP$. We also consider the problem of approximating the global maximum of a quadratic program (QP) subject to bound and (simple) quadratic onstraints. Based on several early results, we show that a 4/7-approximate solution can be obtained in polynomial time.

The rest of the paper is organized as follows. We begin in Section 13.4.1 with the most well known problem in this area, the Max-Cut problem. We present several different relaxations. Surprisingly, following our theme, all these bounds, including the SDP bound, end up being equivalent to the Lagrangian relaxation; see Section 13.4.1. We then present a strengthened SDP bound based on a second lifting procedure.

We discuss the SDP relaxation for general Q^2P in Section 13.4.2. This includes descriptions of the relationships between the SDP relaxation and the Lagrangian relaxation via convex quadratic valid inequalities, following [260, 442]. Several applications, including QAP and GP, are presented in Section 13.4.2.

Occurrences of strong duality for nonconvex quadratic programs is studied in Section 13.4.3. In every instance where one has a tractable bound, we find a Q^2P such that the bound is attained by the Lagrangian relaxation. This follows the work in [41, 37].

13.1.1 Lagrange Multipliers for Q^2P

We now define the (inequality constrained) $\mathrm{Q^2P}$ in x. (Though our notation does differ slightly in the separate parts (sections) of this chapter.)

$$(\mathrm{Q^2P}_x) \qquad \begin{array}{rll} q^* := & \min & q_0(x) := x^T q_0 x + 2g_0^T x + \alpha_0 \\ & \text{subject to} & q_k(x) := x^T Q_k x + 2g_k^T x + \alpha_k \leq 0 \\ & & k \in \mathcal{I} := \{1, \ldots, m\} \\ & & x \in \Re^n, \end{array}$$

where the matrices Q_k are symmetric. The Lagrangian of $\mathrm{Q^2P}_x$ is

$$L(x, \lambda) := q_0(x) + \sum_{k \in \mathcal{I}} \lambda_k q_k(x),$$

where $\lambda = (\lambda_k) \geq 0$ are nonnegative Lagrange multipliers.

Lagrange multipliers can be used in two ways. First, if a constraint qualification holds for $\mathrm{Q^2P}$ at the optimum $\bar{x}$ (e.g. the Mangasarian-Fromovitz constraint qualification), then the Karush-Kuhn-Tucker necessary conditions for optimality hold, i.e.

$$\nabla L(x, \lambda) = 0, \text{ and } \lambda_k q_k(x) = 0, \forall k \in \mathcal{I}.$$

Therefore, the optimum $\bar{x}$ can be searched among the points satisfying stationarity of the Lagrangian and complementary slackness. Moreover, if the Lagrangian is also convex, then this is a sufficient condition for optimality.

Lagrange multipliers can also be used to derive the Lagrangian dual (or relaxation) of $\mathrm{Q^2P}_x$

$$(DQ^2P_x) \qquad q^* \geq d^* := \max_{\substack{\mu \\ \lambda \geq 0}} \min_x q_0(x) + \sum_{k \in \mathcal{I}} \lambda_k q_k(x).$$

A zero duality gap holds if $q^* = d^*$. This can fail in the nonconvex case. Strong duality holds if $q^* = d^*$ and also d^* is attained. Moreover, d^* can be efficiently evaluated using SDP.

13.2 GLOBAL QUADRATIC OPTIMIZATION VIA CONIC RELAXATION

Yuri Nesterov

Starting from the pioneering paper [285], there were obtained several results [572, 859, 575], which show that a solution of an indefinite quadratic maximization problem with some linear constraints on the squared variables can be approximated with a constant relative accuracy. In this Section 13.2 we present some improvements and extensions of the results [575].

In Section 13.2.1 we consider a problem of maximizing an indefinite quadratic form subject to arbitrary convex constraints on the squared variables. For convenience of the dual description we use a conic representation of these constraints. We introduce a convex conic relaxation for that problem and prove that it provides us at least with an approximation of $\frac{\pi-2}{6-\pi}$ relative accuracy. In Section 13.2.2 we show how to improve this approximation using the diagonal elements of the quadratic objective function. The relative accuracy, which we can get in this case, is $\frac{12}{37}$. In Section 13.2.3 we extend the results of Section 13.2.1, 13.2.2 onto the case of general convex constraints on squared variables. We conclude the first part of the section with a discussion of the difficulties which arise in the problems with linear equality constraints on the initial variables (Section 13.2.4).

In the second part of the section, which starts from Section 13.2.5, we study another way of deriving the conic relaxation. This approach can be applied only to a small number of sets (l_p-balls, $p \geq 2$), but it allows to treat also the linear equations. We prove that for such problems the conic relaxation gives $(1-\frac{2}{p})$ relative accuracy. In Section 13.2.6 we apply these results to a problem of maximizing a quadratic function over a unit box subject to a system of homogeneous linear equalities. We show that it is possible to compute in polynomial time a $(1-\frac{1}{e\ln n})$-solution of that problem. We conclude the section with a discussion of the results.

We first recall some of the notation we use. For two vectors x, $y \in R^n$ we denote $\langle x, y\rangle$ the standard inner product:

$$\langle x, y\rangle = \sum_{i=1}^{n} x^{(i)} y^{(i)}.$$

Then $\| x \| = \langle x, x\rangle^{1/2}$. Since we work in several finite-dimensional spaces, the meaning of this notation is defined by the spaces of the arguments. For example, for two $(m \times n)$-matrices X and Y we have

$$\langle X, Y\rangle = \sum_{i=1}^{m}\sum_{j=1}^{n} X_{ij} Y_{ij}.$$

We use the standard notation for l_p-norms:

$$\| x \|_p = \left[\sum_{i=1}^{n} | x^{(i)} |^p\right]^{1/p}, \quad x \in R^n, \; p \geq 1.$$

Again, the meaning of the notation depends on the dimension of space of the argument. Recall, that for $p = \infty$ we have $\| x \|_\infty = \max\limits_{1\leq i\leq n} | x^{(i)} |$. The norm dual to $\| \cdot \|_p$ is $\| \cdot \|_{p^*}$ with $p^* = \frac{p}{p-1}$:

$$\| y \|_{p^*} = \max\{\langle y, x\rangle : \| x \|_p \leq 1\}.$$

For a symmetric matrix A we write $A \succeq 0$ if A is positive semidefinite. Notation $B \succeq A$ means that $B - A \succeq 0$. For $x \in R^n$ we denote $\mathrm{Diag}\,(x)$ the diagonal $(n \times n)$-matrix with diagonal entries $x^{(i)}$. Conversely, $\mathrm{diag}(X) \in R^n$ denotes the diagonal of an $(n \times n)$-matrix X. Notation e_i is used for the ith coordinate vector of R^n and $1_n \in R^n$ stands for the vector of all ones. Thus, $I_n = \mathrm{Diag}\,(1_n)$ is a unit matrix. Notation 0_n is used for the zero vector in R^n.

We use square brackets in order to indicate the component-wise operations with the vectors. For example, notation $[x \cdot y]$ stands for the vector with components $x^{(i)}y^{(i)}$, x, $y \in R^n$. Notation $[x]^2$ is used for the vector with the components $(x^{(i)})^2$. If $f(\tau)$ is a univariate function, we denote $f[x]$ the vector with the components $f(x^{(i)})$. In order to indicate the partial ordering in R^n we use the usual inequality signs. Thus, $x \geq y$ for x and y from R^n means that $x^{(i)} \geq y^{(i)}$, $i = 1, \ldots, n$.

Finally, $[\alpha, \beta]^n$ denotes a continuous box in R^n, that is $\{x \in R^n : \alpha 1_n \leq x \leq \beta 1_n\}$. For a boolean box $\{x \in R^n : x^{(i)} = (\alpha \text{ or } \beta)\}$ we use notation $\{\alpha, \beta\}^n$.

13.2.1 Convex conic constraints on squared variables

Let Q be an arbitrary symmetric $(n \times n)$-matrix. Consider the following pair of optimization problems:

$$\begin{aligned} \text{find } \phi^* = &\ \max\{\langle Qx, x\rangle : [x]^2 \in \mathcal{F}\}, \\ \text{find } \phi_* = &\ \min\{\langle Qx, x\rangle : [x]^2 \in \mathcal{F}\}. \end{aligned} \tag{13.2.1}$$

where $\mathcal{F}$ is a closed convex set. Our main assumption on the problem (13.2.1) is as follows.

Assumption 13.2.1 *1). The set $\mathcal{F}$ is bounded. 2). There exists a strictly positive $v \in \mathcal{F}$.*

In order to simplify the dual analysis, in this section we assume that the feasible set $\mathcal{F}$ is presented in a conic form:

$$\mathcal{F} = \{v \in K : Av = b\}, \tag{13.2.2}$$

where K is a convex closed pointed cone in R^n with non-empty interior, A is an $(m \times n)$-matrix and $b \neq 0_m$. Our additional assumption on the set $\mathcal{F}$ is as follows.

Assumption 13.2.2 $\{v \in \mathrm{int}\, K : Av = b\} \neq \emptyset$.

Note that the form (13.2.2) is quite general, since any bounded convex set can be written in this way (see [583] for details). At the same time, in Section

13.2.3 we will show how to transform our result on the case of a general convex feasible set $\mathcal{F}$.

Using the same technique as in [575], we can rewrite the pair of problems (13.2.1) in a trigonometric form.

Lemma 13.2.1

$$\begin{array}{lcl} \phi^* = & \max\limits_{\substack{X \succeq 0,\, \mathrm{diag}(X) = 1_n, \\ d \geq 0,\, [d]^2 \in \mathcal{F},}} & \frac{2}{\pi}\langle Q, \mathrm{Diag}\,(d) \arcsin[X] \mathrm{Diag}\,(d)\rangle, \\ \phi_* = & \min\limits_{\substack{X \succeq 0,\, \mathrm{diag}(X) = 1_n, \\ d \geq 0,\, [d]^2 \in \mathcal{F}.}} & \frac{2}{\pi}\langle Q, \mathrm{Diag}\,(d) \arcsin[X] \mathrm{Diag}\,(d)\rangle, \end{array} \tag{13.2.3}$$

Proof.
Indeed, let us represent a vector $x \in R^n$ as follows:

$$x = [d \cdot \sigma], \quad d \geq 0 \in R^n, \; \sigma \in \{-1, 1\}^n.$$

Note that $[x]^2 = [d]^2$. Therefore $\phi^* = \max\limits_{d}\{\Phi(d) : d \geq 0, \; [d]^2 \in \mathcal{F}\}$ with

$$\Phi(d) = \max\{\langle \mathrm{Diag}\,(d) Q \mathrm{Diag}\,(d)\sigma, \sigma\rangle : \sigma \in \{-1, 1\}^n\}.$$

Using Theorem 2.3 [575], we can represent $\Phi(d)$ in the following form:

$$\Phi(d) = \max\{\tfrac{2}{\pi}\langle \mathrm{Diag}\,(d) Q \mathrm{Diag}\,(d), \arcsin[X]\rangle : \; X \succeq 0, \; \mathrm{diag}(X) = 1_n\}.$$

Inserting this representation in the above expression for ϕ^* we get the first statement of the lemma. The second one can be obtained in a similar way. ∎

Note that in general none of the problems (13.2.1) is convex in x. Therefore, in order to estimate their optimal values, we need to use a kind of convex relaxation. Let us define the *conic relaxation* of problems (13.2.1):

$$\begin{array}{rcl} \psi^* & = & \max\{\langle Q, X\rangle : \; \mathrm{diag}(X) \in \mathcal{F}, \; X \succeq 0\}, \\ \psi_* & = & \min\{\langle Q, X\rangle : \; \mathrm{diag}(X) \in \mathcal{F}, \; X \succeq 0\}. \end{array} \tag{13.2.4}$$

Sometimes it is convenient to use a dual form of these relaxations. Recall that for a convex cone $K \subseteq R^n$ the dual cone K^* is defined as follows:

$$K^* = \{u \in R^n : \; \langle u, v\rangle \geq 0, \; \forall v \in K\}.$$

Lemma 13.2.2

$$\begin{array}{lcl} \psi^* = & \min\limits_{y \in R^m, u \in R^n} & \{\langle b, y\rangle : \; Q + \mathrm{Diag}\,(u) \preceq \mathrm{Diag}\,(A^T y), \; u \in K^*\}, \\ \psi_* = & \max\limits_{y \in R^m, u \in R^n} & \{\langle b, y\rangle : \; Q \succeq \mathrm{Diag}\,(u) + \mathrm{Diag}\,(A^T y), \; u \in K^*\}. \end{array} \tag{13.2.5}$$

Proof.
In view of Assumptions 13.2.1, 13.2.2, we can get a dual representation of the upper relaxation ψ^* as follows:

$$\psi^* = \max_{X,v}\{\langle Q, X\rangle : \ A\mathrm{diag}(X) = b, \ \mathrm{diag}(X) = v, \ X \succeq 0, \ v \in K\}$$

$$= \max_{X\succeq 0, v\in K} \ \min_{y\in R^m, u\in R^n} \{\langle Q, X\rangle + \langle y, b - A\mathrm{diag}(X)\rangle + \langle u, \mathrm{diag}(X) - v\rangle\}$$

$$= \min_{y\in R^m, u\in R^n} \ \max_{X,v}\{\langle Q + \mathrm{Diag}\,(u - A^T y), X\rangle + \langle b, y\rangle - \langle u, v\rangle : \|X \succeq 0, \ v\in K\}$$

$$= \min_{y\in R^m, u\in R^n} \{\langle b, y\rangle : \ Q + \mathrm{Diag}\,(u) \preceq \mathrm{Diag}\,(A^T y), \ u \in K^*\}.$$

Similarly, for the lower relaxation we get the following:

$$\psi \min_{X,v}\{\langle Q, X\rangle : \ A\mathrm{diag}(X) = b, \ \mathrm{diag}(X) = v, \ X \succeq 0, \ v \in K\}$$

$$= \min_{X\succeq 0, v\in K} \ \max_{y\in R^m, u\in R^n} \{\langle Q, X\rangle + \langle y, b - A\mathrm{diag}(X)\rangle + \langle u, v - \mathrm{diag}(X)\rangle\}$$

$$= \max_{y\in R^m, u\in R^n} \ \min_{X,v}\{\langle Q - \mathrm{Diag}\,(u + A^T y), X\rangle + \langle b, y\rangle + \langle u, v\rangle : \|X \succeq 0, \ v \in K\}$$

$$= \max_{y\in R^m, u\in R^n} \{\langle b, y\rangle : \ Q \succeq \mathrm{Diag}\,(u) + \mathrm{Diag}\,(A^T y), \ u \in K^*\}.$$

■

Let us establish some relations between the relaxations (13.2.4) and the optimal values of the problems (13.2.1). Denote

$$\psi(\alpha) = \alpha\psi^* + (1-\alpha)\psi_*. \tag{13.2.6}$$

The proof of the following theorem is similar to that of Theorem 3.3 [575].

Theorem 13.2.1

$$\psi_* \le \phi_* \le \psi(1 - \tfrac{2}{\pi}) \le \psi(\tfrac{2}{\pi}) \le \phi^* \le \psi^*. \tag{13.2.7}$$

Proof.
Note that $\psi_* \le \psi^*$ by definition. So, the middle inequality in (13.2.7) is correct. Further, if $[x]^2 \in \mathcal{F}$ then the matrix $X = xx^T$ is feasible for both relaxation problems (13.2.4) since $\mathrm{diag}(X) = [x]^2$. Moreover, $\langle Q, X\rangle = \langle Qx, x\rangle$. Thus, both bounding inequalities in the chain (13.2.7) are valid. Let us prove now two remaining inequalities.

Let us choose arbitrary $u \in K$ and $y \in R^m$, which satisfy the constraints of the dual form (13.2.5) of the lower relaxation ψ_*:

$$(u, y) \in \mathcal{F}_d = \{(u, y) \in K^* \times R^m : Q \succeq \text{Diag}\,(u) + \text{Diag}\,(A^T y)\}. \quad (13.2.8)$$

Consider a pair (X, d), which satisfies the constraints of the trigonometric representation (13.2.3) for ϕ^*:

$$X \succeq 0, \quad \text{diag}(X) = 1_n, \quad d \geq 0, \quad A[d]^2 = b, \quad [d]^2 \in K. \quad (13.2.9)$$

Since $X \succeq 0$ and $| X_{ij} | \leq 1$ we have $\arcsin[X] \succeq X$ in view of Corollary 3.2 [575]. Therefore, using Lemma 13.2.1 we get the following:

$$\begin{aligned}
\phi^* &\geq \tfrac{2}{\pi}\langle \text{Diag}\,(d)Q\text{Diag}\,(d), \arcsin[X]\rangle \\
&= \tfrac{2}{\pi}\langle \text{Diag}\,(d)(Q - \text{Diag}\,(u) - \text{Diag}\,(A^T y))\text{Diag}\,(d), \arcsin[X]\rangle \\
&\qquad + \langle u + A^T y, [d]^2\rangle \\
&\geq \tfrac{2}{\pi}\langle \text{Diag}\,(d)(Q - \text{Diag}\,(u) - \text{Diag}\,(A^T y))\text{Diag}\,(d), X\rangle + \langle u + A^T y, [d]^2\rangle \\
&= \tfrac{2}{\pi}\langle Q, \text{Diag}\,(d)X\text{Diag}\,(d)\rangle + (1 - \tfrac{2}{\pi})\langle u + A^T y, [d]^2\rangle.
\end{aligned}$$

Note that $u \in K^*$ and $[d]^2 \in K$. Therefore $\langle u, [d]^2\rangle \geq 0$. In view of (13.2.9) we have

$$\langle A^T y, [d]^2\rangle = \langle A[d]^2, y\rangle = \langle b, y\rangle.$$

Finally, for any pair (X, d), which satisfy (13.2.9) we have $Y = \text{Diag}\,(d)X\text{Diag}\,(d)$ feasible for the primal relaxation problems:

$$Y \in \mathcal{F}_p = \{Y : Y \succeq 0, \; A\text{diag}(Y) = b, \; \text{diag}(Y) \in K\}.$$

On the other hand, any $Y \in \mathcal{F}_p$ can be represented as $Y = \text{Diag}\,(d)X\text{Diag}\,(d)$ with X and d, which satisfy (13.2.9). Therefore, we conclude that

$$\phi^* \geq \tfrac{2}{\pi}\langle Q, Y\rangle + (1 - \tfrac{2}{\pi})\langle b, y\rangle, \quad \forall Y \in \mathcal{F}_p, \; (u, y) \in \mathcal{F}_d.$$

This proves the forth inequality in the chain (13.2.7). The remaining inequality can be proved in a similar way. ■

Definition 13.2.1 *We say that the value ψ approximates ϕ^* with a relative accuracy $\mu \in [0, 1]$ if $| \psi - \phi^* | \leq \mu(\phi^* - \phi_*)$. We call this approximation implementable if $\psi \leq \phi^*$.*

Corollary 13.2.1 *1. Let $\alpha = \frac{2}{\pi}$. Then the value $\psi(\alpha)$ is an implementable approximation of ϕ^* with the relative accuracy $\mu = \frac{\pi}{2} - 1 < \frac{4}{7}$.*

2. Let $\beta = \frac{(1+\alpha)^2 - 2}{3\alpha - 1}$. Then the value $\psi(\beta)$ approximates ϕ^* with the relative accuracy $\mu = \frac{\pi - 2}{6 - \pi} < \frac{2}{5}$.

The proof of that statement is exactly the same as that of Corollary 3.4 in [575].

13.2.2 Using additional information

In this section it is shown how to improve the quality of our bounds by taking into account some additional information. Define

$$\begin{aligned}\tau^* &= \max\{\langle \mathrm{diag}(Q), v\rangle : \ v \geq 0, \ v \in \mathcal{F}\},\\ \tau_* &= \min\{\langle \mathrm{diag}(Q), v\rangle : \ v \geq 0, \ v \in \mathcal{F}\}.\end{aligned} \tag{13.2.10}$$

Note that these values are computable in polynomial time. In view of Lemma 13.2.1 we have

$$\phi_* \leq \tau_* \leq \tau^* \leq \phi^*. \tag{13.2.11}$$

Hence,

$$\beta^* \equiv \frac{\psi^* - \tau^*}{\psi^* - \psi_*} \in [0,1], \quad \beta_* \equiv \frac{\tau_* - \psi_*}{\psi^* - \psi_*} \in [0,1].$$

Using these values we can express τ^* and τ_* as follows:

$$\begin{aligned}\tau^* &= \psi^* - \beta^*(\psi^* - \psi_*) &= \psi(1-\beta^*),\\ \tau_* &= \psi_* + \beta_*(\psi^* - \psi_*) &= \psi(\beta_*).\end{aligned}$$

Denote $\omega(\beta) = \beta \arcsin(\beta) + \sqrt{1-\beta^2} \equiv 1 + \int_0^\beta \arcsin(\tau) d\tau$, $\beta \in [0,1]$. This function is increasing and convex with $\omega(0) = 1$ and $\omega(1) = \frac{\pi}{2}$. In what follows we denote $\bar{\beta}$ the unique root of the following equation:

$$\tfrac{2}{\pi}\omega(\beta) = 1 - \beta, \quad \beta \in [0,1].$$

It can be shown that $\frac{23}{70} < \bar{\beta} < \frac{24}{73}$.

Theorem 13.2.2 *1. Denote*

$$\begin{aligned}\alpha^* &= \max\{\tfrac{2}{\pi}\omega(\beta_*), 1-\beta^*\},\\ \alpha_* &= \min\{1 - \tfrac{2}{\pi}\omega(\beta^*), \beta_*\}.\end{aligned}$$

The optimal values of the problems (13.2.1) satisfy the following relations:

$$\psi^* \geq \ \phi^* \ \geq \psi(\alpha^*), \tag{13.2.12}$$

$$\psi_* \leq \ \phi_* \ \leq \psi(\alpha_*). \tag{13.2.13}$$

2. The value $\psi(\alpha^*)$ is an implementable approximation of ϕ^* with relative accuracy

$$\mu = \frac{1-\alpha^*}{1-\alpha_*} \leq \frac{\bar{\beta}}{1-\bar{\beta}} < \frac{24}{49}.$$

3. Denote $\bar{\alpha} = \frac{\alpha^*(2-\alpha_*)-\alpha_*}{1+\alpha^*-2\alpha_*}$. The value $\psi(\bar{\alpha})$ is a μ-approximation of ϕ^* with

$$\mu = \frac{1-\alpha^*}{1+\alpha^*-2\alpha_*} \leq \frac{\bar{\beta}}{2-3\bar{\beta}} < \frac{12}{37}.$$

In Items 2 and 3 the upper bounds are achieved for $\beta^* = \beta_* = \bar{\beta}$.

Proof.
Let $X \succeq 0$ and $d \geq 0$ be feasible for the trigonometric form of the upper relaxation (13.2.3):

$$\mathrm{diag}(X) = 1_n, \quad A[d]^2 = b, \quad [d]^2 \in K.$$

Consider the matrices $X_\gamma = \gamma X + (1-\gamma)I_n$, $\gamma \in [0,1]$. Then

$$\arcsin[X_\gamma] = \arcsin[\gamma X] + (\tfrac{\pi}{2} - \arcsin(\gamma))I_n.$$

Therefore

$$\begin{aligned}\phi^* &\geq \tfrac{2}{\pi}\langle Q, \mathrm{Diag}\,(d)\arcsin[X_\gamma]\mathrm{Diag}\,(d)\rangle \\ &= \tfrac{2}{\pi}\langle Q, \mathrm{Diag}\,(d)\arcsin[\gamma X]\mathrm{Diag}\,(d)\rangle + \left(1 - \tfrac{2}{\pi}\arcsin(\gamma)\right)\langle \mathrm{diag}(Q), [d]^2\rangle.\end{aligned} \tag{13.2.14}$$

Let us choose now arbitrary $u \in K$ and $y \in R^m$ which satisfy the constraints of the dual form (13.2.5) of the lower relaxation ψ_*:

$$(u,y) \in \mathcal{F}_d = \{(u,y) \in K^* \times R^m : Q \succeq \mathrm{Diag}\,(u) + \mathrm{Diag}\,(A^T y)\}.$$

Then, in view of Corollary 3.2 [575] we have:

$$\begin{aligned}&\langle Q, \mathrm{Diag}\,(d)\arcsin[\gamma X]\mathrm{Diag}\,(d)\rangle \\ &= \langle Q - \mathrm{Diag}\,(u) - \mathrm{Diag}\,(A^T y), \mathrm{Diag}\,(d)\arcsin[\gamma X]\mathrm{Diag}\,(d)\rangle \\ &\quad + \langle \mathrm{Diag}\,(u + A^T y), \mathrm{Diag}\,(d)\arcsin[\gamma X]\mathrm{Diag}\,(d)\rangle \\ &\geq \gamma\langle Q - \mathrm{Diag}\,(u) - \mathrm{Diag}\,(A^T y), \mathrm{Diag}\,(d) X \mathrm{Diag}\,(d)\rangle \\ &\quad + \arcsin(\gamma)\langle u + A^T y, [d]^2\rangle \\ &= \gamma\langle Q, \mathrm{Diag}\,(d) X \mathrm{Diag}\,(d)\rangle + (\arcsin(\gamma) - \gamma)\langle u + A^T y, [d]^2\rangle.\end{aligned}$$

Note that $\arcsin(\gamma) \geq \gamma$ for $\gamma \in [0,1]$. At the same time $u \in K^*$ and $[d]^2 \in K$. Therefore $\langle u, [d]^2\rangle \geq 0$. Finally, $\langle A^T y, [d]^2\rangle = \langle A[d]^2, y\rangle = \langle b, y\rangle$. Thus,

$$\begin{aligned}\langle Q, \mathrm{Diag}\,(d)\arcsin[\gamma X]\mathrm{Diag}\,(d)\rangle &\geq \gamma\langle Q, \mathrm{Diag}\,(d) X \mathrm{Diag}\,(d)\rangle \\ &\quad + (\arcsin(\gamma) - \gamma)\langle b, y\rangle.\end{aligned}$$

Substituting this inequality in (13.2.14) we get the following:

$$\begin{aligned}\phi^* &\geq \tfrac{2}{\pi}\left(\gamma\langle Q, \mathrm{Diag}\,(d) X \mathrm{Diag}\,(d)\rangle + (\arcsin(\gamma) - \gamma)\langle b, y\rangle\right) \\ &\quad + \left(1 - \tfrac{2}{\pi}\arcsin(\gamma)\right)\langle \mathrm{diag}(Q), [d]^2\rangle \\ &\geq \tfrac{2}{\pi}\left(\gamma\langle Q, \mathrm{Diag}\,(d) X \mathrm{Diag}\,(d)\rangle + (\arcsin(\gamma) - \gamma)\langle b, y\rangle\right) \\ &\quad + \left(1 - \tfrac{2}{\pi}\arcsin(\gamma)\right)\tau_*.\end{aligned}$$

Using the same reasoning as in Theorem 13.2.1, we conclude that

$$\begin{aligned}\phi^* &\geq \tfrac{2}{\pi}\gamma\psi^* + \tfrac{2}{\pi}(\arcsin(\gamma) - \gamma)\psi_* + \left(1 - \tfrac{2}{\pi}\arcsin(\gamma)\right)\tau_* \\ &= \tfrac{2}{\pi}\arcsin(\gamma)\psi\left(\tfrac{\gamma}{\arcsin(\gamma)}\right) + \left(1 - \tfrac{2}{\pi}\arcsin(\gamma)\right)\psi(\beta_*) \\ &= \psi\left(\tfrac{2\gamma}{\pi} + \left(1 - \tfrac{2}{\pi}\arcsin(\gamma)\right)\beta_*\right).\end{aligned}$$

The right-hand side of the above inequality is maximal for $\gamma^* = \sqrt{1-\beta_*^2}$. Then

$$\tfrac{2\gamma^*}{\pi} + \left(1 - \tfrac{2}{\pi}\arcsin(\gamma^*)\right)\beta_* = \tfrac{2}{\pi}\left(\sqrt{1-\beta_*^2} + \beta_*\arccos(\gamma^*)\right) = \tfrac{2}{\pi}\omega(\beta_*).$$

Thus, $\phi^* \geq \psi(\frac{2}{\pi}\omega(\beta_*))$. Combining this inequality with (13.2.11), we get the lower bound in (13.2.12). The relations (13.2.13) for ψ_* can be obtained in a similar way.

Let us prove now Item 2 of the theorem. In view of (13.2.12) and (13.2.13) we have

$$0 \leq \frac{\phi^* - \psi(\alpha^*)}{\phi^* - \phi_*} \leq \frac{\psi^* - \psi(\alpha^*)}{\psi^* - \phi_*} \leq \frac{\psi^* - \psi(\alpha^*)}{\psi^* - \psi(\alpha_*)} = \frac{\psi(1) - \psi(\alpha^*)}{\psi(1) - \psi(\alpha_*)} = \frac{1-\alpha^*}{1-\alpha_*}. \tag{13.2.15}$$

Note that

$$\begin{aligned}1 - \alpha^* &= 1 - \max\{\tfrac{2}{\pi}\omega(\beta_*), 1 - \beta^*\} &&= \min\{1 - \tfrac{2}{\pi}\omega(\beta_*), \beta^*\}, \\ 1 - \alpha_* &= 1 - \min\{1 - \tfrac{2}{\pi}\omega(\beta^*), \beta_*\} &&= \max\{\tfrac{2}{\pi}\omega(\beta^*), 1 - \beta_*\}.\end{aligned}$$

Thus, we need to find an upper bound for the ratio

$$\rho(\beta_1, \beta_2) = \frac{\min\{1 - \frac{2}{\pi}\omega(\beta_2), \beta_1\}}{\max\{\frac{2}{\pi}\omega(\beta_1), 1 - \beta_2\}}, \quad 0 \leq \beta_1, \beta_2 \leq 1.$$

Lemma 13.2.3

$$\max\{\rho(\beta_1, \beta_2) : 0 \leq \beta_1, \beta_2 \leq 1\} = \frac{\bar{\beta}}{1 - \bar{\beta}}.$$

Proof.
We need to prove that

$$(1 - \bar{\beta})\min\{1 - \tfrac{2}{\pi}\omega(\beta_2), \beta_1\} \leq \bar{\beta}\max\{\tfrac{2}{\pi}\omega(\beta_1), 1 - \beta_2\}, \quad 0 \leq \beta_1, \beta_2 \leq 1.$$

This is equivalent to the statement that the convex function

$$g(\beta_1, \beta_2) = \bar{\beta}\max\{\tfrac{2}{\pi}\omega(\beta_1), 1 - \beta_2\} + (1 - \bar{\beta})\max\{\tfrac{2}{\pi}\omega(\beta_2) - 1, -\beta_1\}$$

is non-negative for $0 \leq \beta_1, \beta_2 \leq 1$.

Note that $g(\bar\beta,\bar\beta) = 0$ in view of the definition of $\bar\beta$. The subdifferential of the function $g(\cdot)$ at this point is as follows:

$$\partial g(\bar\beta,\bar\beta) = \bar\beta \mathrm{Conv}\,\{(\tfrac{2}{\pi}\omega'(\bar\beta),0);\ (0,-1)\} + (1-\bar\beta)\mathrm{Conv}\,\{(0,\tfrac{2}{\pi}\omega'(\bar\beta));(-1,0)\}.$$

Thus, this set contains the following points:

$$(\tfrac{2}{\pi}\bar\beta\omega'(\bar\beta) - 1 + \bar\beta, 0), \quad (0, \tfrac{2}{\pi}(1-\bar\beta)\omega'(\bar\beta) - \bar\beta), \quad (\tfrac{2}{\pi}\bar\beta\omega'(\bar\beta), \tfrac{2}{\pi}(1-\bar\beta)\omega'(\bar\beta)).$$

Note that

$$\tfrac{2}{\pi}\omega'(\bar\beta) < \omega'(\bar\beta) < \tfrac{\bar\beta}{1-\bar\beta} < \tfrac{1-\bar\beta}{\bar\beta}.$$

Therefore the first coordinate of the first point and the second coordinate of the second point are negative. Since both coordinates of the third point are positive, we conclude that $0 \in \mathrm{int}\,\partial g(\bar\beta,\bar\beta)$. ■

Applying Lemma 13.2.3 to (13.2.15) we prove the statement of Item 2.

In order to prove Item 3 note that in view of inequalities (13.2.12) and (13.2.13), for any $\alpha \in [0,1]$ we have

$$\begin{aligned}\frac{|\psi(\alpha)-\phi^*|}{\phi^*-\phi_*} &\le \frac{|\psi(\alpha)-\phi^*|}{\phi^*-\psi(\alpha_*)} = \max\left\{\frac{\psi(\alpha)-\phi^*}{\phi^*-\psi(\alpha_*)}, \frac{\phi^*-\psi(\alpha)}{\phi^*-\psi(\alpha_*)}\right\}\\ &\le \max\left\{\frac{\psi(\alpha)-\psi(\alpha^*)}{\psi(\alpha^*)-\psi(\alpha_*)}, \frac{\psi^*-\psi(\alpha)}{\psi^*-\psi(\alpha_*)}\right\} = \max\left\{\frac{\alpha-\alpha^*}{\alpha^*-\alpha_*}, \frac{1-\alpha}{1-\alpha_*}\right\} \equiv r(\alpha).\end{aligned}$$

The minimum $\bar\alpha$ of the function $r(\alpha)$ is a solution of the following equation:

$$(\alpha-\alpha^*)(1-\alpha_*) = (1-\alpha)(\alpha^*-\alpha_*).$$

That is $\bar\alpha = \frac{\alpha^*(2-\alpha_*)-\alpha_*}{1+\alpha^*-2\alpha_*}$. Using Lemma 13.2.3 we can estimate the optimal value $r(\bar\alpha)$ as follows:

$$r(\bar\alpha) = \frac{1}{1-\alpha_*}\left(1 - \frac{\alpha^*(2-\alpha_*)-\alpha_*}{1+\alpha^*-2\alpha_*}\right) = \frac{1-\alpha^*}{1+\alpha^*-2\alpha_*} = \frac{\rho(\beta^*,\beta_*)}{2-\rho(\beta^*,\beta_*)} \le \frac{\bar\beta}{2-3\bar\beta}.$$

■

13.2.3 General constraints on squared variables

Let us consider now the quadratic optimization problems in the following form:

$$\begin{aligned}\text{find } \phi^* &= \max\{\langle Qx,x\rangle : [x]^2 \in \mathcal{F}\},\\ \text{find } \phi_* &= \min\{\langle Qx,x\rangle : [x]^2 \in \mathcal{F}\}.\end{aligned} \tag{13.2.16}$$

where $\mathcal{F}$ is a closed convex set, which satisfies Assumption 13.2.1. Let us show that all results of Sections 13.2.1, 13.2.2 can be easily applied to the problem (13.2.16). Denote by $\xi(u)$ the support function of the set $\mathcal{F}$:

$$\xi(u) = \max\{\langle u,v\rangle : v \in \mathcal{F}\}.$$

Theorem 13.2.3 *The statements of Theorems 13.2.1, 13.2.2 are valid for the problem (13.2.16) with the relaxation values ψ^*, ψ_* and τ^*, τ_* defined as follows:*

$$\begin{aligned} \psi^* &= \min_u \{\xi(u) : \ \mathrm{Diag}\,(u) \succeq Q\}, \\ \psi_* &= \max_u \{-\xi(u) : \ Q + \mathrm{Diag}\,(u) \succeq 0)\}, \\ \tau^* &= \xi(\mathrm{diag}(Q)), \quad \tau_* = -\xi(-\mathrm{diag}(Q)). \end{aligned} \tag{13.2.17}$$

Proof.
In order to prove the theorem we need to rewrite the problem (13.2.16) in a conic form. Note that in view of Assumption 13.2.1 the set $\mathcal{F}$ can be represented in the following form:

$$\mathcal{F} = \{v \in S : \ Bv = d\},$$

where S is a bounded convex set with non-empty interior, B is a non-degenerate $(m \times n)$-matrix and $d \in R^m$. Without loss of generality we can assume that

$$\{v \in \mathrm{int}\, S : \ Bv = d\} \neq \emptyset.$$

We allow also $B = 0$; in this case $d = 0$.

Let us consider a conic hull of the set S:

$$K = \{(v, \tau) : \ \tau > 0, \ \tfrac{1}{\tau} v \in S\} \bigcup \{0\}.$$

In view of our assumptions K is a closed convex cone with non-empty interior. The cone dual to K can be represented as follows (see, for example, [345]):

$$K^* = \{\hat{u} = (u, \mu) : \ \mu \geq \xi_S(-u)\},$$

where $\xi_S(\cdot)$ is a support function of the set S.

Now we can rewrite the problem (13.2.16) in the following form:

$$\begin{aligned} \text{find } \phi^* = &\ \max\{\langle \hat{Q}\hat{x}, \hat{x}\rangle : \ [\hat{x}]^2 \in \hat{\mathcal{F}}\}, \\ \text{find } \phi_* = &\ \min\{\langle \hat{Q}\hat{x}, \hat{x}\rangle : \ [\hat{x}]^2 \in \hat{\mathcal{F}}\}, \end{aligned} \tag{13.2.18}$$

where $\hat{x} \in R^{n+1}$, $\hat{Q} = \begin{pmatrix} Q & 0_n \\ 0_n^T & 0 \end{pmatrix}$, $\hat{\mathcal{F}} = \{z = (v, \tau) \in K : \ \hat{A}z = b\}$ and

$$\hat{A} = \begin{pmatrix} B & -d \\ 0_n^T & 1 \end{pmatrix}, \quad b = \begin{pmatrix} 0_m \\ 1 \end{pmatrix}.$$

Note that the problems in (13.2.18) satisfy Assumptions 13.2.1, 13.2.2. Therefore for their relaxation values ψ^* and ψ_* all statements of Theorems

13.2.1, 13.2.2 are valid. Let us find the expressions for ψ^*, ψ_*, τ^* and τ_* in terms of the initial objects of the problem (13.2.16). It is clear that

$$\begin{aligned}
\tau^* &= \max\{\langle \operatorname{diag}(\hat{Q}), z\rangle : \ z \in \hat{\mathcal{F}}\} \\
&= \max\{\langle \operatorname{diag}(Q), v\rangle : \ Bv = \tau d, \ \tau = 1, \ v/\tau \in S\} = \xi(\operatorname{diag}(Q)), \\
\tau_* &= \min\{\langle \operatorname{diag}(\hat{Q}), z\rangle : \ z \in \hat{\mathcal{F}}\} \\
&= \min\{\langle \operatorname{diag}(Q), v\rangle : \ Bv = \tau d, \ \tau = 1, \ v/\tau \in S\} = -\xi(-\operatorname{diag}(Q)).
\end{aligned}$$

Further, in view of Lemma 13.2.2 the upper relaxation value ψ^* can be represented as follows:

$$\begin{aligned}
\psi^* &= \min_{\hat{y}\in R^{m+1}, \hat{u}\in R^{n+1}} \{\langle b, \hat{y}\rangle : \ \hat{Q} + \operatorname{Diag}(\hat{u}) \preceq \operatorname{Diag}(\hat{A}^T\hat{y}), \ \hat{u} \in K^*\} \\
&= \min_{(y,\gamma)\in R^{m+1}, (u,\mu)\in R^{n+1}} \{\gamma : \ Q + \operatorname{Diag}(u) \preceq \operatorname{Diag}(B^T y), \ \mu \\
&\quad \leq \gamma - \langle d, y\rangle, \ \mu \geq \xi_S(-u)\} \\
&= \min_{u,y}\{\xi_S(-u) + \langle d, y\rangle : \ \operatorname{Diag}(B^T y - u) \succeq Q\} \\
&= \min_{u,y}\{\xi_S(u - B^T y) + \langle d, y\rangle : \ \operatorname{Diag}(u) \succeq Q\}.
\end{aligned}$$

Note that in the last expression y does not enter the constraints. Therefore we can replace the objective function of this problem by its minimum in y. That is

$$\begin{aligned}
\min_y\{\xi_S(u - B^T y) + \langle d, y\rangle\} &= \min_y \max_{v\in S}\{\langle u - B^T y, v\rangle + \langle d, y\rangle\} \\
&= \max_{v\in S} \min_y\{\langle u, v\rangle + \langle d - Bv, y\rangle\} \\
&= \max_{v\in S}\{\langle u, v\rangle : \ Bv = d\} = \xi(u).
\end{aligned}$$

Thus, we get the representation (13.2.17) for ψ^*. The representation of ψ_* can be obtained in a similar way:

$$\begin{aligned}\psi^* &= \max_{\hat{y}\in R^{m+1},\hat{u}\in R^{n+1}} \{\langle b,\hat{y}\rangle : \ \hat{Q} \succeq \operatorname{Diag}(\hat{u}) + \operatorname{Diag}(\hat{A}^T\hat{y}),\ \hat{u}\in K^*\}\\
&= \max_{(y,\gamma)\in R^{m+1},(u,\mu)\in R^{n+1}} \{\gamma : \ Q \succeq \operatorname{Diag}(u) + \operatorname{Diag}(B^T y),\\
&\qquad 0 \geq \mu + \gamma - \langle d, y\rangle,\ \mu \geq \xi_S(-u)\}\\
&= \max_{u,y}\{-\xi_S(-u) + \langle d,y\rangle : \ Q \succeq \operatorname{Diag}(u + B^T y)\}\\
&= \max_{u,y}\{-\xi_S(B^T y + u) + \langle d,y\rangle : \ Q + \operatorname{Diag}(u) \succeq 0\}\\
&= \max_{u}\{-\min_{y}\{\xi_S(B^T y + u) - \langle d,y\rangle\} : \ Q + \operatorname{Diag}(u) \succeq 0\}\\
&= \max_{u}\{-\xi(u) : \ Q + \operatorname{Diag}(u) \succeq 0\}.\end{aligned}$$

■

Let us present an example of application of Theorems 13.2.3, 13.2.2. Consider the following problem:

$$\begin{aligned}\text{find } \phi^* = &\quad \max\{\langle Qx_1, x_2\rangle : \ [(x_1,x_2)]^2 \in \mathcal{F}\},\\
\text{find } \phi_* = &\quad \min\{\langle Qx_1, x_2\rangle : \ [(x_1,x_2)]^2 \in \mathcal{F}\},\end{aligned} \qquad (13.2.19)$$

where Q is a $(k\times n)$-matrix, $x_1 \in R^k$, $x_2 \in R^n$ and $\mathcal{F}$ is a closed convex set, which satisfies Assumption 13.2.1. Since the quadratic objective function in this problem is bilinear, we conclude that $\phi_* = -\phi^*$ and $\tau^* = \tau_* = 0$.

The conic relaxation for this problem is defined as follows:

$$\begin{aligned}\psi^* &= \min_{u=(u_1,u_2)} \left\{\xi(u) : \ \begin{pmatrix} \operatorname{Diag}(u_1) & -Q^T \\ -Q & \operatorname{Diag}(u_2)\end{pmatrix} \succeq 0\right\},\\
\psi_* &= \max_{u=(u_1,u_2)} \left\{-\xi(u) : \ \begin{pmatrix} \operatorname{Diag}(u_1) & Q^T \\ Q & \operatorname{Diag}(u_2)\end{pmatrix} \succeq 0\right\}.\end{aligned}$$

It is clear that $\psi_* = -\psi^*$. At the same time, $\beta^* = \beta_* = \frac{1}{2}$. Therefore,

$$\alpha^* = \max\{\tfrac{2}{\pi}\omega(\beta_*), 1-\beta^*\} = \tfrac{2}{\pi}\omega(\tfrac{1}{2}).$$

Therefore, in view of Theorem 13.2.2 we have:

$$\psi^* \geq \phi^* \geq \psi(\alpha^*) = (2\alpha^* - 1)\psi^*.$$

Note that $\alpha^* = \frac{2}{\pi}(\frac{1}{2}\arcsin\frac{1}{2} + \frac{\sqrt{3}}{2}) = \frac{\sqrt{3}}{\pi} + \frac{1}{6}$. Thus, we have proved the following theorem.

Theorem 13.2.4 *In the problem (13.2.19) the optimal and relaxation values are related as follows:*

$$\psi^* \geq \phi^* \geq \gamma\psi^*$$

with $\gamma = \frac{2\sqrt{3}}{\pi} - \frac{2}{3} > 0.43$.

13.2.4 Why the linear constraints are difficult?

In the previous sections we have got a constant relative accuracy estimates for a quadratic maximization problem with convex constraints on *squared* variables. Such type of constraints are rather specific. Therefore it is natural to try to extend the results onto the problems with convex constraints on the variables of the quadratic form. However, it appears that this is not trivial. In this section we show that even a single linear constraint can make a quadratic problem completely intractable by the presented technique.

Consider the following optimization problem:

$$\begin{aligned} \phi^* = \quad & \max \quad \langle Qx, x\rangle, \\ & \text{s.t.} \quad x \in \{-1,1\}^n, \\ & \langle c, x\rangle = \beta, \end{aligned} \tag{13.2.20}$$

where Q is an $(n \times n)$-matrix, $c \in R^n$ and $\beta > 0$. Define ϕ_* as a minimal value of the objective function in (13.2.20). A natural relaxation for this problem is as follows:

$$\psi^* = \max\{\langle Q, X\rangle : \langle Xc, c\rangle = \beta^2, \ \text{diag}(X) = 1_n, \ X \succeq 0\}. \tag{13.2.21}$$

Let us show that this relaxation can be arbitrary bad in terms of relative accuracy.

Denote by v_i, $i = 1,\ldots,2^n$ the nodes of the boolean unit box $\{-1,1\}^n$. Let us assume that there exists only one node v_*, which satisfies the linear constraint of the problem (13.2.20). Moreover, let us assume that there are two other nodes, v_+ and v_- such that

$$0 < \langle c, v_-\rangle < \beta < \langle c, v_+\rangle. \tag{13.2.22}$$

Note that in view of our assumption we have $\phi^* = \phi_*$ independently on our choice of the matrix Q.

Let us define a convex polytope $\mathcal{P}_n$ of positive semidefinite $(n \times n)$-matrices:

$$\mathcal{P}_n = \text{Conv}\,\{V_i = v_i v_i^T, \ i = 1,\ldots,2^n\}.$$

Lemma 13.2.4 *Any* V_i *is an extreme point of* $\mathcal{P}_n$. *Any pair of nodes* V_i, V_j *is connected by an exposed edge.*

Proof.

Since V_i is a rank-one matrix, the first statement is evident. In order to prove the second statement note that the edge $[V_i, V_j]$ is not exposed if and only if there exist some coefficients $\lambda_k > 0$, $k \in \mathcal{I}$, $i, j \notin \mathcal{I}$ such that

$$\alpha V_i + (1-\alpha)V_j = \sum_{k \in \mathcal{I}} \lambda_k V_k, \quad \sum_{k \in \mathcal{I}} \lambda_k = 1,$$

for some $\alpha \in (0,1)$. Since all nodes of $\mathcal{P}_n$ are positive semidefinite rank-one matrices, we conclude that

$$v_k \in \{v : \ v = \alpha v_i + \beta v_j, \ (\alpha, \beta) \in R^2\}, \quad \forall k \in \mathcal{I}.$$

A simple calculation shows that it is possible only for $v_k = \pm v_i$ or $v_k = \pm v_j$. ■

Note that in view of our assumption (13.2.22) there exists a matrix $\widetilde{V} \in \mathcal{P}_n$ such that

$$\widetilde{V} = \alpha v_- v_-^T + (1-\alpha) v_+ v_+^T, \ \alpha \in (0,1), \quad \langle \widetilde{V} c, c \rangle = \beta^2.$$

Let us choose now $Q = \widetilde{V} - v_* v_*^T$. Note that the feasible set of the relaxation problem (13.2.21) contains $\mathcal{P}_n$. Therefore

$$\psi^* \geq \langle Q, \widetilde{V} \rangle > \langle Q, v_* v_*^T \rangle = \phi^*.$$

The lower relaxation value ψ_* never exceed $\phi_* = \phi^*$. Therefore, for our example the value $\psi^* - \psi_*$ is strictly positive. This means that the relative accuracy of the value ψ^* is infinitely bad.

Note that the main source of our troubles in the above example is that the linear constraint $\langle Xc, c \rangle = \beta^2$ intersects an edge of the matrix polytope $\mathcal{P}_n$. That can happen with any value of β except $\beta = 0$. Thus, we still can hope that for the problems with homogeneous linear constraints the conic relaxation can work. In the next sections we will see some problems, for which it is true.

13.2.5 Maximization with a smooth constraint

In the previous section we have established some constant bounds on relative accuracy of the conic relaxations (13.2.4) for a quadratic maximization problem with convex constraints for the squared variables. At the same time, in Section 13.2.4 we have seen that some linear constraints on the initial variables can make the problem intractable in terms of relative accuracy. In this section we present another approach for deriving the conic relaxations. This approach is based on the standard second order optimality conditions and it allows to treat the quadratic maximization problems over l_p-boxes, $p \geq 2$, with homogeneous linear equality constraints (see Section 13.2.6). However, the quality of relaxation in this framework becomes dependent on p.

Let $f(y)$, $y \in R^m$, be a homogeneous function of degree p:

$$f(\tau y) = \tau^p f(y), \quad y \in R^m, \ \tau \geq 0. \qquad (13.2.23)$$

We assume that $f(y)$ is non-negative and twice continuously differentiable at any non-zero point of R^m (notation $f \in H_p$). Recall, that for homogeneous functions we have the following simple relations.

Lemma 13.2.5 *If $f(y)$ is homogeneous of degree p then for any $y \in R^m$ and $\tau \geq 0$ we have*

$$\begin{aligned} f'(\tau y) &= \tau^{p-1} f'(y), & (13.2.24)\\ f''(y)y &= (p-1)f'(y), & (13.2.25)\\ \langle f'(y), y\rangle &= pf(y), & (13.2.26)\\ \langle f''(y)y, y\rangle &= p(p-1)f(y). & (13.2.27) \end{aligned}$$

Proof.
Indeed, if we differentiate (13.2.23) in y we get (13.2.24). If we differentiate (13.2.24) in τ and take $\tau = 1$ we get (13.2.25). In order to get (13.2.26) we differentiate (13.2.23) in τ and take $\tau = 1$. Finally, (13.2.25) and (13.2.26) give (13.2.27). ∎

Let Q be a symmetric $(m \times m)$-matrix. Consider the following maximization problem:

$$\text{find } \phi^*(Q) = \max\{\langle Qy, y\rangle : \; f(y) \leq 1\}. \qquad (13.2.28)$$

If $Q \preceq 0$ then (13.2.28) is a concave maximization problem and $\phi^*(Q) = 0$. In the other cases we need some necessary conditions to characterize the local solutions of the problem (13.2.28).

Lemma 13.2.6 *Let $f \in H_p$ with $p > 0$. Then for any local maximum y_* of the problem (13.2.28) with $\langle Qy_*, y_*\rangle > 0$ we have $f(y_*) = 1$. Moreover, there exists a value $\lambda = \lambda(y_*) > 0$ such that*

$$\begin{aligned} \langle Qy_*, y_*\rangle &= p\lambda, & (13.2.29)\\ Qy_* &= \lambda f'(y_*), & (13.2.30)\\ Q &\preceq \lambda\left[f''(y_*) - \frac{p-2}{p} f'(y_*) f'(y_*)^T\right]. & (13.2.31) \end{aligned}$$

Proof.
Since $\langle Qy_*, y_*\rangle > 0$ and $f(y)$ is a homogeneous function of positive degree, we necessarily have $f(y_*) = 1$. Let us write down a Lagrangean for this problem:

$$\mathcal{L}(y, \lambda) = \tfrac{1}{2}\langle Qy, y\rangle - \lambda[f(y) - 1].$$

Then, the second order necessary conditions for the problem (13.2.28) can be written as follows:

$$\begin{aligned} \mathcal{L}'_y(y_*, \lambda) &= 0, & (13.2.32)\\ \langle \mathcal{L}''_{yy}(y_*, \lambda)h, h\rangle &\preceq 0, \quad \forall h : \langle f'(y_*), h\rangle = 0, & (13.2.33) \end{aligned}$$

with some $\lambda \in R$. Equation (13.2.32) is exactly (13.2.30). Multiplying (13.2.30) by y_* and using (13.2.26) we get

$$\langle Qy_*, y_*\rangle = \lambda\langle f'(y_*), y_*\rangle = \lambda p f(y_*)\lambda p > 0,$$

and that is (13.2.29). Finally, since $\langle f'(y_*), y_*\rangle = p > 0$, any $h \in R^m$ such that $\langle f'(y_*), h\rangle = 0$ can be represented in the form

$$h = \left(I - \tfrac{1}{p} y_* f'(y_*)^T\right) u, \quad u \in R^n.$$

Therefore the condition (13.2.33) can be rewritten as

$$(I - \tfrac{1}{p} f'(y_*) y_*^T) \mathcal{L}''_{yy}(y_*, \lambda)(I - \tfrac{1}{p} y_* f'(y_*)^T) \preceq 0. \qquad (13.2.34)$$

Note that $\mathcal{L}''_{yy}(y_*, \lambda) = Q - \lambda f''(y_*)$ and

$$\begin{aligned}
&(I - \tfrac{1}{p} f'(y_*) y_*^T) Q (I - \tfrac{1}{p} y_* f'(y_*)^T) \\
&= Q - \tfrac{1}{p} f'(y_*) y_*^T Q - \tfrac{1}{p} Q y_* f'(y_*)^T + \tfrac{1}{p^2}\langle Qy_*, y_*\rangle f'(y_*) f'(y_*)^T \\
&= Q - \tfrac{\lambda}{p} f'(y_*) f'(y_*)^T
\end{aligned}$$

in view of (13.2.30) and (13.2.29). Similarly, since $f(y_*) = 1$ we have

$$\begin{aligned}
(I- &\tfrac{1}{p} f'(y_*) y_*^T) f''(y_*)(I - \tfrac{1}{p} y_* f'(y_*)^T) \\
&= f''(y_*) - \tfrac{1}{p} f'(y_*) y_*^T f''(y_*) - \tfrac{1}{p} f''(y_*) y_* f'(y_*)^T \\
&+ \tfrac{1}{p^2}\langle f''(y_*) y_*, y_*\rangle f'(y_*) f'(y_*)^T \\
&= f''(y_*) - 2\tfrac{p-1}{p} f'(y_*) f'(y_*)^T + \tfrac{p(p-1)}{p^2} f'(y_*) f'(y_*)^T \\
&= f''(y_*) - \tfrac{p-1}{p} f'(y_*) f'(y_*)^T,
\end{aligned}$$

in view of (13.2.25) and (13.2.27). Substituting these expressions in (13.2.34) we get

$$\begin{aligned}
Q &\preceq \lambda f''(y_*) + \tfrac{\lambda}{p} f'(y_*) f'(y_*)^T - \lambda \tfrac{p-1}{p} f'(y_*) f'(y_*)^T \\
&= \lambda\left[f''(y_*) - \tfrac{p-2}{p} f'(y_*) f'(y_*)^T\right].
\end{aligned}$$

■

We will use Lemma 13.2.6 in order to estimate the quality of relaxations for some non-convex maximization problems. Let $A = (a_1, \ldots, a_n) \in R^{m\times n}$ be a non-degenerate $(m \times n)$-matrix. Consider the following function:

$$f_A(y) = \sum_{i=1}^{n} |\langle a_i, y\rangle|^p,$$

where $p \geq 2$. The problem we are going to address now is as follows:

$$\text{find } \phi^*(Q,A) = \max\{\langle Qy, y\rangle : \ f_A(y) \leq 1\}. \qquad (13.2.35)$$

For this problem we can introduce the following relaxation:

$$\psi_p^*(Q,A) = \min_u\{\| u \|_q\colon \ A\text{Diag}\,(u)A^T \succeq Q\}, \qquad (13.2.36)$$

where $q = (\frac{p}{2})^* = \frac{p}{p-2}$ (compare with (13.2.17)). Now we can prove the main result of this section.

Theorem 13.2.5 *Let the feasible set of the problem (13.2.35) be bounded. Then*

$$\frac{1}{p-1}\psi_p^*(Q,A) \leq \phi^*(Q,A) \leq \psi_p^*(Q,A). \qquad (13.2.37)$$

Moreover, any local maximum y_* of the problem (13.2.35) with positive value of the objective function satisfies inequality $\langle Qy_*, y_*\rangle \geq \frac{1}{p-1}\psi_p^*(Q,A)$.

Proof.
Indeed, let u be feasible for the problem (13.2.36). Then for any $y \in R^m$ with $f_A(y) \leq 1$ we have

$$\langle Qy, y\rangle \leq \langle A\text{Diag}\,(u)A^T y, y\rangle = \langle u, [A^T y]^2\rangle.$$

At the same time,

$$\| [A^T y]^2 \|_{p/2}^{p/2} = \sum_{i=1}^n \mid \langle a_i, y\rangle \mid^p = f_A(y) \leq 1.$$

Therefore, for any feasible y we have

$$\langle Qy, y\rangle \leq \langle u, [A^T y]^2\rangle \leq \| u \|_q \cdot \| [A^T y]^2 \|_{p/2} \leq \| u \|_q \, .$$

Hence, $\phi^*(Q,A) \leq \psi_p^*(Q,A)$.

On the other hand, let y_* be a local maximum of (13.2.35) with $\langle Qy_*, y_*\rangle > 0$. Then, in view of Lemma 13.2.6 (13.2.31) for $\lambda = \lambda(y_*)$ we have:

$$Q \preceq \lambda f''(y_*) = p(p-1)\lambda \sum_{i=1}^n \mid \langle a_i, y_*\rangle \mid^{p-2} a_i a_i^T$$

(we have used the condition $p \geq 2$). Thus, the vector $u \in R^n$ with the components

$$u^{(i)} = p(p-1)\lambda \mid \langle a_i, y_*\rangle \mid^{p-2}, \ i = 1, \ldots, n,$$

is feasible for the problem (13.2.36). Note that

$$\begin{aligned} \| u \|_q &= p(p-1)\lambda \left[\sum_{i=1}^n \mid \langle a_i, y_*\rangle \mid^{(p-2)q}\right]^{1/q} \\ &= p(p-1)\lambda \left[\sum_{i=1}^n \mid \langle a_i, y_*\rangle \mid^{p}\right]^{1/q} \\ &= p(p-1)\lambda [f_A(y_*)]^{1/q} = p(p-1)\lambda. \end{aligned}$$

Hence, in view of (13.2.29) we have

$$\langle Qy_*, y_* \rangle = p\lambda = \frac{\| u \|_q}{p-1} \geq \frac{1}{p-1} \psi_p^*(Q, A).$$

Note that the above proof shows that under assumptions of the theorem the function $\psi_p^*(Q, A)$ is well defined.

Finally, if there is no local maximum of the problem (13.2.35) with $\langle Qy_*, y_* \rangle > 0$, then $Q \preceq 0$ and in this case we have $\psi_p^*(Q, A) = \phi^*(Q, A) = 0$. ■

Let us estimate the relative accuracy of the relaxation (13.2.36). First, we need the following trivial result.

Lemma 13.2.7 *Let for some non-negative values ϕ, ψ and γ we have the following relations:*

$$\gamma\psi \leq \phi \leq \psi.$$

Then, for $\beta = \frac{2\gamma}{1+\gamma}$ we have: $\mid \beta\psi - \phi \mid \leq (1-\beta)\phi$.

Define $\phi_*(Q, A) = \min\{\langle Qy, y \rangle : \; f_A(y) \leq 1\}$.

Theorem 13.2.6 *Let $\psi^* = \frac{2}{p}\psi_p^*(Q, A)$. Then*

$$\mid \phi^*(Q, A) - \psi^* \mid \leq (1 - \tfrac{2}{p})(\phi^*(Q, A) - \phi_*(Q, A)). \qquad (13.2.38)$$

Proof.
Note that $\phi_*(Q, A) \leq 0$. Therefore it is sufficient to prove

$$\mid \phi^*(Q, A) - \psi^* \mid \leq (1 - \tfrac{2}{p})\phi^*(Q, A).$$

Let us choose $\gamma = \frac{1}{p-1}$ and $\beta = \frac{2\gamma}{1+\gamma} = \frac{2}{p}$. Then the above inequality follows from Theorem 13.2.5 and Lemma 13.2.7. ■

Let us compare now the relaxation (13.2.36) with the conic relaxation (13.2.17). Of course, we have to choose a problem which can be treated by both approaches. Consider the problem

$$\max\{\langle Qx, x \rangle : \| x \|_p \leq 1\}, \quad p \geq 2.$$

This problem can be presented in the form (13.2.35) with $A = I_n$. On the other hand, it can be written in the form (13.2.16) with

$$\mathcal{F} = \{v : \| v \|_{p/2} \leq 1\}.$$

In this case $\xi(u) = \| u \|_q$ and we can see that (13.2.36) coincides with (13.2.17).

13.2.6 Some applications

Let us show that the results of the previous section can be extended onto the problems with linear equality constraints. Consider the following quadratic maximization problem:

$$\begin{aligned} \text{find } \phi_p^* = \quad & \max_{x \in R^n} \quad \langle Cx, x \rangle, \\ \text{s.t.} \quad & \| x \|_p \leq 1, \\ & Bx = 0, \end{aligned} \tag{13.2.39}$$

where C is an arbitrary $(n \times n)$-matrix, $p \geq 2$ and B is a non-degenerate $((n-m) \times n)$-matrix with $n > m$. Let the rows of some $(m \times n)$-matrix A span the null space of the matrix B:

$$BA^T y = 0, \quad \forall y \in R^m.$$

Then we can change variables $x = A^T y$ and obtain a problem, which is equivalent to (13.2.39):

$$\phi_p^* = \max_{y \in R^m} \{ \langle ACA^T y, y \rangle : \; f_A(y) \leq 1 \} = \phi^*(ACA^T, A).$$

Thus, in view of Theorem 13.2.5 and Lemma 13.2.7 we get the following result.

Theorem 13.2.7 *For any $p \geq 2$ we have*

$$\frac{1}{p-1} \psi_p^*(ACA^T, A) \leq \phi_p^* \leq \psi_p^*(ACA^T, A).$$

The value $\psi^* = \frac{2}{p} \psi_p^*(ACA^T, A)$ approximates the solution of the problem (13.2.39) with $(1 - \frac{2}{p})$ relative accuracy.

Now, let us consider the case when the objective function of the problem (13.2.39) has a non-zero linear term:

$$\begin{aligned} \text{find } \hat{\phi}_p^* = \quad & \max_{x \in R^n} \quad \langle Cx, x \rangle + 2\langle c, x \rangle, \\ \text{s.t.} \quad & \| x \|_p \leq 1, \\ & Bx = 0. \end{aligned} \tag{13.2.40}$$

This problem can be homogenized in a standard way:

$$\begin{aligned} & \max_{(x,\tau) \in R^{n+1}} \quad \langle Cx, x \rangle + 2\tau \langle c, x \rangle, \\ \text{s.t.} \quad & \| x \|_p \leq 1, \; | \tau | \leq 1, \\ & Bx = 0. \end{aligned} \tag{13.2.41}$$

Clearly, the optimal value of this problem is $\hat{\phi}_p^*$. However, this problem has two separate constraints for x and τ. Therefore, in order to apply the results of Section 13.2.5 we need to replace them by a single functional inequality. Consider the following problem:

$$\begin{aligned} \text{find } \bar{\phi}_p^* = \quad & \max_{(x,\tau)\in R^{n+1}} \quad \langle Cx, x\rangle + 2\tau\langle c, x\rangle, \\ \text{s.t.} \quad & \| (x,\tau) \|_p \leq 1, \\ & Bx = 0, \end{aligned} \tag{13.2.42}$$

Denote by $\bar{\psi}_p^*$ the value of the conic relaxation for the last problem.

Theorem 13.2.8 *Let $p \geq 2$. For $\psi_a^* = 2^{2/p}\bar{\psi}_p^*$ we have:*

$$\frac{1}{2^{2/p}(p-1)}\psi_a^* \leq \hat{\phi}_p^* \leq \psi_a^*.$$

The value $\psi_r^* = \frac{2}{p+2^{-2/p}-1}\bar{\psi}_p^*$ has at least $(1-\frac{1}{2p})$ relative accuracy.

Proof.
Note that the problems (13.2.41) and (13.2.42) have the same objective function and the same system of linear equations. Denote by $\mathcal{F}_0$ the feasible set of the problem (13.2.42) and by $\mathcal{F}_1$ the feasible set of the problem (13.2.41). Clearly, $\mathcal{F}_0 \subset \mathcal{F}_1 \subset 2^{1/p}\mathcal{F}_0$. Therefore

$$\bar{\phi}_p^* \leq \hat{\phi}_p^* \leq 2^{2/p}\bar{\phi}_p^*.$$

On the other hand, in view of Theorem 13.2.7, we have:

$$\frac{1}{p-1}\bar{\psi}_p^* \leq \bar{\phi}_p^* \leq \bar{\psi}_p^*.$$

Hence, for $\psi_a^* = 2^{2/p}\bar{\psi}_p^*$ we obtain:

$$\psi_a^* = 2^{2/p}\bar{\psi}_p^* \geq 2^{2/p}\bar{\phi}_p^* \geq \hat{\phi}_p^* \geq \bar{\phi}_p^* \geq \tfrac{1}{p-1}\bar{\psi}_p^* = \tfrac{1}{2^{2/p}(p-1)}\psi_a^*.$$

In order to get the statement on the relative accuracy, we take $\psi = \psi_a^*$, $\phi = \hat{\phi}_p^*$, $\gamma = \frac{1}{2^{2/p}(p-1)}$ and apply Lemma 13.2.7. Then the values β and ψ_r^* can be obtained as follow:

$$\begin{aligned} \beta &= \tfrac{2\gamma}{1+\gamma} = \tfrac{2}{1+2^{2/p}(p-1)} \geq \tfrac{1}{2p}, \\ \psi_r^* &= \beta\psi_a^* = \tfrac{2}{p+2^{-2/p}-1}\bar{\psi}_p^*. \end{aligned}$$

■

We see that the quality of conic relaxation decreases as p increase. Therefore, we cannot directly apply the results of Section 13.2.5 to a problem with box constraints. However, at the same time, when p increase the shape of l_p balls becomes very close to the shape of the n-dimensional unit box. Therefore, we can use the values $\psi_p^*(ACA^T, A)$ with p large enough in order to get some bounds for ϕ_∞^*.

Theorem 13.2.9 *Let* $p = 2\ln n$, $\psi_a^* = e\psi_p^*(ACA^T, A)$ *and* $\gamma = \frac{1}{e(2\ln n - 1)}$. *Then*

$$\gamma\psi_a^* \leq \psi_\infty^* \leq \psi_a^*.$$

The value $\psi_r^* = \frac{2\gamma}{1+\gamma}\psi_a^*$ *has at least* $(1 - \frac{1}{e\ln n})$ *relative accuracy.*

Proof.
It is well known that for any two values $p \geq 2$ we have:

$$\frac{1}{n^{1/p}} \| x \|_p \leq \| x \|_\infty \leq \| x \|_p, \quad x \in R^n.$$

Therefore

$$\{x \in R^n : \| x \|_p \leq 1\} \subset \{x \in R^n : \| x \|_\infty \leq 1\} \subset \{s \in R^n : \| x \|_p \leq n^{1/p}\}.$$

Since the objective function of the problem (13.2.39) is homogeneous of degree two, this implies that $\phi_p^* \leq \phi_\infty^* \leq n^{2/p}\phi_p^*$. Thus, using Theorem 13.2.7 we obtain the following:

$$\begin{aligned} \psi_\infty^* &= e\psi_p^*(ACA^T, A) = n^{2/p}\psi_p^*(ACA^T, A) \geq n^{2/p}\phi_p^* \\ &\geq \phi_\infty^* \geq \phi_p^* \geq \tfrac{1}{p-1}\psi_p^*(ACA^T, A) = \tfrac{1}{e(2\ln n - 1)}\psi^*. \end{aligned}$$

In order to get the statement on relative accuracy we apply Lemma 13.2.7 with

$$\beta = \frac{2\gamma}{1+\gamma} = \frac{2}{1 + e(2\ln n - 1)} > \frac{1}{e\ln n}.$$

■

13.2.7 Discussion

In the previous sections we have presented some estimates for the quality of the conic relaxation for different non-convex quadratic maximization problems. The constant bounds of Sections 13.2.1, 13.2.2 can be applied to a quite large class of non-convex problems and we can expect that they can be used in many practical applications. The bounds we get in Section 13.2.5 are not so good. Indeed, they can be applied only to a rather special feasible set, that is an intersection of an l_p-ball, $p \geq 2$, with a linear subspace. Moreover, the quality of these bounds decrease as p increase.

Nevertheless, the results of Section 13.2.5 suggest some interesting conclusions. Firstly, the relative accuracy we get from the relaxation (13.2.36) is $(1 - \frac{2}{p})$. Thus, the accuracy goes to zero as p approaches two. For p small enough the results of Theorem 13.2.5 become even better than the bounds of Section 13.2.1. An important advantage of the estimates (13.2.37) is that we get the separate bounds for the minimal and the maximal value of the problem. The lower estimate for the maximal value remains positive even if the minimal value of the problems is a large negative value.

Secondly, Theorem 13.2.5 tells us that the value of the objective function of the problem (13.2.35) at *any* local solution is not worse than the lower bound we get from the conic relaxation. In fact, this statement is a kind of surprise. Indeed, if we measure a hardness of a problem as a largest ratio of the values of the objective function at the global and a local maximum, it appears that the problem (13.2.35) is not so difficult, at least for p small enough. Usually the general methods of nonlinear optimization are quite efficient in finding a local solution. Since the computational cost of such schemes is much less than that of the schemes of semidefinite programming, we can conclude that for practical applications the traditional schemes look quite attractive.[1]

Finally, in Section 13.2.6 we have shown that the results of Theorem 13.2.5 provides us with some bounds for very difficult problems. Indeed, during last years there were obtained many negative results related to the possibilities to find an approximate solution of an NP-hard problem under hypothesis that $P \neq NP$. The results relevant to the topic of our section can be found in [79]:

> Consider a quadratic optimization problem in the following form:
>
> $$\max\{\langle Cx, x\rangle : Bx \leq b,\ 0 \leq x \leq 1_n\}. \qquad (13.2.43)$$
>
> Denote by $\widetilde{P}$ the class of languages recognizable in quasi-polynomial time.
> **Theorem 1.2.** *Assume $NP \not\subseteq \widetilde{P}$. Then there is a constant $\delta > 0$ such that the problem (13.2.43) has no polynomial time, $(1 - 2^{-\log^{\delta} n})$-approximation algorithm.*
> **Theorem 1.3.** *Assume $P \neq NP$. Then there is a constant $\mu \in (0, \frac{1}{3})$ such that a μ-approximation of the problem (13.2.43) cannot be found in polynomial time.*

In these statements the μ-approximation is understood in a weak sense. We need to compute an estimate for the value of the objective function only.

Note that using Theorems 13.2.8 and 13.2.9, we can approximate in polynomial time the optimal value of the problem

$$\max\{\langle Cx, x\rangle : Bx = \tfrac{1}{2}1_n,\ 0 \leq x \leq 1_n\}. \qquad (13.2.44)$$

with $(1 - O(\frac{1}{\ln n}))$ relative accuracy. This result is better than the limiting bound of Theorem 1.2 [79]. At the same time, the optimization problem, which

[1] Of course, in non-convex case we cannot prove any global efficiency estimates. Moreover, in general we cannot guarantee a convergence to a point, which satisfies the necessary second order optimality conditions. This negative result is valid even for the second order methods.

is used in the proof of Theorems 1.2, 1.3 [79], has, in fact, only linear equalities constraint:

$$\max\{\langle Cx, x\rangle : \; Bx = b, \; 0 \leq x \leq 1_n\}. \tag{13.2.45}$$

Thus, the difference in the formulations (13.2.44) and (13.2.45) looks very minor. Indeed, any system of linear equations $Bx = b$ can be rewritten in the following form:

$$\bar{B}x = \tfrac{1}{2}1_n, \quad \langle a, x\rangle = 1,$$

with some matrix $\bar{B}$ and a vector $a \in R^n$. Hence, the feasible set of the problem (13.2.45) differs from the feasible set of the problem (13.2.44) just by a single linear equation, which does not pass through the center of the box. However, it appears that this linear equation makes the problem (13.2.45) completely different.

Let us look at the concrete form of the problem (13.2.45) ([79], p.438). Denote by X and Y two $(n \times n)$-matrices. And let $\phi(X, Y)$ be a bilinear form in X and Y with all non-negative coefficients. Then the problem (13.2.45) is as follows:

$$\begin{aligned} &\max \; \phi(X, Y), \\ \text{s.t.} \quad & X1_n = 1_n, \; Y1_n = 1_n, \\ & 0 \leq X, Y \leq 1_{n\times n}. \end{aligned} \tag{13.2.46}$$

Now we can see the source of our troubles. Indeed, the technique of Section 13.2.5 can be applied only to l_p boxes with $p \geq 2$. However, if we will try to approximate the feasible set of the problem (13.2.46) with the boxes $\mathcal{B}_p = \{x : \| x - \frac{1}{2}1_n \|_p \leq \frac{1}{2}\}$, we need to choose p very large. It is necessary to take $p = O(n \ln n)$ just to have a non-empty intersection of the box $\mathcal{B}_p$ with the system of linear constraints in (13.2.46).

Thus, we conclude that the feasible set of the problem (13.2.46) is too far from the center of the box. On the other hand, it is clear that the box structure in (13.2.46) is quite artificial: the constraint X, $Y \leq 1_{n\times n}$ can be eliminated without changing the feasible set of the problem. Note that we can easily rewrite the problem (13.2.46) in a more symmetric form:

$$\begin{aligned} &\max \; \phi(X, Y), \\ \text{s.t.} \quad & \| Xe_i \|_1 \leq 1, \; i = 1, \ldots n, \\ & \| Ye_i \|_1 \leq 1, \; i = 1, \ldots n. \end{aligned} \tag{13.2.47}$$

Since the coefficients of the form $\phi(X, Y)$ are non-negative, the optimal value of the problem (13.2.47) is the same as that of (13.2.43). The polyhedral structure of the feasible set in (13.2.47) can be seen as a combination of l_∞-structure with l_1-structure. However, it appears the latter structure is exactly that one, for which no reasonable bounds for quadratic problems are known.

Thus, the above discussion highlights the following unsolved problem:

Find some bounds for the optimal value of the following quadratic problem:

$$\phi^* = \max\{\langle Qx, x\rangle : \| x \|_p \leq 1,\ x \in R^n\}, \quad 1 \leq p < 2. \tag{13.2.48}$$

For an indefinite Q a trivial bound for ϕ^* is given by its maximal eigenvalue $\lambda_{\max}(Q)$:

$$\lambda_{\max}(Q) \geq \phi^* \geq \lambda_{\max}(Q) \cdot n^{1-\frac{2}{p}}, \quad 1 \leq p \leq 2.$$

For $p = 1$ we can suggest for the problem (13.2.48) a kind of semidefinite relaxation:

$$\begin{aligned} \psi^* &= \max_{X,u}\{\langle Q, X\rangle : \ \mathrm{Diag}\,(u) \succeq X,\ \langle 1_n, u\rangle \leq 1,\ X \succeq 0\} \\ &= \min_{S,\lambda}\{\lambda : \ \lambda 1_n = \mathrm{diag}(S),\ S \succeq Q,\ S \succeq 0\}. \end{aligned} \tag{13.2.49}$$

Note that for any x, $\| x \|_1 \leq 1$, the pair $(X = xx^T, u = \mathrm{abs}[x])$ is feasible for the primal form of the relaxation (13.2.49). Therefore we can guarantee that $\psi^* \geq \phi^*$. However, the relative accuracy of such a bound is not known.

13.3 QUADRATIC CONSTRAINTS

Yinyu Ye

Consider the quadratic programming (QP) problem with diagonally quadratic equality and inequality constraints

$$(QP) \quad \begin{array}{rll} \bar{q}(Q) := & \text{Maximize} & q(x) := x^T Q x \\ & \text{Subject to} & \sum_{j=1}^n a_{ij} x_j^2 = b_i,\ i = 1, \ldots, m, \\ & & \sum_{j=1}^n c_{ij} x_j^2 \leq d_i,\ i = 1, \ldots, p \end{array}$$

where the symmetric matrix $Q \in \mathcal{S}^n$, $A = \{a_{ij}\} \in \mathcal{M}_{m,n}$, $C = \{c_{ij}\} \in \mathcal{M}_{p,n}$, $b \in \Re^m$, and $d \in \Re^p$ are given. We assume that the QP problem is feasible and its feasible set is bounded (this can be checked by a linear program considering x_j^2 as nonnegative variables). Let $\bar{x}(Q)$ be a maximizer of the problem.

The (QP) problem has applications in combinatorial and global optimization problems, see, e.g., Gibbons et al. [273]. Note that this quadratic problem includes the max-cut problem by letting $x_j^2 = 1$, $j = 1, ..., n$, be the quadratic constraints. Also note that perturbing the diagonal of Q may change the objective function on the feasible set of the problem.

Normally, there is a linear term in the objective function:

$$\begin{array}{ll} \text{Maximize} & x^T Q x + c^T x \\ \\ \text{Subject to} & \sum_{j=1}^n a_{ij} x_j^2 = b_i, \ i = 1, \ldots, m, \\ & \sum_{j=1}^n c_{ij} x_j^2 \leq d_i, \ i = 1, \ldots, p \end{array}$$

However, the problem can be homogenized as

$$\begin{array}{ll} \text{Maximize} & x^T Q x + t c^T x \\ \\ \text{Subject to} & \sum_{j=1}^n a_{ij} x_j^2 = b_i, \ i = 1, \ldots, m, \ t^2 = 1, \\ & \sum_{j=1}^n c_{ij} x_j^2 \leq d_i, \ i = 1, \ldots, p \end{array}$$

by adding a scalar variable t. There always is an optimal solution $(\bar{x}, \bar{t})$ for this problem in which $\bar{t} = 1$ or $\bar{t} = -1$. If $\bar{t} = 1$, then $\bar{x}$ is also optimal for the non-homogeneous problem; if $\bar{t} = -1$, then $-\bar{x}$ is optimal for the non-homogeneous problem. Thus, without loss of generality, we can let $q(x) = x^T Q x$ throughout this Section 13.3.

The function $q(x)$ has a minimizer and a maximizer over the bounded feasible set

$$\mathcal{F} := \{x \in \Re^n : \sum_{j=1}^n a_{ij} x_j^2 = b_i, \ i = 1, \ldots, m, \ \sum_{j=1}^n c_{ij} x_j^2 \leq d_i, \ i = 1, \ldots, p\}.$$

Let $\underline{q} := -\bar{q}(-Q)$ and $\bar{q} := \bar{q}(Q)$ denote their minimal and maximal objective values, respectively. An ϵ-maximal solution or ϵ-maximizer, $\epsilon \in [0, 1]$, for (QP) is defined as an $x \in \mathcal{F}$ such that

$$\frac{\bar{q} - q(x)}{\bar{q} - \underline{q}} \leq \epsilon.$$

Recently, there were several significant results on approximating specific quadratic problems. Goemans and Williamson [285] (also see Frieze and Jerrum [255]) proved an approximation result for the Maxcut problem where $\epsilon \leq 1 - 0.878$ when all arc weights are nonnegative. Nesterov [572] generalized their result to approximating a boolean QP problem

$$\begin{array}{ll} \text{Maximize} & q(x) = x^T Q x \\ \\ \text{Subject to} & x_j^2 = 1, \ j = 1, \ldots, n, \end{array}$$

where $\epsilon \leq 4/7$. Ye [859] extended the 4/7 result to solving continuous nonconvex QP problems, such as,

$$\text{Maximize} \quad q(x) = x^T Q x$$

$$\text{Subject to} \quad x_j^2 \leq 1, \; j = 1, \ldots, n.$$

Note that some negative results on this problem were given by Bellare and Rogaway [79]. Other results can be found in Fu, Luo and Ye [258], Pardalos and Rosen [619], Vavasis [818], and Ye [855].

In this Section 13.3, we, based on the analyses of Ye [859] and Nesterov [575], further generalize the 4/7 result to approximating (QP) containing (diagonally) quadratic constraints. These constraints have added a few difficulties in analyzing the problem, and they frequently appear in some practical applications.

13.3.1 Positive Semi-Definite Relaxation

The approximation algorithm for (QP) is to solve a positive semi-definite programming (SDP) relaxation problem

$$(SDP) \quad \begin{array}{lll} \bar{p}(Q) := & \text{Maximize} & \langle Q, X \rangle \\ & \text{Subject to} & \langle D(a_i), X \rangle = b_i, \; i = 1, \ldots, m, \\ & & \langle D(c_i), X \rangle \leq d_i, \; i = 1, \ldots, p. \end{array} \qquad (13.3.50)$$

Here, $a_i = (a_{i1}, \ldots, a_{in}) \in \Re^n$, $c_i = (c_{i1}, \ldots, c_{in}) \in \Re^n$, and unknown $X \in \Re^{n \times n}$ is a symmetric matrix. Furthermore, $\langle \cdot, \cdot \rangle$ is the matrix inner product $\langle Q, X \rangle = \text{trace}(Q^T X)$, $D(a)$ is the diagonal matrix of vector a, and $X \succeq Z$ means that $X - Z$ is positive semi-definite. Since the original QP problem is feasible and bounded, so is the SDP relaxation.

The dual of the problem is

$$\begin{array}{lll} \bar{p}(Q) = & \text{Minimize} & d^T z + b^T y \\ & \text{Subject to} & \sum_{i=1}^p z_i D(c_i) + \sum_{i=1}^m y_i D(a_i) \succeq Q, \; z \geq 0. \end{array} \qquad (13.3.51)$$

Note that the primal is feasible and bounded and the dual has an interior so that there is no duality gap between the primal and dual. Denote by $\bar{X}(Q)$ and $(\bar{y}(Q), \bar{z}(Q))$ an optimal solution pair for the primal (13.3.50) and dual (13.3.51).

The positive semi-definite relaxation was first proposed by Lovász and Shrijver [497], also see recent papers by Alizadeh [17], Fujie and Kojima [260] and Polijak, Rendl and Wolkowicz [635]. This relaxation problem pair can be solved in polynomial time, e.g., see Nesterov and Nemirovskii [583] and Alizadeh [17].

We have the following relations between (QP) and (SDP) from Ye [859].

Proposition 13.3.1 *Let* $\bar{q} = \bar{q}(Q)$, $\underline{q} = -\bar{q}(-Q)$, $\bar{p} = \bar{p}(Q)$, $\underline{p} = -\bar{p}(-Q)$, *and*
$(\underline{y}, \underline{z}) = (-\bar{y}(-Q), -\bar{z}(-Q))$. *Then,* $\underline{q}$ *is the minimal objective value of* $x^T Qx$ *in the feasible set of (QP) and* $\underline{p} = d^T \underline{z} + b^T \underline{y}$ *is the minimal objective value of* $\langle Q, X\rangle$ *in the feasible set of (SDP). Furthermore,*

$$\underline{p} = -\bar{p}(-Q) \leq \underline{q} = -\bar{q}(-Q) \leq \bar{q}(Q) = \bar{q} \leq \bar{p}(Q) = \bar{p}.$$

In what follows, we let $\bar{x} = \bar{x}(Q)$, $\bar{X} = \bar{X}(Q)$. Since $\bar{X}$ is positive semidefinite, there is a factorization matrix $\bar{V} = (\bar{v}_1, \ldots, \bar{v}_n) \in \Re^{n\times n}$, i.e., $\bar{v}_j$ is the jth column of $\bar{V}$, such that $\bar{X} = \bar{V}^T\bar{V}$. The algorithm (Goemans and Williamson [285], Nesterov [572], and Ye [859]) generates a random vector u uniformly distributed on the n-dimensional unit ball and then assigns

$$\hat{x} = \bar{D}\sigma(\bar{V}^T u), \tag{13.3.52}$$

where

$$\bar{D} = \text{diag}(\|\bar{v}_1\|, \ldots, \|\bar{v}_n\|) = \text{diag}(\sqrt{\bar{x}_{11}}, \ldots, \sqrt{\bar{x}_{nn}}),$$

and for any $x \in \Re^n$, $\sigma(x)$ is the vector whose components are $\text{sign}(x_j)$, $j = 1, \ldots, n$, that is,

$$\text{sign}(x_j) = \begin{cases} 1 & \text{if } x_j \geq 0 \\ -1 & \text{otherwise.} \end{cases}$$

It is easily seen that $\hat{x}$ is a feasible point for (QP) and we will show later that the expected objective value, $\mathrm{E}_u q(\hat{x})$, satisfies

$$\frac{\bar{q} - \mathrm{E}_u q(\hat{x})}{\bar{q} - \underline{q}} \leq \frac{\pi}{2} - 1 \leq \frac{4}{7}.$$

13.3.2 Approximation Analysis

The following lemma is an analogue to the lemma of Nesterov [572] and Ye [859].

Lemma 13.3.1 *Let* u *be uniformly distributed on the* n*-dimensional unit ball. Then,*

$$\begin{aligned} \bar{q}(Q) = \quad & \textit{Maximize} \quad \mathrm{E}_u(\sigma(V^T u)^T DQD\sigma(V^T u)) \\ & \textit{Subject to} \quad \langle D(a_i), V^T V\rangle = b_i, \ i = 1, \ldots, m, \\ & \qquad\qquad\quad \langle D(c_i), V^T V\rangle \leq d_i, \ i = 1, \ldots, p, \end{aligned}$$

where

$$D = \textit{diag}(\|v_1\|, \ldots, \|v_n\|).$$

Proof. Since, for any feasible V, $D\sigma(V^T u)$ is a feasible point for (QP), we have

$$\bar{q}(Q) \geq \mathrm{E}_u(\sigma(V^T u)^T DQD\sigma(V^T u)).$$

On the other hand, for any fixed u with $||u|| = 1$, we have

$$\mathrm{E}_u(\sigma(V^T u)^T DQD\sigma(V^T u)) = \sum_{i=1}^{n}\sum_{j=1}^{n} q_{ij}||v_i||||v_j||\mathrm{E}_u(\sigma(v_i^T u)\sigma(v_j^T u)). \tag{13.3.53}$$

Let us choose $v_i = \frac{\bar{x}_i}{||\bar{x}||}\bar{x}$, $i = 1, \ldots, n$. (Note that V is feasible for the problem above.) Then

$$\mathrm{E}_u(\sigma(v_i^T u)\sigma(v_j^T u)) = \begin{cases} 1 & \text{if } \sigma(\bar{x}_i) = \sigma(\bar{x}_j) \\ -1 & \text{otherwise.} \end{cases}$$

Thus,

$$||v_i||||v_j||\mathrm{E}_u(\sigma(v_i^T u)\sigma(v_j^T u)) = \bar{x}_i\bar{x}_j$$

which implies that for this particular feasible V

$$\bar{q}(Q) = q(\bar{x}) \leq \mathrm{E}_u(\sigma(V^T u)^T DQD\sigma(V^T u)).$$

These two relations give the desired result. ■

For any function of one variable $f(t)$ and $X \in \Re^{n\times n}$, let $f[X] \in \Re^{n\times n}$ be the matrix with the components $f(x_{ij})$. Nesterov [572] has proved the next technical lemma.

Lemma 13.3.2 *Let* $X \succeq 0$ *and* $d(X) \leq 1$. *Then* $\arcsin[X] \succeq X$. ■

Now we are ready to prove the following theorem.

Theorem 13.3.1

$$\begin{aligned} \bar{q}(Q) = \quad \sup \quad & \tfrac{2}{\pi}\langle Q, D\arcsin[D^{-1}XD^{-1}]D\rangle \\ \textit{Subject to} \quad & \langle D(a_i), X\rangle = b_i,\ i = 1, \ldots, m, \\ & \langle D(c_i), X\rangle \leq d_i,\ i = 1, \ldots, p, \\ & X \succ 0, \end{aligned}$$

where

$$D = \mathrm{Diag}\,(\sqrt{x_{11}}, \ldots, \sqrt{x_{nn}}).$$

Proof. For any $X = V^T V \succ 0$, we have

$$\begin{aligned} \mathrm{E}_u(\sigma(v_i^T u)\sigma(v_j^T u)) &= 1 - 2\Pr\{\sigma(v_i^T u) \neq \sigma(v_j^T u)\} \\ &= 1 - 2\Pr\{\sigma(\tfrac{v_i^T u}{||v_i||}) \neq \sigma(\tfrac{v_j^T u}{||v_j||})\}. \end{aligned}$$

From Lemma 1.2 of Goemans and Williamson [285], we have

$$\Pr\{\sigma(\frac{v_i^T u}{||v_i||}) \neq \sigma(\frac{v_j^T u}{||v_j||})\} = \frac{1}{\pi}\arccos(\frac{v_i^T v_j}{||v_i||||v_j||}).$$

Using the above lemma and equality (13.3.53) and noting $\arcsin(t)+\arccos(t) = \frac{\pi}{2}$ give the desired result. ■

We have used *Supremum* and $X \succ 0$ in the problem above merely for the technical presentation of D^{-1}. The feasible set of this problem can be closed if we rewrite it in terms of variable $Y = D^{-1}XD^{-1}$

Theorem 13.3.1 leads to our main result.

Theorem 13.3.2 *We have*

1.

$$\bar{q} - \underline{p} \geq \frac{2}{\pi}(\bar{p} - \underline{p}).$$

2.

$$\bar{p} - \underline{q} \geq \frac{2}{\pi}(\bar{p} - \underline{p}).$$

3.

$$\bar{p} - \underline{p} \geq \bar{q} - \underline{q} \geq \frac{4-\pi}{\pi}(\bar{p} - \underline{p}).$$

Proof. Recall $\underline{z} = -\bar{z}(-Q) \leq 0$, $\underline{p} = -\bar{p}(-Q) = d^T\underline{z} + b^T\underline{y}$, and

$$Q - \sum_{i=1}^{p} \underline{z}_i D(c_i) - \sum_{i=1}^{m} \underline{y}_i D(a_i) \succeq 0.$$

Thus, for any $X \succ 0$ feasible for (SDP), and $D = \operatorname{diag}(\sqrt{x_{11}}, \ldots, \sqrt{x_{nn}})$, we have from Theorem 13.3.1

$$\begin{aligned} \frac{\pi}{2}\bar{q} &= \frac{\pi}{2}\bar{q}(Q) \\ &\geq \langle Q, D\arcsin[D^{-1}XD^{-1}]D\rangle \\ &= \left\langle Q - \sum_{i=1}^{p} \underline{z}_i D(c_i) - \sum_{i=1}^{m} \underline{y}_i D(a_i) + \sum_{i=1}^{p} \underline{z}_i D(c_i) \right. \\ &\qquad \left. + \sum_{i=1}^{m} \underline{y}_i D(a_i), D\arcsin[D^{-1}XD^{-1}]D \right\rangle \end{aligned}$$

$$= \left\langle Q - \sum_{i=1}^{p} \underline{z}_i D(c_i) - \sum_{i=1}^{m} \underline{y}_i D(a_i), D \arcsin[D^{-1}XD^{-1}]D \right\rangle$$
$$+ \left\langle \sum_{i=1}^{p} \underline{z}_i D(c_i) + \sum_{i=1}^{m} \underline{y}_i D(a_i), D \arcsin[D^{-1}XD^{-1}]D \right\rangle$$
$$\geq \left\langle Q - \sum_{i=1}^{p} \underline{z}_i D(c_i) - \sum_{i=1}^{m} \underline{y}_i D(a_i), DD^{-1}XD^{-1}D \right\rangle$$
$$+ \left\langle \sum_{i=1}^{p} \underline{z}_i D(c_i) + \sum_{i=1}^{m} \underline{y}_i D(a_i), D \arcsin[D^{-1}XD^{-1}]D \right\rangle$$
$$\left(\text{since } Q - \sum_{i=1}^{p} \underline{z}_i D(c_i) - \sum_{i=1}^{m} \underline{y}_i D(a_i) \succeq 0 \right.$$
$$\left. \text{and } \arcsin[D^{-1}XD^{-1}] \succeq D^{-1}XD^{-1} \right)$$
$$= \left\langle Q - \sum_{i=1}^{p} \underline{z}_i D(c_i) - \sum_{i=1}^{m} \underline{y}_i D(a_i), X \right\rangle$$
$$+ \left\langle \sum_{i=1}^{p} \underline{z}_i D(c_i) + \sum_{i=1}^{m} \underline{y}_i D(a_i), D \arcsin[D^{-1}XD^{-1}]D \right\rangle$$
$$= \langle Q, X \rangle - \left\langle \sum_{i=1}^{p} \underline{z}_i D(c_i) + \sum_{i=1}^{m} \underline{y}_i D(a_i), X \right\rangle$$
$$+ \left\langle \sum_{i=1}^{p} \underline{z}_i D(c_i) + \sum_{i=1}^{m} \underline{y}_i D(a_i), D \arcsin[D^{-1}XD^{-1}]D \right\rangle$$
$$= \langle Q, X \rangle - \sum_{i=1}^{p} \underline{z}_i \langle D(c_i), X \rangle - \sum_{i=1}^{m} \underline{y}_i \langle D(a_i), X \rangle$$
$$+ \sum_{i=1}^{p} \underline{z}_i \langle D(c_i), D \arcsin[D^{-1}XD^{-1}]D \rangle$$
$$+ \sum_{i=1}^{m} \underline{y}_i \langle D(a_i), D \arcsin[D^{-1}XD^{-1}]D \rangle$$
$$= \langle Q, X \rangle - \sum_{i=1}^{p} \underline{z}_i \langle D(c_i), X \rangle - \underline{y}^T b + \sum_{i=1}^{p} \underline{z}_i \left(\frac{\pi}{2} \langle D(c_i), X \rangle\right) + \underline{y}^T \left(\frac{\pi}{2} b\right)$$
$$= \langle Q, X \rangle + (\tfrac{\pi}{2} - 1) \sum_{i=1}^{p} \underline{z}_i \langle D(c_i), X \rangle + (\frac{\pi}{2} - 1) \underline{y}^T b$$
$$\geq \langle Q, X \rangle + (\tfrac{\pi}{2} - 1)(\underline{z}^T d + \underline{y}^T b)$$
$$(\text{since } \langle D(c_i), X \rangle \leq d_i \;\; i = 1, ..., p, \text{ and } \underline{z} \leq 0)$$
$$= \langle Q, X \rangle + (\tfrac{\pi}{2} - 1)\underline{p}.$$

Let X converge to $\bar{X}$, then $\langle Q, X \rangle \to \bar{p}$ and we have the desired first inequality.
Replacing Q with $-Q$ proves the second inequality in the theorem.

Adding the first two inequalities gives the third statement in the theorem. ■

The result indicates that the positive semi-definite relaxation value $\bar{p} - \underline{p}$ is a constant approximation of $\bar{q} - \underline{q}$.

Similarly, the following corollary can be devised.

Corollary 13.3.1 *Let $X = V^T V \succ 0$, $\langle D(a_i), X\rangle \le d_i$ $(i = 1, \ldots, p)$, $\langle D(a_i), X\rangle = b_i$ $(i = 1, \ldots, m)$, $D = diag(\sqrt{x_{11}}, \ldots, \sqrt{x_{nn}})$, and $\hat{x} = D\sigma(V^T u)$ where u with $\|u\| = 1$ is a random vector uniformly distributed on the unit ball. Moreover, let $X \to \bar{X}$. Then,*

$$\lim_{X\to\bar{X}} \mathrm{E}_u(q(\hat{x})) = \lim_{X\to\bar{X}} \frac{2}{\pi}\langle Q, D \arcsin[D^{-1}XD^{-1}]D\rangle \ge \frac{2}{\pi}\bar{p} + (1 - \frac{2}{\pi})\underline{p}.$$

Finally, we have

Theorem 13.3.3 *Let $\hat{x}$ be generated above from $X = \bar{X}$. Then*

$$\frac{\bar{q} - \mathrm{E}_u q(\hat{x})}{\bar{q} - \underline{q}} \le \frac{\pi}{2} - 1.$$

Proof. The proof is similar to that in Nesterov [572] and Ye [859]. We include it here for completeness. Since

$$\bar{p} \ge \bar{q} \ge \frac{2}{\pi}\bar{p} + (1 - \frac{2}{\pi})\underline{p} \ge (1 - \frac{2}{\pi})\bar{p} + \frac{2}{\pi}\underline{p} \ge \underline{q} \ge \underline{p}$$

we have

$$\begin{aligned}
\frac{\bar{q} - \mathrm{E}_u q(\hat{x})}{\bar{q} - \underline{q}} &\le \frac{\bar{q} - \frac{2}{\pi}\bar{p} - (1 - \frac{2}{\pi})\underline{p}}{\bar{q} - \underline{q}} \\
&\le \frac{\bar{q} - \frac{2}{\pi}\bar{p} - (1 - \frac{2}{\pi})\underline{p}}{\bar{q} - (1 - \frac{2}{\pi})\bar{p} - \frac{2}{\pi}\underline{p}} \\
&\le \frac{\bar{p} - \frac{2}{\pi}\bar{p} - (1 - \frac{2}{\pi})\underline{p}}{\bar{p} - (1 - \frac{2}{\pi})\bar{p} - \frac{2}{\pi}\underline{p}} \\
&= \frac{(1 - \frac{2}{\pi})(\bar{p} - \underline{p})}{\frac{2}{\pi}(\bar{p} - \underline{p})} \\
&= \frac{(1 - \frac{2}{\pi})}{\frac{2}{\pi}} = \frac{\pi}{2} - 1.
\end{aligned}$$

■

13.3.3 *Results for Other Quadratic Problems*

Consider now another nonconvex QP problem:

$$\begin{aligned} &\text{Maximize} \quad x^T Q x + c^T x \\ &\text{Subject to} \quad x^T A_i x + c_i^T x \leq b_i, \ i = 1, \ldots, m, \end{aligned}$$

where given symmetric matrices $A_i \in \Re^{n \times n}$. We summarize approximation results for solving this problem.

- If $m = 1$, $A_1 = I$, the identity matrix, and $c_1 = 0$, then the problem is polynomially solvable. That is, there is an algorithm to generate an ϵ-solution for any $\epsilon > 0$, and its running time is polynomial in n and $\log(1/\epsilon)$, see an early proof by Vavasis [818] and Ye [855] and a later by Rendl and Wolkowicz [661]. (Ye [856] further reduced the complexity time dependency on ϵ to $\log\log(1/\epsilon)$.)
- If all A_i are mutually commutative (they can be simultaneously diagonalized) and all $c_i = 0$, then the problem can be transformed into a problem with only diagonally quadratic constraints, and thus can be approximated for $\epsilon = 4/7$ according to our early analysis, also see Ye [859] and Nesterov [575].
- If all A_i are positive semidefinite, then the problem can be approximated for $\epsilon = 1 - \frac{constant}{m^2}$ by Fu et al. [258]; and in addition, if all $c_i = 0$, then it can be approximated for $\epsilon = 1 - \frac{constant}{\log(mn)}$ by Nemirovskii et al. [568].

13.4 RELAXATIONS OF Q^2P

Henry Wolkowicz

In this part of the chapter we look at several different instances of Q^2P. In particular, we start with several different tractable relaxations for the max-cut problem and show that, surprisingly, they are all equal to the Lagrangian (**and** SDP) relaxation.

We then illustrate a recipe for constructing relaxations for QQPs by finding a strengthened SDP bound for the max-cut problem.

Other instances discussed are the quadratic assignment and graph partitioning problems.

We then consider trust region type problems and discuss when strong duality holds. This includes problems where orthogonal constraints arise, e.g. orthogonal relaxations of the quadratic assignment and graph partitioning problems. In particular, this part of the chapter emphasizes the theme about the strength of the Lagrangian relaxation.

13.4.1 Relaxations for the Max-cut Problem

The success of the SDP relaxation (equivalently Lagrangian relaxation) over the last few years is exemplified by the success on the *Max-Cut Problem.* Let $G = (V, E)$ be an undirected graph with edge set $V = \{v_i\}_{i=1}^n$ and weights w_{ij} on the edges $(v_i, v_j) \in E$. We want to find the index set $\mathcal{I} \subset \{1, 2, \dots n\}$, to maximize the weight of the edges with one end point with index in $\mathcal{I}$ and the other in the complement. This is equivalent to

$$(MC) \qquad \max\ \tfrac{1}{2} \textstyle\sum_{i<j} w_{ij}(1 - x_i x_j), \quad x \in \mathcal{F},$$

where $\mathcal{F} := \{\pm 1\}^n$, and $x_i = 1$ if $i \in \mathcal{I}$ and -1 otherwise. The objective function is a (homogeneous) quadratic form, $x^T Q x$.

Several *Different* Relaxations. We now look at several different tractable relaxations of MCQ, (13.4.54). These have different motivations. For example, one bound relaxes the constraints to the unit ball of radius $\sqrt{n}$, while another relaxes the constraints to the convex hull, i.e. to the unit cube. Following [638, 635], we observe that several quadratic type bounds considered in the literature are actually equal. The key to the simple proofs is the strong duality result for the trust region subproblem, see [747]. A similar phenomenon occurs for linearizations of (P), such as in *roof duality,* see e.g. [324], where many bounds obtained from various linearizations have been shown to be equal and, in fact, they have been shown to be equal to the Lagrangian dual of a linearized problem, see [4]. (The quality of the SDP bounds is the main topic in the first two parts of this chapter; see above.)

We allow a more general objective function, i.e. we consider the ± 1 constrained quadratic program

$$(MCQ) \qquad \mu^* := \max_{x \in \mathcal{F}} q_0(x) \quad (:= x^T Q x - 2c^T x). \tag{13.4.54}$$

The bounds are derived using the fact that we can perturb the objective function q_0 and exploit the fact that $x_i^2 = 1$ on the feasible set $\mathcal{F}$. Note that

$$\begin{aligned} q_u(x) &:= x^T(Q + \mathrm{Diag}\,(u))x - 2c^T x - u^T e \\ &= q_0(x), \quad \forall x \in \mathcal{F}. \end{aligned} \tag{13.4.55}$$

For each u we get a trivial upper bound obtained from ignoring the constraints and allowing the diagonal perturbations, i.e. we have

$$\mu^* \leq f_0(u) := \max_x q_u(x). \tag{13.4.56}$$

But, the function f_0 can take on the value $+\infty$. Let

$$S := \left\{u : u^T e = 0, Q + \text{Diag}\,(u) \preceq 0\right\}.$$

We then get the following trivial bound.

$$\mu^* \leq B_0 := \min_u f_0(u) \qquad \left(= \min_{u^T e=0} f_0(u), \text{ if } S \neq \emptyset\right). \tag{13.4.57}$$

Note that if the set S is not empty, then we can minimize over the unconstrained parameter u or add the restriction to $u^T e = 0$. This can be seen from the optimality conditions for min-max problems. This comment is true for the following bounds as well. (Details can be found in [638].)

In addition we can restrict the parameters and avoid infinite values for the inner maximization problem by adding the hidden semidefinite constraint, i.e. we use the fact that a quadratic function is unbounded above if the Hessian is not negative semidefinite. (Note that a quadratic function is bounded above if and only if the Hessian is negative semidefinite and the stationarity equation is consistent.) The following is a tractable bound since we minimize a convex function over a convex set.

$$\mu^* \leq B_0 = \min_{Q+\text{Diag}\,(u)\preceq 0} f_0(u). \tag{13.4.58}$$

Next we relax the feasible set to the sphere of radius $\sqrt{n}$. We get

$$\mu^* \leq f_1(u) := \max_{||x||^2=n} q_u(x). \tag{13.4.59}$$

And our next bound is

$$\mu^* \leq B_1 := \min_u f_1(u). \tag{13.4.60}$$

The inner maximization problem is the trust region subproblem and is tractable, see e.g. Section 13.4.3 below. Thus we have our second tractable bound.

We can replace the spherical constraint with the box constraint.

$$\mu^* \leq f_2(u) := \max_{|x_i|\leq 1} q_u(x). \tag{13.4.61}$$

After adding the semidefinite constraint to make the bound tractable, i.e. to make the calculation of f_2 tractable, we get our next bounds.

$$\mu^* \leq \min_u f_2(u) \tag{13.4.62}$$

and

$$\mu^* \leq B_2 := \min_{Q+\text{Diag}\,(u)\preceq 0} f_2(u). \tag{13.4.63}$$

Given Q and c, define the $(n+1)\times(n+1)$-matrix Q^c by adding a $0-th$ row and column, so that

$$\begin{array}{ll} q^c_{00} = 0 & \\ q^c_{0i} = q^c_{i0} = -c_i & \text{for } i > 0 \\ q^c_{ij} = q_{ij} & \text{for } i,j > 0, \end{array}$$

i.e.

$$Q^c := \begin{bmatrix} 0 & -c^T \\ -c & Q \end{bmatrix}. \tag{13.4.64}$$

In order to have analogous functions $q^c_u(y)$ and $f_i(u)$ as in the previous cases, let us introduce

$$q^c_u(y) := y^T(Q^c + \mathrm{diag}(u))y - u^T e. \tag{13.4.65}$$

Note that q^c_u reduces to q_u if the first component y_0 is ± 1. The equivalent relaxed problem is

$$\mu^* \leq f^c_1(u) := \max_{||y||^2=n+1} q^c_u(y) = (n+1)\lambda_{\max}(Q^c + \mathrm{diag}(u)) - u^T e, \tag{13.4.66}$$

where $\lambda_{\max}$ denotes the maximum eigenvalue. Now another bound is

$$\mu^* \leq B^c_1 := \min_u f^c_1(u). \tag{13.4.67}$$

Similarly, we get equivalent bounds B^c_0 and homogenized bounds for the other models.

The above argument shows that we can homogenize the problem by moving into a higher dimension. Therefore, we can consider the special case that $c = 0$. We now look at the SDP bound, see also Section 13.2 above for the performance guarantees. The relaxation comes from the fact that the trace is commutative, i.e.

$$x^T Q x = \mathrm{Trace}\, x^T Q x = \mathrm{Trace}\, Q x x^T$$

and, for $x \in \mathcal{F}$, $y_{ij} = x_i x_j$ defines a symmetric, rank one, positive semidefinite matrix Y with diagonal elements 1. Therefore, we can *lift* the problem into the higher dimensional space of symmetric matrices and relax the rank one constraint. This yields the following relaxation and our bound 3.

$$\begin{array}{llc} B_3 := & \max & \mathrm{Trace}\, QY \\ & \text{subject to} & \mathrm{diag}(Y) = e \\ & & Y \succeq 0. \end{array} \tag{13.4.68}$$

This SDP is a convex programming problem and is tractable.

Now we replace the ± 1 constraints with $x_i^2 = 1, \forall i$. This does not change the feasible set of the original problem. In [638, 635] it is shown that all the above relaxations and bounds for MC come from the Lagrangian dual of (P_E), the following equivalent problem to MCQ. Thus we enforce our theme about the strength of the Lagrangian relaxation. The strong duality result for the trust region subproblem is the key to the proofs.

$$(P_E) \qquad \begin{array}{ll} \max & q_0(x) = x^T Qx - 2c^T x \\ \text{subject to} & x_i^2 = 1, \quad i = 1, \cdots, n. \end{array} \qquad (13.4.69)$$

Note that the Lagrangian dual of P_E yields precisely our trivial first bound B_0 in (13.4.57).

Theorem 13.4.1 *All the bounds for MCQ discussed above are equal to the optimal value of the Lagrangian dual of the equivalent program P_E.*

A Strengthened Bound for MC. From the results above, it would appear that we might have the strongest possible tractable bound. However, adding redundant constraints can strengthen bounds. The following bound is motivated by the strong duality results presented in Section 13.4.3 below and is presented in [41]. The SDP bound (13.4.68) for MCQ arises from a lifting procedure, i.e. identifying

$$0 \preceq X = xx^T \text{ and } x^T Qx = \operatorname{Trace} X.$$

Discarding the rank one condition on X results in the tractable SDP bound. It is not clear what constraints one can add to P_E in order to strengthen the Lagrangian relaxation, i.e. linear combinations of the constraints will not help since they are already included in the Lagrangian. But, in the space of matrices, it is also true that

$$X^2 = xx^T xx^T = nX.$$

Therefore we can use the following equivalent quadratic matrix model for MCQ.

$$\begin{array}{rll} \mu^* := & \max & \operatorname{Trace} QX \\ & \text{s.t.} & \operatorname{diag}(X) = e \\ & & X^2 - nX = 0, \end{array}$$

where X is a symmetric matrix. This problem is equivalent to P_E since $X^2 = nX$ and $\operatorname{Trace} X = n$ implies X is rank one. Therefore we are including the rank one information from the original problem. However, this problem is a nonconvex problem and cannot be solved in general. Note that if $X^2 = nX$, then $\operatorname{Trace} QX = (1/n)\operatorname{Trace} QX^2$, and $\operatorname{diag}(X^2) = ne$. As a result, the above quadratic model is equivalent to the model:

$$\begin{array}{rll} \mu^* = & \max & \frac{1}{n}\operatorname{Trace} QX^2 \\ & \text{s.t.} & x_i^T x_i = n, \quad i = 1, \ldots, n \\ & & X^2 - nx_0 X = 0 \\ & & x_0^2 = 1, \end{array} \qquad (13.4.70)$$

where x_i^T, $i = 1, \ldots, n$ denotes the ith row of X, and x_0 is a scalar. Having a quadratic objective is an advantage only if it results in a larger class of available

Lagrange multipliers. Therefore, the sign of the eigenvalues of Q will determine whether this objective or $\frac{1}{n}\text{Trace}\, QXx_0$ is better. (Note that if $x_0 = -1$ then changing x_0 to 1 and replacing X with $-X$ leaves the objective and constraints in (13.4.70) unchanged.) We will obtain an upper bound $\mu_2 \geq \mu^*$ by applying a Lagrangian procedure to all of the constraints in (13.4.70). Using multipliers u_i for the constraints $x_i^T x_i = n$, $i = 1, \dots, n$, u_0 for the constraint $x_0^2 = 1$, and a symmetric matrix S for the matrix equality $X^2 - nX = 0$, we obtain a Lagrangian problem

$$\begin{array}{rl}\mu_2 := \min_{u_0,u,S} & u_0 + nu^T e + \max_{x_0,X} \frac{1}{n}\text{Trace}\, QX^2 - \text{Trace}\, UX^2 \\ & \quad + \text{Trace}\, SX^2 - nx_0\text{Trace}\, SX - u_0 x_0^2,\end{array}$$

where $U = \text{Diag}\,(u)$. Letting $\bar{x}^T = (x_0, \text{vec}\,(X)^T)$, this problem can be written in Kronecker product form as

$$\mu_2 = \min_{u_0,u,S} \ u_0 + ne^T u + \max_{\bar{x}} \ \bar{x}^T \bar{Q} \bar{x},$$

where

$$\bar{Q} = \begin{pmatrix} -u_0 & -\frac{n}{2}\text{vec}\,(S)^T \\ -\frac{n}{2}\text{vec}\,(S) & I \otimes \left(\frac{1}{n}Q - U + S\right) \end{pmatrix}.$$

Applying the hidden semidefinite constraint $\bar{Q} \preceq 0$, we obtain an equivalent problem

$$\begin{array}{rll}\mu_2 = & \min & u_0 + ne^T u \\ & \text{s.t.} & \begin{pmatrix} u_0 & \frac{n}{2}\text{vec}\,(S)^T \\ \frac{n}{2}\text{vec}\,(S) & I \otimes \left(-\frac{1}{n}Q + U - S\right) \end{pmatrix} \succeq 0 \\ & & S = S^T.\end{array} \qquad (13.4.71)$$

Note that if we take $S = 0$ in (13.4.71), then $u_0 = 0$ is clearly optimal, and the problem reduces to

$$\begin{array}{rcl}\min & e^T u & \\ -Q + U & \succeq & 0,\end{array}$$

which is exactly the dual of the usual SDP relaxation for MC. It follows that we have obtained an upper bound μ_2 which is a strengthening of the usual SDP bound, i.e. $\mu_2 \leq B_0$.

Alternative Strengthened Relaxation. This presents an alternative strengthened SDP relaxation for the max-cut problem, i.e. this continues from the above Section 13.4.1 but tries to fully exploit the rank-one condition in the Lagrangian.

We use the notation: For $S \in \mathcal{S}^n$, the vector $s = \text{svec}\,(S) \in \Re^{t(n)}$, is formed (columnwise) from S while ignoring the strictly lower triangular part of S. Its inverse is the operator $S = \text{sMat}\,(s)$. The adjoint of svec is the operator hMat (v) which forms a symmetric matrix where the off-diagonal terms are multiplied by a half, i.e. this satisfies

$$\operatorname{svec}(S)^T v = \operatorname{Trace} S \operatorname{hMat}(v), \qquad \forall S \in \mathcal{S}^n, v \in \Re^{t(n)},$$

where $t(n) = n(n+1)/2$. The adjoint of sMat is the operator dsvec (S) which works like svec except that the off diagonal elements are multiplied by 2, i.e. this satisfies

$$\operatorname{dsvec}(S)^T v = \operatorname{Trace} S \operatorname{sMat}(v), \qquad \forall S \in \mathcal{S}^n, v \in \Re^{t(n)}.$$

For notational convenience, we define the vectors sdiag (s) := diag(sMat (s)) and vsMat (s) = vec (sMat (s)); the adjoint of vsMat is then given by

$$\operatorname{vsMat}^*(s) = \operatorname{dsvec}\left(\left(\operatorname{Mat}(v) + \operatorname{Mat}(v)^T\right)/2\right).$$

As above we can start with the following equivalent program

$$\text{MC}_\text{O} \qquad \begin{array}{rll} \mu^* = & \max & \frac{1}{2}\operatorname{Trace} QX \\ & \text{s.t.;} & \operatorname{diag}(X) = e \\ & & X \circ X = e \\ & & X^2 - nX = 0. \end{array} \tag{13.4.72}$$

There are many redundant constraints. However, it is uncertain which of these become redundant in the SDP relaxation. The recipe is to throw in redundant constraints; then take the Lagrangian dual twice and delete redundant constraints at the end. At the end one has an SDP with linear constraints and one can often remove the redundancy using the structure of the problem. This illustrates the strength of the Lagrangian relaxation approach. This is done in [34]. (See also [635] and more recently [473].) The result after deleting redundant constraints is the simplified SDP relaxation (see [34]):

$$\text{MCPSDP2} \qquad \begin{array}{rll} \nu_2^* = & \max & \operatorname{Trace} H_c Y \\ & \text{s.t.} & \operatorname{diag}(Y) = e \\ & & Y_{0,t(i)} = 1, \quad \forall i = 1,\ldots,n \\ & & \sum\limits_{k=1}^{i} Y_{t(i-1)+k,t(j-1)+k} + \sum\limits_{k=i+1}^{j} Y_{t(k-1)+i,t(j-1)+i} \\ & & \quad + \sum_{k=j+1}^{n} Y_{t(i-1)+i,t(k-1)+j} - nY_{0,t(j-1)+i} = 0 \\ & & \quad \forall 1 \le i < j \le n \\ & & Y \succeq 0, Y \in \mathcal{S}^{t(n)+1}. \end{array} \tag{13.4.73}$$

This problem has $2t(n)-1$ constraints. In fact, there is still some redundancy as it can be shown that Slater's constraint qualification fails for this problem. This can be further exploited by projecting the problem onto the space determined by the *minimal face* of the problem, see [34].

13.4.2 General Q^2P

We now move on to applying the Lagrangian relaxation to *general* quadratic constrained quadratic problems, denoted Q^2P; and, we apply it to several specific instances: the quadratic assignment, graph partitioning, max-clique problems. The general Q^2P problem is also studied in e.g. [260, 442] and [652, 451, 449, 510].

Quadratic bounds using a Lagrangian relaxation have been extensively studied and applied in the literature, for example in [444] and, more recently, in [445]. The latter calls the Lagrangian relaxation the "best convex bound". Discussions on Lagrangian relaxation for nonconvex programs also appear in [245]. More references are given throughout this chapter.

Remark 13.4.1 *Any equality constraints are written as two inequality constraints; any linear equality constraints, $Ax = b$, is transformed to a quadratic constraint via $||Ax - b||^2 = 0$. The reason for these transformations for linear equality constraints is discussed in [635], i.e. the Lagrangian dual essentially ignores linear constraints as can be seen from: $-\infty = \max_\lambda \min_x -x^2 + \lambda x$, which is the dual of the problem* $\min\{-x^2 : x = 0\}$.

We now recall the Q^2P in x.

$$(Q^2P_x) \qquad \begin{array}{rll} q^* := & \min & q_0(x) := x^T Q_0 x + 2g_0^T x + \alpha_0 \\ & \text{subject to} & q_k(x) := x^T Q_k x + 2g_k^T x + \alpha_k \leq 0 \\ & & k \in \mathcal{I} := \{1, \ldots, m\} \\ & & x \in \Re^n, \end{array} \qquad (13.4.74)$$

where the matrices Q_k are symmetric. The *feasible set* is

$$F_x := \{x \in \Re^n : q_k(x) \leq 0, \forall k \in \mathcal{I}\}.$$

(Note that though the feasible set F_x may be empty, the feasible set of the relaxation may not be.) The objective function and the constraints are not convex, necessarily. Therefore the feasible set can be a very "nasty" set. This problem is a very hard problem to solve in general, see e.g. [614].

Let

$$P_k := \begin{bmatrix} \alpha_k & g_k^T \\ g_k & Q_k \end{bmatrix} \qquad (13.4.75)$$

and, by abuse of notation, define

$$q_k(y) := y^T P_k y, \; k = 0, 1, \ldots, m.$$

Then an equivalent homogenized formulation to (Q^2P_x) is

$$(Q^2P_y) \qquad \begin{array}{rll} q^* = & \min & q_0(y) \\ & \text{subject to} & q_k(y) \leq 0, k \in \mathcal{I} \\ & & y_0^2 = 1 \\ & & y = \begin{pmatrix} y_0 \\ x \end{pmatrix} \in \Re^{n+1}. \end{array}$$

It is clear that the optimal values of the two equivalent formulations are equal. In fact, if $y_0 = -1$ is optimal, then we can replace y by $-y$. This is because the objective function and all but the last constraint are homogeneous.

We will refer to both equivalent formulations of Q^2P in the sequel. The correct reference will be clear from the context.

Remark 13.4.2 *Note that we could replace the constraint $y_0^2 = 1$ by $y_0 = 1$. (The constraint $y_0 = 1$ is used in [260].) In the latter case, the feasible sets of the two formulations coincide exactly, while in the former case they can differ by a sign, i.e. $x \in F_x$ implies that both $\begin{pmatrix} -1 \\ -x \end{pmatrix}$ and $\begin{pmatrix} 1 \\ x \end{pmatrix}$ are in F_y, i.e. are feasible for the homogenized problem Q^2P_y.*

The Lagrangian Relaxation of a General $\mathbf{Q^2P}$. The Lagrangian relaxation of the homogenized problem Q^2P_y provides a simple technique for obtaining the SDP relaxation. In addition, an application of the strong duality result for the trust region subproblem shows that both the SDP and Lagrangian relaxation are equal. The Lagrangian of Q^2P_y is

$$L(y, \mu, \lambda) := y^T P_0 y - \mu(y_0^2 - 1) + \sum_{k \in \mathcal{I}} \lambda_k y^T P_k y.$$

The Lagrangian relaxation of Q^2P_y is

$$(DQ^2P_y) \qquad d^* := \max_{\substack{\mu \\ \lambda \geq 0}} \min_y y^T P_0 y - \mu(y_0^2 - 1) + \sum_{k \in \mathcal{I}} \lambda_k y^T P_k y.$$

Note that

$$\begin{aligned} d^* &= \max_{\lambda \geq 0} \max_{\mu} \min_y y^T P_0 y - \mu(y_0^2 - 1) + \sum_{k \in \mathcal{I}} \lambda_k y^T P_k y \\ &= \max_{\lambda \geq 0} \min_{y_0^2 = 1} y^T P_0 y + \sum_{k \in \mathcal{I}} \lambda_k y^T P_k y, \end{aligned}$$

from strong duality of the trust region subproblem, see [747]. Therefore, we get equivalence of the dual values for the problems in x and in y. (This is similar to the approaches in [849, 733].)

$$(DQ^2P_x) \qquad d^* = \max_{\lambda \geq 0} \min_x q_0(x) + \sum_{k \in \mathcal{I}} \lambda_k q_k(x).$$

We immediately conclude that *weak duality* holds

$$d^* \leq q^* = \min_y \max_{\substack{\mu \\ \lambda \geq 0}} y^T P_0 y - \mu(y_0^2 - 1) + \sum_{k \in \mathcal{I}} \lambda_k y^T P_k y.$$

Therefore, if the optimal μ^*, λ^* can be found, we have found a single quadratic function whose minimal value approximates the original minimal value q^*, i.e.

$$q^* \geq d^* = \min_y y^T P_0 y - \mu^*(y_0^2 - 1) + \sum_{k \in \mathcal{I}} \lambda_k^* y^T P_k y. \tag{13.4.76}$$

Moreover, in the dual program, the Lagrangian is a quadratic function of y. Therefore, the outer maximization problem has the nonnegativity and an additional hidden semidefinite constraint

$$P_0 - \mu E_{00} + \sum_{k \in \mathcal{I}} \lambda_k P_k \succeq 0, \quad \lambda \geq 0, \tag{13.4.77}$$

where E_{00} is the zero matrix with 1 in the top left corner and $M \succeq 0$ denotes the Löwner partial order, i.e. that the symmetric matrix M is positive semidefinite. The minimum of the minimization subproblem, in this case, is attained by $y = 0$. Therefore the Lagrangian dual is equivalent to the SDP problem

$$(DSDP) \qquad \begin{array}{rll} d^* := & \max & \mu \\ & \text{subject to} & \mu E_{00} - \sum_{k \in \mathcal{I}} \lambda_k P_k \preceq P_0 \\ & & \lambda \geq 0. \end{array}$$

Valid Inequalities. Using the above approach we see that more constraints $q_k(y)$ means that we have a stronger dual. This can be phrased as adding redundant constraints to get new valid inequalities to strengthen the relaxation. We will see how this occurs when we look at orthogonally constrained problems below. Another approach is also specified in detail in Kojima and Tuncel [442, 440].

For problems that also have linear equality constraints, one can use the notion of copositivity to strengthen the SDP relaxation. However, this does not result in a tractable relaxation in general, see [649].

Specific Instances of SDP Relaxation. We now study four specific instances and show how to apply the recipe for relaxations. In each case we derive a min-max eigenvalue problem from the Lagrangian dual of an appropriately chosen quadratic constrained program. The dual of this dual problem provides a semidefinite relaxation for the original problem. Adding redundant constraints at the start helps in reducing the duality gap. These redundant constraints are automatically deleted at the end, i.e. in the SDP relaxation, by ensuring full row rank and Slater's condition. We do this for: the quadratic assignment problem; graph partitioning; max-clique problem; and the stable set problem.

Quadratic Assignment Problem

Typical relaxations for QAP, see the definition in Section 13.1, try to exploit the trace formulation and use perturbations on A, B separately. Current approaches have two serious drawbacks. They completely discard the nonnegativity constraints and then they derive a bound from the sum of two bounds obtained by treating the quadratic and linear parts of the objective function separately, see e.g. [610]. However, the Lagrangian relaxations and homogenization for the special case $S = \Re^n$ shows that we should consider more general perturbations and, in particular, we should consider perturbations that arise from Lagrangian quadratic relaxations. This approach does not have the two drawbacks mentioned above.

We now use the fact that the set of permutation matrices is equal to the intersection of the orthogonal matrices with the 0,1 matrices. We get the following equivalent program to QAP.

$$
(QAP_E) \qquad \begin{array}{lll} \mu^* := & \max & q(X) = \operatorname{Trace}(AXB - 2C)X^T \\ & \text{subject to} & XX^T = I \\ & & X_{ij}^2 - X_{ij} = 0, \quad \forall i,j. \end{array} \tag{13.4.78}
$$

We could also consider the square of the norm of the residual of the (redundant) linear constraints

$$
Xe = e, \; X^T e = e.
$$

Other relaxations and bounds can be obtained by adding redundant constraints such as

$$
\operatorname{Trace} XX^T = n, \quad X^T X = I,
$$

or

$$
0 \leq X_{ij} \leq 1, \; \forall i,j.
$$

We now devote our attention to homogenization since that results in a min-max eigenvalue problem and an equivalent semidefinite programming problem. We have seen that we can homogenize by increasing the dimension of the problem by 1. We first add the 0,1 constraints to the objective function using Lagrange multipliers W_{ij}.

$$
\min_{W} \max_{XX^T=I} \operatorname{Trace}(AXB - 2C)X^T + \sum_{ij} W_{ij}(X_{ij}^2 - X_{ij}). \tag{13.4.79}
$$

We now homogenize the objective function by multiplying by a constrained scalar x.

$$
\min_{W} \max_{XX^T=I, x^2=1} \operatorname{Trace}\left[AXBX^T + W(X \circ X)^T - x(2C + W)X^T\right]. \tag{13.4.80}
$$

We can now use Lagrange multipliers to get a parametrized min-max eigenvalue problem in dimension $n^2 + 1$. We get the following bound. The parameters are: the symmetric $n \times n$ matrix $\Lambda = \Lambda^T$, the general $n \times n$ matrix W and the scalar α.

$$
\begin{array}{rcl} B_{QAP} & := & \min_{\Lambda, W, \alpha} \max_{X} \operatorname{Trace}\,[\\ & & AXBX^T + \Lambda XX^T + W^T(X \circ X) + \alpha x^2 \\ & & -x(2C + W)X^T\,] - \alpha - \operatorname{Trace}\Lambda. \end{array} \tag{13.4.81}
$$

We have grouped the quadratic, original linear, and constant terms together. The hidden semidefinite constraint now yields a semidefinite programming problem.

$$
\begin{array}{ll} \min & -\operatorname{Trace}\Lambda - \alpha \\ \text{subject to} & L_Q + \operatorname{Arrow}(\alpha, \operatorname{vec}(W)) + \mathrm{B}^0\operatorname{Diag}(\Lambda) \preceq 0, \end{array} \tag{13.4.82}
$$

where we define the matrix

$$L_Q := \begin{bmatrix} 0 & -\text{vec}\,(C)^T \\ -\text{vec}\,(C) & B \otimes A \end{bmatrix}, \tag{13.4.83}$$

and the linear operators

$$\text{Arrow}\,(\alpha, \text{vec}\,(W)) := \begin{bmatrix} \alpha & -\frac{1}{2}\text{vec}\,(W)^T \\ -\frac{1}{2}\text{vec}\,(W) & \text{Diag}\,(\text{vec}\,(W)) \end{bmatrix}, \tag{13.4.84}$$

$$\text{B}^0\text{Diag}\,(\Lambda) := \begin{bmatrix} 0 & 0 \\ 0 & I \otimes \Lambda \end{bmatrix}. \tag{13.4.85}$$

We can now introduce the $(n^2+1) \times (n^2+1)$ dual variable matrix $Y \succeq 0$ and derive the dual program to this min-max eigenvalue problem, i.e.

$$\max_{Y \succeq 0} \min_{\Lambda, W, \alpha} -\text{Trace}\,\Lambda - \alpha + \text{Trace}\,Y(L_Q + \text{Arrow}\,(\alpha, \text{vec}\,(W)) + \text{B}^0\text{Diag}\,(\Lambda)).$$

The inner minimization problem is unconstrained and linear in the variables. Therefore, after reorganizing the variables, we can differentiate to get the dual problem to this dual problem, or the semidefinite relaxation to the original QAP. (Recall that $Y_{i,j:k}$ refers to the i-th row and columns j to k of the matrix Y; and $\text{b}^0\text{diag}\,(Y)$ is the block diagonal sum of Y which ignores the first row.) The derivatives with respect to α and W yields the first constraint and the derivative with respect to Λ yields the second constraint in the following program. Equivalently, the constraints are the adjoints of the linear operators Arrow and B^0Diag.

$$\begin{array}{ll} \max & \text{Trace}\,L_Q Y \\ \text{subject to} & \text{diag}(Y) = (1, Y_{0,1:n^2})^T \\ & \text{b}^0\text{diag}\,(Y) = I \\ & Y \succeq 0. \end{array} \tag{13.4.86}$$

Another primal-dual pair can be obtained using a trust region subproblem as the inner maximization problem, rather than homogenizing to an eigenvalue problem. This is done by adding the redundant trust region constraint $\text{Trace}\,XX^T = n$. Also, as mentioned above, we can add the redundant constraint

$$||Xe - e||^2 + ||X^T e - e||^2 = 0.$$

This type of constraint is discussed below for the graph partitioning problem. A primal-dual interior point method based on the these types of dual pairs of programs, such as (13.4.86),(13.4.82), are being tested and studied in [870].

Graph Partitioning

Let $G = (V, E)$ be an undirected graph as in the description for (MC). The graph partitioning problem is the problem of partitioning the node set V into k disjoint subsets of specified sizes so as to minimize the total weight of the edges connecting nodes in distinct subsets of the partition. Let $A = (a_{ij})$ be the weighted adjacency matrix of G, i.e.

$$a_{ij} = \begin{cases} w_{ij} & ij \in E \\ 0 & \text{otherwise.} \end{cases}$$

The graph partitioning problem can be described by the following (0,1)-quadratic program see e.g. [660].

$$(\mathbf{GP}) \quad \begin{array}{rc} w(E_{uncut}) = \max & \frac{1}{2}\text{Trace}\, X^t AX \\ \text{subject to} & Xe_k = e_n \\ & X^T e_n = m \\ & X_{ij} \in \{0,1\},\ \forall ij, \end{array}$$

where e_k is the vector of ones of appropriate size and m is the vector of ordered set sizes

$$m_1 \geq \ldots \geq m_k \geq 1 \quad \text{and} \quad k < n.$$

The columns of the 0,1 $n \times k$ matrices X are the indicator vectors for the sets. We can replace the 0,1 constraints by quadratic and also change the linear constraints to quadratic by squaring. We get the following equivalent program.

$$\begin{array}{rc} w(E_{uncut}) = \max & \frac{1}{2}\text{Trace}\, X^t AX \\ \text{subject to} & ||Xe_k - e_n||^2 + ||X^T e_n - m||^2 = 0 \\ & X_{ij}^2 - X_{ij} = 0,\ \forall ij. \end{array}$$

The Lagrangian relaxation yields the following bound.

$$\begin{array}{rcl} B_{GP} & := & \min_{\alpha,W} \max_{X} \text{Trace}\, [\\ & & \frac{1}{2}X^T AX + \alpha(e_k e_k^T X^T X + X^T e_n e_n^T X) + W^T(X \circ X) \\ & & -2\alpha(e_k e_n^T X + m e_n^T X) - W^T X\,] \\ & & +\alpha(n + \sum_i m_i^2). \end{array} \tag{13.4.87}$$

We can now homogenize the problem by adding a variable x.

$$\begin{array}{rcl} B_{GP} & := & \min_{\alpha,W} \max_{\substack{X \\ x^2=1}} \text{Trace}\, [\\ & & \frac{1}{2}X^T AX + \alpha(e_k e_k^T X^T X + X^T e_n e_n^T X) + W^T(X \circ X) \\ & & +x(-2\alpha(e_k e_n^T X + m e_n^T X) - W^T X)\,] \\ & & +\alpha(n + \sum_i m_i^2). \end{array}$$

We now lift the variable x into the Lagrangian to get a min-max eigenvalue problem.

$$\begin{array}{rcl} B_{GP} & := & \min_{\alpha,W,\delta} \max_{X,x} \text{Trace}\, [\\ & & \frac{1}{2}X^T AX + \alpha(e_k e_k^T X^T X + X^T e_n e_n^T X) + W^T(X \circ X) + \delta x^2 \\ & & +x(-2\alpha(e_k e_n^T X + m e_n^T X) - W^T X)\,] \\ & & +\alpha(n + \sum_i m_i^2) - \delta. \end{array}$$

The above has a hidden semidefinite constraint.

$$\begin{array}{cc} \min & \alpha(n+\sum_i m_i^2)-\delta \\ \text{subject to} & L_A + \text{Arrow}\,(\delta, \text{vec}\,(W)) + \alpha L_\alpha \preceq 0, \end{array} \tag{13.4.88}$$

where we define the matrices

$$L_A := \begin{bmatrix} 0 & 0 \\ 0 & \frac{1}{2} I \otimes A \end{bmatrix}, \tag{13.4.89}$$

$$v = \text{vec}\, e_n m^T,$$

$$L_\alpha := \begin{bmatrix} 0 & -(e+v)^T \\ -(e+v) & (e_k e_k^T I \otimes I + I \otimes e_n e_n^T) \end{bmatrix}, \tag{13.4.90}$$

and the linear operator

$$\text{Arrow}\,(\delta, \text{vec}\,(W)) := \begin{bmatrix} \delta & -\frac{1}{2}(\text{vec}\,(W))^T \\ -\frac{1}{2}(\text{vec}\,(W)) & \text{Diag}\,(\text{vec}\,(W)) \end{bmatrix}. \tag{13.4.91}$$

The dual program yields the semidefinite relaxation of (GP).

$$\begin{array}{cc} \max & \text{Trace}\, L_A Y \\ \text{subject to} & \text{diag}(Y) = (1, Y_{0,1:n})^T \\ & \text{Trace}\, Y L_\alpha = 0 \\ & Y \succeq 0. \end{array} \tag{13.4.92}$$

Max-Clique and Stable Set

Consider again the undirected graph $G = (E, V)$ defined above. The *max-clique* problem consists in finding the largest connected subgraph. We let $\omega(G)$ denote the size of the largest clique in G. A *stable set* is a subset of vertices of V such that no two vertices are adjacent. We denote the size of the largest stable set in $\bar{G}$, the complement of G, by $\alpha(\bar{G})$. Clearly

$$\alpha(\bar{G}) = \omega(G).$$

Bounds for these problems and relationships to the theta function, or Lovász number of the graph, are described in the expository paper e.g. [425]; see also [701].

In this section we show that the Lovasz bound on $\omega(G)$ can be alternatively obtained from two distinct 01-programs (13.4.93) and (13.4.96) by Lagrangian relaxations. Let A be the incidence matrix of the graph, i.e. $A = (a_{ij})$ with $a_{ij} = 1$ if $ij \in E$ and 0 otherwise. If x is the indicator vector for the largest clique in G of size k, A then $x^T(I+A)x/x^Tx = k^2/k = k$. A quadratic formulation of the max-clique problem is the following (0,1)-quadratic program.

$$\begin{array}{ccc} \omega(G) = & \max & \frac{x^T(I+A)x}{x^Tx} \\ & \text{subject to} & x_i x_j = 0, \text{ if } ij \notin E,\ i \neq j \\ & & x_i \in \{0,1\},\ \forall i. \end{array} \tag{13.4.93}$$

Therefore, a quadratic relaxation of the max-clique problem is the following quadratic constrained program.

$$\omega(G) \leq \omega_1^* := \quad \begin{array}{ll} \max & x^T(I+A)x \\ \text{subject to} & x_i x_j = 0, \text{ if } ij \notin E, \ i \neq j \\ & x^T x = 1. \end{array} \tag{13.4.94}$$

The Lagrangian relaxation for this problem is the perturbed min-max eigenvalue problem and the equivalent semidefinite program

$$\begin{array}{rcll} \omega_1^* & \leq & \min\limits_{W_{ij}=0, \text{ if } ij \in E, \text{ or } i=j} & \lambda_{\max}(I+A+W) - \alpha x^T x + \alpha \\ & = & \min\limits_{w,\alpha} \max\limits_{x} x^T(I+A)x + \sum\limits_{ij \notin E, \ i \neq j} w_{ij} x_i x_j - \alpha x^T x + \alpha & \\ & = & \min\limits_{\substack{I+A+W \preceq \alpha I \\ W_{ij}=0, \text{ if } ij \in E, \text{ or } i=j}} & \alpha \end{array}$$

i.e. minimize the max eigenvalue over perturbations in the off-diagonal elements corresponding to disjoint nodes. This bound is equal to the Lovasz theta function on the complementary graph.

$$\vartheta(\bar{G}) = \min_{A \in \mathcal{A}} \lambda_{\max}(A), \tag{13.4.95}$$

where $\mathcal{A} = \{A : A$ symmetric $n \times n$ matrix with $A_{ij} = 1$, if $ij \in E$, or $i = j\}$.

By considering the (optimal) indicator vector for the largest clique, we see that a (0,1)-quadratic program that describes the max-clique problem exactly is the following one. Note that if node i is not in the largest clique, then necessarily, $x_i x_j = 0$ for some j with node j in the clique, i.e. necessarily $x_i = 0$ in the indicator vector.

$$\omega(G) = \quad \begin{array}{ll} \max & x^T x \\ \text{subject to} & x_i x_j = 0, \text{ if } ij \notin E, \ i \neq j \\ & x_i^2 - x_i = 0, \ \forall i. \end{array} \tag{13.4.96}$$

The Lagrangian relaxation yields the bound

$$B_{\text{clique}} := \min_{W,\lambda} \max_{x} x^T x + \sum_{ij \notin E, \ i \neq j} w_{ij} x_i x_j + \sum_i \lambda_i (x_i^2 - x_i).$$

We let W be an $n \times n$ matrix with zeros in positions where $ij \in E$. We can homogenize by adding the constraint $y^2 = 1$ and then lifting it into the Lagrangian.

$$\min_{\alpha,W,\lambda} \max_{x,y} x^T x + \sum_{ij \notin E} w_{ij} x_i x_j + \sum_i \lambda_i x_i^2 + \alpha y^2 - y \sum_i \lambda_i x_i - \alpha.$$

We now exploit the hidden semidefinite constraint to get the semidefinite program.

$$B_{\text{clique}} = \quad \begin{array}{ll} \min\limits_{W,\lambda,\alpha} & -\alpha \\ \text{subject to} & L_A + L_W(W) + \text{Arrow}(\alpha, \lambda) \preceq 0 \\ & W_{ij} = 0, \ \forall ij \in E, \text{ or } i = j, \end{array} \tag{13.4.97}$$

where the matrix

$$L_A := \begin{bmatrix} 0 & 0 \\ 0 & I \end{bmatrix}, \tag{13.4.98}$$

and the linear operators

$$L_W(W) := \begin{bmatrix} 0 & 0 \\ 0 & W \end{bmatrix}, \tag{13.4.99}$$

$$\text{Arrow}\,(\alpha, \lambda) := \begin{bmatrix} \alpha & -\frac{1}{2}\lambda^T \\ -\frac{1}{2}\lambda & \text{Diag}\,(\lambda) \end{bmatrix}. \tag{13.4.100}$$

The dual of the above min-max eigenvalue problem yields the semidefinite relaxation for the max-clique problem with $Y \in \mathcal{S}_{n+1}$.

$$\begin{array}{ll} \max & \text{Trace}\, L_A Y \\ \text{subject to} & \text{diag}(Y) = (1, Y_{0,1:n})^T \\ & Y_{ij} = 0, \ \forall ij \notin E \\ & Y \succeq 0. \end{array} \tag{13.4.101}$$

The equivalence of the bounds (13.4.95) and (13.4.101) was shown in lemma 2.17 of [497].

Consider the program (13.4.93) with an additional redundant constraint

$$x_i x_j \geq 0 \text{ for } ij \in E \tag{13.4.102}$$

That is

$$\begin{array}{lll} \omega(G) = & \max & \frac{x^T(I+A)x}{x^T x} \\ & \text{subject to} & x_i x_j = 0, \ \text{if } ij \notin E, \ i \neq j \\ & & x_i x_j \geq 0, \ \text{if } ij \in E, \\ & & x_i \in \{0, 1\}, \ \forall i. \end{array} \tag{13.4.103}$$

A quadratic relaxation of the max-clique problem is the following quadratic constrained program.

$$\begin{array}{lll} \omega(G) \leq \omega_1^* := & \max & x^T(I+A)x \\ & \text{subject to} & x_i x_j = 0, \ \text{if } ij \notin E, \ i \neq j \\ & & x_i x_j \geq 0, \ \text{if } ij \in E, \\ & & x^T x = 1. \end{array} \tag{13.4.104}$$

The Lagrangian relaxation for this problem is equal to the Schrijver's improvement [701] of the theta function on the complementary graph.

$$\vartheta'(\bar{G}) = \min_{A \in \mathcal{A}'} \lambda_{\max}(A),$$

where $\mathcal{A}' = \{A : A \text{ symmetric } n \times n \text{ matrix with } A_{ij} \geq 1, \text{ if } ij \in E, \text{ or } i = j\}$. Haemmers [321] constructed graphs where $\vartheta'(\bar{G})$ is strictly smaller than $\vartheta(\bar{G})$.

Analogously, it is possible to modify the program (13.4.96) by adding the constraint (13.4.102).

13.4.3 Strong Duality

In the case of strong duality (zero duality gap and dual attainment), our bounds are exact. As expected, this holds (generically) in the convex case. Surprisingly, there are several cases on nonconvex quadratic programs where this holds as well. In this Section 13.4.3 we amplify on our theme that illustrates the strength of the Lagrangian relaxation, i.e. that a tractable bound implies a Lagrangian relaxation is at work.

Recall the general quadratically constrained quadratic program (13.4.74). For simplicity we have replaced each equality constraint by two inequality constraints. We will use equality constraints when absolutely required. We let $\mathcal{F}$ denote the feasible set.

We define the Lagrangian

$$L(x,\lambda) := q_0(x) + \sum_{k=1}^{m} \lambda_k q_k(x),$$

and the dual functional

$$\phi(\lambda) := \min_x L(x,\lambda).$$

The Lagrangian is linear in λ and so the dual function is a minimum of linear functions, i.e. it is a concave function of λ. Thus the maximum of this concave function is a tractable problem if the dual functional can be evaluated efficiently. For each $\lambda \geq 0$, we have the lower bound

$$\begin{aligned} \mu^* &= \min_{x \in \mathcal{F}} q_0(x) \\ &\geq \min_{x \in \mathcal{F}} L(x,\lambda) \\ &\geq \min_x L(x,\lambda) \\ &\geq \nu^* := \max_{\lambda \geq 0} \phi(\lambda). \end{aligned}$$

Thus we have defined our dual problem

$$\mu^* \geq \nu^* = \max_{\lambda \geq 0} \phi(\lambda),$$

which provides a lower bound for our primal problem. If, in addition, we have found the feasible $\bar{x} \in \mathcal{F}$ with attainment in the Lagrangian $\bar{x} \in \operatorname{argmin}_x L(x,\bar{\lambda})$ and with complementary slackness $\sum_k \bar{\lambda}_k q_k(\bar{x}) = 0$, then

$$\begin{aligned} \mu^* &\geq \nu^* = L(\bar{x},\bar{\lambda}) \\ &= q_0(\bar{x}) \\ &\geq \mu^*, \end{aligned}$$

i.e. we have found an optimum $\bar{x}$ and have a zero duality gap when these sufficiency conditions (feasibility, attainment, complementary slackness) hold. Note that since we are dealing with an unconstrained minimum of a quadratic

Lagrangian, we obtain the interesting statement: *necessary conditions for the sufficiency conditions to hold*, i.e. we need stationarity of the Lagrangian and positive semidefiniteness of the Hessian of the Lagrangian. Thus, when these two conditions are incompatible we lose strong duality; we can even expect a duality gap.

We now present several Q^2P problems where the Lagrangian relaxation is important and well known. In all these cases, the Lagrangian dual provides an important theoretical tool for algorithmic development, even where the duality gap may be nonzero. We continue to emphasize our theme that illustrates that the Lagrangian relaxation is best.

Convex Quadratic Programs. We start with the easy case; consider the convex quadratic program

$$\text{CQP} \qquad \mu^* := \begin{array}{ll} \min & q_0(x) \\ \text{s.t.} & q_k(x) \leq 0, \quad k = 1, \dots m, \end{array}$$

where all $q_i(x)$ are convex quadratic functions. We now see that Lagrangian duality can always solve this problem.

The dual is

$$\text{DCQP} \qquad \nu^* := \max_{\lambda \geq 0} \min_x \; q_0(x) + \sum_{k=1}^m \lambda_k q_k(x).$$

If ν^* is attained at λ^*, x^*, then a *sufficient* condition for x^* to be optimal for CQP is primal feasibility and complementary slackness, i.e.

$$\sum_{k=1}^m \lambda_k^* q_k(x^*) = 0.$$

In addition, it is well known that the Karush-Kuhn-Tucker (KKT) conditions are sufficient for global optimality, and under an appropriate constraint qualification the KKT conditions are also necessary. Therefore strong duality holds if a constraint qualification is satisfied, i.e. in this case there is no duality gap and the dual is attained.

However, surprisingly, *if the primal value of CQP is bounded then it is attained and there is no duality gap*, see e.g. [776, 630, 631, 629]. (This can be considered to be an extension of the Frank-Wolfe Theorem, [510].) However, the dual may not be attained, e.g. consider the convex program

$$0 = \min\{x : x^2 \leq 0\}$$

and its (unattained) dual

$$0 = \max_{\lambda \geq 0} \min_x x + \lambda x^2 = \max_{\lambda > 0} \min_x x + \lambda x^2.$$

Algorithmic approaches based on Lagrangian duality appear in e.g. [363, 509, 583].

Nonconvex Quadratic Programs.

Rayleigh Quotient. Suppose that $A = A^T \in \mathcal{S}^n$. It is well known that the smallest eigenvalue λ_1 of A is obtained from the Rayleigh quotient, i.e.

$$\lambda_1 = \min\{x^T Ax : x^T x = 1\}. \tag{13.4.105}$$

Since A is not necessarily positive semidefinite, this is the minimization of a nonconvex function on a nonconvex set. However, the Rayleigh quotient forms the basis for many algorithms for finding the smallest eigenvalue, and these algorithms are very efficient. In fact, it is easy to see that there is no duality gap for this nonconvex problem, i.e.

$$\lambda_1 = \max_{\lambda} \min_{x} x^T Ax - \lambda(x^T x - 1) = \max_{A-\lambda I \succeq 0} \lambda. \tag{13.4.106}$$

To see this note that the inner minimization problem in (13.4.106) is unconstrained. This implies that the outer maximization problem has the hidden semidefinite constraint (an ongoing theme in the chapter)

$$A - \lambda I \succeq 0,$$

i.e. λ is at most the smallest eigenvalue of A. With λ set to the smallest eigenvalue, the inner minimization yields the eigenvector corresponding to λ_1. Thus, we have an example of a *nonconvex problem for which strong duality holds.* Note that the problem (13.4.105) has the special norm constraint, and a homogeneous quadratic objective.

Trust Region Subproblem. We will next see that strong duality holds for a larger class of seemingly nonconvex problems. The trust region subproblem, TRS, is the minimization of a quadratic function subject to a norm constraint. No convexity or homogeneity of the objective function is assumed. We allow for a further extension, i.e. we do not assume convexity of the constraint and allow indefinite quadratic functions for both objective and constraint. (See e.g. [155] for applications of indefinite quadratic forms.) This problem is important in nonlinear programming, e.g. [552, 551].

$$\begin{array}{lll} \text{TRS} & \mu^* := \min & q_0(x) = x^T Q_0 x - 2c_0^t x \\ & \text{s.t.} & x^T x - \delta^2 \leq 0 \text{ (or } = 0). \end{array}$$

or the generalized trust region subproblem [747, 549].

$$\begin{array}{lll} \text{GTRS} & \mu^* := \min & q_0(x) = x^T Q_0 x - 2c_0^t x \\ & \text{s.t.} & q_1(x) \leq 0 \text{ (or } = 0), \end{array}$$

where q_1 is another quadratic function. In addition, one can have two sided constraints $\alpha \leq q_1(x) \leq \beta$, which are used in trust region algorithms as well.

For TRS, assuming that the constraint is written "$\leq$," the Lagrangian dual is:

$$\text{DTRS} \qquad \nu^* := \max_{\lambda \geq 0} \min_x \; q_0(x) + \lambda(x^T x - \delta^2).$$

This is equivalent to (see [747]) the (concave) nonlinear semidefinite program

$$\begin{array}{llll} \text{DTRS} & \nu^* := \max & c_0^T (Q_0 + \lambda I)^\dagger c_0 - \lambda \delta^2 \\ & \text{s.t.} & Q_0 + \lambda I \succeq 0 \\ & & \lambda \geq 0. \end{array}$$

where $\cdot^\dagger$ denotes Moore-Penrose inverse. It is shown in [747] that strong duality holds for TRS, i.e. there is a zero duality gap $\mu^* = \nu^*$, and the dual is attained. (The primal is also attained.) Thus, as in the eigenvalue case, we see that this is an example of a nonconvex program where strong duality holds. In addition, this implies that this problem can be solved efficiently; polynomial time results are presented in [854].

Proof.

We include a short proof of strong duality, for the inequality constrained case, based on the outline in [478], i.e. we fall back on the convex case after a perturbation. Note that the key to the proof is being able to pass between the inequality and equality constraints.

Without loss of generality, we can assume that TRS is nonconvex. (Otherwise, we apply the convex results discussed above.) Therefore μ^* is attained on the boundary of the feasible set and the smallest eigenvalue of Q_0, denoted γ, is negative. Then TRS is equivalent to

$$\begin{array}{rll} \mu^* & = & \min\limits_{x^T x \leq \delta^2} \; x^T(Q_0 - \gamma I)x - 2c_0^t x + \gamma x^T x \\ & = & \min\limits_{x^T x = \delta^2} \; x^T(Q_0 - \gamma I)x - 2c_0^t x + \gamma x^t x, \quad (Q_0 \text{ is indefinite}) \\ & = & \min\limits_{x^T x = \delta^2} \; x^T(Q_0 - \gamma I)x - 2c_0^t x + \gamma \delta^2 \\ & = & \min\limits_{x^T x \leq \delta^2} \; x^T(Q_0 - \gamma I)x - 2c_0^t x + \gamma \delta^2, \quad (Q_0 - \gamma I \text{ is singular}) \\ & = & \max\limits_{\lambda \geq 0} \min\limits_x \; x^T(Q_0 - \gamma I)x - 2c_0^t x + \lambda(x^T x - \delta^2) + \gamma \delta^2 \text{ (convex case)} \\ & = & \max\limits_{\lambda \geq 0} \min\limits_x \; x^T Q_0 x - 2c_0^t x + (\lambda - \gamma)(x^T x - \delta^2) \\ & \leq & \max\limits_{\lambda \geq \gamma} \min\limits_x \; x^T Q_0 x - 2c_0^t x + (\lambda - \gamma)(x^T x - \delta^2) \quad (\gamma < 0) \\ & = & \nu^* \leq \mu^*. \end{array} \tag{13.4.107}$$

■

As mentioned above, extensions of this result to a two-sided general, possibly nonconvex, constraint are discussed in [747, 549]. An algorithm based on Lagrangian duality appears in [661] and (implicitly) in [551, 691]. These algorithms are extremely efficient for the TRS problem, i.e. they solve this problem almost as quickly as an eigenvalue problem.

The fact that we can solve the TRS efficiently even though the objective and constraint may be nonconvex is surprising. In fact, in [524] Martinez shows that the TRS can have at most one local and nonglobal optimum, and the Lagrangian at this point has one negative eigenvalue. Therefore, it is even more surprising that the Lagrangian dual (relaxation) allows one to find the global minimum without ever getting trapped near the local minimum.

In fact, for GTRS we still have a 0 duality gap, though strong duality may fail, e.g. consider the simple program $\min x$ s.t. $x^2 \leq 0$. The results in [747] provide strong duality for GTRS with a two sided constraint using the constraint qualification that $\alpha < \beta$. In [549], necessary and sufficient optimality conditions are presented for GTRS using the constraint qualification that $\min q_0(x) < \max q_0(x)$. Using these results in combination with the extension of the Frank-Wolfe result (e.g. [510]) gives us the following.

Theorem 13.4.2 *Consider GTRS: a zero duality gap always holds and, moreover, if the optimal value is finite, then it is attained.* ■

Two Trust Region Subproblem. The two trust region subproblem, TTRS, consists in minimizing a (possibly nonconvex) quadratic function subject to a norm and a least squares constraint, i.e. two convex quadratic constraints. This problem arises in solving general nonlinear programs using a sequential quadratic programming approach, and is often called the CDT problem, see [154].

In contrast to the above single TRS, the TTRS can have a nonzero duality gap, see e.g. [626, 862, 863, 864]. This is closely related to quadratic theorems of the alternative, e.g. [177]. In addition, if the constraints are not convex, then the primal may not be attained, see e.g. [510].

As mentioned above, Martinez [524] shows that the TRS can have at most one local and nonglobal optimum, and the Lagrangian at this point has one negative eigenvalue. Therefore, if we have such a case and add another ball constraint that contains the local, nonglobal, optimum in its interior and also makes this point the global optimum, we obtain a TTRS where we cannot have a zero duality gap due to the negative eigenvalue. It is uncertain what constraints could be added to close this duality gap. In fact, it is still an open problem whether TTRS is an NP-hard or a polynomial time problem.

General Q^2P. The general, possibly nonconvex, Q^2P has many applications in modeling and approximation theory, see e.g. the applications to SQP methods in [451]. Examples of approximations to Q^2P also appear in [258].

The Lagrangian relaxation of a Q^2P is equivalent to the SDP relaxation, and is sometimes referred to as the Shor relaxation, see [733]. The Lagrangian relaxation can be written as an SDP if one takes into the account the hidden semidefinite constraint, i.e. a quadratic function is bounded below only if the Hessian is positive semidefinite. The SDP relaxation is then the Lagrangian dual of this semidefinite program. It can also be obtained directly by *lifting* the problem into matrix space using the fact that $x^TQx = \operatorname{Trace} x^TQx = \operatorname{Trace} Qxx^T$, and relaxing xx^T to a semidefinite matrix X.

One can relate the geometry of the original feasible set of Q^2P with the feasible set of the SDP relaxation. The connection is through *valid quadratic inequalities*, i.e. nonnegative (convex) combinations of the quadratic functions; see [260, 442] and our Section 13.4.2.

Orthogonally Constrained Programs with Zero Duality Gaps. We now follow the approach in [41, 37, 36] and consider the *orthonormal type constraints*

$$X^TX = I, \qquad X \in \mathcal{M}_{m,n}$$

(sometimes known as the Stiefel manifold, e.g. [203]) and the trust region type constraint

$$X^TX \preceq I, \qquad X \in \mathcal{M}_{m,n}.$$

Applications and algorithms for optimization on orthonormal sets of matrices are discussed in [203].) In this section we will show that for $m = n$, strong duality holds for a certain (nonconvex) quadratic program defined over orthonormal matrices. Because of the similarity of the orthonormality constraint to the norm constraint $x^Tx = 1$, the results of this section can be viewed as a matrix generalization of the strong duality result for the Rayleigh Quotient problem (13.4.105).

Let A and B be $n \times n$ symmetric matrices, and consider the orthonormally constrained homogeneous Q^2P

$$\text{QQP}_\text{O} \qquad \mu^O := \begin{array}{ll} \min & \operatorname{Trace} AXBX^T \\ \text{s.t.} & XX^T = I. \end{array} \tag{13.4.108}$$

This problem can be solved exactly using Lagrange multipliers, see e.g. [318], or using the classical Hoffman-Wielandt inequality, e.g. [112].

Proposition 13.4.1 *Suppose that the orthogonal diagonalizations of A, B are $A = V\Sigma V^T$ and $B = U\Lambda U^T$, respectively, where the eigenvalues in Σ are ordered nonincreasing, and the eigenvalues in Λ are ordered nondecreasing. Then the optimal value of* QQP_O *is $\mu^O = \operatorname{Trace} \Sigma\Lambda$, and the optimal solution is obtained using the orthogonal matrices that yield the diagonalizations, i.e. $X^* = VU^T$.* ■

The Lagrangian dual of QQP_O is

$$\max_{S=S^T} \min_X \operatorname{Trace} AXBX^T - \operatorname{Trace} S(XX^T - I). \tag{13.4.109}$$

However, there can be a nonzero duality gap for the Lagrangian dual, see [870] for an example. The inner minimization in the dual problem (13.4.109) is an unconstrained quadratic minimization in the variables $\mathrm{vec}\,(X)$, with hidden constraint on the Hessian

$$B \otimes A - I \otimes S \succeq 0.$$

The first order stationarity conditions are equivalent to $AXB = SX$ or $AXBX^T = S$. Once can easily construct examples where the semidefinite condition and the stationarity are in conflict and result in a duality gap. In order to close the duality gap, we need a larger class of quadratic functions.

Note that in $\mathrm{QQP_O}$ the constraints $XX^T = I$ and $X^TX = I$ are equivalent. Adding the redundant constraints $X^TX = I$, we arrive at

$$\begin{array}{llll} \mathrm{QQP_{OO}} & \mu^O := & \min & \mathrm{Trace}\, AXBX^T \\ & & \text{s.t.} & XX^T = I,\ X^TX = I. \end{array}$$

Using symmetric matrices S and T to relax the constraints $XX^T = I$ and $X^TX = I$, respectively, we obtain a dual problem

$$\begin{array}{llll} \mathrm{DQQP_{OO}} & \mu^O \geq \mu^D := & \max & \mathrm{Trace}\, S + \mathrm{Trace}\, T \\ & & \text{s.t.} & (I \otimes S) + (T \otimes I) \preceq (B \otimes A) \\ & & & S = S^T,\ T = T^T. \end{array}$$

Theorem 13.4.3 *Strong duality holds for* $\mathrm{QQP_{OO}}$ *and* $\mathrm{DQQP_{OO}}$, *i.e.,* $\mu^D = \mu^O$ *and both primal and dual are attained.* ■

A further relaxation of the above orthogonal relaxation is the trust region relaxation studied in [398]

$$\begin{array}{lll} \mu^*_{QAPT} := & \min & \mathrm{Trace}\, AXBX^T \\ & \text{s.t.} & XX^T \preceq I. \end{array}$$

The constraints are convex with respect to the Löwner partial order and so it is hoped that solving this problem would be useful. Also, this problem is visually similar to the TRS discussed above. And so we would like to find a characterization of optimality.

The set

$$\{X : W = XX^T \preceq I\}$$

is studied separately in [604, 233] and is useful in eigenvalue variational principles.

We now study the matrix trust-region relaxation of QAP:

$$\begin{array}{lll} \mu^*_{SDPT} = & \min & \mathrm{Trace}\, AXBX^T \\ & \text{s.t.} & XX^T \preceq I. \end{array}$$

The following generalization of the Hoffman-Wielandt inequality holds.

Theorem 13.4.4 *For any $XX^T \leq I$, we have*

$$\sum_{i=1}^n \min\{\lambda_i\mu_{n-i+1}, 0\} \leq trAXBX^T \leq \sum_{i=1}^n \max\{\lambda_i\mu_i, 0\}$$

And, the upper bound is attained if

$$X = PDiag(\epsilon_1, \epsilon_2, \cdots, \epsilon_n)Q^T, \qquad (13.4.110)$$

where

$$\varepsilon_i = \begin{cases} 1, & \lambda_i\mu_i > 0, \\ \alpha \in [0,1], & \lambda_i\mu_i = 0, \\ 0, & \lambda_i\mu_i < 0; \end{cases} \qquad (13.4.111)$$

The lower bound is attained if

$$X = PDiag(\epsilon_1, \epsilon_2, \cdots, \epsilon_n)JQ^T, \qquad (13.4.112)$$

where

$$\varepsilon_i = \begin{cases} 1, & \lambda_i\mu_{n-i+1} < 0, \\ \alpha \in [0,1], & \lambda_i\mu_{n-i+1} = 0, \\ 0, & \lambda_i\mu_{n-i+1} > 0. \end{cases} \qquad (13.4.113)$$

■

For a scalar ξ, let $\xi^- := \min\{0, \xi\}$. The lower bound in the above theorem states that $\mu^*_{SDPT} = \sum_{i=1}^n [\lambda_i\mu_i]^-$. Since the Theorem provides the feasible point of attainment, i.e. an upper bound for the relaxation problem, we will prove the theorem by proving another theorem that shows that the value μ^*_{SDPT} is also attained by a Lagrangian dual program. Note that since XX^T and X^TX have the same eigenvalues, $XX^T \preceq I$ if and only if $X^TX \preceq I$. Explicitly using both sets of constraints, as in [41], we obtain

$$\text{QAPTR} \qquad \mu^*_{QAPT} := \begin{array}{ll} \min & \text{Trace}\, AXBX^T \\ \text{s.t.} & XX^T \preceq I, \quad X^TX \preceq I. \end{array}$$

Next we apply Lagrangian relaxation to QAPTR, using matrices $S \succeq 0$ and $T \succeq$ to relax the constraints $XX^T \preceq I$ and $X^TX \preceq I$, respectively. This results in the dual problem

$$\text{DQAPTR} \qquad \mu^*_{QAPT} \geq \mu^D_{QAPT} := \begin{array}{ll} \max & -\text{Trace}\, S - \text{Trace}\, T \\ \text{s.t.} & (B \otimes A) + (I \otimes S) + (T \otimes I) \succeq 0 \\ & S \succeq 0, \ T \succeq 0. \end{array}$$

To prove that $\mu^*_{QAPT} = \mu^D_{QAPT}$ we will use the following simple result.

Lemma 13.4.1 *Let $\lambda \in \Re^n$, $\lambda_1 \leq \lambda_2 \leq \ldots \leq \lambda_n$. For $\gamma \in \Re^n$ consider the problem*

$$\min \quad z_\pi := \sum_{i=1}^n [\lambda_i \gamma_{\pi(i)}]^-,$$

where $\pi(\cdot)$ is a permutation of $\{1, \ldots, n\}$, Then the permutation that minimizes z_π satisfies $\gamma_{\pi(1)} \geq \gamma_{\pi(2)} \geq \cdots \gamma_{\pi(n)}$. ■

Theorem 13.4.5 *Strong duality holds for QAPTR and DQAPTR, i.e., $\mu^D_{QAPT} = \mu^*_{QAPT}$ and both primal and dual are attained.* ■

The above results illustrate the theme about the strength of the Lagrangian relaxation, i.e. that tractable problems can be solved using Lagrangian duality in some form.

14 SEMIDEFINITE PROGRAMMING IN SYSTEMS AND CONTROL THEORY

Venkataramanan Balakrishnan, Fan Wang

14.1 INTRODUCTION

It has been long recognized that SDP constraints, i.e., Linear Matrix Inequalities (LMIs), arise naturally and frequently in the analysis of the solution of finite-dimensional differential equations that model control systems. The earliest LMI for systems and control is the "Lyapunov" LMI [512, p277]

$$P > 0, \quad A^T P + PA < 0. \tag{14.1.1}$$

This LMI is feasible if and only if every solution of

$$\frac{d}{dt}x(t) = Ax(t) \tag{14.1.2}$$

satisfies $\lim_{t\to\infty} x(t) = 0$. It turns out that a suitable P satisfying LMI (14.1.1) can be found simply by solving a Lyapunov equation, say $A^T P_0 + P_0 A + I = 0$. Then, (14.1.1) is feasible if and only if $P_0 > 0$.

Another important instance where LMIs arise in control theory is in absolute stability theory. In the 1940s, Lur'e, Postnikov, and others in the Soviet Union applied Lyapunov's methods to some specific practical problems in control engineering, especially, the problem of stability of a control system with a nonlinearity in the actuator [511]. Their stability criteria were expressed as LMIs. However, as numerical algorithms for checking the feasibility of these LMIs were unavailable then, the LMIs were reduced to polynomial inequalities which were then checked "by hand". Connections between the LMIs that arise in absolute stability theory and certain frequency-domain inequalities were derived in the 1960s by Yakubovich, Popov, Kalman; these are known by various names as the Kalman-Yakubovich-Popov (KYP) lemmas, or the positive- and bounded-real lemmas [30]. These connections enabled the graphical verification of the LMI conditions from absolute stability theory, and resulted in the celebrated Popov criterion, Circle criterion, Tsypkin criterion, and many variations.

In the 1960s and 1970s, the important role of LMIs in control theory was already recognized, especially in [847] and in [830]. Similar observations were explicitly made by several researchers, in [648] and [351], to name just a few. Thus, it can be said that by the mid-eighties, system and control theory were ripe for the application of SDP. Thus, with Nesterov and Nemirovskii's seminal work on interior point methods that apply directly to convex problems involving matrix inequalities [576, 583], there was a spurt in research efforts directed towards the numerical solution of systems and control problems using SDP; this has continued into this decade as well.

A number of publications can be found in the control literature that survey applications of SDP to the solution of system and control problems. Perhaps the most comprehensive list can be found in the book [137]. Since its publication, a number of papers have appeared chronicling further applications of SDP in control; we cite for instance the survey article [804], and the special issue of the International Journal of Robust and Nonlinear Control on *Linear Matrix Inequalities in Control Theory and Applications*, published in November-December, 1996. The growing popularity of LMI methods for control is also evidenced by the large number of publications at recent control conferences.

Our objective, in this chapter, is to describe the application of SDP towards the solution of problems from systems and control. All these applications fall under the topic of Robust Control, that of analysis of and design for control systems for which only inexact models are available. The list of references that we cite is by no means complete. In most cases, we have attempted to refer to the most relevant or up-to-date citation, which should serve as a starting point for a more careful literature search for interested readers.

14.2 CONTROL SYSTEM ANALYSIS AND DESIGN: AN INTRODUCTION

A number of control systems are well-modeled by finite-dimensional differential and/or difference equations. Systems modeled by differential equations are usually referred to as "continuous-time" systems, those modeled by difference equations are referred to as "discrete-time" systems, and those modeled by a mixture of differential and difference equations are called "hybrid systems". For simplicity, we will henceforth focus on continuous-time systems, noting that all the problems we discuss in this chapter have a counterpart in discrete-time systems.

It is customary to represent the differential equations modeling continuous-time systems as a single first-order vector differential equation:

$$\frac{d}{dt}x(t) = f(x, w, u, t), \quad z(t) = g(x, w, u, t), \quad y(t) = h(x, w, u, t), \tag{14.2.3}$$

where $x(t) \in \mathbf{R}^n$, $w(t) \in \mathbf{R}^{n_w}$, $u(t) \in \mathbf{R}^{n_u}$, $y(t) \in \mathbf{R}^{n_y}$ and $z(t) \in \mathbf{R}^{n_z}$. The function x is called the "state" of the system, while w and u are "inputs", and z and y are "outputs". w consists of exogenous inputs, i.e., inputs that we have no control over, such as noises, reference inputs etc. u consists of control inputs; we may set $u(t)$ to any value we wish, for every t. The outputs z are those of interest; these may consist, for instance, of components of x or even those of u. y consists of outputs that can be measured. f, g and h are either fixed functions, or are known only to satisfy some properties. The latter situation arises when the model only approximates the system; in this case, equations (14.2.3) are said to describe an "uncertain" system[1]. (We will give specific examples shortly.)

Control system analysis problems consist of the study of the solutions of equations (14.2.3). Typical questions that arise in this context are "Are the solutions x of equations (14.2.3) bounded?" or "With $x(0) = 0$, how large can $\int_0^\infty z(t)^T z(t)\, dt$ be, over all w with $\int_0^\infty w(t)^T w(t)\, dt \leq 1$?" Control system design problems consist of designing control laws $u(t) = \mathcal{K}(y, t)$, so that with the control law in place, desired answers are obtained for the analysis questions. Figure 14.1 shows a block diagram of the control system model (14.2.3) with the controller, i.e., the control law, in place. In this chapter, we present some examples of the application of SDP towards solving analysis and design problems in uncertain control systems.

14.2.1 Linear fractional representation of uncertain systems

We now focus on a special instance of system (14.2.3), consisting of an interconnection of a linear time-invariant system and an "uncertainty" or "perturbation" in the feedback loop. This model has found wide applicability in the analysis and design of control systems for which only imperfect models are

[1] Control system models must often explicitly incorporate in them "uncertainties", which model a number of factors, including: dynamics that are neglected to make the model tractable, as with large scale structures; nonlinearities that are either hard to model or too complicated; and parameters that are not known exactly, either because they are hard to measure or because of varying manufacturing conditions. Robust control deals with the analysis of and design for such control system models.

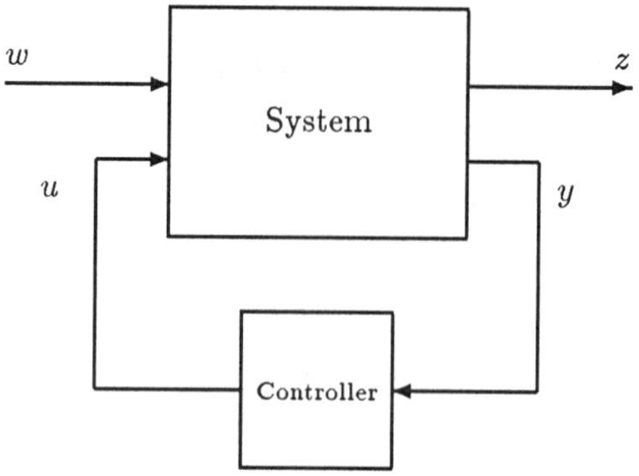

Figure 14.1 A standard controller design framework.

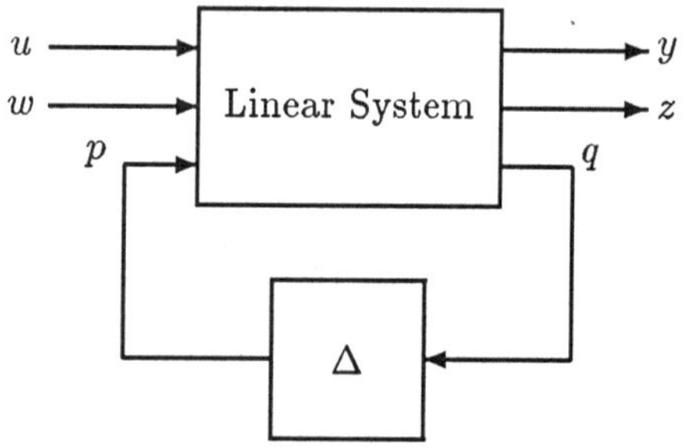

Figure 14.2 A common framework for robustness analysis and robust synthesis.

available; see for example, [296]. The model is described by

$$
\begin{aligned}
\frac{d}{dt}x(t) &= Ax(t) + B_p p(t) + B_u u(t) + B_w w(t), \\
q(t) &= C_q x(t) + D_{qp} p(t) + D_{qu} u(t) + D_{qw} w(t), \\
y(t) &= C_y x(t) + D_{yp} p(t) + D_{yu} u(t) + D_{yw} w(t), \\
z(t) &= C_z x(t) + D_{zp} p(t) + D_{zu} u(t) + D_{zw} w(t), \\
p(t) &= \Delta(q,t),
\end{aligned}
\tag{14.2.4}
$$

where $p \in \mathbf{R}^m$, $q \in \mathbf{R}^m$, A, B_p, B_u, B_w, C_q, C_y, C_z, D_{yp}, D_{yu}, D_{yw}, D_{qp}, D_{qu}, D_{qw}, D_{zp}, D_{zu} and D_{zw} are real matrices of appropriate sizes. $\Delta : \mathbf{L}_2^m[0,\infty) \to \mathbf{L}_2^m[0,\infty)$ is in general a nonlinear operator representing the "uncertainty" in modeling, and is known or assumed to lie in some set $\boldsymbol{\Delta}$. Often $\boldsymbol{\Delta}$ contains the origin, i.e., $\Delta = 0$; the linear time-invariant system that results is called the "nominal model". A block diagram of this system model is shown in Figure 14.2.

Many commonly encountered systems with structured and/or parametric uncertainties can be represented by the system model (14.2.4) [198, 385]. In the control literature, model (14.2.4) is also known as the "Linear Fractional Representation" of the uncertain system, or simply an LFR system. Usually, additional information about the size of the uncertainty (typically some bound on the norm of $\Delta \in \mathbf{\Delta}$), its structure (i.e., diagonal or block-diagonal), and nature (for instance, sector-bounded memoryless, linear time-invariant (LTI) or parametric, etc) is available.

14.2.2 Polytopic systems

Polytopic systems form a special class of LFR systems. For these systems, there exists an extensive body of work on analysis and synthesis using quadratic Lyapunov functions and SDP [137]. These systems are described by

$$\begin{aligned} \frac{d}{dt}x(t) &= A(t)x(t) + B_u(t)u(t) + B_w(t)w(t), \\ y(t) &= C_y(t)x(t) + D_{yu}(t)u(t) + D_{yw}(t)w(t), \\ z(t) &= C_z(t)x(t) + D_{zu}(t)u(t) + D_{zw}(t)w(t), \end{aligned} \tag{14.2.5}$$

$$\Sigma(t) = \begin{bmatrix} A(t) & B_u(t) & B_w(t) \\ C_y(t) & D_{yu}(t) & D_{yw}(t) \\ C_z(t) & D_{zu}(t) & D_{zw}(t) \end{bmatrix} \in \Xi$$

where

$$\Xi = \mathbf{Co}\left\{ \begin{bmatrix} A_1 & B_{u,1} & B_{w,1} \\ C_{y,1} & D_{yu,1} & D_{yw,1} \\ C_{z,1} & D_{zu,1} & D_{zw,1} \end{bmatrix}, \ldots, \begin{bmatrix} A_L & B_{u,L} & B_{w,L} \\ C_{y,L} & D_{yu,L} & D_{yw,L} \\ C_{z,L} & D_{zu,L} & D_{zw,L} \end{bmatrix} \right\}, \tag{14.2.6}$$

where $\mathbf{Co}$ denotes the convex hull. (The matrices $\begin{bmatrix} A_i & B_{u,i} & B_{w,i} \\ C_{y,i} & D_{yu,i} & D_{yw,i} \\ C_{z,i} & D_{zu,i} & D_{zw,i} \end{bmatrix}$, $i = 1, \ldots, L$ are given.)

14.2.3 Robust stability analysis and design problems

We consider questions of stability analysis and stabilizing controller synthesis for both polytopic systems (14.2.5) and the more general LFR systems (14.2.4):

(P1) With w and u identically zero, does the state x of system (14.2.5) (respectively system (14.2.4)) satisfy $\lim_{t\to\infty} x(t) = 0$ for every initial condition $x(0)$? If so, we say that the system (14.2.5) (respectively system (14.2.4)) is "robustly stable over Ξ (respectively $\mathbf{\Delta}$)".

(P2) With w identically zero, does there exist a control law u such that the state x of system (14.2.5) (respectively system (14.2.4)) satisfies $\lim_{t\to\infty} x(t) = 0$ for every initial condition $x(0)$? If so, we say that the system (14.2.5) (respectively system (14.2.4)) is "robustly stabilizable over Ξ (respectively $\mathbf{\Delta}$)".

Each of these "robust stability" questions has a "robust performance" counterpart: For a robustly stable system, measures of performance—usually with smaller values being better—can be defined that quantify how good the map from the exogenous inputs w to the outputs of interest z is. Robust performance analysis questions then ask how large these performance measures can be over Ξ or $\mathbf{\Delta}$. Robust performance design questions concern the design of control laws that minimize the largest values of the performance measures over Ξ or $\mathbf{\Delta}$. However, for simplicity, we will not consider robust performance problems further in the sequel.

One approach towards answering question (P1) uses the notion of *quadratic stability*. A system is said to be quadratically stable if there exists a single positive-definite quadratic Lyapunov function $V(\zeta) = \zeta^T P \zeta$ that decreases along every trajectory of the system. For system (14.2.5), a sufficient and necessary condition for quadratic stability can be directly formulated in terms of a finite number of LMIs [137]. For system (14.2.4), in general, only sufficient conditions for quadratic stability are known; these are stated in terms of a finite number of LMIs.

A system can be robustly stable without being quadratically stable, and more general Lyapunov functions can be employed to derive weaker sufficient conditions for robust stability. For instance, when the state-space matrices of the polytopic system (14.2.5) vary slowly with time, stability analysis using parameter-dependent Lyapunov functions usually leads to less conservative robust stability conditions than the analysis based on quadratic Lyapunov functions [266]. For the LFR system (14.2.4), the framework of integral quadratic constraints (IQCs) [529] provides a systematic method for deriving sufficient conditions for robust stability that are weaker than quadratic stability. In many cases, this framework can be interpreted as searching for more general Lyapunov functions.

In addressing the problem of controller synthesis (P2), there are several possibilities for generating the control input $u(t)$. Perhaps the simplest control law is that of constant state-feedback, $u(t) = Kx(t)$, where K is a real matrix. Of course, in order to implement a state-feedback scheme, the state $x(t)$ has to be measurable at every time t. If only the measured output y is available for generating u, output feedback control laws of the form $u = \mathcal{K}(y, t)$ can be envisioned; a simple example of such a control law is constant output feedback $u = Ky(t)$. If in addition to the measured output, the uncertainty Σ in a polytopic system (or Δ in an LFR system) is measurable in real time [606, 44], a control law $u = \mathcal{K}(y, \Sigma, t)$ (or $u = \mathcal{K}(y, \Delta, t)$) that explicitly depends on the uncertainty can be implemented. This is the so-called *gain-scheduled controller*[2].

[2] The use of the term "gain-scheduled" in the context of this chapter refers to the framework of LMI-based gain-scheduling techniques [78, 606, 265, 842, 44, 158, 272, 695, 42]. This is to be contrasted from the classical gain-scheduling controller synthesis where several controllers are designed for the system under different operating conditions, with the actual control law

The problem of synthesizing robustly stabilizing constant state-feedback for both polytopic and LFR systems can be formulated as LMI feasibility problems [137]. However, no convex reformulation is known for the problem of even constant output feedback synthesis for even polytopic systems. However, gain-scheduled controllers appear to hold promise: Designing gain-scheduled output feedback controllers for polytopic systems using quadratic Lyapunov functions can be reduced to the solution of an optimization problem with a finite number of LMIs [78, 44, 42]. For LFR systems, conditions for the existence of robustly stabilizing gain-scheduled output feedback controllers, derived using quadratic Lyapunov functions, result in a finite number of LMIs [272, 78, 43, 696]. As for the stability criteria derived using parameter-dependent Lyapunov functions or in the IQC framework, although they may yield less conservative conditions for robust stability, the corresponding conditions for the existence of robustly stabilizing controllers (gain-scheduled or otherwise) turn out to be nonconvex.

We next present, in detail, some of the SDP-based robust stability analysis and controller synthesis techniques that we have summarized so far. The specific problems that we consider are:

- For polytopic systems (14.2.5), robust stability analysis, state-feedback synthesis, and gain-scheduled controller synthesis, using quadratic Lyapunov functions.
- For LFR systems (14.2.4), robust stability analysis in the IQC framework.
- For a class of LFR systems (14.2.4), state-feedback synthesis and gain-scheduled controller synthesis, using quadratic Lyapunov functions.

14.3 ROBUSTNESS ANALYSIS AND DESIGN FOR LINEAR POLYTOPIC SYSTEMS USING QUADRATIC LYAPUNOV FUNCTIONS

For the polytopic system (14.2.5), we derive a necessary and sufficient condition for quadratic stability. This stability condition in turn is used to design state feedback and gain-scheduled output feedback controllers that stabilize system (14.2.5). Conditions for the existence of these controllers are formulated in terms of a finite number of LMIs.

14.3.1 Robust stability analysis

Setting u and w to be identically zero in equations (14.2.5) yields the following state equation

$$\frac{dx}{dt} = A(t)x(t), \quad A(t) \in \mathbf{Co}\{A_1, \ldots, A_L\}. \tag{14.3.7}$$

"switching" between the locally designed controllers using some "scheduling" scheme [704, 705].

This autonomous system is quadratically stable if there exists a quadratic Lyapunov function $V(\psi) = \psi^T P \psi$, with $P > 0$ such that $dV(x(t))/dt < 0$ for every nonzero solution x. Since

$$\frac{d}{dt}V(x(t)) = x(t)^T \left(A(t)^T P + PA(t)\right) x(t), \tag{14.3.8}$$

system (14.3.7) is quadratically stable if and only if there exists P such that

$$P = P^T > 0, \qquad A(t)^T P + PA(t) < 0, \; A(t) \in \mathbf{Co}\{A_1, \ldots, A_L\}, \tag{14.3.9}$$

or equivalently

$$P = P^T > 0, \quad A_i^T P + PA_i < 0, \; i = 1, \ldots, L. \tag{14.3.10}$$

V is sometimes called a "simultaneous quadratic Lyapunov function" since it proves the stability of every element of $\mathbf{Co}\{A_1, \ldots, A_L\}$. Thus, determining quadratic stability is an SDP feasibility problem.

14.3.2 Stabilizing state-feedback controller synthesis

Consider the system (14.3.7) with a control input u that is generated via state-feedback:

$$\frac{d}{dt}x(t) = A(t)x(t) + B_u(t)u(t), \;\; u(t) = Kx(t), \tag{14.3.11}$$

where

$$[A(t) \;\; B_u(t)] \in \mathbf{Co}\{[A_1 \;\; B_{u,1}], \ldots, [A_L \;\; B_{u,L}]\}. \tag{14.3.12}$$

Our objective is to design the matrix K such that (14.3.11) is quadratically stable. This is a "quadratic stabilizability" problem.

System (14.3.11) is quadratically stable with some state-feedback gain K if there exist P and K such that

$$P = P^T > 0, \quad (A_i + B_{u,i}K)^T P + P(A_i + B_{u,i}K) < 0, \;\; i = 1, \ldots, L. \tag{14.3.13}$$

Note that the matrix inequality (14.3.13) is *not* jointly convex in P and K. However, with the bijective transformation $Q \triangleq P^{-1}$, $Y \triangleq KP^{-1}$, we may rewrite (14.3.13) as

$$Q = Q^T > 0, \quad (A_i + B_{u,i}YQ^{-1})^T Q^{-1} + Q^{-1}(A_i + B_{u,i}YQ^{-1}) < 0, \; i = 1, \ldots, L. \tag{14.3.14}$$

Multiplying the second inequality on the left and right by Q (such a congruence preserves the inequality) we get an LMI in Q and Y:

$$Q = Q^T > 0, \quad QA_i^T + Y^T B_{u,i}^T + A_i Q + B_{u,i}Y < 0, \; i = 1, \ldots, L. \tag{14.3.15}$$

If this LMI problem is feasible and Q and Y are feasible solutions, then the Lyapunov function $V(\psi) = \psi^T Q^{-1} \psi$ proves quadratic stability of the closed-loop system, with state-feedback $u(t) = YQ^{-1}x(t)$. In other words, we can synthesize a constant state-feedback controller that (quadratically) stabilizes the polytopic system (14.3.7) by solving an LMI feasibility problem.

14.3.3 *Gain-scheduled output feedback controller synthesis*

Consider the system (14.3.7) with a control input u and a measured output y:

$$\frac{d}{dt}x(t) = A(t)x(t) + B_u u(t), \quad y(t) = C_y x(t), \tag{14.3.16}$$

where $A(t) \in \mathbf{Co}\{A_1, \ldots, A_L\}$. (For simplicity, we have set $D_{yu} = 0$. It turns out that this assumption entails no loss of generality.)

When attempting to synthesize a simple constant output feedback control $u(t) = Ky(t)$, one arrives at the following condition for quadratic stability of the closed-loop system:

$$P = P^T > 0, \quad (A(t) + B_u K C_y)^T P + P(A(t) + B_u K C_y) < 0, \tag{14.3.17}$$

and there is no way known to derive equivalent LMI conditions (unlike with state-feedback). However, in many cases, although the uncertainties are not known a priori, they can be measured in real time, so that the control law can be allowed to "schedule" itself according to the measured uncertainties. Then, it turns out that conditions for the existence of so-called gain-scheduled control laws that stabilize the system as well as expressions for the control laws themselves can be derived using SDP. We therefore first pose the gain-scheduled controller design problem below, and describe its solution.

Consider system (14.3.16), where

$$A(t) = \sum_{i=1}^{L} \theta_i(t) A_i, \quad \sum_{i=1}^{L} \theta_i(t) = 1.$$

$\theta(t) = [\theta_1(t), \cdots, \theta_L(t)]$ is unknown a priori, but can be measured in real time. We will design a gain-scheduled output feedback controller

$$\frac{d}{dt}x_k(t) = A_k(\theta(t))x_k(t) + B_k(\theta(t))y(t), \quad u(t) = C_k(\theta(t))x_k(t) + D_k(\theta(t))y(t),$$

where $x_k(t) \in \mathbf{R}^{n_k}$, and $A_k(\cdot)$, $B_k(\cdot)$, $C_k(\cdot)$ and $D_k(\cdot)$ are real valued functions of $\theta(t)$.

The state space representation of the closed loop system is

$$\frac{d}{dt}\begin{bmatrix} x(t) \\ x_k(t) \end{bmatrix} = A_{\text{cl}}(t)\begin{bmatrix} x(t) \\ x_k(t) \end{bmatrix},$$

where

$$A_{\text{cl}}(t) = A_0(t) + \mathcal{B}\Omega(\theta(t))\mathcal{C},$$

$$\Omega(\theta(t)) = \begin{bmatrix} A_k(\theta(t)) & B_k(\theta(t)) \\ C_k(\theta(t)) & D_k(\theta(t)) \end{bmatrix}, \quad A_0(t) = \begin{bmatrix} A(t) & 0 \\ 0 & 0 \end{bmatrix},$$

$$\mathcal{B} = \begin{bmatrix} 0 & B_u \\ I & 0 \end{bmatrix} \quad \text{and} \quad \mathcal{C} = \begin{bmatrix} 0 & I \\ C_y & 0 \end{bmatrix}. \tag{14.3.18}$$

Now, the closed loop system is quadratically stable if there exists $P = P^T > 0$ such that

$$A_{\text{cl}}(t)^T P + P A_{\text{cl}}(t) < 0 \tag{14.3.19}$$

for all possible values of $\theta(t)$.

The following two lemmas play a central role in the derivation of LMI conditions that are equivalent to (14.3.19). The first lemma is called the Elimination lemma (see for example, [137]). The second lemma is called the Completion lemma [606].

Lemma 14.3.1 *Given matrices $G \in \mathbf{R}^{n\times n}$, $U \in \mathbf{R}^{n\times p}$, and $V \in \mathbf{R}^{n\times q}$, there exists $\Omega \in \mathbf{R}^{p\times q}$ such that*

$$G + U\Omega V^T + V\Omega^T U^T > 0$$

if and only if

$$U_\perp^T G U_\perp > 0 \quad \text{and} \quad V_\perp^T G V_\perp > 0,$$

where $U_\perp$ and $V_\perp$ are the orthogonal complements of U and V respectively.

Lemma 14.3.2 *Let $X = X^T \in \mathbf{R}^{n\times n}$ and $Y = Y^T \in \mathbf{R}^{n\times n}$ be positive-definite matrices. There exist $X_2 \in \mathbf{R}^{n\times r}$, $X_3 \in \mathbf{R}^{r\times r}$, $Y_2 \in \mathbf{R}^{n\times r}$ and $Y_3 \in \mathbf{R}^{r\times r}$ such that*

$$\begin{bmatrix} X & X_2 \\ X_2^T & X_3 \end{bmatrix} > 0 \quad \text{and} \quad \begin{bmatrix} X & X_2 \\ X_2^T & X_3 \end{bmatrix}^{-1} = \begin{bmatrix} Y & Y_2 \\ Y_2^T & Y_3 \end{bmatrix}$$

if and only if

$$\begin{bmatrix} X & I_n \\ I_n & Y \end{bmatrix} \geq 0 \quad \text{and} \quad \text{rank} \begin{bmatrix} X & I_n \\ I_n & Y \end{bmatrix} \leq n + r,$$

where I_n denotes an $n \times n$ identity matrix.

Using the Elimination Lemma 14.3.1 and Completion Lemma 14.3.2 and some straightforward matrix manipulations, it can be shown that there exists a full order stabilizing gain-scheduled output feedback controller, if there exist $R = R^T \in \mathbf{R}^{n\times n}$ and $S = S^T \in \mathbf{R}^{n\times n}$ such that the following matrix inequalities hold:

$$\begin{array}{l} N_R^T (A(t)R + RA(t)^T) N_R < 0, \\ N_S^T (A(t)^T S + SA(t)) N_S < 0, \end{array} \quad \begin{bmatrix} S & I \\ I & R \end{bmatrix} \geq 0, \tag{14.3.20}$$

where N_R and N_S are matrices whose columns comprise the bases of the null spaces of B_u^T and C_y respectively. Since $A(t)$ lies in a polytope with vertices $A_1, \ldots, A_L$, condition (14.3.20) is equivalent to

$$\begin{bmatrix} S & I \\ I & R \end{bmatrix} \geq 0, \quad \begin{array}{l} N_R^T (A_i R + R A_i^T) N_R < 0, \\ N_S^T (A_i^T S + S A_i) N_S < 0, \quad i = 1, \ldots, L. \end{array} \tag{14.3.21}$$

Thus, we have an LMI condition that is necessary and sufficient for the existence of a quadratically stabilizing gain-scheduled output feedback controller for the polytopic system (14.3.16). We next describe an algorithm for explicitly constructing a family of gain-scheduled output feedback controllers that are guaranteed to stabilize the system.

Step 1. Design controllers at each vertex of the polytope $\mathbf{Co}\{A_1, \ldots, A_L\}$.
Let $(R,\ S)$ be a feasible solution to (14.3.21). With $S > R^{-1}$, define $P_{12} = (S - R^{-1})^{1/2}$, $Q_{12} = -RP_{12}$ and

$$P = \begin{bmatrix} S & P_{12} \\ P_{12}^T & I \end{bmatrix}.$$

It is easy to check that $P > 0$ and

$$P^{-1} = \begin{bmatrix} R & Q_{12} \\ Q_{12}^T & I - P_{12}^T Q_{12} \end{bmatrix} > 0.$$

Let $A(\theta(t)) = A_i$, $i = 1, \ldots, L$. With P defined above, LMI (14.3.19) is feasible (from the feasibility of (14.3.21)). Solve (14.3.19) for Ω_i, which comprises the state space matrices of the controller corresponding to the vertex A_i.

Step 2. Design the gain-scheduled controller.
For any $A(t) \in \mathbf{Co}\{A_1, \ldots, A_L\}$, solve the set of linear equations

$$A(t) = \sum_{i=1}^{L} \theta_i(t) A_i, \quad \sum_{i=1}^{L} \theta_i(t) = 1,$$

to get $\theta_i(t)$. Inequality (14.3.19) is affine on $\Omega(\theta(t))$. Therefore,

$$\Omega(\theta(t)) = \sum_{i=1}^{L} \theta_i(t) \Omega_i, \tag{14.3.22}$$

comprises the state space matrices of a stabilizing gain-scheduled output feedback controller.

14.4 ROBUST STABILITY ANALYSIS OF LFR SYSTEMS IN THE IQC FRAMEWORK

We next focus on the robust stability analysis of system (14.2.4). With the inputs u and w identically zero, the autonomous uncertain system is described by

$$\frac{d}{dt}x(t) = Ax(t) + B_p p(t), \quad q(t) = C_q x(t) + D_{qp} p(t), \quad p = \Delta(q, t). \tag{14.4.23}$$

Let $H(s) = C_q(sI - A)^{-1}B_p + D_{qp}$.

Recall that the uncertainty Δ is known to lie in some set $\mathbf{\Delta}$. This information can be represented in an elegant and mathematically tractable way using the framework of Integral Quadratic Constraints[3] or IQCs [529].

We first provide a brief description of the IQC framework (for notation, terminology and details, we refer the reader to [529]). Two signals $p \in \mathbf{L}_2^m[0,\infty)$ and $q \in \mathbf{L}_2^m[0,\infty)$, with Fourier Transforms $\hat{p}$ and $\hat{q}$ respectively (assuming that the Fourier Transforms exist), are said to "satisfy the IQC defined by Π", if

$$\int_{-\infty}^{\infty} \begin{bmatrix} \hat{p}(j\omega) \\ \hat{q}(j\omega) \end{bmatrix}^* \Pi(j\omega) \begin{bmatrix} \hat{p}(j\omega) \\ \hat{q}(j\omega) \end{bmatrix} d\omega \geq 0, \tag{14.4.24}$$

where $\Pi : j\mathbf{R} \to \mathbf{C}^{2m\times 2m}$ is a measurable Hermitian function, bounded on the imaginary axis. We also say that $\Delta : \mathbf{L}_2^m[0,\infty) \to \mathbf{L}_2^m[0,\infty)$ "satisfies the IQC Π", if for every $q \in \mathbf{L}_2^m[0,\infty)$, q and Δq satisfy the IQC defined by Π. With this terminology, we assume that Δ lies in the set

$$\mathbf{\Delta} = \left\{ \Delta \;\middle|\; \begin{array}{l} \text{For every } \Pi \in \mathbf{\Pi}, \text{ for every } \tau \in [0,1], \\ \tau\Delta \text{ satisfies the IQC defined by } \Pi \end{array} \right\},$$

where $\mathbf{\Pi}$ is some specified set that can be thought of as summarizing the information known about Δ. In all the examples that we will consider shortly, the elements of $\mathbf{\Pi}$ have the following property:

$$\begin{array}{l} \text{Partitioning any } \Pi \in \mathbf{\Pi} \text{ as } \Pi = \begin{bmatrix} \Pi_{11} & \Pi_{12} \\ \Pi_{12}^* & \Pi_{22} \end{bmatrix}, \\ \text{for some } \epsilon > 0, \text{ for all } \omega \in \mathbf{R},\ \Pi_{11}(j\omega) \geq 2\epsilon \text{ and } \Pi_{22}(j\omega) \leq -2\epsilon. \end{array} \tag{14.4.25}$$

We first review a sufficient condition [529, Theorem 1] for the robust stability of system (14.4.23) over $\mathbf{\Delta}$.

Theorem 14.4.1 *Suppose that the system represented by the equations*

$$\frac{d}{dt}x(t) = Ax(t) + B_p p(t), \quad q(t) = C_q x(t) + D_{qp} p(t), \quad p(t) = \tau\Delta(q,t),$$

[3]The framework of integral quadratic constraints helps unify a number of sufficient conditions for the robust stability of system (14.4.23) over $\mathbf{\Delta}$. When Δ can be *any* operator satisfying an $\mathbf{L}_2$-gain bound, the small-gain theorem provides a necessary and sufficient condition for robust stability [190]. When Δ is structured—say diagonal—the small gain condition is no longer necessary for stability; diagonal scalings can then be used to derive less conservative robust stability conditions [198, 690]. In addition, if Δ is a memoryless time-invariant sector-bounded nonlinearity, the celebrated Popov criterion yields a sufficient condition for robust stability (see for example, [190]). When Δ is LTI or parametric, the well-known μ analysis and K_m analysis methods provide sufficient conditions for robust stability [61, 157, 214]. Besides enabling the rederivation and theoretical analysis of all these sufficient conditions, the IQC framework also lends itself to the derivation of other, new, suffcient conditions for robust stability [529].

is well-posed for any $\tau \in [0,1]$ *and any* $\Delta \in \mathbf{\Delta}$. *Then, if there exist* $\Pi \in \mathbf{\Pi}$ *and* $\epsilon > 0$ *such that*

$$\begin{bmatrix} H(j\omega) \\ I \end{bmatrix}^* \Pi(j\omega) \begin{bmatrix} H(j\omega) \\ I \end{bmatrix} \leq -2\epsilon I, \text{ for all } \omega \in \mathbf{R}, \qquad (14.4.26)$$

then system (14.4.23) is robustly stable over $\mathbf{\Delta}$.

In general, $\mathbf{\Pi}$—the set defining the IQCs corresponding to $\mathbf{\Delta}$—is not described by a finite number of variables. In order to reduce the number of optimization variables to a finite number, a subset of $\mathbf{\Pi}$ is defined as

$$\mathbf{\Pi}_{\text{fin}} = \left\{ \Pi \;\middle|\; \begin{array}{l} \Pi(j\omega) = \begin{bmatrix} \Pi_{11} & \Pi_{12} \\ \Pi_{12}^* & \Pi_{22} \end{bmatrix}; \; \Pi_{ij} = W(j\omega)^* T_{ij} W(j\omega); \\ W(j\omega) = \begin{bmatrix} C_W(j\omega I - A_W)^{-1} B_W \\ D_W \end{bmatrix}; \\ \begin{bmatrix} T_{11} & T_{12} \\ T_{12}^T & -T_{22} \end{bmatrix} \in \mathbf{T}; \text{ for some } \epsilon > 0, \text{ for all } \omega \in \mathbf{R}, \\ W(j\omega)^* T_{11} W(j\omega) \geq 2\epsilon I, \; W(j\omega)^* T_{22} W(j\omega) \geq 2\epsilon I. \end{array} \right\}, \qquad (14.4.27)$$

where $A_W \in \mathbf{R}^{n_W \times n_W}$, $B_W \in \mathbf{R}^{n_W \times m}$, $C_W \in \mathbf{R}^{N_1 \times n_W}$, $D_W \in \mathbf{R}^{N_2 \times m}$, and $\mathbf{T}$ is an appropriately chosen subspace of $\mathbf{R}^{2(N_1+N_2)\times 2(N_1+N_2)}$. (We will give specific examples later.)

The "frequency-domain" condition (14.4.26) plays a key role in the robust stability analysis. The celebrated positive-real lemma[4](or Kalman-Yakubovich-Popov lemma), which provides a connection between frequency-domain conditions on transfer functions and the underlying state-space matrices, enables us to derive an LMI that is equivalent to condition (14.4.26).

Lemma 14.4.1 (Positive Real Lemma [656]) *Let* $A \in \mathbf{R}^{n\times n}$, $B \in \mathbf{R}^{n\times m}$ *and* $M = M^T \in \mathbf{R}^{(m+n)\times(m+n)}$, *with* A *having no eigenvalues on the imaginary axis. Then, the following statements are equivalent.*

1. *For some* $\epsilon > 0$,

$$\begin{bmatrix} (j\omega I - A)^{-1}B \\ I \end{bmatrix}^* M \begin{bmatrix} (j\omega I - A)^{-1}B \\ I \end{bmatrix} \geq 2\epsilon I, \qquad \text{for all } \omega \in \mathbf{R}.$$

2. *There exists a symmetric matrix* $P = P^T$ *such that*

$$\begin{bmatrix} A^T P + PA & PB \\ B^T P & 0 \end{bmatrix} < M.$$

[4] A number of different versions of the positive-real lemma can be found in the literature; see for example [845, 846, 388, 389, 30, 137, 656].

Using this lemma, checking if condition (14.4.26) holds can be reduced to an LMI.

Theorem 14.4.2 *Let*

$$\tilde{A} = \begin{bmatrix} A_W & B_W C_q & 0 \\ 0 & A & 0 \\ 0 & 0 & A_W \end{bmatrix}, \quad \tilde{C} = \begin{bmatrix} I & 0 & 0 \\ 0 & C_q & 0 \\ 0 & 0 & I \\ 0 & 0 & 0 \end{bmatrix},$$

$$\tilde{B} = \begin{bmatrix} B_W D_{qp} \\ B_p \\ B_W \end{bmatrix}, \quad \tilde{D} = \begin{bmatrix} 0 \\ D_{qp} \\ 0 \\ I \end{bmatrix}, \quad E = \begin{bmatrix} C_W & 0 \\ 0 & D_W \end{bmatrix}.$$

Then, condition (14.4.26) holds with $\Pi(j\omega) \in \mathbf{\Pi}_{\text{fin}}$, *if and only if the LMIs*

$$M_1 = E^T T_{11} E - \begin{bmatrix} A_W^T Q_1 + Q_1 A_W & Q_1 B_W \\ B_W^T Q_1 & 0 \end{bmatrix} > 0,$$

$$M_2 = E^T T_{22} E - \begin{bmatrix} A_W^T Q_2 + Q_2 A_W & Q_2 B_W \\ B_W^T Q_2 & 0 \end{bmatrix} > 0,$$

$$\begin{bmatrix} \tilde{A}^T P + P\tilde{A} & P\tilde{B} \\ \tilde{B}^T P & 0 \end{bmatrix} + \begin{bmatrix} \tilde{C}^T \\ \tilde{D}^T \end{bmatrix} \begin{bmatrix} M_1 & E^T T_{12} E \\ E^T T_{12}^T E & -M_2 \end{bmatrix} \begin{bmatrix} \tilde{C} & \tilde{D} \end{bmatrix} < 0,$$

$$P = P^T,\ Q_1 = Q_1^T,\ Q_2 = Q_2^T,\quad \begin{bmatrix} T_{11} & T_{12} \\ T_{12}^T & -T_{22} \end{bmatrix} \in \mathbf{T}$$

are feasible.

We now describe the results that can be obtained from an application of Theorem 14.4.2 in several commonly-encountered situations.

14.4.1 Diagonal nonlinearities

Consider the special case when Δ is a "diagonal" uncertainty, i.e., if $p(t) = \Delta(q,t)$, then $p_i(t) = \delta_i(q_i, t)$, or the ith component of p is purely a function of the ith component of q. Moreover, suppose that the $\mathbf{L}_2$ gain of Δ does not exceed one, i.e., if $p(t) = \Delta(q,t)$, then

$$\int_0^T p(t)^T p(t)\, dt \le \int_0^T q(t)^T q(t)\, dt, \text{ for all } T > 0. \tag{14.4.28}$$

Then, it turns out that Δ satisfies every IQC from the set

$$\mathbf{\Pi}^{\text{DNL}} = \left\{ \Pi \;\middle|\; \begin{array}{l} \Pi(j\omega) = \begin{bmatrix} W & 0 \\ 0 & -W \end{bmatrix} \text{ for all } \omega \in \mathbf{R}, \\ W \in \mathbf{R}^{m\times m},\ W > 0 \text{ and diagonal} \end{array} \right\}. \tag{14.4.29}$$

Note that $\mathbf{\Pi}^{\mathrm{DNL}}$ is already described by a finite number of variables so that $\mathbf{\Pi}^{\mathrm{DNL}} = \mathbf{\Pi}^{\mathrm{DNL}}_{\mathrm{fin}}$ and is defined by (14.4.27), where A_W, B_W and C_W are vacuous, $D_W = I$, and $\mathbf{T} = \mathbf{\Pi}^{\mathrm{DNL}}$.

From Theorem 14.4.2, condition (14.4.26) is equivalent to the LMI

$$P = P^T, \; W \in \mathbf{R}^{m\times m} \text{ and diagonal}, W > 0,$$

$$\begin{bmatrix} A^TP + PA & PB_p \\ B_p^TP & 0 \end{bmatrix} + \begin{bmatrix} C_q^T & 0 \\ D_{qp}^T & I \end{bmatrix} \begin{bmatrix} W & 0 \\ 0 & -W \end{bmatrix} \begin{bmatrix} C_q & D_{qp} \\ 0 & I \end{bmatrix} < 0. \tag{14.4.30}$$

It turns out that this LMI is equivalent to the condition that a diagonally scaled $\mathbf{H}_\infty$ norm of $H(s) = C_q(sI - A)^{-1}B_p + D_{qp}$ is less than one; see [206, 137].

14.4.2 Parametric uncertainties

Suppose Δ is a constant real matrix with a specified block-diagonal structure, and with a spectral norm that does not exceed one:

$$\Delta = \operatorname{diag}\left(D_1, \ldots, D_M, d_1 I_{\ell_1}, \ldots, d_N I_{\ell_N}\right),$$

$D_i \in \mathbf{R}^{k_i \times k_i}$, $i = 1, \ldots, M$, $d_i \in \mathbf{R}$, $i = 1, \ldots, N$, with $\sigma_{\max}(\Delta) \leq 1$. Then, Δ satisfies every IQC from

$$\mathbf{\Pi}^{\mathrm{par}} = \left\{ \Pi \;\middle|\; \begin{array}{l} \Pi(j\omega) = \begin{bmatrix} X(j\omega) & Y(j\omega) \\ -Y(j\omega) & -X(j\omega) \end{bmatrix}, \\ X(j\omega) = X(j\omega)^* \text{ and } Y(j\omega)^* = -Y(j\omega) \text{ in } \mathcal{W}, \\ \text{for some } \epsilon > 0,\; X(j\omega) \geq 2\epsilon I, \text{ for all } \omega \in \mathbf{R} \end{array} \right\},$$

where

$$\mathcal{W} = \left\{ \begin{array}{l} \operatorname{diag}\left(w_1 I_{k_1}, \ldots, w_M I_{k_M},\right. \\ \qquad\qquad \left. W_1, \ldots, W_N\right) \end{array} \;\middle|\; \begin{array}{l} w_i \in \mathbf{C},\; i = 1, \ldots, M \\ W_i \in \mathbf{C}^{\ell_i \times \ell_i},\; i = 1, \ldots, N \end{array} \right\}. \tag{14.4.31}$$

(See for example [51, 529].)

For such uncertainties, Theorem 14.4.2 can be immediately used to obtain sufficient condition that guarantees the stability of the closed loop system. Let $\mathcal{P}$ denote the subset of real matrices that lie in $\mathcal{W}$, i.e., $\mathcal{P} = \mathcal{W} \bigcap \mathbf{R}^{m\times m}$. Then, a subset $\mathbf{\Pi}^{\mathrm{par}}_{\mathrm{fin}}$ of $\mathbf{\Pi}^{\mathrm{par}}$, that is described by a finite number of variables, can be defined as follows. Let $W^{(1)}, \ldots, W^{(N-1)}$ be strictly proper, stable $m \times m$ transfer functions, with each $W^{(i)}$ satisfying $W^{(i)}(j\omega) \in \mathcal{W}$ for every $\omega \in \mathbf{R}$. Let

$$\mathbf{\Theta}^{\mathrm{par}} = \left\{ \begin{bmatrix} \theta_{11} & \theta_{12} & \cdots & \theta_{1N} \\ \theta_{21} & \theta_{22} & \cdots & \theta_{2N} \\ \vdots & \vdots & \ddots & \vdots \\ \theta_{N1} & \theta_{N2} & \cdots & \theta_{NN} \end{bmatrix} \;\middle|\; \theta_{ij} \in \mathcal{P} \right\}. \tag{14.4.32}$$

Then, a subset $\mathbf{\Pi}_{\text{fin}}^{\text{par}}$ of $\mathbf{\Pi}^{\text{par}}$, that is described by a finite number of variables is given by (14.4.27), where (A_W, B_W, C_W) is any state space realization of $[W^{(1)}(s)^T \cdots W^{(N-1)}(s)^T]^T$, $D_W = I$, and

$$\mathbf{T}^{\text{par}} = \left\{ \begin{bmatrix} \Theta + \Theta^T & \Phi - \Phi^T \\ \Phi^T - \Phi & -(\Theta + \Theta^T) \end{bmatrix} \,\middle|\, \Theta, \Phi \in \mathbf{\Theta}^{\text{par}} \right\}. \tag{14.4.33}$$

Note that the choice of $W^{(i)}$ is *ad hoc*, and the robust analysis result will certainly depend on this choice. However, it can be shown (see [162]) that the actual choice of the $W^{(i)}$ is immaterial, provided the set of $W^{(i)}$s is chosen to be "rich enough".

14.4.3 *Structured dynamic uncertainties*

Suppose that Δ is a linear time-invariant operator, such that for all $\omega \in \mathbf{R}$, $\Delta(j\omega) = \operatorname{diag}(D_1, \ldots, D_M, d_1 I_{\ell_1}, \ldots, d_N I_{\ell_N})$, $D_i \in \mathbf{C}^{k_i \times k_i}$, $i = 1, \ldots, M$, $d_i \in \mathbf{C}$, $i = 1, \ldots, N$, with $\sigma_{\max}(\Delta(j\omega)) \leq 1$. Thus, Δ is a dynamic block-structured uncertainty, with an $\mathbf{L}_2$-gain that does not exceed one.

Then, Δ satisfies every IQC from

$$\mathbf{\Pi}^{\text{LTI}} = \left\{ \Pi \,\middle|\, \begin{array}{l} \Pi(j\omega) = \begin{bmatrix} X(j\omega) & 0 \\ 0 & -X(j\omega) \end{bmatrix}, \; X(j\omega) = X(j\omega)^* \in \mathcal{W}, \\ \text{for some } \epsilon > 0, \; X(j\omega) \geq 2\epsilon I, \text{ for all } \omega \in \mathbf{R} \end{array} \right\},$$

where $\mathcal{W}$ is defined in (14.4.31) (see for example [51, 529]). Similarly to the case of parametric uncertainty, $\mathbf{\Pi}_{\text{fin}}^{\text{LTI}} \subset \mathbf{\Pi}^{\text{LTI}}$ can be described by a finite number of variables. Let $W^{(1)}, \ldots, W^{(N-1)}$ be strictly proper, stable $m \times m$ transfer functions, with each $W^{(i)}$ satisfying $W^{(i)}(j\omega) \in \mathcal{W}$ for every $\omega \in \mathbf{R}$. Then, a subset $\mathbf{\Pi}_{\text{fin}}^{\text{LTI}}$ of $\mathbf{\Pi}^{\text{LTI}}$ described by a finite number of variables is given by (14.4.27), where (A_W, B_W, C_W) is any state space realization of $\begin{bmatrix} W^{(1)}(s)^T & \cdots & W^{(N-1)}(s)^T \end{bmatrix}^T$, $D_W = I$, and

$$\mathbf{T}^{\text{LTI}} = \left\{ \begin{bmatrix} \Theta + \Theta^T & 0 \\ 0 & -(\Theta + \Theta^T) \end{bmatrix} \,\middle|\, \Theta \in \mathbf{\Theta}^{\text{par}} \right\}. \tag{14.4.34}$$

14.5 STABILIZING CONTROLLER DESIGN FOR LFR SYSTEMS

Finally, we consider problem (P2) for system (14.2.4), i.e., the problem of synthesizing control laws that stabilize system (14.2.4). It is yet unknown how to gracefully extend the IQC-based robust stability analysis that we described in the previous section to synthesizing even the simplest of control laws, for instance state-feedback. We will therefore focus on the special case of an uncertain system described by

$$
\begin{aligned}
\frac{d}{dt}x(t) &= Ax(t) + B_p p(t) + B_u u(t), \\
q(t) &= C_q x(t) + D_{qp} p(t) + D_{qu} u(t), \\
y(t) &= C_y x(t) + D_{yp} p(t), \\
p(t) &= \Delta(t) q(t),
\end{aligned}
\tag{14.5.35}
$$

where:

- The uncertainty $\Delta(t)$ is measurable in real-time.
- The "$\mathbf{L}_2$ gain" of $\Delta(t)$ does not exceed one (see (14.4.28)).

For this case, we will describe how state-feedback and gain-scheduled output feedback control laws can be designed using quadratic Lyapunov functions and SDP.

14.5.1 *Quadratic stability analysis of LFR systems*

For the robust stability analysis problem, we set the control input u to be identically zero. Then, the Lyapunov function $V(x(t)) = x(t)^T P x(t)$ with $P = P^T > 0$ decreases along the trajectories of (14.5.35) for all Δ if

$$x(t)^T \left(A^T P + PA\right) x(t) + 2x(t)^T P B_p p(t) < 0 \text{ for nonzero } x(t),$$

whenever

$$\int_0^T p(t)^T p(t)\, dt \le \int_0^T q(t)^T q(t)\, dt.$$

It can be shown that this holds if

$$
\begin{bmatrix} x(t) \\ p(t) \end{bmatrix}^T \begin{bmatrix} PA + A^T P & PB_p \\ B_p^T P & 0 \end{bmatrix} \begin{bmatrix} x(t) \\ p(t) \end{bmatrix} < 0,
$$

for every x and p that satisfy (14.5.36)

$$
\begin{bmatrix} x(t) \\ p(t) \end{bmatrix}^T \begin{bmatrix} C_q^T C_q & C_q^T D_{qp} \\ D_{qp}^T C_q & D_{qp}^T D_{qp} - I \end{bmatrix} \begin{bmatrix} x(t) \\ p(t) \end{bmatrix} \ge 0.
$$

There exists P such that condition (14.5.36) holds if there exists P such that the following LMIs are feasible:

$$
P = P^T > 0, \quad \begin{bmatrix} PA + A^T P + C_q^T C_q & PB_p + C_q^T D_{qp} \\ B_p^T P + D_{qp}^T C_q & D_{qp}^T D_{qp} - I \end{bmatrix} < 0. \tag{14.5.37}
$$

Thus, we have an LMI condition that is sufficient for the quadratic stability of system (14.5.35).

14.5.2 *State feedback controller design for LFR systems*

Consider system (14.5.35) with constant state feedback $u(t) = Kx(t)$. The closed loop system is

$$\begin{aligned} \frac{d}{dt}x(t) &= (A + B_u K)x(t) + B_p p(t), \\ q(t) &= (C_q + D_{qu}K)x(t) + D_{qp}p(t), \\ p(t) &= \Delta(t)q(t), \end{aligned} \tag{14.5.38}$$

Using the sufficient condition for robust stability (14.5.37), and following a line of argument similar to that in the derivation of stabilizing state feedback laws for polytopic systems, we conclude that there exists a stabilizing constant state feedback controller for system (14.5.35) if there exist $Q = Q^T > 0$ and Y such that the following LMI holds.

$$\begin{bmatrix} AQ + QA^T + B_p B_p^T + B_u Y + Y^T B_u^T & B_p D_{qp}^T + QC_q^T + Y^T D_{qu}^T \\ D_{qp} B_p^T + C_q Q + D_{qu} Y & D_{qp} D_{qp}^T - I \end{bmatrix} < 0. \tag{14.5.39}$$

$V(\psi) = \psi^T Q^{-1} \psi$ is a Lyapunov function that proves the quadratic stability of the closed loop system (14.5.38), and $K = YQ^{-1}$ is the corresponding stabilizing state feedback gain.

14.5.3 *Gain-scheduled output feedback controller design*

As with polytopic systems, the general output feedback controller synthesis problem for LFR systems is not a convex feasibility problem. However, if the uncertainty $\Delta(t)$ can be measured in real time, the design of a gain-scheduled controller, i.e., one that depends on $\Delta(t)$, turns out to be an SDP problem.

The closed-loop system, with the gain-scheduled controller enclosed in dotted lines, is shown in Figure 14.3. The controller consists of an LTI system $\mathcal{K}_{\text{LTI}}$, with the uncertainty $\Delta(t)$ appearing in a feedback configuration. Let the state space representation of $\mathcal{K}_{\text{LTI}}$ be

$$\begin{aligned} \frac{d}{dt}x_k(t) &= A_k x_k(t) + B_{k1} y(t) + B_{k2} v(t), \\ u(t) &= C_{k1} x_k(t) + D_{k11} y(t) + D_{k12} v(t), \\ f(t) &= C_{k2} x_k(t) + D_{k21} y(t) + D_{k22} v(t). \end{aligned} \tag{14.5.40}$$

Then, the equations governing the closed loop system in Figure 14.3 are

$$\frac{d}{dt}\begin{bmatrix} x(t) \\ x_k(t) \end{bmatrix} = A_{\text{cl}} \begin{bmatrix} x(t) \\ x_k(t) \end{bmatrix} + B_{\text{cl}} \begin{bmatrix} v(t) \\ p(t) \end{bmatrix},$$

$$\begin{bmatrix} f(t) \\ q(t) \end{bmatrix} = C_{\text{cl}} \begin{bmatrix} x(t) \\ x_k(t) \end{bmatrix} + D_{\text{cl}} \begin{bmatrix} v(t) \\ p(t) \end{bmatrix},$$

$$\begin{bmatrix} v(t) \\ p(t) \end{bmatrix} = \begin{bmatrix} \Delta(t) & \\ & \Delta(t) \end{bmatrix} \begin{bmatrix} f(t) \\ q(t) \end{bmatrix},$$

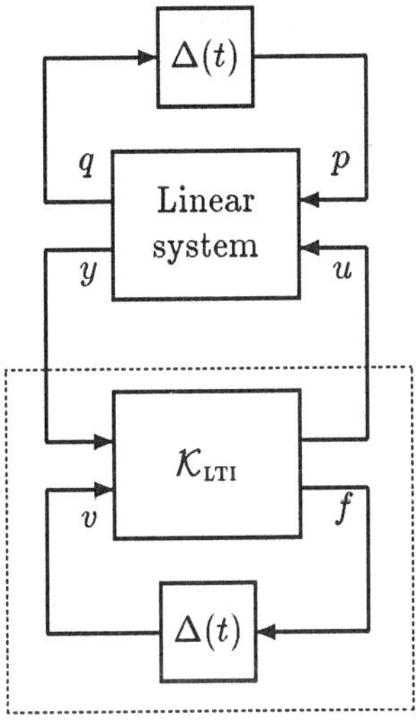

Figure 14.3 Gain scheduled output feedback control framework.

where

$$A_{\rm cl} = A_0 + \mathcal{B}\Omega\mathcal{C}, \quad B_{\rm cl} = B_0 + \mathcal{B}\Omega\mathcal{D}_{yp},$$

$$C_{\rm cl} = C_0 + \mathcal{D}_{qu}\Omega\mathcal{C}, \quad D_{\rm cl} = D_0 + \mathcal{D}_{qu}\Omega\mathcal{D}_{yp},$$

with

$$\Omega = \begin{bmatrix} A_k & B_{k1} & B_{k2} \\ C_{k1} & D_{k11} & D_{k12} \\ C_{k2} & D_{k21} & D_{k22} \end{bmatrix} \tag{14.5.41}$$

and

$$\begin{gathered} A_0 = \begin{bmatrix} A & 0 \\ 0 & 0 \end{bmatrix}, \; B_0 = \begin{bmatrix} 0 & B_p \\ 0 & 0 \end{bmatrix}, \; \mathcal{C} = \begin{bmatrix} 0 & I \\ C_y & 0 \\ 0 & 0 \end{bmatrix}, \\ C_0 = \begin{bmatrix} 0 & 0 \\ C_q & 0 \end{bmatrix}, \; D_0 = \begin{bmatrix} 0 & 0 \\ 0 & D_{qp} \end{bmatrix}, \; \mathcal{D}_{yp} = \begin{bmatrix} 0 & 0 \\ D_{yp} & 0 \\ I & 0 \end{bmatrix}, \\ \mathcal{B} = \begin{bmatrix} 0 & B_u & 0 \\ I & 0 & 0 \end{bmatrix}, \quad \mathcal{D}_{qu} = \begin{bmatrix} 0 & 0 & I \\ 0 & D_{qu} & 0 \end{bmatrix}. \end{gathered} \tag{14.5.42}$$

The sufficient condition for robust stability (14.5.37) implies that there exists a stabilizing gain-scheduled output feedback controller for system (14.5.35) if there exist $P = P^T > 0$ and Ω such that

$$\begin{bmatrix} PA_{\rm cl} + A_{\rm cl}^T P + C_{\rm cl}^T C_{\rm cl} & PB_{\rm cl} + C_{\rm cl}^T D_{\rm cl} \\ B_{\rm cl}^T P + D_{\rm cl}^T C_{\rm cl} & D_{\rm cl}^T D_{\rm cl} - I \end{bmatrix} < 0, \tag{14.5.43}$$

Proceeding along the lines of the derivation of a gain-scheduled output feedback controller for polytopic systems, we conclude that there exists a full order stabilizing gain-scheduled controller such that condition (14.5.43) holds for some $P = P^T > 0$ if and only if there exist $R = R^T \in \mathbf{R}^{n\times n}$ and $S = S^T \in \mathbf{R}^{n\times n}$ such that the following LMIs hold:

$$\begin{bmatrix} N_R & 0 \\ 0 & I \end{bmatrix}^T \left[\begin{array}{cc|c} AR + RA^T & RC_q^T & B_p \\ C_q R & -I & D_{qp} \\ \hline B_p^T & D_{qp}^T & -I \end{array}\right] \begin{bmatrix} N_R & 0 \\ 0 & I \end{bmatrix} < 0, \qquad (14.5.44)$$

$$\begin{bmatrix} N_S & 0 \\ 0 & I \end{bmatrix}^T \left[\begin{array}{cc|c} A^T S + SA & SB_p & C_q^T \\ B_p^T S & -I & D_{qp}^T \\ \hline C_q & D_{qp} & -I \end{array}\right] \begin{bmatrix} N_S & 0 \\ 0 & I \end{bmatrix} < 0, \qquad (14.5.45)$$

$$\begin{bmatrix} S & I \\ I & R \end{bmatrix} \geq 0, \qquad (14.5.46)$$

where N_R and N_S are matrices whose columns comprise the bases of the null spaces of $[B_u^T \;\; D_{qu}^T]$ and $[C_y \;\; D_{yp}]$ respectively.

Next, suppose there exist R and S such that the synthesis conditions are feasible. A gain-scheduled output feedback controller can then be designed as follows.

Step 1. Define $P_{12} = (S - R^{-1})^{1/2}$ and $Q_{12} = -RP_{12}$. It is easy to check that

$$P = \begin{bmatrix} S & P_{12} \\ P_{12}^T & I \end{bmatrix} > 0 \quad \text{and} \quad P^{-1} = \begin{bmatrix} R & Q_{12} \\ Q_{12}^T & I - P_{12}^T Q_{12} \end{bmatrix} > 0.$$

Step 2. Solve the following LMI for Ω.

$$X + U^T \Omega V + V^T \Omega^T U < 0, \qquad (14.5.47)$$

where

$$X = \begin{bmatrix} A_0^T P + PA_0 & PB_0 & C_0^T \\ B_0^T P & -I & D_0^T \\ C_0 & D_0 & -I \end{bmatrix},$$

$$U = \begin{bmatrix} \mathcal{B}^T P & 0 & \mathcal{D}_{qu}^T \end{bmatrix} \text{ and } V = \begin{bmatrix} \mathcal{C} & \mathcal{D}_{yp} & 0 \end{bmatrix}.$$

The feasible solution Ω in (14.5.47) comprises the state space matrices of the LTI part $\mathcal{K}_{\text{LTI}}$ of a stabilizing gain-scheduled output feedback controller (see (14.5.40)).

14.6 CONCLUSION

We have described some examples of the application of SDP in robust control, specifically the problems of robust stability analysis and stabilizing controller design of system models that incorporate model uncertainties. The numerical solution of engineering problems via an SDP reformulation continues to be a popular solution technique not only in robust control, but also in other related areas of system theory, such as communications and signal processing.

15 STRUCTURAL DESIGN

Aharon Ben-Tal, Arkadi Nemirovski

Mathematics and it symbols are an enormous reservoir for unexploited forms.

Robert le Ricolais (1894-1977) French Engineer/Artist/Sculptor/Poet

15.1 STRUCTURAL DESIGN: GENERAL SETTING

Consider a mechanical construction $\mathcal{C}$ with $M < \infty$ degrees of freedom; thus, virtual displacements of $\mathcal{C}$ are specified by vectors $w \in \mathbb{R}^M$. The potential energy capacitated by $\mathcal{C}$ under a displacement w is assumed to be a nonnegative quadratic function $\frac{1}{2}w^T \mathcal{Q} w$ of the displacement ("linearly elastic construction"), $\mathcal{Q}$ being a symmetric positive semidefinite matrix characterizing $\mathcal{C}$; this matrix is assumed to depend linearly on the vector t of design variables:

$$\mathcal{Q} = \mathcal{Q}(t).$$

The displacement vectors w are restricted to reside in a set $W \subset \mathbb{R}^M$ of *kinematically admissible* displacements.

The construction can be subject to an external *load*; mathematically, a load is a vector $f \in \mathbb{R}^M$. The displacement caused by the load maximizes the function

$$2f^T w - w^T \mathcal{Q}(t) w$$

over $w \in W$; the corresponding optimal value

$$\text{compl}(f; t) = \sup_{w \in W} \left[2f^T w - w^T \mathcal{Q}(t) w\right]$$

is twice the *compliance* of the construction under load f; the less is the compliance, the better are the rigidity properties of the construction with respect to the load.

The problem of *optimal structural design* is as follows: given a set $\mathcal{F} \subset \mathbb{R}^M$ of tentative loads and a set T of admissible values of the design vector t, find $t \in T$ which minimizes the worst-case, w.r.t. loads from $\mathcal{F}$, compliance of the construction, i.e., solve the optimization problem

$$\min_{t \in T} \left\{ \mathrm{compl}_{\mathcal{F}}(t) \equiv \sup_{f \in \mathcal{F}} \mathrm{compl}(f;t) \right\} \qquad (P_{\mathrm{ini}})$$

This general setting has two particular cases which are of especial interest.

Truss design. A *truss* is a construction, like an electric mast or the Eifel Tower, comprised of thin elastic *bars* linked with each other at *nodes*. In the standard Truss Topology Design problem the nodes form a given finite set in $\mathbb{R}^d$, where $d = 2$ for planar and $d = 3$ for spatial trusses. It is assumed that all pair connections of nodes by bars are allowed. For every node, its virtual displacements form a given linear subspace of $\mathbb{R}^d$, and the space $\mathbb{R}^M$ of virtual displacements of the truss is just the direct product of these spaces over all nodes. The set W of admissible displacements is a subset of $\mathbb{R}^M$ given by a number of inequality constraints (typically linear) representing *obstacles* – restrictions on the displacements of the nodes coming from absolutely rigid partial supports.

The entries of the design vector $t = (t_1, ..., t_N)$ represent volumes of tentative bars, and the matrix $Q(t)$ ("the bar-stiffness matrix") is of the form

$$Q(t) = \sum_{\ell=1}^{N} b_\ell b_\ell^T t_\ell,$$

where the vectors $b_\ell \in \mathbb{R}^M$ are readily given by positions of the nodes linked by the corresponding bars. The set T of feasible design vectors is always a subset of $\mathbb{R}^N_+$ – bar volumes must be nonnegative – satisfying the *resource constraint*

$$\sum_{\ell=1}^{N} t_\ell \le v$$

(upper bound on the volume of the construction), and, possibly, some other (normally linear) constraints. The most important case is as follows:

$$\begin{aligned} T = \{t \in \mathbb{R}^N_+ \mid \underline{\rho}_\ell \le t_\ell \le \overline{\rho}_\ell, \ell = 1, ..., N; \sum_{\ell=1}^{N} t_\ell \le v\} \\ \left[0 \le \underline{\rho}_\ell \le \overline{\rho}_\ell < \infty, \ell = 1, ..., N\right] \end{aligned} \qquad (15.1.1)$$

Shape design. In this case the construction in question occupies a given 2D or 3D domain Ω and is comprised of material with mechanical properties varying continuously from point to point; thus, in fact we are speaking about a "distributed" mechanical system with infinitely many degrees of freedom. However, in order to get a computationally tractable model, we apply from the very beginning the finite element method in order to pass from the actual model to its finite-dimensional approximation. Namely, we replace the infinite-dimensional space of displacements of the actual construction (which is the space of vector fields on Ω) by its finite-dimensional subspace $\mathbb{R}^M$. Similarly, we partition Ω into N cells C_ℓ, $\ell = 1, ..., N$, and assume that the mechanical properties of the material are constant within every cell. With this approximation, the potential energy capacitated by the construction under displacement w is

$$\frac{1}{2}\sum_{\ell=1}^{N} \nu_\ell^{-1}\mathrm{Trace}(t_\ell \int_{C_\ell} e_P(w)e_P^T(w)dP), \tag{15.1.2}$$

where

• ν_ℓ is the d-dimensional volume of cell C_ℓ ($d = 2$ for planar and $d = 3$ for spatial shapes);

• $e_P(w)$ is the "strain tensor" associated with displacement w at a point $P \in \Omega$; all which is important for us is that $e_P(w)$ is an L_∞ function of P taking values in the Euclidean space $\mathbb{R}^D$ ($D = d(d+1)/2$), and that $e_P(w)$ is linear in $w \in \mathbb{R}^M$;

• $\nu_\ell^{-1}t_\ell$ is the "rigidity tensor of the material" in the cell C_ℓ; all which is important for us is that t_ℓ is a symmetric positive semidefinite $D \times D$ matrix. The set T of feasible design vectors is always a subset of $(\mathcal{P}^D)^N$, where $\mathcal{P}^D$ is the cone of positive semidefinite symmetric $D \times D$ matrices – the rigidity tensors must be positive semidefinite. Typical additional restrictions defining T are "resource constraints" imposed on the quantities $\mathrm{Trace}(t_\ell)$ – these quantities in a sense measure densities of the material in the cells. The most important case is

$$T = \{t \in (\mathcal{P}^D)^N \mid \underline{\rho}_\ell \le \mathrm{Trace}(t_\ell) \le \overline{\rho}_\ell, \ell = 1, ..., N; \sum_{\ell=1}^{N} \mathrm{Trace}(t_\ell) \le v\}$$
$$\left[0 \le \underline{\rho}_\ell \le \overline{\rho}_\ell < \infty, \ell = 1, ..., N\right] \tag{15.1.3}$$

The "standard" case. In fact the Truss and the Shape problems can be covered by a single particular case of our general setting – the one where the "design variables" t_ℓ are positive semidefinite symmetric matrices of certain row dimension D, the constraints defining T are restrictions on vectors comprised of traces of these matrices, and $\mathcal{Q}(t)$ has the following structure:

$$\mathcal{Q}(t) = \sum_{\ell=1}^{N}\sum_{j=1}^{S} b_{\ell j} t_\ell b_{\ell j}^T, \tag{15.1.4}$$

where $b_{\ell j}$ are given $M \times D$ matrices.

Indeed, the Truss problem clearly fits the indicated scheme (with $D = S = 1$). To see that the Shape problem also fits it, note that there exists S, "quadrature grids" $\{P_{\ell j}\}_{j=1}^{S}$ and "quadrature weights" $\{\gamma_{\ell j}^2\}_{j=1}^{S}$, $\ell = 1, ..., N$, such that

$$\frac{1}{\nu_\ell}\int_{C_\ell} e_P(w)e_P^T(w) = \sum_{j=1}^{S} \gamma_{\ell j}^2 e_{P_{\ell j}}(w)e_{P_{\ell j}}^T(w)$$

identically in $w \in \mathbb{R}^M$. Defining matrices $b_{\ell j}$ by the relation

$$b_{\ell j}^T w = \gamma_{\ell j} e_{P_{\ell j}}(w), \quad w \in \mathbb{R}^M,$$

we represent the potential energy (15.1.2) in the form of

$$\frac{1}{2} w^T Q(t) w$$

with $Q(t)$ given by (15.1.4).

In what follows we refer to our original setting as to the *general case*, and specify its particular *standard case* as the one where

S.1. The space of design vectors is the direct product of N copies of the space $\mathcal{S}^D$ of symmetric $D \times D$ matrices, so that the design vector is

$$t = (t_1, ..., t_N) : t_\ell \in \mathcal{S}^D, l = 1, ..., N;$$

S.2. The mapping $t \mapsto Q(t)$ is

$$Q(t) = \sum_{\ell=1}^{N}\sum_{j=1}^{S} b_{\ell j} t_\ell b_{\ell j}^T,$$

where $b_{\ell j}$ are given $M \times D$ matrices;

S.3 The set W of kinematically admissible displacements is a polytope:

$$W = \{w \in \mathbb{R}^M \mid Rw \le r\},$$

and the system of constraints $Rw \le r$ satisfies the Slater condition: there exists $\hat{w}$ such that $R\hat{w} < r$;

S.4. The set T of admissible design vectors is

$$T = \{t \in (\mathcal{P}^D)^N \mid \underline{\rho}_\ell \le \text{Trace}(t_\ell) \le \overline{\rho}_\ell, \ell = 1, ..., N; \sum_{\ell=1}^{N} \text{Trace}(t_\ell) \le v\}$$

with

$$0 \le \underline{\rho}_\ell < \overline{\rho}_\ell < \infty, \ell = 1, ..., N; \sum_{\ell=1}^{N} \underline{\rho}_\ell < v.$$

S.5. The matrix

$$\sum_{\ell=1}^{N}\sum_{j=1}^{S} b_{\ell j} b_{\ell j}^T$$

is positive definite.

The set of loads. In the traditional literature on structural design (see [3, 82, 85, 90, 91, 89] and references therein) $\mathcal{F}$ is either a singleton $\{f\}$ ("single-load case") or a finite set $\{f_1, f_2, ..., f_k\}$ ("multi-load case"). Recently, a *robust* setting of the problem was proposed and motivated [86], where $\mathcal{F}$ is an ellipsoid:

$$\mathcal{F} = \{f = Qu \mid u \in \mathbb{R}^k, u^T u \leq 1\}.$$

The rest of the chapter is organized as follows. In Section 15.2 we demonstrate that under reasonable assumptions the general structural design problem can be posed as a semidefinite program; in particular, this is true for the situations we are actually interested in, namely, for

(i) The standard case of the multi-load structural design;

(ii) The obstacle-free (i.e., with $W = \mathbb{R}^M$) standard case of the robust structural design.

Moreover, we present a semidefinite program (P) which, on one hand, is specific enough to allow for instructive processing, and on the other hand, is general enough to cover problems of interest (i) – (ii). Sections 15.3 and 15.4 are devoted to mathematical processing of (P). In Section 15.3 we demonstrate that the (Fenchel) dual problem of (P) admits analytical elimination of "most" of the variables, so that the resulting problem (D) is much better suited for numerical solution than the original problem (P). In Section 15.4 we build the Fenchel dual to (D), thus coming to an instructive equivalent reformulation (P^+) of (P). Explicit forms of problems (P), (D), (P^+) for the situations of interest (i), (ii) are listed in Section 15.5. The concluding Section 15.6 deals with computational issues, in particular, with those of recovering the design parameters of a (nearly) optimal construction from a (nearly) optimal solution to (D).

15.2 SEMIDEFINITE REFORMULATION OF (P_{INI})

In this section we demonstrate that the standard case of (P_{ini}) can be posed as a semidefinite program.

Our initial observation is as follows:

Proposition 15.2.1 *Let $Q \in \mathcal{P}^M$, $f \in \mathbb{R}^M$, $\tau \in \mathbb{R}$, and let $W \subset \mathbb{R}^M$ be a closed convex set with a nonempty interior, and*

$$W^* = \{(\nu, \mu) \in \mathbb{R}^M \times \mathbb{R} : \nu^T w \leq \mu \quad \forall w \in W\}$$

be the polar cone of W.

The inequality

$$\sup_{w \in W} [2f^T w - w^T Q w] \leq \tau$$

is satisfied if and only if there exists $(\mu, \nu) \in W^*$ such that the matrix

$$\mathcal{C}[f, Q, \tau, \nu, \mu] \equiv \begin{pmatrix} \tau - \mu & f^T + \frac{1}{2}\nu^T \\ f + \frac{1}{2}\nu & Q \end{pmatrix}$$

is positive semidefinite.

Proof. We have

$$\sup_{w\in W}[2f^T w - w^T Q w] \le \tau$$

$$\Updownarrow$$

$$g(w) \equiv w^T Q w - 2f^T w \ge -\tau \forall w \in W$$

$$\Updownarrow$$

$$\inf_{w\in \mathbb{R}^M}[g(w) + \delta(w)] \ge -\tau,$$

where $\delta(w) = \begin{cases} 0, & w \in W \\ +\infty, & w \notin W \end{cases}$ is the indicator function of W. Now, $g(w)$ is convex due to $Q \succeq 0$, and W is a closed convex set with a nonempty interior. Thus, we can apply the Fenchel Duality Theorem to get the equivalence

$$\left\{\sup_{w\in W}[2f^T w - w^T Q w] \le \tau\right\}_1 \Leftrightarrow \{\exists \nu \in \mathbb{R}^M : g^*(\nu) + \delta^*(-\nu) \ge -\tau\}_2, \tag{15.2.5}$$

where h^* denotes the function conjugate to a function h:

$$h^*(\pi) = \inf_{w\in\mathbb{R}^M}\{h(w) - \pi^T w\}.$$

Now assume that $\{\cdot\}_1$ holds. Then there exists ν such that

$$g^*(\nu) \ge -\tau - \delta^*(-\nu). \tag{15.2.6}$$

Setting

$$\mu = \delta^*(-\nu) \equiv \inf_{w\in W}(-\nu)^T w,$$

we get $(\nu, \mu) \in W^*$. Now, (15.2.6) means exactly that

$$w^T \mathcal{Q}(t) w - 2f^T w - \nu^T w \ge -\tau + \mu \quad \forall w \in \mathbb{R}^M, \tag{15.2.7}$$

or, which is the same, that the matrix $\mathcal{C}[f, Q, \tau, \nu, \mu]$ is positive semidefinite.

Vice versa, if there exists $(\nu, \mu) \in W^*$ such that the matrix $\mathcal{C}[f, Q, \tau, \nu, \mu]$ is positive semidefinite, then, inverting step by step the above reasoning, we conclude that the predicate $\{\cdot\}_2$ holds true; by (15.2.5), $\{\cdot\}_1$ holds true as well. ■

In view of Proposition 15.2.1, whenever the design vector t is such that $\mathcal{Q}(t) \succeq 0$, the predicate

$$\mathcal{R}(t, \tau): \quad \forall (f \in \mathcal{F}) \exists ((\nu, \mu) \in W^*): \quad \mathcal{C}[f, \mathcal{Q}(t), \tau, \nu, \mu] \succeq 0,$$

says exactly that $\text{compl}_{\mathcal{F}}(t) \le \tau$. It follows that in order to convert (P_{ini}) into a semidefinite program, it suffices

(I) To find a semidefinite representation of the predicate $I\!R$, i.e., to find a symmetric matrix $\mathcal{A}(t,\tau,\xi)$ affinely depending on t,τ and a vector ξ of additional variables in such a way that

$$\{I\!R(t,\tau) \text{ is true }\} \Leftrightarrow \{\exists\xi : \mathcal{A}(t,\tau,\xi) \succeq 0\};$$

(II) To find a semidefinite representation of T, i.e., to find a symmetric matrix $\mathcal{B}(t,\eta)$ affinely depending on t and a vector η of additional variables in such a way that

$$\{t \in T\} \Leftrightarrow \{\exists\eta : \mathcal{B}(t,\eta) \succeq 0\}.$$

Indeed, given $\mathcal{A},\mathcal{B}$ satisfying (I) and (II), we can write down (P_{ini}) equivalently as the semidefinite program

$$\begin{array}{rrcl} & \multicolumn{3}{c}{\text{minimize } \tau} \\ \text{s.t.} & & & \\ & \mathcal{A}(t,\tau,\xi) & \succeq & 0, \\ & \mathcal{B}(t,\eta) & \succeq & 0 \end{array} \tag{15.2.8}$$

in variables t,τ,ξ,η.

Let us look at two generic examples where we can easily satisfy the requirements (I), (II).

Example A. Multi-load structural design with polyhedral W and "simple" T. Here

$$\mathcal{F} = \{f_1,...,f_k\}, W = \{w \in \mathbb{R}^M \mid Rw \le r\},$$
$$T = \{(t_1,...,t_N) \mid t_\ell \in \mathcal{P}^D, \ell = 1,...,N, Pt \le p\}.$$

In this case the semidefinite reformulation of (P_{ini}) is

$$\begin{array}{ll} & \text{minimize } \tau \\ \text{s.t.} & \\ (a.1) & \mathcal{C}[f_i, \mathcal{Q}(t), \tau, R^T y_i, \mu_i] \succeq 0, i = 1,...,k; \\ (a.2) & y_i \ge 0, r^T y_i \le \mu_i, i = 1,...,k; \\ (b.1) & t_\ell \succeq 0, \ell = 1,...,N; \\ (b.2) & Pt \le p, \ell = 1,...,N \end{array} \tag{15.2.9}$$

in variables $t,\tau,\{y_i,\mu_i\}$. Indeed, $W = \{w \mid Rw \le r\}$, whence $W^* = \{(R^T y_i, \mu_i) \mid y_i \ge 0, \mu_i \ge r^T y_i\}$. Thus, for every $i \le k$ the corresponding constraints in $(a.1-2)$ say exactly that there exist $(\nu_i,\mu_i) \in W^*$ such that $\mathcal{C}[f_i,\mathcal{Q}(t),\tau,\nu_i,\mu_i] \succeq 0$, and consequently the constraints $(a.1-2)$ represent the predicate $I\!R$. The constraints $(b.1-2)$ clearly represent T.

Note that in the standard case (15.2.9) becomes the problem

$$\begin{array}{c}\text{minimize } \tau \\ \text{s.t.} \\ \begin{pmatrix} \tau - \mu_i & f_i^T + \frac{1}{2} y_i^T R \\ f_i + \frac{1}{2} R^T y_i & \sum_{\ell=1}^{N} \sum_{j=1}^{S} b_{\ell j} t_\ell b_{\ell j}^T \end{pmatrix} \succeq 0, i = 1, ..., k; \\ y_i \geq 0, r^T y_i \leq \mu_i, i = 1, ..., k; \\ t_\ell \succeq 0, \ell = 1, ..., N; \\ \underline{\rho}_\ell \leq \mathrm{Trace}(t_\ell) \leq \overline{\rho}_\ell, \ell = 1, ..., N; \\ \sum_{\ell=1}^{N} \mathrm{Trace}(t_\ell) \leq v. \end{array} \tag{15.2.10}$$

Example B. Robust structural design without obstacles and with "simple" T.. Here

$$\begin{array}{l} \mathcal{F} = \{f = Qu \mid u \in \mathbb{R}^k, u^T u \leq 1\}; \quad W = \mathbb{R}^M; \\ T = \{(t_1, ..., t_N) \mid t_\ell \in \mathcal{P}^D, \ell = 1, ..., N, Pt \leq p\}. \end{array} \tag{15.2.11}$$

In this case $W^* = \{(0, \mu) \mid \mu \geq 0\}$, and we have the following chain of equivalences:

$$\begin{array}{c} \forall (f \in \mathcal{F}) \exists ((\nu, \mu) \in W^*) : \begin{pmatrix} \tau - \mu & f^T + \frac{1}{2}\nu^T \\ f + \frac{1}{2}\nu & \mathcal{Q}(t) \end{pmatrix} \geq 0 \\ \Updownarrow \\ \forall (f \in \mathcal{F}) : \begin{pmatrix} \tau & f^T \\ f & \mathcal{Q}(t) \end{pmatrix} \geq 0 \\ \Updownarrow \\ \forall (u : u^T u \leq 1) : \begin{pmatrix} \tau & u^T Q^T \\ Qu & \mathcal{Q}(t) \end{pmatrix} \geq 0 \\ \Updownarrow \\ \forall (u : u^T u \leq 1, w, p) : \tau p^2 + 2w^T Q(pu) + w^T \mathcal{Q}(t) w \geq 0 \\ \Updownarrow \\ \forall (u : u^T u = 1, w, p) : \tau p^2 + 2w^T Q(pu) + w^T \mathcal{Q}(t) w \geq 0 \\ \Updownarrow \\ \forall (u' = pu, u^T u = 1, w) : \tau (u')^T u' + 2w^T Q u' + w^T \mathcal{Q}(t) w \geq 0 \\ \Updownarrow \\ \begin{pmatrix} \tau I_k & Q^T \\ Q & \mathcal{Q}(t) \end{pmatrix} \geq 0, \\ \Updownarrow \\ \exists (\mu \geq 0) : \begin{pmatrix} (\tau - \mu) I_k & Q^T \\ Q & \mathcal{Q}(t) \end{pmatrix} \geq 0, \end{array}$$

I_q being the $q \times q$ unit matrix.

Thus, here a semidefinite representation of the predicate $\mathbb{R}$ does exist, e.g., the one given by the system

$$\begin{pmatrix} (\tau-\mu)I_k & Q^T \\ Q & \mathcal{Q}(t) \end{pmatrix} \geq 0, \mu \geq 0,$$

and (P_{ini}) can be posed as the semidefinite problem (cf. (15.2.8))

$$\begin{array}{c} \text{minimize } \tau \\ \text{s.t.} \\ \begin{pmatrix} (\tau-\mu)I_k & Q^T \\ Q & \mathcal{Q}(t) \end{pmatrix} \succeq 0; \\ \mu \geq 0; \\ t_\ell \succeq 0, \ell = 1, ..., N; \\ Pt \leq p, \ell = 1, ..., N \end{array} \tag{15.2.12}$$

in variables t, τ, μ.

Note that in the standard case (15.2.12) becomes the problem

$$\begin{array}{c} \text{minimize } \tau \\ \text{s.t.} \\ \begin{pmatrix} (\tau-\mu)I_k & Q^T \\ Q & \sum\limits_{\ell=1}^{N}\sum\limits_{j=1}^{S} b_{\ell j} t_\ell b_{\ell j}^T \end{pmatrix} \succeq 0; \\ \mu \geq 0; \\ t_\ell \succeq 0, \ell = 1, ..., N; \\ \underline{\rho}_\ell \leq \text{Trace}(t_\ell) \leq \overline{\rho}_\ell, \ell = 1, ..., N; \\ \sum\limits_{\ell=1}^{N} \text{Trace}(t_\ell) \leq v. \end{array} \tag{15.2.13}$$

Intermediate summary and further goals. We have demonstrated that in two important cases (Examples A, B above) the structural design problem (P_{ini}) can be posed as a semidefinite program. Note that in the situations of actual interest these programs are extremely large-scale. Indeed, consider the standard cases (i), (ii) of Truss and Shape design. Assume, as it is normally the case, that the number of obstacles $\dim(r)$, the parameter k (number of loading scenarios in the multi-load design or the dimension of the ellipsoid of loads in the robust design) and S are small integers. Under this assumption, the sizes of the respective programs (15.2.10), (15.2.13) are as follows:

- the design dimension is $O(N)$;
- among the matrix inequalities in the program there is a small number of "large" ones – those involving matrices of order $O(M)$, and $O(N)$ "small" ones, with matrices of order of $O(1)$.

Now, in problems of actual interest M is at least thousands. As about N, it is larger than M by a moderate constant factor (shape design) or in order (truss design, where $N = O(M^2)$).

To get an impression of the resulting sizes, note that already a "toy" single-load design of a planar truss ("a cantilever arm") on 15×15 nodal grid results in semidefinite program (15.2.10) with $N = 25,096$ design variables. We see that the semidefinite programs arising from structural design are extremely demanding computationally; in their original forms (15.2.10), (15.2.13), they are too big for the existing SDP software. It turns out, however, that (15.2.10), (15.2.13) admit *analytical* processing which results in computationally less demanding equivalent reformulations of the original problems. Specifically, one can eliminate analytically "most" of the variables in the problems *dual* to (15.2.10), (15.2.13), thus reducing (sometimes – dramatically) the computational price of finding the optimal solution. E.g., in the aforementioned "cantilever arm" example, the dual problem has only 420 design variables. Our goal in the rest of the chapter is to carry out the aforementioned analytical processing. When achieving this goal, it makes sense to deal with a *single* semidefinite program which, on one hand, is enough general to cover problems (15.2.10), (15.2.13) we actually are interested in, and, on the other hand, is enough specialized to allow for the processing we are interested in. This "unified setting" is as follows:

$$
\begin{array}{llcl}
(P): & & \text{minimize } \tau & \\
\text{s.t.} & & & \\
\text{(i)} & \begin{pmatrix} \tau I_{p_i} + D_i z + d_i & [E_i z + e_i]^T \\ [E_i z + e_i] & \sum\limits_{\ell=1}^{N}\sum\limits_{j=1}^{S} b_{\ell j} t_\ell b_{\ell j}^T \end{pmatrix} & \succeq & 0, i = 1, ..., k; \\
\text{(ii)} & t_\ell & \succeq & 0, l = 1, ..., N; \\
\text{(iii)} & \underline{\rho}_\ell \leq \mathrm{Trace}(t_\ell) & \leq & \overline{\rho}_\ell, l = 1, ..., N; \\
\text{(iv)} & \sum\limits_{\ell=1}^{N} \mathrm{Trace}(t_\ell) & \leq & v; \\
\text{(v)} & z & \geq & 0,
\end{array}
$$

the design variables in the problem being $\tau \in \mathbb{R}$, $\{t_\ell \in \mathcal{S}^D\}_{\ell=1}^N$, $z \in \mathbb{R}^q$.

It is immediately seen that with an appropriate setup (P) indeed covers both (15.2.10) and (15.2.13). To transform (15.2.10) to (P), it suffices to specify the data of (P) as

$$
\begin{array}{c}
z = \{z_{2i-1} = y_i, z_{2i} = \mu_i - r^T y_i\}_{i=1}^k; \\
p_i = 1, D_i z + d_i = -z_{2i} - r^T z_{2i} [= -\mu_i], \\
E_i z + e_i = f_i + \frac{1}{2} R^T z_{2i-1} [= f_i + \frac{1}{2} y_i], i = 1, ..., k.
\end{array}
\tag{15.2.14}
$$

To get (15.2.13), it suffices to specify the data of (P) as

$$
z = \mu, D_1 z + d_1 = -\mu, E_1 z + c_1 \equiv Q \tag{15.2.15}
$$

and to set $k = 1$ in (P).

From now on we focus on problem (P); the results related to the "problems of actual interest" (i) – (ii) will be obtained by specifying the data of (P) accordingly.

When processing (P), we always make the following assumptions:

A. The objective function of (P) is bounded below on the feasible set of the problem.

B. (P) is strictly feasible, i.e., there exists a feasible solution at which all the inequality constraints of the problem are satisfied strictly.

C. For every $i = 1, ..., k$ there exist $w_i \in \mathbb{M}^{M \times p_i}$ and a positive definite $p_i \times p_i$ matrix α_i such that the q-dimensional vector

$$\sum_{i=1}^{k}[D_i^* \alpha_i + 2E_i^* w_i]$$

is negative.

Note that in the cases when (P) is obtained from problems (i), (ii) and the data of these latter problems satisfy **S.1** – **S.5**, the assumptions **A** – **C** are satisfied. Indeed, **A** immediately follows from the fact that in the cases in question the τ-component of a feasible solution to (P) is an upper bound on the compliance of certain construction (given by a design vector $t \in T$) w.r.t. a fixed load f; since T is compact by **S.4**, the compliances, taken w.r.t. f, W, of all constructions coming from T are uniformly bounded from below. To verify **B**, note that by **S.4** there exists $t^* = \{t_\ell^*\}_{\ell=1}^N$ satisfying strictly the constraints (P.ii-iv). By **S.2** and **S.5**, the matrix $\mathcal{Q}(t^*)$ is positive definite. Now let us choose somehow a positive vector z^*; since $\mathcal{Q}(t^*)$ is positive definite, the collection (τ^*, t^*, z^*) is, for all large enough values of τ^*, a strictly feasible solution to (P). It remains to verify **C**. This assumption is clearly true in the case of (ii), since here by (15.2.15) one has $E_1 = 0$ and $D_1 z = -zI_{p_1}$, whence $D_1^*\alpha = -\operatorname{Tr}(\alpha)$. Thus, **C** is satisfied with, say, $\alpha = I_{p_1}$. Now consider the case when (P) comes from problem (i). Taking into account (15.2.14), it is immediately seen that in order to verify **C** it suffices to set $\alpha_i = 1$ and $w_i = w$, $i = 1, ..., k$, where w satisfies $Rw < r$; such a w exists due to **S.5**.

In what follows we refer to (P) as to the *primal* form of the structural design problem.

15.3 FROM PRIMAL TO DUAL

Here we derive the Fenchel dual to (P). Recall that the Fenchel dual of a generic semidefinite program

$$\text{minimize } c^T x \text{ s.t. } A_0 + \sum_j x_j A_j \succeq 0$$

is the semidefinite program

$$\text{minimize Trace}(A_0 X) \text{ s.t. Trace}(A_j X) = c_j, j = 1, ..., \dim(x), X \succeq 0$$

(see Chapter 3). Applying this construction to (P), we get the following dual problem:

$$
\begin{array}{lrcl}
(D_{\rm ini}): & \text{minimize } \phi & \equiv & \sum_{i=1}^{k} \mathrm{Trace}(d_i\alpha_i + 2e_i^T w_i) \\
& & & + \sum_{\ell=1}^{N} [\overline{\rho}_\ell \sigma_\ell^+ - \underline{\rho}_\ell \sigma_\ell^-] + v\gamma \\
\text{s.t.} & & & \\
& \begin{pmatrix} \alpha_i & w_i^T \\ w_i & \mu_i \end{pmatrix} & \succeq & 0, i = 1, \ldots, k; \\
& \zeta_\ell & \succeq & 0, \ell = 1, \ldots, N; \\
& \sigma_\ell^+, \sigma_\ell^- & \geq & 0, \ell = 1, \ldots, N; \\
& \gamma & \geq & 0; \\
& \eta & \succeq & 0; \\
& \sum_{i=1}^{k} \mathrm{Trace}(\alpha_i) & = & 1; \\
& \sum_{i=1}^{k} [D_i^* \alpha_i + 2E_i^* w_i] + \eta & = & 0; \\
& \sum_{i=1}^{k} \sum_{j=1}^{S} b_{\ell j}^T \mu_i b_{\ell j} + \tau_\ell & & \\
& +[\sigma_\ell^- - \sigma_\ell^+ - \gamma] I_D & = & 0, l = 1, \ldots, N,
\end{array}
$$

the design variables being $\alpha_i \in \mathcal{S}^{p_i}, \mu_i \in \mathcal{S}^M, w_i \in \mathbb{R}^{p_i \times M}, \tau_\ell \in \mathcal{S}^D, \eta \in \mathbb{R}^q$ and reals $\sigma_\ell^\pm, \gamma$.

Since (P) is below bounded and strictly feasible by **A**, **B**, the standard results on Fenchel duality imply the following

Proposition 15.3.1 *Problem $(D_{\rm ini})$ is solvable, and its optimal value D^* and the optimal value P^* of (P) satisfy the relation*

$$P^* + D^* = 0. \qquad (15.3.16)$$

Moreover, every level set of the problem $(D_{\rm ini})$ – the intersection of the feasible set of the problem and the set where its objective is $\leq a$ for a given $a < \infty$ – is bounded.

Our local goal is to convert $(D_{\rm ini})$ into an equivalent simpler form.

Step 1. Eliminating η and $\{\tau_\ell\}_{\ell=1}^N$. $(D_{\rm ini})$ clearly is equivalent to the problem

$$
\begin{array}{lrcl}
(D_1): & \text{minimize } \phi & \equiv & \sum_{i=1}^k \mathrm{Trace}(d_i\alpha_i + 2e_i^T w_i) \\
 & & & +\sum_{\ell=1}^N [\overline{\rho}_\ell \sigma_\ell^+ - \underline{\rho}_\ell \sigma_\ell^-] + v\gamma \\
\text{s.t.} & & & \\
(a) & \begin{pmatrix} \alpha_i & w_i^T \\ w_i & \mu_i \end{pmatrix} & \succeq & 0, i = 1, ..., k; \\
(b) & \sigma_\ell^+, \sigma_\ell^- & \geq & 0, \ell = 1, ..., N; \\
(c) & \gamma & \geq & 0; \\
(d) & \sum_{i=1}^k \mathrm{Trace}(\alpha_i) & = & 1; \\
(e) & \sum_{i=1}^k [D_i^* \alpha_i + 2E_i^* w_i] & \leq & 0; \\
(f) & \sum_{i=1}^k \sum_{j=1}^S b_{\ell j}^T \mu_i b_{\ell j} & \preceq & [\gamma + \sigma_\ell^+ - \sigma_\ell^-] I_D, \ell = 1, ..., N.
\end{array}
$$

Step 2. Eliminating $\{\mu_i\}_{i=1}^k$. We start with the observation that (D_1) is strictly feasible. Indeed, let us set, say, $\sigma_\ell^\pm = 1$. By **C** there exist positive definite matrices α_i and rectangular matrices w_i of appropriate sizes satisfying $(D_1.e)$ strictly; by normalization, we may enforce these matrices to satisfy $(D_1.d)$. Given the indicated α_i, w_i and choosing "large enough" μ_i, we can enforce the validity of $(D_1.a)$ as strict matrix inequalities. Finally, choosing large enough $\gamma > 0$, we can satisfy $(D_1.c)$ and $(D_1.f)$ as strict inequalities.

Since (D_1) is strictly feasible, its optimal value D^* is the same as in problem (D_2) obtained from (D_1) by adding to the set of constraints the restrictions

$$(D_1.g) \quad \alpha_i \succ 0, i = 1, ..., k.$$

Now note that if a collection

$$(\alpha = \{\alpha_i\}_{i=1}^k, w = \{w_i\}_{i=1}^k, \mu = \{\mu_i\}_{i=1}^k, \sigma = \{\sigma_\ell^\pm\}_{\ell=1}^N, \gamma)$$

is a feasible solution to (D_2), then the collection

$$(\alpha, w, \mu(\alpha, w) = \{\mu_i(\alpha, w) = w_i \alpha_i^{-1} w_i^T\}_{i=1}^k, \sigma, \gamma)$$

is a feasible solution to (D_2) with the same value of the objective; indeed, from the matrix inequalities $(D_1.a)$ it follows that $\mu_i(\alpha, w) \preceq \mu_i$, so that, replacing μ_i with $\mu_i(\alpha, w)$, we preserve validity of the constraints $(D_1.f)$, $(D_1.a)$.

Consequently, (D_2) is equivalent to the problem

$$\begin{array}{llcl}
(D_3): & \text{minimize } \phi & \equiv & \sum\limits_{i=1}^{k}\mathrm{Trace}(d_i\alpha_i + 2e_i^T w_i) \\
 & & & +\sum\limits_{\ell=1}^{N}[\overline{\rho}_\ell\sigma_\ell^+ - \underline{\rho}_\ell\sigma_\ell^-] + v\gamma \\
\text{s.t.} & & & \\
(b) & \sigma_\ell^+, \sigma_\ell^- & \geq & 0, l = 1, ..., N; \\
(c) & \gamma & \geq & 0; \\
(d) & \sum\limits_{i=1}^{k}\mathrm{Trace}(\alpha_i) & = & 1; \\
(e) & \sum\limits_{i=1}^{k}[D_i^*\alpha_i + 2E_i^* w_i] & \leq & 0; \\
(f') & \sum\limits_{i=1}^{k}\sum\limits_{j=1}^{S} b_{\ell j}^T w_i\alpha_i^{-1}w_i^T b_{\ell j} & \preceq & [\gamma + \sigma_\ell^+ - \sigma_\ell^-]I_D, \ell = 1, ..., N, \\
(g) & \alpha_i & \succ & 0, i = 1, ..., k.
\end{array}$$

Now note that $(D_3.g)$ and $(D_3.f')$ together clearly are equivalent to the system of semidefinite constraints

$$\begin{array}{rcl}
\begin{pmatrix} A(\alpha) & B_\ell^T(w) \\ B_\ell(w) & (\gamma + \sigma_\ell^+ - \sigma_\ell^-)I_D \end{pmatrix} & \succeq & 0, \ell = 1, ..., N, \\
A(\alpha) & \succ & 0,
\end{array} \tag{15.3.17}$$

where

$$\alpha = \{\alpha_i\}_{i=1}^k, w = \{w_i\}_{i=1}^k,$$

$$A(\alpha) = \mathrm{Diag}(\overbrace{\alpha_1, ..., \alpha_1}^{S \text{ times}}, \overbrace{\alpha_2, ..., \alpha_2}^{S \text{ times}}, ..., \overbrace{\alpha_k, ..., \alpha_k}^{S \text{ times}}),$$
$$B_\ell(w) = [b_{\ell 1}^T w_1, b_{\ell 2}^T w_1, ..., b_{\ell S}^T w_1; b_{\ell 1}^T w_2, b_{\ell 2}^T w_2, ..., b_{\ell S}^T w_2; ...; b_{\ell 1}^T w_k, b_{\ell 2}^T w_k, ..., b_{\ell S}^T w_k]; \tag{15.3.18}$$

indeed, the left-hand side of $(D_3.f')$ is the Schur complement of the South-Eastern block in the left-hand side matrix in (15.3.17). Consequently, (D_3) is equivalent to the problem

$$
\begin{array}{llcl}
(D_4): & \text{minimize}\,\phi & \equiv & \sum\limits_{i=1}^{k} \mathrm{Trace}(d_i\alpha_i + 2e_i^T w_i) \\
& & & +\sum\limits_{\ell=1}^{N} [\overline{\rho}_\ell \sigma_\ell^+ - \underline{\rho}_\ell \sigma_\ell^-] + v\gamma \\
\text{s.t.} & & & \\
(a) & \begin{pmatrix} A(\alpha) & B_\ell^T(w) \\ B_\ell(w) & (\gamma + \sigma_\ell^+ - \sigma_\ell^-) I_D \end{pmatrix} & \succeq & 0, \ell = 1, ..., N, \\
(b) & \sigma_\ell^+, \sigma_\ell^- & \geq & 0, l = 1, ..., N; \\
(c) & \gamma & \geq & 0; \\
(d) & \sum\limits_{i=1}^{k} \mathrm{Trace}(\alpha_i) & = & 1; \\
(e) & \alpha_i & \succ & 0, i = 1, ..., k; \\
(f) & \sum\limits_{i=1}^{k} [D_i^* \alpha_i + 2E_i^* w_i] & \leq & 0.
\end{array}
$$

Problem (D_4) is strictly feasible along with (D_1), so that its optimal value remains unchanged when we replace in (D_4) the strict inequalities $\alpha_i \succ 0$ with the nonstrict ones $\alpha_i \succeq 0$. The latter inequalities are in fact redundant, since they are consequences of the inequality $A(\alpha) \succeq 0$, which in turn is a consequence of $(D_4.a)$. We have arrived at the following final form of the problem dual to (P):

$$
\begin{array}{llcl}
(D): & \text{minimize}\ \phi & \equiv & \sum\limits_{i=1}^{k} \mathrm{Trace}(d_i\alpha_i + 2e_i^T w_i) \\
& & & +\sum\limits_{\ell=1}^{N} [\overline{\rho}_\ell \sigma_\ell^+ - \underline{\rho}_\ell \sigma_\ell^-] + v\gamma \\
\text{s.t.} & & & \\
& \begin{pmatrix} A(\alpha) & B_\ell^T(w) \\ B_\ell(w) & (\gamma + \sigma_\ell^+ - \sigma_\ell^-) I_{D_\ell} \end{pmatrix} & \succeq & 0, l = 1, ..., N, \\
& \sigma_\ell^+, \sigma_\ell^- & \geq & 0, l = 1, ..., N; \\
& \gamma & \geq & 0; \\
& \sum\limits_{i=1}^{k} \mathrm{Trace}(\alpha_i) & = & 1; \\
& \sum\limits_{i=1}^{k} [D_i^* \alpha_i + 2E_i^* w_i] & \leq & 0,
\end{array}
$$

the design variables of the problem being

$$
\alpha = \{\alpha_i \in \mathcal{S}^{p_i}\}_{i=1}^k, w = \{w_i \in \mathbb{M}^{M \times p_i}\}_{i=1}^k, \sigma = \{\sigma_\ell^\pm \in \mathbb{R}\}_{\ell=1}^N, \gamma \in \mathbb{R}.
$$

Due to its connection with (D_{ini}) and by Proposition 15.3.1, (D) is solvable, and the level sets of the problem are bounded; the optimal value of (D) is minus the one of (P). Besides this, we have seen that (D) is strictly feasible along with (D_1).

15.4 FROM DUAL TO PRIMAL

Problem (D) is not the Fenchel dual to (P), it is obtained from this dual by eliminating part of the variables. What happens when we pass from (D) to *its* Fenchel dual? It turns out that we end up with a nontrivial (and instructive) equivalent reformulation of (P), namely, with the following problem

(P^+): minimize τ

s.t.

$$\text{(i)}\quad \left(\begin{array}{c|ccc|c|ccc}
\tau I_{p_i} + D_i z + d_i & [q^i_{11}]^T & \cdots & [q^i_{1S}]^T & \cdots & [q^i_{N1}]^T & \cdots & [q^i_{NS}]^T \\
\hline
q^i_{11} & t_1 & & & & & & \\
\cdots & & \ddots & & & & & \\
q^i_{1S} & & & t_1 & & & & \\
\hline
\cdots & & & & \ddots & & & \\
\hline
q^i_{N1} & & & & & t_N & & \\
\cdots & & & & & & \ddots & \\
q^i_{NS} & & & & & & & t_N
\end{array}\right) \succeq 0, \quad i = 1, ..., k;$$

(ii) $t_\ell \succeq 0, \ell = 1, ..., N;$

(iii) $\underline{\rho}_\ell \leq \mathrm{Trace}(t_\ell) \leq \overline{\rho}_\ell, \ell = 1, ..., N;$

(iv) $\sum\limits_{\ell=1}^{N} \mathrm{Trace}(t_\ell) \leq v;$

(v) $\sum\limits_{\ell=1}^{N} \sum\limits_{j=1}^{S} b_{\ell j} q^i_{\ell j} = e_i + E_i z, i = 1, ..., k;$

(vi) $z \geq 0,$

the design variables in the problem being symmetric $D \times D$ matrices t_ℓ, $\ell = 1, ..., N$, $D \times p_i$ matrices $q^i_{\ell j}$, $i = 1, ..., k, \ell = 1, ..., N, j = 1, ..., S$, real τ and $z \in \mathbb{R}^q$. Problem (P^+) is not the straightforward Fenchel dual of (D); it is obtained from this dual by eliminating part of the variables. Instead of deriving (P^+) via Fenchel duality, we prefer to give a direct proof of the equivalence between (P) and (P^+):

Theorem 15.4.1 *A collection $(\{t_\ell\}_{\ell=1}^N, z, \tau)$ is a feasible solution to (P) if and only if it can be extended, by properly chosen $\{q^i_{\ell j} | i = 1, ..., k, \ell = 1, ..., N, s = 1, ..., S\}$, to a feasible solution of (P^+).*

Proof. "if" part: let a collection

$$\Theta = (\{t_\ell\}_{\ell=1}^N, z, \tau, \{q^i_{\ell j} | i = 1, ..., k, \ell = 1, ..., N, s = 1, ..., S\})$$

be a feasible solution to (P^+); all we have to prove is the validity of the constraint $(P.\text{(i)})$. Let us fix $i \leq k$; we should prove that for every pair (x, y) of vectors of appropriate dimensions we have

$$x^T[\tau I_{p_i} + D_i z + d_i]x + 2x^T[e_i + E_i z]^T y + y^T[\sum_{\ell=1}^{N} \sum_{j=1}^{S} b_{\ell j}^T t_\ell b_{\ell j}^T]y \geq 0. \quad (15.4.19)$$

Indeed, in view of $(P^+.\text{v})$, the left hand side of (15.4.19) is equal to

$$\begin{array}{rl} & x^T[\tau I_{p_i} + D_i z + d_i]x + 2x^T[\sum_{\ell=1}^{N}\sum_{j=1}^{S} b_{\ell j}q^i_{\ell j}]^T y + y^T[\sum_{\ell=1}^{N}\sum_{j=1}^{S} b_{\ell j}t_\ell b_{\ell j}^T]y \\ = & x^T[\tau I_{p_i} + D_i z + d_i]x + 2\sum_{\ell=1}^{N}\sum_{j=1}^{S} x^T[q^i_{\ell j}]^T y_{\ell j} + \sum_{\ell=1}^{N}\sum_{j=1}^{s} y_{\ell j}^T t_\ell y_{\ell j}, \\ & y_{\ell j} = b_{\ell j}^T y. \end{array}$$

The resulting expression is nothing but the value of the quadratic form of the variables $(x, \{y_{\ell j}\})$ with the matrix from the left hand side of the constraint $(P^+.\text{i})$; since Θ is feasible for (P^+), this value is nonnegative, as claimed.

"only if" part: let

$$\Theta = (\{t_\ell\}_{\ell=1}^N, z, \tau)$$

be a feasible solution to (P). Let us fix i, $1 \le i \le k$, and let us set $f_i = E_i z + e_i$. By the constraint $(P.\text{i})$, for every $x \in \mathbb{R}^{p_i}$ the quadratic form

$$x^T[\tau I_{p_i} + D_i z + d_i]x + 2x^T f_i^T y + y^T \mathcal{Q}(t)y$$

of $y \in \mathbb{R}^M$ is nonnegative, hence the equation

$$\mathcal{Q}(t)y = f_i x$$

is solvable for every x; of course, we can choose its solution to be linear in x:

$$y = R_i x,$$

in which case

$$\mathcal{Q}(t)R_i x = f_i x \quad \forall x \Leftrightarrow \mathcal{Q}(t)R_i = f_i.$$

Let us now set

$$[q^i_{\ell j}]^T = R_i^T b_{\ell j} t_\ell; \tag{15.4.20}$$

then

$$\sum_{\ell=1}^{N}\sum_{j=1}^{S} b_{\ell j}q^i_{\ell j} = \sum_{\ell=1}^{N}\sum_{j=1}^{S} b_{\ell j}t_\ell b_{\ell j}^T R_i = \mathcal{Q}(t)R_i = f_i \qquad [= E_i z + e_i].$$

Thus, extending $(\{t_\ell\}, z, \tau)$ by the variables $\{q^i_{\ell j}\}$ given by (15.4.20), we ensure the validity of constraints $(P^+.\text{v})$. It remains to verify that the indicated extension ensures the validity of constraint $(P^+.\text{i})$. The latter requires verifying that for every collection $\{y_{\ell j}\}$ of vectors of appropriate dimension, and for every $x \in \mathbb{R}^{p_i}$ we have

$$F(x, \{y_{\ell j}\}) \equiv x^T[\tau I_{p_i} + D_i z + d_i]x + 2x^T\sum_{\ell=1}^{N}\sum_{j=1}^{S}[q^i_{\ell j}]^T y_{\ell j} + \sum_{\ell=1}^{N}\sum_{j=1}^{S} y_{\ell j}^T t_\ell y_{\ell j} \ge 0. \tag{15.4.21}$$

Given x, let us set

$$y^*_{\ell j} = -b^T_{\ell j} R_i x,$$

and let us prove that the collection $\{y^*_{\ell j}\}$ minimizes $F(x,\cdot)$. This is immediate, since $F(x,\cdot)$ is a convex quadratic form, and its partial derivative w.r.t. $y_{\ell j}$ at the point $\{y^*_{\ell j}\}$ is equal to

$$2q^i_{\ell j}x + 2t_\ell y^*_{\ell j} = 2[t_\ell b^T_{\ell j} R_i x - t_\ell b^T_{\ell j} R_i x] \equiv 0.$$

It remains to note that

$$\begin{array}{rl} & F(x,\{y^*_{\ell j}\}) \\ = & x^T[\tau I_{p_i} + D_i z + d_i]x - 2x^T \sum\limits_{\ell=1}^{N}\sum\limits_{j=1}^{S}[q^i_{\ell j}]^T b^T_{\ell j} R_i x + \sum\limits_{\ell=1}^{N}\sum\limits_{j=1}^{S} x^T R_i^T b_{\ell j} t_\ell b^T_{\ell j} R_i x \\ = & x^T[\tau I_{p_i} + D_i z + d_i]x - 2x^T[\sum\limits_{\ell=1}^{N}\sum\limits_{j=1}^{S} b_{\ell j} q^i_{\ell j}]^T R_i x + x^T R_i^T \mathcal{Q}(t) R_i x \\ & = x^T[\tau I_{p_i} + D_i z + d_i]x - 2x^T[E_i z + e_i]^T R_i x + x^T R_i^T \mathcal{Q}(t) R_i x \\ & \qquad\qquad [\text{due to already proved inequality } (P^+.\mathrm{v})] \\ = & \begin{pmatrix} x \\ -R_i x\end{pmatrix}^T \begin{pmatrix} \tau I_{p_i} + D_i z + d_i & [E_i z + e_i]^T \\ E_i z + e_i & \mathcal{Q}(t)\end{pmatrix} \begin{pmatrix} x \\ -R_i x\end{pmatrix} \\ \geq & 0 \\ & \qquad\qquad [\text{since } (\{t_\ell\}, z, \tau) \text{ is feasible for } (P)] \end{array}$$

Thus, the minimum of $F(x,\{y_{\ell j}\})$ in $\{y_{\ell j}\}$ is nonnegative, and therefore (15.4.21) is valid. ■

The relations between problems (P), (D), (P^+) are summarized in the following

Theorem 15.4.2 *Under assumptions* **A**, **B**, **C** *all three problems* (P), (D), (P^+) *are strictly feasible and solvable, and their level sets are bounded. The optimal values in* (P) *and* (P^+) *are equal to each other and are minus the optimal value in* (D).

15.5 EXPLICIT FORMS OF THE STANDARD TRUSS AND SHAPE PROBLEMS

Let us list the explicit forms of problems (P), (D), (P^+) for the standard cases of multi-load and robust truss/shape design (the case of single-load design is obtained from the multi-load one by specifying the number of loads as 1).

Multi-load truss design. Here $\mathcal{F} = \{f_1, ..., f_k\}$ and

$$\begin{array}{c} S = 1; \quad D = 1, l = 1, ..., N; \quad W = \{w \in \mathbb{R}^M \mid Rw \leq r\} \quad [\dim(r) = q]; \\ T = \{t \in \mathbb{R}^N \mid [0 \leq]\underline{\rho}_\ell \leq t_\ell \leq \overline{\rho}_\ell, \ell = 1, ..., N, \sum\limits_{\ell=1}^{N} t_\ell \leq v\}. \end{array} \tag{15.5.22}$$

The settings are

$$
\begin{array}{lc}
(P): & \text{minimize } \tau \\
\text{s.t.} & \\
& \begin{pmatrix} \tau + r^T y_i - \mu_i & f_i^T + \frac{1}{2} y_i^T R \\ f_i + \frac{1}{2} R^T y_i & \sum_{\ell=1}^{N} b_\ell b_\ell^T t_\ell \end{pmatrix} \succeq 0, i = 1, ..., k; \\
& \underline{\rho}_\ell \leq t_\ell \leq \overline{\rho}_\ell, \ell = 1, ..., N; \\
& \sum_{\ell=1}^{N} t_\ell \leq v; \\
& y_i \geq 0, i = 1, ..., k; \\
& \mu_i \geq r^T y_i, i = 1, ..., k. \\
& [\tau, t_\ell, \mu_i \in \mathbb{R}, y_i \in \mathbb{R}^q]
\end{array}
$$

$$
\begin{array}{lc}
(D): & \text{minimize } -2 \sum_{i=1}^{k} f_i^T w_i + \sum_{\ell=1}^{N} [\overline{\rho}_\ell \sigma_\ell^+ - \underline{\rho}_\ell \sigma_\ell^-] + v\gamma \\
\text{s.t.} & \\
& Z_\ell \equiv \left(\begin{array}{ccc|c} \alpha_1 & & & b_\ell^T w_1 \\ & \ddots & & \vdots \\ & & \alpha_k & b_\ell^T w_k \\ \hline b_\ell^T w_1 & \cdots & b_\ell^T w_k & \gamma + \sigma_\ell^+ - \sigma_\ell^- \end{array} \right) \succeq 0, \ell = 1, ..., N; \\
& \sigma_\ell^\pm \geq 0, \ell = 1, ..., N; \\
& \gamma \geq 0; \\
& R w_i \leq \alpha_i r_i, i = 1, ..., k; \\
& \sum_{i=1}^{k} \alpha_i = 1. \\
& [\alpha_i, \sigma_\ell^\pm, \gamma \in \mathbb{R}, w_i \in \mathbb{R}^M]
\end{array}
$$

$$
\begin{array}{lc}
(P^+): & \text{minimize } \tau \\
\text{s.t.} & \\
& \left(\begin{array}{c|ccc} \tau + r^T y_i - \mu_i & q_1^i & \cdots & q_N^i \\ \hline q_1^i & t_1 & & \\ \vdots & & \ddots & \\ q_N^i & & & t_N \end{array} \right) \succeq 0, i = 1, ..., k; \\
& \underline{\rho}_\ell \leq t_\ell \leq \overline{\rho}_\ell, \ell = 1, ..., N; \\
& \sum_{\ell=1}^{N} t_\ell \leq v; \\
& \sum_{\ell=1}^{N} q_\ell^i b_\ell = f_i + \frac{1}{2} R^T y_i, i = 1, ..., k; \\
& y_i \geq 0, i = 1, ..., k; \\
& \mu_i \geq r^T y_i, i = 1, ..., k. \\
& [\tau, t_\ell, q_\ell^i, \mu_i \in \mathbb{R}, y_i \in \mathbb{R}^q]
\end{array}
$$

Note that in the simplest case of obstacle-free single-load truss design with simple bounds on bar volumes, i.e., in the case of

$$k = 1; R = 0; \underline{\rho}_\ell = 0, \overline{\rho}_\ell = v, \ell = 1, ..., N,$$

problem (P^+) becomes the problem

$$\text{minimize} \sum_{\ell=1}^{N} \frac{q_\ell^2}{t_\ell} \text{ s.t. } \sum_\ell q_\ell b_\ell = f, t \geq 0, \sum_\ell t_\ell \leq v.$$

One can further eliminate analytically the variables t_ℓ, thus coming to the problem

$$\text{minimize} \left(\sum_\ell |q_\ell|\right)^2 \text{ s.t. } \sum_\ell q_\ell b_\ell = f,$$

which is equivalent to a Linear Programming program.

Multi-load shape design. Here $\mathcal{F} = \{f_1, ..., f_k\}$ and

$$\begin{array}{c} D_\ell = D, l = 1, ..., N [D = 3 \text{ in the planar and } D = 6 \text{ in the spatial case}]; \\ W = \{w \in \mathbb{R}^M \mid Rw \leq r\} \quad [\dim(r) = q]; \\ T = \{t \in (\mathcal{P}^D)^N \mid [0 \leq] \underline{\rho}_\ell \leq \mathrm{Trace}(t_\ell) \leq \overline{\rho}_\ell, 1 \leq l \leq N, \sum_{\ell=1}^{N} \mathrm{Trace}(t_\ell) \leq v\}. \end{array} \tag{15.5.23}$$

The settings are

$$\begin{array}{ll} (P): & \text{minimize } \tau \\ \text{s.t.} & \\ & \begin{pmatrix} \tau + r^T y_i - \mu_i & f_i^T + \frac{1}{2} y_i^T R \\ f_i + \frac{1}{2} R^T y_i & \sum_{\ell=1}^{N} \sum_{j=1}^{S} b_{\ell j} t_\ell b_{\ell j}^T \end{pmatrix} \succeq 0, i = 1, ..., k; \\ & t_\ell \succeq 0, \ell = 1, ..., N; \\ & \underline{\rho}_\ell \leq \mathrm{Trace}(t_\ell) \leq \overline{\rho}_\ell, \ell = 1, ..., N; \\ & \sum_{\ell=1}^{N} \mathrm{Trace}(t_\ell) \leq v; \\ & y_i \geq 0, i = 1, ..., k; \\ & \mu_i \geq r^T y_i, i = 1, ..., k. \\ & [\tau, \mu_i \in \mathbb{R}, t_\ell \in \mathcal{S}^D, y_i \in \mathbb{R}^q] \end{array}$$

$$(D): \quad \text{minimize } -2\sum_{i=1}^{k} f_i^T w_i + \sum_{\ell=1}^{N}[\overline{\rho}_\ell \sigma_\ell^+ - \underline{\rho}_\ell \sigma_\ell^-] + v\gamma$$

s.t.

$$Z_\ell \equiv \left(\begin{array}{ccc|c|ccc|c}
\alpha_1 & & & & & & & w_1^T b_{l1} \\
 & \ddots & & & & & & \dots \\
 & & \alpha_1 & & & & & w_1^T b_{lS} \\
\hline
 & & & \ddots & & & & \vdots \\
\hline
 & & & & \alpha_k & & & w_k^T b_{l1} \\
 & & & & & \ddots & & \dots \\
 & & & & & & \alpha_k & w_k^T b_{lS} \\
\hline
b_{l1}^T w_1 & \cdots & b_{lS}^T w_1 & \cdots & b_{l1}^T w_k & \cdots & b_{lS}^T w_k & (\gamma + \sigma_\ell^+ - \sigma_\ell^-) I_D
\end{array}\right) \succeq 0, \quad \ell = 1, ..., N;$$

$$\sigma_\ell^\pm \geq 0, \ell = 1, ..., N;$$
$$\gamma \geq 0;$$
$$Rw_i \leq \alpha_i r_i, i = 1, ..., k;$$
$$\sum_{i=1}^{k} \alpha_i = 1.$$
$$[\alpha_i, \sigma_\ell^\pm, \gamma \in \mathbb{R}, w_i \in \mathbb{R}^M]$$

$$(P^+): \quad \text{minimize } \tau$$

s.t.

$$\left(\begin{array}{c|ccc|c|ccc}
\tau + r^T y_i - \mu_i & [q_{11}^i]^T & \cdots & [q_{1S}^i]^T & \cdots & [q_{N1}^i]^T & \cdots & [q_{NS}^i]^T \\
q_{11}^i & t_1 & & & & & & \\
\vdots & & \ddots & & & & & \\
q_{1S}^i & & & t_1 & & & & \\
\hline
\vdots & & & & \ddots & & & \\
\hline
q_{N1}^i & & & & & t_N & & \\
\vdots & & & & & & \ddots & \\
q_{NS}^i & & & & & & & t_N
\end{array}\right) \succeq 0;$$

$$\underline{\rho}_\ell \leq \mathrm{Trace}(t_\ell) \leq \overline{\rho}_\ell, \ell = 1, ..., N;$$
$$\sum_{\ell=1}^{N} \mathrm{Trace}(t_\ell) \leq v;$$
$$\sum_{\ell=1}^{N}\sum_{j=1}^{S} b_{\ell j} q_{\ell j}^i = f_i + \tfrac{1}{2} R^T y_i, i = 1, ..., k;$$
$$y_i \geq 0, i = 1, ..., k;$$
$$\mu_i \geq r^T y_i, i = 1, ..., k.$$
$$[\tau, \mu_i \in \mathbb{R}, t_\ell \in \mathcal{S}^D, q_{\ell j}^i \in \mathbb{R}^D, y_i \in \mathbb{R}^q]$$

Robust truss design without obstacles. Here S, D_ℓ, W, T are given by (15.5.22) and

$$\mathcal{F} = \{f = Qu \mid u \in \mathbb{R}^k, u^T u \leq 1\}. \tag{15.5.24}$$

The settings are

$$
\begin{array}{ll}
(P): & \text{minimize } \tau \\
\text{s.t.} & \\
& \begin{pmatrix} \tau I_k & Q^T \\ Q & \sum_{\ell=1}^{N} t_\ell b_\ell b_\ell^T \end{pmatrix} \succeq; \\
& \underline{\rho}_\ell \le t_\ell \le \overline{\rho}_\ell, \ell = 1, ..., N; \\
& \sum_{\ell=1}^{N} t_\ell \le v. \\
& [\tau, t_\ell \in \mathbb{R}]
\end{array}
$$

$$
\begin{array}{ll}
(D): & \text{minimize } -2\text{Trace}(Q^T w) + \sum_{\ell=1}^{N} [\overline{\rho}_\ell \sigma_\ell^+ - \underline{\rho}_\ell \sigma_\ell^-] + v\gamma \\
\text{s.t.} & \\
& Z_\ell \equiv \begin{pmatrix} \alpha & w^T b_\ell \\ b_\ell^T w & \gamma + \sigma_\ell^+ - \sigma_\ell^- \end{pmatrix} \succeq 0, \ell = 1, ..., N; \\
& \sigma_\ell^\pm \ge 0, \ell = 1, ..., N; \\
& \gamma \ge 0; \\
& \text{Trace}(\alpha) = 1. \\
& [\alpha \in \mathcal{S}^k, \sigma_\ell^\pm, \gamma \in \mathbb{R}, w \in \mathbb{M}^{M \times k}]
\end{array}
$$

$$
\begin{array}{ll}
(P^+): & \text{minimize } \tau \\
\text{s.t.} & \\
& \left(\begin{array}{c|ccc} \tau I_k & q_1^T & \cdots & q_N^T \\ \hline q_1 & t_1 & & \\ \cdots & & \ddots & \\ q_N & & & t_N \end{array}\right) \succeq 0; \\
& \underline{\rho}_\ell \le t_\ell \le \overline{\rho}_\ell, \ell = 1, ..., N; \\
& \sum_{\ell=1}^{N} t_\ell \le v; \\
& \sum_{\ell=1}^{N} b_\ell q_\ell = Q; \\
& [\tau, t_\ell \in \mathbb{R}, q_\ell^T \in \mathbb{R}^k]
\end{array}
$$

Robust shape design without obstacles. Here D_ℓ, W, T are given by (15.5.23) and $\mathcal{F}$ is given by (15.5.24). The settings are

$$
\begin{array}{ll}
(P): & \text{minimize } \tau \\
\text{s.t.} & \\
& \begin{pmatrix} \tau I_k & Q^T \\ Q & \sum_{\ell=1}^{N} \sum_{j=1}^{S} b_{\ell j} t_\ell b_{\ell j}^T \end{pmatrix} \succeq 0; \\
& t_\ell \succeq 0, \ell = 1, ..., N; \\
& \underline{\rho}_\ell \le \text{Trace}(t_\ell) \le \overline{\rho}_\ell, \ell = 1, ..., N; \\
& \sum_{\ell=1}^{N} \text{Trace}(t_\ell) \le v. \\
& [\tau \in \mathbb{R}, t_\ell \in \mathcal{S}^D]
\end{array}
$$

$$(D): \quad \text{minimize} \; -2\text{Trace}(Q^T w) + \sum_{\ell=1}^{N}[\overline{\rho}_\ell \sigma_\ell^+ - \underline{\rho}_\ell \sigma_\ell^-] + v\gamma$$

s.t.

$$Z_\ell \equiv \left(\begin{array}{ccc|c} \alpha & & & w^T b_{\ell 1} \\ & \ddots & & \cdots \\ & & \alpha & w^T b_{\ell S} \\ \hline b_{\ell 1}^T w & \cdots & b_{\ell S}^T w & (\gamma + \sigma_\ell^+ - \sigma_\ell^-) I_D \end{array}\right) \succeq 0, \ell = 1, ..., N;$$

$$\sigma_\ell^\pm \geq 0, \ell = 1, ..., N;$$

$$\gamma \geq 0;$$

$$\text{Trace}(\alpha) = 1.$$

$$[\alpha \in \mathcal{S}^k, \sigma_\ell^\pm, \gamma \in \mathbb{R}, w \in \mathbb{M}^{M\times k}]$$

$$(P^+): \quad \text{minimize} \; \tau$$

s.t.

$$\left(\begin{array}{c|ccc|c|ccc} \tau I_k & [q_{11}]^T & \cdots & [q_{1S}]^T & \cdots & [q_{N1}]^T & \cdots & [q_{NS}]^T \\ \hline q_{11} & t_1 & & & & & & \\ \cdots & & \ddots & & & & & \\ q_{1S} & & & t_1 & & & & \\ \hline \cdots & & & & \ddots & & & \\ \hline q_{N1} & & & & & t_N & & \\ \cdots & & & & & & \ddots & \\ q_{NS} & & & & & & & t_N \end{array}\right) \succeq 0;$$

$$\underline{\rho}_\ell \leq \text{Trace}(t_\ell) \leq \overline{\rho}_\ell, l = 1, ..., N;$$

$$\sum_{\ell=1}^{N} \text{Trace}(t_\ell) \leq v;$$

$$\sum_{\ell=1}^{N} \sum_{j=1}^{S} b_{\ell j} q_{\ell j} = Q;$$

$$[\tau \in \mathbb{R}, t_\ell \in \mathcal{S}^D, q_{\ell j} \in \mathbb{M}^{D\times k}]$$

Note that in the case of "simple bounds" $\underline{\rho}_\ell = 0, \overline{\rho}_\ell > v$ on the material densities the formulations in question can be slightly simplified. Namely, in problems (P) and (P^+) one should just skip the constraints $\underline{\rho}_\ell \leq \cdots \leq \overline{\rho}_\ell$; in (D), one should skip the variables $\sigma_\ell^\pm$ (formally speaking, should set them to 0).

15.6 CONCLUDING REMARKS

Which formulation to choose. We have presented several SDP settings of the (standard cases of the) truss and the shape problems. A natural question is which formulation is better suited for a straightforward numerical processing by interior point methods (which are the best known so far optimization techniques for semidefinite programming). The answer, of course, depends on the particular problem we are interested in, and it turns out that in many cases the dual problem (D) is much better suited for numerical processing than (P), (P'). Consider, e.g., multi-load truss design; for the sake of simplicity, let us restrict ourselves with the case of simple bounds on bar volumes ($\underline{\rho}_\ell = 0, \overline{\rho}_\ell = v$ for all ℓ). Assume also, as it normally is the case, that the number of obstacles q and the number of loading scenarios k are once for ever fixed small integers, like 3 or 5, the number M of degrees of freedom is large, and the number of tentative bars is $N \approx \frac{M^2}{2d^2}$, where $d = 2$ for planar and $d = 3$ for spatial trusses (the latter assumption means that the space of virtual displacements of the major part of the nodes is the entire "physical" space $\mathbb{R}^d$, so that the total number of nodes is approximately M/d, and all pair connections of these nodes by bars are allowed). In this situation the principal, as $M \to \infty$, terms in the sizes of (P), (D), (P^+) are as follows:

	(P)	(D)	(P^+)
# of design variables	$\frac{M^2}{2d^2}$	kM	$\frac{(k+1)M^2}{2d^2}$
# and sizes of semidefinite constraints	$k \times (M+1 \times M+1)$	$\frac{M^2}{2d^2} \times (k+1 \times k+1)$	$k \times \left(\frac{M^2}{2d^2} \times \frac{M^2}{2d^2}\right)$
# of linear constraints	$\frac{M^2}{2d^2}$	$kq+2$	kM

We see that for large M problem (D) is incomparably better suited for interior point methods than (P), (P^+). Indeed, it can be easily verified that computational effort per iteration of an interior point method as applied to every one of our three problems is dominated by the necessity to solve the corresponding Newton system, i.e., is proportional to the cube of the design dimension of the problem. And for k fixed, the design dimensions of (P), (P^+) are of order of *squared* design dimension of (D)! An additional computational advantage of (D) as compared to (P) is that instead of a small number k of *large* $(M+1 \times M+1)$ semidefinite constraints in (P), in (D) we have a large number $N \approx \frac{M^2}{2d^2}$ of *small* $(k+1 \times k+1)$ semidefinite constraints. It follows that typical auxiliary operations in interior point methods, like checking feasibility of the semidefinite constraints at a given point, are much cheaper for (D) ($O(N(k+1)^3) = O(M^2k^3)$ arithmetic operations) than for (P) ($O(kM^3)$ arithmetic operations).

In shape design $N = O(M)$, and the advantages of (D), if any, are not that dramatic. However, in many shape design problems (D) still is by far more preferable than (P) and (P^+).

Recovering primal solutions from the dual ones. We have seen that in some important cases the dual form of the structural design problem is by far better suited for direct numerical processing than other SDP formulations of the problem. There is, however, an evident drawback of the dual form: the actual design variables t_ℓ do not appear in (D) at all. Thus, we should resolve the question of how to recover a "good" approximate solution to the original problem (P) from a good approximate solution to (D). This question can be easily answered when the dual approximate solution in question is a *central* one. The latter notion is defined as follows. (D) is of the generic form

$$d^T\xi \to \min \mid \mathcal{G}\xi + g \succeq 0, \tag{SDP}$$

where the design vector ξ varies in certain Euclidean space X, and $\xi \mapsto \mathcal{G}\xi$ is a linear mapping from X to the space $\mathcal{S}$ of block-diagonal symmetric matrices of a given block-diagonal structure[1)]. We say that a strictly feasible solution $\xi(\rho)$ to the problem is the central (or the "central path") solution corresponding to a value $\rho > 0$ of the penalty parameter, if $\xi(\rho)$ minimizes the aggregate

$$\rho d^T\xi - \ln \mathrm{Det}(\mathcal{G}\xi + g)$$

over the set of all strictly feasible solutions to the problem. From the general theory of interior point polynomial time methods it is known that

1. If (D) is strictly feasible and has bounded level sets, then central solutions exist for every value $\rho > 0$ of the penalty parameter;
2. If $\xi(\rho)$ is a central solution, then the point

   $$\Xi(\rho) = \rho^{-1}[\mathcal{G}\xi(\rho) + g]^{-1}$$

   is a strictly feasible solution of the Fenchel dual of (SDP) – the problem

   $$\mathrm{Trace}(g\Xi) \to \min \mid \mathcal{G}^*\Xi = d, \Xi \succeq 0, \tag{SDP*}$$

   and this solution is $O(\rho^{-1})$-optimal, namely,

   $$\mathrm{Trace}(g\Xi) - \mathrm{Opt}(\mathrm{SDP}^*) \leq \frac{m}{\rho},$$

   where Opt(SDP*) is the optimal value of (SDP*), and m is the total row size of matrices from $\mathcal{S}$.

In our context, (D) indeed is strictly feasible with bounded level sets (see Proposition 15.3.1). Furthermore, the Fenchel dual to (D), after eliminating part of the variables, is exactly the problem (P^+). Combining 2) and Theorem 15.4.1, we conclude that a central solution $\xi(\rho)$ to (D) can be immediately converted to a $(m\rho^{-1})$-optimal solution to (P). Note also that when (D) is solved by path-following interior point methods, the iterates can be easily converted to central solutions to (D), and the corresponding values of the penalty parameter grow linearly with the rate $(1 + O(1)m^{-1/2})$.

In the situations considered in Section 15.5, the outlined scheme for recovering nearly optimal solutions of the primal problem from nearly optimal central solutions of the dual one becomes extremely simple: ρt_ℓ is just the South-Eastern $D \times D$ block in the matrix Z_ℓ^{-1}, the matrix being evaluated at a central solution $\xi(\rho)$ of (D) (see explicit forms of (D) in Section 15.5).

[1)] Note that in order to represent (D) in this form, one should include the linear equality constraints of the formulation as given above into the definition of X

16 MOMENT PROBLEMS AND SEMIDEFINITE OPTIMIZATION

Dimitris Bertsimas, Jay Sethuraman

Chance favors the prepared mind.

Quoted in H. Eves Return to Mathematical Circles, Prindle, Wever and Schmidt, Boston, 1988.

16.1 INTRODUCTION

Problems involving moments of random variables arise naturally in many areas of mathematics, economics, and operations research. Let us give some examples that motivate the present paper.

Moment problems in probability theory

The problem of deriving bounds on the probability that a certain random variable belongs in a set, given information on some of the moments of this random variable, has a rich history, which is very much connected with the development of probability theory in the twentieth century. The inequalities due to Markov, Chebyshev and Chernoff are some of the classical and widely used results of modern probability theory. Natural questions arise, however:

(a) *Are such bounds "best possible," i.e., do there exist distributions that match them? A concrete and simple question in the univariate case: Is the Chebyshev inequality "best possible"?*

(b) *Can such bounds be generalized in multivariate settings?*

(c) *Can we develop a general theory based on optimization methods to address moment problems in probability theory?*

Moment problems in finance

A central question in financial economics is to find the price of a derivative security given information on the underlying asset. This is exactly the area of the 1997 Nobel prize in economics to Robert Merton and Myron Scholes. Under the assumption that the price of the underlying asset follows a geometric Brownian motion and using the no-arbitrage assumption, the Black-Scholes formula provides an explicit and insightful answer to this question. Natural questions arise, however. Making no assumptions on the underlying price dynamics, but only using the no-arbitrage assumption:

(a) *What are the best possible bounds for the price of a derivative security based on the first and second moments of the price of the underlying asset?*

(b) *How can we derive optimal bounds on derivative securities that are based on multiple underlying assets, given the first two moments of the asset prices and their correlations?*

(c) *Conversely, given observable option prices, what are the best bounds that we can derive on the moments of the underlying asset?*

(d) *Finally, given observable option prices, what are the best bounds that we can derive on prices of other derivatives on the same asset?*

Moment problems in stochastic optimization

Scheduling a multiclass queueing network is a central problem in stochastic optimization. Queueing networks represent dynamic and stochastic generalizations of job shops, and have been used in the last thirty years to model communication, computer, and manufacturing systems. The central optimization problem is to find a scheduling policy that optimizes a performance cost function $\mathbf{c}'\mathbf{x} + \mathbf{d}'\mathbf{y}$, where $\mathbf{x} = (x_1, \ldots, x_N)$, x_j is the mean number of jobs of class j, $\mathbf{y} = (y_1, \ldots, y_N)$, and y_j is the second moment of the number of jobs of class j, and $\mathbf{c}$, $\mathbf{d}$ are N-vectors of nonnegative constants. The design of optimal policies is $EXPTIME$-hard (Papadimitiou and Tsitsiklis [608]), i.e., it provably requires exponential time, as $P \neq EXPTIME$. A natural question that arises:
Can we find strong lower bounds efficiently, exploiting the fact that the performance vectors represent moments of random variables?

Moment problems in discrete optimization

The development of semidefinite relaxations in recent years represents an important advance in discrete optimization. In several problems, semidefinite relaxations are provably closer to the discrete optimization solution value (see Goemans and Williamson [285] for the max-cut problem for example) than linear ones. The proof of closeness of the semidefinite relaxation to the discrete optimization solution value involves a randomized argument that exploits the geometry of the semidefinite relaxation. A key question arises:

Is there a general method of generating near optimal integer solutions starting from an optimal solution of the semidefinite relaxation?
We will see that the interpretation of the solution of the semidefinite relaxation as a covariance matrix for a collection of random variables leads to such a method and connects moment problems and discrete optimization.

Our goal in this paper is to demonstrate that convex and, in particular, semidefinite optimization methods give interesting and often unexpected answers to moment problems arising in probability, economics, and operations research. We also report new computational results in the area of stochastic optimization that show the effectiveness of semidefinite relaxations.

The key connection

The key connection between moment problems and semidefinite optimization is centered in the notion of a *feasible moment vector.* Let $\mathbf{k} = (k_1, \ldots, k_n)$ be a vector of nonnegative integers.

Definition 16.1.1 *A vector $\boldsymbol{\sigma} : (\sigma_{\mathbf{k}})_{k_1+\cdots+k_n \leq k}$ is a feasible (n, k, Ω)-moment vector, if there is a multivariate random variable $\boldsymbol{X} = (X_1, \ldots, X_n)$ with domain $\Omega \subseteq R^n$, whose moments are given by $\boldsymbol{\sigma}$, that is $\sigma_{\mathbf{k}} = E\big[X_1^{k_1} \cdots X_n^{k_n}\big]$, $\forall\, k_1 + \cdots + k_n \leq k$. We say that any such multivariate random variable $\boldsymbol{X}$ has a $\boldsymbol{\sigma}$-feasible distribution and denote this as $\boldsymbol{X} \sim \boldsymbol{\sigma}$.*

We denote by $\mathcal{M} = \mathcal{M}(n, k, \Omega)$ the set of feasible (n, k, Ω)-moment vectors.

The univariate case

For the univariate case ($n=1$), the problem of deciding if $\boldsymbol{\sigma} = (M_1, M_2, \ldots, M_k)$ is a feasible $(1, k, \Omega)$-moment vector is the classical moment problem, which has been completely characterized by necessary and sufficient conditions (see Karlin and Shapley [399], Akhiezer [10], Siu, Sengupta and Lind [703] and Kemperman [409]).

Theorem 16.1.1 (a) (Nonnegative random variables)
The vector $(M_1, \ldots, M_{2n+1})$ is a feasible $(1, 2n+1, R^+)$-moment vector if and only if the following matrices are positive semidefinite:

$$\mathbf{R}_{2n} = \begin{pmatrix} 1 & M_1 & \ldots & M_n \\ M_1 & M_2 & \ldots & M_{n+1} \\ \vdots & \vdots & \ddots & \vdots \\ M_n & M_{n+1} & \ldots & M_{2n} \end{pmatrix} \succeq \mathbf{0},$$

$$\mathbf{R}_{2n+1} = \begin{pmatrix} M_1 & M_2 & \ldots & M_{n+1} \\ M_2 & M_3 & \ldots & M_{n+2} \\ \vdots & \vdots & \ddots & \vdots \\ M_{n+1} & M_{n+2} & \ldots & M_{2n+1} \end{pmatrix} \succeq \mathbf{0}.$$

(b) (Arbitrary random variables) *The vector* $(M_1, M_2, \ldots, M_{2n})$ *is a feasible* $(1, 2n, R)$*-moment vector if and only if* $\mathbf{R}_{2n} \succeq \mathbf{0}$.

Proof:
We will only show the necessity of part (b). If $(M_1, M_2, \ldots, M_{2n})$ is a feasible $(1, 2n, R)$-moment vector, then there exists a probability measure $f(x)$ such that

$$\int_{-\infty}^{\infty} x^k f(x)dx = M_k, \qquad k = 0, 1, \ldots, 2n,$$

where $M_0 = 1$. Consider the vector $\mathbf{x} = (1, x, x^2, \ldots, x^{2n})'$, and the positive semidefinite matrix $\mathbf{x}\mathbf{x}'$. Since $f(x)$ is nonnegative, the matrix

$$\mathbf{R}_{2n} = \int_{-\infty}^{\infty} \mathbf{x}\mathbf{x}' f(x)dx$$

should be positive semidefinite. ■

The multivariate case

We consider the question of whether a vector $\boldsymbol{\sigma} = (\mathrm{M}, \boldsymbol{\Gamma})$ is a feasible $(n, 2, R^n)$-moment vector, i.e., whether there exists a random vector $\boldsymbol{X}$ such that $E[\boldsymbol{X}] = \mathrm{M}$, $E[\boldsymbol{X}\boldsymbol{X}'] = \boldsymbol{\Gamma}$.

Theorem 16.1.2 *A vector* $\boldsymbol{\sigma} = (\mathrm{M}, \boldsymbol{\Gamma})$ *is a feasible* $(n, 2, R^n)$*-moment vector if and only if the following matrix is positive semidefinite:*

$$\boldsymbol{\Sigma} = \begin{bmatrix} 1 & \mathrm{M}' \\ \mathrm{M} & \boldsymbol{\Gamma} \end{bmatrix} \succeq \mathbf{0}.$$

Proof:
Suppose $(\mathrm{M}, \boldsymbol{\Gamma})$ is a feasible $(n, 2, R^n)$-moment vector. Then, there exists a random variable $\boldsymbol{X}$ such that $E[\boldsymbol{X}] = \mathrm{M}$, $E[\boldsymbol{X}\boldsymbol{X}'] = \boldsymbol{\Gamma}$. The matrix $(\boldsymbol{X} - \mathrm{M})(\boldsymbol{X} - \mathrm{M})'$ is positive semidefinite. Taking expectations, we obtain that

$$E[(\boldsymbol{X} - \mathrm{M})(\boldsymbol{X} - \mathrm{M})'] = \boldsymbol{\Gamma} - \mathrm{M}\mathrm{M}' \succeq \mathbf{0},$$

which expresses the fact that a covariance matrix needs to be positive semidefinite. It is easy to see that $\boldsymbol{\Gamma} - \mathrm{M}\mathrm{M}' \succeq \mathbf{0}$ if and only if $\boldsymbol{\Sigma} \succeq \mathbf{0}$.

Conversely, if $\boldsymbol{\Sigma} \succeq \mathbf{0}$, then $\boldsymbol{\Gamma} - \mathrm{M}\mathrm{M}' \succeq \mathbf{0}$. Let $\boldsymbol{X}$ be a multivariate normal distribution with mean M and covariance matrix $\boldsymbol{\Gamma} - \mathrm{M}\mathrm{M}'$. This shows that the vector $\boldsymbol{\sigma} = (\mathrm{M}, \boldsymbol{\Gamma})$ is a feasible $(n, 2, R^n)$-moment vector. ■

There are known necessary conditions for a vector $\boldsymbol{\sigma}$ to be a feasible (n, k, R^n)-moment vector for $k \geq 3$, that also involve the semidefiniteness of a matrix derived from the vector $\boldsymbol{\sigma}$, but these conditions are not known to be sufficient. To the best of our knowledge the complexity of deciding whether a vector $\boldsymbol{\sigma}$ is a feasible (n, k, R^n)-moment vector has not been resolved.

Structure of the paper

The structure of the paper is as follows. In Section 16.2, we outline the application of semidefinite optimization to stochastic optimization problems. In Section 16.3, we derive explicit and often surprising optimal bounds in probability theory using convex and semidefinite optimization methods. In Section 16.4, we apply convex and semidefinite optimization methods to problems in finance. In Section 16.5, we illustrate a connection between moment problems and semidefinite relaxations in discrete optimization. Section 16.6 contains some concluding remarks.

16.2 SEMIDEFINITE RELAXATIONS FOR STOCHASTIC OPTIMIZATION PROBLEMS

The development of semidefinite relaxations represents an important advance in discrete optimization. In this section, we review a theory for deriving semidefinite relaxations for classical stochastic optimization problems. The idea of deriving semidefinite relaxations for this class of problems is due to Bertsimas [98]. Our development in this paper follows Bertsimas and Niño-Mora [101], [102]; the interested reader is referred to these papers for further details. We demonstrate the central ideas for the problem of optimizing a multiclass queueing network that represents a stochastic and dynamic generalization of a job shop.

16.2.1 Model description

We consider a network of queues composed of K single-server stations and populated by N job classes. The set of job classes $\mathcal{N} = \{1, \dots, N\}$ is partitioned into subsets $\mathcal{C}_1, \dots, \mathcal{C}_K$, so that station $m \in \mathcal{K} = \{1, \dots, K\}$ only serves classes in its *constituency* $\mathcal{C}_m$. We refer to jobs of class i as i-jobs, and we let $s(i)$ be the station that serves i-jobs. The network is *open*, so that jobs arrive from outside, follow a Markovian route through the network (i-jobs wait for service at the *i-queue*) and eventually exit. External arrivals of i-jobs follow a Poisson process with rate α_i (if class i does not have external arrivals $\alpha_i = 0$). The service times of i-jobs are independent and identically distributed, having an exponential distribution with mean $\beta_i = 1/\mu_i$. Upon completion of service at station $s(i)$, an i-job becomes a j-job (and hence is routed to the j-queue), with probability p_{ij}, or leaves the system, with probability $p_{i0} = 1 - \sum_{j \in \mathcal{N}} p_{ij}$. We assume that the routing matrix $\mathbf{P} = (p_{ij})_{i,j \in \mathcal{N}}$ is such that a single job moving through the network eventually exits, i.e., the matrix $\mathbf{I} - \mathbf{P}$ is invertible. We further assume that all service times and arrival processes are mutually independent.

The network is controlled by a *scheduling policy*, which specifies dynamically how each server is allocated to waiting jobs. Scheduling policies can be either *dynamic* or *static*. In a *dynamic* policy, scheduling decisions may depend on the current or past states of all queues; in a *static* policy, the scheduling decisions of each server are independent of the queue lengths of the job classes. A scheduling policy is *stable* if the queue-length vector process has an equilibrium distribution with finite mean. We allow policies to be *preemptive*, i.e., a job's service may be interrupted and resumed later. Finally, a scheduling policy is *nonidling* if a server cannot idle whenever there is a job waiting for service at that station.

Next, we define other model parameters of interest. The *effective arrival rate* of j-jobs, denoted by λ_j, is the total rate at which both external and internal jobs arrive to the j-queue. The λ_j's are computed by solving the system

$$\lambda_j = \alpha_j + \sum_{i \in \mathcal{N}} p_{ij} \lambda_i, \qquad \text{for } j \in \mathcal{N}.$$

The *traffic intensity* of j-jobs, denoted by $\rho_j = \lambda_j \beta_j$, is the time-stationary probability that a j-job is in service. The *total traffic intensity* at station m is $\rho(\mathcal{C}_m) = \sum_{j \in \mathcal{C}_m} \rho_j$, and is the time-stationary probability that server m is busy. We note that the condition

$$\rho(\mathcal{C}_m) < 1, \qquad \text{for } m \in \mathcal{K}$$

is necessary but not sufficient for guaranteeing the stability of any nonidling policy.

We assume that the system operates in a steady-state regime (under a stable policy), and introduce the following variables:

- $L_i(t)$ = number of i-jobs in system at time t.
- $B_i(t) = 1$ if an i-job is in service at time t; 0 otherwise.
- $B^m(t) = 1$ if server m is busy at time t; 0 otherwise; notice that $B^m(t) = \sum_{i \in \mathcal{C}_m} B_i(t)$.

In what follows we write, for convenience of notation, $L_i = L_i(0)$, $B_i = B_i(0)$ and $B^m = B^m(0)$.

16.2.2 The performance optimization problem

The *performance measures* we are interested in are $\mathbf{x} = (x_j)_{j \in \mathcal{N}}$, and $\mathbf{y} = (y_j)_{j \in \mathcal{N}}$, where

$$x_j = E\left[L_j\right], \qquad y_j = E\left[L_j^2\right], \qquad \text{for } j \in \mathcal{N},$$

i.e., the vectors whose components are the time-stationary mean and second moment of the number of jobs from each class in the system.

Given a *performance cost function* $\mathbf{c}'\mathbf{x} + \mathbf{d}'\mathbf{y}$, we investigate the following *performance optimization problem*: compute a lower bound $\underline{Z} \leq \mathbf{c}'\mathbf{x} + \mathbf{d}'\mathbf{y}$ that is valid under a given class of admissible policies, and design a policy which nearly minimizes the cost $\mathbf{c}'\mathbf{x} + \mathbf{d}'\mathbf{y}$. For our purposes, any preemptive, nonidling policy is admissible. In this paper, we restrict our attention to the question of computing strong lower bounds. As we mentioned in the Introduction, the design of optimal policies is $EXPTIME$-hard (Papadimitiou and Tsitsiklis [608]), i.e., it is provably requires exponential time, as $P \neq EXPTIME$. In recent years, progress has been made in designing near-optimal scheduling policies based on the idea of fluid control, in which discrete jobs are replaced by the flow of a fluid; we refer the interested reader to the papers by Avram, Bertsimas and Ricard [47], Weiss [825], Luo and Bertsimas [504], and the references cited therein for details.

We study the problem of computing good lower bounds via the *achievable region* approach. The achievable region (equivalently, performance region) $\mathcal{X}$ is defined as the set of all performance vectors $(\mathbf{x}, \mathbf{y})$ that can be achieved under admissible policies. Our goal is to derive constraints on the performance vector $(\mathbf{x}, \mathbf{y})$ that define a relaxation of performance region $\mathcal{X}$. Since it is not obvious how to derive such constraints directly, we pursue the following plan: (a) identify system *equilibrium relations* and formulate them as constraints involving *auxiliary performance variables*; (b) formulate additional constraints (both *linear* and *positive semidefinite*) on the auxiliary performance variables; (c) formulate constraints that express the original performance vector, $(\mathbf{x}, \mathbf{y})$, in terms of the auxiliary variables.

Notice that this approach is fairly standard in the mathematical programming literature and has a clear geometric interpretation: It corresponds to constructing a relaxation of the performance region of the natural variables, $(\mathbf{x}, \mathbf{y})$, by (a) *lifting* this region into a higher dimensional space, by means of auxiliary variables, (b) bounding the lifted region through constraints on the auxiliary variables, and (c) *projecting* back into the original space. *Lift and project* techniques have proven powerful tools for constructing tight relaxations for hard discrete optimization problems (see, e.g., Lovász and Schrijver [497]). We have summarized the performance measures considered in this paper (including auxiliary ones) in Table 16.1.

The rest of this section is organized as follows. In Section 16.2.3, we include linear constraints that relate the natural performance measures in terms of auxiliary performance variables. Using the fact that our performance measures are expectations of random variables, we describe a set of positive semidefinite constraints in Section 16.2.4. In Section 16.2.5, we introduce a linear and a semidefinite relaxation using the constraints of the previous sections. We further present computational results that illustrate that the semidefinite relaxation is substantially stronger than the linear programming relaxation.

16.2.3 Linear constraints

In this section, we present several sets of linear constraints that express natural performance measures in terms of auxiliary ones. The first set of constraints follow from elementary arguments.

Theorem 16.2.1 (Elementary constraints) *Under any stable policy, the following equations hold:*

Performance variables	Interpretation
x_j; $\mathbf{x} = (x_j)_{j\in\mathcal{N}}$	$E[L_j]$
x_j^i; $X = (x_j^i)_{i,j\in\mathcal{N}}$; $\mathbf{x}^i = (x_j^i)_{j\in\mathcal{N}}$	$E[L_j \mid B_i = 1]$
x_j^{0m}; $X^0 = (x_j^{0m})_{m\in\mathcal{K},j\in\mathcal{N}}$; $\mathbf{x}^{0m} = (x_j^{0m})_{j\in\mathcal{N}}$	$E[L_j \mid B^m = 0]$
r_{ij}; $\mathbf{R} = (r_{ij})_{i,j\in\mathcal{N}}$	$E[B_iB_j]$
r_{ij}^k; $\mathbf{R}^k = (r_{ij}^k)_{i,j\in\mathcal{N}}$	$E[B_iB_j \mid B_k = 1]$
r_{ij}^{0m}; $\mathbf{R}^{0m} = (r_{ij}^{0m})_{i,j\in\mathcal{N}}$	$E[B_iB_j \mid B^m = 0]$
y_{ij}; $\mathbf{Y} = (y_{ij})_{i,j\in\mathcal{N}}$	$E[L_iL_j]$
y_{ij}^k; $\mathbf{Y}^k = (y_{ij}^k)_{i,j\in\mathcal{N}}$	$E[L_iL_j \mid B_k = 1]$
y_{ij}^{0m}; $\mathbf{Y}^{0m} = (y_{ij}^{0m})_{i,j\in\mathcal{N}}$	$E[L_iL_j \mid B^m = 0]$

Table 16.1 Network performance measures.

(a) *Projection Constraints:*

$$x_j = \sum_{i\in\mathcal{C}_m} \rho_i x_j^i + (1-\rho(\mathcal{C}_m))\, x_j^{0m}, \qquad j\in\mathcal{N},\ m\in\mathcal{K}, \qquad (16.2.1)$$

$$r_{ij} = \sum_{k\in\mathcal{C}_m} \rho_k r_{ij}^k + (1-\rho(\mathcal{C}_m))\, r_{ij}^{0m}, \qquad i,j\in\mathcal{N},\ m\in\mathcal{K}, \qquad (16.2.2)$$

$$y_{ij} = \sum_{k\in\mathcal{C}_m} \rho_k y_{ij}^k + (1-\rho(\mathcal{C}_m))\, y_{ij}^{0m}, \qquad i,j\in\mathcal{N},\ m\in\mathcal{K}. \qquad (16.2.3)$$

(b) *Definitional Constraints:*

$$r_{ij} = \rho_j r_{ii}^j, \qquad i,j\in\mathcal{N}, \qquad (16.2.4)$$

$$r_{ii} = \rho_i,\ r_{ii}^i = 1, \qquad i\in\mathcal{N}, \qquad (16.2.5)$$

$$r_{ij} = 0,\ r_{ij}^k = 0, \qquad i,j\in\mathcal{C}_m, \qquad (16.2.6)$$

$$r_{ij}^k = 0, \qquad i,k\in\mathcal{C}_m, \text{ or } j,k\in\mathcal{C}_m, \qquad (16.2.7)$$

$$r_{ij}^{0m} = 0, \qquad i \text{ or } j\in\mathcal{C}_m. \qquad (16.2.8)$$

(c) *Lower bound constraints:*

$$r_{ij} \ge \max(0, \rho_i + \rho_j - 1), \qquad i,j\in\mathcal{N}, \qquad (16.2.9)$$

$$x_j^i \ge \frac{r_{ij}}{\rho_i}, \qquad i,j\in\mathcal{N}, \qquad (16.2.10)$$

$$x_j^{0m} \ge \max\left(0, \frac{\rho_j - \rho(\mathcal{C}_m)}{1-\rho(\mathcal{C}_m)}\right), \quad m\in\mathcal{K},\ j\in\mathcal{N}, \qquad (16.2.11)$$

$$r_{ij}^k \ge \max\left(0, \frac{r_{ki}+r_{kj}}{\rho_k} - 1\right), \qquad i,j,k\in\mathcal{N}, \qquad (16.2.12)$$

$$r_{ij}^{0m} \ge \max\left(0, \frac{\max(0,\rho_i-\rho(\mathcal{C}_m)) + \max(0,\rho_j-\rho(\mathcal{C}_m))}{1-\rho(\mathcal{C}_m))} - 1\right),$$
$$i,j\in\mathcal{N},\ m\in\mathcal{K}, \qquad (16.2.13)$$

$$y_{ij} \geq r_{ij}, \qquad i,j \in \mathcal{N}, \tag{16.2.14}$$
$$y_{ij}^k \geq r_{ij}^k, \qquad i,j,k \in \mathcal{N}, \tag{16.2.15}$$
$$y_{ij}^{0m} \geq r_{ij}^{0m}, \quad i,j \in \mathcal{N}, \ m \in \mathcal{K}. \tag{16.2.16}$$

Proof
The constraints in (a) follow by a simple conditioning argument, by noticing that at each time instant, a server is either serving some job class in its constituency or idling. The constraints in (b), (c) follow from elementary arguments. ■

Flow conservation constraints

We next present a set of linear constraints on performance measures using the classical *flow conservation law* of queueing theory, $L^- = L^+$. We first provide a brief discussion of flow conservation in stochastic systems, and then show how to use these ideas to derive linear relations between time-stationary moments of queue lengths.

The classical flow conservation law of queueing systems states that, under mild restrictions, the stationary state probabilities of the number in system at arrival epochs and that at departure epochs are equal. The key assumption is that jobs arrive to the system and depart from the system one at a time, so that the queue size can change only by unit steps.

Consider a multiclass queueing network operating in a steady state regime, with the number in system process $\{L(t)\}$. We assume that the process $\{L(t)\}$ has right-continuous sample paths, and we use $L(t^-)$ to denote the left limit of the process at time t. The corresponding right limit $L(t^+) = L(t)$ because of right-continuity of sample paths. Let $A = \{\tau_k^a\}$ and $D = \{\tau_k^d\}$ be the sequences of arrival and departure epochs of jobs respectively. Let $L(\tau_k^{a-})$ be the number of jobs in the system seen by the kth arriving job just before its arrival; similarly, let $L(\tau_k^d)$ be the number of jobs in the system seen by the kth departing job just after its departure. We define

$$L^- = L(\tau_0^{a-}),$$

and

$$L^+ = L(\tau_0^d).$$

Since we assumed the system to be in steady-state, L^- may be interpreted as the number of jobs in the system seen by a typical arrival, while L^+ may be interpreted as the number of jobs in the system seen by a typical departure. By considering any realization, we see that for every upward transition for the number in system from i to $(i+1)$, there is a corresponding downward transition from $(i+1)$ to i; thus every $L(\tau_k^{a-})$ is equal to a distinct $L(\tau_k^d)$ in a sample-path sense. In particular, we have $L^- = L^+$, yielding the following theorem.

Theorem 16.2.2 (Flow Conservation Law) *If jobs enter and leave the system one at a time, then*

$$L^- = L^+$$

holds in distribution.

In what follows, we apply the law $L^- = L^+$ to a family of queues obtained by aggregating job classes, as explained next. Let $S \subseteq \mathcal{N}$.

Definition 16.2.1 (S-queue) *The S-queue is the queueing system obtained by aggregating job classes in S. The number in system at time t in the S-queue is denoted by $L_S(t) = \sum_{j \in S} L_j(t)$.*

As usual we write $L_S = L_S(0)$, $L_S^- = L_S(0-)$, $L_S^+ = L_S(0+) = L_S(0)$. For convenience of notation we also write

$$p(i, S) = \sum_{j \in S} p_{ij}$$

and

$$\alpha(S) = \sum_{j \in S} \alpha_j.$$

The next theorem formulates the law $L^- = L^+$ as it applies to the S-queue.

Theorem 16.2.3 (The law $L^- = L^+$ in MQNETs) *Under any dynamic stable policy, and for any subset of job classes $S \subseteq \mathcal{N}$ and nonnegative integer l:*

$$\begin{aligned} \alpha(S)P\left(L_S = l\right) + \sum_{i \in S^c} \lambda_i p(i, S) P\left(L_S = l \mid B_i = 1\right) \\ = \sum_{i \in S} \lambda_i \left(1 - p(i, S)\right) P\left(L_S = l + 1 \mid B_i = 1\right). \end{aligned} \qquad (16.2.17)$$

Proof

By applying Theorem 16.2.2 to the S-queue, we have that

$$P(L_S^- = l) = P(L_S^+ = l).$$

An arrival epoch to the S-queue is either an arrival from the outside world (external arrival) that happens with rate $\alpha(S)$, or an internal movement from a class i in S^c to a class in S (internal arrival) that happens only if $B_i = 1$, for $i \in S^c$ with rate

$$\mu_i p(i, S) P(B_i = 1) = \mu_i p(i, S) \rho_i = \lambda_i p(i, S).$$

The total arrival rate to S-queue is

$$\lambda_S = \alpha(S) + \sum_{i \in S^c} \lambda_i p(i, S).$$

Therefore,

$$P\left(L_S^- = l\right) = \frac{\alpha(S)}{\lambda_S} P\left(L_S = l\right) + \sum_{i \in S^c} \frac{\lambda_i p(i,S)}{\lambda_S} P\left(L_S = l \mid B_i = 1\right).$$

A departure epoch from S-queue happens with rate

$$\mu_i(1 - p(i,S))P(B_i = 1) = \lambda_i(1 - p(i,S),$$

for all $i \in S$. The total departure rate is:

$$\mu_S = \sum_{i \in S} \lambda_i(1 - p(i,S).$$

It can be easily checked that the total arrival rate to the S-queue and the total departure rate from the S-queue are equal, i.e., $\lambda_S = \mu_S$. Therefore,

$$P\left(L_S^+ = l\right) = \sum_{i \in S} \frac{\lambda_i\left(1 - p(i,S)\right)}{\mu_S} P\left(L_S = l + 1 \mid B_i = 1\right).$$

By applying $P(L_S^- = l) = P(L_S^+ = l)$, Eq. (16.2.17) follows. ■

Taking expectations in identity (16.2.17) we obtain:

Corollary 16.2.1 *Under any stable policy, and for any subset of job classes $S \subseteq \mathcal{N}$ and positive integer K for which $E\left[(L_1 + \cdots + L_N)^K\right] < \infty$,*

$$\begin{aligned}\alpha(S)E\left[L_S^K\right] + \textstyle\sum_{i \in S^c} \lambda_i p(i,S) E\left[L_S^K \mid B_i = 1\right] \\ = \textstyle\sum_{i \in S} \lambda_i\left(1 - p(i,S)\right) E\left[(L_S - 1)^K \mid B_i = 1\right].\end{aligned} \quad (16.2.18)$$

Note that Corollary 16.2.1 formulates a linear relation between time-stationary moments of queue lengths. The equilibrium equations in Corollary 16.2.1 corresponding to $K = 1, 2$ and $S = \{i\}, \{i,j\}$, for $i, j \in \mathcal{N}$, yield directly the system of linear constraints on performance variables shown next. Let $\mathbf{\Lambda} = \mathrm{Diag}(\boldsymbol{\lambda})$.

Corollary 16.2.2 (Flow conservation constraints) *Under any dynamic stable policy, the following linear constraints hold:*

(a)

$$-\alpha \mathbf{x}' - \mathbf{x}\alpha' + (\mathbf{I} - \mathbf{P})'\mathbf{\Lambda} X + X'\mathbf{\Lambda}(\mathbf{I} - \mathbf{P}) = (\mathbf{I} - \mathbf{P})'\mathbf{\Lambda} + \mathbf{\Lambda}(\mathbf{I} - \mathbf{P}). \quad (16.2.19)$$

(b) *If $E\left[(L_1 + \cdots + L_N)^2\right] < \infty$, then for $i, j \in \mathcal{N}$,*

$$\alpha_j y_{jj} + \sum_{r \in \mathcal{N}} \lambda_r p_{rj} y_{jj}^r - \lambda_j y_{jj}^j + 2\lambda_j(1 - p_{jj})x_j^j = \lambda_j(1 - p_{jj}), \quad (16.2.20)$$

$$\alpha_i y_{jj} + \alpha_j y_{ii} + 2(\alpha_i + \alpha_j) y_{ij} + \sum_{r \in \mathcal{N}} \lambda_r p_{ri} y^r_{jj} + \sum_{r \in \mathcal{N}} \lambda_r p_{rj} y^r_{ii}$$
$$+ \sum_{r \in \mathcal{N}} 2\lambda_r (p_{ri} + p_{rj}) y^r_{ij} - \lambda_i y^i_{jj} - \lambda_j y^j_{ii} - 2\lambda_i y^i_{ij} - 2\lambda_j y^j_{ij}$$
$$-2\lambda_i p_{ij} x^i_i - 2\lambda_j p_{ji} x^j_j$$
$$+2\lambda_i (1 - p_{ii} - p_{ij}) x^i_j + 2\lambda_j (1 - p_{ji} - p_{jj}) x^j_i = -\lambda_i p_{ij} - \lambda_j p_{ji}, \qquad . \qquad (16.2.21)$$

The equilibrium equations in Corollary 16.2.1 corresponding to $K = 2$ and $S = \{i, j, k\}$, for $i, j, k \in \mathcal{N}$, yield the following linear constraints on performance variables.

Corollary 16.2.3 (Flow conservation constraints (continued)) *Let* $A_{ijk} = \alpha_i + \alpha_j + \alpha_k$, *and* $Y_{ijk} = y_{ii} + y_{jj} + y_{kk} + 2y_{ik} + 2y_{jk} + 2y_{ik}$. *Also, for* $r \in \mathcal{N}$, *let* $Y^r_{ijk} = y^r_{ii} + y^r_{jj} + y^r_{kk} + 2y^r_{ij} + 2y^r_{jk} + 2y^r_{ik}$, $W^r_{ijk} = x^r_i + x^r_j + x^r_k$, *and* $P^r_{ijk} = p_{ri} + p_{rj} + p_{rk}$. *Under any dynamic stable policy, the following linear constraints hold:*

$$A_{ijk} Y_{ijk} + \sum_{r \in \mathcal{N}} \lambda_r P^r_{ijk} Y^r_{ijk} - \lambda_i Y^i_{ijk} - \lambda_j Y^j_{ijk} - \lambda_k Y^k_{ijk}$$
$$+2\lambda_i (1 - P^i_{ijk}) W^i_{ijk} + 2\lambda_j (1 - P^j_{ijk}) W^j_{ijk} + 2\lambda_k (1 - P^k_{ijk}) W^k_{ijk}$$
$$= \lambda_i (1 - P^i_{ijk}) + \lambda_j (1 - P^j_{ijk}) + \lambda_k (1 - P^k_{ijk}), \quad i, j, k \in \mathcal{N}. \qquad (16.2.22)$$

Remark: We note that for a fixed K in Corollary 16.2.1, it is enough to consider subsets of classes, S, such that $|S| \leq (K + 1)$. Thus, the constraints we present in Corollaries 16.2.2 and 16.2.3 suffice for our purposes because we shall restrict our performance variables to first and second moments only.

The flow conservation constraints were first derived for multi-station MQNETs by Bertsimas, Paschalidis and Tsitsiklis [105], and by Kumar and Kumar [455], using a potential function approach. The derivation and the interpretation we presented is from Bertsimas and Niño-Mora [102].

16.2.4 Positive semidefinite constraints

We present in this section, a set of *positive semidefinite constraints* that strengthen the formulations obtained through equilibrium relations. Recall that the performance measures $\mathbf{x}, \mathbf{y}$ are moments of random variables. Applying Theorem 16.1.2 to the performance variables introduced in Table 16.1 yields directly the following result.

Theorem 16.2.4 *Under any dynamic stable policy, the following semidefinite constraints hold:*

(a) *Let* $\mathbf{r}^k = (r_{ii}^k)_{i\in\mathcal{N}}$ *and* $\mathbf{r}^{0m} = (r_{ii}^{0m})_{i\in\mathcal{N}}$.

$$\begin{bmatrix} 1 & \rho' \\ \rho & \mathbf{R} \end{bmatrix} \succeq \mathbf{0}, \tag{16.2.23}$$

$$\begin{bmatrix} 1 & \mathbf{r}^{k\prime} \\ \mathbf{r}^k & \mathbf{R}^k \end{bmatrix} \succeq \mathbf{0}, \qquad k \in \mathcal{N}, \tag{16.2.24}$$

$$\begin{bmatrix} 1 & \mathbf{r}^{0m\prime} \\ \mathbf{r}^{0m} & \mathbf{R}^{0m} \end{bmatrix} \succeq \mathbf{0}, \qquad m \in \mathcal{K}. \tag{16.2.25}$$

(b) *If* $E\left[\left(\sum_{j\in\mathcal{N}} L_j\right)^2\right] < \infty$, *then*

$$\begin{bmatrix} 1 & \mathbf{x}' \\ \mathbf{x} & \mathbf{Y} \end{bmatrix} \succeq \mathbf{0}, \tag{16.2.26}$$

$$\begin{bmatrix} 1 & \mathbf{x}^{k\prime} \\ \mathbf{x}^k & \mathbf{Y}^k \end{bmatrix} \succeq \mathbf{0}, \qquad k \in \mathcal{N}, \tag{16.2.27}$$

$$\begin{bmatrix} 1 & \mathbf{x}^{0m\prime} \\ \mathbf{x}^{0m} & \mathbf{Y}^{0m} \end{bmatrix} \succeq \mathbf{0}, \qquad m \in \mathcal{K}. \tag{16.2.28}$$

16.2.5 On the power of the semidefinite relaxation

Our objective in this section is to compare computationally the linear and semidefinite relaxations of the multiclass queueing network performance optimization problem. The linear programming relaxation is defined as follows:

$$Z_{LP} = \text{minimize} \quad \mathbf{c}'\mathbf{x} + \mathbf{d}'\mathbf{y}$$

$$\begin{aligned} \text{subject to} \quad & \text{Projection constraints : (16.2.1) } - \text{ (16.2.3)}, \\ & \text{Definitional constraints : (16.2.4) } - \text{ (16.2.8)}, \\ & \text{Lower bound constraints :(16.2.9) } - \text{ (16.2.16)}, \\ & \text{Flow} - \text{conservation constraints : (16.2.19) } - \text{ (16.2.22)}, \\ & \mathbf{x} \geq \mathbf{0}, \mathbf{y} \geq \mathbf{0}. \end{aligned}$$

The semidefinite relaxation Z_{SD} is obtained by adding the constraints (16.2.23) - (16.2.28).

A multiclass network

We consider the network of Figure 16.1. In this network external arrivals come into either class 1 or class 3, and so $\alpha_2 = \alpha_4 = 0$. In our computations we fix the service times as shown in the figure, and vary only the arrival rates. We maintain the symmetry between classes, and so we set $\alpha_1 = \alpha_3 = \alpha$, where α varies from 0.1 to 1.18. We select $c_i = 1$ and $d_i = 0$, i.e., we are interested in minimizing the expected number of jobs in the system in steady-state. We present below the optimal values Z_{LP} and Z_{SD}. The SDP relaxation

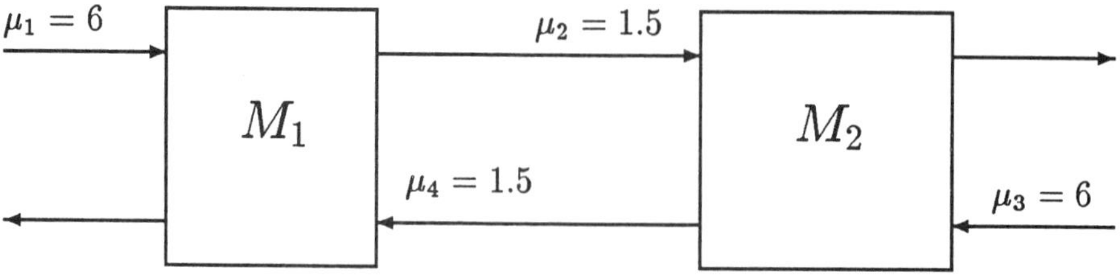

Figure 16.1 A Multiclass Network.

ρ	Z_{LP}	Z_{SD}	$E[Z_{LBFS-B}]$	Best B
0.083	0.170	0.170	0.179 ± 0.001	0
0.167	0.347	0.347	0.390 ± 0.002	0
0.250	0.538	0.538	0.643 ± 0.003	0
0.333	0.793	0.794	0.951 ± 0.004	1
0.417	1.113	1.113	1.339 ± 0.006	1
0.500	1.530	1.530	1.842 ± 0.008	1
0.583	2.102	2.103	2.527 ± 0.012	1
0.667	2.947	2.976	3.526 ± 0.017	1
0.750	4.360	4.416	5.131 ± 0.027	1
0.833	7.167	7.220	8.203 ± 0.045	2
0.875	9.930	9.980	11.227 ± 0.065	2
0.917	15.413	15.497	17.078 ± 0.106	2
0.958	31.777	31.832	34.414 ± 0.246	2
0.983	80.766	81.093	85.139 ± 0.893	3

Table 16.2 Comparison of LP and SDP relaxations for the network of Figure 16.1.

has 283 variables (including slack variables) and 259 constraints. We solve the semidefinite relaxation using the package SDPA developed by Fujisawa, Kojima and Nakata [264]. All of these instances were solved in less than one minute by SDPA on a Pentium II workstation as the SDP relaxation has a simple block structure.

For comparison purposes, we also report simulation results for a particular policy that was derived from fluid optimal control (see Avram et. al. [47]): When both $L_4(t)$, $L_2(t) > B$, the first station gives preemptive priority to class 4 and the second station gives preemptive priority to class 2. When $L_4(t) \leq B$, class 3 has preemptive priority over class 2. Similarly, when $L_2(t) \leq B$, class 1 has preemptive priority over class 4. We call this policy last-buffer-first-served with a threshold B, denoted by $LBFS - B$. We let $E[Z_{LBFS-B}]$ denote the expected number of jobs under this policy. We select the value of B optimally using simulation.

In Table 16.2, we report the values Z_{LP}, Z_{SD}, the simulation value $E[Z_{LBFS-B}]$, and the value of the threshold B that gives the optimal performance. The computational results suggest the following:

(a) The semidefinite relaxation strictly improves the linear programming relaxation, although the improvement is typically small.

(b) The value of the semidefinite relaxation is close to the expected value of the policy $LBFS - B$. This shows that not only the semidefinite relaxation produces near optimal bounds, but the particular policy we constructed is a near optimal.

A multiclass single queue

We consider a single station network with four classes. Our objective here is to minimize $\sum_{i=1}^{4}(x_i + y_{ii})$. For the case that we do not include terms involving y_{ii} in the objective function, the LP relaxation is exact (see Bertsimas and Niño-Mora [100]).

We assume that the arrival rate for each class is the same, and that the mean service times for the job classes are 0.05, 0.1, 0.2, and 0.4 respectively. The results of the LP and SDP relaxations are tabulated in Table 16.3. In this experiment the SDP relaxation has 234 variables and 220 constraints. All of these instances were solved in less than two minutes by SDPA on a Pentium II workstation.

For comparison purposes we have simulated the following dynamic priority policy P: At every service completion time t, we give priority to the class that has the highest index $\mu_i L_i(t)$. The policy was derived from fluid optimal control (see Avram et. al. [47]).

We observe that the semidefinite relaxation provides a substantial improvement over the value of the LP relaxation often by an order of magnitude.

Both computational experiments demonstrate that, unlike the LP relaxation, the semidefinite relaxation provides practically useful suboptimality guarantees that can be used to asscss the closeness to optimality of heuristic policies. We believe that the combination of fluid optimal control methods to generate near optimal policies for large scale problems (see Luo and Bertsimas [504]), and semidefinite relaxations to provide near optimal bounds is perhaps the most promising methodology to address the multiclass queueing network optimization problem.

16.3 OPTIMAL BOUNDS IN PROBABILITY

In this section, we review the work of Bertsimas and Popescu [107]. Suppose that $\boldsymbol{\sigma}$ is a feasible moment vector and $\boldsymbol{X}$ has a $\boldsymbol{\sigma}$-feasible distribution. We now define the central problem we address in this section:

ρ	Z_{LP}	Z_{SD}	$E[Z_P]$
0.075	0.162	0.162	0.165 ± 0.001
0.150	0.352	0.358	0.368 ± 0.002
0.225	0.578	0.598	0.620 ± 0.005
0.300	0.854	0.901	0.946 ± 0.008
0.375	1.198	1.302	1.379 ± 0.013
0.450	1.639	1.857	1.982 ± 0.022
0.525	2.227	2.676	2.869 ± 0.038
0.600	3.047	3.982	4.255 ± 0.066
0.675	4.270	6.287	6.671 ± 0.129
0.750	6.269	10.991	11.588 ± 0.289
0.825	10.072	22.314	24.495 ± 0.267
0.900	19.811	60.948	75.429 ± 0.394
0.975	89.332	725.855	1203.778 ± 12.864

Table 16.3 Comparison of LP and SDP relaxations for a multiclass queue.

The (n, k, Ω)-bound problem

Given a vector $\boldsymbol{\sigma}$ of up to kth order moments

$$\sigma_{\mathbf{k}} = E[X_1^{k_1} X_2^{k_2} \cdots X_n^{k_n}], \qquad k_1 + k_2 + \cdots + k_n \le k,$$

of a multivariate random variable $\boldsymbol{X} = (X_1, X_2, \ldots, X_n)$ on $\Omega \subseteq R^n$, find the "best possible" or "tight" upper and lower bounds on $P(\boldsymbol{X} \in S)$, for arbitrary events $S \subseteq \Omega$.

The term "best possible" or "tight" upper (and by analogy lower) bound above is defined as follows.

Definition 16.3.1 *We say that α is a tight upper bound on $P(\boldsymbol{X} \in S)$, and we will denote it by $\sup_{\boldsymbol{X} \sim \boldsymbol{\sigma}} P(\boldsymbol{X} \in S)$ if :*

(a) *it is an upper bound, i.e., $P(\boldsymbol{X} \in S) \le \alpha$ for all random variables $\boldsymbol{X} \sim \boldsymbol{\sigma}$;*

(b) *it cannot be improved, i.e., for any $\epsilon > 0$ there is a random variable $\boldsymbol{X}_\epsilon \sim \boldsymbol{\sigma}$ for which $P(\boldsymbol{X}_\epsilon \in S) > \alpha - \epsilon$.*

The well known inequalities due to Markov, Chebyshev and Chernoff, which are widely used if we know the first moment, the first two moments, and all moments (i.e., the generating function) of a random variable, respectively, are feasible but not necessarily optimal solutions to the $(1, k, \Omega)$-bound problem, i.e., they are not necessarily tight bounds.

The history of the developments in the area of (n, k, Ω)-bound problems, sometimes referred to as Chebyshev type inequalities, can be traced back to the work of Gauss, Cauchy, Chebyshev, and Markov, and has witnessed an unexpected evolution. The problem of finding bounds on univariate distributions under moment constraints has actually been proposed and formulated without proof initially by Chebychev [156] in 1874 and resolved ten years later by his student Markov [518] in his PhD thesis, using continued fractions techniques. In the 1950s and 1960s there has been a revival of the interest in this area, that resulted in a large literature on the topic of generalized Chebyshev inequalities. Surveys of early literature can be found in Shohat and Tamarkin [730] and Godwin [279], [280].

The idea that optimization methods and duality theory can be used to address moment-type inequalities in probability first appeared in 1960, and is due independently and simultaneously to Isii [354] and Karlin (lecture notes at Stanford, see [400], p.472), who show that certain types of Chebyshev inequalities for univariate random variables are sharp, via strong duality results. Isii [355] extends these results for multivariate random variables. Marshall and Olkin in 1961 [522] give a game theoretic proof of the sharpness of Chebyshev type inequalities with first and second order moment constraints, as well as with trigonometric moments. The same authors [520], [521] were the first to actually compute tight, explicit bounds on probabilities given first and second order moments (the $(n, 2, \Omega)$ problem in our context), thus generalizing Chebyshev's inequality to a multivariate setting. A detailed, unified account of the evolution of *Chebyshev Systems* is given by Karlin and Studden [400] in their 1966 monograph (see in particular chapters 12 and 13, that deal with (n, k, Ω)-type bounds).

Not for the first time in its history, "the problem of moments lay dormant for more than 20 years" (Shohat and Tamarkin [730], p.10.) It revives briefly in the 1980s, with the book on *Probability Inequalities and Multivariate Distributions* of Y.L. Tong [785] in 1980, who also publishes a monograph on probability inequalities in 1984. The latter notably contains, among others, a generalization of Markov's inequality for multivariate tails, due to Marshall [519], and an application of moment inequalities for computing error bounds in stochastic programming, by Birge and Wets [114]. A volume on *Moments in Mathematics* edited by Landau in 1987 includes a background survey by the same author [458], as well as relevant papers of Kemperman [410] and Diaconis [192]. Thirty two years after Isii's [355] original multivariate proof, Smith [741] rederived the same duality results and proposed new interesting applications in decision analysis, dynamic programming, statistics and finance. He also introduces a computational procedure for the (n, k, R^n)-bound problem, although he does not refer to it in this way. Unfortunately, the procedure is far from an actual algorithm, as there is no proof of convergence, and no investigation (theoretical or experimental) of its efficiency. Bertsimas and Popescu [107], whose work we survey in this paper, have proposed convex and, in particular semidefinite, optimization algorithms to address the (n, k, Ω)-bound problem.

We examine the existence of an algorithm that on input $\langle n, k, \boldsymbol{\sigma}, \epsilon\rangle$ computes a value $\alpha \in [\alpha^* - \epsilon, \alpha^* + \epsilon]$, where $\alpha^* = \sup_{\boldsymbol{X} \sim \boldsymbol{\sigma}} P(\boldsymbol{X} \in S)$, and runs in time polynomial in $n, k, \log \sigma_{max}$ and $\log \frac{1}{\epsilon}$, where $\sigma_{max} = \max(\sigma_{\mathbf{k}})$. We assume the availability of an oracle to test membership in $S \subset \Omega$, and we allow our algorithm to make oracle queries of the type "is $\mathbf{x}$ in S ?".

Primal and dual formulations

The (n, k, Ω)-upper bound problem can be formulated as the following optimization problem (P):

$$\begin{aligned}(P) \quad Z_P = \text{maximize} \quad & \int_S f(\mathbf{z})d\mathbf{z} \\ \text{subject to} \quad & \int_\Omega z_1^{k_1} \cdots z_n^{k_n} f(\mathbf{z})d\mathbf{z} = \sigma_{\mathbf{k}}, \quad \forall\, k_1 + \cdots + k_n \leq k, \\ & f(\mathbf{z}) = f(z_1, \ldots, z_n) \geq 0, \quad \forall\, \mathbf{z} = (z_1, \ldots, z_n) \in \Omega.\end{aligned}$$

Notice that if Problem (P) is feasible, then $\boldsymbol{\sigma}$ is a feasible moment vector, and any feasible distribution $f(\mathbf{z})$ is a $\boldsymbol{\sigma}$-feasible distribution.

In the spirit of linear programming duality theory, we associate a dual variable $u_{\mathbf{k}}$ with each equality constraint of the primal. We can identify the vector of dual variables with a k-degree, n-variate dual polynomial:

$$g(x_1, \ldots, x_n) = \sum_{k_1 + \cdots + k_n \leq k} u_{\mathbf{k}} x_1^{k_1} \cdots x_n^{k_n}.$$

We refer to such a polynomial as a k-degree, n-variate polynomial. The dual objective translates to finding the smallest value of:

$$\sum_{\mathbf{k}} u_{\mathbf{k}} \sigma_{\mathbf{k}} = \sum_{\mathbf{k}} u_{\mathbf{k}} E[X_1^{k_1} \cdots X_n^{k_n}] = E[g(\boldsymbol{X})],$$

where the expected value is taken over any $\boldsymbol{\sigma}$-feasible distribution. In this framework, the Dual Problem (D) corresponding to Problem (P) can be written as:

$$\begin{aligned}(D) \quad Z_D = \text{minimize} \quad & E[g(\boldsymbol{X})] \\ \text{subject to} \quad & g(\mathbf{x}) \ k\text{-degree, } n\text{-variate polynomial,} \\ & g(\mathbf{x}) \geq \chi_S(\mathbf{x}), \forall\, \mathbf{x} \in \Omega,\end{aligned}$$

where $\chi_S(\mathbf{x})$ is the indicator function of the set S, defined by:

$$\chi_S(\mathbf{x}) = \begin{cases} 1, & \text{if } \mathbf{x} \in S, \\ 0, & \text{otherwise.} \end{cases}$$

Notice that in general the optimum may not be achievable. Whenever the primal optimum is achieved, we call the corresponding distribution **an extremal distribution.** We next establish weak duality.

Theorem 16.3.1 (Weak duality) $Z_P \leq Z_D$.

Proof
Let $f(\mathbf{z})$ be a primal optimal solution and let $g(\mathbf{z})$ be any dual feasible solution. Then:

$$Z_P = \int_S f(\mathbf{z})d\mathbf{z} = \int_\Omega \chi_S(\mathbf{z})f(\mathbf{z})d\mathbf{z} \leq \int_\Omega g(\mathbf{z})f(\mathbf{z})d\mathbf{z} = E[g(\boldsymbol{X})],$$

and hence $Z_P \leq \inf_{g(\cdot)\leq\chi_S(\cdot)} E[g(\boldsymbol{X})] = Z_D$. ■

Theorem 16.3.1 indicates that by solving the Dual Problem (D) we obtain an upper bound on the primal objective and hence on the probability we are trying to bound. Under some mild restrictions on the moment vector $\boldsymbol{\sigma}$, the dual bound turns out to be tight. This strong duality result follows from a univariate result due to Karlin and Isii in 1960 (see Karlin and Studden [400], p.472), and generalized by Isii in 1963 [355] for the multivariate case. The following theorem holds for arbitrary distributions and is a consequence of their work:

Theorem 16.3.2 (Strong Duality and Complementary Slackness) *If the moment vector $\boldsymbol{\sigma}$ is an interior point of the set $\mathcal{M}$ of feasible moment vectors, then the following results hold:*

(a) *Strong Duality: $Z_P = Z_D$.*

(b) *Complementary Slackness: If the dual is bounded, there exists a dual optimal solution $g_{opt}(\cdot)$ and a discrete extremal distribution concentrated on points $\mathbf{x}$, where $g_{opt}(\mathbf{x}) = \chi_S(\mathbf{x})$, that achieves the bound.*

It can also be shown that if the dual is unbounded, then the primal is infeasible, i.e., the multidimensional moment problem is infeasible. Moreover, if $\boldsymbol{\sigma}$ is a boundary point of $\mathcal{M}$, then it can be shown that the $\boldsymbol{\sigma}$-feasible distributions are concentrated on a subset Ω_0 of Ω, and strong duality holds provided we relax the dual to Ω_0 (see Isii [355], p.190 or Smith [741], p. 824). These authors also prove that it is equivalent to optimize only over distributions that are concentrated on $m + 2$ points, where m is the number of moment constraints (in our case $m = \frac{n(n+1)}{2}$). Little is known, however, about the uniqueness of such extremal distributions. In the univariate case, Isii [355] proves that if $\boldsymbol{\sigma}$ is a boundary point of $\mathcal{M}$, then exactly one $\boldsymbol{\sigma}$-feasible distribution exists.

If strong duality holds, then by optimizing over Problem (D) we obtain a **tight** bound on $P(X \in S)$. On the other hand, solving Problem (D) is equivalent to solving the corresponding separation problem, under certain technical conditions (see Grötschel, Lovász and Schrijver [304]).

16.3.1 Optimal bounds for the univariate case using semidifinite optimization

In this section, we restrict our attention to univariate random variables. Given the first k moments $M_1, \ldots, M_k$ (we let $M_0 = 1$) of a real random variable X with domain Ω, we are interested in deriving tight bounds on $P(X \in S)$. The main result in this section is that optimal bounds can be derived as a solution to a single semidefinite optimization problem.

Given the first k moments of X with domain Ω, we can use the dual formulation (D) to find a tight bound for $P(X \in S)$ by solving the following problem

$$\begin{array}{lll} \text{minimize} & \sum_{r=0}^{k} y_r M_r & \\ \text{subject to} & \sum_{r=0}^{k} y_r x^r \geq 1, & \forall\, x \in S \\ & \sum_{r=0}^{k} y_r x^r \geq 0, & \forall\, x \in \Omega. \end{array} \tag{16.3.29}$$

Since S and Ω are intervals in the real line we show in the next proposition that the feasible region of Problem (16.3.29) can be expressed using semidefinite constraints. The results and the proofs in the following proposition are due to Bertsimas and Popescu [107] and they are inspired by Nesterov [573].

Proposition 16.3.1 **(a)** *The polynomial* $g(x) = \sum_{r=0}^{2k} y_r x^r$ *satisfies* $g(x) \geq 0$ *if and only if there exists a positive semidefinite matrix* $\boldsymbol{X} = [x_{ij}]_{i,j=0,\dots,k}$, *such that*

$$y_r = \sum_{i,j:\ i+j=r} x_{ij}, \qquad r = 0, \dots, 2k, \qquad \boldsymbol{X} \succeq \boldsymbol{0}. \tag{16.3.30}$$

(b) *The polynomial* $g(x) = \sum_{r=0}^{k} y_r x^r$ *satisfies* $g(x) \geq 0$ *for all* $x \geq 0$ *if and only if there exists a positive semidefinite matrix* $\boldsymbol{X} = [x_{ij}]_{i,j=0,\dots,k}$, *such that*

$$\begin{array}{rcll} 0 & = & \sum_{i,j:\ i+j=2l-1} x_{ij}, & l = 1, \dots, k, \\ y_l & = & \sum_{i,j:\ i+j=2l} x_{ij}, & l = 0, \dots, k, \\ & & \boldsymbol{X} \succeq \boldsymbol{0}. & \end{array} \tag{16.3.31}$$

(c) *The polynomial* $g(x) = \sum_{r=0}^{k} y_r x^r$ *satisfies* $g(x) \geq 0$ *for all* $x \in [0, a]$ *if and only if there exists a positive semidefinite matrix* $\boldsymbol{X} = [x_{ij}]_{i,j=0,\dots,k}$, *such* *that*

$$
\begin{aligned}
0 &= \sum_{i,j:\ i+j=2l-1} x_{ij}, && l = 1, \ldots, k,\\
\sum_{r=0}^{l} y_r \binom{k-r}{l-r} a^r &= \sum_{i,j:\ i+j=2l} x_{ij}, && l = 0, \ldots, k, \qquad (16.3.32)\\
\boldsymbol{X} &\succeq \mathbf{0}.
\end{aligned}
$$

(d) *The polynomial* $g(x) = \sum_{r=0}^{k} y_r x^r$ *satisfies* $g(x) \geq 0$ *for all* $x \in [a, \infty)$ *if and only if there exists a positive semidefinite matrix* $\boldsymbol{X} = [x_{ij}]_{i,j=0,\ldots,k}$, *such that*

$$
\begin{aligned}
0 &= \sum_{i,j:\ i+j=2l-1} x_{ij}, && l = 1, \ldots, k,\\
\sum_{r=l}^{k} y_r \binom{r}{l} a^r &= \sum_{i,j:\ i+j=2l} x_{ij}, && l = 0, \ldots, k, \qquad (16.3.33)\\
\boldsymbol{X} &\succeq \mathbf{0}.
\end{aligned}
$$

(e) *The polynomial* $g(x) = \sum_{r=0}^{k} y_r x^r$ *satisfies* $g(x) \geq 0$ *for all* $x \in (-\infty, a]$ *if and only if there exists a positive semidefinite matrix* $\boldsymbol{X} = [x_{ij}]_{i,j=0,\ldots,k}$, *such that*

$$
\begin{aligned}
0 &= \sum_{i,j:\ i+j=2l-1} x_{ij}, && l = 1, \ldots, k,\\
\sum_{r=0}^{k-l} y_r \binom{k-r}{l} a^r &= \sum_{i,j:\ i+j=2l} x_{ij}, && l = 0, \ldots, k, \qquad (16.3.34)\\
\boldsymbol{X} &\succeq \mathbf{0}.
\end{aligned}
$$

(f) *The polynomial* $g(x) = \sum_{r=0}^{k} y_r x^r$ *satisfies* $g(x) \geq 0$ *for all* $x \in [a, b]$ *if and only if there exists a positive semidefinite matrix* $\boldsymbol{X} = [x_{ij}]_{i,j=0,\ldots,k}$, *such that*

$$
\begin{aligned}
0 &= \sum_{i,j:\ i+j=2l-1} x_{ij}, && l = 1, \ldots, k,\\
\sum_{m=0}^{l} \sum_{r=m}^{k+m-l} y_r \binom{r}{m} \binom{k-r}{l-m} a^{r-m} b^m &= \sum_{i,j:\ i+j=2l} x_{ij}, && l = 0, \ldots, k,\\
\boldsymbol{X} &\succeq \mathbf{0}.
\end{aligned}
\qquad (16.3.35)
$$

Proof:
(a) Suppose (16.3.30) holds. Let $\boldsymbol{e}_x = (1, x, x^2, \ldots, x^k)'$. Then

$$\begin{aligned} g(x) &= \sum_{r=0}^{2k} \sum_{i+j=r} x_{ij} x^r \\ &= \sum_{i=0}^{k} \sum_{j=0}^{k} x_{ij} x^i x^j \\ &= \boldsymbol{e}_x' \boldsymbol{X} \boldsymbol{e}_x \\ &\geq 0, \end{aligned}$$

since $\boldsymbol{X} \succeq \boldsymbol{0}$.

Conversely, suppose that the polynomial $g(x)$ of degree $2k$ is nonnegative for all x. Then, the real roots of $g(x)$ should have even multiplicity, otherwise $g(x)$ would alter its sign in a neighborhood of a root. Let λ_i, $i = 1, \ldots, r$ be its real roots with corresponding multiplicity $2m_i$. Its complex roots can be arranged in conjugate pairs, $a + ib_j$, $a_j - ib_j$, $j = 1 \ldots, h$. Then,

$$g(x) = y_{2k} \prod_{i=1}^{r} (x - \lambda_i)^{2m_i} \prod_{j=1}^{h} \left((x - a_j)^2 + b_j^2 \right).$$

Note that the leading coefficient y_{2k} needs to be positive. Thus, by expanding the terms in the products, we see that $g(x)$ can be written as a sum of squares of polynomials, of the form

$$\begin{aligned} g(x) &= \sum_{i=0}^{k} \left(\sum_{j=0}^{k} x_{ij} x^j \right)^2 \\ &= \boldsymbol{e}_x' \boldsymbol{X} \boldsymbol{e}_x, \end{aligned}$$

with $\boldsymbol{X}$ positive semidefinite, from where Equation (16.3.30) follows.
(b) We observe that $g(x) \geq 0$ for $x \geq 0$ if and only if $g(t^2) \geq 0$ for all t. Since

$$g(t^2) = y_0 + 0 \cdot t + y_1 t^2 + 0 \cdot t^3 + y_2 t^4 + \cdots + y_k t^{2k},$$

we obtain (16.3.31) by applying part (a).
(c) We observe that $g(x) \geq 0$ for $x \in [0, a]$ if and only if

$$(1 + t^2)^k g\left(\frac{at^2}{1 + t^2} \right) \geq 0, \qquad \text{for all } t.$$

Since

$$\begin{aligned}(1+t^2)^k g\left(\frac{at^2}{1+t^2}\right) &= \sum_{r=0}^{k} y_r a^r t^{2r}(1+t^2)^{k-r} \\ &= \sum_{r=0}^{k} y_r a^r \sum_{l=0}^{k-r} \binom{k-r}{l} t^{2(l+r)} \\ &= \sum_{j=0}^{k} t^{2j} \left(\sum_{r=0}^{j} y_r \binom{k-r}{j-r} a^r \right),\end{aligned}$$

by applying part (a) we obtain (16.3.32).
(d) We observe that $g(x) \geq 0$ for $x \in [a, \infty)$ if and only if

$$g(a(1+t^2)) \geq 0, \qquad \text{for all } t.$$

Since

$$\begin{aligned}g(a(1+t^2)) &= \sum_{r=0}^{k} y_r a^r (1+t^2)^r \\ &= \sum_{r=0}^{k} y_r a^r \sum_{l=0}^{r} \binom{r}{l} t^{2l} \\ &= \sum_{l=0}^{k} t^{2l} \left(\sum_{r=l}^{k} y_r \binom{r}{l} a^r \right),\end{aligned}$$

by applying part (a) we obtain (16.3.33).
(e) We observe that $g(x) \geq 0$ for $x \in (-\infty, a]$ if and only if

$$(1+t^2)^k g\left(\frac{a}{1+t^2}\right) \geq 0, \qquad \text{for all } t.$$

Since

$$\begin{aligned}(1+t^2)^k g\left(\frac{a}{1+t^2}\right) &= \sum_{r=0}^{k} y_r a^r (1+t^2)^{k-r} \\ &= \sum_{r=0}^{k} y_r a^r \sum_{l=0}^{k-r} \binom{k-r}{l} t^{2l} \\ &= \sum_{l=0}^{k} t^{2l} \left(\sum_{r=0}^{k-l} y_r \binom{k-r}{l} a^r \right),\end{aligned}$$

by applying part (a) we obtain (16.3.34).
(f) We observe that $g(x) \geq 0$ for $x \in [a, b]$ if and only if

$$(1+t^2)^k g\left(a + (b-a)\frac{t^2}{1+t^2}\right) \geq 0, \qquad \text{for all } t.$$

Since

$$
\begin{aligned}
(1+t^2)^k g\left(a+(b-a)\frac{t^2}{1+t^2}\right) &= \sum_{r=0}^{k} y_r (a+bt^2)^r (1+t^2)^{k-r} \\
&= \sum_{r=0}^{k} y_r \sum_{m=0}^{r} \binom{r}{m} a^{r-m} b^m t^{2m} \sum_{j=0}^{k-r} \binom{k-r}{j} t^{2j} \\
&= \sum_{l=0}^{k} t^{2l} \left(\sum_{m=0}^{l} \sum_{r=m}^{k+m-l} y_r \binom{r}{m} \binom{k-r}{l-m} a^{r-m} b^m \right),
\end{aligned}
$$

by applying part (a) we obtain (16.3.35). ■

We next show that Problem (16.3.29) can be written as a semidefinite optimization problem.

Theorem 16.3.3 *Given the first k moments $(M_1, \ldots, M_k)$ (we let $M_0 = 1$) of a random variable X defined on Ω we obtain the following tight upper bounds:* **(a)** *If $\Omega = R^+$, the tight upper bound on $P(X \geq a)$ is given as the solution of the semidefinite optimization problem*

$$
\begin{aligned}
\text{minimize} \quad & \sum_{r=0}^{k} y_r M_r \\
\text{subject to} \quad & 0 = \sum_{i,j:\ i+j=2l-1} x_{ij}, && l = 1, \ldots, k, \\
& (y_0 - 1) + \sum_{r=1}^{k} y_r \binom{r}{l} a^r = x_{00}, \\
& \sum_{r=l}^{k} y_r \binom{r}{l} a^r = \sum_{i,j:\ i+j=2l} x_{ij}, && l = 1, \ldots, k, \qquad (16.3.36) \\
& 0 = \sum_{i,j:\ i+j=2l-1} z_{ij}, && l = 1, \ldots, k, \\
& \sum_{r=0}^{l} y_r \binom{k-r}{l-r} a^r = \sum_{i,j:\ i+j=2l} z_{ij}, && l = 0, \ldots, k, \\
& \boldsymbol{X}, \boldsymbol{Z} \succeq \boldsymbol{0}.
\end{aligned}
$$

If $\Omega = R$, then the tight bound on $P(X \geq a)$ is as above with the next to last equation in (16.3.36) replaced by

$$
\sum_{r=0}^{k-l} y_r \binom{k-r}{l} a^r = \sum_{i,j:\ i+j=2l} z_{ij}, \qquad l = 0, \ldots, k.
$$

(b) *If $\Omega = R^+$, the tight upper bound on $P(a \leq X \leq b)$ is given as the solution of the semidefinite optimization problem*

$$
\begin{aligned}
\min \quad & \sum_{r=0}^{k} y_r M_r \\
\text{s.t.} \quad & 0 = \sum_{i,j:\ i+j=2l-1} x_{ij}, \qquad l = 1, \ldots, k, \\
& \sum_{m=0}^{l} \sum_{r=m}^{k+m-l} y_r \binom{r}{m} \binom{k-r}{l-m} a^{r-m} b^m = \binom{k}{l} + \sum_{i,j:\ i+j=2l} x_{ij}, \quad l = 0, \ldots, k, \\
& 0 = \sum_{i,j:\ i+j=2l-1} z_{ij}, \qquad l = 1, \ldots, k, \\
& y_l = \sum_{i,j:\ i+j=2l} z_{ij}, \qquad l = 0, \ldots, k, \\
& \boldsymbol{X}, \boldsymbol{Z} \succeq \boldsymbol{0}.
\end{aligned}
\tag{16.3.37}
$$

If $\Omega = R$, then the tight upper bound on $P(a \leq X \leq b)$ is as above with the next to last equation in (16.3.37) replaced by

$$
\sum_{r=0}^{k-l} y_r \binom{k-r}{l} a^r = \sum_{i,j:\ i+j=2l} z_{ij}, \qquad l = 0, \ldots, k,
$$

and the following equations added

$$
\begin{aligned}
0 &= \sum_{i,j:\ i+j=2l-1} u_{ij}, \qquad & l = 1, \ldots, k, \\
\sum_{r=l}^{k} y_r \binom{r}{l} b^r &= \sum_{i,j:\ i+j=2l} u_{ij}, \qquad & l = 0, \ldots, k, \\
& \boldsymbol{U} \succeq \boldsymbol{0}.
\end{aligned}
$$

Proof:

(a) The feasible region of Problem (16.3.29) for $S = [a, \infty)$ and $\Omega = R_+$, becomes:

$$
g(x) = \sum_{r=0}^{k} y_r x^r \geq 1, \ \forall\, x \in [a, \infty), \qquad \text{and } g(x) \geq 0, \ \forall\, x \in [0, a).
$$

By applying Proposition 16.3.1(c),(d) we obtain (16.3.36). If $\Omega = R$, we apply Proposition 16.3.1(d),(e).

(b) The feasible region of Problem (16.3.29) for $S = [a, b]$ and $\Omega = R_+$, becomes:

$$
g(x) = \sum_{r=0}^{k} y_r x^r \geq 1, \ \forall\, x \in [a, b], \qquad \text{and } g(x) \geq 0, \ \forall\, x \in [0, \infty).
$$

By applying Proposition 16.3.1(b),(f) we obtain (16.3.37). If $\Omega = R$, we apply Proposition 16.3.1(c),(d),(f). ■

16.3.2 Explicit bounds for the $(n, 1, \Omega)$, $(n, 2, R^n)$-bound problems

In this section, we present tight bounds as solutions to n convex optimization problems for the $(n, 1, R^n_+)$-bound problems, and as a solution to a single convex optimization problem for the $(n, 2, R^n)$-bound problem for the case when the event S is a convex set. Marshall [519] derived a tight bound for the case that $S = \{x_i > (1+\delta_i)M_i,\ i = 1, \ldots, n\}$. For general convex sets S, the following result is due to Bertsimas and Popescu [107].

Theorem 16.3.4 **(a)** *The tight $(n, 1, R^n_+)$-upper bound for an arbitrary convex event S is given by:*

$$\sup_{\boldsymbol{X}\sim\mathbf{M}} P(\boldsymbol{X} \in S) = \min\left(1, \max_{i=1,\ldots,n} \frac{M_i}{\inf_{x\in S_i} x_i}\right), \tag{16.3.38}$$

where $S_i = S \cap (\cap_{j\neq i}\{\mathrm{x} |\ M_i x_j - M_j x_i \leq 0\})$.
(b) *If the Bound (16.3.38) is achievable, then there is an extremal distribution that exactly achieves it; otherwise, there is a sequence of distributions with mean* $\mathbf{M}$, *that asymptotically achieve it.*

Theorem 16.3.4 constitutes a multivariate generalization of Markov's inequality. We denote by

$$P(\boldsymbol{X} > \mathbf{M}_{\mathbf{e}+\boldsymbol{\delta}}) = P(X_i > (1+\delta_i)M_i\ ,\ \forall i = 1, \ldots, n),$$

where $\boldsymbol{\delta} = (\delta_1, \ldots, \delta_n)'$, $\mathbf{e} = (1, 1, \ldots, 1)'$ and $\mathbf{M}_{\boldsymbol{\delta}} = (\delta_1 M_1, \ldots, \delta_n M_n)'$. Then, applying Theorem 16.3.4 leads to:

$$\sup_{\boldsymbol{X}\sim\mathbf{M}} P(\boldsymbol{X} > \mathbf{M}_{\mathbf{e}+\boldsymbol{\delta}}) = \min_{i=1,\ldots,n} \frac{1}{1+\delta_i}.$$

The $(n, 2, R^n)$-bound problem for convex sets

We first rewrite the $(n, 2, R^n)$-bound problem in a more convenient form. Rather than assuming that $E[\boldsymbol{X}]$ and $E[\boldsymbol{X}\boldsymbol{X}']$ are known, we assume equivalently that the vector $\mathrm{M} = E[\boldsymbol{X}]$ and the covariance matrix $\boldsymbol{\Gamma} = E[(\boldsymbol{X} - \mathrm{M})(\boldsymbol{X} - \mathrm{M})']$ are known. Given a set $S \subset R^n$, we find tight upper bounds, denoted by $\sup_{\boldsymbol{X}\sim(\mathbf{M},\boldsymbol{\Gamma})} P(\boldsymbol{X} \in S)$, on the probability $P(\boldsymbol{X} \in S)$ for all multivariate random variables $\boldsymbol{X}$ defined on R^n with mean $\mathrm{M} = E[\boldsymbol{X}]$ and covariance matrix $\boldsymbol{\Gamma} = E[(\boldsymbol{X} - \mathrm{M})(\boldsymbol{X} - \mathrm{M})']$.

First, notice that a necessary and sufficient condition for the existence of such a random variable $\boldsymbol{X}$, is that the covariance matrix $\boldsymbol{\Gamma}$ is symmetric and positive semidefinite. Indeed, given $\boldsymbol{X}$, for an arbitrary vector $\mathbf{a}$ we have:

$$0 \leq E[(\mathbf{a}'(\boldsymbol{X} - \mathrm{M}))^2] = \mathbf{a}'E[(\boldsymbol{X} - \mathrm{M})(\boldsymbol{X} - \mathrm{M})']\mathbf{a} = \mathbf{a}'\boldsymbol{\Gamma}\mathbf{a},$$

so $\mathbf{\Gamma}$ must be positive semidefinite. Conversely, given a symmetric semidefinite matrix Γ and a mean vector $\mathbf{M}$, we can define a multivariate normal distribution with mean $\mathbf{M}$ and covariance $\mathbf{\Gamma}$. Moreover, notice that $\mathbf{\Gamma}$ is positive definite if and only if the components of $\boldsymbol{X} - \mathrm{M}$ are linearly independent. Indeed, the only way that $0 = \mathbf{a}'\mathbf{\Gamma}\mathbf{a} = E[(\mathbf{a}'(X - \mathrm{M}))^2]$ for a nonzero vector $\mathbf{a}$ is that $\mathbf{a}'(\boldsymbol{X} - \mathrm{M}) = 0$.

We assume that $\mathbf{\Gamma}$ has full rank and is positive definite. This does not reduce the generality of the problem, it just eliminates redundant constraints, and thereby insures that Theorem 16.3.2 holds. Indeed, the tightness of the bound is guaranteed by Theorem 16.3.2 whenever the moment vector is interior to $\mathcal{M}$. If the moment vector is on the boundary, it means that the covariance matrix of X is not of full rank, implying that the components of X are linearly dependent. By eliminating the dependent components, we reduce without loss of generality the problem to one of smaller dimension for which strong duality holds. Hence, the primal and the dual problems (P) and (D) satisfy $Z_P = Z_D$. The following theorem is due to Marshall and Olkin [520] (see also Bertsimas and Popescu [107] for an alternative proof).

Theorem 16.3.5 **(a)** *The tight $(n, 2, R^n)$-upper bound for an arbitrary convex event S is given by:*

$$\sup_{\boldsymbol{X} \sim (\boldsymbol{M}, \mathbf{\Gamma})} P(\boldsymbol{X} \in S) = \frac{1}{1 + d^2}, \qquad (16.3.39)$$

where $d^2 = \inf_{\mathbf{x} \in S} (\mathbf{x} - \mathrm{M})'\mathbf{\Gamma}^{-1}(\mathbf{x} - \mathrm{M})$, *is the squared distance from* M *to the set S, under the norm induced by the matrix $\mathbf{\Gamma}^{-1}$.*

(b) *If* $\mathrm{M} \notin S$ *and if* $d^2 = \inf_{\mathbf{x} \in S} (\mathbf{x} - \mathrm{M})'\mathbf{\Gamma}^{-1}(\mathbf{x} - \mathrm{M})$ *is achievable, then there is an extremal distribution that exactly achieves the Bound (16.3.39); otherwise, if* $\mathrm{M} \in S$ *or if* d^2 *is not achievable, then there is a sequence of* $(\mathrm{M}, \mathbf{\Gamma})$*-feasible distributions that asymptotically approach the Bound (16.3.39).*

Theorem 16.3.5 constitutes a multivariate generalization of Chebyshev's inequality. The tight multivariate one-sided Chebyshev bound is

$$\sup_{\boldsymbol{X} \sim (\mathrm{M}, \mathbf{\Gamma})} P(\boldsymbol{X} > \mathrm{M}_{\mathbf{e}+\boldsymbol{\delta}}) = \frac{1}{1 + d^2}, \qquad (16.3.40)$$

where d^2 is given by:

$$d^2 = \text{minimize} \quad \mathbf{x}'\mathbf{\Gamma}^{-1}\mathbf{x} \qquad (16.3.41)$$
$$\text{subject to} \quad \mathbf{x} \geq \mathrm{M}_{\boldsymbol{\delta}},$$

or alternatively d^2 is given by the Gauge dual problem of (16.3.41):

$$\frac{1}{d^2} = \text{minimize} \quad \mathbf{x}'\mathbf{\Gamma}\mathbf{x} \qquad (16.3.42)$$
$$\text{subject to} \quad \mathbf{x}'\mathrm{M}_{\boldsymbol{\delta}} = 1$$
$$\mathbf{x} \geq \mathbf{0}.$$

If $\mathbf{\Gamma}^{-1}\mathbf{M}_{\delta} \geq \mathbf{0}$, then the tight bound is expressible in closed form:

$$\sup_{X\sim(\mathbf{M},\mathbf{\Gamma})} P(X > \mathbf{M}_{\mathbf{e}+\delta}) = \frac{1}{1+\mathbf{M}'_{\delta}\mathbf{\Gamma}^{-1}\mathbf{M}_{\delta}}. \qquad (16.3.43)$$

Surprisingly, the bound (16.3.43) improves upon the Chebyshev's inequality for scalar random variables. In order to express Chebyshev's inequality we define the squared coefficient of variation: $C_M^2 = \dfrac{M_2 - M_1^2}{M_1^2}$. Chebyshev's inequality is given by:

$$P(X > (1+\delta)M_1) \leq \frac{C_M^2}{\delta^2},$$

where as bound (16.3.43) is stronger:

$$P(X > (1+\delta)M_1) \leq \frac{C_M^2}{C_M^2 + \delta^2}.$$

16.3.3 The complexity of the $(n, 2, R_+^n)$, (n, k, R^n)-bound problems

Bertsimas and Popescu [107] show that the separation problem associated with Problem (D) for the cases $(n, 2, R_+^n)$, (n, k, R^n)-bound problems are NP-hard for $k \geq 3$. By the equivalence of optimization and separation (see Grötschel, Lovász and Schrijver [304]), solving Problem (D) is NP-hard as well. Finally, because of Theorem 2, solving the $(n, 2, R_+^n)$, (n, k, R^n)-bound problems with $k \geq 3$ is NP-hard.

The complexity of the $(n, 2, R_+^n)$-bound problem

The separation problem can be formulated as follows in this case:
Problem 2SEP: Given a multivariate polynomial $g(\mathbf{x}) = \mathbf{x}'\mathbf{H}\mathbf{x} + \mathbf{c}'\mathbf{x} + d$, and a set $S \subseteq R_+^n$, does there exist $\mathbf{x} \in S$ such that $g(\mathbf{x}) < 0$?

If we consider the special case $\mathbf{c} = \mathbf{0}$, $\mathbf{d} = \mathbf{0}$, and $S = R_+^n$, Problem 2SEP reduces to the question whether a given matrix H is co-positive, which is NP-hard (see Murty and Kabadi [560]).

The complexity of the (n, k, R^n)-bound problem for $k \geq 3$

For $k \geq 3$, the separation problem can be formulated as follows:
Problem 3SEP: Given a multivariate polynomial $g(\mathbf{x})$ of degree $k \geq 3$, and a set $S \subseteq R^n$, does there exist $\mathbf{x} \in S$ such that $g(\mathbf{x}) < 0$?

Bertsimas and Popescu [107] show that problem $3SEP$ is NP-hard by performing a reduction from $3SAT$.

16.4 MOMENT PROBLEMS IN FINANCE

The idea of investigating the relation of option and stock prices just based on the no-arbitrage assumption, but without assuming any model for the underlying price dynamics has a long history in the financial economics literature. Cox and Ross [169] and Harrison and Kreps [328] show that the no-arbitrage assumption is equivalent with the existence of a probability distribution π (the so-called martingale measure) such that that option prices become martingales under π. In this section, we survey some recent work of Bertsimas and Popescu [106] that sheds new light to the relation of option and stock prices, and shows that the natural way to address this relation, without making distributional assumptions for the underlying price dynamics, but only using the no-arbitrage assumption, is the use of convex, and in particular semidefinite, optimization methods.

In order to motivate the overall approach we formulate the problem of deriving optimal bounds on the price of a European call option given the mean and variance of the underlying stock price solved by Lo [492]. A call option on a certain stock with maturity T and strike k gives the owner of the option the right to buy the underlying stock at time T at price k. If X is the price of the stock at time T, then the payoff of such an option is zero if $X < k$ (the owner will not exercise the option), and $X - k$ if $X \geq k$, i.e., it is $\max(0, X - k)$. Following Cox and Ross [169] and Harrison and Kreps [328], the no-arbitrage assumption is equivalent with the existence of a probability distribution π of the stock price X, such that the price of any European call option with strike price k is given by

$$q(k) = E[\max(0, X - k)],$$

where the expectation is taken over the unknown distribution π. Note that we have assumed, without loss of generality, that the risk free interest rate is zero. Moreover, given that the mean and variance of the underlying asset are observable:

$$E[X] = \mu, \qquad \text{and} \qquad Var[X] = \sigma^2,$$

the problem of finding the best possible upper bound on the call price, written as

$$Z^* = \sup_{X \sim (\mu,\sigma^2)^+} E[\max(0, X - k)],$$

can be formulated as follows:

$$\begin{aligned} Z^* = \sup \quad & E[\max(0, X - k)] \\ \text{subject to} \quad & E[X] = \mu \\ & Var[X] = \sigma^2. \end{aligned}$$

Conversely, the problem of finding sharp upper bounds on the moments of the stock price using known option prices, can be formulated as follows:

$$\begin{aligned} \sup \quad & E[X]\text{ , or } E[X^2]\text{ , or } E[\max(0, X - k)] \\ \text{subject to} \quad & E[\max(0, X - k_i)] = q_i, \; i = 1, \ldots, n. \end{aligned}$$

These formulations naturally lead us to the following general optimization problem:

$$(P)\quad Z_P = \sup \quad E[\phi(\boldsymbol{X})]$$
$$\text{subject to} \quad E[f_i(\boldsymbol{X})] = q_i,\ i = 0, 1, \ldots, n\,, \tag{16.4.44}$$

where $\boldsymbol{X} = (X_1, \ldots, X_m)$ is a multivariate random variable, and $\phi : R^m \to R$ is a real-valued objective function, $f_i : R^m \to R$, $i = 1, \ldots, n$ are also real-valued, so-called *moment functions* whose expectations $q_i \in R$, referred to as *moments*, are known and finite. We assume that $f_0(\mathbf{x}) = 1$ and $q_0 = E[f_0(\boldsymbol{X})] = 1$, corresponding to the implied probability-mass constraint. Note that the problem of finding optimal bounds on $P(\boldsymbol{X} \in S)$ of the previous section given moments of $\boldsymbol{X}$ can be formulated as a special case of Problem (16.4.44) with $\phi(\mathbf{x}) = \chi_S(\mathbf{x})$.

The dual problem can be written as

$$(D)\quad Z_D = \inf \quad E[\boldsymbol{y}'f(\boldsymbol{X})] \;=\; \inf\ \boldsymbol{y}'\boldsymbol{q}$$
$$\text{subject to} \quad \boldsymbol{y}'f(\mathbf{x}) \geq \phi(\mathbf{x}), \quad \forall \mathbf{x} \in R^m.$$

Smith [741] shows that if the vector of moments $\mathbf{q}$ is interior to the feasible moment set $\mathcal{M} = \{E[f(\boldsymbol{X})] \mid \boldsymbol{X} \text{ arbitrary multivariate distribution}\}$, then strong duality holds: $Z_P = Z_D$. Thus by solving Problem (D) we obtain the desired sharp bounds.

16.4.1 Bounds in one dimension

We are given the n first moments $(q_1, q_2, \ldots, q_n)$, (we let $q_0 = 1$) of the price of an asset, and we are interested in finding the best possible bounds on the price of an option with payoff $\phi(x)$. An example is a European call option with payoff $\phi(x) = \max(0, x-k)$. Problems (P) and (D) are as follows:

$$(P)\quad \text{maximize} \quad E_\pi[\max(0, X-k)] = \int_0^\infty \max(0, x-k)\pi(x)dx$$
$$\text{subject to} \quad E_\pi[X^i] = \int_0^\infty x^i\pi(x)dx = q_i, \quad i = 0, 1, \ldots, n, \tag{16.4.45}$$
$$\pi(x) \geq 0.$$

$$(D)\quad \text{minimize} \quad \sum_{r=0}^{n} y_i q_i$$
$$\text{subject to} \quad \sum_{r=0}^{n} y_r x^r \geq \max(0, x-k), \quad \forall x \in R_+. \tag{16.4.46}$$

The next theorem shows that Problem (16.4.45) can be solved as a semidefinite optimization problem.

Theorem 16.4.1 *The best upper bound on the price of a European call option with strike k given the n first moments $(q_1, \ldots, q_n)$ $(q_0 = 1)$ of the underlying stock is given by the solution of the following semidefinite optimization problem:*

$$
\begin{aligned}
\text{minimize} \quad & \sum_{r=0}^{n} y_i q_i \\
\text{subject to} \quad & 0 = \sum_{i,j:\ i+j=2l-1} x_{ij}, && l = 1, \ldots, n, \\
& \sum_{r=0}^{l} y_r \binom{k-r}{l-r} k^r = \sum_{i,j:\ i+j=2l} x_{ij}, && l = 0, \ldots, n, \\
& 0 = \sum_{i,j:\ i+j=2l-1} z_{ij}, && l = 1, \ldots, n, \\
& (y_0 + k) + (y_1 - 1)k + \sum_{r=2}^{k} y_r k^r = x_{00}, \\
& (y_1 - 1)k + \sum_{r=2}^{k} y_r r k^r = \sum_{i,j:\ i+j=2} x_{ij}, \\
& \sum_{r=l}^{k} y_r \binom{r}{l} k^r = \sum_{i,j:\ i+j=2l} x_{ij}, && l = 2, \ldots, n, \\
& \boldsymbol{X}, \boldsymbol{Z} \succeq \boldsymbol{0}.
\end{aligned}
\qquad (16.4.47)
$$

Proof: We note that the feasible region of Problem (16.4.46) can be written as

$$
\sum_{r=0}^{n} y_r x^r \geq 0 \quad \text{for all } x \in [0, k],
$$

$$
(y_0 + k) + (y_1 - 1)x + \sum_{r=2}^{n} y_r x^r \geq 0 \quad \text{for all } x \in [k, \infty).
$$

by applying Proposition 16.3.1 (a), (b) we reformulate Problem (16.4.46) as the semidefinite optimization Problem (16.4.47). ∎

We next consider an option with payoff function given as follows:

$$\phi(x) = \begin{cases} \phi_0(x), & x \in [0, k_1], \\ \phi_1(x), & x \in [k_1, k_2], \\ \vdots & \vdots \\ \phi_{d-1}(x), & x \in [k_{d-1}, k_d], \\ \phi_d(x), & x \in [k_d, \infty), \end{cases} \tag{16.4.48}$$

where the functions $\phi_r(x)$, $r = 0, 1, \ldots, d$ are polynomials. Given the generality of the payoff function (16.4.48), we can approximate the payoff of any option using the payoff function (16.4.48). In this case the dual problem becomes:

$$\begin{aligned} \text{minimize} \quad & \sum_{r=0}^{n} y_i q_i \\ \text{subject to} \quad & \sum_{r=0}^{n} y_r x^r \geq \begin{cases} \phi_0(x), & x \in [0, k_1], \\ \phi_1(x), & x \in [k_1, k_2], \\ \vdots & \vdots \\ \phi_{d-1}(x), & x \in [k_{d-1}, k_d], \\ \phi_d(x), & x \in [k_d, \infty), \end{cases} \end{aligned} \tag{16.4.49}$$

The next theorem shows that the problem of finding best possible bounds on an option with a general piecewise polynomial payoff function can be formulated as a semidefinite optimization problem.

Theorem 16.4.2 *The best possible bounds for the price of an option with a piecewise polynomial payoff function $\phi(x)$ shown in (16.4.48), given moments of the underlying asset, can be formulated as a semidefinite optimization problem.*

Proof:

The constraint set for Problem (16.4.49) can be written as follows:

$$\sum_{r=0}^{n} y_r x^r \geq \phi_i(x), \qquad x \in [k_{i-1}, k_i], \;\; i = 1, \ldots, d+1,$$

with $k_0 = 0$, $k_{d+1} = \infty$. Let $\phi_i(x) = \sum_{r=0,\ldots,m_i} a_{ir} x^r$, and assume without loss of generality that $m_i \leq n$. Then, the constraint set for Problem (16.4.49) can be equivalently written as

$$\sum_{r=0}^{m_i} (y_r - a_{ir}) x^r + \sum_{r=m_i+1}^{n} y_r x^r \geq 0, \qquad x \in [k_{i-1}, k_i], \;\; i = 1, \ldots, d+1.$$

For the the interval $[k_0, k_1]$ we apply Proposition 16.3.1(a), for the intervals $[k_{i-1}, k_i]$, $i = 2, \ldots, d$, we apply Proposition 16.3.1(c), and for the interval

$[k_d, \infty)$, we apply Proposition 16.3.1(b), to express Problem (16.4.49) as a semidefinite optimization problem. ■

In the special case that we are only given the mean μ and variance σ^2, Lo [492] found a tight upper bound in closed form. The proof below follows Bertsimas and Popescu [106].

Theorem 16.4.3 (Tight upper bound on option prices) *The tight upper bound on the price of an option with strike k, on a stock whose price at maturity has a known mean μ and variance σ^2, is computed by:*

$$\max_{X \sim (\mu, \sigma^2)^+} E[\max(0, X-k)] = \begin{cases} \frac{1}{2}\left[(\mu-k)+\sqrt{\sigma^2+(\mu-k)^2}\right], & \text{if } k \geq \frac{\mu^2+\sigma^2}{2\mu}, \\ \mu - k + k\frac{\sigma^2}{\mu^2+\sigma^2}, & \text{if } k < \frac{\mu^2+\sigma^2}{2\mu}. \end{cases}$$

Proof:
The dual in this case can be formulated by associating dual variables y_0, y_1, y_2 with the probability-mass, mean and respectively, variance constraints. We obtain the following dual formulation:

$$\begin{aligned} Z_D = \text{minimize} \quad & (\mu^2+\sigma^2)y_2 + \mu y_1 + y_0 \\ \text{subject to} \quad & g(x) = y_2 x^2 + y_1 x + y_0 \geq \max(0, x-k), \quad \forall x \geq 0. \end{aligned}$$

A dual feasible function $g(\cdot)$ is any quadratic function that, on the positive orthant, is nonnegative and lies above the line $(x-k)$. In an optimal solution, such a quadratic should be tangent to the line $(x-k)$, so we can write $g(x) - (x-k) = a(x-b)^2$, for some $a \geq 0$. The nonnegativity constraint on $g(\cdot)$ can be expressed as $a(x-b)^2 + x - k \geq 0$, $\forall x \geq 0$. Let $x_0 = b - \frac{1}{2a}$ be the point of minimum of this quadratic. Depending whether x_0 is nonnegative or not, either the inequality at $x = x_0$ or at $x = 0$ is binding in an optimal solution. We have two cases:

(a) If $b \geq \frac{1}{2a}$, then $-\frac{1}{4a} + b - k = 0$ (binding constraint at x_0);
Substituting $a = \frac{1}{4(b-k)}$ in the objective, we obtain:

$$Z_D = \min_b \frac{((\mu-k)+(b-k))^2+\sigma^2}{4(b-k)} = \frac{1}{2}\left[(\mu-k)+\sqrt{\sigma^2+(\mu-k)^2}\right],$$

achieved at $b_0 = \frac{\mu^2+\sigma^2}{\mu}$.
This bound is valid whenever $b_0 \geq \frac{1}{2a_0} = 2(b_0 - k)$, that is $\frac{\mu^2+\sigma^2}{2\mu} \leq k$.

(b) If $b < \frac{1}{2a}$, then $ab^2 - k = 0$ (binding constraint at $x = 0$).

Substituting $a = \frac{k}{b^2}$ in the objective, we obtain:

$$Z_D = \min_b \frac{k}{b^2}(\mu^2 + \sigma^2) - 2\frac{k}{b}\mu + \mu = \mu - k\frac{\mu^2}{\mu^2 + \sigma^2},$$

achieved at $b_0 = \dfrac{\mu^2 + \sigma^2}{\mu}$.

This bound is valid whenever $b_0 < \frac{1}{2a_0} = \frac{b_0^2}{2k}$, that is $\frac{\mu^2 + \sigma^2}{2\mu} > k$. ■

16.4.2 Bounds in multiple dimensions

In this section, we provide bounds for the case that we have information about a set of m different stocks. In particular, we have an option with payoff function $\phi(\boldsymbol{x})$, $\phi : R_+^m \to R$, and a vector of n moment functions $\boldsymbol{f} = (f_1, \ldots, f_n)$ (we let $f_0(\boldsymbol{x}) = 1$), $f_i : R_+^m \to R$, $i = 0, 1, \ldots, n$, and the corresponding vector of moments $\boldsymbol{q} = (q_1, \ldots, q_n)$ (we let $q_0 = 1$). We address in this section the upper bound problem (16.4.44):

$$\begin{aligned} \text{maximize} \quad & E_\pi[\phi(\boldsymbol{X})] \\ \text{subject to} \quad & E_\pi[f_i(\boldsymbol{X})] = q_i, \; i = 1, \ldots, n. \\ & \int_0^\infty \pi(\boldsymbol{x}) d\boldsymbol{x} = 1 \\ & \pi(\boldsymbol{x}) \geq 0, \quad \boldsymbol{x} \in R_+^m, \end{aligned} \qquad (16.4.50)$$

where the expectation is taken over all martingale measures defined on R_+^m. We can solve the lower bound problem by changing the sign of the objective function ϕ in Problem (16.4.50).

Bertsimas and Popescu [106] show that solving Problem (16.4.50) is NP-hard. For this reason, we find a weaker bound by optimizing over all martingale measures defined on R^m as opposed to R_+^m. For this reason we consider the following problem:

$$\begin{aligned} \text{maximize} \quad & E_\pi[\phi(\boldsymbol{X})] \\ \text{subject to} \quad & E_\pi[f_i(\boldsymbol{X})] = q_i, \; i = 1, \ldots, n. \\ & \int_{-\infty}^\infty \pi(\boldsymbol{x}) d\boldsymbol{x} = 1 \\ & \pi(\boldsymbol{x}) \geq 0, \quad \boldsymbol{x} \in R^m, \end{aligned} \qquad (16.4.51)$$

and its dual:

$$\begin{aligned} \text{minimize} \quad & y_0 + \sum_{i=1}^{n} y_i q_i \\ \text{subject to} \quad & y_0 + \sum_{i=1}^{n} y_i f_i(\boldsymbol{x}) \geq \phi(\boldsymbol{x}), \quad \forall \boldsymbol{x} \in R^m. \end{aligned} \tag{16.4.52}$$

The best possible upper bound corresponds to the optimal solution value of Problem (16.4.50). Since Problem (16.4.51) is a relaxation of Problem (16.4.50), we obtain an upper bound, although not necessarily the optimal one, by solving Problem (16.4.51), and by strong duality, Problem (16.4.52). In the next theorem we identify cases under which we can solve Problem (16.4.52), efficiently.

Theorem 16.4.4 *An upper bound on Problem (16.4.50) can be found in polynomial time in the following cases:*

(a) *If ϕ and f_i, $i = 1, \ldots, n$ are quadratic or linear functions of the form*

$$\begin{aligned} \phi(\boldsymbol{x}) &= \boldsymbol{x}'\boldsymbol{A}\boldsymbol{x} + \boldsymbol{b}'\boldsymbol{x} + c \\ f_i(\boldsymbol{x}) &= \boldsymbol{x}'\boldsymbol{A}_i\boldsymbol{x} + \boldsymbol{b}_i'\boldsymbol{x} + c_i, \qquad i = 1, \ldots, n. \end{aligned} \tag{16.4.53}$$

then Problem (16.4.52), and thus Problem (16.4.51), can be solved in polynomial time by solving the following semidefinite optimization problem:

$$\begin{aligned} \text{minimize} \quad & \sum_{i=1}^{n} y_i q_i \\ \textit{subject to} \quad & \begin{bmatrix} \sum_{i=1}^{n} y_i c_i + y_0 - c & \left(\sum_{i=1}^{n} y_i \boldsymbol{b}_i - \boldsymbol{b}\right)' /2 \\ \left(\sum_{i=1}^{n} y_i \boldsymbol{b}_i - \boldsymbol{b}\right) /2 & \sum_{i=1}^{n} y_i \boldsymbol{A}_i - \boldsymbol{A} \end{bmatrix} \succeq \boldsymbol{0}. \end{aligned} \tag{16.4.54}$$

(b) *If ϕ and f_i, $i = 1, \ldots, n$, are quadratic or piecewise linear functions of the form*

$$\begin{aligned} \phi(x) &= \boldsymbol{x}'\boldsymbol{A}\boldsymbol{x} + \boldsymbol{b}_k'\boldsymbol{x} + c_k \qquad \boldsymbol{x} \in D_k, \quad k = 1, \ldots, d, \\ f_i(\boldsymbol{x}) &= \boldsymbol{x}'\boldsymbol{A}_i\boldsymbol{x} + \boldsymbol{b}_{ik}'\boldsymbol{x} + c_{ik}, \quad \boldsymbol{x} \in D_k, \quad i = 1, \ldots, n, \ k = 1, \ldots, d, \end{aligned} \tag{16.4.55}$$

over the d disjoint polyhedra $D_1, \ldots, D_d$ that form a partition of R^m, and d is a polynomial in n, m, then Problem (16.4.52), and thus Problem (16.4.51), can be solved in polynomial time.

Proof:

(a) We consider first the case when all the functions ϕ and f_i are are quadratic or linear as in Eq. (16.4.53). In this case, Problem (16.4.52) becomes:

$$\begin{array}{ll} \text{minimize} & y_0 + \sum_{i=1}^{n} y_i q_i \\ \text{subject to} & g(\boldsymbol{x}) \geq 0, \quad \forall \boldsymbol{x} \in R^m, \end{array}$$

where

$$g(\boldsymbol{x}) = y_0 + \sum_{i=1}^{n} y_i f_i(\boldsymbol{x}) - \phi(\boldsymbol{x}) = \boldsymbol{x}'\hat{\boldsymbol{A}}\boldsymbol{x} + \hat{\boldsymbol{b}}'\boldsymbol{x} + \hat{c},$$

with

$$\hat{\boldsymbol{A}} = \sum_{i=1}^{n} y_i \boldsymbol{A}_i - \boldsymbol{A}, \qquad \hat{\boldsymbol{b}} = \sum_{i=1}^{n} y_i \boldsymbol{b}_i - \boldsymbol{b}, \qquad \hat{c} = \sum_{i=1}^{n} y_i c_i + y_0 - c.$$

Thus, the constraints $g(\boldsymbol{x}) \geq 0$ are equivalent to

$$\boldsymbol{x}'\hat{\boldsymbol{A}}\boldsymbol{x} + \hat{\boldsymbol{b}}'\boldsymbol{x} + \hat{c} \geq 0, \quad \forall \boldsymbol{x} \in R^m,$$

or equivalently

$$\begin{pmatrix} 1 \\ \boldsymbol{x} \end{pmatrix}' \begin{bmatrix} \hat{c} & \hat{\boldsymbol{b}}/2 \\ \hat{\boldsymbol{b}}/2 & \hat{\boldsymbol{A}} \end{bmatrix} \begin{pmatrix} 1 \\ \boldsymbol{x} \end{pmatrix} \geq 0, \quad \forall \boldsymbol{x} \in R^m. \tag{16.4.56}$$

Eq. (16.4.56) holds if and only if

$$\begin{bmatrix} \hat{c} & \hat{\boldsymbol{b}}'/2 \\ \hat{\boldsymbol{b}}/2 & \hat{\boldsymbol{A}} \end{bmatrix} \succeq \boldsymbol{0},$$

i.e., the matrix $\begin{bmatrix} \hat{c} & \hat{\boldsymbol{b}}'/2 \\ \hat{\boldsymbol{b}}/2 & \hat{\boldsymbol{A}} \end{bmatrix}$ is positive semidefinte. Thus, Problem (16.4.52) is equivalent to the semidefinite optimization problem (16.4.54).

(b) If the functions ϕ or $f_i, i = 1, \ldots, n$ are as in (16.4.55), then Problem (16.4.52) can be expressed as

$$\begin{array}{lll} \text{minimize} & y_0 + \sum_{i=1}^{n} y_i q_i & \\ \text{subject to} & g_k(\boldsymbol{x}) = \boldsymbol{x}'\hat{\boldsymbol{A}}\boldsymbol{x} + \hat{\boldsymbol{b}}_k'\boldsymbol{x} + \hat{c}_k \geq 0, & \forall\, \boldsymbol{x} \in D_k,\ k = 1, \ldots, d, \end{array} \tag{16.4.57}$$

where

$$\hat{\boldsymbol{A}} = \sum_{i=1}^{n} y_i \boldsymbol{A}_i - \boldsymbol{A}, \qquad \hat{\boldsymbol{b}}_k = \sum_{i=1}^{n} y_i \boldsymbol{b}_{ik} - \boldsymbol{b}_k, \qquad \hat{c}_k = \sum_{i=1}^{n} y_i c_{ik} + y_0 - c_k.$$

By the equivalence of separation and optimization (see Grötschel, Lovász and Schrijver [302]), Problem (16.4.57) can be solved in polynomial time if and only if the following separation problem can be solved in polynomial time.

The Separation Problem:

Given an arbitrary $\boldsymbol{y} = (y_0, y_1, \ldots, y_n)$, check whether $g_k(\boldsymbol{x}) \geq 0$, for all $\boldsymbol{x} \in D_k$, $k = 1, \ldots, n$ and if not, find a violated inequality.

We show next that solving the separation problem reduces to checking whether the matrix $\hat{\boldsymbol{A}}$ is positive semidefinite, and in this case solving the convex quadratic problems

$$\min_{\boldsymbol{x} \in D_k} g_k(\boldsymbol{x}), \qquad k = 1, \ldots, d.$$

This can be done in polynomial time using the ellipsoid algorithm (see Grötschel, Lovász and Schrijver [302]). The following algorithm solves the separation problem in polynomial time:

Algorithm A:

1. If $\hat{\boldsymbol{A}}$ is not positive semidefinite, we construct a vector $\boldsymbol{x_0}$ so that $g_k(\boldsymbol{x_0}) < 0$ for some $k = 1, \ldots, n$. We decompose $\hat{\boldsymbol{A}} = \boldsymbol{Q}'\boldsymbol{\Lambda}\boldsymbol{Q}$, where $\boldsymbol{\Lambda} = \text{diag}(\lambda_1, \ldots, \lambda_n)$ is the diagonal matrix of eigenvalues of $\hat{\boldsymbol{A}}$. Let $\lambda_i < 0$ be a negative eigenvalue of $\hat{\boldsymbol{A}}$. Let $\boldsymbol{u}$ be a vector with $u_j = 0$, for all $j \neq i$, and u_i selected as follows: Let v_k be the largest root of each polynomial if it exists. Let $u_i = \max_k v_k + 1$. If all the polynomials do not have real roots, then u_i can be chosen arbitrarily. Then

$$\lambda_i u_i^2 + (\boldsymbol{Q}\hat{\boldsymbol{b}}_k)_i u_i + \hat{c}_k < 0, \qquad \forall\ k = 1, \ldots, d.$$

   Let $\boldsymbol{x}_0 = \boldsymbol{Q}'\boldsymbol{u}$. Since the polyhedra D_k form a partition of R^m, then $\boldsymbol{x}_0 \in D_{k_0}$ for some k_0. Then,

$$\begin{aligned} g_{k_0}(\boldsymbol{x}_0) &= \boldsymbol{x}_0'\hat{\boldsymbol{A}}\boldsymbol{x}_0 + \hat{\boldsymbol{b}}_{k_0}'\boldsymbol{x}_0 + \hat{c}_{k_0} \\ &= \boldsymbol{u}'\boldsymbol{Q}\boldsymbol{Q}'\boldsymbol{\Lambda}\boldsymbol{Q}\boldsymbol{Q}'\boldsymbol{u} + \hat{\boldsymbol{b}}_{k_0}'\boldsymbol{Q}'\boldsymbol{u} + \hat{c}_{k_0} \\ &= \boldsymbol{u}'\boldsymbol{\Lambda}\boldsymbol{u} + (\boldsymbol{Q}\hat{\boldsymbol{b}}_{k_0})'\boldsymbol{u} + \hat{c}_{k_0} \\ &= \sum_{j=1}^{n} \lambda_j u_j^2 + \sum_{j=1}^{n} (\boldsymbol{Q}\hat{\boldsymbol{b}}_{k_0})_j u_j + \hat{c}_{k_0} \\ &= \lambda_i u_i^2 + (\boldsymbol{Q}\hat{\boldsymbol{b}}_{k_0})_i u_i + \hat{c}_{k_0} < 0. \end{aligned}$$

   This produces a violated inequality.

2. Otherwise, if $\hat{\boldsymbol{A}}$ is positive semidefinite, then we test if $g_k(\boldsymbol{x}) \geq 0$, $\forall\ \boldsymbol{x} \in D_k$ by solving d convex quadratic optimization problems:

$$\min_{\boldsymbol{x} \in D_k} \boldsymbol{x}'\hat{\boldsymbol{A}}\boldsymbol{x} + \hat{\boldsymbol{b}}_k'\boldsymbol{x} + \hat{c}_k, \quad \text{for } k = 1, \ldots, d. \tag{16.4.58}$$

   We denote by $\boldsymbol{x}_k^*$ an optimal solution of Problem (16.4.58), and $z_k = g_k(\boldsymbol{x}_k^*)$ the optimal value of Problem (16.4.58). If $z_k \geq 0$ for all $k = 1, \ldots, d$, then there is no violated inequality. Otherwise, if $z_{k_0} < 0$ for some k_0, then we find x_k^* such that $g(x_k^*) < 0$, which represents a violated inequality.

Thus, Algorithm A solves the separation problem in polynomial time. It follows that Problem (16.4.52), and hence Problem (16.4.51), can be solved in polynomial time. ∎

Examples

Suppose we have observed the price q_1 of a European call option with strike k_1 for stock 1, and the price q_2 of a European call option with strike k_2 for stock 2. In addition, we have estimated the means μ_1, μ_2, the variances σ_1^2, σ_2^2 and the covariance σ_{12}^2 of the prices of the two underlying stocks. Suppose, in addition, we are interested in obtaining an upper bound on the price of a European call option with strike k for stock 1. Intuition suggests that since the prices of the two stocks are correlated, the price of a call option on stock 1 with strike k might be affected by the available information regarding stock 2. We can find an upper bound on the price of a call option on stock 1 with strike k, by solving Problem (16.4.50), with $m = 2$, $n = 7$. ¿From Theorem 16.4.4(b), Problem (16.4.51) can be solved efficiently. In this case, there are six sets D_k as follows:

$$\begin{aligned}
D_1 &= \{(x_1, x_2) \mid x_1 \geq k, x_2 \geq k_2\},\\
D_2 &= \{(x_1, x_2) \mid x_1 \geq k, x_2 \leq k_2\},\\
D_3 &= \{(x_1, x_2) \mid k_1 \leq x_1 \geq k, x_2 \geq k_2\},\\
D_4 &= \{(x_1, x_2) \mid k_1 \leq x_1 \geq k, x_2 \leq k_2\},\\
D_5 &= \{(x_1, x_2) \mid x_1 \leq k, x_2 \geq k_2\},\\
D_6 &= \{(x_1, x_2) \mid x_1 \leq k, x_2 \leq k_2\}.
\end{aligned}$$

As another example, suppose we are interested to find an upper bound on the price of an option with payoff

$$\phi(\boldsymbol{x}) = \max(0, \boldsymbol{a}_1'\boldsymbol{x} - k_1, \boldsymbol{a}_2'\boldsymbol{x} - k_2).$$

This option allows its holder to buy at maturity two stock indices: the first one (given by the vector $\boldsymbol{a}_1$) at price k_1, and the second one (given by the vector $\boldsymbol{a}_2$) at price k_2. Suppose we have estimated the mean and covariance matrix of the underlying securities. Again, Theorem 16.4.4(b) applies. In this case there are three sets D_k that form a polyhedral partition of R^m:

$$\begin{aligned}
D_1 &= \{\boldsymbol{x} \in R^m \mid \boldsymbol{a}_1'\boldsymbol{x} - k_1 \leq 0, \boldsymbol{a}_2'\boldsymbol{x} - k_2 \leq 0\},\\
D_2 &= \{\boldsymbol{x} \in R^m \mid 0 \leq \boldsymbol{a}_1'\boldsymbol{x} - k_1, \boldsymbol{a}_2'\boldsymbol{x} - k_2 \leq \boldsymbol{a}_1'\boldsymbol{x} - k_1\},\\
D_3 &= \{\boldsymbol{x} \in R^m \mid 0 \leq \boldsymbol{a}_2'\boldsymbol{x} - k_1, \boldsymbol{a}_1'\boldsymbol{x} - k_2 \leq \boldsymbol{a}_2'\boldsymbol{x} - k_1\}.
\end{aligned}$$

Note that if $\boldsymbol{x} \in D_1$, $\phi(\boldsymbol{x}) = 0$, while if $\boldsymbol{x} \in D_2$, $\phi(\boldsymbol{x}) = \boldsymbol{a}_1'\boldsymbol{x} - k_1$. Finally, if $\boldsymbol{x} \in D_3$, $\phi(\boldsymbol{x}) = \boldsymbol{a}_2'\boldsymbol{x} - k_2$.

16.5 MOMENT PROBLEMS IN DISCRETE OPTIMIZATION

In this section, we explore the connection of moment problems and discrete optimization. We consider the maximum $s-t$ cut problem. Goemans and Williamson [285] showed that a natural semidefinite relaxation is within 0.878 of the value of the maximum $s-t$ cut. Bertsimas and Ye [111] provide an alternative interpretation of their method that makes the connection of moment problems and discrete optimization explicit. We review this development in this section.

Given an undirected graph on n nodes, and weights c_{ij} on the edges we would like to find an $s-t$ cut of maximum weight. We formulate the problem as follows:

$$\begin{aligned} Z_{IP} = \text{maximize} \quad & \frac{1}{2}\sum_{i,j} c_{ij}(1 - x_i x_j) \\ \text{subject to} \quad & x_s + x_t = 0, \\ & x_j^2 = 1, \; j = 1, \ldots, n, \end{aligned} \tag{16.5.59}$$

and consider the semidefinite relaxation

$$\begin{aligned} Z_{SD} = \text{maximize} \quad & \frac{1}{2}\sum_{i,j} c_{ij}(1 - y_{ij}) \\ \text{subject to} \quad & y_{st} = 0, \\ & y_{jj} = 1, \; j = 1, \ldots, n, \\ & \mathbf{Y} \succeq \mathbf{0}. \end{aligned} \tag{16.5.60}$$

We solve the semidefinite relaxation (16.5.60) and obtain the semidefinite matrix $\mathbf{Y}$. In order to understand the closeness of Z_{SD} and Z_{IP}, we create a feasible solution to the $s-t$ maximum cut problem by interpreting the matrix $\mathbf{Y}$ as a covariance matrix.

Randomized heuristic H:

1. We generate a vector $\overline{\mathbf{x}}$ from **a multivariate normal distribution** with 0 mean and covariance matrix $\mathbf{Y}$, that is,

$$\overline{\mathbf{x}} \sim N(0, \mathbf{Y}).$$

2. We create a vector $\hat{\mathbf{x}}$ with components equal to 1 or -1:

$$\hat{\mathbf{x}} = \text{sign}(\overline{\mathbf{x}}), \tag{16.5.61}$$

i.e., $\hat{x}_j = 1$ if $\overline{x}_j > 0$, and $\hat{x}_j = -1$ if $\overline{x}_j \leq 0$.

Notice that instead of using a multivariate normal distribution for $\overline{\mathbf{x}}$ we can use any distribution that has covariance $Cov(\overline{\mathbf{x}}) = \mathbf{Y}$. What is interesting is that we show the degree of closeness of Z_{SD} and Z_{IP} by considering results regarding the normal distributed that were known in 1900.

Proposition 16.5.1

$$\mathrm{E}[\hat{x}_j] = 0, \quad \mathrm{E}[\hat{x}_j^2] = 1, \quad j = 1, 2, \ldots, n,$$

$$\mathrm{E}[\hat{x}_i\hat{x}_j] = \frac{2}{\pi}\arcsin(y_{ij}), \quad i,j = 1,2,\ldots,n.$$

Proof:
The marginal distribution of $\overline{x}_i$ is $N(0,1)$, and thus $\Pr(\hat{x}_i = 1) = \Pr(\hat{x}_i = -1) = 1/2$. Thus, $\mathrm{E}[\hat{x}_i] = 0$ and $\mathrm{E}[\hat{x}_i^2] = 1$. Furthermore,

$$\begin{aligned}
&\mathrm{E}[\hat{x}_i\hat{x}_j] \\
&\quad = P(\hat{x}_i = 1, \hat{x}_j = 1) + P(\hat{x}_i = -1, \hat{x}_j = -1) \\
&\quad\quad -P(\hat{x}_i = 1, \hat{x}_j = -1) - P(\hat{x}_i = -1, \hat{x}_j = 1) \\
&\quad = P(x_i \geq 0, x_j \geq 0) + P(x_i < 0, x_j < 0) \\
&\quad\quad -P(x_i \geq 0, x_j < 0) - P(x_i < 0, x_j \geq 0).
\end{aligned}$$

The tail probabilities of a multivariate normal distribution is a problem that has been studied in the last 100 years. Sheppard [717] shows (see Johnson and Kotz [381], p. 95) that

$$\begin{aligned}
P(x_i \geq 0, x_j \geq 0) = P(x_i < 0, x_j < 0) &= \frac{1}{4} + \frac{1}{2\pi}\arcsin(y_{ij}) \\
P(x_i \geq 0, x_j < 0) = P(x_i < 0, x_j \geq 0) &= \frac{1}{4} - \frac{1}{2\pi}\arcsin(y_{ij}).
\end{aligned}$$

This leads to

$$\mathrm{E}[\hat{x}_i\hat{x}_j] = \frac{2}{\pi}\arcsin(y_{ij}). \tag{16.5.62}$$

■

Theorem 16.5.1 *Heuristic H provides a feasible solution for the $s-t$ maximum cut problem with objective value Z_H:*

$$\mathrm{E}[Z_H] \geq 0.878 Z_{SD}.$$

Proof
First, we notice that $\mathrm{E}[\hat{x}_s + \hat{x}_t] = 0$, and

$$\mathrm{E}[(\hat{x}_s + \hat{x}_t)^2] = 2 + 2\mathrm{E}[\hat{x}_s\hat{x}_t] = 2 + \frac{4}{\pi}\arcsin(-1) = 2 - 2 = 0,$$

i.e., $\hat{x}_s + \hat{x}_t = 0$ with probability 1, i.e., the solution is feasible.

Moreover, the value of the heuristic solution is

$$\begin{aligned}
\mathrm{E}[Z_H] &= \sum_{ij} c_{ij}\Big(P(\hat{x}_i = 1, \hat{x}_j = -1) + P(\hat{x}_i = 1, \hat{x}_j = -1)\Big) \\
&= \sum_{ij} c_{ij}\left(\frac{1}{2} - \frac{1}{\pi}\arcsin(y_{ij})\right) \\
&\geq 0.878\frac{1}{2}\sum_{ij} c_{ij}(1 - y_{ij}) \\
&= 0.878 Z_{SD},
\end{aligned}$$

where we used the inequality $\frac{1}{2} - \frac{1}{\pi}\arcsin(y_{ij}) \geq (0.878)\frac{1}{2}(1 - y_{ij})$ from Goemans and Williamson [285]. ■

Bertsimas and Ye [111] show that the interpretation of the matrix $\mathbf{Y}$ as a covariance matrix leads to interesting bounds for several problems, such as the graph bisection problem, the s, t, u maximum cut problem, and constrained quadratic maximization. If the distribution used is the multivariate normal distribution the method is equivalent to the Goemans and Williamson [285] method. However, it would be interesting to explore the generation of random variables $\overline{\mathbf{x}}$ using a distribution other than the multivariate normal distribution. For the max cut problem, the bound 0.878 is not known to be tight. Generating a random vector using a distribution other than the normal, might lead to a sharper bound.

16.6 CONCLUDING REMARKS

We would like to leave the reader with the following closing thoughts:

(a) Convex optimization has a central role to play in moment problems arising in probability and finance. Optimization not only offers a natural way to formulate and study such problems, it leads to unexpected improvements of classical results.

(b) Semidefinite optimization represents a promising direction for further research in the area of stochastic optimization. The two major areas in which semidefinite optimization has had a very positive impact are discrete optimization and control theory. Paralleling the development in discrete optimization, we believe that semidefinite optimization can play an important role in stochastic optimization by increasing our ability to find better lower bounds.

17 DESIGN OF EXPERIMENTS IN STATISTICS

Valerii Fedorov and Jon Lee

17.1 DESIGN OF REGRESSION EXPERIMENTS

Valerii Fedorov

17.1.1 Main Optimization Problem

Models and Information Matrix. Most of the results in experimental design theory are related to the linear regression models:

$$\mathrm{E}\{y|x\} = \eta(\theta^T, x) \text{ and } \mathrm{Var}\{y|x\} = \sigma^2(x), \tag{17.1.1}$$

where the observation y is a random variable, E and Var stand for expectation and variance, respectively. The transform $\sigma^{-1}(x)y$ allows us to confine ourselves to the case with $\sigma(x) \equiv 1$. Observations y and y' are assumed to be uncorrelated, i.e., $\mathrm{Cov}\{y, y'|x\} = 0$, where Cov stands for the covariance. In most cases we assume that the response function $\eta(\theta, x)$ is linear with respect to θ, i.e., $\eta(\theta, x) = \theta^T f(x)$. The components of the vector $f^T(x) = (f_1(x), \ldots, f_m(x))$ are called "basis" functions and assumed to be known. The variables $x^T = (x_1, \ldots, x_k)$ are called independent, or control, or design variables. The parameters $\theta^T = (\theta_1, \ldots, \theta_m)$ are typically unknown and must be estimated. Whenever it is not confusing the same symbol will be used for a random variable and its observed value.

Let r_i, $i = 1, \ldots, n$, be the number of observations y_{ij}, $j = 1, \ldots, r_i$, that were made at the design or support point x_i. The collection of variables

$$\xi_N = \left\{ \begin{array}{c} x_1, \ldots, x_n \\ p_1, \ldots, p_n \end{array} \right\} = \left\{ \begin{array}{c} x_i \\ p_i \end{array} \right\}_1^n \tag{17.1.2}$$

with $p_i = r_i/N$ and $N = \sum_{i=1}^n r_i$ is called the design (of the experiment). If an experimenter can select any x from X then the latter is called the design region.

Based on the observations

$$y_{11}, \ldots, y_{1r_1}, \ldots, y_{n1}, \ldots, y_{nr_n}$$

the least squares estimator (l.s.e.)/best linear unbiased estimator (b.l.u.e.) [cf. the Gauss-Markov Theorem; see Rao (1973), Chpt. 4a] $\hat{\theta}$ for θ is defined to be

$$\hat{\theta} = \arg\min_{\theta} \sum_{i=1}^{n} p_i \left(\overline{y}_i - \theta^T f(x_i)\right)^2, \tag{17.1.3}$$

where

$$\overline{y}_i = r_i^{-1} \sum_{j=1}^{r_i} y_{ij}.$$

For any $\hat{\theta}$ satisfying (17.1.3)

$$\mathrm{E}\{\hat{\theta}\} = \theta_t \text{ and } \mathrm{Var}\{\hat{\theta}\} \leq \mathrm{Var}\{\widetilde{\theta}\}, \tag{17.1.4}$$

where $\widetilde{\theta}$ is any other linear unbiased estimator and t indicates the true value of the estimated parameters. The above inequality is understood in the sense of the positive semidefinite matrix ordering.

It is known that

$$N\mathrm{Var}\{\hat{\theta}\} = D(\xi_N) = M^{-1}(\xi), \tag{17.1.5}$$

where the information matrix

$$M(\xi) = \sum_{i=1}^{n} p_i f(x_i) f^T(x_i) \tag{17.1.6}$$

is assumed regular, i.e., rank$M(\xi) = m$. If the information matrix is regular, then (17.1.3) has an unique solution. If the linear combination $\gamma = L^T\theta$ must be estimated, then $\hat{\gamma} = L^T\hat{\theta}$ is a b.l.u.e. of γ, and

$$\mathrm{Var}\{\hat{\gamma}\} = L^T M^-(\xi_N)L, \tag{17.1.7}$$

where $M^-(\xi_N)$ can be any pseudoinverse matrix of $M(\xi_N)$.

The information matrix $M(\xi_N)$ can be considered as the information about parameters θ accumulated in the experiment made accordingly to the design ξ_N. It plays a key role in experimental design. The most interesting results are developed for the case in which the set of information matrices defined by (17.1.2) and (17.1.6) is extended and defined in the following manner:

$$\mathcal{M}(X) = \{M : M(\xi) = \int_X f(x)f^T(x)\xi(dx)\}, \tag{17.1.8}$$

where

$$\int_X \xi(dx) = 1\,. \tag{17.1.9}$$

The set of all probability measures [continuous designs, cf. Fedorov and Hackl (1997), Chpt.2] defined by (17.1.9) will be denoted by $\Xi(X)$.

The information matrix can be also presented as

$$M(\xi) = \int_X M(x)\xi(dx), \tag{17.1.10}$$

where the matrix $M(x) = f(x)f^T(x)$ is the information matrix of a single observation made at point x.

Let X be compact, and the basis functions $f(x)$ be continuous in X (cf. Fedorov and Hackl (1997), Chpt. 2.3). Then the following theorem holds.

Theorem 17.1.1

- *For any design ξ the information matrix $M(\xi)$ is symmetric and positive semidefinite.*
- *The set $\mathcal{M}(X)$ is compact and convex.*
- *For any matrix M from $\mathcal{M}(X)$ there exists a design ξ that contains not more than $n_0 = 1+m(m+1)/2$ points with non-zero weights and $M(\xi) = M$. If M is a boundary point of $\mathcal{M}(X)$, $n_0 = m(m+1)/2$.*

Equation (17.1.10) allows us to extend the standard results derived for linear regression (17.1.1) to more general or complicated models. For instance, let us assume that the observation has the density function $p(y|x,\theta)$. Given sufficient regularity of this function and independence of the observations [cf. Rao (1973), Chpt. 5f] the maximum likelihood estimator

$$\hat{\theta} = \arg\min_{\theta\in\Omega} \prod_{i=1}^{n}\prod_{j=1}^{r_i} p(y_{ij}|x_i,\theta)$$

is a popular choice of many practitioners. The information matrix for $\hat{\theta}$ (it is called the Fisher information matrix in this particular setting) is

$$NM(\xi, \theta_t) = \sum_{i=1}^{n} p_i M(x_i, \theta_t),$$

or, considering a design as a probability measure, one can introduce

$$M(\xi, \theta_t) = \int M(x_i, \theta_t)\xi(dx),$$

where

$$M(x, \theta) = \mathrm{E}\{f(y|x, \theta) f^T(y|x, \theta)\}, \quad f(y|x, \theta) = \frac{\partial}{\partial \theta} \log p(y|x, \theta).$$

Thus, the Fisher information matrix has exactly the same structure as the information matrix from (17.1.10). As a consequence, all results that are given in this text and mainly based on (17.1.10), are valid for the maximum likelihood estimators. However, in general the Fisher information matrix depends on θ_t, that are not previously known. The corresponding optimal designs are called "locally optimal". That dependence usually is eliminated by the use of some average value of objective function (the notorious Bayesian approach) or by maximization over some previously given set of possible values for θ_t [see more in Fedorov and Hackl (1997). Chpts. 1 and 5].

Characterization of Optimal Designs. *The Basic Optimization Problem.* In the context of continuous designs we define an optimal design as a solution of the optimization problem

$$\xi^* = \arg \min_{\xi \in \Xi(X)} \Psi\left[M(\xi)\right]. \tag{17.1.11}$$

The above problem and its various modifications are the main content of this chapter. Many experimental design problems can relatively easily be solved if they can be approximated by (17.1.11). The word "approximated" should draw attention to the fact that a solution of (17.1.11) does generally not give an exact solution of a practical design problem in which the design measure is defined by (17.1.2). However, it is often acceptable as an approximate solution, and, in particular, when the total number of possible observations N is large.
Alternative Optimization Problems. Instead of the previous optimization problem we can consider the closely related problem

$$M^* = \arg \min_{M \in \mathcal{M}(X)} \Psi(M). \tag{17.1.12}$$

In theory, problem (17.1.12) is easier to solve than (17.1.11). It may be treated as a special case of minimization in the space of positive semidefinite matrices. There exists rather extensive literature [see, for instance, Lewis (1996),

Vandenberghe et al. (1996)]exploring this topic including chapters 2, 4, 16 and 20 from this book. Of course, the problem can be also embedded in convex analysis [cf. Hiriart-Urruty J.-B. and C. Lemaréchal (1993)]. The approach based on (17.1.12) was applied in experimental design by Pukelsheim (1980), (1993).

In case of (17.1.12) one has to work not with probability measures but with information matrices or vectors that are elements of an $m(m+1)/2$-dimensional space. However, in most practical situations it is more difficult to construct $\mathcal{M}(X)$ numerically than to solve (17.1.11). In addition, experience shows that practitioners preferably think in terms of the design variables x , but seldom in terms of the mapping of $\Xi(X)$ to $\mathcal{M}(X)$.

This text is based on (17.1.11) and all results are formulated in terms (17.1.10). Nevertheless we would like to emphasize that most of them are direct corollaries of more general results from convex analysis.

Popular Optimality Criteria We shall use notation M for $M(\xi)$, $\Psi(\xi)$ for $\Psi[M(\xi)]$, Ψ^* for $\Psi(\xi^*)$ and $\min_x$, $\min_\xi$, $\int$, and so on, instead of $\min_{x\epsilon X}$, $\min_{\xi\epsilon\Xi}$, $\int_X$, respectively, if it does not lead to ambiguity. The equation

$$(\theta-\hat{\theta})^T M(\theta-\hat{\theta}) = m \tag{17.1.13}$$

defines the ellipsoid of concentration. The "larger" the matrix M (the "smaller" D) is, the "smaller" is the ellipsoid of concentration. A measure of the "size" of the ellipsoid is, for instance, its volume which is proportional to $|D|^{1/2} = |M|^{-1/2}$. In this text, $|A|$ stands for the determinant of matrix A. Another measure is the length of the largest principal axis of the ellipsoid which is $\lambda_{\min}^{-1/2}(M)$; here, $\lambda_{\min}(M)$ stands for the minimal eigenvalue of matrix M. Thus, we may introduce the D-criterion

$$\Psi(M) = \ln|M|^{-1} = \ln|D| \tag{17.1.14}$$

and the E-criterion

$$\Psi(M) = \lambda_{\min}(M) = \lambda_{\max}(D). \tag{17.1.15}$$

Both criteria are limit cases of the Kiefer criterion

$$\Psi_\gamma(M) = \left(m^{-1}\text{Trace}M^{-\gamma}\right)^{1/\gamma} = (m^{-1}\text{Trace}D^{\gamma})^{1/\gamma}; \tag{17.1.16}$$

$\Psi_\gamma(M)$ results in (17.1.14) if $\gamma \to 0$, and in (17.1.15) if $\gamma \to \infty$. Another popular choice is $\gamma = 1$:

$$\Psi(M) = m^{-1}\,\text{Trace}M^{-1} = m^{-1}\text{Trace}D. \tag{17.1.17}$$

Using Liapunov's Inequality [see, for instance, Wang and Zhang (1995)]

$$(\mathrm{E}\{|z|^\gamma\})^{1/\gamma} \le (E\{|z|^{\gamma'}\})^{1/\gamma'} \text{ for } 0<\gamma\le\gamma',$$

we find that

$$\begin{aligned} |D|^{1/m} &= \lim_{\gamma\to 0}(m^{-1}\text{Trace}D^{\gamma})^{1/\gamma} \le (m^{-1}\text{Trace}D^{\gamma})^{1/\gamma} \\ &\le \lim_{\gamma\to\infty}(m^{-1}\text{Trace}D^{\gamma})^{1/\gamma} = \lambda_{\max}(D). \end{aligned}$$

Thus, the three criteria may be ordered for any design:

$$|D|^{1/m} \leq m^{-1}\text{Trace}D \leq \lambda_{\max}(D)$$

If we are interested in the linear combination $\zeta = L^T\theta$ of the unknown parameters where L is a (known) matrix of order $m \times s$, $s \leq m$, then

$$N\text{Var}\{\hat{\zeta}\} = L^T M^{-1} L = D_\zeta \tag{17.1.18}$$

Although the vector ζ may be estimable even if for some singular M, we always assume that for optimal designs the information matrices are regular if it is not stated otherwise. All the above-mentioned criteria work for $\hat{\zeta}$ if D is replaced by D_ζ. Note, that

$$\Psi(M) = \max_L \frac{L^T M^{-1} L}{L^T L} = \lambda_{\min}(M).$$

Criteria based on (17.1.17) include the linear criterion

$$\Psi(M) = \text{Trace}AM^{-1},$$

where $A = LL^T$ is a positive semidefinite matrix that is usually called the loss matrix.

The variance of the estimated response at x, i.e., $\hat{\eta}(x,\theta) = \hat{\theta}^T f(x)$, is

$$N\text{Var}[\hat{\theta}^T f(x)] = d(x) = f^T(x) M^{-1} f(x). \tag{17.1.19}$$

If we are interested in the response function at a particular point x_0, then

$$\Psi(M) = d(x_0) \tag{17.1.20}$$

is an obvious optimality criterion. If we want to know the response function sufficiently well on some set Z, two popular criteria are

$$\Psi(M) = \int_Z w(x) d(x) dx \tag{17.1.21}$$

and

$$\Psi(M) = \max_{x \in Z} w(x) d(x); \tag{17.1.22}$$

the function $w(x)$ describes the relative importance of a response at x.

All the above introduced criteria are sublinear functions of M, i.e. they are convex, subadditive and positively homogeneous; see definitions, for instance, in Hiriart-Urruty and Lemaréchal (1993), Chpt. V.

Equivalence Theorems. The optimization problem (17.1.12) has been intensively studied since Kiefer's pioneering paper (1959). In this section we summarize the major properties of the traditional optimal designs.

Let us assume that $\Psi(M)$ is a convex function of $M \in \mathcal{M}(X)$, and let

$$\Psi'(M^*, M - M^*) = \lim_{\alpha \to 0} \frac{\Psi[(1-\alpha)M^* + \alpha M] - \Psi[M^*]}{\alpha} \tag{17.1.23}$$

be a directional derivative of $\Psi(M)$. The existence of this derivative is assured by convexity of $\Psi(M)$; see Hiriart-Urruty and Lemaréchal (1993), Chpts. I.4 and VI.1. Using Theorem 1.1.1 from Chpt. VI.1 of the above book we may conclude that:

Theorem 17.1.2 *There exists a solution of (17.1.12) , the set of all solutions is compact and the following statements are equivalent when $M \in \mathcal{M}(X)$:*

- M^* *minimizes* Ψ *over* $\mathcal{M}(X)$;
- $\Psi'(M^*, M - M^*) \geq 0$ *for all* $M \in \mathcal{M}(X)$.

The result is simple and shows how closely the experimental design related to convex analysis. However, it is not very constructive until a simple form of a directional derivative and a simple presentation of $\mathcal{M}(X)$ are found. More constructive version of the same result can be derived if we move from the information matrix space to the space of all possible designs Ξ. Let us note that every information matrix $M(\xi)$ is uniquely defined by the design ξ. That allows to introduce the function $\Psi'(\xi^*, \xi - \xi^*) = \Psi'[M(\xi^*), M(\xi) - M(\xi^*)]$, that can be seen as a directional derivative in $\Xi(X)$. For all the criteria which have been introduced in this section (see Table 1; we still assume regularity of optimal designs) the directional derivative can be presented as:

$$\Psi'(\xi^*, \xi - \xi^*) = \int_X \psi(x, \xi^*)\xi(dx). \qquad (17.1.24)$$

It is worthwhile to emphasize that (17.1.24) takes place for the reported criteria because of the special structure of the information matrices. The use of this structure, and (17.1.24) in particular, helps to enhance theorem 17.1.2:

Theorem 17.1.3 *There exists at least one optimal design containing no more than $n = m(m+1)/2$ supporting points. The set of optimal designs is convex and the following three statements are equivalent:*

- ξ^* *minimizes* $\Psi(\xi)$ *over* $\Xi(X)$;
- $\psi(x, \xi^*) \geq 0$ *for all* $x \in X$;
- ξ^* *maximizes* $\min_{x \in X} \psi(x, \xi)$ *over* $\Xi(X)$.

In addition, the function $\psi(x, \xi^*) = 0$ *for all* $x \in \mathrm{supp}\xi^*$.

Transition from Theorem 17.1.2 to Theorem 17.1.3 is based on the fact that due to (17.1.24) and to the definition of the information matrix [see (17.1.8) and (17.1.9)]:

$$\min_{M \in \mathcal{M}(X)} \Psi'[M^*, M - M^*] = \min_{x \in X} \psi(x, \xi^*) . \qquad (17.1.25)$$

Theorem 17.1.3 is the basic one in convex design theory and its various modifications have been extensively discussed in the statistical literature: see Fedorov

Table 17.1 Sensitivity function for various optimality criteria.

$\Psi(\xi)$	$\phi(x,\xi) = C - \psi(x,\xi)$	C
$\ln\|D\|$	$d(x) = f^T(x)Df(x)$	m
$\text{Trace}D^\gamma$	$f^T(x)D^{\gamma+1}f(x)$	$\text{Trace}D^\gamma$
$\text{Trace}AD,\ A \geq 0$	$f^T(x)DADf(x)$	$\text{Trace}AD$
$\ln\|D_\ell\|$ *	$d(x) - d_k(x)$ $d_k(x) = f_k^T(x)D_k f_k(x)$ $f^T(x) = (f_\ell^T(x), f_k^T(x))$	k
$\delta = \lambda_{\max}(D)$	$\delta^2 f^T(x)(\sum_{i=1}^{\alpha} \pi_i P_i P_i^T)f(x)$ α is a multiple of δ $\sum_{i=1}^{\alpha} \pi_i = 1,\ 0 \leq \pi_i \leq 1$	δ
$d(x_0)$	$d^2(x, x_0)$ $d(x, x_0) = f^T(x)Df(x_0)$	$d(x_0)$
$\int_Z d(x)dx$	$\int_Z d^2(x,z)dz$	$\int_Z d(x)dx$

* **D_ℓ is a submatrix of D corresponding to ℓ parameters, $k = m - \ell$.**

(1972), Fedorov and Hackl (1997), Fedorov and Malyutov (1972), Karlin and Studden (1966), Chpt. X, Pukelsheim (1993), Silvey (1980), Whittle (1973). In all cited publications, except Pukelsheim (1993), the direct proofs were used and they are transparent and simple. Pukelsheim works within the framework of (17.1.12) and it allows him to overcome some difficulties related to cases in which optimal designs may have singular information matrices. In practice to cope with a possible singularity we may introduce

$$\Psi_\gamma(\xi) = \Psi\left[(1-\gamma)M(\xi) + \gamma M(\xi_0)\right],$$

where $0 < \gamma < 1$ and $\text{rank}M(\xi_0) = m$. If

$$\xi_\gamma^* = \arg\min_\xi \Psi_\gamma(\xi),$$

then from the convexity of $\Psi_\gamma(\xi)$, it immediately follows that

$$\Psi(\xi_\gamma^*) - \Psi(\xi^*) \leq \gamma\left[\Psi(\xi_0) - \Psi(\xi^*)\right]. \tag{17.1.26}$$

The proper choice of γ and ξ_0 may assure "practical" optimality of ξ_γ^*.

In Table 17.1.1 we use the sensitivity function $\phi(x,\xi) = \psi(x,\xi) - C(\xi)$. The reason is that in many cases this function is statistically meaningful. For instance, for the D-criterion $ln|D|$ this function coincides with the standardized variance of the estimated response $d(x,\xi)$ [see (17.1.19)] and $C(\xi) = m$. Theorem 17.1.3 can be reformulated as Kiefer's celebrated equivalence theorem; see Kiefer (1959).

Theorem 17.1.4 *The following problems:*

- $\max_\xi \ln |M(\xi)|$;
- $\arg\min_\xi \max_{x\in X} d(x,\xi)$;
- $\max_{x\in X} d(x,\xi) = m$

are equivalent.

It is named "the equivalence theorem" because it shows that in the homogeneous case, i.e., $\sigma^2(x) \equiv const$, two criteria: $\ln |D|$ and $\max_{x\in X} d(x,\xi)$ have the same optimal designs. It is expedient to emphasize that only the use of specific structure of the information matrix (not only the fact that it is positive semidefinite) allows us to get the more constructive results than those provided by the general theory.

17.1.2 Constraints Imposed on Designs

Linear Constraints. The location of support points may define not only the informative nature of an experiment but also its cost. For instance, let the functions $\zeta(x)$, be the costs when observation is taken at point x. Assume that the total cost can not exceed $\mathcal{C}$. Then

$$\Sigma r_i \zeta(x_i) \le \mathcal{C},$$

where r_i is the number of observations at point x_i. For continuous designs the latter inequality takes the form:

$$\int \varphi(x)\xi(dx) = \Phi(\xi) \le 0,$$

where $\varphi(x) = \zeta(x) - \mathcal{C}/N$. Thus, we arrive at the following optimization problem [see Cook and Fedorov (1995) for the "statistical" details]:

$$\xi^* = \arg\min_\xi \Psi(\xi), \tag{17.1.27}$$

$$\int \xi(dx) = 1, \qquad \Phi(\xi) = \int \varphi(x)\xi(dx) \le 0, \tag{17.1.28}$$

where, in general, we can have several constraints, i.e., $\varphi = (\varphi_1, \ldots, \varphi_l)^T$. From now on all results will be formulated in the space of probability (design) measures. In other words we will work within the framework of (17.1.11) staying closer to the imposed constraints and trying to elaborate the possible structure of admissible information matrices.

To avoid pure technical complications we assume that all components of $\varphi(x)$ are continuous in X.

Theorem 17.1.5 *There exists at least one optimal design containing no more than $n = \ell + m(m+1)/2$ supporting points. The set of optimal designs is convex and the following two statements are equivalent:*

- *ξ^* minimizes $\Psi(\xi)$ over $\Xi(X)$ subject to constraints (17.1.28);*
- *for all $x \in X$ $\quad q(x, u^*, \xi^*) \geq 0$, where $q(x, u, \xi) = \psi(x, \xi) + u^T \varphi(x)$, $u^* = \arg\max_{u \in U} \min_x \, q(x, u, \xi^*)$, $U = \{u : u \in R^l, u_\alpha \geq 0\}$.*

In addition, the function $q(x, u^*, \xi^*) = 0$ for all $x \in \mathrm{supp}\xi^*$.

The first part of the theorem follows from the fact that any pair $M(\xi), \Phi(\xi)$ belongs to the convex hull of

$$\{m(x), \varphi(x)\} \in R^{m(m+1)/2+l}, x \in X.$$

Then apply Caratheodory's theorem.

To prove the second part of the theorem let us notice (see Theorem 17.1.2) that

$$\min_{M \in \mathcal{M}(X)} \Psi'[M^*, M - M^*] = \min_{\xi} \int_X \psi(x, \xi^*)\xi(dx) \geq 0, \tag{17.1.29}$$

where the minimum is taken over all designs such that $\int \varphi(x)\xi(dx) \leq 0$. Unlike the standard case (see comments to Theorem 17.1.3) there is generally no single point design that can be a solution of the above problem. The Lagrangian technique [cf. Laurent (1972), Ch. 7, see also details in Fedorov and Gaivoronski (1984), Gaivoronsky (1986)] leads to duality of the optimization problem

$$\min_{\xi} \int_X \psi(x, \xi^*)\xi(dx), \quad \Phi(\xi) \leq 0,$$

with the following max–min problem:

$$\max_{u \in U} \min_{\xi} \int [\psi(x, \xi^*) + u^T \varphi(x)]\xi(dx),$$

or equivalently

$$\max_{u \in U} \min_{x} q(x, u, \xi^*).$$

Combining the above duality with (17.1.29) we verify the second part of the theorem.

The proof of the rest of the theorem continues as in Theorem 17.1.3.

Linearization of Nonlinear Convex Constraints. Now let

$$\xi^* = \arg\min_{\xi} \Psi(\xi), \quad \Phi(\xi) \leq 0, \tag{17.1.30}$$

where all ℓ components of the vector function Φ are convex in $\Xi(X)$. Let, in addition, the vector of directional derivatives be presented [compare with (17.1.24)] as

$$\Phi'(\xi^*, \xi - \xi^*) = \int_X \varphi(x, \xi^*)\xi(dx) \tag{17.1.31}$$

for any ξ from $\Xi(X)$, and suppose there exists at least one design satisfying all the constraints.

Theorem 17.1.6 *There exists at least one optimal design containing no more than $n = \ell+m(m+1)/2$ supporting points. The set of optimal designs is convex and the following statements are equivalent:*

- ξ^* *is a solution to problem (17.1.30);*
- *for all* $x \in X$ $\quad q(x,u^*,\xi^*) \geq 0$, *where* $q(x,u,\xi) = \psi(x,\xi) + u^T\varphi(x,\xi)$, $u^* = \arg\ \max_{u\in U}\ \min_x\ q(x,u,\xi^*)$, $U = \{u : u \in R^l, u_\alpha \geq 0\}$.

The function $q(x,u^*,\xi^*) = 0$ *for all* $x \in \text{supp}\xi^*$.

The possibility to linearize $\Phi(\xi)$ near an optimal design [compare with Gaivoronski, (1984)] allows to use Theorem 17.1.5 to verify the above statements.

Examples of application of Theorem 17.1.6 to various statistical problems such as minimization of the main optimality criterion with bounded auxiliary criteria, model testing and robust optimal design for nonlinear regression can be found in Cook and Fedorov (1995), Dette (1995), Lee (1988).

Directly Constrained Design Measures. Optimization of sampling experiments, sensors allocation, selection of optimal sites for observing stations in meteorology or seismology is frequently performed under condition that "experimental points" must not be placed too close to each other. The corresponding design problem may be reasonably approximated in many cases [see Fedorov (1986, 1989, 1996), Wynn (1982)] by the following optimization problem:

$$\xi^* = \arg\min_{\xi} \Psi(\xi), \quad \xi(dx) \leq \varrho(dx), \quad \int_X \varrho(dx) = K \geq 1. \qquad (17.1.32)$$

Similar to moment space theory [see Fedorov (1989), Karlin and Studden (1966), Chapt. VIII, Krein and Nudelman (1973) we assume that $\varrho(dx)$ is atomless, i.e., $\lim_{n\to\infty} \int_{\Delta X_n} \varrho(dx) = 0$, for any diminishing sequence ΔX_n. Let also $\Psi(\xi_\varrho) \leq \infty$, where $\xi_\varrho = \varrho/K$. We denote $\bar{\Xi}$ a set of design measures such that $\xi(\Delta x) = \varrho(\Delta x)$ for any $\Delta X \subset \text{supp}\xi$.

A function $\psi(x,\xi)$ is said to separate sets X_1 and X_2 with respect to the measure $\varrho(dx)$ if for any two subsets $\Delta X_1 \in X_1$ and $\Delta X_2 \in X_2$ with equal nonzero measures the following inequality holds:

$$\int_{\Delta X_1} \phi(x,\xi)\varrho(dx) \leq \int_{\Delta X_2} \phi(x,\xi)\varrho(dx).$$

Theorem 17.1.7 *There exists an optimal design* $\xi^* \in \bar{\Xi}$. *A necessary and sufficient condition for this design to be optimal is that* $\psi(x,\xi^*)$ *separates* $X^* = \text{supp}\xi^*$ *and* $X \setminus X^*$.

The results of the theorem are strongly related to the moment spaces theory, and the proof is based on the corresponding ideas and, in particular, on Liapunov's Theorem on the range of a vector measure [see Karlin and Studden, (1966), Ch. VIII]. Various modifications of that proof can be found in Cook and Fedorov (1995), Fedorov (1989), Wynn (1982).

Marginal Design Measures. Another interesting example of constraints directly implied in $\Xi(X)$ appears when the vector of independent variables x in two groups: the first one, x, includes variables that can be controlled by an experimenter, and the second group, t, consists of uncontrolled (but predetermined!) variables. The typical example of uncontrolled but predetermined variable is time. We replace x and X by (x, t) and $X \otimes T$ correspondingly to remind the reader that one of uncontrolled variables frequently is associated with time. The design problem can be described as the following optimization problem with constrained imposed in $\Xi(X \otimes T)$, where the latter is a set of all probability measures on $X \otimes T$:

$$\xi^* = \arg\min_{\xi} \Psi(\xi), \quad \int_X \xi(dx, dt) = \tau(dt), \tag{17.1.33}$$

where $\tau(dt)$ is a given measure and $\int_T \tau(dt) = 1$.

Similar to the conditional distribution we may introduce

$$\xi(dx|t) = \xi(dx, dt)/\tau(dt)$$

and assume a special form of the conditional directional derivatives

$$\Psi'(\xi^*(dx|t), \xi(dx|t) - \xi^*(dx|t) = \int_X \psi(x, t; \xi^*)\xi(dx|t).$$

When t is associated with time, a time dependent measure can be considered as the generalized trajectory and is closely related to generalized controls considered in the optimal control theory [see Balakrishnan (1971), Fedorov and Nachtsheim (1995)]. The following theorem is a rather direct extension of the first two sections of Theorem 17.1.3.

Theorem 17.1.8 *For every t there exists at least one optimal conditional measure $\xi^*(dx|t)$ containing no more than $n = m(m+1)/2$ supporting points. The set of optimal designs is convex and the following statements are equivalent:*

- *ξ^* minimizes $\Psi(\xi)$ over $\Xi(X \otimes T)$;*
- *$\psi(x, t; \xi^*) \geq 0$ for all $x \in X$ and $t \in T$.*

The function $\psi(x, t; \xi^*) = 0$ for all $x \in \text{supp}\xi^*(dx|t)$.

The function $\psi(x, t; \xi)$ can be easily found for any criterion listed in Table 1. Actually the sensitivity functions look exactly as they reported in this table if x is replaced by x, t. However, C must be calculated. For instance, in the case of the D–optimality criterion,

$$\psi(x, t; \xi) = C(t, \xi) - \phi(x, t; \xi) = \int_X d(x, t, \xi)\xi(dx|t) - d(x, t, \xi).$$

where

$$d(x, t, \xi) = f^T(x, t, \xi)M^{-1}(\xi)f(x, t, \xi).$$

For comparison, note that in the standard case

$$C(\xi) = \int_X d(x,\xi)\xi(dx) = \int_X f^T(x)M^{-1}(\xi)f(x)\xi(dx)$$

$$= \text{Trace}M^{-1}(\xi)\int_X f(x)f^T(x)\xi(dx) = \text{Trace}M^{-1}(\xi)M(\xi) = m;$$

see the first row in Table 17.1.1.

Various modifications of the above theorem and possible applications can be found in Cook and Thibodeau (1980), Fedorov and Nachtsheim (1995), Huang and Hsu (1993), Schwabe (1996).

17.1.3 Numerical Construction of Optimal Designs

Direct Approaches. If for a particular design problem there exists a simple way to describe the set of admissible information matrices $\mathcal{M}(X)$, then a variety of different numerical methods can be applied to find a solution to (17.1.12). One can use any algorithm developed for convex analysis (cf. Hiriart-Urruty and Lemaréchal (1993), Chpts. II and VII or any other monograph on optimization). Examples of adaptation of such methods to experimental design may be found, for instance, in Gaffke and Heiligers (1996), Gaffke and Mathar (1993). We may expect better results if the positive semidefiniteness of information matrices is taken into account as it was done in Vandenberghe et al. (1996), see also Vandenberghe and Boyd (1996). However, the numerically practical or tractable description of $\mathcal{M}(X)$ is possible only in some trivial cases. Frequently, starting with the algorithms developed in convex analysis researches end up with algorithms which coincide or very close to those elaborated in experimental design theory as soon as they try to incorporate more detailed information about elements of $\mathcal{M}(X)$. The paper by Gaffke and Mathar (1993) is a good example of such evolution. We will discuss only methods which are essentially based on the specific structure of information matrices.

The First Order Algorithms. We continue to work almost entirely in the space of design measures $\Xi(X)$ in belief that it leads to a better understanding of algorithms developed in the experimental design paradigm, and assuming that the reader can get results in the space of information matrices using mapping (17.1.8) or (17.1.10). We start with the simplest case (17.1.11) without any constraints.

The core of the first order iterative procedure is the following natural updating procedure:

$$\xi_{s+1} = (1-\alpha_s)\xi_s + \alpha_s\xi^s = \xi_s + \alpha_s(\xi_s - \xi^s), \qquad (17.1.34)$$

where

$$\xi^s = \arg\min_{\xi} \Psi[(1-\alpha_s)\xi_s + \alpha_s\xi]. \qquad (17.1.35)$$

There is no much gain in using (17.1.34) and (17.1.35) directly because of (17.1.35) is not simpler than the original problem (17.1.11). However, if we recollect (17.1.24), then for the relatively small α the first order approximation

$$\xi^s = \arg\min_{\xi}\{\Psi(\xi_s) + \alpha_s \int \psi(x, \xi_s)\xi(dx)\} \tag{17.1.36}$$

may substitute (17.1.35).

The objective function in (17.1.36) is linear with respect to ξ thus it is easier than the original problem. Moreover, when there are no constraints imposed on ξ then for $\alpha_s > 0$

$$\xi^s = \xi(x_s), \qquad x_s = \arg\min_{x\in X} \psi(x, \xi_s) = \arg\max_{x\in X} \phi(x, \xi_s), \tag{17.1.37}$$

where $\xi(x)$ is a design measure completely atomized at point x. The existence of x_s with a positive decrement of the criterion Ψ is guaranteed by Theorem 17.1.3.

The use of approximations (17.1.36)–(17.1.37) generated a number of the first order algorithms. The simplest of them can be embedded in the following scheme:

- Find $x_s = \arg\min_x \psi(x, \xi_s)$.
- Choose $0 \leq \alpha_s \leq 1$ and construct $\xi_{s+1} = (1 - \alpha_s)\xi_s + \alpha_s\xi(x_s)$.

The choice of a sequence α_s defines a variety of the algorithms (see Atkinson and Donev (1992), Cook and Nachtsheim (1989), Fedorov (1972), Fedorov and Uspensky (1975), Silvey (1980).

The following three sequences α_s are most popular (especially in the theoretical considerations):

- $\lim_{s\to\infty} \alpha_s = 0, \quad \sum_{s=0}^{\infty} \alpha_s = \infty$;
- $\alpha_s = \arg\min_\alpha \Psi[\xi_{s+1}(\alpha)], \quad \xi_{s+1}(\alpha) = (1-\alpha)\xi_s + \alpha\xi(x_s)$;
- $\alpha_s = \begin{cases} \alpha_{s-1}, \text{ if } \Psi[\xi_s(\alpha_{s-1})] < \Psi(\xi_{s-1}); \\ \alpha_{s-1}/\gamma, \quad \gamma > 1 \text{ otherwise.} \end{cases}$

Proving the (weak) convergence (i.e. that $\lim_{s\to\infty} \Psi(\xi_s) = \Psi^*$) of the algorithms corresponding to any of these sequences is rather standard; see Cook and Fedorov (1995), Fedorov and Uspensky(1975), Wu and Wynn (1978). Usually it is assumed that for any ξ, such that $\Psi(\xi) \leq \infty$, and all $\xi' \in \Xi(X)$

$$\Psi[(1-\alpha)\xi + \alpha\xi'] = \Psi(\xi) + \int_X \psi(x, \xi)\xi'(dx) + O(\alpha^2|\xi). \tag{17.1.38}$$

All criteria included in Table 17.1.1 satisfy that assumption.

Note, that for $\alpha_s = (s+1)^{-1}$ the above iterative procedure mimics the sequential design of experiments when every $(s+1) - th$ observation is placed at the point where the sensitivity function is maximal. Historically the sequential design was first applied for D-criterion for which the sensitivity function coincides with the variance of the estimated response [cf. Fedorov (1972), Chpt. 4].

The significant improvement in the rate of convergence is observing when in the iterative procedure together with adding new "good" points corresponding to the maximal values of the directional derivative an opportunity to delete "bad" supporting points with the small values of directional derivative; see Atwood (1973), Fedorov and Uspensky (1975). For instance, one can use the following iterative procedure:

$$x_s = \arg\ \min\ [\psi(x_s^+, \xi_s), -\psi(x_s^-, \xi_s)], \tag{17.1.39}$$

where

$$x_s^+ = \arg \min_{x\in X} \psi(x, \xi_s) \text{ and } x_s^- = \arg \min_{X\backslash X_s} \psi(x, \xi_s),\ X_s = \text{supp}\xi_s.$$

and

$$\alpha_s = \begin{cases} \gamma_s, & x_s J = J x_s^+ \\ -\min[\gamma_s, p_s/(1-p_s)], & x_s = x_s^- \end{cases}$$

where $\{\gamma_s\}$ has to obey one of the rules for the step selection formulated for the simpler iterative procedure and p_s is the measure of a point $x_s^- \in X_s$.

One can introduce "forward" and "backward" excursions when correspondingly n^+ steps operating with "new" points x_s^+ and n^- steps handling points x_s^- are doing subsequently; the numbers n^+ and n^- are called a "length" of an excursion; see Mitchell (1974). All computations are simplified with the following useful recursions:

$$D(\xi_{s+1}) = \frac{D(\xi_s)}{1-\alpha}\left[I - \frac{\alpha f^T(x_s) f(x_s) D(\xi_s)}{1-\alpha+\alpha d(x_s, \xi_s)}\right].$$

and

$$|D^{-1}(\xi_{s+1})| = (1-\alpha)^m \left(1 + \frac{\alpha d(x_s, \xi_s)}{1-\alpha}\right) |D^{-1}(\xi_s)|.$$

To conclude the discussion of the iterative procedure we would like to emphasize that its simplicity has been achieved only by the heavy use of the information matrix special structure.

Second Order Algorithms. The main assumption that is needed in order to develop a first order iterative procedure is the existence of the presentation (17.1.25) for the first directional derivative. Evidently, the idea can be extended in a straight manner if the similar presentations exist for the higher order derivatives. The corresponding algorithms have theoretically a better rate of convergence in the vicinity of a solution, but are more complicated. Thus, we face the usual trade-off between efficiency and complexity of the algorithm. We discuss a second order iterative procedure for the D-criterion. Generalization for other criteria is straightforward. We start again with the updating relation $\xi_{s+1} = (1-\alpha)\xi_s + \alpha\xi$. Some algebra shows that for any ξ_s such, that $|D(\xi_s)| < \infty$, and all $\xi \in \Xi(X)$

$$\begin{aligned}\ln|D(\xi_{s+1})| &= \ln|D(\xi_s)| + \alpha\int(m - d(x,\xi_s))\xi(dx)\\ &+ \frac{\alpha^2}{2}\int\int[d^2(x,x',\xi_s) - d(x,\xi_s) - d(x',\xi_s) + m]\xi(dx)\xi(dx')\\ &+ O(\alpha^3|\xi).\end{aligned}$$

It is not difficult to prove the existence of a two-point design

$$\xi_s = (p_{s1}, x_{s1}, p_{s2}, x_{s2})$$

that maximizes the decrement at the s-th iteration. The weights may be found analytically for given x_{s1} and x_{s2}. Therefore, at every iteration we have to solve an optimization problem in $X \otimes X$. Instead of a simple one-point correction as in the first order procedure we have to use

$$\xi_{s+1} = (1-\alpha)\xi_s + \alpha[p_1\xi(x_{s1}) + p_2\xi(x_{s2})].$$

Linear Constraints. Direct First Order Algorithm.. For the optimization problem (17.1.27) and (17.1.28), an iterative procedure very similar to the procedure from Section 17.1.3 (with all improvements discussed at the end of the previous section be used to construct an optimal design with constraints:

- There is a design $\xi_s \in \Xi_q$, such that $\Psi(\xi_s) < \infty$ and $\int\varphi(x)\xi_s(dx) \leq 0$. Find

$$\xi^s = \arg\min_{\xi}\int\psi(x,\xi_s)\xi(dx), \quad \int\zeta(x)\xi(dx) \leq 0. \qquad (17.1.40)$$

- Choose $0 \leq \alpha_s \leq 1$ and construct $\xi_{s+1} = (1-\alpha_s)\xi_s + \alpha_s\xi^s$.

Assume that (17.1.38) holds. Then [cf. Cook and Fedorov (1995)]

$$\lim_{s\to\infty}\Psi[M(\xi_s)] = \Psi^*.$$

Thus one does not face any difficulties in optimal design construction if optimization problem (17.1.40) can be easily solved. Unlike to the case without constraints, it is not so in the majority of experimental situations. Some simplification can be obtained if one observe (see Theorem 17.1.5) that (17.1.40) can be reduced to a finite dimension problem:

$$\xi^s = \left\{\begin{array}{c} x_1^s,\ldots,x_{l+1}^s \\ p_1^s,\ldots,p_{l+1}^s \end{array}\right\} = \arg\min_{p_j,x_j}\sum_{j=1}^{l+1}p_j\psi(x_j,\xi_s), \qquad (17.1.41)$$

$$\sum_{j=1}^{l+1} p_j \varphi(x_j) \leq 0, \quad \sum_{j=1}^{l+1} p_j = 1.$$

This problem can be considered as practically solvable when the number of constraints and the dimension of X is reasonably small. Numerical methods to solve (17.1.41) based on the cutting–plane technique are considered by Gaivoronski (1986). They assume extra iterating within every s–th iteration.

If the number of constraints is small, and that is true for majority of practical cases, then the Lagrangian approach looks most promising. On an intuitive level it is clear that minimizing function

$$\mathcal{L}(\xi) = \Psi(\xi) + \lambda^T \Phi(\xi) \tag{17.1.42}$$

where $\lambda_\alpha > 0, \alpha = 1, \ldots, l$ and $\Phi(\xi) = \int \varphi(x)\xi(dx)$, one guaranties that $\Phi(\xi) \leq c_\lambda$, where c_λ become known after minimization of (17.1.42). More accurately, the saddle-point theory leads [cf. Hiriart-Urruty and Lemaréchal (1993), Chpt. VII.4, Cook and Fedorov (1995)] to the following simple theorem.

Theorem 17.1.9 *A design:*

$$\xi_\lambda = \arg \min_\xi [\Psi(\xi) + \lambda^T \Phi(\xi)]$$

is a solution of the constrained design problem (17.1.27) with $\Phi(\xi) \leq c_\lambda$*, where* $c_\lambda = \Phi(\xi_\lambda)$.

The theorem provides a basis for the numerical solution of (17.1.27), (17.1.28), if one manages to find out the vector λ such that all components of vector c_λ are small enough, for instance, a squared distance $\rho_\lambda = c_\lambda^T A c_\lambda$, $A > 0$, would be small. Thus, one can consider the "empirical" optimization problem:

$$\lambda^* = \arg \min_\lambda c_\lambda^T A c_\lambda, \quad \lambda_\alpha \geq 0 \ , \ \alpha = 1, \ldots, l.$$

Viewing ρ_λ as a response function depending upon λ, one can apply to the empirical optimization technique which is the famous milestone in the experimental design [cf. Box and Draper (1987)]. It is essential that the approach assumes that for any given λ the design problem can be numerically solved at moderate expense. The likeliness of that is confirmed by existence of many commercial design packages. Some useful details about compound criteria can be found in Läuter (1976).

Nonlinear Constraints. All results from Sections 17.1.3 can be generalized for the case with nonlinear constraints using the idea of linearization. To do this we have to assume that

$$\Phi[\xi(\alpha)] = \Phi(\xi) + \alpha \int \varphi(x, \xi)\xi'(dx) + O(\alpha^2|\xi)],$$

where $\xi(\alpha) = (1 - \alpha)\xi + \alpha\xi'$ and $\Phi(\xi) \leq 0$.

By linearizing $\Phi(\xi)$ in the "vicinity" of design ξ_s (compare with Section 17.1.2) one can check that all results of Sections 17.1.3 stay valid if the above defined $\varphi(x, \xi_s)$ replaces everywhere $\varphi(x)$. The approach proposed at the concluding part of Section 17.1.3 looks less elegant from a mathematical point of view then the approach based on (17.1.40), but it is more practical. Again, as in Section 17.1.3, instead of the original problem one has to consider a sequence of optimization problems for a "compound" criterion of optimality:

$$\xi_\lambda = \arg \min_{\xi}[\Psi(\xi) + \lambda^T \Phi(\xi)], \quad \lambda_\alpha > 0, \alpha = 1, \ldots, l.$$

17.2 SEMIDEFINITE PROGRAMMING IN EXPERIMENTAL DESIGN

Jon Lee

There are several problems in the *design of experiments (DOE)* that can be recast as semidefinite programming problems. Below $\mathcal{E}[\cdot]$ denotes expected value, and for a matrix A, $A_{j\cdot}$ (respectively, $A_{\cdot j}$) denotes its row (column) indexed by j . For an $n \times n$ matrix A, $\mathrm{diag}(A)$ denotes the n-vector of diagonal entries from A. Finally, with respect to an n-vector a, $\mathrm{Diag}(a_j)$ denotes the $n \times n$ diagonal matrix having a_j as its jth element.

17.2.1 Covariance Matrices

Some DOE problems are formulated based on a covariance matrix C for a set of n random variables $Z_1, Z_2, \ldots, Z_n$. We have

$$C_{ij} := \mathcal{E}\left[(Z_i - \mathcal{E}[Z_i])\,(Z_j - \mathcal{E}[Z_j])\right].$$

In practice, we do not ordinarily know the matrix C, but we estimate it using m covariate observations Z_{ik} ($1 \le i \le n$, $1 \le k \le m$). Specifically, we make the approximation

$$C_{ij} \approx \frac{1}{m-1} \sum_{k=1}^{m} \left(Z_{ik} - \bar{Z}_i\right) \left(Z_{jk} - \bar{Z}_j\right),$$

where $\bar{Z}_i = \frac{1}{m} \sum_{k=1}^m Z_{ik}$ and $\bar{Z}_j = \frac{1}{m} \sum_{k=1}^m Z_{jk}$. In applications, we assume that m is large enough so that we have estimated C reliably.

Since covariances matrices are symmetric and positive semidefinite, it is not surprising that semidefinite programming has a role to play in the DOE.

Reliability of Test Scores. Fletcher [240] proposed a solution method for the semidefinite program

$$\begin{aligned} &\max \operatorname{Trace}(\operatorname{Diag}(y_j)), \\ &\quad \text{subject to } C - \operatorname{Diag}(y_j) \succeq 0; \\ &\quad\quad \operatorname{Diag}(y_j) \succeq 0. \end{aligned}$$

Fletcher was motivated by a problem of determining the reliability of a student's total test score on a single examination consisting of a number of subject tests. In this setting, Z_{ik} is the score on subject i for student k. So the matrix C is the covariance matrix for the n *subjects*. Underlying the scores Z_{ik} are the hypothetical unknown "true scores" S_{ik} and the associated "errors" $E_{ik} := Z_{ik} - S_{ik}$. We assume that each E_{ik} has mean zero and that the errors are uncorrelated (specifically, we assume that $\mathcal{E}[E_{ik}E_{il}] = 0$ when $k \neq l$, and $\mathcal{E}[E_{ik}S_{il}] = \mathcal{E}[E_{ik}]S_{il}$). The S_{ik} have unknown covariance matrix B defined by

$$B_{ij} := \frac{1}{m-1} \sum_{k=1}^{m} \left(S_{ik} - \bar{S}_i\right)\left(S_{jk} - \bar{S}_j\right),$$

where $\bar{S}_i = \frac{1}{m}\sum_{k=1}^{m} S_{ik}$ and $\bar{S}_j = \frac{1}{m}\sum_{k=1}^{m} S_{jk}$. Our definitions and assumptions imply that $Y := C - B$ is the covariance matrix of the E_{ik}, and that Y is a diagonal matrix.

The variance of the realized scores is $\sigma_Z^2 = e^T Ce$ and the unknown variance of the true scores is $\sigma_S^2 = e^T Be$. So, the objective of maximizing $\operatorname{Trace}(Y) = e^T(C - B)e = \sigma_Z^2 - \sigma_S^2$ reflects the goal of obtaining an upper bound on the amount by which the variance of the realized scores exceeds the variance of the true scores.

Maximum-Entropy Sampling. The *Constrained Maximum-Entropy Sampling Problem* (CMESP) is

$$\begin{aligned} &\max \ln \det\left(C[\underline{x}, \underline{x}]\right), \\ &\quad \text{subject to } Ax \leq b; \\ &\quad\quad e^T x = s; \\ &\quad\quad x \in \{0,1\}^n, \end{aligned}$$

where s is an integer satisfying $0 < s < n$, A is a real $m \times n$ matrix, b is a real m-vector, x is an n-vector of variables, $\underline{x} = \{j \; : 1 \leq j \leq n, \; x_j \neq 0\}$, and $C[S,T]$ is the submatrix of C indexed by $S, T \subset \{1, 2, \ldots, n\}$.

Applications of this NP-Hard problem arise in environmental monitoring (see Ko, Lee and Queyranne [426]). For example, in environmental monitoring, the quantity Z_{ik} is the logarithm of the concentration of a pollutant in wet deposition at monitoring site i collected during time period k. It is common practice to assume that the underlying random variables Z_i have a Gaussian distribution. In this case, $\ln\,\det(C[S,S])$ is, up to constants, the entropy of $\{Z_i\ :\ i \in S\}$. The constraints of the CMESP might reflect budgetary or other logistical considerations.

Short of complete enumeration, all exact methods for the CMESP use the branch-and-bound framework (see Ko et al. [426], Lee [469],and Anstreicher, Fampa, Lee and Williams [38], [39]). Extending the work of Ko et al. [426] who considered only $m = 0$, Lee [469] developed the spectral upper bound

$$\min_{\pi \in \mathbb{R}^m_+} \left\{ \ln \left(\prod_{l=1}^{s} \lambda_l \left(D^\pi C D^\pi\right) \right) + \pi^T b \right\},$$

where D^π is the diagonal matrix having

$$D^\pi_{jj} = -\frac{1}{2}\pi^T A_{\cdot j} \ .$$

Anstreicher et al. [38] and [39] developed an upper bound based on the nonlinear programming relaxation

$$\begin{aligned} &\max\ f(x), \\ &\quad \text{subject to } Ax \le b; \\ &\quad e^T x = s; \\ &\quad 0 \le\ x \le e, \end{aligned}$$

where

$$f(x) := \ln\ \det\left(\mathrm{Diag}(x_j^{p_j}) C \mathrm{Diag}(x_j^{p_j}) + \mathrm{Diag}(d_j^{x_j} - d_j x_j^{p_j})\right),$$

the constants $d_j > 0$ and $p_j \ge 1$ satisfy $p_j \ge 1$, $0 < d_j \le \exp(p_j - \sqrt{p_j})$, and $\mathrm{Diag}(d_j) \succeq C$ (which makes the objective function concave on $0 < x \le e$).

Recently, C. Helmberg has suggested an upper bound based on a semidefinite programming relaxation. His relaxation is based on letting $Y = xx^T$. Since, $(1\ x^T)^T(1\ x^T)$ is positive semidefinite, and since the first row of this matrix is equal to the diagonal for 0/1-vectors x, we are lead to requiring

$$\begin{pmatrix} 1 & \mathrm{diag}(Y)^T \\ \mathrm{diag}(Y) & Y \end{pmatrix} \succeq 0,$$

or, equivalently, $Y - \mathrm{diag}(Y)\mathrm{diag}(Y)^T \succeq 0$. Finally, we observe that for 0/1-vectors x,

$$\det\left(C[\underline{x},\underline{x}]\right) = \det\left((C - I) \circ Y + I\right).$$

So we are led to the semidefinite programming relaxation

$$\begin{aligned} &\max\ \ln\ \det\left((C - I) \circ Y + I\right), \\ &\quad \text{subject to } \mathcal{A}(Y) \le \beta; \\ &\quad Y - \mathrm{diag}(Y)\mathrm{diag}(Y)^T \succeq 0, \end{aligned}$$

where the constraints involving the linear operator $\mathcal{A}$ capture the linear constraints of the CMESP. Helmberg's formulation can be tightened by including further linear inequalities that are valid for the 0/1-solutions to $y_{ii}y_{jj} = y_{ij}$ (see Padberg [607], for example).

17.2.2 Linear Models

Let A be a known $n \times m$ "design matrix". We think of the rows $A_{j\cdot}$ as specifying the levels of each of m "independent variables". We can set the independent variables according to the levels specified by $A_{j\cdot}$ and observe the "dependent (random) response variable" Z_j. Unfortunately, sampling can be rather costly, and running an experiment for each of the n rows may be impractical. Our goal is to choose just s of the n rows (with replication allowed!) according to some criterion.

We assume that our variables follow the *linear model*

$$Z = A\beta + E,$$

where E is an n-vector of "errors". As is usual, we assume that the E_j are independent and identically distributed random variables having mean zero and constant variance σ^2 (i.e, the covariance matrix for E is $\sigma^2 I_n$).

We let x_j denote the number of times $A_{j\cdot}$ is selected divided by s, and we refer to the vector x as a *design.* We relax the assumption that each x_j should be a multiple of $\frac{1}{s}$; if we really care about the discrete nature of the x_j, then we can regard the formulations below as providing bounds for use in a branch-and-bound scheme (note that the discrete versions of most of these problems are NP-Hard (see Welch [826])).

Common optimality criteria relate to the "Fisher information matrix", which in our setting is $F(x) := \sum_{j=1}^{n} x_j A_{j\cdot} A_{j\cdot}^T$. The information matrix appears in the formula for the least-squares estimator of β, based on the design x. The matrix $\sigma^2 F^{-1}(x)$ is the covariance matrix for the least-squares estimator of β, based on the design x. Classical design criteria involve making $\sigma^2 F^{-1}(x)$ "small" (see Pukelsheim [647], for example).

Below, we describe some of the common criterion and describe how these optimization problems can be recast as semidefinite programs. More details can be found in Vandenberghe, Boyd and Wu [813].

E-Optimal Design. An *E-optimal design* x minimizes the maximum eigenvalue of $F^{-1}(x)$. The appeal of the E-optimality criterion is it minimizes the maximum variance of least-square estimators of $z^T\beta$, where the maximum is taken over $\{z \in \mathbb{R}^m : ||z|| = 1\}$. The problem of finding an *E-optimal design* can be recast as the semidefinite program

$$\begin{aligned} \max\ & t, \\ \text{subject to } & F(x) \succeq tI; \\ & e^T x = 1; \\ & x \geq 0, \end{aligned}$$

where t is a scalar variable.

A-Optimal Design. An *A-optimal design* x minimizes the trace of $F^{-1}(x)$. The A-optimality criterion is appealing since it minimizes the average variance of the least-square estimators of the unknown parameter vector β. The problem of finding an *A-optimal design* can be recast as the semidefinite program

$$\begin{aligned} &\min \sum_{i=1}^{m} t_i, \\ &\text{subject to } \begin{pmatrix} F(x) & e_i \\ e_i^T & t_i \end{pmatrix} \succeq 0, \text{ for } i = 1, 2, \ldots, m; \\ &e^T x = 1; \\ &x \geq 0, \end{aligned}$$

where e_i is the ith standard unit vector, and t_i is a scalar variable.

An alternative formulation, using a matrix Y in place of the vector t, is

$$\begin{aligned} &\min \operatorname{Trace}(Y), \\ &\text{subject to } \begin{pmatrix} F(x) & I \\ I & Y \end{pmatrix} \succeq 0; \\ &e^T x = 1; \\ &x \geq 0, \end{aligned}$$

D-Optimal Design. A *D-optimal design* x minimizes the determinant of $F^{-1}(x)$. The D-optimality criterion is particularly appealing since it minimizes the volume of the "confidence ellipsoid" of the unknown parameter vector β. Another appealing quality of the D-optimality criterion is that it is invariant under "reparameterization" (i.e, linear transformation of β). The problem of finding a D-optimal design has the formulation

$$\begin{aligned} &\min \ \ln\det(F^{-1}(x)), \\ &\text{subject to } e^T x = 1; \\ &x \geq 0. \end{aligned} \tag{17.2.43}$$

It can be shown using convex duality (see Titterington [778] and Vandenberghe, Boyd and Wu [813]) that the dual of (17.2.43) is equivalent to the program

$$\begin{aligned} &\max \ \ln\det(Y), \\ &\text{subject to } Y \succ 0; \\ &A_{j\cdot}^T Y A_{j\cdot} \leq m, \text{ for } j = 1, 2, \ldots, n. \end{aligned}$$

Welch [827] worked with the discrete version of the D-optimal design problem. He used a bound based on Hadamard's inequality and also worked directly with the formulation (17.2.43). Ko, Lee and Wayne [427] also worked with the discrete version of the problem and developed bounds based on singular values.

18 MATRIX COMPLETION PROBLEMS

Abdo Alfakih, Henry Wolkowicz

18.1 INTRODUCTION

In the matrix completion problem we are given a partial symmetric real matrix A with certain elements specified or *fixed* and the rest are unspecified or *free*; and, we are asked whether A can be completed to satisfy a given property (P) by assigning certain values to its free elements. In this chapter, we are interested in the following two completion problems: the *positive semidefinite matrix completion problem* corresponding to (P) being the positive semi-definiteness (PSD) property; and the *Euclidean distance matrix completion problem* corresponding to (P) being the Euclidean distance (ED) property. (We concentrate on the latter problem. A survey of the former is given in [373]. The relationships between the two is discussed in e.g. [465, 464, 380].) These problems can be generalized to allow for approximate completion by formulating the completion problems as weighted closest matrix problems:

$$\min \{f(X) = || H \circ (A - X) ||_F^2 : \text{for all } X \text{ satisfying (P)}\}, \qquad (18.1.1)$$

where $||\,.\,||_F$ denotes the *Frobenius* norm defined as $|| A ||_F = \sqrt{\text{Trace} A^T A}$, $(\circ)$ denotes the *Hadamard* or the element-wise product, and the weight matrix H is symmetric with nonnegative elements. If each element of the weight matrix H is set to 0 or 1 depending respectively on whether the corresponding element of A is free or fixed, then matrix A has an exact completion if and only if the optimal objective value of problem (18.1.1) is zero. Problem (18.1.1) will be referred to as CPM when (P) is the PSD property, and as CDM when (P) is the

ED property. Both CPM and CDM are examples of semidefinite programming with the former problem being a special case of the latter.

An $n \times n$ symmetric matrix D with nonnegative elements and zero diagonal is called a *dissimilarity matrix* or a *pre-distance matrix*. A dissimilarity matrix D is said to be a *Euclidean distance matrix* (EDM) if there exist points $x^1, x^2, \ldots, x^n$ in some Euclidean space $\Re^r$ such that

$$D_{ij} = \|x^i - x^j\|^2, \qquad i, j = 1, 2, \ldots, n, \tag{18.1.2}$$

where $\| \,.\, \|$ denotes the usual Euclidean norm. The *embedding dimension* of D is given by r where r is the dimension of the smallest space containing $x^1, x^2, \ldots, x^n$. Note that r is always less than or equal to $n - 1$.

Euclidean distance matrices have received much attention in recent years both for their elegance and for their many important applications. Applications of EDMs abound: e.g. molecular conformation problems in chemistry [175, 613]; multidimensional scaling and multivariate analysis problems in statistics [470, 471]; genetics, geography, and others [11]. In many of these applications, a low embedding dimension, e.g. $r = 3$, is required.

Theoretical properties of EDMs can be found in e.g. [50, 217, 295, 331, 380, 465, 699]. This includes characterizations as well as graph theoretic conditions for existence of completions. More information can be found in the recent survey article by Laurent [465].

18.2 WEIGHTED CLOSEST EUCLIDEAN DISTANCE MATRIX

Consider the objective function

$$f(D) := \| H \circ (A - D) \|_F^2.$$

The weighted, *closest Euclidean distance matrix problem* is

$$(CDM_0) \quad \mu^* := \begin{array}{cc} \min & f(D) \\ \text{subject to} & D \in \mathcal{E}, \end{array}$$

where $\mathcal{E}$ is the set of all EDMs. In the following section, it is shown that $\mathcal{E}$ is in fact a convex cone whose structure is similar to that of the cone of positive semidefinite matrices $\mathcal{P}$. This relationship between $\mathcal{E}$ and $\mathcal{P}$ will allow us to formulate (CDM_0) as a semidefinite program.

18.2.1 Distance Geometry

It is well known, e.g. [295, 699, 331], that a pre-distance matrix D is a EDM if and only if D is negative semidefinite on

$$M := \left\{ x \in \Re^n : x^T e = 0 \right\},$$

the orthogonal complement of e where e is the vector of all ones. Thus $\mathcal{E}$ is a convex cone. We exploit this result and the fact that faces of $\mathcal{P}_n$, the cone of positive semidefinite matrices of order n, are isomorphic to $\mathcal{P}_r$, for some $0 \le r \le n$, to translate the cone $\mathcal{E}$ to the cone $\mathcal{P}_{n-1}$.

Let V be the $n \times (n-1)$ matrix whose columns form an orthonormal basis for the subspace M. i.e., V satisfies:

$$V^T e = 0 \ , \text{ and } V^T V = I. \tag{18.2.3}$$

Then,

$$J := VV^T = I - \frac{ee^T}{n} \tag{18.2.4}$$

is the orthogonal projection onto M. Now define the *centered* and *hollow* subspaces of $\mathcal{S}^n$, the space of $n \times n$ symmetric matrices, as

$$\begin{aligned} \mathcal{S}_C &:= \{B \in \mathcal{S}^n : Be = 0\}, \\ \mathcal{S}_H &:= \{D \in \mathcal{S}^n : \operatorname{diag}(D) = 0\}, \end{aligned} \tag{18.2.5}$$

where $\operatorname{diag}(D)$ denotes the column vector formed from the diagonal of D. Following [176], we define the two linear operators $\mathcal{K} : \mathcal{S}_C \to \mathcal{S}_H$ and $\mathcal{T} : \mathcal{S}_H \to \mathcal{S}_C$ such that

$$\mathcal{K}(B) \ := \ \operatorname{diag}(B)e^T + e\operatorname{diag}(B)^T - 2B, \tag{18.2.6}$$

and

$$\mathcal{T}(D) \ := \ -\tfrac{1}{2} JDJ. \tag{18.2.7}$$

Theorem 18.2.1 *The linear operators $\mathcal{K}$ and $\mathcal{T}$ are inverses of each other.*

Proof. See [295, 380]. ■

It can be easily verified that

$$\mathcal{K}^*(D) \ = \ 2\left(\operatorname{Diag}(De) - D\right) \tag{18.2.8}$$

is the adjoint operator of $\mathcal{K}$, where $\operatorname{Diag}(De)$ denotes the diagonal matrix formed from the vector De. In addition, a hollow matrix D is EDM if and only if $B = \mathcal{T}(D) \succeq 0$. Equivalently, D is EDM if and only if $D = \mathcal{K}(B)$, for some B with $Be = 0$ and $B \succeq 0$. In this case the embedding dimension r is given by the rank of B. If $B = XX^T$ for some $n \times r$ matrix X, then the rows of X are the coordinates of the points $x^1, x^2, \ldots, x^n$ that generate D. Furthermore, since $Be = 0$, it follows that the origin coincides with the centroid of these points. For these and other basic results on EDM see e.g. [295, 331, 380, 465, 699].

We now introduce the composite operators $\mathcal{K}_V : \mathcal{S}_{n-1} \to \mathcal{S}_H$ and $\mathcal{T}_V : \mathcal{S}_H \to \mathcal{S}_{n-1}$ such that

$$\mathcal{K}_V(X) \ := \ \mathcal{K}(VXV^T), \tag{18.2.9}$$

and

$$\mathcal{T}_V(D) \ := \ V^T \mathcal{T}(D) V = -\tfrac{1}{2} V^T D V, \tag{18.2.10}$$

where V is defined in (18.2.3).

Lemma 18.2.1 *The linear operators $\mathcal{K}_V$ and $\mathcal{T}_V$ are inverses of each other.*

Proof. This immediately follows from Theorem 18.2.1 and the definition of V. ■

From (18.2.8), we get that

$$\mathcal{K}_V^*(D) \quad = \quad V^T\mathcal{K}^*(D)V \tag{18.2.11}$$

is the adjoint operator of $\mathcal{K}_V$. The following corollary summarizes the relationships between the cones $\mathcal{E}$ and $\mathcal{P}_{n-1}$.

Corollary 18.2.1 *Suppose that V is defined as in (18.2.3). Then:*

$$\begin{aligned} \mathcal{K}_V(\mathcal{P}_{n-1}) &= \mathcal{E}, \\ \mathcal{T}_V(\mathcal{E}) &= \mathcal{P}_{n-1}. \end{aligned}$$

Proof. We saw earlier that D is EDM if and only if $D = \mathcal{K}(B)$ with $Be = 0$ and $B \succeq 0$. Let $X = V^TBV$, then since $Be = 0$ we have $B = VXV^T$. Therefore, $VXV^T \succeq 0$ if and only if $X \succeq 0$; and the result follows using (18.2.9) and Lemma 18.2.1. ■

18.2.2 Program Formulations

It can be assumed without loss of generality that $\text{diag}(H) = 0$ since both $\text{diag}(A)$ and $\text{diag}(D)$ are 0. Also, all unspecified elements of A can be originally set to 0. Elements D_{ij} corresponding to $H_{ij} = 0$ are free, while elements D_{ij} corresponding to $H_{ij} > 0$ are approximately fixed since the weight $H_{ij} > 0$ forces D_{ij} to be approximately equal to A_{ij}. Hence, if the optimal value of the objective function $f(\bar{D})$ is 0, then an exact completion of A is found. i.e., $\bar{D}$ differs from A only in free elements i.e., elements with zero weight. If D_{ij} is fixed i.e., it is required to be equal to A_{ij} exactly, then a linear constraint can be added to the program. (See below.)

Recall that the graph of H is *connected* if for all indices $i \neq j$ there is a *path* of indices $i_1, i_2, ..., i_k$ such that $H_{i,i_1} \neq 0, H_{i_1,i_2} \neq 0, \dots, H_{i_{k-1},i_k} \neq 0, H_{i_k,j} \neq 0$, see e.g. [143]. Thus, it is assumed that the graph of H is connected or the problem can be solved more simply as two or more smaller problems one for each component of H, see Lemma 18.2.3 below. In particular, it is assumed that H does not have a row (hence a column) of all zeros; otherwise the corresponding row and column in A and D consists of free (independent) elements and the problem can be posed in a lower dimensional space.

By abuse of notation, let the function

$$f(X) := \| H \circ (A - \mathcal{K}_V(X)) \|_F^2 = \| H \circ \mathcal{K}_V(B - X) \|_F^2,$$

where $B = \mathcal{T}_V(A)$. Using Corollary 18.2.1, (CDM_0) is equivalent to the following problem

$$\begin{array}{llll} & \mu^* := & \min & f(X) \\ (CDM) & & \text{subject to} & \mathcal{A}(X) = b \\ & & & X \succeq 0. \end{array}$$

An additional constraint using the linear operator $\mathcal{A} : \mathcal{S}_{n-1} \longrightarrow \Re^m$ is allowed here. The addition of this linear constraint could represent some of the fixed elements in the given matrix A, e.g. adding the constraint $(K_V(X))_{ij} = A_{ij}$ fixes the ij element of D. Also, note that $X \in \mathcal{S}_{n-1}$. It is in this lower dimensional space that the problem is solved. The optimal distance matrix is recovered by using the optimal X and the relation

$$D = \mathcal{K}_V(X).$$

Using finite precision, one can never solve the approximation problem exactly. In addition, one needs to calculate the embedding dimension. The following lemma shows that little is lost in the objective function if a small embedding dimension is chosen using a numerical rank approach i.e., if only very small eigenvalues are discarded, then the objective function changes very little.

Lemma 18.2.2 *Suppose that X^* solves (CDM). Let $\bar{X}$ be the closest symmetric matrix to X^* with rank k, i.e. we set the smallest $n-k$ eigenvalues of X^* to 0, $\lambda_{k+1} = \ldots \lambda_n = 0$. Then,*

$$\sqrt{f(\bar{X})} \leq \sqrt{f(X^*)} + 2\gamma\left(\sqrt{n}+1\right)\sqrt{\sum_{i=k+1}^{n} \lambda_i^2}, \tag{18.2.12}$$

where $\gamma := \max_{ij} H_{ij}$.

Proof. See [14] ∎

18.2.3 Duality and Optimality

Duality theory and optimality conditions needed for a primal-dual interior-point approach are derived in this section.

For $\Lambda \in \mathcal{S}_{n-1}$ and $y \in \Re^m$, let

$$L(X, y, \Lambda) = f(X) + \langle y, b - \mathcal{A}(X)\rangle - \operatorname{trace}\Lambda X \tag{18.2.13}$$

denote the *Lagrangian* of (CDM), where $\langle , \rangle$ denotes the usual inner product in $\Re^m$. It is easy to see that the primal program (CDM) is equivalent to

$$\mu^* = \min_{X} \max_{\substack{y \\ \Lambda \succeq 0}} L(X, y, \Lambda) = \min_{X \succeq 0} \max_{\substack{y \\ \Lambda \succeq 0}} L(X, y, \Lambda). \tag{18.2.14}$$

We assume that the generalized Slater's constraint qualification,

$$\exists\, X \succ 0 \text{ with } \mathcal{A}(X) = b,$$

holds for (CDM). Slater's condition implies that strong duality holds, i.e.,

$$\mu^* = \max_{\substack{y \\ \Lambda \succeq 0}} \min_{X} L(X, y, \Lambda) = \max_{\substack{y \\ \Lambda \succeq 0}} \min_{X \succeq 0} L(X, y, \Lambda), \tag{18.2.15}$$

and μ^* is attained for some y and $\Lambda \succeq 0$, see e.g. [833]. Since the semidefinite constraint on X can be treated as redundant, the inner minimization of the convex, in X, Lagrangian is unconstrained and we can differentiate to get the equivalent problem

$$\mu^* = \max_{\substack{\nabla f(X) - \mathcal{A}^* y = \Lambda \\ \Lambda \succeq 0}} f(X) + \langle y, b - \mathcal{A}(X) \rangle - \text{trace}\Lambda X. \tag{18.2.16}$$

We can now state the dual problem

$$\begin{array}{llll} \text{(DCDM)} & \mu^* := & \max & f(X) + \langle y, b - \mathcal{A}(X)\rangle - \text{ trace } \Lambda X \\ & & \text{subject to} & \nabla f(X) - \mathcal{A}^* y - \Lambda = 0 \\ & & & \Lambda \succeq 0, (X \succeq 0). \end{array} \tag{18.2.17}$$

The semidefinite constraint on X is kept in brackets to emphasize that it is a hidden constraint in the dual, though it is a given constraint in the primal. The above pair of dual programs, (CDM) and (DCDM), provides an optimality criteria in terms of feasibility and complementary slackness. This provides the basis for many algorithms including primal-dual interior-point algorithms. In particular, we see that the duality gap, in the case of primal and dual feasibility, is given by the complementary slackness condition:

$$\text{trace } X(\mathcal{K}_V^*(H^{(2)} \circ \mathcal{K}_V(\bar{X} - B)) - \mathcal{A}^* \bar{y}) = 0, \tag{18.2.18}$$

or equivalently

$$X(\mathcal{K}_V^*(H^{(2)} \circ \mathcal{K}_V(\bar{X} - B)) - \mathcal{A}^* \bar{y}) = 0,$$

where $H^(2) = H \circ H$.

Theorem 18.2.2 *Suppose that Slater's condition holds. Then $\bar{X} \succeq 0$, and $\bar{y}$, $\bar{\Lambda} \succeq 0$ solve (CDM) and (DCDM), respectively, if and only if the following three equations hold.*

$$\begin{array}{cc} \mathcal{A}(\bar{X}) = b & \textit{primal feasibility} \\ 2\mathcal{K}_V^*(H^{(2)} \circ \mathcal{K}_V(\bar{X} - B)) - \mathcal{A}^* \bar{y} - \bar{\Lambda} = 0 & \textit{dual feasibility} \\ \textit{trace } \bar{\Lambda}\bar{X} = 0 & \textit{complementary slackness} \end{array}$$

18.2.4 Primal-Dual Interior-Point Algorithm

In this section, a primal-dual interior point method is derived using the log-barrier approach [339]. This is an alternative way of deriving the optimality conditions in Theorem 18.2.2. For simplicity, the unconstrained problem i.e., (CDM) without the linear equality constraint $\mathcal{A}X = b$, is treated. In this case, the Slater constraint qualification holds for both primal and dual problems, as can be seen from the following lemma. (In the case that some of the elements of A are definitely fixed, then one needs to re-derive the algorithm and include the constraint $\mathcal{A}X = b$.)

Lemma 18.2.3 *Let H be an $n \times n$ symmetric matrix with nonnegative elements and 0 diagonal such that the graph of H is connected. Then*

$$\mathcal{K}_V^*(H^{(2)} \circ \mathcal{K}_V(I)) \succ 0,$$

where $I \in \mathcal{S}_{n-1}$ is the identity matrix.

Proof. See [14] ∎

Strict primal feasibility follows since every $X \succeq 0$ is feasible. Strict dual feasibility, i.e. $\Lambda \succ 0$, follows in the case of a connected graph, since a positive scalar α can be chosen such that

$$X = B + \alpha I \succ 0; \tag{18.2.19}$$

one can then apply Lemma 18.2.3 to get

$$\mathcal{K}_V^*(H^{(2)} \circ \mathcal{K}_V(X - B)) \succ 0. \tag{18.2.20}$$

Corollary 18.2.2 *Slater's constraint qualification holds for the dual problem (18.2.17) if and only if the graph of H is connected.*

Proof. See [14] ∎

The log-barrier problem for (CDM) is

$$\min_{X \succ 0} B_\mu(X) := f(X) - \mu \log\det(X),$$

where $\mu \downarrow 0$. For each $\mu > 0$, we take one Newton step for solving the stationarity condition

$$\nabla B_\mu(X) = 2\mathcal{K}_V^*(H^{(2)} \circ \mathcal{K}_V(X - B)) - \mu X^{-1} = 0. \tag{18.2.21}$$

Let

$$C := 2\mathcal{K}_V^*(H^{(2)} \circ \mathcal{K}_V(B)) = 2\mathcal{K}_V^*(H^{(2)} \circ A). \tag{18.2.22}$$

Then the stationarity condition is equivalent to

$$\nabla B_\mu(X) = 2\mathcal{K}_V^*\left(H^{(2)} \circ \mathcal{K}_V(X)\right) - C - \mu X^{-1} = 0. \tag{18.2.23}$$

By equating $\Lambda = \mu X^{-1}$ and multiplying through by X, one gets the optimality conditions, $F := \begin{pmatrix} F_d \\ F_c \end{pmatrix} = 0$,

$$\begin{array}{llcll} F_d & := 2\mathcal{K}_V^* \left(H^{(2)} \circ \mathcal{K}_V(X)\right) - C - \Lambda & = & 0 & \text{dual feas.} \\ F_c & := \Lambda X - \mu I & = & 0 & \text{perturbed compl. slack.,} \end{array} \tag{18.2.24}$$

and the estimate of the barrier parameter

$$\mu = \frac{1}{n-1} \text{ trace } \Lambda X. \tag{18.2.25}$$

(Recall that X and Λ are $(n-1) \times (n-1)$ matrices.) The linearization of the optimality conditions (18.2.24) yield

$$\begin{array}{rcl} 2\mathcal{K}_V^* \left(H^{(2)} \circ \mathcal{K}_V(\delta X)\right) - \delta\Lambda & = & -F_d \\ \Lambda\ \delta X + \delta\Lambda\ X & = & -F_c. \end{array} \tag{18.2.26}$$

which is written in block form as

$$F' \begin{pmatrix} \delta X \\ \delta\Lambda \end{pmatrix} = \begin{bmatrix} F'_{u1} & F'_{u2} \\ F'_{l1} & F'_{l2} \end{bmatrix} \begin{pmatrix} \delta X \\ \delta\Lambda \end{pmatrix} = -F = \begin{pmatrix} -F_d \\ -F_c \end{pmatrix}, \tag{18.2.27}$$

where F' denotes the derivative of F. Following, we present the p-d i-p framework used in [14]. This is a common framework for both linear and semidefinite programming, see e.g. [841]. A centering parameter σ_k is included rather than the customary predictor-corrector approach. Let $\mathcal{F}^0$ denote the set of strictly feasible primal-dual points.

Algorithm 18.2.1 *(p-d i-p framework:)*
Given $(X^0, \Lambda^0) \in \mathcal{F}^0$
for $k = 0, 1, 2 \ldots$
solve *for the search direction (in a least squares sense)*

$$F'(X^k, \Lambda^k) \begin{pmatrix} \delta X^k \\ \delta\Lambda^k \end{pmatrix} = \begin{pmatrix} -F_d \\ \Lambda^k X^k + \sigma_k \mu_k I \end{pmatrix}$$

where σ_k *centering,* $\mu_k = \text{Trace} X^k \Lambda^k / (n-1)$

$$(X^{k+1}, \Lambda^{k+1}) = (X^k, \Lambda^k) + \alpha_k (\delta X^k, \delta\Lambda^k), \quad \alpha_k > 0,$$

so that $(X^{k+1}, \Lambda^{k+1}) \succ 0$
end (for).

The linear system for the search direction in Algorithm 18.2.1 is over-determined. Therefore, it is not clear what is meant by solving this system. There are many different search directions that have been used for SDP, see e.g. [780]. Two approaches to find the least squares solution of the over-determined linear system (18.2.27) can be used. First, one can solve the large least squares problem. This is called the Gauss-Newton (GN) method. Second, one can restrict dual feasibility and substitute for $\delta\Lambda^k$ in the second equation. This is called the restricted Gauss-Newton (RGN) method. The Gauss-Newton direction, first introduced in [450], always exists and it has many other good properties. See [450] for more details.

There does not seem to be an easy way to exploit the sparsity of H without using an iterative solution technique, e.g. a Lanczos approach to solving the least squares problem. In fact, such an approach (with a pre-conditioner) does exploit sparsity and can be very effective. Alternatively, one can exploit the fact that the sparsity pattern does not change and use a predetermined pivot order from a symbolic factorization.

Numerical tests were done on randomly generated problems. The program was written in MATLAB and ran on sun SPARC 20 with 64 megs of RAM and SPECint 65.3, SPECfp 53.1. Table 1 shows the results of problems with matrices up to dimension $n = 42$. the Algorithm has proven to be efficient and robust, i.e., it has not failed on any of the test problems. An important observation is that the ranks of the optimal solution X, for sparse problems where no completion existed, were typically small. i.e., a small embedding dimension was obtained without any original rank restriction. However, when a completion existed, the typically the embedding dimension was very high. For more details see [14].

dim	toler	H dens.	$rank(X)$	iterations	lss cpu-time	
					GN	RGN
8	10^{-13}	.8	2	25	.16	.1
9	10^{-13}	.8	2	23	.24	.13
10	10^{-13}	.8	3	25	.34	.18
12	10^{-9}	.5	3	17	.73	.32
15	10^{-9}	.5	2	20	2.13	.79
18	10^{-9}	.5	4	20	6.15	1.9
20	10^{-9}	.3	2	20	11.35	3.3
24	10^{-9}	.3	2	20	34.45	8.4
30	10^{-9}	.3	4	20	138.	31.5
35	10^{-9}	.2	3	19	373.	77.
38	10^{-9}	.2	3	19	634.	127
40	10^{-8}	.1	2	20	845.9	181.7
42	10^{-8}	.1	4	18	1118.	232.02

Table 18.1 data for closest distance matrix: dimension; tolerance for duality gap; density of nonzeros in H; rank of optimal X; number of iterations; cpu-time for one least squares solution of the GN and restricted GN directions.

18.3 WEIGHTED CLOSEST POSITIVE SEMIDEFINITE MATRIX

Consider the objective function

$$f(P) := \|H \circ (A - P)\|_F^2,$$

The weighted, *closest positive semidefinite matrix problem* is

$$\begin{array}{llll} & \mu^* := & \min & f(P) \\ (CPM) & & \text{subject to} & \mathcal{A}P = b \\ & & & P \succeq 0. \end{array}$$

As in the case of CDM, the linear constraint $\mathcal{A}P = b$ is added in case some elements of P are required to be exactly equal to the corresponding elements of A. However, unlike CDM, the weight matrix H here is assumed to have positive diagonal elements since otherwise if $H_{kk} = 0$, then P_{kk} is free and by setting $P_{kj} = A_{kj}$ for all j, and by taking P_{kk} large enough, P will maintain its positive semi-definiteness. Also note that CDM reduces to CPM if the linear operator $\mathcal{K}_V$ in CDM is defined as the identity operator. Thus, optimality conditions for the CPM follows from (18.2.2) by replacing $\mathcal{K}_V$ and $\mathcal{K}_V^*$ with the identity operator.

Theorem 18.3.1 *Suppose that Slater's condition holds. Then $\bar{P} \succeq 0$, and $\bar{y}$, $\bar{\Lambda} \succeq 0$ solve (CPM) and (DCPM), respectively, if and only if the following three equations hold.*

$$\begin{array}{rc} \mathcal{A}(\bar{P}) = b & \textit{primal feasibility} \\ 2H^{(2)} \circ (\bar{P} - A) - \mathcal{A}^*\bar{y} - \bar{\Lambda} = 0 & \textit{dual feasibility} \\ \textit{trace } \bar{\Lambda}\bar{P} = 0 & \textit{complementary slackness} \end{array}$$

In the case that there is no linear operator $\mathcal{A}$, and in some special cases, explicit solutions can be found for CPM as can be shown by the following theorem and its corollaries.

Theorem 18.3.2 *Suppose that (CPM) has no linear constraint $\mathcal{A}$. The matrix $\bar{P} \succeq 0$ solves (CPM) if and only if*

$$\textit{trace } (H^{(2)} \circ (\bar{P} - A))\,(P - \bar{P}) \geq 0, \quad \forall P \succeq 0, \tag{18.3.28}$$

where $H^{(2)} = H \circ H$.

Proof. See [375].

Let $A_{\succ}$ denote the positive semidefinite part of A.

Corollary 18.3.1 *If $H = E$, the matrix of ones, then the (unique) optimal solution of CPM is $\bar{P} = A_{\succ}$.*

Proof. See [375].

Corollary 18.3.2 *If $H \succ 0$, then the (unique) optimal solution of CPM is the solution of*

$$H \circ \bar{P} = (H \circ A)_{\succ}.$$

Proof. See [375].

Whereas a theorem similar to Theorem 18.3.2 is known for CDM [14], no explicit solutions, similar to those in the previous two corollaries, are known for the CDM.

18.3.1 Primal-Dual Interior-Point Algorithms

As in the case of CDM, the optimality conditions of CPM (Theorem 18.3.1) can be derived using the log-barrier approach. The stationarity condition for the CPM is obtained from (18.2.21) by setting both $\mathcal{K}_V$ and $\mathcal{K}_V^*$ to the identity operator.

$$\nabla B_\mu(X) = 2H^{(2)} \circ (P - A) - \mu P^{-1} = 0. \tag{18.3.29}$$

By equating $\Lambda = \mu P^{-1}$ we get the optimality conditions,

$$\begin{array}{rcll} 2H^{(2)} \circ (P - A) - \Lambda & = & 0 & \text{dual feasibility} \\ -P + \mu\Lambda^{-1} & = & 0 & \text{perturbed complementary slackness,} \end{array} \tag{18.3.30}$$

and the estimate of the barrier parameter

$$\mu = \frac{2}{n} \text{ trace } PH^{(2)} \circ (P - A). \tag{18.3.31}$$

Note that the perturbed complementary slackness in (18.3.30) could have been written as $\Lambda P - \mu I = 0$ as was done in the CDM case. However, unlike the CDM case where there was no easy way for exploiting the sparsity of H, the current form (18.3.30) for CPM allows one to take full advantage of the sparsity of H. Furthermore, both dual feasibility and perturbed complementary slackness maintain symmetry in (18.3.30). Thus, (18.3.30) is a square system which can be solved using Newton's method.

The linearization of equations (18.3.30) yield

$$\begin{array}{rcl} 2H^{(2)} \circ \delta P - \delta\Lambda & = & -(2H^{(2)} \circ (P - A) - \Lambda) \\ P + \delta P & = & \mu\Lambda^{-1} - \mu\Lambda^{-1}\ \delta\Lambda\ \Lambda^{-1}. \end{array} \tag{18.3.32}$$

From the second equation in (18.3.32) it follows that

$$\delta P = \mu\Lambda^{-1} - \mu\Lambda^{-1}\ \delta\Lambda\ \Lambda^{-1} - P \tag{18.3.33}$$

and

$$\delta\Lambda = \frac{1}{\mu}(-\Lambda(P + \delta P)\Lambda) + \Lambda. \tag{18.3.34}$$

Substituting (18.3.33) into the first equation in (18.3.32), one arrives at Newton equation in terms of the dual step $\delta\Lambda$

$$2H^{(2)} \circ (\mu\Lambda^{-1}\ \delta\Lambda\ \Lambda^{-1}) + \delta\Lambda = 2H^{(2)} \circ (\mu\Lambda^{-1} - A) - \Lambda. \tag{18.3.35}$$

One can solve (18.3.35) for $\delta\Lambda$ and substitute in (18.3.33) to recover δP (*the dual-step-first algorithm*). Note that (18.3.35) is equivalent to

$$\begin{array}{c} [\ 2\text{Diag}\ (\text{vec}(H^{(2)}))\ \mu\ (\Lambda^{-1} \otimes \Lambda^{-1}) + I\ ]\ \ \text{vec}(\delta\Lambda) \\ = \text{vec}(2H^{(2)} \circ (\mu\Lambda^{-1} - A) - \Lambda). \end{array} \tag{18.3.36}$$

In this final form possible sparsity of H can be exploited. In particular, $H_{ij} = 0$ implies that $\delta\Lambda_{ij} = 0$. Therefore, the corresponding rows and columns of the Kronecker product can be deleted when solving (18.3.36). This simplifies the construction of the linear system (18.3.36) considerably when H is sparse. On the other hand, if H has many significantly large elements. i.e., many elements of A are specified (or fixed). Then it is more efficient to substitute (18.3.34) into the first equation in (18.3.32) to obtain the following Newton equation in terms of the primal step δP

$$2H^{(2)} \circ \delta P + \frac{1}{\mu}\Lambda \ \delta P \ \Lambda = \Lambda - \frac{1}{\mu}\Lambda P \Lambda - (2H^{(2)} \circ (P - A) - \Lambda) \qquad (18.3.37)$$

or equivalently,

$$\begin{array}{l} [\,2\mathrm{Diag}\,(\mathrm{vec}(H^{(2)})) + \frac{1}{\mu}(\Lambda \otimes \Lambda)\,] \ \mathrm{vec}(\delta P) \\ = \mathrm{vec}(\Lambda - \frac{1}{\mu}\Lambda P \Lambda - (2H^{(2)} \circ (P - A) - \Lambda)). \end{array} \qquad (18.3.38)$$

Then, this system is solved for δP and substitute in (18.3.34) to recover $\delta\Lambda$ (*the primal-step-first algorithm).* Note that if P_{ij} is fixed, i.e., if H_{ij} is sufficiently large, then $\delta P_{ij} = 0$. Therefore, the system (18.3.38) can be very small if H has many elements which are sufficiently large.

In [375], numerical tests on randomly generated dense and sparse problems were done on a SUN SPARC 1 with 16megs RAM and with SPECint18.2 and SPECfp 17.9. The program was written in MATLAB. Tables 2 and 3 summarize the results of these test for the dual-step-first and for the primal-step-first algorithms respectively. For more details see [375].

dim	toler	H dens. / infty	$A \succeq 0$	cond(A)	$H \succ 0$	min/max	iters
60	10^{-6}	.01 / .001	yes	79.738	no	15/23	16.8
65	10^{-6}	.015 / .001	yes	49.8611	yes	18/24	21.25
83	10^{-6}	.007 / .001	no	235.1547	no	24/29	25.45
85	10^{-5}	.008 / .001	yes	94.6955	no	11/17	13.05
85	10^{-6}	.0075 / .001	no	299.8517	no	23/27	25.25
87	10^{-6}	.006 / .001	yes	74.163	yes	14/19	16.85
89	10^{-6}	.006 / .001	no	179.33	no	23/28	15.2
110	10^{-6}	.007 / .001	yes	172.255	yes	15/20	17.8
155	10^{-6}	.01 / 0	yes	643.9619	yes	14/18	15.3

Table 18.2 data for dual-step-first (20 problems per test); dimension; tolerance for duality gap; density of nonzeros in H/ density of infinite values in H; positive semidefiniteness of A; positive definiteness of H; min and max number of iterations; average number of iterations.

18.4 OTHER COMPLETION PROBLEMS

The list of papers on completion problems and applications is large and varied. The different types of completion problems include: maximum rank completions; M-matrices and inverse M-matrices; partially upper triangular matrices;

dim	toler	H dens. / infty	$A \succeq 0$	cond(A)	$H \succ 0$	min/max	iters
15	10^{-5}	.0751 / .02	yes	222.52	no	8/17	10.25
15	10^{-6}	.1 / .95	yes	19.6102	no	10/23	15.55
15	10^{-6}	.01 / .95	yes	21.0473	no	10/20	13.2
19	10^{-6}	.005 / .1	yes	14.811	no	10/18	12.95
21	10^{-6}	.005 / .1	yes	20.3077	no	8/24	15
38	10^{-6}	1 / .99	yes	49.6068	yes	14/24	16.3
45	10^{-6}	1 / .99	yes	46.8396	yes	15/22	17.1
55	10^{-6}	1 / .99	yes	37.1883	yes	15/30	17.45
85	10^{-5}	.0219 / .02	yes	1374.54	no	16/23	18.95
95	10^{-5}	.0206 / .02	yes	2.6983	no	8/14	11.05
95	10^{-6}	1 / .999	yes	196.0130	yes	14/18	16.8
145	10^{-6}	.01 / .997	yes	658.5103	yes	13/17	14.95

Table 18.3 data for primal-step-first (20 problems per test): dimension; tolerance for duality gap; density of nonzeros in H/ density of infinite values in H; positive semidefiniteness of A; positive definiteness of H; min and max number of iterations; average number of iterations.

partial Jordan matrices; completions to normal matrices and to P-matrices. One can see the vast assortment of further problems and applications by searching MathSciNet with the keywords *matrix* and *completion.*

19 EIGENVALUE PROBLEMS AND NONCONVEX MINIMIZATION

Florian Jarre

AMS 1991 subject classification. Primary: 90C.

Key words. Eigenvalue, predictor corrector method, semidefinite program.

19.1 INTRODUCTION

Many semidefinite optimization problems are reformulations of problems that are initially derived from certain eigenvalue bounds. As examples we discuss the problems of minimizing the largest eigenvalue of a symmetric matrix, and, more generally, a weighted sum of the k largest eigenvalues of a symmetric matrix, or the largest generalized eigenvalue of a symmetric matrix pencil. In extension to the theory in Chapter 1 of this book, we summarize some known theoretical results on eigenvalues of symmetric matrices. These results can be used to reformulate eigenvalue problems such that the problems involve smooth functions only—for the constraints as well as for the objective function—and that convexity is preserved if possible.

Based on such a reformulation we discuss a generalization of a primal predictor corrector interior-point method for smooth constrained minimization. In recent years, a number of approaches were proposed to generalize interior-point methods to nonconvex programs, see for example Argaez and Tapia [45], Byrd, Gilbert, and Nocedal [150], Coleman and Li [165], Conn, Gould, and Toint [167], Forsgren and Gill [248], Gay, Overton, and Wright [269], Sargent and Zhang [692], Urban, Tits, and Lawrence [801], Vanderbei and Shanno [815]. Most closely related to the approach presented here is the article by Kruk and Wolkowicz [451]. The main difference lies in the "distribution" of computational effort. For the method presented here, the individual subproblems (certain QQP's) are simpler, but we have to be prepared to solve more subproblems than are necessary in [451].

Some references for nonsmooth reformulations and nonsmooth optimization involving eigenvalue problems can be found in [483].

19.2 SELECTED EIGENVALUE PROBLEMS

Eigenvalue optimization has many origins. A very prominent one is based on the Lyapunov stability theorem, [513].

Theorem 19.2.1 (Lyapunov) *Let A be an $n \times n$-matrix with real or complex entries. The following statements are equivalent:*

i) All eigenvalues of A have strictly positive real part.

ii) A is similar to some matrix B such that $B+B^*$ is positive definite. (Here, $B^* = \bar{B}^T$ is the transpose of the complex conjugate of B.)

iii) There exists a positive definite matrix W such that $WA+A^*W$ is positive definite.

A matrix satisfying *i)*—and thus also *ii)* and *iii)*—is called positive stable.

For several applications arising in control theory with a real symmetric matrix A one would like to impose certain constraints on the matrix W in *iii)*. Finding such a restrained matrix W can be done by minimizing the maximum eigenvalue λ of $WA+A^TW$ i.e. by solving a semidefinite optimization problem of the form

$$\begin{array}{ll} \text{minimize} & \lambda \\ \text{subject to} & \lambda I + WA + A^*W \succeq 0 \\ & F_1(W) = 0 \\ & F_2(W) \leq 0, \\ & W \succeq 0 \end{array} \qquad (19.2.1)$$

see e.g. [27, 137, 139, 364, 583, 810] and the references therein. When F_1 is linear and F_2 is convex, problem (19.2.1) is convex, and the techniques discussed in

Another interesting example is the minimization of the weighted sum of the eigenvalues of some symmetric $n \times n$-matrix variable X. By

$$\lambda_1(X) \geq \lambda_2(X) \geq \ldots \geq \lambda_n(X)$$

we denote the eigenvalues of X in nonincreasing order. Let $\Lambda_k := \sum_{i=1}^k \lambda_i(X)$ be the sum of the k largest eigenvalues of X.

Theorem 19.2.2 (Nesterov and Nemirovskii) *Let $t \in \mathbb{R}$ and $X \in \mathcal{S}^n$ be given. The following statements are equivalent* *i)*

$$t \geq \Lambda_k(X),$$

ii)

$$\exists\ s \in \mathbb{R},\ Z \in \mathcal{S}^n: \quad t - ks - \text{Trace}(Z) \geq 0 \qquad (19.2.2)$$

$$Z \succeq 0 \qquad (19.2.3)$$

$$Z - X + sI \succeq 0. \qquad (19.2.4)$$

A proof is given in [583, Chapter 6.4.3]. Minimizing Λ_k can therefore be reduced to the problem of minimizing t subject to the convex (linear and semidefinite) constraints (19.2.2)–(19.2.4). We note that the dual to the problem of minimizing $\Lambda_k(X)$ subject to linear equality constraints on X can also be written as a semidefinite program, see, e.g. [483].

When $d_1 \geq d_2 \geq \ldots \geq d_k > 0$, the problem of minimizing the weighted sum

$$\Lambda_d(X) := d_1\lambda_1(X) + d_2\lambda_2(X) + \ldots + d_k\lambda_k(X)$$

can be reformulated as

$$\begin{array}{lll} \text{minimize} & t & \\ \text{subject to} & t & \geq d_k t_k + \sum_{i=1}^{k-1}(d_i - d_{i+1})t_i \qquad (19.2.5) \\ & t_1 I - X \succeq 0 & \\ & t_i & \geq \Lambda_i(X) \ \text{ for } 2 \leq i \leq k, \\ & t, t_1, \ldots, t_k \in \mathbb{R} & \end{array}$$

and the constraints $t_i \geq \Lambda_i(X)$ can be modelled using (19.2.2)–(19.2.4) of Theorem 19.2.2. For small values of k, the optimization problem (19.2.5) (along with Theorem 19.2.2) is computationally better tractable than the original formulation; it is a smooth problem with convex inequalities, while the function Λ_d is only piecewise smooth.

As was shown in [213] (see also [483]) the function Λ_d is convex, and there are explicit formulas to evaluate its subgradient, allowing for example, the use of bundle-methods for minimizing Λ_d. A formula for the subgradient of the largest eigenvalue $\lambda_1(X)$ and nonsmooth methods for large scale applications are discussed in Chapter 11 of this Handbook, see also [338]. In [604], the case of minimizing the sum of the k largest absolute values of the eigenvalues of A is analyzed. (This function is convex as well.) In [605], a Newton type method for eigenvalue optimization is discussed. See also [483] for eigenvalue problems with nonsymmetric matrices.

A minor generalization to problem (19.2.1), also the problem of minimizing the largest generalized eigenvalue of a symmetric matrix pencil, arises in applications of control theory, see e.g. [136, 320]. Let $A(x)$ and $B(x)$ be smooth functions: $x \in \mathbb{R}^n \to \mathcal{S}^k$ such that $B(x)$ is positive definite for all x. Let

$A_i(x)$ denote the derivative of $A(\ .\ )$ at the point x with respect to the i-th component of x, and likewise for $B_i(x)$. The optimality conditions that ω is the largest generalized eigenvalue of $(A(x), B(x))$ (i.e. the largest value such that $A(x) - \omega B(x)$ is singular) are given in [320]:

$$\begin{aligned} \exists\, H, U \in \mathcal{S}^n,\ \ U \succeq 0: \quad & U \bullet H(\omega B_i(x) - A_i(x))H = 0 \quad \text{for } 1 \le i \le n \\ & \text{Trace}(U) = 1 \\ & HB(x)H = I \\ & U(\omega I - HA(x)H) = 0. \end{aligned} \tag{19.2.6}$$

(The conditions are stated in a slightly more general form in [320].) Based on these conditions, a locally superlinearly convergent algorithm is proposed in [320]. For the case that A, B depend linearly on x, a very powerful polynomial time method for solving (19.2.6) up to a specified accuracy was developed in [584].

An interesting result relating the eigenvalues of two symmetric matrices is the following theorem [659, 235].

Theorem 19.2.3 (Rendl,Wolkowicz;Finke,Burkard,Rendl) *Let A, B be real symmetric matrices. Then,*

$$\begin{aligned} \max\{\ \text{Trace}(AUBU^T)\ \mid\ U^TU = I\ \} &= \lambda(A)^T\lambda(B), \\ \min\{\ \text{Trace}(AUBU^T)\ \mid\ U^TU = I\ \} &= -\lambda(-A)^T\lambda(B). \end{aligned}$$

The most general form of this theorem is given in [660], where $A \in \mathcal{S}^p$ and $B \in \mathcal{S}^q$ may be of different dimension $q \ge p$, (and $U \in \mathbb{R}^{p\times q}$ is a rectangular matrix). The proof of Theorem 19.2.3 given in [659] also implies the following well-known result [590].

Theorem 19.2.4 (Neumann) *For real symmetric matrices $X, Y \in \mathcal{S}^n$ the inequality*

$$\text{Trace}(XY) \le \lambda(X)^T\lambda(Y)$$

is true, and equality holds if and only if X and Y have a simultaneous ordered eigenvalue decomposition.

For completeness, we mention the following corollary:

Theorem 19.2.5 (Neumann) *For real square matrices X and Y, the inequality*

$$\text{Trace}(X^TY) \le \sigma(X)^T\sigma(Y)$$

is true, where $\sigma(X)$ is the vector of singular values of X in nonincreasing order.

(Theorem 19.2.5 follows when applying Theorem 19.2.4 to the block matrices $\begin{pmatrix} 0 & X^T \\ X & 0 \end{pmatrix}$ and $\begin{pmatrix} 0 & Y^T \\ Y & 0 \end{pmatrix}$).

Graph partitioning problems have long been associated with eigenvalue problems. In [197], a relation bounding the partition of a graph in terms of a weighted sum of the eigenvalues of a symmetric matrix with variable diagonal entries was established, and in [196, 178] early implementations are presented. In [660], Theorem 19.2.3 is used to reprove the result of [197] in an elegant way, and to derive various semidefinite relaxations for the graph partitioning problem. Another surprising graph-theoretical eigenvalue bound was established in [496, 497]. In recent years, with the development of interior point methods, these eigenvalue bounds have regained public interest, see e.g. the survey [483].

Depending on the specifications of the particular problem, (additional convex or nonconvex constraints) some of the above examples (19.2.1), (19.2.2)–(19.2.4), (19.2.5), (19.2.6) can be reformulated as convex optimization problems and can be solved by interior-point methods. There are also important applications for which a convex reformulation is not known. These applications may still contain a semidefiniteness constraint, and our general problem formulation (19.2.7) distinguishes between semidefinite and other nonlinear constraints.

$$\begin{array}{ll} \text{minimize} & c^T x \\ \text{subject to} & \mathcal{A}(x) \succeq 0, \\ & f_i(x) \leq 0 \quad \text{for} \quad 1 \leq i \leq p, \\ & f_j(x) = 0 \quad \text{for} \quad p+1 \leq j \leq m, \end{array} \tag{19.2.7}$$

where $\mathcal{A} : \mathbb{R}^n \to \mathcal{S}^k$, $f_i : \mathbb{R}^n \to \mathbb{R}$, and $f_j : \mathbb{R}^n \to \mathbb{R}$ are smooth mappings[1].

In the remaining part of this chapter, a method for finding a local minimizer of problem (19.2.7) is derived. The method is motivated by a primal predictor corrector method as discussed in [367]. The usual linear systems of the Newton step and of the predictor step are replaced by quadratically constrained quadratic programs. These QQP's are of a special structure and can be solved directly. They also allow for a special line search which effects that the algorithm is applicable to nonconvex programs.

To explain the concept of our method we proceed in two steps. We first discuss a method for smooth unconstrained nonconvex minimization that is based on merging trust-region and line search methods. This method is then generalized to constrained nonlinear programs.

19.3 GENERALIZATION OF NEWTONS METHOD

In the framework of interior methods for convex programs, the two basic elements are the existence of a *central path*, and, as a main tool, the use of *Newton's method* to approximate the path, see e.g. [742, 583].

For the nonconvex case, there may be bifurcations in the path (so, at best, we may expect to find a local minimizer when following a branch of the path), and we need to replace Newton's method by a new tool for nonconvex minimization.

[1]The linearity of the objective function is no restriction of generality; a nonlinear objective function $f(x)$ can be included as an additional constraint "$f(x) \leq x_{n+1}$", and then, the linear quantity x_{n+1} can be minimized.

19.3.1 An algorithm for unconstrained minimization

We first describe a tool for smooth nonconvex minimization that may be regarded as a modification of the well known Levenberg-Marquardt type methods, or, as a modification of a class of trust region algorithms described in Shultz et.al. [737].

We start by recalling an important feature of Newton's method for minimizing a smooth convex function: Its affine invariance. For descent methods for nonconvex minimization this property is lost. The approach below is intended to preserve the independence of affine transformations "as much as possible". In particular, locally, near an optimal solution the method behaves like Newton's method.

Let a smooth function $f : \mathbb{R}^n \to \mathbb{R}$ be given. Let $g = g(x) := \nabla f(x)$, and $H := H(x) := \nabla^2 f(x)$ be the gradient and the Hessian. To minimize f (without any constraints on x) we consider the problem of finding a search direction

$$\Delta x(r) \in \arg\min\{g^T s + \frac{1}{2} s^T H s \mid s^T s \leq 2r^2\}. \tag{19.3.8}$$

Evaluating the KKT-conditions, it is straightforward to derive (and well known) that, in general, there exists a monotone function $\lambda = \lambda(r)$ such that the following three properties hold:

1. $\lim_{r\to 0} \lambda(r) = \infty$,
2. $\lim_{r\to\infty} \lambda(r) = \underline{\nu} := \max\{0, -\nu_{min}\}$ where ν_{min} is the smallest eigenvalue of H,
3. and

$$s(r) := -(H + \lambda(r)I)^{-1} g \tag{19.3.9}$$

   is a KKT-point for (19.3.8).

One can construct degenerate examples where property 2. fails to hold[2]. However, when r is sufficiently small and $g \neq 0$, then $s(r)$ is always a global minimizer. In fact it is obvious, that

- $\lim_{r\to 0} \frac{s(r)}{\|s(r)\|} = -\frac{\nabla f(x)}{\|\nabla f(x)\|}$ (steepest descent), and
- $\lim_{r\to\infty} s(r) = -(\nabla^2 f(x))^{-1} \nabla f(x)$ (Newton step) if $H \succ 0$.

We now do a line search along r, i.e. we find the point where $f(x + s(r))$ is minimized for $r > 0$. Since there is a one-to-one correspondence of r and λ for $\lambda > \underline{\nu}$, we will also write $s(\lambda)$ instead of $s(r)$, and minimize $f(x + s(\lambda))$ with respect to λ. The "line search" is done as follows: First compute

$$H = UDU^T,$$

where U is unitary, and D is a *tri*diagonal matrix[3]

[2]This situation is usually referred to as the "hard case" [550]. Here, we can ignore the hard case. Instead, we secure the descent properties by a line search.

[3]We thank Prof. M. Powell for pointing us to tridiagonal systems which are much cheaper to compute than diagonal decompositions, see also [644, 645].

If H is dense, the tridiagonal factorization proposed here is somewhat more expensive than a Cholesky factorization of H (provided H is positive definite). For some standard (convex) semidefinite program, the factorization is about the same computational effort as computing H. As a technical detail we remark that we need **not** compute U explicitly, but rather store the Householder vectors w_i that are used to tridiagonalize H. The knowledge of the w_i is sufficient to compute the multiplications Uz and $U^T z$ for a given vector z.

Now,

$$s(\lambda) = -U(D + \lambda I)^{-1} U^T g.$$

can easily be computed for any value of $\lambda > \underline{\nu}$. (Solving a tridiagonal system takes $O(n)$ operations.) Moreover, the derivatives of the 'line search function'

$$l(\lambda) := f(x + s(\lambda)) \tag{19.3.10}$$

are readily computed,

$$\begin{aligned} l'(\lambda) &= \nabla f(x + s(\lambda))^T \dot{s}(\lambda), \qquad (19.3.11) \\ l''(\lambda) &= \nabla f(x + s(\lambda))^T \ddot{s}(\lambda) + \dot{s}(\lambda)^T \nabla^2 f(x + s(\lambda)) \dot{s}(\lambda). \end{aligned}$$

Here, the vectors $\dot{s}$ and $\ddot{s}$ are obtained from systems as in (19.3.9)

$$\begin{aligned} (H + \lambda I) s(\lambda) &= -g \\ (H + \lambda I) \dot{s}(\lambda) &= -s(\lambda) \\ (H + \lambda I) \ddot{s}(\lambda) &= -\dot{s}(\lambda). \end{aligned}$$

Note that directional derivatives like $\dot{s}^T \nabla^2 f(x) \dot{s}$ are typically much cheaper to evaluate than the full derivatives e.g. than $\nabla^2 f(x)$.

The function l can be minimized by some Newton method with Armijo globalization. The minimization of l results in a (curved) line search for f.

Trust region problems of the form (19.3.8) have been considered in numerous papers in connection with various strategies for doubling or halving r depending on how well a certain prediction matches the actual function value. Here, we propose to do a line search which helps to avoid issues related to the "hard case" when the KKT-point of (19.3.8) is not a global (or even local) minimizer. The following result is shown in [365].

Proposition 19.3.1 *A repeated application of the search step* (19.3.9) *with exact line search for the first local minimizer with respect to* (19.3.10) *either returns a (curved) line $s(\lambda)$ along which f is monotonically decreasing (but possibly bounded), or it yields an infinite sequence of iterates along which $f(x)$ is monotonically decreasing, and each accumulation point of this sequence is a zero of the gradient of f.*

If the Hessian of f is positive definite at one of the accumulation points then the iterates converge superlinearly to this accumulation point (in particular, then there exists only one accumulation point).

Discussion. The global convergence behaviour of the above scheme is not fully satisfactory in practice. It may be slow in spite of the high costs (the second derivative of f and its UDU^T-factorization) of evaluating the search step s. In numerical experiments we have tested numerous variations of this scheme in the spirit of Mehrotra's corrector, but at present we did not find any improvements that are comparable to the effectiveness of Mehrotra's corrector in primal-dual space. It appears that as long as there is negative curvature, the second order information can only be locally useful, and any search step derived from local information may exhibit poor global convergence properties. (This contrasts the performance of the Mehrotra corrector observed in implementations for linear and quadratically constrained convex programs.) In fact, it is easy to show that there does not exist any affine invariant method that generates a descent direction for a smooth but nonconvex function f based on knowing a fixed number of points and the associated function values and first and second derivatives.

We found in particular, that for barrier sub-problems with a poorly chosen barrier parameter, the rate of convergence of the qqp-steps may be very slow: This happens, for example, when the Hessian is positive definite, but the full Newton step is "much too long", i.e. when a line search along the Newton direction returns a very small step length, say $\alpha << 1$. In this case, the qqp-step cannot result in the Newton direction; instead it will generate a very small multiple of the steepest descent direction. Obviously, a small multiple of the Newton direction would have been a better search step in this case. To avoid this situation, we propose to do an approximate plane search along α and λ for the case that λ is large compared to the eigenvalues of the Hessian (i.e. when the information of H in $(H + \lambda I)^{-1}$ is nearly wiped out by a large value of λ). Typically, a line search with respect to α for very few fixed values of λ suffices; we omitt the details.

With this modification, our observation for some (unconstrained) nonconvex example problems (e.g. various modifications of Rosenbrock's function) was that the qqp scheme converges *very reliably*.

For sparse large scale applications we refer to the techniques discussed in [185].

We now apply the new main tool to constrained optimization problems.

19.4 A METHOD FOR CONSTRAINED PROBLEMS

19.4.1 The constrained problem

We now return to problem (19.2.7), namely

$$\begin{array}{ll} \text{minimize} & c^T x \\ \text{subject to} & \mathcal{A}(x) \succeq 0, \\ & f_i(x) \leq 0 \quad \text{for} \quad 1 \leq i \leq p, \\ & f_j(x) = 0 \quad \text{for} \quad p+1 \leq j \leq m. \end{array}$$

For compactness of notation we write

$$f_0(x) := -\det(\mathcal{A}(x)). \tag{19.4.12}$$

Further we denote

$$F_1(x) := \begin{pmatrix} f_0(x) \\ \vdots \\ f_p(x) \end{pmatrix}, \qquad F_2(x) := \begin{pmatrix} f_{p+1}(x) \\ \vdots \\ f_m(x) \end{pmatrix}. \tag{19.4.13}$$

We assume that a point $\bar{x}^{(0)}$ is known such that $\mathcal{A}(\bar{x}^{(0)}) \succ 0$, and $f_i(\bar{x}^{(0)}) < 0$. The point $\bar{x}^{(0)}$ can be found by solving a Phase 1 problem of the form

$$\text{minimize}\{\lambda \mid \mathcal{A}(x) + \lambda I \succeq 0, \quad f_i(x + \frac{\lambda}{\lambda_0}(\bar{x}^{(i)} - x^{(0)})) \leq 0, \quad \lambda \geq 0\}. \tag{19.4.14}$$

In (19.4.14) we assume that points $\bar{x}^{(i)}$ are known such that $f_i(\bar{x}^{(i)}) < 0$. If the inequalities in (19.4.14) are modelled by a barrier function, the algorithm below applied to this problem will—under certain conditions—return a strictly feasible starting point.

19.4.2 Outline of the method

For the inequalities of (19.2.7) (including $f_0(x) < 0$) we define the barrier function

$$\phi(x, \mu) := \frac{c^T x}{\mu} - \sum_{i=0}^{p} \log(-f_i(x)). \tag{19.4.15}$$

By the above assumption, $\bar{x}^{(0)}$ is strictly feasible for $\phi(\;.\;, \mu)$, but does not necessarily satisfy $F_2(\bar{x}(0)) = 0$.

The method consists of two steps. First, a barrier problem for some fixed barrier parameter $\mu > 0$ is solved, and then a predictor step is performed to reduce μ. Here, the role of μ is twofold. It is used as a bound for the residual of the equation $F_2(x) = 0$, and as a weight to the logarithmic barrier terms of the inequalities. (For convenience we divided by μ in the definition of ϕ, so that in (19.4.15), μ seems to be a weight for the objective function rather than for the barrier terms.)

To initialize the algorithm we define some quantities

$$x^{(0)} := \bar{x}^{(0)}, \qquad \rho_0 > \left\| F_2(x^{(0)}) \right\|, \qquad \mu_0 := 1.$$

Given $x^{(k)}$ and $\mu = \mu_k$, we consider the barrier problem

$$\text{minimize} \quad \phi(x, \mu) \quad \text{subject to} \quad F_2(x) = F_2(x^{(k)}). \tag{19.4.16}$$

The barrier subproblem (19.4.16) does not aim at reducing $||F_2(x)||$, and in fact, we will replace the condition $F_2(x) = F_2(x^{(k)})$ by a weaker condition

$$||F_2(x)|| \leq 2\mu_k \rho_0, \tag{19.4.17}$$

which in the case $k = 0$, for example, allows an increase of $||F_2(x)||$ by a factor of more than 2.

19.4.3 *Solving the barrier subproblem ("centering step")*

The barrier subproblem will be solved with a generalization of the method from the previous section, namely as follows:

The KKT-conditions for (19.4.16) are

$$\nabla_x \phi(x,\mu) + \nabla F_2(x) y = 0 \tag{19.4.18}$$

and $F_2(x) = F_2(x^{(k)})$. Given an iterate x we find the "best" y associated to x by solving a linear least squares problem to approximate (19.4.18). (The QR-factorization of $\nabla F_2(x)$ will be needed below, so that y can be obtained at little extra cost.) We then define

$$H := \nabla_x^2 \phi(x,\mu) + \sum_{j=p+1}^m y_j \nabla^2 f_j(x), \qquad g := \nabla_x \phi(x,\mu),$$
$$A := \nabla F_2(x)^T, \qquad b := -F_2(x).$$

Let

$$A^T = Q \begin{bmatrix} R \\ 0 \end{bmatrix} \quad \text{with} \quad Q^T Q = I \in \mathbb{R}^{n \times n} \tag{19.4.19}$$

and with a full row rank upper triangular matrix R be a QR-factorization of the transpose of the Jacobian of F_2. Here, $R \in \mathbb{R}^{q \times (m-p)}$ with $m-p \geq q$, and $0 \in \mathbb{R}^{(n-q)\times(m-p)}$. When A has full rank then $q = m-p$, else $q < m-p$. Note that we referred to this QR-factorization after equation (19.4.18) where a multiplyer y was to be found. We divide $Q = (Q_1, Q_2)$ into the first q columns Q_1, and the remaining $n-q$ columns Q_2 that form a basis of the null space of A. Let

$$\tilde{s} := Q^T s = \begin{pmatrix} \tilde{s}_1 \\ \tilde{s}_2 \end{pmatrix} \tag{19.4.20}$$

be partitioned analogously to the partition of Q (i.e. $\tilde{s}_1 = Q_1^T s \in \mathbb{R}^q$), and likewise for $\tilde{g} := Q^T g$. Let $\tilde{H} := Q^T H Q$ be divided in four blocks $\tilde{H}_{ij} = Q_i^T H Q_j$ for $i,j \in \{1,2\}$. By I_1 we denote the $q \times q$ identity matrix, and by I_2 we denote the $(n-q) \times (n-q)$ identity matrix. With this notation we define

$$\tilde{s}_1 = \theta(RR^T + \lambda^{3/2} I_1)^{-1} Rb, \tag{19.4.21}$$
$$\tilde{s}_2 = (\tilde{H}_{22} + \lambda I_2)^{-1}(-\tilde{g}_2 - \tilde{H}_{21}\tilde{s}_1),. \tag{19.4.22}$$

(There is no need to use the same number λ in (19.4.21) and (19.4.22); instead, one might consider using parameters λ_1 in (19.4.21) and λ_2 in (19.4.22).)

Note that $\tilde{s}_1$ in (19.4.21) is a KKT point—and hence an optimal solution—of the convex problem

$$\text{minimize} \quad \left\|R^T s_1 - \theta b\right\|^2 \quad \text{subject to} \quad \|s_1\|^2 \leq \hat{r}^2$$

for some $\hat{r} > 0$. System (19.4.21), (19.4.22) is derived from the problem

$$\text{minimize}\{g^T s + \frac{1}{2} s^T H s \mid \|As - \theta b\| \leq \epsilon(r)\,\|\theta b\|, \;\; s^T s \leq 2r^2\}, \tag{19.4.23}$$

with some number $\epsilon(r) \le 1$, see [365]. A positive value of $\theta \le 1$ may be necessary in order to avoid that $\|F_2(x)\|$ grows too large when iterating the process $x \to x + s$, When $\theta > 0$ is positive, also a positive value of $\epsilon(r)$ may be necessary in order to guarantee a descent property of the search step s[4]. The step length along the curved line $s(r)$ will be chosen as to minimize $\phi(\ .\ ,\mu)$ subject to the constraint (19.4.17), i.e. we compute

$$\lambda^* := \text{argmin}\ \ \phi(x + s(\lambda), \mu)\ \ \text{subject to}\ \ \|F_2(x + s(\lambda))\| \le 2\mu\rho_0, \quad (19.4.24)$$

We hope, of course, that the constraint (19.4.17) will not become active, and that the step length λ^* is solely determined by minimizing ϕ. To this end we set $\theta = \bar{\theta}$ in (19.4.23) if $\|F_2(x)\| \ge \mu\rho_0$, and $\theta = 0$ if $\|F_2(x)\| \le 0.5\mu\rho_0$, and we leave θ at its value from the previous iteration if $0.5\mu\rho_0 < \|F_2(x)\| < \mu\rho_0$.

In [365] it is shown that for large λ the search direction $s(\lambda)$ reduces both, ϕ as well as $\|F_2\|$ provided that $F_2 \ne 0$ and $\nabla\phi \notin \text{range}(A^T)$.

The idea underlying (19.4.21), (19.4.22) is very simple: The power $\lambda^{3/2}$ in (19.4.21) guarantees for large λ that $\|F_2(x + s(\lambda))\|$ is reduced by $O(\lambda^{-3/2})$, while the increase of $\|F_2\|$ due to its nonlinearity is $O(\lambda^{-2})$. Likewise, the factor λI in (19.4.22) guarantees that $\phi(x+s(\lambda))$ is reduced by $O(\lambda^{-1})$ while the perturbation resulting from (19.4.21) (which could possibly increase $\phi(x+s(\lambda))$) is $O(\lambda^{-3/2})$.

For the systems (19.4.21) and (19.4.22) the UDU^T-decomposition of Section 19.3 can be applied again. Here, the computation of $\widetilde{H}_{22}$ typically is more expensive than the cost of its UDU^T factorization.

Proposition 19.4.1 *Let $\mu > 0$ be fixed, and $x^{(k)}$ satisfy the condition* (19.4.17) *and be feasible for $\phi(\ .\ ,\mu)$. If the above procedure* (19.4.21), (19.4.22) *and* (19.4.24) *to minimize $\phi(\ .\ ,\mu)$ for a fixed value $\mu > 0$ is iterated, then each accumulation point x^* of the sequence of iterates is either*

1. *a stationary point in the sense that*

$$\|F_2(x^*)\| \le 2\mu\rho_0, \quad F_1(x^*) < 0, \quad A(x^*) \succ 0,$$

and there is a vector y with $\nabla\phi(x^*, \mu) + \nabla F_2(x^*)y = 0$.

2. *or a singular point such that $\nabla F_2(x^*)$ is rank deficient.*

For a proof see [365].

The stationary point of $\phi(\ .\ ,\mu_k)$ found at iteration k will be denoted by $x^{(k^+)}$, and the residual for the equality constraints by

$$\rho_k := \left\| F_2(x^{(k^+)}) \right\| \le 2\mu_k\rho_0 \quad (19.4.25)$$

(because of (19.4.17)).

[4]In using the Hessian H of the Lagrangian of (19.4.16), problem (19.4.23) is very similar to an SQP subproblem. As in the unconstrained case (Section 19.3.1), the approximate solution (19.4.21), (19.4.22) of (19.4.23) allows a continuous adjustment of the trust region radius r. In view of the similarity to the SQP method, and since we solve a simple **Q**uadratically constrained **QP** to determine the search direction, we call this method "QQP-method".

19.4.4 The predictor step

In principle, after an approximate local minimizer $x^{(k^+)}$ of the barrier problem has been computed, one could simply reduce the barrier parameter and continue using Newtons method. We propose a special predictor step with the aim of reducing $||F_2||$ if it is large, and moving towards optimality when $||F_2||$ is small. Let $\sigma = \max\{1, ||F_2(x^{(k^+)})||/(\mu_k||F_2(x^{(0)})||)\}$ measure the size of $||F_2||$ relative to μ.

To define a predictor step we approximate the feasible set by linearizing all constraints and by using the Dikin-ellipsoid of the linearized set. Then, the infeasibility (i.e. $||F_2||$) is reduced over the Dikin ellipsoid using the approach of Section 3.

To this end let $\tilde{c} := Q^T c$ and $\tilde{x} = Q^T x$ and define

$$\phi_x^{lin}(s,\mu) := \frac{c^T(x+s)}{\mu} - \log\det \mathcal{A}_x^{lin}(s) - \sum_{i=1}^{p} \log(-f_i(x) - \nabla f_i(x)^T s) \quad (19.4.26)$$

as the barrier function of problem (19.2.7) linearized at x. Here, $\mathcal{A}_x^{lin}(s)$ is the linearization of $\mathcal{A}$ at x. Define a positive definite approximation $H^+ := I + \nabla_s^2 \phi_x^{lin}(s,\mu)$ to H. (When H is sufficiently positive definite, we may use H in place of H^+.) The above definition of ϕ^{lin} and hence of H^+ is not unique; if we treat the constraint $-\det \mathcal{A}(x) < 0$ like the remaining constraints $f_i(x) < 0$, we obtain a different matrix H^+. In this case also the convergence result will be weaker. We require that ϕ^{lin} is a self-concordant barrier function (which is automatically true if we define ϕ^{lin} as in (19.4.26)).

We propose to reduce $||F_2||$ by exploiting the barrier property of ϕ^{lin},

$$\text{minimize} \quad ||As - b|| \quad \text{subject to} \quad s^T H^+ s \le \check{r}^2. \quad (19.4.27)$$

As before for H we also denote $\widetilde{H}^+ = Q^T H^+ Q$ and $\tilde{\hat{s}} = Q^T \hat{s}$. The predictor step is given by

$$\tilde{\hat{s}}_1 = \sigma(RR^T + \lambda \widetilde{H}_{11}^+)^{-1} Rb, \quad (19.4.28)$$

$$\tilde{\hat{s}}_2 = (\widetilde{H}_{22}^+)^{-1}\left(\frac{-\tilde{c}_2}{\mu(1+\lambda)} - \widetilde{H}_{21}^+ \tilde{\hat{s}}_1\right). \quad (19.4.29)$$

Here, (19.4.28) is based on the same ideas as the search step in the previous discussion, it minimizes the norm of the linearization of F_2 over a trust region given by the Hessian of ϕ^{lin}. When $H = H^+$, the term in (19.4.29) reduces to the tangent of the central path, damped by a factor $\mu/(1+\lambda)$. Again, different values of λ in (19.4.28) and (19.4.29) may be efficient in an implementation.

As for the search step s in the trust region problem, also $\hat{s}$ depends on the choice of λ; occasionally we will write $\hat{s}(\lambda)$ to emphasize this dependence.

The step lengths $\Delta\mu_k$ and λ are determined by two different cases.

When $\sigma > 1$ define

$$\Delta\mu = \Delta\mu(\lambda) = \mu_k\left(1 - \frac{\left\|F_2(x^{(k^+)} - \hat{s}(\lambda))\right\|}{\max\{\frac{3}{2}\rho_0\mu_k, \ \left\|F_2(x^{(k^+)})\right\|\}}\right) \quad (19.4.30)$$

and find $\lambda > 0$ that maximizes $\Delta\mu$ subject to the condition

$$\phi(x^{(k^+)} - t\hat{s}(\lambda)) \quad \text{is defined for} \quad t \in [0,\ 2 - \frac{\Delta\mu}{\mu}]. \tag{19.4.31}$$

When $\sigma = 1$ we define $\Delta\mu = \frac{\mu}{1+\lambda}$ and minimize λ subject to

$$\frac{\left\| F_2(x^{(k^+)} - \hat{s}(\lambda)) \right\|}{\mu_k - \Delta\mu} \leq \max\{\frac{3}{2}\rho_0, \frac{\left\| F_2(x^{(k^+)}) \right\|}{\mu_k}\} \tag{19.4.32}$$

and (19.4.31). (Note that (19.4.30) implies (19.4.32).)

If the step length for the predictor step lies below a certain preset threshold ϵ and if $\sigma > 1$, we will replace the right hand side term $\tilde{c}_2/\mu$ in (19.4.29) by 0 and use the rules (19.4.32), (19.4.31). This effects that the only purpose of the predictor step is to reduce $||F_2||$ while staying feasible with respect to ϕ, and the goal to head towards optimality is "suspended". (In this case one might limit the reduction of μ to at most, say 50%; if μ is reduced substantially while $c^T x$ is not being "updated", it is likely that many corrector steps will be needed for the next centering.)

19.4.5 The overall algorithm

We now summarize the derivations of the previous chapters and state the framework of a prototype algorithm.

Let us recall the problem and the notation.

$$\begin{array}{ll} \text{minimize} & c^T x \\ \text{subject to} & \mathcal{A}(x) \succeq 0, \\ & f_i(x) \leq 0 \ \text{ for } \ 1 \leq i \leq p, \\ & f_j(x) = 0 \ \text{ for } \ p+1 \leq j \leq m. \end{array}$$

We assume that $\bar{x}^{(0)}$ is given such that $\mathcal{A}(\bar{x}^{(0)}) \succ 0$, and $f_i(\bar{x}^{(0)}) < 0$. As before we treat the inequality $f_0(x) := -\det \mathcal{A}(x) < 0$ like the other f_i, $(i \in 1, \ldots, p)$. As in (19.4.13) let F_1 be the vector formed by the f_i, and F_2 by the f_j. The logarithmic barrier function ϕ for the inequalities is defined in (19.4.15), and the barrier function ϕ^{lin} for the linearized problem is given in (19.4.26). For a given triple x, y, μ we denote as in (19.4.18)

$$H = H(x,y) = \nabla_x^2 \phi(x,\mu) + \sum_{j=p+1}^{m} y_j \nabla^2 f_j(x), \qquad g = g(x,\mu) = \nabla_x \phi(x,\mu),$$

$$A = A(x) = \nabla F_2(x)^T, \qquad b = b(x) = -F_2(x)$$

Note that H and the matrix H^+ given by $H^+ = \nabla^2 \phi^{lin}(x,\mu)$ do not depend on μ. Whenever we use the letters H, g, A or b without their arguments we imply that these quantities are evaluated at a "current" point which is given in the context. We also assume that $Q = Q(x)$ is given by $A^T = Q \begin{bmatrix} R \\ 0 \end{bmatrix}$ as in

(19.4.19). Here, $R \in \mathbb{R}^{q\times(m-p)}$. (Note that this notation hides the fact that Q is not unique.) We partition $Q = (Q_1, Q_2)$ into the first q columns Q_1 and the remaining $n-q$ columns Q_2.

Algorithm 19.4.1 (Conceptual Algorithm)
Initialization: Choose $\bar{\theta} \in (0,1)$ *and* $\epsilon \in (0,1)$ *(e.g.* $\bar{\theta} = 0.2$, $\epsilon = 0.1$*). Let*

$$x = \bar{x}^{(0)}, \quad \rho_0 = 1 + \left\|F_2(\bar{x}^{(0)})\right\|, \quad \mu = \mu_0 = 1, \quad \Delta\mu_0 = 1, \quad \theta = \bar{\theta}.$$

For $k = 0, 1, 2, \ldots$ *do (outer loop)*

1. *Do until convergence (Corrector steps, inner loop)*

   (a) *Given* x*, compute a least squares solution* y *to* $g + A^T y = 0$*,*
   i.e. $y = R^{-1}Q_1^T g$ *if* A *has full row rank.*
   Let $\tilde{g}_2 = Q_2^T g$.
   Update $\theta = \bar{\theta}$ *if* $\|F_2(x)\| \geq \mu\rho_0$ *and* $\theta = 0$ *if* $\|F_2(x)\| \leq 0.5\mu\rho_0$.

   (b) *Let* $\tilde{s} = \tilde{s}(\lambda)$ *be given by*

   $$\begin{aligned} \tilde{s}_1 &= \theta(RR^T + \lambda^{3/2}I_1)^{-1}Rb \\ \tilde{s}_2 &= (\tilde{H}_{22} + \lambda I_2)^{-1}(-\tilde{g}_2 - \tilde{H}_{21}\tilde{s}_1), \end{aligned}$$

   and $s(\lambda) = Q\tilde{s}$.

   (c) *Find the step length* λ^* *that minimizes* $\phi(x + s(\lambda), \mu)$ *subject to the constraint*

   $$\|F_2(x + s(\lambda))\| \leq 2\mu\rho_0.$$

   (d) *Set* $x = x + s(\lambda^*)$.

2. *(Predictor step)*
   Denote the result of the corrector steps by $x = x^{(k^+)}$*, and set*

   $$\begin{aligned} \rho_k &= \left\|F_2(x^{(k^+)})\right\| \quad (\text{only for } k \geq 1) \\ \sigma &= \max\{1, \rho_k/(\mu\rho_0)\} \\ \tilde{c}_2 &= Q_2^T c \end{aligned}$$

3. *Compute a least squares solution* y *to* $g + A^T y = 0$.

4. *Let* $\tilde{\hat{s}} = \tilde{\hat{s}}(\lambda)$ *be given by*

   $$\begin{aligned} \tilde{\hat{s}}_1 &= \sigma(RR^T + \lambda\tilde{H}_{11}^+)^{-1}Rb \\ \tilde{\hat{s}}_2 &= \begin{cases} (\tilde{H}_{22}^+)^{-1}(-\tilde{H}_{21}^+\tilde{\hat{s}}_1) & \text{if } \Delta\mu_k \leq \epsilon\mu_k \text{ and } \sigma > 1 \\ (\tilde{H}_{22}^+)^{-1}(\frac{-1}{\mu(1+\lambda)}\tilde{c}_2 - \tilde{H}_{21}^+\tilde{\hat{s}}_1) & \text{if } \Delta\mu_k > \epsilon\mu_k \text{ or } \sigma = 1 \end{cases} \end{aligned}$$

   and $\hat{s}(\lambda) = Q\tilde{\hat{s}}$.

5. When $\sigma > 1$ define

$$\Delta\mu = \Delta\mu(\lambda) = \mu_k(1 - \frac{\left\| F_2(x^{(k^+)} + \hat{s}(\lambda)) \right\|}{\max\{\frac{3}{2}\rho_0\mu_k,\ \left\| F_2(x^{(k^+)}) \right\|\}}) \qquad (19.4.33)$$

and find $\lambda > 0$ that maximizes $\Delta\mu$ subject to the condition

$$\phi(x^{(k^+)} + t\hat{s}(\lambda)) \quad \text{is defined for} \quad t \in [0,\ 2 - \frac{\Delta\mu}{\mu}]. \qquad (19.4.34)$$

When $\sigma = 1$ define $\Delta\mu = \frac{\mu}{1+\lambda}$ and minimize λ subject to (19.4.34) *and*

$$\frac{\left\| F_2(x^{(k^+)} + \hat{s}(\lambda)) \right\|}{\mu_k - \Delta\mu} \leq \max\{\frac{3}{2}\rho_0, \frac{\left\| F_2(x^{(k^+)}) \right\|}{\mu_k}\}. \qquad (19.4.35)$$

Update $x = x + \hat{s}(\lambda)$ and $\mu = \mu - \Delta\mu$.

Suppose that at each outer iteration k the barrier subproblem converges to some point $x^{(k^+)}$.

If $\mu \to 0$ then any accumulation point x^* of the $x^{(k^+)}$ is either a feasible stationary point of (19.2.7) in the sense that there exists a multiplyer $z \geq 0$ and y such that

$$c + \nabla F_1(x^*)z + \nabla F_2(x^*)y = 0 \qquad \text{and} \qquad z^T F_1(x^*) = 0,$$

or x^* is a feasible singular point.

Suppose that $\mu^k \to \bar{\mu} > 0$. Then for any accumulation point x^* of the iterates $x^{(k^+)}$ either the implication

If $z \in \mathbb{R}^n$ satisfies $\nabla F_2(x^*)^T z = 0$
then $\exists i \in 0, \ldots, p$ with $f_i(x^*) = 0$ and $\nabla f_i(x^*)^T z \geq 0$

holds or x^* is a singular point such that $\nabla F_2(x^*)$ is rank deficient.

The above statements are shown in [365]. Also some reasons are given why one may conjecture for the case $\bar{\mu} > 0$ that the condition $\nabla F_2(x^*)^T z = 0$ can be replaced by the condition $F_2(x^*)^T \nabla F_2(x^*)^T z < 0$. The latter condition excludes the existence of a direction along which $||F_2||$ can locally be reduced while simultaneously reducing F_1.

The above convergence result assumes that the QQP method converges for each barrier subproblem. If some of the barrier subproblems are unbounded it may be necessary to include additional bounds like $||x|| \leq M$ to the problem description (19.2.7).

In the absence of nonlinear equality constraints in (19.2.7), the convergence results become much stronger and simplify considerably. A detailed analysis along with numerical results is currently in preparation.

19.5 CONCLUSION

We propose an algorithm for nonconvex programming that combines features of trust region methods, SQP methods, and interior-point methods. The method does not use a merit function, i.e. for the barrier subproblems the "merit function" is the same (barrier) function whose linearization was used to compute the search step. Therefore, "incompatibility effects" like the Maratos effect are not expected. Likewise the predictor step along the tangent of the central path is only restricted by feasibility conditions, and when the optimal solution is strictly complementary, the path behaves (under mild regularity assumptions) locally like a linear function. In this case we may expect the predictor step to be effective as well.

Acknowledgement

The author is very grateful for many corrections and comments from the editor, Prof. H. Wolkowicz. He is also thankful to his colleague J. Launer for converting the bibliography to bibtex.

20 SEQUENTIAL, QUADRATIC CONSTRAINED, QUADRATIC PROGRAMMING FOR GENERAL NONLINEAR PROGRAMMING

Serge Kruk and Henry Wolkowicz

20.1 INTRODUCTION

A proven approach for unconstrained minimization of a function, $f(x), x \in \Re^n$, is to build and solve a quadratic model at a local estimate $x^{(k)}$, i.e. apply the *trust region method.* In this paper we propose a *direct* extension of this modeling approach to constrained minimization. A local quadratic model of *both* the objective function and the constraints is built. This model is too hard to solve, so it is relaxed using the Lagrangian dual, which is then solved by semidefinite programming techniques. The key ingredient in this approach is the equivalence between the Lagrangian and semidefinite relaxations.

As illustration, recall the well-known Rayleigh-Ritz program to obtain the smallest eigenvalue of a symmetric matrix A,

$$\lambda_1(A) = \min\left\{x^t Ax \mid x^t x = 1\right\}.$$

One approach to prove this result involves Lagrange multipliers, i.e. the optimal x must be a stationary point of the Lagrangian $L(x, \lambda) = x^t Ax - \lambda(x^t x - 1)$. This shows that the optimal x is an eigenvector; and substitution into the objective function shows that the corresponding eigenvalue is the smallest. But now, consider instead, $x^t Ax = \text{Trace}(x^t Ax) = \text{Trace}(Axx^t)$ and let $X := xx^t$. We can write the above program as

$$\min\left\{\langle A, X\rangle \mid \langle I, X\rangle = 1, X \succeq 0, X = xx^t\right\},$$

where $\langle A, B\rangle = \mathrm{Trace}(AB)$, the trace inner product; $A \succeq 0$ (resp. $A \succ 0$) denotes positive semidefiniteness (resp. positive definiteness); and $A \succeq B$ denotes $A - B \succeq 0$, i.e. the symmetric matrix space $\mathcal{S}^n$ is equipped with the Löwner partial order.

Note that the rank one constraint ($X = xx^t$) is redundant (see Chapter 3). Therefore drop it and construct the dual to obtain

$$\max\left\{\lambda \mid \lambda I \preceq A, \lambda \in \Re\right\},$$

which obviously has $\lambda_1(A)$ as optimal value. Since the dual has a strictly interior point, the primal attains the same value and we get the Rayleigh-Ritz result. Thus we can use SDP to solve Lagrangian relaxations. In fact, throughout this handbook we have seen that we can solve these very efficiently.

In this chapter, we wish to illustrate some of the strengths, both theoretical and practical, of considering semidefinite relaxations of quadratic programs as the tool of choice for solving Lagrangian relaxations that arise from quadratic models of general nonlinear programs. Some of this material comes from Kruk and Wolkowicz [15, 449], where missing details and references can be found.

20.2 THE SIMPLEST CASE

Consider the unconstrained problem

$$UNC \qquad \min\left\{f(x) \mid x \in \Re^n\right\}.$$

When possible, the method of choice for this problem is Newton's method, which solves a quadratic model of the objective function. To ensure a solution (or convexity) of the model, Newton's method is often implemented within a Trust-Region, or Restricted-Step approach. This very efficient variation proceeds from an initial estimate of the solution; develops a second-order model of the objective function deemed valid in a region around the estimate; and finally solves the model (the trust-region subproblem)

$$TRS \qquad \min\left\{q_0(d) = d^tQd + 2b^td \mid q_1(d) = d^td \le \delta^2, d \in \Re^n\right\}.$$

The model is constructed from $Q = \nabla^2 f(x^{(k)})$ (or an approximation of the Hessian), $b = \nabla f(x^{(k)})$ and the parameter δ represents the radius where the model is trusted. The trust-region can be scaled or even arise from a nonconvex quadratic. A solution d is then used as the step to the next estimate $x^{(k+1)} = x^{(k)} + d$.

One of the interesting properties of *TRS*, first shown in Stern and Wolkowicz [747] using semidefinite programming, is that even though generally nonconvex, the problem exhibits no duality gaps. The Lagrangian dual of *TRS* can be written as

$$\text{NonLinDualSDP-TRS} \qquad \max\left\{-b^t(Q+\lambda I)^\dagger b - \lambda\delta^2 \mid Q+\lambda I \succ 0, \lambda \geq 0\right\},$$

a nonlinear (concave) semidefinite program, where $(\cdot)^\dagger$ is the Moore-Penrose generalized inverse. In addition, the Lagrangian dual has been shown to be equivalent to the following linear semidefinite program by Rendl and Wolkowicz [661],

$$\text{DSDP-TRS} \qquad \max\left\{(\delta^2+1)\lambda - t \;\middle|\; \begin{bmatrix} t & b^t \\ b & Q \end{bmatrix} \succeq \lambda I, t \in \Re, \lambda \geq 0\right\}.$$

We can take the dual of the above linear semidefinite program (DSDP-TRS) and get a semidefinite program equivalent to *TRS*.

$$\text{PSDP-TRS} \qquad \min\left\{\langle P_0, Y\rangle \mid \langle E_0, Y\rangle = 1, \langle P_I, Y\rangle \leq \delta^2, Y \succeq 0\right\}.$$

The variable in this program, Y, belongs to the cone of symmetric positive semidefinite matrices of dimension $(n+1)\times(n+1)$. Also,

$$P_0 = \begin{bmatrix} 0 & b^t \\ b & Q \end{bmatrix}, P_I = \begin{bmatrix} 0 & 0 \\ 0 & I \end{bmatrix}, E_0 = \begin{bmatrix} 1 & 0 \\ 0 & \mathbf{0} \end{bmatrix}.$$

The reader will note that PSDP-TRS can be obtained by homogenization of *TRS* followed by the same trick we used for the Rayleigh-Ritz program. (By homogenization we mean that we get a quadratic function without a linear term.)

This pair of linear primal-dual semidefinite programs (PSDP-TRS, DSDP-TRS) have strict interior points. Therefore the optimal values are equal; moreover, they are both attained. Finally, part of the first column of the primal semidefinite solution, the matrix Y, is feasible for *TRS*. And, possibly with an additional displacement, chosen in the nullspace of the Lagrangian, this first column yields the same objective value for *TRS* as its dual optimal. By this procedure, usually known as *lifting*, of *TRS* to the cone of semidefinite matrices, and projecting back (by the first column), we see that there are no duality gaps for *TRS*. This was first shown in [747].

Theorem 20.2.1 *The optimal solution to TRS and to its Lagrangian dual problem (NonLinDualSDP-TRS) are attained and the corresponding objective values are equal.*

An interesting consequence is that polynomial-time interior-point algorithms can be used to solve *TRS* , *even if the objective function and the feasible set are non-convex*. The result has been extended to upper and lower bounded trust-region subproblems but, interestingly, not to a finite number of constraints. With as few as two constraints, a duality gap can appear, see e.g. Yuan [864, 863].

20.3 MULTIPLE TRUST-REGIONS

Consider now a quadratic objective function constrained by multiple quadratics,

$$\text{Q}^2\text{P} \qquad \min\Big\{x^t Q_0 x + 2b_0^t x - a_0 \mid x^t Q_k x + 2b_k^t x - a_k \leq 0, 1 \leq k \leq m, x \in \Re^n\Big\}.$$

As soon as two or more trust-regions are considered, the necessary and sufficient conditions that hold for one trust region may no longer be necessary for Q^2P. This is reflected in the duality gap exhibited by some instances of multiple trust-region programs.

To directly derive the relaxations, we introduce the vector $y = (x_0 \; x)^t$. We then require $x_0^2 = 1$ or, in terms of the new variable, $y^t E_0 y = 1$, to get an homogeneous program equivalent to Q^2P,

$$\text{H}Q^2P \qquad \min\Big\{y^t P_0 y \mid y^t E_0 y = 1, y^t P_k y \leq 0, 1 \leq k \leq m, y \in \Re^{n+1}\Big\},$$

where

$$E_0 = \begin{bmatrix} 1 & 0 \\ 0 & \mathbf{0} \end{bmatrix} \text{ and } P_k = \begin{bmatrix} -a_k & b_k^t \\ b_k & Q_k \end{bmatrix}, 0 \leq k \leq m.$$

The homogenization simplifies the notation and opens the way to the semidefinite relaxation since we can rewrite $\text{H}Q^2P$ using matrix variables.

$$\text{HM}Q^2P \qquad \min\Big\{\langle Y, P_0\rangle \mid \langle Y, E_0\rangle = 1, \langle Y, P_k\rangle \leq 0, 1 \leq k \leq m, Y = yy^t\Big\}.$$

The rank-one constraint can be relaxed to a semidefinite constraint; a procedure we justify by showing its equivalence with the Lagrangian relaxation. After some rearrangement of terms, the Lagrangian dual of $\text{H}Q^2P$ reads

$$\max\left\{\min\{y^t(P_0 + \sum_{k=1}^m \lambda_k P_k + \lambda_0 E_0)y - \lambda_0 \mid y \in \Re^{n+1}\} \mid \lambda \in \Re \times \Re_+^m\right\}.$$

For the inner minimization to be bounded we must now have

$$P_0 + \sum_{k=1}^m \lambda_k P_k + \lambda_0 E_0 \succeq 0, \quad \text{which implies} \quad Q + \sum_{k=1}^m \lambda_k Q_k \succeq 0.$$

Remark 20.3.1 *This, by the way, is where the duality gap arises. The standard necessary optimality conditions for Q^2P do not require the Hessian of the Lagrangian to be semidefinite. But the Lagrangian dual program we are deriving here requires the same Hessian to be semidefinite. We therefore cannot expect the primal variables corresponding to an optimal dual solution to be, in general, optimal for Q^2P.*

To complete the derivation, we note that the minimum over y will be attained at $y = 0$ from which we get the dual program

$$\textit{Dual-}Q^2P \qquad \max\Big\{-\lambda_0 \mid P_0 + \lambda_0 E_0 + \sum_{k=1}^m \lambda_k P_k \succeq 0, \lambda \in \Re \times \Re_+^m\Big\}.$$

We have now justified the claim of equivalence of the Lagrangian and semidefinite relaxations since dropping the rank-one condition on the homogenized primal (HMQ^2P) or taking the semidefinite dual of Dual-Q^2P will result in the following, which we will therefore simply refer to as the relaxation of Q^2P,

$$\text{SDP-Q}^2\text{P} \qquad \min\Big\{\langle P_0, Y\rangle \mid \langle E_0, Y\rangle = 1, \langle P_k, Y\rangle \leq 0, 1 \leq k \leq m, Y \succeq 0\Big\}.$$

This resulting SDP relaxation of HQ^2P is equivalent to the one considered in the literature, e.g. in [733, 137, 635].

The optimal value of the relaxation provides a lower bound for the Q^2P. We now need an approximation for the optimum x. Feasibility properties of the first column of the semidefinite relaxation were first shown by Fujie and Kojima [260] for an equivalent problem with linear objective function. For an alternate view of this result, see [15]. Consider the feasible set of Q^2P,

$$\hat{F} := \{x \in \Re^n \mid x^t Q_k x + 2b_k^t x - a_k \leq 0, 1 \leq k \leq m\};$$

the feasible set of *SDP*-Q^2P,

$$\widetilde{F} := \{Y \succeq 0 \mid \langle E_0, Y\rangle = 1, \langle P_k, Y\rangle \leq 0, 1 \leq k \leq m\};$$

and the projector map,

$$P_R : S^n \mapsto \Re^n, \quad P_R(Y) = P_R\left(\begin{bmatrix} a & x^t \\ x & X \end{bmatrix}\right) = x.$$

Theorem 20.3.1 *Suppose that Y is a feasible solution of SDP-Q^2P. The projected vector, $x = P_R(Y)$, is then feasible for all convex constraints of Q^2P.*

This can be seen as follows. Since we are only concerned with convex constraints, we may consider only those where $Q_k \succeq 0$ and compute

$$\begin{aligned} x^t Q_k x + 2b_k^t x - a_k - \langle P_k, Y\rangle &= x^t Q_k x - \langle Q_k, X\rangle \\ &= -\langle Q, X - xx^t\rangle. \end{aligned}$$

Since $Y \succeq 0$ implies $X - xx^t \succeq 0$, we obtain

$$\begin{aligned} x^t Q_k x + 2b_k^t x - a_k &= \langle P_k, Y\rangle - \langle Q_k, X - xx^t\rangle \\ &\leq \langle P_k, Y\rangle \\ &\leq 0. \end{aligned}$$

And therefore x is feasible for all convex constraints of Q^2P.

This feasibility of the first column is interesting to consider in more detail. First, in the case of a problem where the quadratic constraints are convex (but maybe the objective is not) there is an obvious way to improve this first column solution when it is not optimal.

A optimal pair Y, λ to the semidefinite relaxation, if Y is not rank one, will in general map to a vector x for which complementarity fails but improving the objective value while remaining feasible is then easy.

Lemma 20.3.1 *Consider a Q^2P with convex constraints. If the semidefinite primal optimal solution Y is not rank one, let $\tilde{x} = P_R(Y)$, (part of the first column of Y). Then there is a $\bar{x}$ chosen in $\mathcal{N}(Q_0 + \sum \lambda_k Q_k)$, the nullspace of the Lagrangian, such that $x = \tilde{x} + \bar{x}$, is feasible and will improve the primal objective value of Q^2P.*

The idea is to choose a displacement along the nullspace of the Lagrangian until one or more slack constraints is satisfied with equality. The value of the objective function is lowered since

$$0 = \bar{x}^t(Q_0 + \sum \lambda_k Q_k)\bar{x} \geq \bar{x}^t Q_0 \bar{x}$$

and therefore $(\tilde{x} + \bar{x})^t Q_0 (\tilde{x} + \bar{x}) \leq \tilde{x}^t Q_0 \tilde{x}$.

Consider now a more general case where the constraints may not be convex. Note that Theorem 20.3.1 implies that the projected first column x is feasible for any nonnegative combination of constraints,

$$\sum_{k=1}^{m} \lambda_k (x^t Q_k x + 2b_k^t x - a_k) \leq 0, \lambda \geq 0, \tag{20.3.1}$$

which results in a convex function. Thus we can obtain feasible points for convex combinations of constraints of Q^2P as in (20.3.1) from feasible points Y of the relaxation (*SDP*-Q^2P), even when these are not rank one. Therefore the relaxation provides a convex approximation to the feasible set $\hat{F}$. However, it actually provides a better approximation than this would initially lead us to believe.

Let us define a *valid inequality* for Q^2P as

$$\sum_{k=1}^{m} \lambda_k (x^t Q_k x + 2b_k^t x - a_k) \leq 0, \quad \text{where} \quad Q_0 + \sum_{k=1}^{m} \lambda_k Q_k \succeq 0, \lambda \geq 0. \tag{20.3.2}$$

These inequalities, an infinite number of them, are not, in general, convex. (Simply consider a *TRS* where the objective is strictly convex while the constraint is not.) However, they provide geometric insight into the *SDP* relaxation. The set of vectors satisfying all valid inequalities,

$$\left\{x \mid \sum_{k=1}^{m} \lambda_k (x^t Q_k x + 2b_k^t x - a_k) \leq 0, \quad Q_0 + \sum_{k=1}^{m} \lambda_k Q_k \succeq 0, \lambda \geq 0\right\}$$

establishes a relation between the set of projected columns of *SDP* solutions and some intersection of the original constraints.

We now use the above geometric descriptions to provide an approximate solution to Q^2P from the optimum of *SDP*. We use the first column of the optimum Y but then we use the properties of the valid inequalities (20.3.2) to improve this column by moving onto a boundary of a valid inequality.

In the general case of a nonconvex feasible region, we can obtain a step, which, unlike Lemma 20.3.1, *attains* complementary slackness, though not necessarily feasibility. Again, the value of the objective function is improved. This additional step is a generalization of an idea introduced by Moré and Sorensen [551] to solve *TRS* and there is an explicit expression for the step as there is for *TRS*, given here in Lemma (20.3.2). We give the technical construction of the step in the following lemma and its value in Corollary 20.3.1.

Lemma 20.3.2 *Suppose that* (λ), *and*

$$Y = \begin{bmatrix} 1 & x^t \\ x & X \end{bmatrix}$$

are feasible for the primal-dual pair DSDP and PSDP, respectively. Let

$$y := \begin{bmatrix} 1 \\ x \end{bmatrix},$$

$$Z := P_0 + \lambda_0 E_0 + \sum_{k \in \mathcal{I}} \lambda_k P_k, \quad \mathcal{I} = \{1, \ldots m\}$$

and suppose that they satisfy

$$ZY = 0.$$

Let the matrix Y *be factored as*

$$Y = TT^t,$$

where T *is* $(n+1) \times r$ *and full column rank* $r \geq 2$. *Let the matrix* S *be* $r \times (r-1)$ *and full column rank with* $\mathcal{R}(S) = \mathcal{N}(T_{1,:})$, *i.e. with range space given by the orthogonal complement to the first row of* T. *Define*

$$R := TS, \; \bar{P} := \sum_{k \in \mathcal{I}} \lambda_k P_k, \; c := y^t \bar{P} y,$$

$$K := R^t \left(\bar{P} y y^t \bar{P} - c \bar{P} \right) R.$$

Choose v *such that*

$$Rv \neq 0 \text{ and } v^t K v \geq 0, \tag{20.3.3}$$

and define

$$a := v^t R^t \bar{P} R v, \; b := 2 v^t R^t \bar{P} y.$$

Then, for z *defined as follows, we have*

$$TSv = \begin{bmatrix} 0 \\ z \end{bmatrix} \neq 0, \tag{20.3.4}$$

and

$$b^2 - 4ac \geq 0. \tag{20.3.5}$$

Moreover, if we define

$$\alpha := \begin{cases} \left(-b \pm \sqrt{b^2 - 4ac}\right) / (2a) & \text{if } a \neq 0 \\ \frac{-c}{b} & \text{if } a = 0, \end{cases}$$

and

$$w := y + \alpha \begin{bmatrix} 0 \\ z \end{bmatrix},$$

then

$$w^t \bar{P} w = 0, \text{ and } Zw = 0. \tag{20.3.6}$$

The proof is as follows. That (20.3.4) holds and $Zw = 0$ follows directly from construction of R and the assumption of complementary slackness, $ZY = 0$. Note that $ZY = ZTT^t = 0$ implies $ZT = 0$. We still need to show the equality of the quadratic form in (20.3.6). Now

$$w^t \bar{P} w = y^t \bar{P} y + \alpha 2 v^t R^t \bar{P} y + \alpha^2 v^t R^t \bar{P} R v.$$

(We assume that a w exists to make this quadratic 0. This can be seen from using the ordinary TRS with the given λ defining the single constraint.) The discriminant for this quadratic in α is defined in (20.3.5), where

$$b^2 - 4ac = 4 v^t K v.$$

Therefore, the discriminant is nonnegative, and the quadratic has a real solution α as given by the standard formula.

We can now make explicit the value of the above lemma in finding an approximate solution to Q^2P.

Corollary 20.3.1 *Suppose that Y, Z, λ, w are defined as in Lemma 20.3.2 above. Then the Lagrangian dual bound is attained by w as well as complementary slackness.*

That complementarity is attained is seen directly from the second equation of (20.3.6). And from both equations we obtain

$$0 = w^t Z w - w^t \bar{P} w = w^t (P_0 - \lambda_0 E_0) w.$$

Therefore, $w^t P_0 w = q_0(x + \alpha z) = -\lambda_0$, the dual Lagrangian bound.

20.4 APPROXIMATIONS OF NONLINEAR PROGRAMS

We assume the reader is familiar with *Sequential Quadratic Programming*, denoted *SQP*. We only recall the main features and refer the reader to [753], [115] for details. The usual justifications for the application of *SQP* to the nonlinear program

$$NEP \qquad \min\left\{ f_0(x) \mid f_i(x) = 0, 1 \le i \le m, x \in \Re^n \right\},$$

stem from applying Newton's method to obtain stationarity of its Lagrangian $\mathcal{L}(x, \lambda) := f_0(x) + \sum \lambda_i f_i(x)$,

$$\begin{aligned} \nabla f_0(x^*) + \sum \nabla f_i(x^*) \lambda_i^* &= 0, \\ f(x^*) &= 0. \end{aligned}$$

We will sometimes use the notation $f(x) := [f_1(x) \dots f_m(x)]^t$ and $f'(x)$ for the first derivative. An iterative attempt at the non-linear system above by Newton's method with some simplification involving $d = x^{(k+1)} - x^{(k)}$ and $\delta_\lambda = \lambda^{(k+1)} - \lambda^{(k)}$, will produce the First-Order Newton Step,

$$\textit{FONS} \quad \begin{bmatrix} \nabla^2 \mathcal{L}(x^{(k)}, \lambda^{(k)}) & f'(x^{(k)}) \\ f'(x^{(k)})^t & 0 \end{bmatrix} \begin{bmatrix} d \\ \lambda^{(k+1)} \end{bmatrix} = \begin{bmatrix} -\nabla f_0(x^{(k)}) \\ -f(x^{(k)}) \end{bmatrix}.$$

This system produces a direction d and a new vector of Lagrangian multiplier estimates $\lambda^{(k+1)}$. The key justification for *SQP* is that the system of equations *FONS* can also be derived as the first-order necessary conditions of the quadratic program

$$\begin{array}{lll} QP & \min & f_0(x^{(k)}) + \nabla f_0(x^{(k)})^t d + \frac{1}{2} d^t \nabla^2 \mathcal{L}(x^{(k)}, \lambda^{(k)}) d \\ & \text{s.t.} & f_i(x^{(k)}) + \nabla f_i(x^{(k)})^t d = 0, \qquad 1 \le i \le m. \end{array}$$

Stationarity of the Lagrangian of *QP* yields the first line of *FONS*, and feasibility yields the second line. This is why *SQP* is viewed as an extension of Newton's method to constrained optimization.

It is now standard procedure to extend the above derivation to the inequality constrained program.

$$NLP \qquad \min\left\{ f_0(x) \mid f_i(x) \le 0, 1 \le i \le m, x \in \Re^n \right\},$$

and obtain the subproblem,

$$\begin{array}{lll} NLP\text{-}QP & \min & f_0(x^{(k)}) + \nabla f_0(x^{(k)})^t d + \frac{1}{2} d^t \nabla^2 \mathcal{L}(x^{(k)}, \lambda^{(k)}) d \\ & \text{s.t.} & f_i(x^{(k)}) + \nabla f_i(x^{(k)})^t d \le 0, \qquad 1 \le i \le m, \end{array}$$

In summary, the above derivation, from the Taylor first-order expansion of $\mathcal{L}(x, \lambda)$, obtained the standard *SQP* subproblem, which approximates the objective function to second order yet approximates the constraints only to first order. Consider now a second-order Taylor expansion of $\mathcal{L}(x, \lambda)$,

$$\begin{bmatrix} \sum \lambda_i \nabla f_i(x^{(k)}) + \nabla^2 \mathcal{L}(x^{(k)}, \lambda^{(k)}) d + H_3(\delta_x, \delta_\lambda) \\ f'(x^{(k)}) d + \frac{1}{2} d^t f''(x^{(k)}) d \end{bmatrix} = \begin{bmatrix} -\nabla f_0(x^{(k)}) \\ -f(x^{(k)}) \end{bmatrix},$$

where we have grouped the third-order derivatives under the name H_3 because we intend to neglect them. Consider also replacing $\nabla^2 \mathcal{L}(x^{(k)}, \lambda^{(k)})$ by $\nabla^2 \mathcal{L}(x^{(k)}, \lambda^{(k+1)})$. We then obtain an approximation of the necessary optimality conditions which sits between a first and a second-order expansion and can be obtained by solving

$$\begin{array}{lll} NLP\text{-}Q^2P & \min & f_0(x^{(k)}) + \nabla f_0(x^{(k)})^t d + \frac{1}{2} d^t \nabla^2 f_0(x^{(k)}) d \\ & \text{s.t.} & f_i(x^{(k)}) + \nabla f_i(x^{(k)})^t d + \frac{1}{2} d^t \nabla^2 f_i(x^{(k)}) d \le 0, 1 \le i \le m \\ & & d^t d \le \delta^2, \end{array}$$

without the additional trust-region, which is added to ensure a bounded solution.

Such a straightforward subproblem has often been considered, but has, just as often, been discarded as unsolvable. One notable exception is an algorithm by Maany [514] developed, interestingly enough, because the standard *SQP* approach failed on the highly nonlinear orbital trajectory problems they were studying. (See [194].) Because *NLP-Q^2P* is a closer approximation to the original problem (*NLP*) than the quadratic program, we expect it to be a better subproblem to solve in a sequential programming approach and, in fact we have the following,

Lemma 20.4.1 *Assume that $x^{(k)}$ is feasible for NLP. If the NLP-Q^2P subproblem is solved by $d = 0$ with multipliers λ, then the pair of vectors $x^{(k)}$ and λ satisfies the first-order conditions and second-order conditions of NLP. Conversely, if $x^{(k)}$ and λ satisfy the first and second-order necessary conditions of NLP, then the pair of vectors $d = 0$, λ satisfy the first and second-order conditions of NLP-Q^2P.*

This implies that the Q^2P subproblem does better than the *QP* subproblem since they both solve the first-order conditions but only the former guarantees second-order optimality conditions. This is expected of a trust-region approach.

Is also does better by providing second-order multiplier estimates in the sense that the multipliers $\lambda^{(k+1)}$ obtained from *NLP-Q^2P* satisfy

$$\min\left\{ ||\nabla f_0(x^{(k)}) + \nabla^2 \mathcal{L}(x^{(k)}, \eta)d + \sum \eta_i \nabla f_i(x^{(k)})||_2^2 \mid \eta \in \Re^m \right\}.$$

If we are close to the solution we therefore obtain, directly from the solution of the subproblem, not only a good search direction in primal space, but better multiplier estimates than provided by the standard *QP* subproblem. (For more details on second-order multiplier estimates, see [274].)

20.5 QUADRATICALLY CONSTRAINED QUADRATIC PROGRAMMING

Note that, for simplicity, we assume that our constraints are nonlinear. Linear constraints have to be treated differently, essentially squared, see [635]. Equivalently, linear constraints can be eliminated or mapped to a linear constraint in matrix space.

Homogenization of *NLP-Q^2P*, obtained by adding a component d_0 to the vector d, together with the constraint $d_0^2 = 1$, yield the semidefinite relaxation,

$$\textit{PSDP} \qquad \min\left\{ \langle P_0, Y\rangle \mid \langle E_0, Y\rangle = 1, \langle P_i, Y\rangle \le 0, 1 \le i \le m, \langle P_I, Y\rangle \le \delta^2 \right\},$$

where

$$P_i = \begin{bmatrix} -a_i & \nabla f_i(x^{(k)})^t & 0 \\ \nabla f_i(x^{(k)}) & \nabla^2 f_i(x^{(k)}) & 0 \\ 0 & 0 & 0 \end{bmatrix}, \quad a_i = -2 f_i(x^{(k)}), \quad 0 \le i \le m,$$

and where E_0 and P_I have their usual definitions,

$$E_0 = \begin{bmatrix} 1 & 0 \\ 0 & 0 \end{bmatrix}, P_I = \begin{bmatrix} 0 & 0 \\ 0 & I \end{bmatrix},$$

and $Y \succeq 0$.

Remark 20.5.1 *But this relaxation, can possibly be infeasible if the current estimate is too far from the feasible region. To overcome this difficulty in SQP, Vardi [817] suggested a heuristic shift of the linear constraints. We can do a related shift of our second-order constraints by allowing the additional component d_0 to take values between zero and one. That is, we change $d_0^2 = 1$ to $d_0^2 \leq 1$. This additional relaxation allows for a feasible subproblem. Of course we would want d_0 to be as close to 1 as possible and examination of the subproblem shows that it automatically tries to make d_0 'large'. We need no heuristic to choose a Vardi-type parameter.*

The dual program to *PSDP* is then

$$\textit{DSDP} \qquad \max\Big\{-\lambda_0 \mid P_0 + \lambda_0 E_0 + \sum_{i=1}^{m} \lambda_i P_i + \lambda_I P_I \succeq 0, \lambda \in \Re \times \Re_+^m\Big\}.$$

Solving the above primal-dual pair *PSDP*,*DSDP*, in the case of gap-free *NLP*, is enough since, as we have seen, the first column is optimal for the quadratic approximation. But, in general, we need an appropriate merit function to ensure sufficient decrease at each step and guarantee global convergence of the algorithm, whether we use a line search or a trust-region strategy.

After solving the Q^{2}P subproblem for a direction $d \neq 0$, the next iterate is obtained by $x^{(k+1)} = x^{(k)} + d$. This new point serves for the expansion of a new problem by second-order polynomials and we iterate until the subproblem yields $d = 0$. As with any trust-region based algorithm, we adjust the trust-region radius according to the ratio of predicted improvement to actual improvement. At the end, we have a solution satisfying both first and second-order conditions of *NLP*. Somewhat more formally, here is the *SQ^{2}P* algorithm.

SEQUENTIAL QUADRATICALLY CONSTRAINED PROGRAMMING

$SQ^2P(f_i, \nabla f_i, \nabla^2 f_i, x^{(0)})$
  $k = 0$
  do
    $Y \in \text{argmin}\{\langle P_0, Y\rangle : \langle P_i, Y\rangle \leq 0, \langle E_0, Y\rangle = 1, Y \succeq 0\}$
    $\lambda^{(k+1)} \in \text{argmax}\{-\lambda_0 : P_0 + \sum \lambda P_i + \lambda_0 E_0 \succeq 0, \lambda \in \Re \times \Re^m_+\}$
    $d = P_R(Y)$
    $x^{(k+1)} = x^{(k)} + d$
    $r^k = \frac{\varphi(x^{(k)}) - \varphi(x^{(k+1)})}{q_0(x^{(k)}) - q_0(x^{(k+1)})}$
    if $(r^k < \frac{1}{4})$
      $\delta = \delta/4$
    elseif $(r^k > \frac{3}{4})$ and $\|x^{(k+1)} - x^{(k)}\| = \delta$
      $\delta = 2\delta$
    fi
    $k = k + 1$
  while $(\|d\| > \varepsilon)$
  Find maximal $d \in \mathcal{N}(\nabla^2 \mathcal{L})$ such that $f(x^{(k)} + d) \leq 0$
  $x^{(k)} = x^{(k)} + d$
return$(x^{(k)}, \lambda^{(k)})$

If the *NLP-Q²P* subproblem is convex, or more generally, if it is an instance without duality gaps, then solving the semidefinite relaxation, which can be done in polynomial time, will be enough since the primal semidefinite solution will be rank one. We will have a pair of primal-dual vectors satisfying the sufficient conditions for optimality of Q^2P.

This takes care of the convex case and of many non-convex cases. In other cases, we can move along the nullspace of the Lagrangian until we hit one of the constraints. This nullspace-restricted step improves the objective value even if it does not lead to an optimal solution.

20.6 CONCLUSION

Efficient approaches to unconstrained optimization based on Newton's method all involve local quadratic models of the objective function. Yet for constrained optimization, the so-called extension of Newton's method, *SQP*, uses mostly linear approximations. Some second-order information is included in the model, but in an aggregate form.

In this chapter we have outlined an approach that deals more closely with the true quadratic model of the problem at hand. One of the key features is the relationship between the Lagrangian and Semidefinite relaxations which leads to what we have called the *SQ²P* algorithm for general nonlinear programs.

This algorithm builds second-order approximations of both the objective function and the constraints and then solves the Lagrangian relaxation of this quadratic model via semidefinite programming. The approach provides a stronger relaxation than the standard quadratic program used in *SQP* methods; at every step it provides better multiplier estimates; it handles potential infeasibility of the subproblem in a straight-forward manner; finally, it aims at solutions satisfying both first and second-order optimality conditions. Many implementation issues still need to be resolved but the recent advances in numerical solutions of large semidefinite programs encourage further study.

References

[1] L. AARTS. Primal-dual search directions in semidefinite optimization. Master's thesis, Delft University of Technology, 1999.

[2] R. ABRAHAM, J.E. MARSDEN, and T. RATIU. *Manifolds, Tensor Analysis, and Applications*, volume 75 of *Applied Mathematical Sciences*. Springer-Verlag, 1988. Second edition.

[3] A. ACHTZIGER, A. BEN-TAL, M. BENDSŒ, and J. ZOWE. Equivalent displacement-based formulations for maximum strength truss topology design. *IMPACT of Computing in Science and Engineering*, 4:315–345, 1992.

[4] W.P. ADAMS and P.M. DEARING. On the equivalence between roof duality and Lagrangian duality for unconstrained $0-1$ quadratic programming problems. *Discrete Appl. Math.*, 48:1–20, 1994.

[5] W.P. ADAMS and H.D. SHERALI. A tight linearization and an algorithm for zero-one quadratic programming problems. *Management Sci.*, 32(10):1274–1290, 1986.

[6] W.P. ADAMS and H.D. SHERALI. Mixed-integer bilinear programming problems. *Math. Programming*, 59(3, Ser. A):279–305, 1993.

[7] I. ADLER and F. ALIZADEH. Primal-dual interior point algorithms for convex quadratically constrained and semidefinite optimization problems. Technical report, Rutcor, Rutgers University, New Brunswick, NJ, 1995.

[8] I. ADLER and R.D.C. MONTEIRO. A geometric view of parametric linear programming. *Algorithmica*, 8:161–176, 1992.

[9] J. AGLER, J.W. HELTON, and S. McCULLOUGH. Positive semidefinite matrices with a given sparsity pattern. *Linear Algebra Appl.*, 107(3):101–149, 1988.

[10] N.I. AKHIEZER. *The Classical Moment Problem.* Hafner, New York, NY, 1965.

[11] S. AL-HOMIDAN. *Hybrid methods for optimization problems with positive semidefinite matrix constraints.* PhD thesis, University of Dundee, 1993.

[12] S. AL-HOMIDAN and R. FLETCHER. Hybrid methods for finding the nearest Euclidean distance matrix. In *Recent advances in nonsmooth optimization*, pages 1–17. World Sci. Publishing, River Edge, NJ, 1995.

[13] F.A. AL-KHAYYAL, R. HORST, and P.M. PARDALOS. Global optimization of concave functions subject to quadratic constraints: an application in nonlinear bilevel programming. *Ann. Oper. Res.*, 34(1-4):125–147, 1992.

[14] A. ALFAKIH, A. KHANDANI, and H. WOLKOWICZ. Solving Euclidean distance matrix completion problems via semidefinite programming. *Comput. Optim. Appl.*, 12(1-3):13–30, 1999. A tribute to Olvi Mangasarian, Part I.

[15] A. ALFAKIH, S. KRUK, and H. WOLKOWICZ. A note on geometry of semidefinite relaxations. Technical Report in progress, University of Waterloo, Waterloo, Canada, 1999.

[16] A. ALFAKIH and H. WOLKOWICZ. On the embeddability of weighted graphs in Euclidean spaces. Technical Report CORR Report 98-12, University of Waterloo, 1998.

[17] F. ALIZADEH. *Combinatorial optimization with interior point methods and semidefinite matrices.* PhD thesis, University of Minnesota, 1991.

[18] F. ALIZADEH. Combinatorial optimization with semidefinite matrices. In *Proceedings of the Second Annual Integer Programming and Combinatorial Optimization Conference*, Carnegie-Mellon University, 1992.

[19] F. ALIZADEH. Optimization over positive semi-definite cone; interior-point methods and combinatorial applications. In P.M. Pardalos, editor, *Advances in Optimization and Parallel Computing*, pages 1–25. North–Holland, 1992.

[20] F. ALIZADEH. Semidefinite programming: duality theory, eigenvalue optimization and combinatorial applications. Technical report, Stanford University, 1992. Presented at the Fourth SIAM Conference on Optimization, 1992.

[21] F. ALIZADEH. Interior point methods in semidefinite programming with applications to combinatorial optimization. *SIAM J. Optim.*, 5:13–51, 1995.

[22] F. ALIZADEH, J.-P. HAEBERLY, M.V. NAYAKKANKUPPAM, M.L. OVERTON, and S. SCHMIETA. SDPpack user's guide – version 0.9 Beta. Technical Report TR1997–737, Courant Institute of Mathematical Sciences, NYU, New York, NY, June 1997.

[23] F. ALIZADEH, J-P.A. HAEBERLY, and M.L. OVERTON. A new primal-dual interior-point method for semidefinite programming. In J.G. Lewis, editor, *Proceedings of the Fifth SIAM Conference on Applied Linear Algebra*, pages 113–117. SIAM, 1994.

[24] F. ALIZADEH, J-P.A. HAEBERLY, and M.L. OVERTON. Complementarity and nondegeneracy in semidefinite programming. *Math. Programming*, 77:111–128, 1997.

[25] F. ALIZADEH, J-P.A. HAEBERLY, and M.L. OVERTON. Primal-dual interior-point methods for semidefinite programming: Convergence rates, stability and numerical results. *SIAM J. Optim.*, 8:746–768, 1998.

[26] F. ALIZADEH and S. SCHMIETA. Optimization with semidefinite, quadratic and linear constraints. Technical Report RRR23–97, RUTCOR, Rutgers University, New Brunswick, NJ, November 1997.

[27] J.C. ALLWRIGHT. On maximizing the minimum eigenvalue of a linear combination of symmetric matrices. *SIAM J. Matrix Anal. Appl.*, 10(3):347–382, 1989.

[28] N. ALON and B. SUDAKOV. Bipartite subgraphs and the smallest eigenvalue. Technical report, 1998.

[29] E.D. ANDERSEN, J. GONDZIO, C. MÉSZÁROS, and X. XU. Implementation of interior-point methods for large scale linear programs. In T. Terlaky, editor, *Interior point methods of mathematical programming*, pages 189–252. Kluwer, Dordrecht, The Netherlands, 1996.

[30] B. ANDERSON and S. VONGPANITLERD. *Network analysis and synthesis: a modern systems theory approach.* Prentice-Hall, 1973.

[31] T.W. ANDERSON and I. OLKIN. An extremal problem for positive definite matrices. *Linear and Multilinear Algebra*, 6:257–262, 1978.

[32] T. ANDO. Concavity of certain maps and positive definite matrices and applications to Hadamard products. *Linear Algebra Appl.*, 26:203–241, 1979.

[33] T. ANDO and F. HIAI. Inequality between powers of positive semidefinite matrices. *Linear Algebra Appl.*, 208/209:65–71, 1994.

[34] M. ANJOS and H. WOLKOWICZ. A strengthened SDP relaxation via a second lifting for the max-cut problem. Research report, corr 99-55, University of Waterloo, Waterloo, Ontario, 1999.

[35] K.M. ANSTREICHER. On the equivalence of convex programming bounds for Boolean quadratic programming. Technical report, University of Iowa, Iowa City, IA, 1998.

[36] K.M. ANSTREICHER. Eigenvalue bounds versus semidefinite relaxations for the quadratic assignment problem. Technical report, University of Iowa, Iowa City, IA, 1999.

[37] K.M. ANSTREICHER, X. CHEN, H. WOLKOWICZ, and Y. YUAN. Strong duality for a trust-region type relaxation of QAP. *Linear Algebra Appl.*, To appear, 1999.

[38] K.M. ANSTREICHER, M. FAMPA, J. LEE, and J. WILLIAMS. Continuous relaxations for constrained maximum-entropy sampling. In W.H. Cunningham, T. McCormick, and M. Queyranne, editors, *Integer Programming and Combinatorial Optimization (Vancouver, BC, 1996)*, volume 1084 of *Lecture Notes in Comput. Sci.*, pages 234–248. Springer, Berlin, 1996.

[39] K.M. ANSTREICHER, M. FAMPA, J. LEE, and J. WILLIAMS. Using continuous nonlinear relaxations to solve constrained maximum-entropy sampling problems. *Math. Programming*, Ser. A, 85(2):221–240, 1999.

[40] K.M. ANSTREICHER and J.-P. VIAL. On the convergence of an infeasible primal-dual interior-point method for convex programming. *Optim. Methods Softw.*, 3:273–283, 1994.

[41] K.M. ANSTREICHER and H. WOLKOWICZ. On Lagrangian relaxation of quadratic matrix constraints. *SIAM J. Matrix Anal. Appl.*, To appear, 1999.

[42] P. APKARIAN and R.J. ADAMS. Advanced gain-scheduled techniques for uncertain systems. *IEEE Trans. Control Sys. Tech.*, 6(1):21–32, January 1998.

[43] P. APKARIAN and P. GAHINET. A convex characterization of gain-scheduled $\mathcal{H}_\infty$ controllers. *IEEE Trans. Aut. Control*, AC-40(5):853–864, 1995.

[44] P. APKARIAN, P. GAHINET, and G. BECKER. Self-scheduled $\mathcal{H}_\infty$ control of linear parameter-varying systems: A design example. *Automatica*, 31(9):1251–1261, 1995.

[45] M. ARGAEZ and R.A. TAPIA. On the global convergence of a modified augmented Lagrangian linesearch interior-point Newton method for nonlinear programming. Technical Report 95-38, Dept. of Comp. and Appl. Math., Rice University, 1997.

[46] T. ASANO and D.P. WILLIAMSON. Improved approximation algorithms for MAX SAT. In *Proc.11th ACM-SIAM Symp. on Disc. Algs.*, 2000.

[47] F. AVRAM, D. BERTSIMAS, and M. RICARD. Optimization of multiclass queueing networks: a linear control approach. In *Stochastic Networks, Proceedings of the IMA*, pages 199–234. F. Kelly and R. Williams eds., 1995.

[48] F. BACCELLI and P. BRÉMAUD. *Elements of Queueing Theory: Palm-Martingale Calculus and Stochastic Recurrences.* Springer-Verlag, Berlin, 1994.

[49] R. BAČ´IK and S. MAHAJAN. Semidefinite programming and its applications to NP problems. In *Computing and combinatorics (Xi'an, 1995)*, volume 959 of *Lecture Notes in Comput. Sci.*, pages 566–575. Springer, Berlin, 1995.

[50] M. BAKONYI and C. R. JOHNSON. The Euclidean distance matrix completion problem. *SIAM J. Matrix Anal. Appl.*, 16(2):646–654, 1995.

[51] V. BALAKRISHNAN. Linear matrix inequalities in robustness analysis with multipliers. *Syst. Control Letters*, 25(4):265–272, 1995.

[52] V. BALAKRISHNAN and E. FERON, editors. *Linear matrix inequalities in control theory and applications.* John Wiley & Sons Ltd., Chichester, 1996. Internat. J. Robust Nonlinear Control **6** (1996), no. 9-10.

[53] E. BALAS. Disjunctive programming. *Ann. Discrete Math.*, 5:3–51, 1979. Discrete optimization (Proc. Adv. Res. Inst. Discrete Optimization and Systems Appl., Banff, Alta., 1977), II.

[54] E. BALAS. A modified lift-and-project procedure. *Math. Programming*, 79(1-3, Ser. B):19–31, 1997. Lectures on Mathematical Programming (ismp97) (Lausanne, 1997).

[55] E. BALAS. Disjunctive programming: properties of the convex hull of feasible points. *Discrete Appl. Math.*, 89(1-3):3–44, 1998.

[56] E. BALAS, S. CERIA, and G. CORNUEJOLS. A lift-and-project cutting plane algorithm for mixed 0-1 programs. *Math. Programming*, 58:295–324, 1993.

[57] E. BALAS, S. CERIA, and G. CORNUÉJOLS. Solving mixed 0-1 programs by a lift-and-project method. In *Proceedings of the Fourth Annual ACM-SIAM Symposium on Discrete Algorithms (Austin, TX, 1993)*, pages 232–242, New York, 1993. ACM.

[58] E. BALAS and M. FISCHETTI. A lifting procedure for the asymmetric traveling salesman polytope and a large new class of facets. *Math. Programming*, 58(3, Ser. A):325–352, 1993.

[59] E. BALAS and S. M. NG. On the set covering polytope. II. Lifting the facets with coefficients in $\{0, 1, 2\}$. *Math. Programming*, 45(1 (Ser. B)):1–20, 1989.

[60] E. BALAS and E. ZEMEL. Lifting and complementing yields all the facets of positive zero-one programming polytopes. In *Mathematical programming (Rio de Janeiro, 1981)*, pages 13–24. North-Holland, Amsterdam, 1984.

[61] G. BALAS, J. C. DOYLE, K. GLOVER, A. PACKARD, and R. SMITH. *μ-analysis and synthesis.* MUSYN, inc., and The Mathworks, Inc., 1991.

[62] R. BALDICK. A unified approach to polynomially solvable cases of integer "non-separable" quadratic optimization. *Discrete Appl. Math.*, 61(3):195–212, 1995.

[63] J.R. BAR-ON and K.A. GRASSE. Global optimization of a quadratic functional with quadratic equality constaints. *J. Optim. Theory Appl.*, 82(2):379–386, 1994.

[64] J.R. BAR-ON and K.A. GRASSE. Global optimization of a quadratic functional with quadratic equality constraints. II. *J. Optim. Theory Appl.*, 93(3):547–556, 1997.

[65] F. BARAHONA. The max-cut problem on graphs not contractible to K_s. *Oper. Res. Lett.*, 2(3):107–111, 1983.

[66] F. BARAHONA, M. GRÖTSCHEL, M. JÜNGER, and G. REINELT. An application of combinatorial optimization to statistical physics and circuit layout design. *Oper. Res.*, 36:493–513, 1988.

[67] G. P. BARKER, M. LAIDACKER, and G. POOLE. Projectionally exposed cones. *SIAM J. Algebraic Discrete Methods*, 8(1):100–105, 1987.

[68] G.P. BARKER. The lattice of faces of a finite dimensional cone. *Linear Algebra Appl.*, 7:71–82, 1973.

[69] G.P. BARKER. Faces and duality in convex cones. *Linear and Multilinear Algebra*, 6:161–169, 1977.

[70] G.P. BARKER. Theory of cones. *Linear Algebra Appl.*, 39:263–291, 1981.

[71] G.P. BARKER. Automorphism groups of algebras of triangular matrices. *Linear Algebra Appl.*, 121:207–215, 1989.

[72] G.P. BARKER. Automorphism of triangular matrices over graphs. *Linear Algebra Appl.*, 160:63–74, 1992.

[73] G.P. BARKER and D. CARLSON. Cones of diagonally dominant matrices. *Pacific J. of Math.*, 57:15–32, 1975.

[74] G.P. BARKER and B.-S. TAM. Baer semirings and Baer *-semirings of cone-preserving maps. *Linear Algebra Appl.*, 256:165–183, 1997.

[75] E.R. BARNES. Bounds for the largest clique in a graph. Technical report, Georgia Tech., 1997.

[76] A. BARVINOK. Problems of distance geometry and convex properties of quadratic maps. *Discrete Comput. Geom.*, 13(2):189–202, 1995.

[77] A. BARVINOK. On convex properties of the quadratic image of the sphere. Technical report, University of Michigan, 1999.

[78] G. BECKER. *Quadratic Stability and Performance of Linear Parameter Dependent Systems.* PhD thesis, University of California, Berkeley, 1993.

[79] M. BELLARE and P. ROGAWAY. The complexity of approximating a nonlinear program. *Math. Programming*, 69:429–442, 1995.

[80] R. BELLMAN and K. FAN. On systems of linear inequalities in Hermitian matrix variables. In *Proceedings of Symposia in Pure Mathematics, Vol 7, AMS*, 1963.

[81] A. BEN-ISRAEL, A. CHARNES, and K. KORTANEK. Duality and asymptotic solvability over cones. *Bulletin of American Mathematical Society*, 75(2):318–324, 1969.

[82] A. BEN-TAL and M. P. BENDSŒ. A new method for optimal truss topology design. *SIAM Journal of Optimization*, 3:322–358, 1993.

[83] A. BEN-TAL, M. KOCVARA, A.S. NEMIROVSKI, and J. ZOWE. Free material design via semidefinite programming: The multiload case with contact conditions. *SIAM J. Optim.*, 9(4):813 – 832, 1999.

[84] A. BEN-TAL and A.S. NEMIROVSKI. Robust solutions to uncertain linear problems. *Math. Oper. Res.* To appear.

[85] A. BEN-TAL and A.S. NEMIROVSKI. Potential reduction polynomial time method for truss topology design. *SIAM J. Optim.*, 4:596–612, 1994.

[86] A. BEN-TAL and A.S. NEMIROVSKI. Robust truss topology design via semidefinite programming. *SIAM J. Optim.*, 7(4):991–1016, 1997.

[87] A. BEN-TAL and A.S. NEMIROVSKI. Robust convex optimization. *Math. Oper. Res.*, 23(4):769–805, 1998.

[88] A. BEN-TAL and J. ZOWE. A unified theory of first and second order conditions for extremum problems in topological vector spaces. *Math. Programming Stud.*, 19:39–76, 1982.

[89] M. P. BENDSŒ. *Optimization of Structural Topology, Shape and Material.* Springer, Heidelberg, 1995.

[90] M. P. BENDSŒ, A. BEN-TAL, and J. ZOWE. Optimization methods for truss geometry and topology design. *Structural Optimization*. To appear.

[91] M.P. BENDSŒ, J.M. GUEDES, R.B. HABER, P. REDERSEN, and J.E. TAYLOR. An analytical model to predict optimal material properties in the context of optimal structural design. Technical Report # 453, Technical University of Denmark, Lyngby, 1992.

[92] M. BENGTSSON and B. OTTERSTEN. Optimal transmit beam forming using convex optimization. Technical report, Royal Institute of Technology, Stockholm, Sweden, 1999.

[93] S. J. BENSON, Y. YE, and X. ZHANG. Solving large-scale sparse semidefinite programs for combinatorial optimization. Technical report, The University of Iowa, 1997.

[94] Y. BERGMAN, B. GRUNDY, and Z. WIENER. General properties of option prices. *J. Finance*, 51(5):1573–1610, December 1996.

[95] A.B. BERKELAAR, B. JANSEN, K. ROOS, and T. TERLAKY. Basis- and tripartition identification for quadratic programming and linear complementarity problems. from an interior point solution to an optimal basis and vice versa. Technical report, Delft University of Technology, Delft University, Delft, The Netherlands, 1996.

[96] A.B. BERKELAAR, B. JANSEN, K. ROOS, and T. TERLAKY. Sensitivity analysis in (degenerate) quadratic programming. Technical report, Delft University of Technology, Delft, The Netherlands, 1996.

[97] A. BERMAN. *Cones, Matrices and Mathematical Programming*. Springer-Verlag, Berlin, New York, 1973.

[98] D. BERTSIMAS. The achievable region method in the optimal control of queueing systems; formulations, bounds and policies. *Queueing Systems*, 21:337–389, 1995.

[99] D. BERTSIMAS and D. GAMARNIK. Asymptotically optimal algorithms for job shop scheduling and packet routing. Submitted for publication, 1998.

[100] D. BERTSIMAS and J. NINO-MORA. Conservation laws, extended polymatroids and multi-armed bandit problems; a polyhedral approach to indexable systems. *Math. Oper. Res.*, 21:257–306, 1996.

[101] D. BERTSIMAS and J. NINO-MORA. Optimization of multiclass queueing networks with changeover times via the achievable region approach: Part i, the single-station case. *Math. Oper. Res.*, To appear.

[102] D. BERTSIMAS and J. NINO-MORA. Optimization of multiclass queueing networks with changeover times via the achievable region approach: Part ii, the multi-station case. *Math. Oper. Res.*, To appear.

[103] D. BERTSIMAS and J. NINO-MORA. Restless bandits, linear programming relaxations and a primal-dual heuristic. *Oper. Res.*, To appear.

[104] D. BERTSIMAS, I. PASCHALIDIS, and J. TSITSIKLIS. Branching bandits and Klimov's problem: achievable region and side constraints. *IEEE Trans. Automat. Control*, 40:2063–2075, 1995.

[105] D. BERTSIMAS, I. Ch. PASCHALIDIS, and J. N. TSITSIKLIS. Optimization of multiclass queueing networks: polyhedral and nonlinear characterizations of achievable performance. *Ann. Appl. Probab.*, 4(1):43–75, 1994.

[106] D. BERTSIMAS and I. POPESCU. On the relation between option and stock prices: an optimization approach. Submitted for publication, 1998.

[107] D. BERTSIMAS and I. POPESCU. Optimal inequalities in probability: a convex programming approach. Submitted for publication, 1998.

[108] D. BERTSIMAS and J. TSITSIKLIS. *Introduction to Linear Optimization.* Athena Scientific, Belmont, MA, 1997.

[109] D. BERTSIMAS and R. VOHRA. Rounding algorithms for covering problems. *Math. Progr.*, 80:63–89, 1998.

[110] D. BERTSIMAS and H. XU. Optimization of polling systems and dynamic vehicle routing problems on networks. Working paper, Operations Research Center, MIT, 1993.

[111] D. BERTSIMAS and Y. YE. Semidefinite relaxations, multivariate normal distributions, and order statistics. In D.-Z. Du and P.M. Pardalos, editors, *Handbook of Combinatorial Optimization*, volume 3, pages 1–19. Kluwer Academic Publishers, 1998.

[112] R. BHATIA. *Perturbation Bounds for Matrix Eigenvalues : Pitman Research Notes in Mathematics Series 162.* Longman, New York, 1987.

[113] R. BHATIA. *Matrix analysis.* Springer-Verlag, New York, 1997.

[114] J.R. BIRGE and R.J.-B. WETS. Approximations and error bounds in stochastic programming. In *Inequalities in statistics and probability (Lincoln, Neb., 1982)*, pages 178–186. Inst. Math. Statist., Hayward, Calif., 1984.

[115] P.T. BOGGS and J.W. TOLLE. Sequential quadratic programming. *Acta Numerica*, 4:1–50, 1995.

[116] F. BOHNENBLUST. Joint positiveness of matrices. Manuscript, 1948.

[117] I. M. BOMZE. On standard quadratic optimization problems. *J. Global Optim.*, 13(4):369–387, 1998. Workshop on Global Optimization (Trier, 1997).

[118] I. M. BOMZE and G. DANNINGER. A global optimization algorithm for concave quadratic programming problems. *SIAM J. Optim.*, 3(4):826–842, 1993.

[119] J.F. BONNANS, R. COMINETTI, and A. SHAPIRO. Sensitivity analysis of optimization problems under second order regular constraints. *Math. Oper. Res.*, 23:806–831, 1998.

[120] J.F. BONNANS, R. COMINETTI, and A. SHAPIRO. Second order optimality conditions based on parabolic second order tangent sets. *SIAM J. Optim.*, 9:466–492, 1999.

[121] J.F. BONNANS and A. SHAPIRO. Nondegeneracy and quantitative stability of parameterized optimization problems with multiple solutions. *SIAM J. Optim.*, 8(4):940–946 (electronic), 1998.

[122] J.F. BONNANS and A. SHAPIRO. Optimization problems with perturbations, a guided tour. *SIAM Review*, 40(2):228–264 (electronic), 1998.

[123] R.B. BOPPANA. Eigenvalues and graph bisection: An average case analysis. In *Proceedings of the 28th Annual Symposium on Foundations of Computer Science*, pages 280–285, Los Angeles, California, 1987. IEEE.

[124] B. BORCHERS. Csdp, a C library for semidefinite programming. Technical report, New Mexico Tech, Soccorrow, NM, 1999. To appear in Optim. Methods Softw.

[125] B. BORCHERS. SDPLIB 1.2, a library of semidefinite programming test problems. Technical report, New Mexico Tech, Soccorrow, NM, 1999. To appear in Optim. Methods Softw.

[126] E. BOROS and P. L. HAMMER. The max-cut problem and quadratic 0-1 optimization; polyhedral aspects, relaxations and bounds. *Ann. Oper. Res.*, 33(1-4):151–180, 1991. Topological network design (Copenhagen, 1989).

[127] J.M. BORWEIN and H. WOLKOWICZ. Characterization of optimality for the abstract convex program with finite-dimensional range. *J. Austral. Math. Soc. Ser. A*, 30(4):390–411, 1980/81.

[128] J.M. BORWEIN and H. WOLKOWICZ. Facial reduction for a cone-convex programming problem. *J. Austral. Math. Soc. Ser. A*, 30(3):369–380, 1980/81.

[129] J.M. BORWEIN and H. WOLKOWICZ. Cone-convex programming stability and affine constraint functions. In *Generalized Concavity in Optimization and Economics*, pages 379–397. NATO conference, Academic Press, 1981. invited paper.

[130] J.M. BORWEIN and H. WOLKOWICZ. Regularizing the abstract convex program. *J. Math. Anal. Appl.*, 83(2):495–530, 1981.

[131] J.M. BORWEIN and H. WOLKOWICZ. Characterizations of optimality without constraint qualification for the abstract convex program. *Math. Programming Stud.*, 19:77–100, 1982. Optimality and stability in mathematical programming.

[132] J.M. BORWEIN and H. WOLKOWICZ. A simple constraint qualification in infinite-dimensional programming. *Math. Programming*, 35(1):83–96, 1986.

[133] O.J. BOXMA. Workloads and waiting times in single-server systems with multiple customer classes. *Queueing Systems*, 5:185–214, 1989.

[134] S. BOYD, V. BALAKRISHNAN, E. FERON, and L. El GHAOUI. Control system analysis and synthesis via linear matrix inequalities. *Proc. ACC*, pages 2147–2154, 1993.

[135] S. BOYD and C. H. BARRATT. *Linear Controller Design. Limits of Performance.* Prentice-Hall, 1991.

[136] S. BOYD and L. EL GHAOUI. Method of centers for minimizing generalized eigenvalues. *Linear Algebra Appl.*, 188:63–111, 1993.

[137] S. BOYD, L. El GHAOUI, E. FERON, and V. BALAKRISHNAN. *Linear Matrix Inequalities in System and Control Theory*, volume 15 of *Studies in Applied Mathematics.* SIAM, Philadelphia, PA, June 1994.

[138] S. BOYD and S. WU. *SDPSOL: A Parser/Solver for Semidefinite Programs with Matrix Structure. User's Guide, Version Alpha.* Stanford University, March 1995.

[139] S. BOYD and Q. YANG. Structured and simultaneous Lyapunov functions for system stability problems. *Int. J. Control*, 49(6):2215–2240, 1989.

[140] N. BRIXIUS, F.A. POTRA, and R. SHENG. SDPHA: a matlab implementation of homogeneous interior-point algorithms for semidefinite programming. Technical report, University of Iowa, Iowa City, IA, 1998.

[141] N. BRIXIUS, F.A. POTRA, and R. SHENG. Nonsymmetric search directions for semidefinite programming. *SIAM J. Optim.*, 9(4):863 – 876, 1999.

[142] A. BRONDSTED. *An Introduction to Convex Polytopes.* Springer Verlag, Berlin, 1983.

[143] R. A. BRUALDI and H. J. RYSER. *Combinatorial Matrix Theory.* Cambridge University Press, New York, 1991.

[144] L. BRUNETTA, M. CONFORTI, and G. RINALDI. A branch-and-cut algorithm for the equicut problem. *Math. Programming*, 78:243–263, 1997.

[145] S. BURER and R.D.C. MONTEIRO. An efficient algorithm for solving the MAXCUT SDP relaxation. Technical report, Georgia Tech, Atlanta, GA, 1999.

[146] S. BURER and R.D.C. MONTEIRO. A general framework for establishing polynomial convergence of long-step methods for semidefinite programming. Technical report, Georgia Tech, Atlanta, GA, 1999.

[147] S. BURER, R.D.C. MONTEIRO, and Y. ZHANG. Solving semidefinite programs via nonlinear programming part i: Transformations and derivatives. Technical report, Department of Computational and Applied Mathematics, Rice University, Houston, Texas, 1999.

[148] P.J. BURKE. The output of a queueing system. *Oper. Res.*, 4:699–704, 1956.

[149] J.A. BUZACOTT and J.G. SHANTHIKUMAR. *Stochastic Models of Manufacturing Systems.* Prentice Hall, Englewood Cliffs, NJ, 1993.

[150] R.H. BYRD, J.C. GILBERT, and R.H. NOCEDAL. A trust region method based on interior point techniques for nonlinear programming. Technical Report Report OTC 96/02, Optimization Technology Center, Northwestern University, Evanston Il 60208, 1996.

[151] R.H. BYRD, D.C. LIU, and R.H. NOCEDAL. On the behaviour of Broyden's class of quasi-Newton methods. *SIAM J. Optim.*, 2, 1992.

[152] J.A. CALVIN and DYKSTRA. A note on maximizing a special concave function subject to simultaneous Löwner order constraints. *Linear Algebra Appl.*, 176:37–44, 1992.

[153] F. CELA. *The Quadratic Assignment Problem: Theory and Algorithms.* Kluwer, Massachessets, USA, 1998.

[154] M.R. CELIS, J.E. DENNIS Jr., and R.A. TAPIA. A trust region strategy for nonlinear equality constrained optimization. In *Proceedings of the SIAM Conference on Numerical Optimization, Boulder, CO*, 1984. Also available as Technical Report TR84-1, Department of Mathematical Sciences, Rice University, Houston, TX.

[155] S. CHANDRASEKARAN, M. GU, and A. H. SAYED. The sparse basis problem and multilinear algebra. *SIAM J. Matrix Anal. Appl.*, 20(2):354–362, 1999.

[156] P. CHEBYCHEV. Sur les valeurs limites des intégrales. *Jour. Math. Pur. Appl.*, 19(2):157–160, 1874.

[157] R. CHIANG and M.G. SAFONOV. *Robust Control Toolbox.* The Mathworks, Inc., 1992.

[158] M. CHILALI and P. GAHINET. H_∞ design with pole placement constraints: An lmi approach. *IEEE Trans. Aut. Control*, 40(3):358–367, March 1995.

[159] C.C. CHOI and Y. YE. Application of semidefinite programming to circuit partitioning. Technical report, The University of Iowa, Iowa City, Iowa, 1999.

[160] S. CHOPRA and M.R. RAO. The partition problem. *Math. Programming*, 49:87–115, 1993.

[161] B. CHOR and M. SUDAN. A geometric approach to betweenness. In *Proc. of 3rd Europ. Symp. on Algs.*, volume LNCS 979 of *Lecture Notes in Computer Science*, pages 227–237, 1995.

[162] Y.S. CHOU, A.L. TITS, and V. BALAKRISHNAN. Absolute stability theory, theory, and state-space verification of frequency-domain conditions: Connections and implications for computation. Technical Report TR 97-23, Institute for Systems Research, University of Maryland, 1997. Accepted for publication in *IEEE Trans. Aut. Control*, 1998.

[163] J.P.R. CHRISTENSEN and F. VESTERSTROM. A note on extreme positive definite matrices. *Math. Ann.*, 244:65–68, 1979.

[164] E.G. COFFMAN Jr and I MITRANI. A characterization of waiting time performance realizable by single server queues. *Oper. Res.*, 28:810–821, 1980.

[165] T.F. COLEMAN and Y.Y. LI. A primal dual trust region algorithm for nonconvex programming using a l_1 penalty function. Report, Dept. of Computer Science, Cornell University, 1998.

[166] R. COMINETTI. Metric regularity, tangent sets and second order optimality conditions. *Appl. Math. Optim.*, 21:265–287, 1990.

[167] A.R. CONN, N. GOULD, and Ph.L. TOINT. A primal-dual algorithm for minimizing a nonconvex function subject to bounds and nonlinear constraints. Report RC 20639, IBM T.J. Watson Center, Yorktown Heights, NY, 1996.

[168] R. COURANT and D. HILBERT. *Methods of Mathematical Physics, I.* Interscience, New York, NY, 1953.

[169] J. COX and S. ROSS. The valuation of options for alternative stochastic processes. *Journal of Financial Economics*, 3:145–166, 1976.

[170] B. CRAVEN. Modified Kuhn-Tucker conditions when a minimum is not attained. *Operations Research Letters*, 3:47–52, 1984.

[171] B. CRAVEN and J.J. KOLIHA. Generalizations of Farkas' thoerem. *SIAM J. Matrix Anal. Appl.*, 8:983–997, 1977.

[172] B. CRAVEN and B. MOND. Real and complex Fritz John theorems. *J. Math. Anal. Appl.*, 44:773–779, 1973.

[173] B. CRAVEN and B. MOND. Complementarity for arbitrary cones. *Z. fur Oper. Res.*, 21:143–150, 1977.

[174] B. CRAVEN and B. MOND. Linear programming with matrix variables. *Linear Algebra Appl.*, 38:73–80, 1981.

[175] G.M. CRIPPEN and T.F. HAVEL. *Distance Geometry and Molecular Conformation.* Wiley, New York, 1988.

[176] F. CRITCHLEY. On certain linear mappings between inner-product and squared distance matrices. *Linear Algebra Appl.*, 105:91–107, 1988.

[177] J.-P. CROUZEIX, J.-E. MARTINEZ-LEGAZ, and A. SEEGER. An alternative theorem for quadratic forms and extensions. *Linear Algebra Appl.*, 215:121–134, 1995.

[178] J. CULLUM, W.E. DONATH, and P. WOLFE. The minimization of certain nondifferentiable sums of eigenvalues of symmetric matrices. *Mathematical Programming Study*, 3:35–55, 1975.

[179] D. CVETKOVIĆ, M. CANGALOVIĆ, and V. KOVAČEVIČ-VUJČIĆ. Semidefinite programming methods for the symmetric traveling salesman problem. In *Proceedings of the 7th International IPCO Conference, Graz, Austria*, pages 126–136, 1999.

[180] C. DAVIS. All convex invariant functions of Hermitian matrices. *Archiv der Mathematik*, 8:276–278, 1957.

[181] C. DAVIS. The Toeplitz-Hausdorff theorem explained. *Canad. Math. Bull.*, 14:245–246, 1971.

[182] J. de PILLIS. Linear transformations which preserve Hermitian and positive semidefinite operators. *Pacific J. of Math.*, 23:129–137, 1967.

[183] G. M. del CORSO. Estimating an eigenvector by the power method with a random star t. *SIAM J. Matrix Anal. Appl.*, 18(4):913–937, October 1997.

[184] C. DELORME and S. POLJAK. Laplacian eigenvalues and the maximum cut problem. *Math. Programming*, 62(3):557–574, 1993.

[185] R.S. DEMBO and T. STEIHAUG. Truncated Newton algorithms for large scale unconstrained optimization. *Math. Programming*, 26(72):190–212, 1983.

[186] D. den HERTOG. *Interior point approach to linear, quadratic and convex programming*, volume 277 of *Mathematics and its Applications*. Kluwer Academic Publishers Group, Dordrecht, 1994. Algorithms and complexity.

[187] S. DENG and H. HU. Computable error bounds for semidefinite programming. Technical report, Department of Mathematical Sciences, Northern Illinois University, DeKalb, Illinois, USA, 1996.

[188] J. DENNIS and H. WOLKOWICZ. Sizing and least-change secant methods. *SIAM J. Numer. Anal.*, 30(5):1291–1314, 1993.

[189] E. DERMAN and I. KANI. Riding on the smile. *RISK*, 7:32–39, 1994.

[190] C.A. DESOER and M. VIDYASAGAR. *Feedback Systems: Input-Output Properties*. Academic Press, New York, 1975.

[191] M.M. DEZA and M. LAURENT. *Geometry of cuts and metrics*. Springer-Verlag, Berlin, 1997.

[192] P. DIACONIS. Application of the method of moments in probability and statistics. In *Proc. Sympos. Appl. Math.*, volume 37, pages 125–139. Amer. Math. Soc., 1987.

[193] L.L. DINES. On the mapping of quadratic forms. *Bull. of AMS*, 47:494–498, 1941.

[194] L.C.W. DIXON, S.E. HERSOM, and Z.A. MAANY. Initial experience obtained solving the low thrust satellite trajectory optimisation problem. Technical Report T.R. 152, The Hatfield Polytechnic Numerical Optimization Centre, 1984.

[195] M. DOLJANSKY and M. TEBOULLE. An interior proximal algorithm and the exponential multiplier method for semidefinite programming. *SIAM J. Optim.*, 9(1):1–13, 1998.

[196] W.E. DONATH and A.J. HOFFMAN. Algorithms for partitioning graphs and computer logic based on eigenvectors of connection matrices. *IBM Technical Disclosure Bulletin*, 15(3):938–944, 1972.

[197] W.E. DONATH and A.J. HOFFMAN. Lower bounds for the partitioning of graphs. *IBM J. of Research and Developement*, 17:420–425, 1973.

[198] J. DOYLE. Analysis of feedback systems with structured uncertainties. *IEE Proc.*, 129-D(6):242–250, November 1982.

[199] D.-Z. DU and P.M. PARDALOS. Global minimax approaches for solving discrete problems. In P. Gritzmann, R. Horst, E. Sachs, and R. Tichatschke, editors, *Recent Advances in Optimization*, pages 34–48. Springer-Verlag, 1997.

[200] L.E. DUBINS. On extreme points of convex sets. *J. Math. Anal. Appl.*, 5:237–244, 1962.

[201] B. DUPIRE. Pricing with a smile. *RISK*, 7:18–20, 1994.

[202] H. DYM and I. GOHBERG. Extensions of band matrices with band inverses. *Linear Algebra Appl.*, 36:1–24, 1981.

[203] A. EDELMAN, T. ARIAS, and S.T. SMITH. The geometry of algorithms with orthogonality constraints. *SIAM J. Matrix Anal. Appl.*, 20(2):303–353 (electronic), 1999.

[204] A. EDELMAN and S.T. SMITH. On conjugate gradient-like methods for eigen-like problems. *BIT*, 36(3):494–508, 1996. International Linear Algebra Year (Toulouse, 1995).

[205] L. EL GHAOUI and V. BALAKRISHNAN. Synthesis of fixed-structure controllers via numerical optimization. In *Proc. IEEE Conf. on Decision and Control*, pages 2678–2683, December 1994.

[206] L. EL GHAOUI, V. BALAKRISHNAN, E. FERON, and S. BOYD. On maximizing a robustness measure for structured nonlinear perturbations. In *Proc. American Control Conf.*, volume 4, pages 2923–2924, Chicago, June 1992.

[207] L. EL GHAOUI and G. CALAFIORE. Confidence ellipsoids for uncertain linear equations with structure. In *Proc. IEEE Conf. on Decision and Control*, 1999.

[208] L. EL GHAOUI and G. CALAFIORE. Deterministic state prediction under structured uncertainty. In *Proc. American Control Conf.*, 1999.

[209] L. EL GHAOUI and G. CALAFIORE. Worst-case simulation of uncertain systems. In A. Tesi A. Garulli and A. Vicino, editors, *Robustness in Identification and Control*, Lecture Notes in Control and Information Sciences. Springer, 1999. To appear.

[210] L. EL GHAOUI and H. LEBRET. Robust solutions to least-squares problems with uncertain data. *SIAM J. Matrix Anal. Appl.*, 18(4):1035–1064, 1997.

[211] L. EL GHAOUI and H. LEBRET. Robust solutions to least squares problems with uncertain data. In *Recent advances in total least squares techniques and errors-in-variables modeling (Leuven, 1996)*, pages 161–170. SIAM, Philadelphia, PA, 1997.

[212] L. EL GHAOUI, F. OUSTRY, and H. LEBRET. Robust solutions to uncertain semidefinite programs. *SIAM J. of Optim.*, 9(1):33–52, 1999.

[213] J. FALKNER, F. RENDL, and H. WOLKOWICZ. A computational study of graph partitioning. *Math. Programming*, 66(2, Ser. A):211–239, 1994.

[214] M.K. H. FAN, A.L. TITS, and J.C. DOYLE. Robustness in the presence of mixed parametric uncertainty and unmodeled dynamics. *IEEE Trans. Aut. Control*, AC-36(1):25–38, January 1991.

[215] M.K.H. FAN and Y. GONG. Eigenvalue multiplicity estimate in semidefinite programming. *J. Optim. Theory Appl.*, 94(1):55–72, 1997.

[216] J. FARAUT and A. KORANYI. *Analysis on Symmetric Cones.* Oxford University Press, NY, USA, 1994.

[217] R.W. FAREBROTHER. Three theorems with applications to Euclidean distance matrices. *Linear Algebra Appl.*, 95:11–16, 1987.

[218] L. FAYBUSOVICH. Jordan algebras, symmetric cones and interior-point methods. Technical report, Dept. of Mathematics, University of Notre Dame, Notre Dame, IN 46556-5683, 1995.

[219] L. FAYBUSOVICH. On a matrix generalization of affine-scaling vector fields. *SIAM J. Matrix Anal. Appl.*, 16(3):886–897, 1995.

[220] L. FAYBUSOVICH. Infinite-dimensional semidefinite programming: regularized determinants and self-concordant barriers. Technical report, Dept. of Mathematics, University of Notre Dame, Notre Dame, IN 46556-5683, 1996.

[221] L. FAYBUSOVICH. Semi-definite programming: a path-following algorithm for a linear-quadratic functional. *SIAM J. Optim.*, 6(4):1007–1024, 1996.

[222] L. FAYBUSOVICH. Euclidean Jordan algebras and interior-point algorithms. *Positivity*, 1(4):331–357, 1997.

[223] L. FAYBUSOVICH. Euclidean Jordan algebras and interior-point algorithms. *Journal of Computational and Applied Mathematics*, 86:149–175, 1997.

[224] L. FAYBUSOVICH. A Jordan-algebraic approach to potential-reduction algorithms. Technical report, Dept. of Mathematics, University of Notre Dame, Notre Dame, IN 46556-5683, April 1998.

[225] L. FAYBUSOVICH and J.B. MOORE. Infinite-dimensional quadratic optimization: interior-point methods and control applications. *Appl. Math. Optim.*, 36(1):43–66, 1997.

[226] U. FEIGE. Randomized graph products, chromatic numbers, and the Lovász ϑ-function. *Combinatorica*, 17(1):79–90, 1997.

[227] U. FEIGE and M.X. GOEMANS. Approximating the value of two proper proof systems, with applications to MAX-2SAT and MAX-DICUT. In *Proceeding of the Third Israel Symposium on Theory of Computing and Systems*, pages 182–189, 1995.

[228] E. FERON. Nonconvex quadratic programming, semidefinite relaxations and randomization algorithms in information and decision systems. Technical report, Dept. of Aeronautics and Astronautics, MIT, Cambridge, MA 02139, 1999.

[229] A.V. FIACCO and G.P. McCORMICK. *Nonlinear Programming: Sequential Unconstrained Minimization Techniques.* John Wiley & Sons, New York, NY, 1968.

[230] A.V. FIACCO and G.P. McCORMICK. *Nonlinear programming sequential unconstrained minimization techniques.* Classics in Applied Mathematics. SIAM, Philadelphia, PA, USA, 1990.

[231] M. FIEDLER. Bounds for eigenvalues of doubly stochastic matrices. *Linear Algebra Appl.*, 5:299–310, 1972.

[232] M. FIEDLER. Algebraic connectivity of graphs. *Czechoslovak Math. J.*, 23:298–305, 1973.

[233] P.A. FILLMORE and J.P. WILLIAMS. Some convexity theorems for matrices. *Glasgow Mathematical Journal*, 10:110–117, 1971.

[234] P.D. FINCH. On the distribution of queue size in queueing problems. *Acta Math. Hungar.*, 10:327–336, 1959.

[235] G. FINKE, R.E. BURKARD, and F. RENDL. Quadratic assignment problems. *Ann. Discrete Math.*, 31:61–82, 1987.

[236] A. FISCHER. A Newton-type method for positive-semidefinite linear complementarity problems. *J. Optim. Theory Appl.*, 86(3):585–608, 1995.

[237] C.H. FITZGERALD, C.A. MICCHELLI, and A. PINKUS. Functions that preserve families of positive semidefinite matrices. *Linear Algebra Appl.*, 221:83–102, 1995.

[238] S.D. FLAM and A. SEEGER. Solving cone-constrained convex programs by differential inclusions. *Math. Programming*, 64(1):107–121, 1994.

[239] H. FLANDERS. An extremal problem in the space of positive definite matrices. *Linear and Multilinear Algebra*, 3:33–39, 1975.

[240] R. FLETCHER. A nonlinear programming problem in statistics (educational testing). *SIAM J. Sci. Stat. Comput.*, 2:257–267, 1981.

[241] R. FLETCHER. Semi-definite matrix constraints in optimization. *SIAM J. Control Optim.*, 23:493–513, 1985.

[242] R. FLETCHER. A new variational result for quasi-Newton formulae. *SIAM J. Optim.*, 1:18–21, 1991.

[243] R. FLETCHER. An optimal positive definite update for sparse Hessian matrices. *SIAM J. Optim.*, 5:192–218, 1995.

[244] C.A. FLOUDAS and V. VISWESWARAN. Quadratic optimization. In *Handbook of global optimization*, pages 217–269. Kluwer Acad. Publ., Dordrecht, 1995.

[245] F. FORGÓ. *Nonconvex Programming.* Akadémiai Kiadó, Budapest, 1988.

[246] A. FORSGREN. On linear least-squares problems with diagonally dominant weight matrices. *SIAM J. Matrix Anal. Appl.*, 17:763–788, 1995.

[247] A. FORSGREN. Optimality conditions for nonconvex semidefinite programming. Technical report, Dept. of Mathematics, Royal Institute of Technology, Stockholm, Sweden, 1998.

[248] A. FORSGREN and P.E. GILL. Primal-dual interior methods for nonconvex nonlinear programming. Technical Report NA 96-3, Dept. of Mathematics, University of California, San Diego, 1996.

[249] R.M. FREUND. Dual gauge programs, with applications to quadratic programming and the minimum-norm problem. *Math. Programming*, 38:47–67, 1987.

[250] R.M. FREUND. Complexity of an algorithm for finding an approximate solution of a semi-definite program with no regularity assumption. Technical Report OR 302-94, MIT, Cambridge, MA, 1994.

[251] R.M. FREUND and S. MIZUNO. Interior point methods: current status and future directions. *OPTIMA*, 51, 1996.

[252] R.M. FREUND and J.R. VERA. Some characterizations and properties of the "distance to ill-posedness" and the condition measure of a conic linear system. Technical report, MIT, Cambridge, MA, 1997.

[253] S. FRIEDLAND. Convex spectral functions. *Linear and Multilinear Algebra*, 9:293–316, 1981.

[254] S. FRIEDLAND, J. NOCEDAL, and M.L. OVERTON. The formulation and analysis of numerical methods for inverse eigenvalue problems. *SIAM J. Numer. Anal.*, 24(3):634–667, 1987.

[255] A. FRIEZE and M. JERRUM. Improved approximation algorithms for MAX k-CUT and MAX BISECTION. In *Integer programming and combinatorial optimization (Copenhagen, 1995)*, volume 920 of *Lecture Notes in Comput. Sci.*, pages 1–13. Springer, Berlin, 1995.

[256] A. FRIEZE and M. JERRUM. Improved approximation algorithms for MAX k-CUT and MAX BISECTION. *Algorithmica*, 18:67–81, 1997.

[257] K.R. FRISCH. The logarithmic potential method of convex programming. Technical report, Institute of Economics, Oslo University, Oslo, Norway, 1955.

[258] M. FU, Z. LUO, and Y. YE. Approximation algorithms for quadratic programming. *J. Comb. Optim.*, 2(1):29–50, 1998.

[259] S.W. FUHRMANN and R.B. COOPER. Stochastic decompositions in the $m/g/1$ queue with generalized vacations. *Oper. Res.*, 33:1117–1129, 1985.

[260] T. FUJIE and M. KOJIMA. Semidefinite programming relaxation for nonconvex quadratic programs. *J. Global Optim.*, 10(4):367–380, 1997.

[261] K. FUJISAWA, M. KOJIMA, and K. NAKATA. Numerical evaluation of SDPA (semidefinite programming algorithm). In *Proceedings of the Second Workshop on High Performance Optimization Techniques, Rotterdam.* To appear.

[262] K. FUJISAWA, M. KOJIMA, and K. NAKATA. SDPA semidefinite programming algorithm. Technical report, Dept. of Information Sciences, Tokyo Institute of Technology, Tokyo, Japan, 1995.

[263] K. FUJISAWA, M. KOJIMA, and K. NAKATA. Exploiting sparsity in primal-dual interior-point methods for semidefinite programming. *Math. Programming*, 79, 1997.

[264] K. FUJISAWA, M. KOJIMA, and K. NAKATA. SDPA (semidefinite programming algorithm user's manual, version 4.10. Research Report on Mathematical and Computing Sciences, Tokyo Institute of Technology, Tokyo, Japan, 1998.

[265] P. GAHINET and P. APKARIAN. A linear matrix inequality approach to H^∞ control. *Int. J. Robust and Nonlinear Control*, 4:421–448, 1994.

[266] P. GAHINET, P. APKARIAN, and M. CHILALI. Affine parameter-dependent Lyapunov functions for real parameter uncertainty. *IEEE Trans. Aut. Control*, 41(3):436–442, March 1996.

[267] P. GAHINET and A.S. NEMIROVSKI. The projective method for solving linear matrix inequalities. *Math. Programming*, 77(2, Ser. B):163–190, 1997.

[268] G. GALLEGO. A minmax distribution free procedure for the (q,r) inventory model. *Oper. Res. Letters*, 11:55–60, 1992.

[269] D.M. GAY, M.L. OVERTON, and M.H. WRIGHT. A primal-dual interior method for nonconvex nonlinear programming. Technical Report 97-4-08, Bell Laboratories, Murray Hill, NJ, 1997.

[270] J. F. GEELEN. Maximum rank matrix completion. *Linear Algebra Appl.*, 288(1-3):211–217, 1999.

[271] E. GELENBE and I. MITRANI. *Analysis and Synthesis of Computer Systems.* Academic Press, London, 1980.

[272] L. EL GHAOUI and G. SCORLETTI. Control of rational systems using Linear-Fractional Representations and Linear Matrix Inequalities. *Automatica*, 32(9), September 1996.

[273] L.E. GIBBONS, D.W. HEARN, and P.M. PARDALOS. A continuous based heuristic for the maximum clique problem. Technical report, DIMACS Series in Discrete Mathematics and Theoretical Computer Science, 1996.

[274] P.E. GILL, W. MURRAY, and M.H. WRIGHT. *Practical Optimization.* Academic Press, New York, London, Toronto, Sydney and San Francisco, 1981.

[275] W. GLUNT, T.L. HAYDEN, S. HONG, and J. WELLS. An alternating projection algorithm for computing the nearest Euclidean distance matrix. *SIAM J. Matrix Anal. Appl.*, 11(4):589–600, 1990.

[276] W. GLUNT, T.L. HAYDEN, C.R. JOHNSON, and P. TARAZAGA. Maximum determinant completions. Technical report, Dept. of Mathematics, College of William and Mary, Williamsburg, Va, 1994.

[277] W. GLUNT, T.L. HAYDEN, C.R. JOHNSON, and P. TARAZAGA. Positive definite completions and determinant maximization. *Linear Algebra Appl.*, 288(1-3):1–10, 1999.

[278] C. GODSIL. Comment on D. E. Knuth: "The sandwich theorem" [Electron. J. Combin. **1** (1994), Article 1, approx. 48 pp. (electronic); MR 95g:05048a]. *Electron. J. Combin.*, 1:Article 1, Comment 1, approx. 1 p. (electronic), 1994.

[279] H.J. GODWIN. On generalizations of Tchebycheff's inequality. *Am. Stat. Assoc. J.*, 50:923–945, 1955.

[280] H.J. GODWIN. Inequalities on distribution functions. In ed. M.G. Kendall, editor, *Griffin's Statistical Monographs and Courses*, number 16. Charles Griffin and Co., London, 1964.

[281] M.X. GOEMANS. Semidefinite programming in combinatorial optimization. *Math. Programming*, 79:143–162, 1997.

[282] M.X. GOEMANS. Semidefinite programming and combinatorial optimization. *Documenta Mathematica*, Extra Volume ICM 1998:657–666, 1998. Invited talk at the International Congress of Mathematicians, Berlin, 1998.

[283] M.X. GOEMANS and F. RENDL. Semidefinite programs and association schemes. *Computing*, 1999. To appear.

[284] M.X. GOEMANS and D.P. WILLIAMSON. .878-approximation algorithms for MAX CUT and MAX 2SAT. In *ACM Symposium on Theory of Computing (STOC)*, 1994.

[285] M.X. GOEMANS and D.P. WILLIAMSON. Improved approximation algorithms for maximum cut and satisfiability problems using semidefinite programming. *J. Assoc. Comput. Mach.*, 42(6):1115–1145, 1995.

[286] D. GOLDFARB and S. LIU. An $O(n^3L)$ primal interior point algorithm for convex quadratic programming. *Math. Programming*, 53, 1991.

[287] D. GOLDFARB and K. SCHEINBERG. On parametric semidefinite programming. In *Proceedings of the Stieltjes Workshop on High Performance Optimization Techniques*, 1996.

[288] D. GOLDFARB and K. SCHEINBERG. Interior point trajectories in semidefinite programming. *SIAM J. Optim.*, 8(4):871–886, 1998.

[289] A. J. GOLDMAN and A. W. TUCKER. Polyhedral convex cones. In *Linear equalities and related systems*, pages 19–40. Princeton University Press, Princeton, N. J., 1956. Annals of Mathematics Studies, no. 38.

[290] A. J. GOLDMAN and A. W. TUCKER. Theory of linear programming. In *Linear inequalities and related systems*, pages 53–97. Princeton University Press, Princeton, N.J., 1956. Annals of Mathematics Studies, no. 38.

[291] E.G. GOL'SHTEIN. *Theory of Convex Programming*, volume 36 of *Translations of Mathematical Monographs*. American Mathematical Society, Providence, RI, 1972.

[292] G.H. GOLUB and C.F. VAN LOAN. *Matrix Computations*. Johns Hopkins University Press, Baltimore, Maryland, 3nd edition, 1996.

[293] C.C. GONZAGA and M.J. TODD. An $O(\sqrt{n}L)$-iteration large-step primal-dual affine algorithm for linear programming. *SIAM J. Optim.*, 2:349–359, 1992.

[294] N. GOULD, S. LUCIDI, M. ROMA, and Ph. L. TOINT. Solving the trust-region subproblem using the Lanczos method. *SIAM J. Optim.*, 9(2):504–525, 1999.

[295] J. C. GOWER. Properties of Euclidean and non-Euclidean distance matrices. *Linear Algebra Appl.*, 67:81–97, 1985.

[296] M. GREEN and D. J. N. LIMEBEER. *Linear Robust Control.* Information and System sciences. Prentice Hall, Englewood Cliffs, NJ, 1995.

[297] P. GRITZMANN, V. KLEE, and B.-S. TAM. Cross-positive matrices revisited. *Linear Algebra Appl.*, 223/224:285–305, 1995. Special issue honoring Miroslav Fiedler and Vlastimil Pták.

[298] B. GRONE, C.R. JOHNSON, E. MARQUES de SA, and H. WOLKOWICZ. Improving Hadamard's inequality. *Linear and Multilinear Algebra*, 16(1-4):305–322, 1984.

[299] B. GRONE, C.R. JOHNSON, E. MARQUES de SA, and H. WOLKOWICZ. Positive definite completions of partial Hermitian matrices. *Linear Algebra Appl.*, 58:109–124, 1984.

[300] B. GRONE, C.R. JOHNSON, E. MARQUES de SA, and H. WOLKOWICZ. A note on maximizing the permanent of a positive definite Hermitian matrix, given the eigenvalues. *Linear and Multilinear Algebra*, 19(4):389–393, 1986.

[301] R. GRONE, S. PIERCE, and W. WATKINS. Extremal correlation matrices. *Linear Algebra Appl.*, 134:63–70, 1990.

[302] M. GRÖTSCHEL, L. LOVÁSZ, and A. SCHRIJVER. The ellipsoid method and its consequences in combinatorial optimization. *Combinatorica*, 1:169–197, 1981.

[303] M. GRÖTSCHEL, L. LOVÁSZ, and A. SCHRIJVER. Relaxations of vertex packing. *J. Comb. Optim.*, pages 330–343, 1986.

[304] M. GRÖTSCHEL, L. LOVÁSZ, and A. SCHRIJVER. *Geometric Algorithms and Combinatorial Optimization.* Springer Verlag, Berlin-Heidelberg, 1988.

[305] M. GRÖTSCHEL and W. R. PULLEYBLANK. Weakly bipartite graphs and the max cut problem. *Oper. Res. Lett.*, 1:23–27, 1981.

[306] G. GRUBER, S. KRUK, F. RENDL, and H. WOLKOWICZ. Presolving for semidefinite programs without constraint qualifications. Research report, corr 98-32, University of Waterloo, Waterloo, Ontario, 1998.

[307] B. GRUNDY. Option prices and the underlying asset's return distribution. *J. Finance*, 46(3):1045–1070, July 1991.

[308] M. GU. On primal-dual interior point methods for semidefinite programming. Technical Report CAM 97-12, Department of Mathematics, University of California, Los Angeles, CA 90095-1555, 1997.

[309] M. GUIGNARD. Generalized Kuhn-Tucker conditions for mathematical programming problems in a Banach space. *SIAM J. of Control*, 7:232–241, 1969.

[310] O. GÜLER. Barrier functions in interior point methods. *Math. Oper. Res.*, 21:860–885, 1996.

[311] O. GÜLER. Hyperbolic polynomials and interior point methods for convex programming. *Math. Oper. Res.*, 22:350–377, 1997.

[312] O. GÜLER. Private communication. Unpublished, January 1998.

[313] O. GÜLER and L. TUNCEL. Manuscript. Unpublished, 1997.

[314] O. GÜLER and L. TUNCEL. Characterization of the barrier parameter of homogeneous convex cones. *Math. Programming*, 81:55–76, 1998.

[315] O. GÜLER and Y. YE. Convergence behavior of some interior point algorithms. *Math. Programming*, 60:215–228, 1993.

[316] S.M. GUU and Y.C. LIOU. On a quadratic optimization problem with equality constraints. *J. Optim. Theory Appl.*, 98(3):733–741, 1998.

[317] S.W. HADLEY. *Continuous Optimization Approaches to the Quadratic Assignment Problem.* PhD thesis, University of Waterloo, 1989.

[318] S.W. HADLEY, F. RENDL, and H. WOLKOWICZ. A new lower bound via projection for the quadratic assignment problem. *Math. Oper. Res.*, 17(3):727–739, 1992.

[319] F. HADLOCK. Finding a maximum cut of a planar graph in polynomial time. *SIAM J. Comput.*, 4(3):221–225, 1975.

[320] J.-P.A. HAEBERLY and M.L. OVERTON. Optimizing eigenvalues of symmetric definite pencils. In *Proceedings of American Control Conference*, Baltimore, July 1994.

[321] W. HAEMMERS. On some problems of Lovász concerning the Shannon capacity of graphs. *IEEE Transactions on Information Theory*, 25:231–232, 1979.

[322] P. L. HAMMER and P. HANSEN. Logical relations in quadratic $0-1$ programming. *Rev. Roumaine Math. Pures Appl.*, 26(3):421–429, 1981.

[323] P. L. HAMMER and B. SIMEONE. Quadratic functions of binary variables. In *Combinatorial Optimization (Como, 1986)*, pages 1–56. Springer, Berlin, 1989.

[324] P.L. HAMMER, P. HANSEN, and B. SIMEONE. Roof duality, complementation and persistency in quadratic 0-1 optimization. *Math. Programming*, 28:121–155, 1984.

[325] P.L. HAMMER and A.A. RUBIN. Some remarks on quadratic programming with 0-1 variables. *R.I.R.O.*, 3:67–79, 1970.

[326] S.-P. HAN and O. L. MANGASARIAN. Conjugate cone characterization of positive definite and semidefinite matrices. *Linear Algebra Appl.*, 56:89–103, 1984.

[327] E. R. HANSEN. *Global optimization using interval analysis.* Marcel Dekker, New-York, 1992.

[328] M. HARRISON and D. KREPS. Martingales and arbitrage in multiperiod securities markets. *Journal of Economic Theory*, 20:381–408, 1979.

[329] J. HASTAD. Some optimal inapproximability results. In *Proc. of the 29th ACM Symp. on Theory Comput.*, 1997.

[330] R. HAUSER. Self-scaled barrier functions: Decomposition and classification. Technical Report DAMTP 1999/NA13, Department of Applied Mathematics and Theoretical Physics, University of Cambridge, Cambridge, England, 1998.

[331] T.L. HAYDEN, J. WELLS, W-M. LIU, and P. TARAZAGA. The cone of distance matrices. *Linear Algebra Appl.*, 144:153–169, 1991.

[332] M. HEINKENSCHLOSS. On the solution of a two ball trust region subproblem. *Math. Programming*, 64(3):249–276, 1994.

[333] C. HELMBERG. *An interior point method for semidefinite programming and max-cut bounds.* PhD thesis, Graz University of Technology, Austria, 1994.

[334] C. HELMBERG. Fixing variables in semidefinite relaxations. *Lecture Notes in Computer Science*, 1284:259–263, 1997.

[335] C. HELMBERG and K.C. KIWIEL. A spectral bundle method with bounds. ZIB preprint, Konrad-Zuse-Zentrum für Informationstechnik Berlin, Takustraße 7, 14195 Berlin, Germany, November 1999. In preparation.

[336] C. HELMBERG, K.C. KIWIEL, and F. RENDL. Incorporating inequality constraints in the spectral bundle method. In R. E. Bixby, E. A. Boyd, and R. Z. Ríos-Mercado, editors, *Integer Programming and Combinatorial Optimization*, volume 1412 of *Lecture Notes in Computer Science*, pages 423–435. Springer, 1998.

[337] C. HELMBERG, S. POLJAK, F. RENDL, and H. WOLKOWICZ. Combining semidefinite and polyhedral relaxations for integer programs. In *Integer Programming and Combinatorial Optimization (Copenhagen, 1995)*, pages 124–134. Springer, Berlin, 1995.

[338] C. HELMBERG and F. RENDL. A spectral bundle method for semidefinite programming. Technical Report ZIB Preprint SC-97-37, Konrad-Zuse-Zentrum Berlin, Berlin, Germany, 1997. To appear in SIAM J. Optim.

[339] C. HELMBERG, F. RENDL, R. J. VANDERBEI, and H. WOLKOWICZ. An interior-point method for semidefinite programming. *SIAM J. Optim.*, 6(2):342–361, 1996.

[340] C. HELMBERG, F. RENDL, and R. WEISMANTEL. Quadratic knapsack relaxations using cutting planes and semidefinite programming. In *Integer Programming and Combinatorial Optimization (Vancouver, BC, 1996)*, volume 1084 of *Lecture Notes in Comput. Sci.*, pages 190–203. Springer, Berlin, 1996.

[341] D. HERSHKOWITZ, M. NEUMANN, and H. SCHNEIDER. Hermitian positive semidefinite matrices whose entries are 0 or 1 in modulus. Technical report, Dept. of Mathematics, Technion, Israel, 1998.

[342] R.D. HILL and S.R. WATERS. On the cone of positive semidefinite matrices. *Linear Algebra Appl.*, 90:81–88, 1987.

[343] J.-B. HIRIART-URRUTY. Conditions for global optimality. In *Handbook of Global Optimization*, volume 2 of *Nonconvex Optim. Appl.*, pages 1–26. Kluwer Acad. Publ., Dordrecht, 1995.

[344] J.-B. HIRIART-URRUTY and D. YE. Sensitivity analysis of all eigenvalues of a symmetric matrix. *Numer. Math.*, 70(1):45–72, 1995.

[345] J.B. HIRIART-URRUTY and C. LEMARECHAL. *Convex Analysis and Minimization Algorithms.* Springer-Verlag, 1993.

[346] J.B. HIRIART-URRUTY and C. LEMARECHAL. *Convex Analysis and Minimization Algorithms II*, volume 306 of *Grundlehren der mathematischen Wissenschaften.* Springer, Berlin, Heidelberg, 1993.

[347] L.T. HOAI, P.D. TAO, and L.D. MUU. A combined D.C. optimization–ellipsoidal branch-and-bound algorithm for solving nonconvex quadratic programming problems. *J. Comb. Optim.*, 2(1):9–28, 1998.

[348] A.S. HODEL, R.B. TENISON, and K. POOLLA. Numerical solution of large Lyapunov equations by Approximate Power Iteration. *Linear Algebra Appl.*, 236:205–230, 1996.

[349] A.J. HOFFMAN. On approximate solutions of systems of linear inequalities. *J. of Research of the National Bureau of Standards*, 49:263–265, 1952.

[350] R.B. HOLMES. *Geometric Functional Analysis and its Applications.* Springer-Verlag, Berlin, 1975.

[351] H. P. HORISBERGER and P. R. BÉLANGER. Regulators for linear, time invariant plants with uncertain parameters. *IEEE Trans. Aut. Control*, AC-21:705–708, 1976.

[352] R.A. HORN and C.R. JOHNSON. *Topics in Matrix Analysis.* Cambridge University Press, New York, 1991.

[353] T. ILLÉS, J. PENG, C. ROOS, , and T. TERLAKY. A strongly polynomial rounding procedure yielding a maximally complementary solution for $P^*(\kappa)$ linear complementarity problems. Technical Report 98–15, Faculty of Technical Mathematics and Computer Science, Delft University of Technology, Delft, The Netherlands, 1998.

[354] K. ISII. The extrema of probability determined by generalized moments (i) bounded random variables. *Ann. Inst. Stat. Math.*, 12:119–133, 1960.

[355] K. ISII. On sharpness of Chebyshev-type inequalities. *Ann. Inst. Stat. Math.*, 14:185–197, 1963.

[356] N. JACOBSON. Structure and representation of Jordan algebras. *Colloquium Publications*, XXXIX, 1968.

[357] J. JAHN. *Mathematical Vector Optimization in Partially Ordered Linear Spaces.* Peter Lang, Frankfurt am Main, 1986.

[358] G. JAMESON. *Ordered linear spaces.* Springer-Verlag, New York, 1970.

[359] B. JANSEN, C. ROOS, and T. TERLAKY. The theory of linear programming: skew symmetric self-dual problems and the central path. *Optimization*, 29(3):225–233, 1994.

[360] B. JANSEN, C. ROOS, and T. TERLAKY. Interior point methods, a decade after Karmarkar—a survey, with application to the smallest eigenvalue problem. *Statist. Neerlandica*, 50(1):146–170, 1996.

[361] B. JANSEN, C. ROOS, and T. TERLAKY. A family of polynomial affine scaling algorithms for positive semidefinite linear complementarity problems. *SIAM J. Optim.*, 7(1):126–140, 1997.

[362] B. JANSEN, K. ROOS, and T. TERLAKY. An interior point method approach to postoptimal and parametric analysis in linear programming. In *Proceeding of the Workshop on Interior Point Methods*, Budapest, Hungary, 1993.

[363] F. JARRE. On the convergence of the method of analytic centers when applied to convex quadratic programs. *Math. Programming*, 49(3):341–358, 1990.

[364] F. JARRE. An interior-point method for minimizing the maximum eigenvalue of a linear combination of matrices. *SIAM J. Control and Optimization*, 31:1360–1377, 1993.

[365] F. JARRE. A QQP-minimization method for semidefinite and smooth nonconvex programs. Technical report, Universitaet Wurzburg, Wurzburg, Germany, 1998.

[366] F. JARRE, M. KOCVARA, and J. ZOWE. Optimal truss design by interior-point methods. *SIAM J. Optim.*, 1999. To appear.

[367] F. JARRE, G. SONNEVEND, and J. STOER. An implementation of the method of analytic centers. In A. Benoussan and J.L. Lions, editors, *Analysis and Optimization of Systems*, volume 111 of *Lecture Notes in Control and Information Sciences*, pages 297–307. Springer, New York, 1988.

[368] J. JIANG. A long-step primal-dual path-following method for semidefinite programming. *Oper. Res. Lett.*, 23(1-2):53–61, 1998.

[369] J JIANG and M. SHI. A logarithmic barrier Newton method for semidefinite programming. *Systems Sci. Math. Sci.*, 10(2):168–175, 1997.

[370] S. JIBRIN. *Redundancy in Semidefinite Programming.* PhD thesis, Carleton University, Ottawa, Ontario, Canada, 1997.

[371] S. JIBRIN and I. PRESSMAN. Probabilistic algorithms for the detection of nonredundant linear matrix inequality constraints. Technical report, Carleton University, Ottawa, Ontario, Canada, 1998.

[372] F. JOHN. Extremum problems with inequalities as subsidiary conditions. In *Studies and Essays, Courant Anniversary Volume*, pages 187–204. Interscience, New York, 1948.

[373] C.R. JOHNSON. Matrix completion problems: a survey. *Proceedings of Symposium in Applied Mathematics*, 40:171–198, 1990.

[374] C.R. JOHNSON and B. KROSCHEL. Principal submatrices, geometric multiplicities, and structured eigenvectors. *SIAM J. Matrix Anal. Appl.*, 16:1004–1012, 1995.

[375] C.R. JOHNSON, B. KROSCHEL, and H. WOLKOWICZ. An interior-point method for approximate positive semidefinite completions. *Comput. Optim. Appl.*, 9(2):175–190, 1998.

[376] C.R. JOHNSON, B.K. KROSCHEL, and M. LUNDQUIST. The totally nonnegative completion problem. In *Topics in Semidefinite and Interior-Point Methods (Toronto, ON, 1996)*, pages 97–107. Amer. Math. Soc., Providence, RI, 1998.

[377] C.R. JOHNSON and L. RODMAN. Inertia possibilities for completions of partial Hermitian matrices. *Linear and Multilinear Algebra*, 16:179–195, 1984.

[378] C.R. JOHNSON and E.A. SCHREINER. The relationship between *AB* and *BA*. *American Math. Monthly*, Aug-Sept:578–582, 1996.

[379] C.R. JOHNSON and P. TARAZAGA. Approximate semidefinite matrices in a subspace. Technical report, Dept. of Mathematics, College of William and Mary, Williamsburg, Va, 1995. To appear in SIMAX.

[380] C.R. JOHNSON and P. TARAZAGA. Connections between the real positive semidefinite and distance matrix completion problems. *Linear Algebra Appl.*, 223/224:375–391, 1995. Special issue honoring Miroslav Fiedler and Vlastimil Pták.

[381] N. JOHNSON and S. KOTZ. *Distributions in Statistics: Continuous Multivariate Distributions.* John Wiley & Sons, New York, NY, 1972.

[382] P. JORDAN, J. VON NEUMANN, and E. WIGNER. On an algebraic generalization of the quantum mechanical formalism. *Annals of Mathematics*, 35:29–64, 1934.

[383] E. JOUINI and H. KALLAL. Martingales and arbitrage in securities markets with transaction costs. *Journal of Economic Theory*, 66:178–197, 1995.

[384] M. JÜNGER, A. MARTIN, G. REINELT, and R. WEISMANTEL. Quadratic 0/1 optimization and a decomposition approach for the placement of electronic circuits. *Math. Programming*, 63(3, Ser. B):257–279, 1994.

[385] K. ZHOU with J. DOYLE and K. GLOVER. *Robust and Optimal Control.* Prentice Hall, 1996.

[386] N. KAHALE. A semidefinite bound for mixing rates of Markov chains. In *Integer programming and combinatorial optimization (Vancouver, BC, 1996)*, volume 1084 of *Lecture Notes in Comput. Sci.*, pages 190–203. Springer, Berlin, 1996.

[387] C. KALLINA and A.C. WILLIAMS. Linear programming in reflexive spaces. *SIAM Rev.*, 13:350–376, 1971.

[388] R.E. KALMAN. Lyapunov functions for the problem of Lur'e in automatic control. *Proc. Nat. Acad. Sci., USA*, 49:201–205, 1963.

[389] R.E. KALMAN. On a new characterization of linear passive systems. In *Proc. First Annual Allerton Conf. on Communication, Control and Computing*, pages 456–470, 1963.

[390] A. KAMATH and N.K. KARMARKAR. A continuous method for computing bounds in integer quadratic optimization problems. *J. Global Optim.*, 2(3):229–241, 1992. Conference on Computational Methods in Global Optimization, II (Princeton, NJ, 1991).

[391] L.V. KANTOROVICH. Functional analysis and applied mathematics. *Uspekhi Mat. Nauk.*, 3:89–185, 1948. Transl. by C. Benster as N.B.S. Rept. 1509, Washington D.C., 1952.

[392] D. L. KARČICKA. Self-conditionally positive semidefinite matrices. *Mat. Fak. Univ. Kiril Metodij Skopje Godišen Zb.*, 33(34):53–57, 1982/83.

[393] D. KARGER, R. MOTWANI, and M. SUDAN. Approximate graph coloring by semidefinite programming. *J. Assoc. Comput. Mach.*, 45:246–265, 1998.

[394] S. KARISCH. *Nonlinear Approaches for Quadratic Assignment and Graph Partition Problems.* PhD thesis, University of Graz, Graz, Austria, 1995.

[395] S.E. KARISCH. Trust regions and the quadratic assignment problem. Master's thesis, University of Waterloo, 1992.

[396] S.E. KARISCH and F. RENDL. Semidefinite programming and graph equipartition. In *Topics in Semidefinite and Interior-Point Methods*, volume 18 of *The Fields Institute for Research in Mathematical Sciences, Communications Series*, Providence, Rhode Island, 1998. American Mathematical Society.

[397] S.E. KARISCH, F. RENDL, and J. CLAUSEN. solving graph bisection problems with semidefinite programming. Technical Report Report DIKU-Tr-97/9, University of Copenhagen, Dept. of Computer Science, 1997.

[398] S.E. KARISCH, F. RENDL, and H. WOLKOWICZ. Trust regions and relaxations for the quadratic assignment problem. In *Quadratic assignment and related problems (New Brunswick, NJ, 1993)*, pages 199–219. Amer. Math. Soc., Providence, RI, 1994.

[399] S. KARLIN and L.S. SHAPLEY. Geometry of moment spaces. *Memoirs Amer. Math. Soc.*, 12, 1953.

[400] S. KARLIN and W.J. STUDDEN. *Tchebycheff Systems: with Applications in Analysis and Statistics.* Pure and Applied Mathematics, A Series of Texts and Monographs. Interscience Publishers, John Wiley and Sons, 1966.

[401] H. KARLOFF. How good is the Goemans-Williamson MAX CUT algorithm? In *Proceedings of the Twenty-eighth Annual ACM Symposium on the Theory of Computing (Philadelphia, PA, 1996)*, pages 427–434, New York, 1996. ACM.

[402] H. KARLOFF. How good is the Goemans-Williamson MAX CUT algorithm? *SIAM J. on Computing*, 29(1):336–350, 1999.

[403] H. KARLOFF and U. ZWICK. A 7/8-approximation algorithm for MAX 3SAT? In *Proc. 38th Symp. on Found. of Comp. Sci.*, pages 406–415, 1997.

[404] N.K. KARMARKAR. A new polynomial time algorithm for linear programming. *Combinatorica*, 4:373–395, 1984.

[405] R. M. KARP. Reducibility among combinatorial problems. In R. E. Miller and J.W. Thatcher, editors, *Complexity of Computer Computation*, pages 85–103. Plenum Press, New York, 1972.

[406] T. KATO. *Perturbation theory for linear operators.* Springer-Verlag, Berlin, second edition, 1976. Grundlehren der Mathematischen Wissenschaften, Band 132.

[407] H. KAWASAKI. An envelope-like effect of infinitely many inequality constraints on second-order necessary conditions for minimization problems. *Math. Programming*, 41:73–96, 1988.

[408] F.P. KELLY. *Reversibility and Stochastic Networks.* John Wiley & Sons, New York, 1979.

[409] J.H.B. KEMPERMANN. On the role of duality in the theory of moments. In *Lecture Notes in Econom. and Math. Systems*, volume 215, pages 63–92, Berlin-New York, 1983. Springer.

[410] J.H.B. KEMPERMANN. Geometry of the moment problem. In *Proc. Sympos. Appl. Math.*, volume 37, pages 16–53, Providence, RI, 1987.

[411] H.A.L. KIERS and J.M.F. TEN BERGE. Optimality conditions for the trace of certain matrix products. *Linear Algebra Appl.*, 126:125–134, 1989.

[412] D. KINDERLEHRER and G. STAMPACCHIA. *An Introduction to Variational Inequalities and Applications.* Academic Press, New York, NY, 1980.

[413] K.C. KIWIEL. An aggregate subgradient method for nonsmooth convex minimization. *Math. Programming*, 27:320–341, 1983.

[414] K.C. KIWIEL. Proximity control in bundle methods for convex nondifferentiable minimization. *Math. Programming*, 46:105–122, 1990.

[415] P. KLEIN and H.-I LU. Efficient approximation algorithms for semidefinite programs arising from MAX CUT and COLORING. In *Proceedings of the Twenty-eighth Annual ACM Symposium on the Theory of Computing (Philadelphia, PA, 1996)*, pages 338–347, New York, 1996. ACM.

[416] E. DE KLERK. *Interior point methods for semidefinite programming.* PhD thesis, Delft University, 1997.

[417] E. DE KLERK, J. PENG, C. ROOS, and T. TERLAKY. A scaled Gauss-Newton primal-dual search direction for semidefinite programming. Technical report, Delft University of Technology, Faculty of Technical Mathematics and Informatics, Delft, The Netherlands, 1999.

[418] E. DE KLERK, C. ROOS, and T. TERLAKY. Initialization in semidefinite programming via a self-dual skew-symmetric embedding. *Oper. Res. Lett.*, 20(5):213–221, 1997.

[419] E. DE KLERK, C. ROOS, and T. TERLAKY. Infeasible–start semidefinite programming algorithms via self–dual embeddings. In *Topics in Semidefinite and Interior-Point Methods*, volume 18 of *The Fields Institute for Research in Mathematical Sciences, Communications Series*, pages 215–236. American Mathematical Society, 1998.

[420] E. DE KLERK, C. ROOS, and T. TERLAKY. On primal–dual path–following algorithms in semidefinite programming. In *New trends in Mathematical Programming*, pages 137–157, Dordrecht, The Netherlands, 1998. Kluwer Academic Publishers.

[421] E. DE KLERK, C. ROOS, and T. TERLAKY. Polynomial primal-dual affine scaling algorithms in semidefinite programming. *J. Comb. Optim.*, 2(1):51–70, 1998.

[422] E. DE KLERK, C. ROOS, and T. TERLAKY. Primal–dual potential reduction methods for semidefinite programming using affine–scaling directions. *Appl. Numer. Math.*, 29:335–360, 1999.

[423] G.P. KLIMOV. Time sharing service systems I. *Theory Probab. Appl.*, 19:532–551, 1974.

[424] G.P. KLIMOV. Time sharing service systems II. *Theory Probab. Appl.*, 23:314–321, 1978.

[425] D.E. KNUTH. The sandwich theorem. *Electronic J. Combinatorics*, 1:48pp, 1994.

[426] C.-W. KO, J. LEE, and M. QUEYRANNE. An exact algorithm for maximum entropy sampling. *Oper. Res.*, 43(4):684–691, 1995.

[427] C.-W. KO, J. LEE, and K. WAYNE. A spectral bound for D-optimality. In A.C. Atkinson, L. Pronzato, and H.P. Wynn, editors, *MODA 5 - Advances in Model-Oriented Data Analysis and Experimental Design (Marseilles, 1998)*, Contributions to Statistics. Springer, Berlin, 1998.

[428] M. KOJIMA. Private communication. 1995.

[429] M. KOJIMA, S. KOJIMA, and S. HARA. Linear algebra for semidefinite programming. Technical Report 1004, Dept. of Information Sciences, Tokyo Institute of Technology, Tokyo, Japan, 1997. Linear matrix

inequalities and positive semidefinite programming (Japanese) (Kyoto, 1996).

[430] M. KOJIMA, N. MEGIDDO, T. NOMA, and A. YOSHISE. *A Unified Approach to Interior Point Algorithms for Linear Complementarity Problems.* Lecture Notes in Computer Science. Springer-Verlag, NY, USA, 1991.

[431] M. KOJIMA, S. MIZUNO, and A. YOSHISE. A polynomial time algorithm for a class of linear complementarity problems. *Math. Programming*, 44:1–26, 1989.

[432] M. KOJIMA, S. MIZUNO, and A. YOSHISE. A primal–dual interior point algorithm for linear programming. In N. Megiddo, editor, *Progress in Mathematical Programming : Interior Point and Related Methods*, pages 29–47. Springer Verlag, New York, 1989.

[433] M. KOJIMA, S. MIZUNO, and A. YOSHISE. An $O(\sqrt{n}L)$ iteration potential reduction algorithm for linear complementarity problems. *Math. Programming*, 50:331–342, 1991.

[434] M. KOJIMA, M. SHIDA, and S. SHINDOH. Reduction of monotone linear complementarity problems over cones to linear programs over cones. Technical report, Dept. of Information Sciences, Tokyo Institute of Technology, Tokyo, Japan, 1995.

[435] M. KOJIMA, M. SHIDA, and S. SHINDOH. Local convergence of predictor-corrector infeasible-interior-point algorithms for SDPs and SDLCPs. Technical report, Dept. of Information Sciences, Tokyo Institute of Technology, Tokyo, Japan, 1996.

[436] M. KOJIMA, M. SHIDA, and S. SHINDOH. A note on the Nesterov-Todd and the Kojima-Shindoh-Hara search directions in semidefinite programming. Technical report, Department of Mathematical and Computing Sciences Tokyo Institute of Technology, Oh-Okayama, Meguro, Tokyo 152-8552, 1999. To appear in Optim. Methods Softw.

[437] M. KOJIMA, M. SHIDA, and S. SHINDOH. A predictor-corrector interior point algorithm for the semidefinite linear complementarity problem using the Alizadeh-Haeberley-Overton search direction. *SIAM J. Optim.*, 9(2):444–465, 1999.

[438] M. KOJIMA, M. SHIDA, and S. SHINDOH. Search directions in the SDP and the monotone SDLCP: generalization and inexact computation. *Math. Programming*, 85:51–80, 1999.

[439] M. KOJIMA, S. SHINDOH, and S. HARA. Interior-point methods for the monotone semidefinite linear complementarity problem in symmetric matrices. *SIAM J. Optim.*, 7(1):86–125, 1997.

[440] M. KOJIMA and L. TUNCEL. Discretization and localization in successive convex relaxation methods for nonconvex quadratic optimization problems. Technical Report CORR98-34, Dept. of Combinatorics and Optimization, University of Waterloo, 1998.

[441] M. KOJIMA and L. TUNCEL. Monotonicity of primal-dual interior-point algorithms for semidefinite programming problems. *Optim. Methods Softw.*, 10:275–296, 1998.

[442] M. KOJIMA and L. TUNCEL. Cones of matrices and successive convex relaxations of nonconvex sets. *SIAM J. Optim.*, 1999. To appear.

[443] M. KOJIMA and L. TUNCEL. On the finite convergence of successive convex relaxation methods. Technical Report CORR99-36, Dept. of Combinatorics and Optimization, University of Waterloo, 1999.

[444] F. KÖRNER. A tight bound for the Boolean quadratic optimization problem and its use in a branch and bound algorithm. *Optimization*, 19(5):711–721, 1988.

[445] F. KÖRNER. Remarks on a difficult test problem for quadratic Boolean programming. *Optimization*, 26:355–357, 1992.

[446] K.O. KORTANEK. Constructing a perfect duality in infinite programming. *Appl. Math. Optim.*, 3(4):357–372, 1977.

[447] K.O. KORTANEK and Q. ZHANG. Prefect duality in semi-infinite and semidefinite programming. Technical report, Department of Management Sciences The University of Iowa, 1998.

[448] M. KOSHELER and V. KREINOVITCH. Interval computations web site. `http://cs.utep.edu/interval-comp/main.html`, 1996.

[449] S. KRUK. Semidefinite programming applied to nonlinear programming. Master's thesis, University of Waterloo, 1996.

[450] S. KRUK, M. MURAMATSU, F. RENDL, R.J. VANDERBEI, and H. WOLKOWICZ. The Gauss-Newton direction in linear and semidefinite programming. Technical Report CORR 98-16, University of Waterloo, Waterloo, Canada, 1998.

[451] S. KRUK and H. WOLKOWICZ. SQ^2P, sequential quadratic constrained quadratic programming. In Ya xiang Yuan, editor, *Advances in Nonlinear Programming*, volume 14 of *Applied Optimization*, pages 177–204. Kluwer, Dordrecht, 1998. Proceedings of Nonlinear Programming Conference in Beijing in honour of Professor M.J.D. Powell.

[452] H.W. KUHN. Nonlinear programming: a historical view. In R.W. Cottle and C.E. Lemke, editors, *Nonlinear Programming*, pages 1–26, Providence, R.I., 1976. AMS.

[453] H.W. KUHN and A.W. TUCKER. Nonlinear programming. In *Proceedings of the Second Berkeley Symposium on Mathematical Statistics and Probability, 1950*, pages 481–492, Berkeley and Los Angeles, 1951. University of California Press.

[454] P.R. KUMAR and S.P. MEYN. Duality and linear programs for stability and performance analysis of queueing networks and scheduling policies. *IEEE Trans. Autom. Control*, 41:4–16, 1996.

[455] S. KUMAR and P.R. KUMAR. Performance bounds for queueing networks and scheduling policies. *IEEE Trans. Autom. Control*, 39:1600–1611, 1994.

[456] S. KURCYUSZ. On the existence and nonexistence of Lagrange multipliers in Banach spaces. *Journal of Optimization Theory and Applications*, 20:81–110, 1976.

[457] M.K. KWONG. Some results on matrix monotone functions. *Linear Algebra Appl.*, 118:129–153, 1989.

[458] H.J. LANDAU. Classical background on the moment problem. In ed. H.J. Landau, editor, *Moments in Mathematics*, number 37 in Proc. Sympos. Appl. Math., Series III, pages 1–15, Providence, RI, 1987. American Mathematical Society.

[459] J. B. LASSERRE. A new Farkas lemma for positive semidefinite matrices. *IEEE Trans. Automat. Control*, 40(6):1131–1133, 1995.

[460] J. B. LASSERRE. Linear programming with positive semi-definite matrices. *MPE*, 2:499–521, 1996.

[461] J. B. LASSERRE. A Farkas lemma without a standard closure condition. *SIAM J. Control Optim.*, 35(1):265–272, 1997.

[462] M. LAURENT. Cuts, matrix completions and graph rigidity. *Math. Programming*, 79:255–284, 1997.

[463] M. LAURENT. The real positive semidefinite completion problem for series-parallel graphs. *Linear Algebra Appl.*, 252:347–366, 1997.

[464] M. LAURENT. A connection between positive semidefinite and Euclidean distance matrix completion problems. *Linear Algebra Appl.*, 273:9–22, 1998.

[465] M. LAURENT. A tour d'horizon on positive semidefinite and Euclidean distance matrix completion problems. In *Topics in Semidefinite and Interior-Point Methods*, volume 18 of *The Fields Institute for Research in Mathematical Sciences, Communications Series*, Providence, Rhode Island, 1998. American Mathematical Society.

[466] M. LAURENT and S. POLJAK. On a positive semidefinite relaxation of the cut polytope. *Linear Algebra Appl.*, 223/224:439–461, 1995.

[467] M. LAURENT and S. POLJAK. On the facial structure of the correlation matrices. *SIAM J. Matrix Anal. Appl.*, 17(3):530–547, 1996.

[468] M. LAURENT, S. POLJAK, and F. RENDL. Connections between semidefinite relaxations of the max-cut and stable set problems. *Math. Programming*, 77:225–246, 1997.

[469] J. LEE. Constrained maximum-entropy sampling. *Oper. Res.*, 46:655–664, 1998.

[470] J. DE LEEUW and W. HEISER. Theory of multidimensional scaling. In P. R. Krishnaiah and L. N. Kanal, editors, *Handbook of Statistics*, volume 2, pages 285–316. North-Holland, 1982.

[471] S. LELE. Euclidean distance matrix analysis (EDMA): estimation of mean form and mean form difference. *Math. Geol.*, 25(5):573–602, 1993.

[472] C. LEMARÉCHAL and F. OUSTRY. Nonsmooth algorithms to solve semidefinite programs. In L. EL Ghaoui and S-I. Niculescu, editors, *Recent Advances on LMI methods in Control*, Advances in Design and Control series. SIAM, 1999. To appear.

[473] C. LEMARÉCHAL and F. OUSTRY. Semidefinite relaxations and Lagrangian duality with application to combinatorial optimization. Technical report, Institut National de Recherche en Informatique et en Automatique, INRIA, St Martin, France, 1999.

[474] F. LEMPIO and H. MAURER. Differential stability in infinite-dimensional nonlinear programming. *Appl. Math. Optim.*, 6:139–152, 1980.

[475] F. LEMPIO and J. ZOWE. Higher order optimality conditions. In *Modern applied mathematics (Bonn, 1979)*, pages 147–193. North-Holland, Amsterdam, 1982.

[476] E.S. LEVITIN. On differential properties of the optimal value of parametric problems of mathematical programming. *Dokl. Akad. Nauk SSSR*, 224:1354–1358, 1975.

[477] H. LEVY and M. SIDI. Polling systems: applications, modeling, and optimization. *IEEE Trans. Comm.*, 38:1750–1760, 1990.

[478] A.S. LEWIS. *Take-home final exam, Course CO663 in Convex Analysis.* University of Waterloo, Ontario, Canada, 1994.

[479] A.S. LEWIS. Convex analysis on the Hermitian matrices. *SIAM J. Optim.*, 6(1):164–177, 1996.

[480] A.S. LEWIS. Derivatives of spectral functions. *Math. Oper. Res.*, 21(3):576–588, 1996.

[481] A.S. LEWIS. Group invariance and convex matrix analysis. *SIAM J. Matrix Anal. Appl.*, 17(4):927–949, 1996.

[482] A.S. LEWIS. Eigenvalue-constrained faces. *Linear Algebra Appl.*, 269:159–181, 1998.

[483] A.S. LEWIS and M.L. OVERTON. Eigenvalue optimization. *Acta Numerica*, 5:149–190, 1996.

[484] C.K. LI and B.S. TAM. A note on extreme correlation matrices. *SIAM J. Matrix Anal. Appl.*, 15(3):903–908, 1994.

[485] M.H. LIM. Linear transformations on symmetric matrices. *Linear and Multilinear Algebra*, 7:47–57, 1979.

[486] C. LIN and R. SAIGAL. An infeasible start predictor corrector method for semi-definite linear programming. Technical report, Dept. of Ind. and Oper. Eng., Univ. of Michigan, Ann Arbor, MI, 1995.

[487] C. LIN and R. SAIGAL. A predictor corrector method for semi-definite linear programming. Technical report, Dept. of Ind. and Oper. Eng., Univ. of Michigan, Ann Arbor, MI, 1995.

[488] C. LIN and R. SAIGAL. On solving large-scale semidefinite programming problems - a case study of quadratic assignment problem. Technical report, Dept. of Ind. and Oper. Eng., Univ. of Michigan, Ann Arbor, MI, 1997.

[489] A. LISSER and F. RENDL. Telecommunication clustering using linear and semidefinite programming. Technical report, Institut fuer Mathematik, Universitaet Klagenfurt, A - 9020 Klagenfurt, Austria, 1999.

[490] J. LIU and L. ZHU. A minimum principle and estimates of the eigenvalues for schur complements positive semidefinite Hermitian matrices. *Linear Algcbra Appl.*, 265:123–147, 1997.

[491] S. LIU. Reverse of a convex matrix inequality. *IMAGE*, 16:32, 1996. Problem Corner.

[492] A. LO. Semiparametric upper bounds for option prices and expected payoffs. *Journal of Financial Economics*, 19:373–388, 1987.

[493] M.S. LOBO, L. VANDENBERGHE, S. BOYD, and H. LEBRET. Applications of second-order cone programming. *Linear Algebra Appl.*, 284(1-3):193–228, 1998. ILAS Symposium on Fast Algorithms for Control, Signals and Image Processing (Winnipeg, MB, 1997).

[494] R. LOEWY. Extreme points of a convex subset of the cone of positive semidefinite matrices. *Math. Ann.*, 253:227–232, 1980.

[495] F. LONGSTAFF. Martingale restriction tests of option pricing models, version 1. Technical report, University of California, Los Angeles, 1990. Working paper.

[496] L. LOVÁSZ. On the Shannon capacity of a graph. *IEEE Transactions on Information Theory*, 25:1–7, 1979.

[497] L. LOVÁSZ and A. SCHRIJVER. Cones of matrices and set-functions and 0-1 optimization. *SIAM J. Optim.*, 1(2):166–190, 1991.

[498] K. LÖWNER. Über monotone matrixfunktionen. *Math. Z.*, 38:177–216, 1934.

[499] L.Z. LU. Matrix bounds and simple iterations for positive semidefinite solutions to discrete-time algebraic Riccati equations. *Xiamen Daxue Xuebao Ziran Kexue Ban*, 34(4):512–516, 1995.

[500] S. LUCIDI, L. PALAGI, and M. ROMA. Quadratic programs with quadratic constraint; characterization of KKT points and equivalence with an unconstrained problem. Technical report, Universita di Roma la Sapienca, 1995.

[501] D.G. LUENBERGER. *Optimization by Vector Space Methods*. John Wiley, 1969.

[502] M. LUNDQUIST and W. BARRETT. Rank inequalities for positive semidefinite matrices. *Linear Algebra Appl.*, 248:91–100, 1996.

[503] M. LUNDQUIST and C.R. JOHNSON. Linearly constrained positive definite completions. *Linear Algebra Appl.*, 150:195–207, 1991.

[504] X. LUO and D. BERTSIMAS. A new algorithm for state constrained separated continuous linear programs. *SIAM J. Contr. Optim.*, 1:177–210, 1998.

[505] X.D. LUO and Z.-Q. LUO. Extensions of Hoffman's error bound to polynomial systems. *SIAM J. Optim.*, 4:383–392, 1994.

[506] Z-Q. LUO, J. F. STURM, and S. ZHANG. Superlinear convergence of a symmetric primal-dual path-following algorithm for semidefinite programming. *SIAM J. Optim.*, 8:59–81, 1998.

[507] Z-Q. LUO, J.F. STURM, and S. ZHANG. Duality and self-duality for conic convex programming. Technical report, Erasmus University Rotterdam, The Netherlands, 1996.

[508] Z-Q. LUO, J.F. STURM, and S. ZHANG. Duality results for conic convex programming. Technical Report Report 9719/A, April, Erasmus University Rotterdam, Econometric Institute EUR, P.O. Box 1738, 3000 DR, The Netherlands, 1997.

[509] Z-Q. LUO and J. SUN. An analytic center based column generation algorithm for convex quadratic feasibility problems. Technical report, McMaster University, Hamilton, Ontario, 1995.

[510] Z-Q. LUO and S. ZHANG. On the extension of Frank-Wolfe theorem. Technical report, Erasmus University Rotterdam, The Netherlands, 1997.

[511] A.I. LUR'E. *Some Nonlinear Problems in the Theory of Automatic Control.* H. M. Stationery Off., London, 1957. In Russian, 1951.

[512] A.M. LYAPUNOV. *Problème général de la stabilité du mouvement*, volume 17 of *Annals of Mathematics Studies.* Princeton University Press, Princeton, 1947.

[513] A.M. LYAPUNOV. The general problem of stability of motion. *Ann. math. Studies*, 11, 1949. Princeton (in Russian: Moscow 1935).

[514] Z.A. MAANY. A new algorithm for highly curved constrained optimization. Technical Report T.R. 161, The Hatfield Polytechnic Numerical Optimization Centre, 1985.

[515] O.L. MANGASARIAN and S. FROMOVITZ. The Fritz John necessary optimality conditions in the presence of equality and inequality constraints. *J. Math. Anal. Appl.*, 17:37–47, 1967.

[516] M. MARCUS and H. MINC. *A Survey of Matrix Theory and Matrix Inequalities.* Dover Publications Inc., New York, 1992. Reprint of the 1969 edition.

[517] M. MARCUS and B.N. MOYLS. Linear transformations on algebras of matrices. *Canadian J. Math.*, 11:61–66, 1959.

[518] A. MARKOV. *On Certain Applications of Algebraic Continued Fractions.* PhD thesis, St. Petersburg, 1884. in Russian.

[519] A. MARSHALL. Markov's inequality for random variables taking values in a linear topological space. In *Inequalities in statistics and probability (Lincoln, Neb., 1982)*, pages 104–108. Inst. Math. Statist., Hayward, Calif., 1984.

[520] A. MARSHALL and I. OLKIN. Multivariate Chebyshev inequalities. *Ann. Math. Stat.*, 31:1001–1014, 1960.

[521] A. MARSHALL and I. OLKIN. A one-sided inequality of the Chebyshev type. *Ann. Math. Stat.*, 31:488–491, 1960.

[522] A. MARSHALL and I. OLKIN. A game theoretic proof that Chebyshev inequalities are sharp. *Pacific. J. Math.*, 11:1421–1429, 1961.

[523] A.W. MARSHALL and I. OLKIN. *Inequalities: Theory of Majorization and its Applications.* Academic Press, 1979.

[524] J.M. MARTINEZ. Local minimizers of quadratic functions on Euclidean balls and spheres. *SIAM J. Optim.*, 4(1):159–176, 1994.

[525] A. MATVEEV and V.A. YAKUBOVICH. Nonconvex problems of global optimization: linear-quadratic control problems with quadratic constraints. *Dynam. Control*, 7(2):99–134, 1997.

[526] H. MAURER and J. ZOWE. Second-order necessary and sufficient optimality conditions for infinite-dimensional programming problems. In *Optimization techniques (Proc. 8th IFIP Conf., Würzburg, 1977), Part 2*, pages 13–21. Lecture Notes in Control and Information Sci., Vol. 7. Springer, Berlin, 1978.

[527] H. MAURER and J. ZOWE. First and second order necessary and sufficient optimality conditions for infinite-dimensional programming problems. *Math. Programming*, 16(1):98–110, 1979.

[528] N. MEGIDDO. Pathways to the optimal set in linear programming. In N. Megiddo, editor, *Progress in Mathematical Programming, Interior Point and Related Methods.* Springer-Verlag, 1988.

[529] A. MEGRETSKI and A. RANTZER. System analysis via integral quadratic constraints. *IEEE Trans. Aut. Control*, 42(6):819–830, June 1997.

[530] M. MESBAHI and G.P. PAPAVASSILOPOULOS. A cone programming approach to the bilinear matrix inequality problem and its geometry. *Math. Programming*, 77(2, Ser. B):247–272, 1997.

[531] M. MESBAHI and G.P. PAPAVASSILOPOULOS. On the rank minimization problem over a positive semidefinite linear matrix inequality. *IEEE Trans. Automat. Control*, 42(2):239–243, 1997.

[532] C. MESZAROS. *An efficient implementation of interior point methods for linear programming and their applications.* PhD thesis, Eotvos Lorand University of Sciences, 1997.

[533] S. MIZUNO, M. KOJIMA, and M.J. TODD. Infeasible-interior-point primal-dual potential-reduction algorithms for linear programming. *SIAM J. Optim.*, 5:52–67, 1995.

[534] S. MIZUNO, M.J. TODD, and Y. YE. On adaptive-step primal-dual interior-point algorithms for linear programming. *Math. Oper. Res.*, 18(4):964–981, 1993.

[535] J. MOCKUS, L. MOCKUS, and A. MOCKUS. Bayesian approach to combinatorial optimization. Technical report, School of Engineering, Purdue University, W. Lafayette, IN, 1996.

[536] R.D.C. MONTEIRO. Primal-dual path-following algorithms for semidefinite programming. *SIAM J. Optim.*, 7(3):663–678, 1997.

[537] R.D.C. MONTEIRO. Polynomial convergence of primal-dual algorithms for semidefinite programming based on the Monteiro and Zhang family of directions. *SIAM J. Optim.*, 8:797–812, 1998.

[538] R.D.C. MONTEIRO and I. ADLER. Interior path following primal-dual algorithms. Part I: Linear programming. *Math. Programming*, 44:27–42, 1989.

[539] R.D.C. MONTEIRO and I. ADLER. Interior path following primal-dual algorithms. Part II: Convex quadratic programming. *Math. Programming*, 44:43–66, 1989.

[540] R.D.C. MONTEIRO and J.-S. PANG. On two interior-point mappings for nonlinear semidefinite complementarity problems. *Math. Oper. Res.*, 23:39–60, 1998.

[541] R.D.C. MONTEIRO and J.-S. PANG. A potential reduction Newton method for constrained equations. *SIAM J. Optim.*, 9:729–754, 1999.

[542] R.D.C. MONTEIRO and T. TSUCHIYA. Polynomial convergence of primal-dual algorithms for the second-order cone program based on the MZ-family of directions. Technical report, The School of ISyE, Georgia Institute of Technology, Atlanta, GA 30332, USA, 1998.

[543] R.D.C. MONTEIRO and T. TSUCHIYA. Polynomial convergence of a new family of primal-dual algorithms for semidefinite programming. *SIAM J. Optim.*, 9:551–577, 1999.

[544] R.D.C. MONTEIRO and T. TSUCHIYA. Polynomiality of primal-dual algorithms for semidefinite linear complementarity problems based on the Kojima-Shindoh-Hara family of directions. *Math. Programming*, 84:39–53, 1999.

[545] R.D.C. MONTEIRO and P.R. ZANJÁCOMO. Implementation of primal-dual methods for semidefinite programming based on Monteiro and Tsuchiya Newton directions and their variants. Technical report, Georgia Tech, Atlanta, GA, 1997.

[546] R.D.C. MONTEIRO and P.R. ZANJÁCOMO. A note on the existence of the Alizadeh-Haeberly-Overton direction for semidefinite programming. *Math. Programming*, 78(3, Ser. A):393–396, 1997.

[547] R.D.C. MONTEIRO and Y. ZHANG. A unified analysis for a class of long-step primal-dual path-following interior-point algorithms for semidefinite programming. *Math. Programming*, 81(3, Ser. A):281–299, 1998.

[548] R.E. MOORE. *Methods and Applications of Interval Analysis.* SIAM, Philadelphia, 1979.

[549] J. J. MORÉ. Generalizations of the trust region problem. *Optim. Methods Software*, 2:189–209, 1993.

[550] J.J. MORÉ and D.C. SORENSEN. On the use of directions of negative curvature in a modified Newton method. *Math. Programming*, 16:1–20, 1979.

[551] J.J. MORÉ and D.C. SORENSEN. Computing a trust region step. *SIAM J. Sci. Statist. Comput.*, 4:553–572, 1983.

[552] J.J. MORÉ and D.C. SORENSEN. Newton's method. In Gene H. Golub, editor, *Studies in Numerical Analysis*, volume 24 of *MAA Studies in Mathematics*. The Mathematical Association of America, 1984.

[553] J.J. MORÉ and Z. WU. Global continuation for distance geometry problems. Technical Report MCS-P505-0395, Applied Mathematics Division, Argonne National Labs, Chicago, Il, 1996.

[554] J.J. MORÉ and Z. WU. Distance geometry optimization for protein structures. Technical Report MCS-P628-1296, Applied Mathematics Division, Argonne National Labs, Chicago, Il, 1997.

[555] L. MOSHEYEV and M. ZIBULEVSKY. Penalty/barrier multiplier algorithm for semidefinite programming. Technical report, Technion, Haifa, Israel, 1999.

[556] J.M. MULVEY, R.J. VANDERBEI, and S.A. ZENIOS. Robust optimization of large-scale systems. *Operations Research*, 43:264–281, 1995.

[557] M. MURAMATSU. Affine scaling algorithm fails for semidefinite programming. *Sūrikaisekikenkyūsho Kōkyūroku*, 1004:128–137, 1997. Linear matrix inequalities and positive semidefinite programming (Japanese) (Kyoto, 1996).

[558] M. MURAMATSU. Affine scaling algorithm fails for semidefinite programming. *Math. Programming*, 83(3, Ser. A):393–406, 1998.

[559] M. MURAMATSU and R.J. VANDERBEI. Primal-dual affine-scaling algorithms fail for semidefinite programming. Technical report, SOR, Princeton University, Princeton, NJ, 1997.

[560] K.G. MURTY and S.N. KABADI. Some NP-complete problems in quadratic and nonlinear programming. *Math. Programming*, 39:117–129, 1987.

[561] G. NÆVDAL and H.J. WOERDEMAN. Cone inclusion numbers. *SIAM J. Matrix Anal. Appl.*, 19(3):613–639, 1998.

[562] M. V. NAYAKKANKUPPAM and M.L. OVERTON. Primal–dual interior–point methods for semidefinite programming. In *Proceedings of the IEEE International Symposium on Computer–Aided Control System Design*, pages 235–239. IEEE Control Systems Society, September 1996.

[563] M. V. NAYAKKANKUPPAM and M.L. OVERTON. Conditioning of semidefinite programs. Technical report, Courant Institute of Mathematical Sciences, NYU, New York, NY, March 1997.

[564] G.L. NEMHAUSER and L.A. WOLSEY. *Integer and Combinatorial Optimization*. John Wiley & Sons, New York, NY, 1988.

[565] A.S. NEMIROVSKI. The long-step method of analytic centers for fractional problems. *Math. Programming*, 77(2, Ser. B):191–224, 1997.

[566] A.S. NEMIROVSKI. On normal self-concordant barriers and long-step interior point methods. Technical report, Faculty of Industrial Engineering, Institute of Technology, Technion, Haifa, ISRAEL, 1997.

[567] A.S. NEMIROVSKI. Private communication. 1997.

[568] A.S. NEMIROVSKI, C. ROOS, and T. TERLAKY. On maximization of quadratic form over intersection of ellipsoids with common center. *Math. Programming*, 1999. To appear, DOI 10.1007/s101079900099.

[569] A.S. NEMIROVSKI and K. SCHEINBERG. Extension of Karmarkar's algorithm onto convex quadratically constrained quadratic programming. *Math. Programming*, 72:273–289, 1996.

[570] Y. NESTEROV. Semidefinite relaxation and nonconvex quadratic optimization. *Optim. Methods Softw.*, 9(1-3):141–160, 1998.

[571] Y.E. NESTEROV. Long-step strategies in interior-point primal-dual methods. *Math. Programming*, 76:47–94, 1997.

[572] Y.E. NESTEROV. Quality of semidefinite relaxation for nonconvex quadratic optimization. Technical report, CORE, Universite Catholique de Louvain, Belgium, 1997.

[573] Y.E. NESTEROV. Structure of non-negative polynomial and optimization problems. Technical report, Louvan-la-Neuve, Belgium, 1997. Preprint DP 9749.

[574] Y.E. NESTEROV. Global quadratic optimization via conic relaxation. Technical report, CORE, Universite Catholique de Louvain, Belgium, 1998.

[575] Y.E. NESTEROV. Semidefinite relaxation and nonconvex quadratic optimization. *Optim. Methods Softw.*, 9:141–160, 1998. Special Issue Celebrating the 60th Birthday of Professor Naum Shor.

[576] Y.E. NESTEROV and A.S. NEMIROVSKI. A general approach to polynomial-time algorithms design for convex programming. Technical report, Centr. Econ. & Math. Inst., USSR Acad. Sci., Moscow, USSR, 1988.

[577] Y.E. NESTEROV and A.S. NEMIROVSKI. Polynomial barrier methods in convex programming. *Èkonom. i Mat. Metody*, 24(6):1084–1091, 1988.

[578] Y.E. NESTEROV and A.S. NEMIROVSKI. Self–concordant functions and polynomial–time methods in convex programming. Book–Preprint, Central Economic and Mathematical Institute, USSR Academy of Science, Moscow, USSR, 1989. Published in Nesterov and Nemirovsky [583].

[579] Y.E. NESTEROV and A.S. NEMIROVSKI. *Optimization over positive semidefinite matrices: Mathematical background and user's manual.* USSR Acad. Sci. Centr. Econ. & Math. Inst., 32 Krasikova St., Moscow 117418 USSR, 1990.

[580] Y.E. NESTEROV and A.S. NEMIROVSKI. Self-concordant functions and polynomial time methods in convex programming. Technical report, Centr. Econ. & Math. Inst., USSR Acad. Sci., Moscow, USSR, April 1990.

[581] Y.E. NESTEROV and A.S. NEMIROVSKI. Conic formulation of a convex programming problem and duality. Technical report, Centr. Econ. & Math. Inst., USSR Academy of Sciences, Moscow USSR, 1991.

[582] Y.E. NESTEROV and A.S. NEMIROVSKI. Conic duality and its applications in convex programming. *Optim. Methods Softw.*, 1:95–115, 1992.

[583] Y.E. NESTEROV and A.S. NEMIROVSKI. *Interior Point Polynomial Algorithms in Convex Programming.* SIAM Publications. SIAM, Philadelphia, USA, 1994.

[584] Y.E. NESTEROV and A.S. NEMIROVSKI. An interior-point method for generalized linear-fractional programming. *Math. Programming*, 69(1):177–204, 1995.

[585] Y.E. NESTEROV and A.S. NEMIROVSKI. Multi-parameter surfaces of analytic centers and long-step surface-following interior point methods. *Math. Oper. Res.*, 23(1):1–38, 1998.

[586] Y.E. NESTEROV and A.S. NEMIROVSKI. On self-concordant convex-concave functions. *Optim. Methods Softw.*, 1999. To appear.

[587] Y.E. NESTEROV and M.J. TODD. Self-scaled barriers and interior-point methods for convex programming. *Math. Oper. Res.*, 22(1):1–42, 1997.

[588] Y.E. NESTEROV and M.J. TODD. Primal-dual interior-point methods for self-scaled cones. *SIAM J. Optim.*, 8:324–364, 1998.

[589] A. NEUMAIER. An optimality criterion for global quadratic optimization. *J. Global Optim.*, 2(2):201–208, 1992.

[590] J. VON NEUMANN. *Collected Works.* Pergamon Press, New York, 1962. A.H. Traub, ed.

[591] L.W. NEUSTADT. *Optimization.* Princeton University Press, Princeton, N. J., 1976. A Theory of Necessary Conditions, With a chapter by H. T. Banks.

[592] J. NINO-MORA. Optimal resource allocation in a dynamic and stochastic environment: A mathematical programming approach. Ph.D. Dissertation, Sloan School of Management, MIT, 1995.

[593] M.A. NUNEZ. *Condition Numbers and Properties of Central Trajectories in Convex Programming.* PhD thesis, Massachusetts Institute of Technology, 1997.

[594] M.A. NUNEZ. A characterization of ill-posed data instances for convex programming. Technical report, Chapman University, School of Business and Economics, Orange, CA, 1998.

[595] I. OLKIN and R. PUKELSHEIM. The distance between two random vectors with given dispersion matrices. *Linear Algebra Appl.*, 48:257–263, 1982.

[596] G.I. ORLOVA and Ya.G. DORFMAN. Finding the maximum cut in a graph. *Engrg. Cybernetics*, 10(3):502–506, 1972.

[597] F. OUSTRY. The $\mathcal{U}$-Lagrangian of the maximum eigenvalue function. *SIAM J. Optim.*, 1998. To appear.

[598] F. OUSTRY. A second-order bundle method to minimize the maximum eigenvalue function. Submitted to Math. Programming, February 1998.

[599] M.L. OVERTON. On minimizing the maximum eigenvalue of a symmetric matrix. *SIAM J. Matrix Anal. Appl.*, 9:256–268, 1988.

[600] M.L. OVERTON. Large-scale optimization of eigenvalues. *SIAM J. Optim.*, 2:88–120, 1992.

[601] M.L. OVERTON and H. WOLKOWICZ, editors. *Semidefinite programming.* North-Holland Publishing Co., Amsterdam, 1997. Dedicated to the memory of Svatopluk Poljak, Math. Programming **77** (1997), no. 2, Ser. B.

[602] M.L. OVERTON and R.S. WOMERSLEY. On minimizing the spectral radius of a nonsymmetric matrix function: optimality conditions and duality theory. *SIAM J. Matrix Anal. Appl.*, 9(4):473–498, 1988.

[603] M.L. OVERTON and R.S. WOMERSLEY. On the sum of the largest eigenvalues of a symmetric matrix. *SIAM J. Matrix Anal. Appl.*, 13(1):41–45, 1992.

[604] M.L. OVERTON and R.S. WOMERSLEY. Optimality conditions and duality theory for minimizing sums of the largest eigenvalues of symmetric matrices. *Math. Programming*, 62(2, Ser. B):321–357, 1993.

[605] M.L. OVERTON and R.S. WOMERSLEY. Second derivatives for optimizing eigenvalues of symmetric matrices. *SIAM J. Matrix Anal. Appl.*, 16(3):697–718, 1995.

[606] A. PACKARD. Gain scheduling via linear fractional transformations. *Syst. Control Letters*, 22:79–92, 1994.

[607] M. W. PADBERG. The Boolean quadric polytope: Some characteristics, facets and relatives. *Math. Programming*, Ser. B, 45(1):139–172, 1989.

[608] C.H. PAPADIMITRIOU and J.N. TSITSIKLIS. The complexity of optimal queueing network control. *Math. Oper. Res.*, 1, 1998.

[609] F. PAPANGELOU. Integrability of expected increments and a related random change of time scale. *Trans. Amer. Math. Soc.*, 165:483–506, 1972.

[610] P. PARDALOS, F. RENDL, and H. WOLKOWICZ. The quadratic assignment problem: a survey and recent developments. In P. Pardalos and H. Wolkowicz, editors, *Quadratic assignment and related problems (New Brunswick, NJ, 1993)*, pages 1–42. Amer. Math. Soc., Providence, RI, 1994.

[611] P. PARDALOS and H. WOLKOWICZ, editors. *Semidefinite programming and interior-point approaches for combinatorial optimization problems.* Kluwer Academic Publishers, Hingham, MA, 1998. Papers from the workshop held at the University of Toronto, Toronto, ON, May 15–17, 1996, J. Comb. Optim. **2** (1998), no. 1.

[612] P. PARDALOS and H. WOLKOWICZ, editors. *Topics in semidefinite and interior-point methods*, The Fields Institute for Research in Mathematical Sciences, Communications Series, Providence, RI, 1998. American Mathematical Society.

[613] P. M. PARDALOS, D. SHALLOWAY, and G. XUE, editors. *Global minimization of nonconvex energy functions: molecular conformation and protein folding*, volume 23 of *DIMACS Series in Discrete Mathematics and Theoretical Computer Science*. American Mathematical Society, Providence, RI, 1996. Papers from the DIMACS Workshop held as part of the DIMACS Special Year on Mathematical Support for Molecular Biology at Rutgers University, New Brunswick, New Jersey, March 20–21, 1995.

[614] P.M. PARDALOS. Quadratic programming with one negative eigenvalue is NP-hard. *J. Global Optim.*, 1:15–22, 1991.

[615] P.M. PARDALOS. On the passage from local to global in optimization. In *Mathematical Programming: State of the Art 1994*, pages 220–247. The University of Michigan, 1994.

[616] P.M. PARDALOS. Continuous approaches to discrete optimization problems. In *Nonlinear Optimization and Applications*, pages 313–328. Plenum Publishing, 1996.

[617] P.M. PARDALOS and M. RAMANA. Semidefinite programming. In *Interior Point Methods of Mathematical Programming*, pages 369–398. Kluwer Academic Publishers, 1997.

[618] P.M. PARDALOS and M.G.C. RESENDE. Interior point methods for global optimization problems. In *Interior Point methods of Mathematical Programming*, pages 467–500. Kluwer Academic Publishers, 1997.

[619] P.M. PARDALOS and J.B. ROSEN. *Constrained global optimization: Algorithms and applications*, volume 268 of *Lecture Notes in Computer Science*. Springer-Verlag, Berlin, 1987.

[620] G. PATAKI. On the facial structure of cone-LP's and semidefinite programs. Technical Report 595, Graduate School of Industrial Administration, Carnegie Mellon University, 1994.

[621] G. PATAKI. Cone-LP's and semidefinite programs: geometry and a simplex-type method. In *Integer programming and combinatorial optimization (Vancouver, BC, 1996)*, volume 1084 of *Lecture Notes in Comput. Sci.*, pages 162–174. Springer, Berlin, 1996.

[622] G. PATAKI. *Cone Programming and Eigenvalue Optimization: Geometry and Algorithms*. PhD thesis, Carnegie Mellon University, Pittsburgh, PA, 1996.

[623] G. PATAKI. On the rank of extreme matrices in semidefinite programs and the multiplicity of optimal eigenvalues. *Math. Oper. Res.*, 23(2):339–358, 1998.

[624] G. PATAKI and L. TUNCEL. On the generic properties of convex optimization problems in conic form. Technical Report CORR 97-16, Department of Combinatorics and Optimization, Waterloo, Ont, 1997.

[625] J. PENA. Computing the distance to infeasibility: theoretical and practical issues. Technical report, Cornell University, Ithaca,, NY, 1997.

[626] J. PENG and Y. YUAN. Optimality conditions for the minimization of a quadratic with two quadratic constraints. *SIAM J. Optim.*, 7(3):579–594, 1997.

[627] A.L. PERESSINI. *Ordered Topological Vector Spaces.* Harper & Row Publishers, New York, 1967.

[628] E.L. PETERSON and J.G. ECKER. Geometric programming: A unified duality theory for quadratically constrained quadratic programs and l_p-constrained l_p-approximation problems. *Bull. Amer. Math. Soc.*, 74:316–321, 1968.

[629] E.L. PETERSON and J.G. ECKER. Geometric programming: duality in quadratic programming and l_p-approximation. II. (Canonical programs). *SIAM J. Appl. Math.*, 17:317–340, 1969.

[630] E.L. PETERSON and J.G. ECKER. Geometric programming: duality in quadratic programming and l_p-approximation. III. Degenerate programs. *J. Math. Anal. Appl.*, 29:365–383, 1970.

[631] E.L. PETERSON and J.G. ECKER. Geometric programming: duality in quadratic programming and l_p-approximation. I. In *Proceedings of the Princeton Symposium on Mathematical Programming (Princeton Univ., 1967)*, pages 445–480, Princeton, N.J., 1970. Princeton Univ. Press.

[632] I. PITOWSKI. Correlation polytopes: their geometry and complexity. *Math. Programming*, 50:395–414, 1991.

[633] I. PITOWSKI. Correlation polytopes II: the geometry of limit laws in probability. preprint, 1996.

[634] S. POLJAK and F. RENDL. Nonpolyhedral relaxations of graph-bisection problems. *SIAM J. Optim.*, 5(3), 1995. 467-487.

[635] S. POLJAK, F. RENDL, and H. WOLKOWICZ. A recipe for semidefinite relaxation for (0, 1)-quadratic programming. *J. Global Optim.*, 7(1):51–73, 1995.

[636] S. POLJAK and Z. TUZA. On the relative error of the polyhedral approximation of the max-cut problem. *Operations Research Letters*, 16:191–198, 1994.

[637] S. POLJAK and Z. TUZA. The max-cut problem- a survey. In *Special Year on Combinatorial Optimization, DIMACS series in Discrete Mathematics and Theoretical Computer Science.* AMS, 1995.

[638] S. POLJAK and H. WOLKOWICZ. Convex relaxations of $(0,1)$-quadratic programming. *Math. Oper. Res.*, 20(3):550–561, 1995.

[639] L. PORKOLAB and L. KHACHIYAN. On the complexity of semidefinite programs. *J. Global Optim.*, 10(4):351–365, 1997.

[640] L. PORKOLAB and L. KHACHIYAN. Testing the feasibility of semidefinite programs. In P. Pardalos and H. Wolkowicz, editors, *Topics in Semidefinite and Interior-Point Methods (Toronto, ON, 1996)*, pages 17–26. Amer. Math. Soc., Providence, RI, 1998.

[641] L. PORTUGAL and J. JUDICE. A hybrid algorithm for the solution of a single commodity spatial equilibrium model. Technical report, Universidade de Coimbra, Coimbra, Portugal, 1994.

[642] F.A. POTRA and R. SHENG. A superlinearly convergent primal-dual infeasible-interior-point algorithm for semidefinite programming. Technical Report Reports on Computational Mathematics, 78, University of Iowa, Iowa City, IA, 1995.

[643] F.A. POTRA and R. SHENG. On homogeneous interior-point algorithms for semidefinite programming. *Optim. Methods Softw.*, 9:161–184, 1998. Special Issue Celebrating the 60th Birthday of Professor Naum Shor.

[644] M.J.D. POWELL. Trust region calculations revisited. Report DAMTP 1997/NA18, Cambridge University, 1997. Presented at the 17th Biennial Conference on Numerical Analysis (Dundee).

[645] M.J.D. POWELL. The use of band matrices for second derivative approximations in trust region algorithms. DAMTP Report 1997/NA12, Department of Applied Mathematics and Theoretical Physics, University of Cambridge, Cambridge, England, 1997.

[646] A. PREKOPA. Bounds on probabilities and expectations using multivariate moments of discrete distributions. RUTCOR Research Report, 1993.

[647] F. PUKELSHEIM. *Optimal Design of Experiments.* Wiley Series in Probability and Mathematical Statistics: Probability and Mathematical Statistics. John Wiley & Sons Inc., New York, 1993. A Wiley-Interscience Publication.

[648] E. S. PYATNITSKII and V. I. SKORODINSKII. Numerical methods of Lyapunov function construction and their application to the absolute stability problem. *Syst. Control Letters*, 2(2):130–135, August 1982.

[649] A.J. QUIST, E. DE KLERK, C. ROOS, and T. TERLAKY. Copositive relaxation for general quadratic programming. *Optim. Methods Softw.*, 9:185–208, 1998. Special Issue Celebrating the 60th Birthday of Professor Naum Shor.

[650] P. RAGHAVAN. Randomized approximation algorithms in combinatorial optimization. In *Foundations of software technology and theoretical computer science (Madras, 1994)*, volume 880 of *Lecture Notes in Comput. Sci.*, pages 300–317. Springer, Berlin, 1994.

[651] M. RAMANA, L. TUNCEL, and H. WOLKOWICZ. Strong duality for semidefinite programming. *SIAM J. Optim.*, 7(3):641–662, 1997.

[652] M.V. RAMANA. *An Algorithmic Analysis of Multiquadratic and Semidefinite Programming Problems.* PhD thesis, Johns Hopkins University, Baltimore, Md, 1993.

[653] M.V. RAMANA. An exact duality theory for semidefinite programming and its complexity implications. *Math. Programming*, 77:129–162, 1997.

[654] M.V. RAMANA and A.J. GOLDMAN. Some geometric results in semidefinite programming. *J. Global Optim.*, 7(1):33–50, 1995.

[655] M.V. RAMANA and P.M. PARDALOS. Semidefinite programming. In T. Terlaky, editor, *Interior Point Methods of Mathematical programming*, pages 369–398, Dordrecht, The Netherlands, 1996. Kluwer.

[656] A. RANTZER. On the Kalman-Yakubovich-Popov Lemma. *Syst. Control Letters*, 28(1):7–10, 1996.

[657] F. RENDL. Semidefinite programming and combinatorial optimization. *Appl. Numer. Math.*, 29:255–281, 1999.

[658] F. RENDL, R. J. VANDERBEI, and H. WOLKOWICZ. Max-min eigenvalue problems, primal-dual interior point algorithms, and trust region subproblems. *Optim. Methods Softw.*, 5:1–16, 1995.

[659] F. RENDL and H. WOLKOWICZ. Applications of parametric programming and eigenvalue maximization to the quadratic assignment problem. *Math. Programming*, 53(1, Ser. A):63–78, 1992.

[660] F. RENDL and H. WOLKOWICZ. A projection technique for partitioning the nodes of a graph. *Ann. Oper. Res.*, 58:155–179, 1995. Applied mathematical programming and modeling, II (APMOD 93) (Budapest, 1993).

[661] F. RENDL and H. WOLKOWICZ. A semidefinite framework for trust region subproblems with applications to large scale minimization. *Math. Programming*, 77(2, Ser. B):273–299, 1997.

[662] J. RENEGAR. A polynomial-time algorithm, based on Newton's method, for linear programming. *Math. Programming*, 40(1 (Ser. A)):59–93, 1988.

[663] J. RENEGAR. On the computational complexity and geometry of the first-order theory of the reals. III. Quantifier elimination. *J. Symbolic Comput.*, 13(3):329–352, 1992.

[664] J. RENEGAR. Ill-posed problem instances. In *From Topology to Computation: Proceedings of the Smalefest (Berkeley, CA, 1990)*, pages 340–358, New York, 1993. Springer.

[665] J. RENEGAR. Is it possible to know a problem instance is ill-posed? Some foundations for a general theory of condition numbers. *J. Complexity*, 10(1):1–56, 1994.

[666] J. RENEGAR. Some perturbation theory for linear programming. *Math. Programming*, 65(1, Ser. A):73–91, 1994.

[667] J. RENEGAR. Incorporating condition measures into the complexity theory of linear programming. *SIAM J. Optim.*, 5(3):506–524, 1995.

[668] J. RENEGAR. Linear programming, complexity theory and elementary functional analysis. *Math. Programming*, 70(3, Ser. A):279–351, 1995.

[669] J. RENEGAR. Condition numbers, the barrier method, and the conjugate-gradient method. *SIAM J. Optim.*, 6(4):879–912, 1996.

[670] J. RENEGAR. Recent progress on the complexity of the decision problem for the reals. In *Quantifier elimination and cylindrical algebraic decomposition (Linz, 1993)*, pages 220–241. Springer, Vienna, 1998.

[671] J. RENEGAR and M. SHUB. Unified complexity analysis for Newton LP methods. *Math. Programming*, 53(1, Ser. A):1–16, 1992.

[672] J. RENEGAR, M. SHUB, and S. SMALE, editors. *The Mathematics of Numerical Analysis*, Providence, RI, 1996. American Mathematical Society.

[673] G. DELLA RICCIA and A. SHAPIRO. Minimum rank and minimum trace of covariance matrices. *Psychometrika*, 47:443–448, 1982.

[674] S. M. ROBINSON. First order conditions for general nonlinear optimization. *SIAM Journal on Applied Mathematics*, 30:597–607, 1976.

[675] S. M. ROBINSON. Regularity and stability for convex multivalued functions. *Math. Oper. Res.*, 1:130–143, 1976.

[676] S. M. ROBINSON. Stability theorems for systems of inequalities, part ii: differentiable nonlinear systems. *SIAM J. Numerical Analysis*, 13:497–513, 1976.

[677] S. M. ROBINSON. Generalized equations and their solutions. II. Applications to nonlinear programming. *Math. Programming Stud.*, 19:200–221, 1982. Optimality and stability in mathematical programming.

[678] S.M. ROBINSON. An application of error bounds for convex programming in a linear space. *SIAM J. Contr.*, 13:271–273, 1975.

[679] R.T. ROCKAFELLAR. *Convex Analysis.* Princeton University Press, Princeton, NJ, 1970.

[680] R.T. ROCKAFELLAR. *Conjugate Duality and Optimization.* SIAM, Philadelphia, PA, 1974. Regional Conference Series in Applied Mathematics.

[681] J. ROHN. Systems of linear interval equations. *Linear Algebra Appl.*, 126:39–78, 1989.

[682] J. ROHN. Overestimations in bounding solutions of perturbed linear equations. *Linear Algebra Appl.*, 262:55–66, September 1997.

[683] K. ROOS, T. TERLAKY, and J.-Ph. VIAL. *Theory and Algorithms for Linear Optimization: An interior Point Approach.* John Wiley & Sons, New York, 1997.

[684] A.M. ROSS. Optimization over cones: SDPpack versus SeDuMi. Technical report, 1998.

[685] S. ROSS. Options and efficiency. *Quarterly Journal of Economics*, 90:75–89, 1976.

[686] G.-C. ROTA. Twelve problems in probability no one likes to bring up. The Fubini Lectures, Torino, June 3-5, 1998.

[687] O.S. ROTHAUS. Domains of positivity. *Abh. Math. Sem. Univ. Hamburg*, 24:189–235, 1960.

[688] M. RUBINSTEIN. Implied binomial trees. *J. Finance*, 49(3):771–819, 1994.

[689] Y. SAAD. *Numerical Methods for Large Eigenvalue Problems.* Halsted Press, New York, 1992.

[690] M.G. SAFONOV. Stability margins of diagonally perturbed multivariable feedback systems. *IEE Proc.*, 129-D(6):251–256, November 1982.

[691] S.A. SANTOS and D.C. SORENSEN. A new matrix-free algorithm for the large-scale trust-region subproblem. Technical Report TR95-20, Rice University, Houston, TX, 1995.

[692] R.W.H. SARGENT and X. ZHANG. An interior-point algorithm for general nonlinear complementarity problems and nonlinear programs. Report, Centre for Process Systems Engineering, Imperial College, London, 1998.

[693] H. SCARF. A min-max solution of an inventory problem. In *Studies in the mathematical theory of inventory and production*, pages 201–209, Stanford University Press, Stanford, CA, 1958. K.J. Arrow and S. Karlin and H. Scarf, eds.

[694] S. SCHAIBLE and W.T. ZIEMBA, editors. *Generalized Concavity in Optimization and Economics*, New York, 1981. Academic Press Inc. [Harcourt Brace Jovanovich Publishers].

[695] C. SCHERER. Mixed H_2/H_∞ control for time-varying and linear parametrically-varying systems. *Int. J. Robust and Nonlinear Control*, 6(9/10):929–952, Nov–Dec 1996.

[696] C. SCHERER. Robust generalized H_2 control for uncertain and LPV systems with general scalings. In *Proc. IEEE Conf. on Decision and Control*, pages 3970–3975, December 1996.

[697] S. H. SCHMIETA and F. ALIZADEH. Associative and Jordan algebras, and polynomial time interior point algorithms for symmetric cones. Technical Report RRR 12-99, RUTCOR, Rutgers University, 640 Bartholomew Road, Piscataway, NJ 08854-8803, March 1999. Available at URL: `ftp://rutcor.rutgers.edu/pub/rrr/reports99/12.ps`.

[698] S. H. SCHMIETA and F. ALIZADEH. Extension of commutative class of primal-dual interior point algorithms to symmetric cones. Technical Report RRR 13-99, RUTCOR, Rutgers University, 640 Bartholomew Road, Piscataway, NJ 08854-8803, July 1999. Available at URL: `ftp://rutcor.rutgers.edu/pub/rrr/reports99/13.ps`.

[699] I.J. SCHOENBERG. Remarks to Maurice Frechet's article: Sur la definition axiomatique d'une classe d'espaces vectoriels distancies applicables vectoriellement sur l'espace de Hilbert. *Ann. Math.*, 36:724–732, 1935.

[700] H. SCHRAMM and J. ZOWE. A version of the bundle idea for minimizing a nonsmooth function: Conceptual idea, convergence analysis, numerical results. *SIAM J. Optim.*, 2:121–152, 1992.

[701] A. SCHRIJVER. A comparison of the Delsarte and Lovász bounds. *IEEE Trans. Infor. Theory*, IT-25:425–429, 1979.

[702] A. SEIFI and L. TUNCEL. A constant-potential infeasible-start interior-point algorithm with computational experiments and applications. *Comput. Optim. Appl.*, 9:107–152, 1998.

[703] S. SENGUPTA, W. SIU, and N.C. LIND. A linear programming formulation of the problem of moments. *Z. Angew. Math. Mecg.*, 10:533–537, 1979.

[704] J. S. SHAMMA and M. ATHANS. Analysis of gain scheduled control for nonlinear plants. *IEEE Trans. Aut. Control*, 35(8):898–907, August 1990.

[705] J. S. SHAMMA and M. ATHANS. Gain scheduling: Potential hazards and possible remedies. *IEEE Control Syst. Mag.*, pages 101–107, June 1992.

[706] J.G. SHANTHIKUMAR and D.D. YAO. Multiclass queueing systems: Polymatroidal structure and optimal scheduling control. *Oper. Res.*, 40:S293–299, 1992.

[707] A. SHAPIRO. Rank-reducibility of a symmetric matrix and sampling theory of minimum trace factor analysis. *Psycometrika*, 47:187–199, 1982.

[708] A. SHAPIRO. Weighted minimum trace factor analysis. *Psycometrika*, 47:243–263, 1982.

[709] A. SHAPIRO. On the unsolvability of inverse eigenvalues problems almost everywhere. *Linear Algebra Appl.*, 49:27–31, 1983.

[710] A. SHAPIRO. Extremal problems on the set of nonnegative definite matrices. *Linear Algebra Appl.*, 67:7–18, 1985.

[711] A. SHAPIRO. First and second order analysis of nonlinear semidefinite programs. *Math. Programming*, 77:301–320, 1997.

[712] A. SHAPIRO. On uniqueness of Lagrange multipliers in optimization problems subject to cone constraints. *SIAM J. Optim.*, 7(2):508–518, 1997.

[713] A. SHAPIRO and J. F. BONNANS. Sensitivity analysis of parametrized programs under cone constraints. *SIAM J. Control Optim.*, 30(6):1409–1422, 1992.

[714] A. SHAPIRO and J.D. BOTHA. Dual algorithms for orthogonal Procrustes rotations. *SIAM J. Matrix Anal. Appl.*, 9:378–383, 1988.

[715] A. SHAPIRO and M.K.H. FAN. On eigenvalue optimization. *SIAM J. Optim.*, 5:552–569, 1995.

[716] R. SHENG and F.A. POTRA. Nonsymmetric search directions for semidefinite programming. Technical report, University of Iowa, Iowa City, IA, 1997.

[717] W.F. SHEPPARD. On the calculation of the double integral expressing normal correlation. *Trans. Cambr. Phil. Soc.*, 19:23–66, 1900.

[718] H.D. SHERALI and W.P. ADAMS. A hierarchy of relaxations between the continuous and convex hull representations for zero-one programming problems. *SIAM J. Discrete Math.*, 3(3):411–430, 1990.

[719] H.D. SHERALI and W.P. ADAMS. A hierarchy of relaxations and convex hull characterizations for mixed-integer zero-one programming problems. *Discrete Appl. Math.*, 52(1):83–106, 1994.

[720] H.D. SHERALI and W.P. ADAMS. Computational advances using the reformulation-linearization technique (rlt) to solve discrete and continuous nonconvex problems. *Optima*, 49:1–6, 1996.

[721] H.D. SHERALI and W.P. ADAMS. *A reformulation-linearization technique for solving discrete and continuous nonconvex problems.* Kluwer Academic Publishers, Dordrecht, 1999.

[722] H.D. SHERALI and A. ALAMEDDINE. A new reformulation-linearization technique for bilinear programming problems. *J. Global Optim.*, 2(4):379–410, 1992.

[723] H.D. SHERALI, R.S. KRISHNAMURTHY, and F.A. AL-KHAYYAl. Enumeration approach for linear complementarity problems based on a reformulation-linearization technique. *J. Optim. Theory Appl.*, 99(2):481–507, 1998.

[724] H.D. SHERALI and Y. LEE. Tighter representations for set partitioning problems. *Discrete Appl. Math.*, 68(1-2):153–167, 1996.

[725] H.D. SHERALI, Y. LEE, and W.P. ADAMS. A simultaneous lifting strategy for identifying new classes of facets for the Boolean quadric polytope. *Oper. Res. Lett.*, 17(1):19–26, 1995.

[726] H.D. SHERALI and C. H. TUNCBILEK. A global optimization algorithm for polynomial programming problems using a reformulation-linearization technique. *J. Global Optim.*, 2(1):101–112, 1992. Conference on Computational Methods in Global Optimization, I (Princeton, NJ, 1991).

[727] H.D. SHERALI and C. H. TUNCBILEK. A reformulation-convexification approach for solving nonconvex quadratic programming problems. *J. Global Optim.*, 7(1):1–31, 1995.

[728] H.D. SHERALI and C. H. TUNCBILEK. New reformulation linearization/convexification relaxations for univariate and multivariate polynomial programming problems. *Oper. Res. Lett.*, 21(1):1–9, 1997.

[729] M. SHIDA, S. SHINDOH, and M. KOJIMA. Existence and uniqueness of search directions in interior-point algorithms for the SDP and monotone SDLCP. *SIAM J. Optim.*, 8:387–396, 1998.

[730] J.A. SHOHAT and J.D. TAMARKIN. *The Problem of Moments.* American Mathematical Society, New York, 1943. American Mathematical Society Mathematical surveys, vol. II.

[731] N. Z. SHOR. Dual quadratic estimates in polynomial and Boolean programming. *Ann. Oper. Res.*, 25(1-4):163–168, 1990. Computational methods in global optimization.

[732] N. Z. SHOR. Dual estimates in multiextremal problems. *J. Global Optim.*, 2(4):411–418, 1992.

[733] N.Z. SHOR. Quadratic optimization problems. *Izv. Akad. Nauk SSSR Tekhn. Kibernet.*, 222(1):128–139, 222, 1987.

[734] N.Z. SHOR and O.A. BEREZOVSKIUI. New algorithms for solving a weighted problem on the maximum cut of a graph. *Kibernet. Sistem. Anal.*, 2:100–106, 187, 1995.

[735] N.Z. SHOR and A.S. DAVYDOV. A method of obtaining estimates in quadratic extremal problems with Boolean variables. *Kibernetika (Kiev)*, 2:48–50, 54, 131, 1985.

[736] N.Z. SHOR and S.I. STETSENKO. Extremal spectral problems on classes of symmetric matrices and combinatorial problems. *Kibernet. i Vychisl. Tekhn.*, 69:8–15, 113, 1986.

[737] G.A. SHULTZ, R.B. SCHNABEL, and R.H. BYRD. A family of trust-region based algorithms for unconstrained optimization with strong global convergence properties. *SIAM J. Numer. Anal.*, 22(1):47–67, 1985.

[738] C. DE SIMONE and G. RINALDI. A cutting plane algorithm for the max cut problem. Technical report, Istituto di Analisi dei Sistemi ed Informatica del CNR, Rome, 1992.

[739] M. SIPSER. *Introduction to the Theory of Computation.* PWS, Boston, MA, 1996.

[740] M. SKUTELLA. Convex quadratic and semidefinite programming relaxations in scheduling. Technical report, Technische Universitat Berlin, Berlin, Germany, 1999. manuscript.

[741] J. SMITH. Generalized Chebyshev inequalities: theory and applications in decision analysis. *Oper. Res.*, 43:807–825, 1995.

[742] G. SONNEVEND. An 'analytical centre' for polyhedrons and new classes of global algorithms for linear (smooth, convex) programming. In *System Modelling and Optimization*, pages 866–875. Springer, Berlin, 1986. Lecture Notes in Control and Information Sciences No. 84.

[743] D.C. SORENSEN. Implicit application of polynomial filters in a k-step Arnoldi method. *SIAM J. Matrix Anal. Appl.*, 13(1):357–385, 1992.

[744] T.A. SPRINGER. *Jordan Algebras and Algebraic Groups.* Springer-Verlag, 1973.

[745] T. STEPHEN and L. TUNCEL. On a representation of the matching polytope via semidefinite liftings. Technical Report CORR 97-11, Department of Combinatorics and Optimization, Waterloo, Ont, 1997.

[746] R. STERN and H. WOLKOWICZ. Trust region problems and nonsymmetric eigenvalue perturbations. *SIAM J. Matrix Anal. Appl.*, 15(3):755–778, 1994.

[747] R. STERN and H. WOLKOWICZ. Indefinite trust region subproblems and nonsymmetric eigenvalue perturbations. *SIAM J. Optim.*, 5(2):286–313, 1995.

[748] R.J. STERN and H. WOLKOWICZ. Exponential nonnegativity on the ice cream cone. *SIAM J. Matrix Anal. Appl.*, 12(1):160–165, 1991.

[749] R.J. STERN and H. WOLKOWICZ. Invariant ellipsoidal cones. *Linear Algebra Appl.*, 150:81–106, 1991.

[750] R.J. STERN and H. WOLKOWICZ. A note on generalized invariant cones and the Kronecker canonical form. *Linear Algebra Appl.*, 147:97–100, 1991.

[751] R.J. STERN and H. WOLKOWICZ. Results on invariant cones. *Linear Algebra Appl.*, 166:1–26, 1991. Proceedings from the Haifa Matrix Theory Conference, June 1990.

[752] G.W. STEWART and J-G. SUN. *Matrix Perturbation Theory.* Academic Press, 1990.

[753] J. STOER. Principles of sequential quadratic programming methods for solving nonlinear programs. In K. Schittkowski, editor, *Computational Mathematical Programming*, pages 165–207. Springer-Verlag, Berlin, 1985.

[754] J. STOER and M. WECHS. Infeasible-interior-point paths for sufficient linear complementarity problems and their analyticity. Technical report, Institut für Angewandte Mathematik und Statistik, Universität Würzburg, Würzburg, Germany, 1996. Manuscript.

[755] J. STOER and C. WITZGALL. *Convexity and Optimization in Finite Dimensions.* Springer, Berlin-Heidelberg-New York, 1970.

[756] J.F. STURM. *Primal-Dual Interior Point Approach to Semidefinite Programming.* PhD thesis, Erasmus University Rotterdam, 1997.

[757] J.F. STURM. Error bounds for linear matrix inequalities. Technical report, Communications Research Laboratory, McMaster University, Hamilton, Canada, 1998.

[758] J.F. STURM. Using SeDuMi 1.02, a MATLAB toolbox for optimization over symmetric cones. Technical report, Communications Research Laboratory, McMaster University, Hamilton, Canada, 1998. To appear in *Optimization Methods and Software.*

[759] J.F. STURM and S. ZHANG. On sensitivity of central solutions in semidefinite programming. Technical report, Econometric Institute, Erasmus University, Rotterdam, The Netherlands, 1998.

[760] J.F. STURM and S. ZHANG. Symmetric primal-dual path following algorithms for semidefinite programming. *Appl. Numer. Math.*, 29:301–315, 1999.

[761] C. H. SUNG and B.-S. TAM. A study of projectionally exposed cones. *Linear Algebra Appl.*, 139:225–252, 1990.

[762] B.-S. TAM. On the duality operator of a convex cone. *Linear Algebra Appl.*, 64:33–56, 1985.

[763] B.-S. TAM. On the distinguished eigenvalues of a cone-preserving map. *Linear Algebra Appl.*, 131:17–37, 1990.

[764] B.-S. TAM. Graphs and irreducible cone preserving maps. *Linear and Multilinear Algebra*, 31:12–25, 1992.

[765] B.-S. TAM. On the structure of the cone of positive operators. *Linear Algebra Appl.*, 167:65–85, 1992. Sixth Haifa Conference on Matrix Theory (Haifa, 1990).

[766] B.-S. TAM. Extreme positive operators on convex cones. In *Five decades as a mathematician and educator*, pages 515–558. World Sci. Publishing, River Edge, NJ, 1995.

[767] B.-S. TAM and P. H. LIOU. Linear transformations which map the class of inverse M-matrices onto itself. *Tamkang J. Math.*, 21(2):159–167, 1990.

[768] B.-S. TAM and H. SCHNEIDER. On the core of a cone-preserving map. *Trans. Amer. Math. Soc.*, 343(2):479–524, 1994.

[769] K. TANABE. Complementarity-enforcing centered Newton method for linear programming: Global methods. Manuscript at the symposium "New Methods for Linear Programming", The Institute of Statistical Mathematics, 4-6-7 Minamiazabu, Minatoku, Tokyo, Japan 106, 1987. Title of revised version is: Complementarity-enforcing centered Newton method for mathematical programming: Global method.

[770] K. TANABE. Centered Newton method for mathematical programming. In *System Modeling and Optimization*, pages 197–206, NY, USA, 1988. Springer-Verlag.

[771] X. TANG and A. BEN-ISRAEL. Two consequences of Minkowski's 2^n Theorem. Technical report, RUTCOR, Rutgers University, 1996.

[772] P. TARAZAGA, T.L. HAYDEN, and J. WELLS. Circum-Euclidean distance matrices and faces. *Linear Algebra Appl.*, 232:77–96, 1996.

[773] P. TARAZAGA and M.W. TROSSET. An optimization problem on subsets of the symmetric positive semidefinite matrices. Technical report, Rice University, Houston, Texas, 1993.

[774] O. TAUSSKY. Positive definite matrices. In O. Shisha, editor, *Inequalities*. Academic Press Inc., New York and London, 1967.

[775] O. TAUSSKY. Positive-definite matrices and their role in the study of the characteristic roots of general matrices. *Advances in Math.*, 2:175–186 (1968), 1968.

[776] T. TERLAKY. On l_p programming. *European J. Oper. Res.*, 22:70–100, 1985.

[777] G.A. TIJSSEN and G. SIERKSMA. Balinski-Tucker simplex tableaus: Dimensions, degeneracy degrees, and interior points of optimal faces. *Math. Programming*, 81:349–372, 1998.

[778] D.M. TITTERINGTON. Optimal design: some geometrical aspects of D-optimality. *Biometrika*, 62(2):313–320, 1975.

[779] M.J. TODD. 'Fat' triangulations, or solving certain nonconvex matrix optimization problems. *Math. Programming*, 31:123–136, 1985.

[780] M.J. TODD. On search directions in interior-point methods for semidefinite programming. *Optim. Methods Softw.*, (TR1205), 1999.

[781] M.J. TODD, K.C. TOH, and R.H. TUTUNCU. A MATLAB software package for semidefinite programming. Technical report, School of OR and IE, Cornell University, Ithaca, NY, 1996.

[782] M.J. TODD, K.C. TOH, and R.H. TUTUNCU. On the Nesterov-Todd direction in semidefinite programming. *SIAM J. Optim.*, 8(3):769–796, 1998.

[783] M.J. TODD and Y. YE. A centered projective algorithm for linear programming. *Math. Oper. Res.*, 15:508–529, 1990.

[784] K.C. TOH. Search directions for primal-dual interior point methods in semidefinite programming. Technical report, Department of Mathematics, National University of Singapore, Singapore, 1997.

[785] Y.L. TONG. *Probability Inequalities in Multivariate Distributions*. New York, Academic Press, 1980.

[786] W. S. TORGERSON. Multidimensional scaling. I. Theory and method. *Psychometrika*, 17:401–419, 1952.

[787] M.W. TROSSET. Computing distances between convex sets and subsets of the positive semidefinite matrices. Technical report, Rice University, Houston, Texas, 1997.

[788] M.W. TROSSET. Distance matrix completion by numerical optimization. Technical report, Rice University, 1997.

[789] M.W. TROSSET. Applications of multidimensional scaling to molecular conformation. *Computing Science and Statistics*, 29:148–152, 1998.

[790] P. TSENG. Search directions and convergence analysis of some infeasible path-following methods for the monotone semi-definite LCP. *Optim. Methods Softw.*, 9:245–268, 1998.

[791] T. TSUCHIYA. A polynomial primal-dual path-following algorithm for second-order cone programming. Technical report, The Institute of Statistical Mathematics, Tokyo, Japan, 1997.

[792] T. TSUCHIYA. A convergence analysis of the scaling-invariant primal-dual path-following algorithms for second-order cone programming. Technical Report Research Memorandum No. 664, The Institute of Statistical Mathematics, Tokyo, Japan, 1998.

[793] L. TUNCEL. Constant potential primal-dual algorithms: A framework. *Math. Programming*, 66:145–159, 1994.

[794] L. TUNCEL. Lecture notes on "Convex optimization: Barrier functions and interior-point methods". Technical Report B-336, Department of Mathematical and Computing Sciences, Tokyo Institute of Technology, Tokyo, Japan, 1998.

[795] L. TUNCEL. Primal-dual symmetry and scale invariance of interior-point algorithms for convex optimization. *Math. Oper. Res.*, 23:708–718, 1998.

[796] L. TUNCEL. Generalization of primal-dual interior-point methods to convex optimization problems in conic form. Technical Report CORR 99-35, Department of Combinatorics and Optimization, Waterloo, Ont, 1999.

[797] L. TUNCEL. Interior-point methods for semidefinite programming. In *Encyclopedia of Optimization*, MA, USA, 1999. Kluwer Academic Publishers.

[798] L. TUNCEL and M.J. TODD. On the interplay among entropy, variable metrics and potential functions in interior-point algorithms. *Comput. Optim. Appl.*, 8:5–19, 1997.

[799] L. TUNCEL and S. XU. Complexity analyses of discretized successive convex relaxation methods. Technical Report CORR 99-37, Department of Combinatorics and Optimization, Waterloo, Ont, 1999.

[800] L. TUNCEL and S. XU. On homogeneous convex cones, Caratheodory number, and duality mapping semidefinite liftings. Technical Report CORR 99-21, Department of Combinatorics and Optimization, Waterloo, Ont, 1999.

[801] T. URBAN, A.L. TITS, and C.T. LAWRENCE. A primal-dual interior-point method for nonconvex optimization with multiple logarithmic barrier parameters and with strong convergence properties. Report, University of Maryland, College Park, 1998.

[802] C. URSESCU. Multifunctions with convex closed graph. *Czechoslovak Math. J.*, 25:438–441, 1975.

[803] S. VAN HUFFEL and J. VANDEWALLE. *The total least squares problem: computational aspects and analysis*, volume 9 of *Frontiers in applied Math.* SIAM, Philadelphia, PA, 1991.

[804] L. VANDENBERGHE and V. BALAKRISHNAN. Algorithms and software for LMI problems in control. *IEEE Control Syst. Mag.*, pages 89–95, October 1997.

[805] L. VANDENBERGHE and V. BALAKRISHNAN. Semidefinite programming duality and linear system theory: connections and implications for computation. Technical report, UCLA, Los Angeles, CA, 1999. To appear in the Proceedings of the 1999 Conference on Decision and Control.

[806] L. VANDENBERGHE and S. BOYD. A polynomial-time algorithm for determining quadratic Lyapunov functions for nonlinear systems. In *Eur. Conf. Circuit Th. and Design*, pages 1065–1068, 1993.

[807] L. VANDENBERGHE and S. BOYD. Positive definite programming. In *Mathematical Programming: State of the Art, 1994*, pages 276–308. The University of Michigan, 1994.

[808] L. VANDENBERGHE and S. BOYD. *SP: Software for Semidefinite Programming. User's Guide, Beta Version.* K.U. Leuven and Stanford University, October 1994.

[809] L. VANDENBERGHE and S. BOYD. A primal-dual potential reduction method for problems involving matrix inequalities. *Math. Programming*, 69(1, Ser. B):205–236, 1995.

[810] L. VANDENBERGHE and S. BOYD. Semidefinite programming. *SIAM Rev.*, 38(1):49–95, 1996.

[811] L. VANDENBERGHE and S. BOYD. Connections between semi-infinite and semidefinite programming. In *Semi-infinite programming*, pages 277–294. Kluwer Acad. Publ., Boston, MA, 1998.

[812] L. VANDENBERGHE and S. BOYD. Applications of semidefinite programming. *Appl. Numer. Math.*, 29(3):283–299, 1999. Proceedings of the Stieltjes Workshop on High Performance Optimization Techniques (HPOPT '96) (Delft).

[813] L. VANDENBERGHE, S. BOYD, and S.-P. WU. Determinant maximization with linear matrix inequality constraints. *SIAM J. Matrix Anal. Appl.*, 19(2):499–533, 1998.

[814] R. J. VANDERBEI. *Linear Programming: Foundations and Extensions.* Kluwer Acad. Publ., Dordrecht, 1998.

[815] R.J. VANDERBEI and D.F. SHANNO. An interior point algorithm for nonconvex nonlinear programming. Report SOR-97-21, Dept. of Statistics and OR, Princeton University, 1997.

[816] R.J. VANDERBEI and B. YANG. The simplest semidefinite programs are trivial. *Math. Oper. Res.*, 20(3):590–596, 1995.

[817] A. VARDI. A trust region algorithm for equality constrained minimization: convergence properties and implementation. *SIAM J. Numer. Anal.*, 22:575–591, 1985.

[818] S.A. VAVASIS. *Nonlinear Optimization: Complexity Issues.* Oxford Science, New York, 1991.

[819] S.A. VAVASIS. A note on efficient computation of the gradient in semidefinite programming. Technical report, Cornell University, Ithaca, NY, 1999.

[820] J.-P. VIAL. Computational experience with a primal-dual interior-point method for smooth convex programming. *Optim. Methods Softw.*, 3:285–316, 1994.

[821] E.B. VINBERG. The theory of homogeneous cones. *Trans. Moscow Math.*, 12:340–403, 1965.

[822] T. WANG and J.S. PANG. Global error bounds for convex quadratic inequality systems. *Optimization*, 31:1–12, 1994.

[823] W.C. WATERHOUSE. Linear transformations preserving symmetric rank one matrices. *Journal of Algebra*, 125:502–518, 1989.

[824] L.M. WEIN. Scheduling networks of queues: Heavy traffic analysis of a two-station network with controllable inputs. *Oper. Res.*, 38:1065–1078, 1990.

[825] G. WEISS. On the optimal draining of re–entrant fluid lines. In *Stochastic Networks, Proceedings of the IMA*, pages 91–104. F. Kelly and R. Williams eds., 1995.

[826] W.J. WELCH. Algorithmic complexity: Three NP-hard problems in computational statistics. *J. Statist. Comput. Simulation*, 15(1):17–25, 1982.

[827] W.J. WELCH. Branch-and-bound search for experimental designs based on D-optimality and other criteria. *Technometrics*, 24(1):41–48, 1982.

[828] C.P. WELLS. *An Improved Method for Sampling of Molecular Conformation Space.* PhD thesis, University of Kentucky, 1995.

[829] J.H. WILKINSON. *The Algebraic Eigenvalue Problem.* Oxford University Press, London, 1965.

[830] J. C. WILLEMS. Least squares stationary optimal control and the algebraic Riccati equation. *IEEE Trans. Aut. Control*, AC-16(6):621–634, December 1971.

[831] H.K. WIMMER. The set of positive semidefinite solutions of the algebraic Riccati equation of discrete-time optimal control. *IEEE Trans. Automat. Control*, 41(5):660–671, 1996.

[832] H. WOLKOWICZ. A constrained matrix optimization problem. *SIAM Review 23*, 101, 1981.

[833] H. WOLKOWICZ. Some applications of optimization in matrix theory. *Linear Algebra Appl.*, 40:101–118, 1981.

[834] H. WOLKOWICZ. Explicit solutions for interval semidefinite linear programs. *Linear Algebra Appl.*, 236:95–104, 1996.

[835] H. WOLKOWICZ. Semidefiniteness of a sum: Problem solution 19-5.5. *IMAGE, The Bulletin of ILAS*, 20:30–31, 1998.

[836] H. WOLKOWICZ. Duality for semidefinite programming. Research report, University of Waterloo, Waterloo, Ontario, 1999. To appear in the Encyclopedia of Optimization.

[837] H. WOLKOWICZ. Semidefinite programming. In P.M. Pardalos and M.G.C. Resende, editors, *Handbook of Applied Optimization.* Oxford University Press, Waterloo, Canada, 1999. To appear.

[838] H. WOLKOWICZ. Semidefinite and Lagrangian relaxations for hard combinatorial problems. In M.J.D. Powell, editor, *Proceedings of 19th IFIP TC7 CONFERENCE ON System Modelling and Optimization, July, 1999, Cambridge.* Kluwer Academic Publishers, Boston, MA, 2000. To appear.

[839] H. WOLKOWICZ and Q. ZHAO. An all-inclusive efficient region of updates for least change secant methods. *SIAM J. Optim.*, 5(1):172–191, 1995.

[840] H. WOLKOWICZ and Q. ZHAO. Semidefinite relaxations for the graph partitioning problem. *Discrete Applied Math.*, 96-97(1-3):461–479, 1999.

[841] S. WRIGHT. *Primal-Dual Interior-Point Methods.* SIAM, Philadelphia, Pa, 1996.

[842] F. WU. *Control of Linear Parameter Varying Systems.* Doctoral Dissertation. Engineering-Mechanical Eng. Dept., University of California Berkeley, 1995.

[843] Z. WU. A subgradient algorithm for nonlinear integer programming. Technical report, Rice University, Houston, Tx, 1998.

[844] X. XU, P.-F. HUNG, and Y. YE. A simplified homogeneous and self-dual linear programming algorithm and its implementation. *Ann. Oper. Res.*, 62:151–171, 1996.

[845] V.A. YAKUBOVICH. The solution of certain matrix inequalities in automatic control theory. *Soviet Math. Dokl.*, 3:620–623, 1962. In Russian, 1961.

[846] V.A. YAKUBOVICH. Solution of certain matrix inequalities encountered in nonlinear control theory. *Soviet Math. Dokl.*, 5:652–656, 1964.

[847] V.A. YAKUBOVICH. The method of matrix inequalities in the stability theory of nonlinear control systems, *I*, *II*, *III*. *Automation and Remote Control*, 25-26(4):905–917, 577–592, 753–763, April 1967.

[848] V.A. YAKUBOVICH. Minimization of quadratic functionals under the quadratic constraints and the necessity of a frequency condition in the quadratic criterion for absolute stability of nonlinear control systems. *Dokl. Akad. Nauk SSSR*, 209(14):1039–1042, 1973.

[849] V.A. YAKUBOVICH. The S-procedure and duality theorems for nonconvex problems of quadratic programming. *Vestnik Leningrad. Univ.*, 1973(1):81–87, 1973.

[850] V.A. YAKUBOVICH. Nonconvex optimization problem: the infinite-horizon linear-quadratic control problem with quadratic constraints. *Systems Control Lett.*, 19(1):13–22, 1992.

[851] M. YANNAKAKIS. On the approximation of maximum satisfiability. *J. Algorithms*, 17(3):475–502, 1994. Third Annual ACM-SIAM Symposium on Discrete Algorithms (Orlando, FL, 1992).

[852] Y. YE. A class of projective transformations for linear programming. *SIAM J. Comput.*, 19(3):457–466, 1990.

[853] Y. YE. An O(n^3L) potential reduction algorithm for linear programming. *Math. Programming*, 50:239–258, 1991.

[854] Y. YE. A new complexity result on minimization of a quadratic function with a sphere constraint. In *Recent Advances in Global Optimization*, pages 19–31. Princeton University Press, 1992.

[855] Y. YE. On affine scaling algorithms for nonconvex quadratic programming. *Math. Programming*, 56:285–300, 1992.

[856] Y. YE. Combining binary search and Newton's method to compute real roots for a class of real functions. *Journal of Complexity*, 10:271–280, 1994.

[857] Y. YE. *Interior Point Algorithms: Theory and Analysis.* Wiley-Interscience series in Discrete Mathematics and Optimization. John Wiley & Sons, New York, 1997.

[858] Y. YE. A .699-approximation algorithm for max-bisection. Technical report, University of Iowa, Iowa City, IA, 1999.

[859] Y. YE. Approximating quadratic programming with bound and quadratic constraints. *Math. Programming*, 84:219–226, 1999.

[860] Y. YE, M.J. TODD, and S. MIZUNO. An $\mathcal{O}(\sqrt{n}L)$–iteration homogeneous and self–dual linear programming algorithm. *Math. Oper. Res.*, 19:53–67, 1994.

[861] E.A. YILDIRIM and M.J. TODD. Sensitivity analysis in linear programming and semidefinite programming using interior-point methods. Technical Report TR1253, School of Operations Research and Industrial Engineering, Ithaca, NY 14853-3801, 1999.

[862] Y. YUAN. Some properties of trust region algorithms for nonsmooth optimization. Technical Report DAMTP 1983/NA14, Department of Applied Mathematics and Theoretical Physics, University of Cambridge, Cambridge, England, 1983.

[863] Y. YUAN. On a subproblem of trust region algorithms for constrained optimization. *Math. Programming*, 47:53–63, 1990.

[864] Y. YUAN. A dual algorithm for minimizing a quadratic function with two quadratic constraints. *Journal of Computational Mathematics*, 9:348–359, 1991.

[865] F. Z. ZHANG. On the best Euclidean fit to a distance matrix. *Beijing Shifan Daxue Xuebao*, 4:21–24, 1987.

[866] Y. ZHANG. Basic equalities and inequalities in primal-dual interior-point methods for semidefinite programming. Technical report, Department of

Mathematics and Statistics, University of Maryland Baltimore County, 1995.

[867] Y. ZHANG. On extending some primal-dual interior-point algorithms from linear programming to semidefinite programming. *SIAM J. Optim.*, 8:365–386, 1998.

[868] Q. ZHAO. Measures for least change secant methods. Master's thesis, University of Waterloo, 1993.

[869] Q. ZHAO. *Semidefinite Programming for Assignment and Partitioning Problems*. PhD thesis, University of Waterloo, 1996.

[870] Q. ZHAO, S.E. KARISCH, F. RENDL, and H. WOLKOWICZ. Semidefinite programming relaxations for the quadratic assignment problem. *J. Comb. Optim.*, 2(1):71–109, 1998.

[871] A.D. ZIEBUR. Chain rules for functions of matrices. *Linear Algebra Appl.*, 283:87–97, 1998.

[872] G.M. ZIEGLER. Lectures on 0/1-polytopes. Technical report, Technische Universitat Berlin, Berling, Germany, 1999.

[873] Z. ZOU, R.H. BYRD, and R.B. SCHNABEL. A stochastic/perturbation global optimization algorithm for distance geometry problems. Technical report, Dept. of Computer Science, University of Colorado, Boulder, Co, 1996.

[874] J. ZOWE and S. KURCYUSZ. Regularity and stability for the mathematical programming problem in Banach spaces. *Appl. Math. Optim.*, 5:49–62, 1979.

[875] J. ZOWE and H. MAURER. Optimality conditions for the programming problem in infinite dimensions. In *Optimization and operations research (Proc. Workshop, Univ. Bonn, Bonn, 1977)*, volume 157 of *Lecture Notes in Econom. and Math. Systems*, pages 261–270. Springer, Berlin, 1978.

[876] U. ZWICK. Finding almost satisfying assignments. In *Proc. of the 30th ACM Symp. on Theory Comput.*, pages 551–560, 1998.

[877] U. ZWICK. Outward rotations: A tool for rounding solutions of semidefinite programming relaxations, with applications to MAX CUT and other problems. In *Proc. of the 31st ACM Symp. on Theory Comput.*, pages 679–687, 1999.

Appendix

A-.1 CONCLUSION AND FURTHER HISTORICAL NOTES

The area of SDP remains a very active area of research. This is clearly evident in the number of conferences and publications devoted to this area. Recent workshops were held at the Fields Institute in Toronto, Canada, May, 1996; in Delft, Netherlands (June, 1998 and August, 1997); in ZIB in Berlin, November, 1998; and at DIMACS in Princeton, January 1999. Many international conferences such as SIAM and ISMP have many sessions on SDP, e.g. ISMP00 in Atlanta, (August, 2000) will have at least 12 sessions devoted to SDP.

The number of references in this handbook provide an indication of the ongoing research (approximately 1000). There are two main reasons for this. First, is the fact that there are so many interesting applications in so many diverse areas. This began with the early engineering applications for LMIs. (Lyapunov over 100 years ago on stability analysis of differential equations; and then more recently Yakubovitch in the 1960's and Boyd and others on convex optimization in control in the 1980's [135]. Chapter 4 presents an introduction to applications in systems and control theory.) Then, the area of matrix completion problems (another name for LMI's) essentially started in the early 1980's and continues to be a very active area of research, e.g. [202, 299, 373, 50, 465, 277]. This was also the time that applications in combinatorial optimization began. The most interesting and exciting applications were the introduction by Lovász of the *theta function*, also known as the *Lovász number of a graph*, see e.g. [496], which gives an upper bound on the size of the largest clique in a graph (see also [425, 75] for more references and details), and more recently the strong approximation results for the max-cut problem by Goemans-Williamson, e.g. [285] and Chapters 12 and 13 in this handbook.

Second is the fact that SDPs are convex problems that can be solved very efficiently, both in theory (with a polynomial worst-case complexity) and in practice. This follows from the seminal work by Nesterov and Nemirovski, e.g. [576, 578]. They developed the notion of *self-concordant barrier* for general convex optimization, which enabled them to formulate polynomial-time algorithms for a wide range of convex optimization problems. Applying their theory to semidefinite programming, they showed that the log det barrier is a self-concordant barrier for the positive semidefinite cone. They also implemented the first interior-point method for SDP in [579]. This work was summarized in their 1994 book [583] (drafts of which had been circulating at Western universities as early as 1991, [580]). See also [576, 578].

Independently, Alizadeh presented a transparent recipe for extending interior-point polynomial time methods from linear programming to SDP [17, 19, 21], and studied applications in discrete optimization [21]. The name semidefinite programming (SDP) also appears to originate with Alizadeh, and has now become standard.

The success of these interior-point methods for SDP attracted many of the principal researchers who contributed to the development of interior-point methods in linear programming. They, and others, have developed an extensive theory as well as solid algorithms for this area. Many of these researchers have contributions in this handbook, see Part II.

We now try and provide an idea of the breadth of the subject and also some of the history that is missing from the chapters in this book. We list the subjects in alphabetical order. (We realize that this has to be a prejudiced view. We regret any omissions and/or discrepancies. We further hope that omissions are covered inside the papers that we reference.)

A-.1.1 Combinatorial Problems

This topic is discussed extensively in this handbook and is one of the main applications of SDP. The exciting results of the Lovász *theta function* and the strong approximation results for the max-cut problem are mentioned above. Many other researchers worked on very closely related relaxations, or lifting, of hard combinatorial problems. For example: for boolean quadratic maximization problems [325, 322]; for disjunctive programming [53, 55]; for reformulation and linearization [720, 719, 5]; and for Lagrangian relaxation [870, 840]. Other applications have appeared in e.g. [58, 56, 54, 745].

A-.1.2 Complementarity Problems

An area that has not been included here is that of complementarity problems. Many of the elegant results for linear complementarity problems can be extended to problems over the cone of semidefinite matrices, see e.g. [236, 361, 439, 434, 544].

A-.1.3 Complexity, Distance to Ill-Posedness, and Condition Numbers

It is now well known that SDP is a convex program and it can be solved to any desired accuracy in polynomial time. As mentioned above, this is mainly due to the seminal work of Nesterov and Nemirovski e.g. [577, 578, 583, 580, 576, 579, 581] who presented barrier methods for more general conic problems and for SDP.

Detailed complexity issues for this area are new. Some discussion is given in Ramana's thesis [652]. Porkolab and Khachian [639] consider testing the feasibility of a system of linear inequalities with semidefinite matrices as variables and present a polynomial time algorithm if the size of the matrices (or the number of inequalities) is fixed.

Another measure of complexity is the distance to ill-posedness, as this can be used to prove convergence, find condition numbers and provide convergence rates, e.g. the work by Renegar [663, 669, 668, 667, 665], and others [250, 252, 594].

A-.1.4 Cone Programming

As mentioned before, this is a generalization of SDP and is an area that has been studied under many different names, e.g. it is called *generalized linear programming* in the paper by Bellman and Fan, [80]. Problems in function spaces with operator constraints and variational principles have an older history, e.g. Kantorovich's book [391] and Neustadt's book [591]; optimization problems over cones (or partial orders, or wedges) are studied in several books dating back to the 60s, e.g. [350, 501, 358, 357, 627]. This has continued to be a very active area of research, e.g. [170, 171, 173, 174, 446, 309, 81, 131, 833] and [475, 88]. More recently, the generalization of SDP to more general cones has been studied by e.g. Güler and Tuncel, [314, 800] and also Hauser [330]; they discuss the barrier parameter and Caratheodory number for general homogeneous cones.

A-.1.5 Eigenvalue Functions

SDP is closely related to min-max eigenvalue problems. And, we have a lot of material on this in the handbook. Early work on this appeared in [178]. Algorithms with quadratic (asymptotic) convergence for min-max problems were given by Overton in e.g. [599, 600]. See also the more recent survey [483].

A-.1.6 Engineering Applications

Here the history is also quite old; in fact, SDP is called LMI which stands for linear matrix inequalities. A brief history is given in the book by Boyd et al [137].

Early results and applications appeared in the classic work by Lyapunov on stability of differential equations, over 100 years ago. Other applications appear in the work of Yakubovich in the 1960's and solutions using ellipsoid methods appeared in the 80's (as discussed above, see also e.g. [135]).

A-.1.7 Financial Applications

Moment problems in finance are dealt with in Chapter 16.

A-.1.8 Generalized Convexity

The cone of semidefinite matrices is a closed convex cone. Therefore it induces a linear partial order on the space of symmetric matrices. We can therefore talk about generalized convex functions with respect to this partial order, called the Löwner partial order, see e.g. [694, 523].

A-.1.9 Geometry

The geometry of SDP was studied as early as 1948 by Bohnenblust [116]. This unpublished paper contains a study on the facial structure of the cone of semidefinite matrices, denoted $\mathcal{P}$. This work was continued in [774, 775] and [73]. These results presented the characterizations of the faces of $\mathcal{P}$ using the null space (and range space) of elements in the relative interior. The fact that the faces are exposed and projectionally exposed appears in [130, 67]. More recent work appears in [765, 484] and [494]. G.F. Voronoi studied the cone $\mathcal{P}$ in connection with problems of the geometry of numbers; C. Davis related geometry of the partial order $\succeq$ and quadratic convexity, e.g. notably the Toeplitz-Hausdorff convexity theorem, see [181]. Fujie and Kojima [260] describe the geometry in terms of quadratic valid inequalities, thus extending the concept of valid inequalities for convex sets. This is continued in [442].

See also our Chapter 3 and, in particular, Section 3.6.

A-.1.10 Implementation

There are many public domain codes for SDP. In addition, benchmarks have been done on several problem sets, see e.g. Section A-.1.19 below. The different codes use different search directions as well as different strategies for handling sparsity. There are several subtleties that arise for SDP that do not arise for linear programming, e.g. possible duality gaps [130, 652, 651, 418, 416]; overdetermined system of equations for the optimality conditions which results in a large choice of Newton search directions [780, 23, 339, 547, 262, 1]; and loss of sparsity in the primal feasible solution [263, 93]. Currently almost all codes are based on primal-dual interior-point techniques. The following is a small sample of papers that discuss and compare different algorithms, [684, 145, 780, 125]. (Many more are cited in Part II of this handbook.)

A-.1.11 Matrix Completion Problems

Matrix completion problems consist in completing a matrix with incomplete data so as to satisfy some criteria. This matrix theory area developed in parallel and independently of the work in engineering and optimization. An early reference is Dym and Gohberg [202], where they studied completions of Toeplitz matrices. The solution of the general positive definite completion problem was characterized in [299] using chordality. In addition, the unique positive definite solution that maximizes the determinant was characterized, and the proof used the function $\log\det$ as a barrier for $\mathcal{P}$. There is an extensive literature in this area, some of which appears in our Chapter 18 and our bibliography. Surveys is given in [465, 50, 373].

A-.1.12 Nonlinear and Nonconvex SDPs

First and second order optimality conditions appear in this handbook, in chapter 4.1. General higher order optimality conditions over cones have been studied in the literature, e.g. [247, 88, 677, 475, 527, 875, 526].

Nonlinear SDPs also arise from relaxations of QAP, e.g. [395, 398], and many other eigenvalue like problems, e.g. [204]. These can have surprisingly strong duality results, e.g. [41, 37, 36]. Other instances of nonlinear SDPs are [540, 541].

A-.1.13 Nonlinear Programming

Chapter 20 shows how SDP can be used to improve the quadratic model used in SQP methods. SDP like problems also arise in quasi-Newton methods where a positive definite closest approximation to the Hessian is sought subject to a linear constraint (the secant equation). Measures such as log det were used in convergence and existence proofs, see [299, 151, 242]. Another such measure is equivalent to Karmarkar's potential function, $\frac{\mathrm{Trace} A}{\det(A)^{\frac{1}{n}}}$, i.e. it is essentially the arithmetic mean divided by the geometric mean of the eigenvalues, e.g. [188],

A-.1.14 Quadratic Constrained Quadratic Programs

This topic is discussed in Chapter 12 on Combinatorial Optimization and also in Chapter 13. This problem, see e.g. problem Q^2P in Section 13.1.1, is basic to modelling hard problems and gives rise to semidefinite relaxations. The problem was studied as the *S-procedure* in the 1970's, e.g. [849]. More references in this engineering context are given in the book [137]. The semidefinite relaxation for Q^2P is often called the Shor relaxation and is presented in [733]. The problem and SDP relaxation is also studied in [635]. Many other references are given in the above mentioned two chapters. A sample of additional, more recent, references are [568, 652, 727, 343, 228]. Many more are included in the bibliography.

A-.1.15 Sensitivity Analysis

This area is treated in Chapter 4. (See also Section 3.5.2. Additional work can be found in e.g. [370, 861, 759, 713].

A-.1.16 Statistics

Many references are included in our Chapters 16 and 17. Work on optimal design can also be found in [647]. A trace maximization problem is studied in 1989 [411]; see also [710] for a number of particular examples of SDP problems (including the minimum trace factor analysis) that were studied in the statistics literature in the seventies and early eighties. An other early study that used SDP is by Fletcher, 1981 [240], on the educational testing problem. (See also Chapter 17.2.)

A-.1.17 Books and Related Material

Several books, special issues of journals, theses, etc... have appeared that are devoted to SDP. We list several here.

1. Stephen Boyd, Venkataramanan Balakrishnan, Eric Feron, and Laurent El Ghaoui [137] study linear matrix inequalities in system and control theory.

2. Y. E. Nesterov and A. S. Nemirovski [583] is the seminal work on interior-point polynomial time algorithms for convex programs.

3. Yinyu Ye [857] provides a very comprehensive theoretical background on cone programming.

4. Stephen Wright [841] mainly deals with LP but has parts on SDP as does [814].

5. Special issues in journals and proceedings are [601, 612, 611].

6. Many theses on SDP and related work have appeared, e.g. [11, 395, 317, 17, 869, 449, 395, 333, 370, 394, 416, 622, 652, 756, 1, 532, 593].

A-.1.18 Review Articles

1. M.X. Goemans [281, 282] concentrates mainly on combinatorial applications.

2. Monique Laurent [465] on Euclidean Distance Matrices.

3. Franz Rendl [657] on combinatorial problems.

4. Lieven Vandenberghe and Stephen Boyd [810, 807] are two of the earlier survey articles on semidefinite programming motivated by their applications in engineering.

5. [837] a general outline of SDP.

A-.1.19 Computer Packages and Test Problems

A collection of test problems is maintained by Brian Borchers at URL: http://www.nmt.edu/~borchers/sdplib.html and also by Hans Mittelmann at ftp://plato.la.asu.edu/pub/sdplib.txt.

Several public domain packages are currently available. These can be found through links at several home pages, e.g. the home page on SDP maintained by Christoph Helmberg at URL http://www.zib.de/helmberg/semidef.html.

A-.2 INDEX

Keywords and Sections